SURFACE POLARITONS

Electromagnetic Waves at Surfaces and Interfaces

MODERN PROBLEMS IN CONDENSED MATTER SCIENCES

Series editors

V.M. AGRANOVICH

Moscow, USSR

A.A. MARADUDIN

Irvine, California, USA

Advisory editorial board

F. Abelès, Paris, France
N. Bloembergen, Cambridge, MA, USA
E. Burstein, Philadelphia, PA, USA
I.L. Fabelinskii, Moscow, USSR
M.D. Galanin, Moscow, USSR
V.L. Ginzburg, Moscow, USSR
H. Haken, Stuttgart, W. Germany
R.M. Hochstrasser, Philadelphia, PA, USA
I.P. Ipatova, Leningrad, USSR
A.A. Kaplyanskii, Leningrad, USSR
L.V. Keldysh, Moscow, USSR
R. Kubo, Tokyo, Japan
I.M. Lifshitz, Moscow, USSR
R. Loudon, Colchester, UK
A.M. Prokhorov, Moscow, USSR
K.K. Rebane, Tallinn, USSR

NORTH-HOLLAND PUBLISHING COMPANY
AMSTERDAM · NEW YORK · OXFORD

SURFACE POLARITONS

Electromagnetic Waves at Surfaces and Interfaces

Volume editors

V. M. AGRANOVICH
Moscow, USSR

D. L. MILLS
Irvine California, USA

1982

NORTH-HOLLAND PUBLISHING COMPANY
AMSTERDAM · NEW YORK · OXFORD

ISBN 0444 86165 3

PUBLISHERS:
NORTH-HOLLAND PUBLISHING COMPANY
AMSTERDAM · NEW YORK · OXFORD

SOLE DISTRIBUTORS FOR THE USA AND CANADA:
ELSEVIER SCIENCE PUBLISHING COMPANY, INC.
52 VANDERBILT AVENUE
NEW YORK, N.Y. 10017

Library of Congress Cataloging in Publication Data
Main entry under title:

Surface polaritons.

(Modern problems in condensed matter sciences)
Includes bibliographies and indexes.
1. Polaritons. I. Agranovich, V. M. (Vladimir Moiseevich), 1929- . II. Mills, D. L. III. Series.
QC174. 8. P6S93 530.1'41 81-22496
ISBN 0-444-86165-3 (Elsevier AACR2 North-Holland)

PRINTED IN THE NETHERLANDS

Other Volumes in this Series:

EXCITONS
E.I. Rashba and M.D. Sturge, *editors*

ELECTRONIC EXCITATION ENERGY TRANSFER IN CONDENSED MATTER
V.M. Agranovich and M.D. Galanin

SPECTROSCOPY AND EXCITATION DYNAMICS IN CONDENSED MOLECULAR SYSTEMS
V.M. Agranovich and R.M. Hochstrasser, *editors*

Oh, how many of them there
are in the fields!
But each flowers in its
own way—
In this is the highest achievement
of a flower!

Matsuo Bashó
1644–1694

PREFACE TO THE SERIES

"Surface Polaritons" is the first volume in a series of contributed volumes and monographs on condensed matter science that the North-Holland Publishing Company is now beginning to publish. This vast area of physics is developing rapidly at the present time, and the numerous fundamental results in it define to a significant degree the face of contemporary science. This being so, it is clear that the most important results and directions for future developments can only be covered by an international group of authors working in cooperation.

Both Soviet and Western scholars are taking part in the series, and each contributed volume has, correspondingly, two editors. Furthermore, it is intended that the volumes in the series will be published subsequently in Russian by the publishing house "Nauka".

The idea for the series and for its present structure was born during discussions that took place in the USSR and the USA between the President of North-Holland Publishing Company, Drs. W.H. Wimmers, and the General Editors.

The establishment of this series of books, which should become a distinguished encyclopedia of condensed matter science, is not the only important outcome of these discussions. A significant development is also the emergence of a rather interesting and fruitful form of collaboration among scholars from different countries. We are deeply convinced that such international collaboration in the spheres of science and art, as well as other socially useful spheres of human activity, will assist in the establishment of a climate of confidence and peace.

The General Editors of the Series,

V.M. Agranovich A.A. Maradudin

PREFACE

The purpose of this volume is to present a sequence of articles which describe the basic properties of surface polaritons, methods of generating these waves in the laboratory at frequencies of interest to condensed matter physicists, and then the physics that may be learned from them. In our view, the collection provides an excellent summary of the activity that resulted from the recent interest in these modes in the solid state physics community.

A surface polariton is simply an electromagnetic wave that propagates along the surface of a medium, or along the interface between two media. The strength of the electromagnetic fields associated with the wave decay in strength exponentially as one moves away from the interface into either medium, but they vary in a wavelike fashion as one moves parallel to it. Despite the exotic name, upon which we comment in more detail below, if both media may be described in a continuum approximation by a dielectric tensor and a magnetic permeability tensor (both are frequency dependent and complex, in general), then these waves emerge as solutions of Maxwell's equations applied to the pair of media. The chapters in this volume explore a large number of realizable configurations that support propagation of these modes.

Despite the fact that the basic properties of surface polaritons may be obtained directly from Maxwell's equations, most physicists are not familiar with them. This is curious, because they have been studied actively since the turn of the present century, when Sommerfeld and his students discussed them within a certain context (Joos 1934). Interest in the waves has arisen from time to time in many subfields of physics and engineering, where in each instance they have attracted the attention of specialists in the area, but not of the general physics community. Largely because of the recent burst of activity in condensed matter physics, and especially in physics of surfaces, we now appreciate the ubiquitous nature of surface polaritons. We hope this volume will show the condensed matter physics community how surface polaritons may be used to probe the near vicinity of the crystal surface or of an interface under a variety of circumstances, in other words, how they may be used to facilitate the development of the spectroscopy of surfaces.

The interaction of electromagnetic waves with an interface between two media is discussed in nearly all textbooks on electromagnetic theory when a derivation is given of Fresnel formulae for the amplitude of waves reflected from or transmitted through an interface. Yet surface polaritons are almost never treated (Portis 1978), even though they emerge as solutions of Maxwell's equations applied to the same geometry. The reason for this may possibly be appreciated by considering a plane interface between an isotropic dielectric with dielectric constant ϵ, and vacuum. As the reader will appreciate upon examining the discussion in several of the chapters (see, for example, the opening chapter by D. N. Mirlin) for the surface polariton to exist, the requirement is that ϵ, assumed real here for simplicity, be a *negative* number. More specifically, the inequality $\epsilon \leqslant -1$ must be satisfied for this particular example. Most textbooks on electromagnetic theory assume, either implicitly or explicitly, that the dielectric constant of matter is always positive when the interaction of electromagnetic waves with dielectrics is examined.

The *static* dielectric constant of any insulating solid is, ignoring spatial dispersion (i.e. for wave vectors $\boldsymbol{k} = 0$, see Agranovich and Ginzburg 1981) required to be positive, from stability considerations in thermodynamics applied to nonconducting media (Callen 1960). No such constraint applies to the finite frequency dielectric function $\epsilon(\Omega)$ that enters the theory of electromagnetic wave propagation. In fact, as the sequence of chapters in this volume illustrate very nicely, in a wide class of solid systems one may have negative values for $\epsilon(\Omega)$ over substantial ranges of frequency. Consider, for example, the nearly free electron picture of simple metals. Then if Ω_p is the electron plasma frequency, we have $\epsilon(\Omega) \cong 1 - \Omega_p^2/\Omega^2$ (with scattering of the electrons ignored), so that the inequality $\epsilon(\Omega) \leqslant -1$ holds for *all* frequencies below $\Omega_p/\sqrt{2}$. For aluminum, $\Omega_p/\sqrt{2}$ is 10.6 eV, so on the surface of aluminum, surface polaritons may propagate at *all* frequencies from the microwave through the visible and well into the ultraviolet. Similar statements apply to a variety of the simple metals. This example shows that surface polaritons are not strange solutions of Maxwell's equations encountered under exotic conditions, but in fact they are rather commonplace.

As several chapters explain, a photon which strikes a perfectly plane, smooth interface between a dielectric and vacuum fails to "see" or interact with surface polaritons on the interface. But if a suitable prism is placed nearby, if roughness is present or a grating is ruled on the surface, one may easily "drive" the surface polariton by linear coupling to the incoming photon. A variety of other methods of excitation are also discussed in the chapters, including the use of nonlinear interactions which allow the generation of surface polaritons on perfectly smooth surfaces or interfaces.

In an isotropic dielectric, we have $\boldsymbol{D} = \boldsymbol{E} + 4\pi\boldsymbol{P} \equiv \epsilon(\Omega)\boldsymbol{E}$. For the dielectric constant to be negative, the external electric field $\boldsymbol{E}$ must excite a polarization density $\boldsymbol{P}$ 180° out of phase with the exciting field $\boldsymbol{E}$. If the medium has *any* sharp absorption line at frequency Ω_0, the excitation of the medium at a frequency just above Ω_0 will result in a negative contribution to $\boldsymbol{P}$ that can be very large, if the absorption line is sharp. That this is so, quite generally is insured by the Kramers–Kronig relations. In more elementary language, if a very lightly damped harmonic oscillator is driven just *above* its resonance frequency, then the response not only has large amplitude, but is 180° out of phase with the driving field. The induced polarization is thus directed anti-parallel to the electric field that excites it. Hence, the condition for surface polariton propagation is realized at the smooth planar interface between vacuum and an isotropic dielectric almost always just above an absorption line, in spectral regions where narrow lines exist. Thus, just above the reststrahl frequency of simple ionic crystals, we encounter the "surface phonon polaritons" discussed by D. N. Mirlin and others. "Surface exciton polaritons" are formed just above sharp exciton absorption lines, as discussed by J. Lagois and B. Fischer. The additional modifier refers to the elementary excitation in the crystal whose contribution to the dipole moment density $\boldsymbol{P}$ drives the dielectric constant negative.

We are now in a position to understand why the surface waves are referred to commonly as surface polaritons, in the condensed matter physics literature. Near sharp absorption bands such as those just described, $\epsilon(\Omega)$ is a strong function of frequency both above and below Ω_0. When $\epsilon(\Omega)$ is positive, just below Ω_0 and sufficiently far above it, Maxwell's equations admit plane wave solutions in the bulk of the material. In the frequency regime where $\epsilon(\Omega)$ displays such sharp resonant structure near an absorption line, one may describe the electromagnetic wave as a coupled mode which is an admixture of the electromagnetic field and the elementary excitation in the medium that produces the resonance in $\epsilon(\Omega)$ (Mills and Burstein 1974). Such electromagnetic waves, when described within this picture, are commonly called polaritons in the literature of condensed matter physics. Hence, the surface electromagnetic waves of interest to the authors in the present volume are frequently referred to as surface polaritons.

As remarked earlier, surface electromagnetic waves have been studied in various subfields of physics, and within each, a specialized terminology for them has evolved. This is true within condensed matter physics, as well as within physics and engineering as a whole. This generated a problem for the editors of the present volume, in that the set of authors has been drawn from a broad spectrum of areas in condensed matter physics. Not all authors used the same terminology, and after initial correspondence and conversation, it became clear that some felt the particular choice made by them had historical precedence over other possibilities. The arguments in each case seem sensible, but it proved difficult to obtain agreement on the terminology among the diverse array of authors who have contributed to the volume. Thus, A. A. Maradudin prefers to use distinctly different language to discuss waves for which the retardation terms in Maxwell's equations enter in an essential fashion, and those for which these may be ignored and an electrostatic description is sufficient. He uses the term surface polariton to describe surface electromagnetic waves of the first class, and with the metal/vacuum interface in mind, the solutions found in the electrostatic limit are described as surface plasmons. In fact, for a smooth surface, the dispersion relation of the surface electromagnetic wave is continuous, with retardation important when the frequency Ω of the wave and its wave vector parallel to the surface $Q_{\|}$ satisfy $\Omega \cong cQ_{\|}$, and the electrostatic limit appropriate when $Q_{\|}$ is so large that $cQ_{\|} \gg \Omega$. Here c is the velocity of light in vacuum. Thus, the terminology used by A. A. Maradudin, not at all uncommon in the theoretical literature, uses different terms to denote quasi-particles obtained in different approximations. In the chapter by V. M. Agranovich, we see the term "Coulomb wave" used to denote modes that were found when retardation is disregarded. H. Raether, along with J. Sipe and G. Stegeman use the term surface plasmon to refer to the surface electromagnetic wave on metals for *all* values of Ω and $Q_{\|}$, including the region where retardation is important. Thus, these authors use the term surface plasmon in the same manner as we employ surface polariton in the present discussion.

We hope the lack of a uniform terminology will not render the volume confusing to the reader not familiar with the field; it should be clear that the same reader will encounter a similar array of terms if he or she consults the general literature, so it may be that by comparing the material in the various chapters within one set of covers, the reader can become acquainted with the range of descriptive terminology used in the field.

The present volume confines its attention to the elucidation of the basic properties of surface polaritons, the methods of generating them in the laboratory, their use as probes of the surface or interface environment and finally to the nonlinear interactions in which they participate. These topics are covered in depth, and we are particularly pleased to see that a large volume of actual experimental data is reproduced in the figures and tables. Space limitations then require us to omit material on more specialized or less developed topics. There is no discussion of the properties of surface polaritons on magnetic crystals; one of the editors will prepare a chapter which includes discussion of such modes for another volume in the present series (Agranovich and Loudon).

Surface spectroscopy and, in particular, spectroscopy of surface polaritons continues its rapid development. In this connection, we shall mention new results that were obtained after this volume had been prepared for publication. These concern nonlinear surface polaritons (NSP) and waves of self-induced transparency (SIT) on surfaces.

Previously, when we discussed surface polaritons, surface plasmons, etc., we were concerned with solutions of linear Maxwell equations. As is known, nonlinearity appears in these equations when in the relation $D = \epsilon E$ the dependence of ϵ on the strength of the electric field is taken into account, and not only its dependence on the frequency and wave vector (the latter corresponds to the taking of spatial dispersion into account, see Agranovich and Ginzburg (1981)). In the simplest case, i.e., in taking the dependence of ϵ on only the magnitude of $|\boldsymbol{E}|$ into consideration, $\epsilon = \epsilon(\Omega) + \alpha|\boldsymbol{E}|^2$. If $\alpha > 0$, this nonlinearity leads to self-focusing and to other nonlinear effects.

In the papers (Tomlinson 1980, Agranovich et al. 1980, Maradudin 1981) it was established that this same nonlinearity leads to the occurrence of NSP, that those waves can exist in spectral regions in which no linear waves are formed, and that the frequency of these waves depends on the strength of the electric field at the surface. In particular, it was shown by Tomlinson (1980) and Maradudin (1981) that at the interface between linear and nonlinear isotropic media NSP exist only in s-polarization. If, however, the nonlinear medium is optically anisotropic (uniaxial), then, as shown by Agranovich et al. (1980), NSP are formed in p-polarization as well (recall that at the interface of the linear media SP are formed only in p-polarization). The frequency of NSP in p-polarization depends on the wave vector so that it is precisely in this case that NSP have nonzero group velocity $d\omega/dk$. This feature of NSP in p-polarization is important because the waves being discussed can exist, in particular, in the transparency spectral region, according to theory. In virtue of the fact that $d\omega/dk \neq 0$, these waves may have considerable lengths of propagation, even along dielectric surfaces on which linear SP are usually quite rapidly damped. In connection with the aforesaid, the search for methods of exciting and detecting NSP would be of especial interest, as would be an analysis of the feasibility of employing NSP for developing new optical methods of investigating surfaces and thin films.

An entirely different type of nonlinear surface waves occurs in the propagation of ultrashort pulses of intensive and resonant radiation along a surface coated with a thin film of a resonant substance. In this case, as shown by Agranovich et al (1981a), surface "2π" pulses of self-induced transparency are found to be stable. The investigation of this problem is only in an early stage (a similar effect in waveguides is discussed in the paper by Agranovich et al. (1981b)) and only the simplest situations have been studied. In particular, there is no rigorous answer to the question of whether stable surface "2π" pulses exist or not under conditions when the whole mass of the dielectric is resonant, rather than a thin film on a surface. In connection with the development of picosecond spectroscopy, problems of SIT on a surface will undoubtedly be a subject of discussion in the future.

It is clear that surface polaritons can participate in a number of physical processes that take place near surfaces and interfaces. For example, the fluorescent yield of an excited atom placed near a metal surface is less than that appropriate to the gas phase, because near the surface the atom may decay in a nonradiative fashion by emission of a surface polariton. Also, in recent years, there has been considerable discussion of the emission of light from tunnel junctions subject to a DC bias voltage. Surface polaritons enter as an intermediate stage in the emission process. Discussion of these two topics, as well as that of a number of other specific phenomena, does not appear in this volume (see Hansma, ed., 1981), since we felt that by spreading out the range covered by the book, we would produce a volume which covered the key basic properties of surface polaritons in a fashion too superficial to be satisfying.

Finally, we wish to thank the authors for the thought and care they have put into their material.

V. M. Agranovich
Moscow
USSR

D. L. Mills
Irvine, California
USA

References

Agranovich, V.M. and R. Loudon, eds., Surface Excitations (North-Holland, Amsterdam) forthcoming.

Agranovich, V.M. and V.L. Ginzburg, 1982, Crystal Optics with Spatial Dispersion and the Theory of Excitons, sec. ed. (Springer-Verlag, Berlin) in press.

Agranovich, V.M., V.S. Babichenko and V.Ya. Chernyak, 1980, Pisma v ZhETF **32**, 532.

Agranovich, V.M., V.I. Rupasov and V.Ya. Chernyak, 1981a, Pisma v ZhETF **33**, 196

Agranovich, V.M., V.I. Rupasov and V.Ya. Chernyak, 1981b, Opt. Commun. **37**, 363.

Callen, H.B., 1960, Thermodynamics (Wiley, New York) ch. 8.

Hansma, P.K., ed., 1981, Tunneling Spectroscopy: Applications and New Techniques (Plenum Press, New York). (One of the editors is preparing a review chapter on this subject to be included in this volume.)

Joos, G., 1934, Theoretical Physics (Stechert, New York) p. 332. (A discussion of this body of work, which explored the behavior of low frequency waves that propagate on conducting media, may be found in the discussion that begins on p. 332.)

Maradudin, A.A., 1981, Z. f. Physik B, to appear.
Mills, D.L. and E. Burstein, 1974, Reports on Progress in Physics **37**, 817. (This gives a discussion of electromagnetic waves from the polariton point of view.)
Portis, A.M., 1978, Electromagnetic Fields; Sources and Media (Wiley, New York) p. 498 and following section. (This textbook presents a discussion of surface electromagnetic waves.)
Tomlinson, W.J., 1980, Opt. Lett. **5**, 323.

CONTENTS

Preface to the series vi
Preface vii
Contents xv

Part I. Basic properties of surface polaritons on surfaces and interfaces; methods of excitation . .

1. Surface phonon polaritons in dielectrics and semiconductors 3
D. N. Mirlin
2. Surface exciton polaritons from an experimental viewpoint 69
J. Lagois and B. Fischer
3. Surface electromagnetic wave propagation on metal surfaces 93
G. N. Zhizhin, M. A. Moskalova, E. V. Shomina and V. A. Yakovlev
4. Thermally stimulated emission of surface polaritons . . . 145
E. A. Vinogradov, G. N. Zhizhin and V. I. Yudson

Part II. Surface polaritons as a probe of surface and interface properties . . . .

5. Effects of the transition layer and spatial dispersion in the spectra of surface polaritons 187
V. M. Agranovich
6. Surface polaritons at metal surfaces and interfaces . . . 239
F. Abelés and T. Lopez-Rios
7. Resonance of transition layer excitations with surface polaritons 275
G. N. Zhizhin and V. A. Yakovlev
8. The study of solid–liquid interfaces by surface plasmon polariton excitation 299
Dieter M. Kolb

9. Surface plasmons and roughness 331
H. Raether
10. Interaction of surface polaritons and plasmons with surface roughness 405
Alexei A. Maradudin
11. Scattering of surface polaritons by order parameter fluctuations near phase transition points 511
V. M. Agranovich, V. E. Kravtsov and T. A. Leskova

Part III. Nonlinear interactions and surface polaritons

12. Raman scattering by surface polaritons 535
S. Ushioda and R. Loudon
13. Three wave nonlinear interactions involving surface polaritons; Raman scattering, light diffraction and parametric mixing . . 587
Y. J. Chen and E. Burstein
14. Nonlinear wave interaction involving surface polaritons . . 629
Y. R. Shen and F. deMartini
15. Nonlinear optical response of metal surfaces 661
J. E. Sipe and G. I. Stegeman

Author Index 703
Subject Index 715

PART I

Basic Properties of Surface Polaritons on Surfaces and Interfaces; Methods of Excitation

B. FISCHER
J. LAGOIS
D. N. MIRLIN
M. A. MOSKALOVA
E. V. SHOMINA
E. A. VINOGRADOV
V. A. YAKOLEV
V. I. YUDSON
G. N. ZHIZHIN

CHAPTER 1

Surface Phonon Polaritons in Dielectrics and Semiconductors

D. N. MIRLIN

A. F. Ioffe Physico-Technical Institute
USSR Academy of Sciences
Leningrad, 194021
USSR

Surface Polaritons
Edited by
V.M. Agranovich and D.L. Mills

Contents

1. Introduction 5
2. Surface vibrations in isotropic crystals – a phenomenological treatment 7
3. Experimental methods for the excitation and study of surface polaritons 17
   3.1. Methods of electron spectroscopy 18
   3.2. Optical methods 19
4. Experimental studies of surface polaritons in cubic crystals – anharmonic effects . 24
5. Evaluation of anharmonic functions from ATR spectra 30
6. Surface vibrations in anisotropic crystals 33
7. Coupled plasmon–phonon surface waves in semiconductors 42
   7.1. Dispersion and damping of plasmon–phonon surface waves 42
   7.2. Plasmon–phonon surface waves in a magnetic field 47
8. Interface polaritons 52
References. 63

1. Introduction

The boundaries present in real crystals are leading to the surface states in the electron and phonon spectra and also in the spectra of other quasi-particles: excitons, plasmons, magnons. Lately, there has been a noticeable growth of interest in the investigation of such excitations owing to the progress in the traditional experimental techniques and the appearance of new ones. The modified method of attenuated total reflection (ATR), first introduced for the investigation of surface plasmons in metals (see sect. 3), has proved to be especially profitable. Later this method was used for studying both surface phonons in ionic dielectrics and semiconductors and also surface plasma vibrations and mixed excitations in semiconductors. The influence of a magnetic field on surface excitation spectra was studied. Surface excitons were discovered and investigated. A natural generalization of these studies were investigations of states on an interface of two "active media" (interface modes) due to the coupling of excitations in two different media. These investigations have provided a large body of information on the properties of surface excitations. Moreover, they have revealed interesting possibilities for applications, in particular, new methods to determine optical properties of thin layers and monocrystals from the spectra of surface modes.

In this chapter some results in research on surface phonons in dielectrics and semiconductors and also on plasmon–phonons in semiconductors are presented. We would like to make some preliminary remarks to specify the problem.

Cyclic boundary conditions of the Born–Karman type are usually used in calculations of electron and vibration spectra in crystals. However, the presence of a surface destroys the translational symmetry of a crystal. In the slab geometry the translational invariance of a semi-infinite crystal is inevitably violated along one dimension, in the cylinder geometry along two dimensions, and for small crystals it is entirely absent. To obtain surface branches in the energy spectrum theoretically one should use other boundary conditions.

Surface modes* are localized in a relatively thin surface layer. The

*In the following, we will mean surface lattice vibrations, i.e. surface phonons, if not otherwise specified. At least the qualitative character of the results also holds for other types of surface excitations.

theoretical approach depends very strongly upon one characteristic parameter of a surface vibration – the decay distance of the surface vibration amplitude. If the decay distance is much greater than the lattice constant, then such surface vibrations can be treated phenomenologically. Among these are, in particular, surface acoustic vibrations, which in the limiting case of continuous medium are known as Rayleigh waves and which have been intensively studied. Our interest will be in the surface optical vibrations, which have no analogue in the theory of elasticity. If such surface vibrations are accompanied by an alternating dipole moment and a macroscopic electric field, as in the case of polar crystals, then they can be treated by the macroscopic electrodynamics using Maxwell equations and the well-known boundary conditions for the tangential and normal components of electric and magnetic fields. In such treatment the properties of the "active" medium supporting the surface waves are described by the components of its frequency dependent dielectric tensor. Thus, the spectrum of the surface vibrations will be related to the parameters of the bulk spectrum of the crystal (to the frequencies and damping of bulk modes). In this way dispersion can be calculated for other types of surface excitations too (surface plasmons or excitons). In each specific case the characteristic features of the particular excitation branch are revealed in the form of the frequency dependent dielectric function $\epsilon(\omega)$ entering into the dispersion relation.

However, it should be noted that surface states with small decay distances, which arise due to the distortion of the short-range forces near the surface, to the disruption of chemical bonds and the adsorption of foreign atoms, are lost in such a macroscopic theory. Obviously, a concrete model of these vibrations can be investigated only in terms of a microscopic theory. This big area of surface science will be beyond the scope of our discussion. Its theoretical aspects were, for instance, discussed in a review by Bryksin et al. (1974).

Now we will recall one significant result of the dynamic theory of a crystal lattice (Born and Huang, 1954). We refer to the case of a simple diatomic cubic crystal of the NaCl or ZnS type. In the $q \approx 0$ limit, i.e., at $qa \ll 1$ (where q is the phonon wave vector and a is the lattice constant) and within the electrostatic approximation the frequencies of longitudinal ω_{LO} and transverse ω_{TO} optical vibrations are q independent. In the electrostatic approximation the effects due to the retardation of the Coulomb interactions are neglected, i.e., the light velocity c formally approaches infinity. However, if the finite value of c is taken into account and the full set of Maxwell equations is used in solving the problem, the result is different. Though the account of the retardation does not change the longitudinal vibrations, solution corresponding to ω_{TO} becomes coupled to the light wave. The dispersion relation $q^2 = \omega^2\epsilon(\omega)/c^2$ obtained when the retardation is taken into account describes the coupled photon–phonon system, where to each q value correspond two frequencies. These are the frequencies of phonon polaritons, which are observed in appropriate experiments (e.g. the small-angle Raman scattering). A perfectly analogous situation occurs in physics of surface excitations. The coupling of photons and surface phonons results in the appearance of mixed surface excitations – surface phonon polaritons. The dispersion relation of these coupled excitations is different from the $\omega(q)$ dependence for purely mechanical vibrations (see sect. 2). Surface plasmon polaritons, surface exciton polaritons and mixed types of excitations arise in a similar way.

Now the subject of this chapter can be formulated precisely. It is concerned with macroscopic surface excitations, phonon polaritons, appearing near the surface of an ionic crystal. The outline of the chapter is as follows:

Section 2 is devoted to the discussion of the results of a phenomenological theory of surface phonon polaritons in simple, diatomic cubic crystals. Both the case of a "semi-infinite" crystal (one boundary) and the case of a thin slab (two boundaries) are considered.

Section 3 discusses the experimental methods used for investigation of surface vibrations. The attention is largely focussed on the ATR method.

The experimental results on surface vibrations in cubic and anisotropic crystals are discussed in sect. 4 and 6, respectively.

In sect. 5 the evaluation of the optical constants from the surface vibration spectra is briefly considered.

Section 7 deals with the coupled surface plasmon–phonon modes in semiconductors (InSb) and the influence of a magnetic field on these modes.

Finally, sect. 8 is devoted to the vast area of interface modes research. The phenomenon of metallic damping of surface polaritons is also discussed.

We avoided reproducing cumbersome mathematical derivations and the details of the experiments. The list of references is not complete, because the number of publications on the problem is very great. A comparatively large number of references to the papers by the author and his coworkers by no means signifies the neglect of the works of his colleagues who have made a much greater contribution to the development of the problem. The above failings can be overcome with the help of a number of exhaustive reviews published recently (e.g. Fischer et al. 1973, Bryksin et al. 1974, Burstein et al. 1974, Otto 1974, 1975, Agranovich 1975, Halevi 1978).

2. *Surface Vibrations in Isotropic Crystals – a Phenomenological Treatment*

In this section we will derive the dispersion relation of surface modes in the simplest case of an isotropic crystal with a single dispersion oscillator.

Under the category of such crystals fall, for instance, alkali halides, $A^{III}B^{V}$ semiconductors with the zincblende structure, such as GaP, InSb, etc. The frequency dependent dielectric function in the form neglecting the damping and spatial dispersion is:

$$\epsilon(\omega) = \epsilon_\infty + \frac{(\epsilon_0 - \epsilon_\infty)\,\omega_{TO}^2}{\omega_{TO}^2 - \omega^2} \tag{1}$$

where ϵ_∞ and ϵ_0 are the high-frequency and static dielectric constants, respectively, and ω_{TO} is the frequency of the transverse $q = 0$ optical phonon. The curve $\epsilon(\omega)$ (fig. 1) has a pole at ω_{TO}. In the frequency range between ω_{TO} and ω_{LO} (longitudinal phonon frequency) $\epsilon(\omega)$ is negative. It is this frequency range that will be of special interest to us henceforth.

We consider first the case of a "semi-infinite" crystal occupying the halfspace $z < 0$. The plane $z = 0$ is the crystal–vacuum interface (the vacuum can be replaced by a different medium whose dielectric function is assumed to be frequency independent and equal to ϵ_1). The cyclic boundary condition in the plane $z = 0$ can be applied since the crystal is taken to be infinite along the x and y directions. Thus its translation invariance is conserved and the two-dimensional wave vector $\boldsymbol{q}$ with q_x and q_y com-

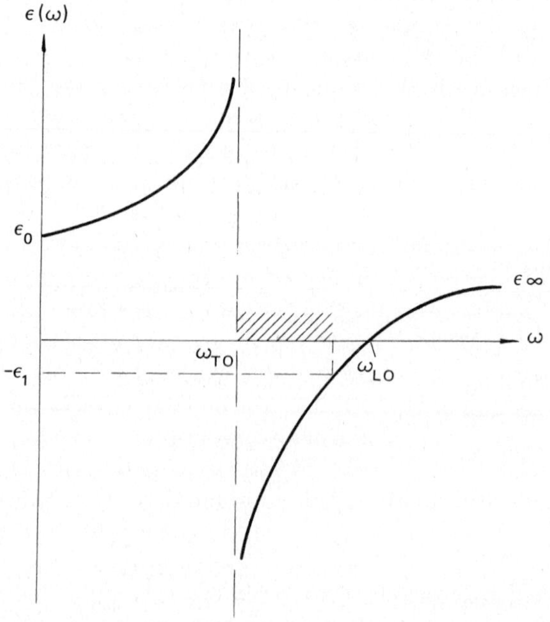

Fig. 1. Frequency dependence of the dielectric function of an isotropic crystal. The pole and zero of $\epsilon(\omega)$ correspond to the transverse and longitudinal $q = 0$ optical vibrations respectively. The frequency range where surface modes exist is indicated by the shaded area.

ponents is valid. It is convenient to direct $\boldsymbol{q}$ along the x-axis, thus $q_y = 0$, $q_x = |\boldsymbol{q}|$. The crystal and the medium are regarded as nonmagnetic ($\mu = 1$).

Maxwell equations for a harmonic wave have the form (see e.g. Landau and Lifshitz, 1960):

$$\text{curl}\, \boldsymbol{E} = (\mathrm{i}\omega/c)\, \boldsymbol{H} \tag{2}$$

$$\text{curl}\, \boldsymbol{H} = -(\mathrm{i}\omega/c)\, \epsilon \boldsymbol{E}. \tag{3}$$

Substituting $\boldsymbol{H}$ from eq. (2) into eq. (3), we get for $\boldsymbol{E}$ the equation:

$$\Delta \boldsymbol{E} + (\epsilon\omega^2/c^2)\, \mathbf{E} - \text{grad div}\, \boldsymbol{E} = 0, \tag{4}$$

and from the vector equation (4) we get for the x component of the electric field:

$$\frac{\partial^2 E_x}{\partial z^2} + \frac{\epsilon\omega^2}{c^2} E_x - \frac{\partial^2 E_z}{\partial x \partial z} = 0. \tag{5}$$

The solutions decaying away from both sides of the interface exist only for the so-called TM waves. The magnetic field of these waves is perpendicular to the plane containing the normal to the surface (z-axis) and the propagation direction, i.e., $\boldsymbol{H} \| \mathrm{y}$ and $\boldsymbol{E}$ lies in the xz-plane. We shall write the solution for $\boldsymbol{E}$ in the form:

$$\boldsymbol{E} = \boldsymbol{E}^{(1)} \exp(\mathrm{i}q_x x - \kappa_1 z) \quad \text{for} \quad z > 0,$$

$$\boldsymbol{E} = \boldsymbol{E}^{(2)} \exp(\mathrm{i}q_x x + \kappa_2 z) \quad \text{for} \quad z < 0, \tag{6}$$

where κ_1 and κ_2 are real and positive.

$$\kappa_i^2 = q_x^2 - (\omega^2/c^2)\, \epsilon_i \qquad i = 1, 2. \tag{7}$$

The eqs. (7) for "decay constants" κ_i follow, for instance, from the equations for the magnetic field:

$$\partial^2 H_y/\partial z^2 - (q_x^2 - \epsilon_i\omega^2/c^2) H_y = 0.$$

This equation, like eq. (5), is derived from eqs. (2) and (3). But unlike eq. (5) it contains for TM waves only z derivatives. Upon substituting eq. (6) into eq. (5), we get the relation between E_x and E_z:

$$\text{for } z > 0, \text{ i.e. in the medium} \quad E_z^{(1)} = (\mathrm{i}q_x/\kappa_1)\, E_x^{(1)}, \tag{8}$$

$$\text{for } z < 0, \text{ i.e. in the crystal} \quad E_z^{(2)} = -(\mathrm{i}q_x/\kappa_2)\, E_x^{(2)}. \tag{9}$$

Due to the continuity of the normal components of the displacement in Maxwell boundary condition, we have $\epsilon_1 E_{z0}^{(1)} = \epsilon_2 E_{z0}^{(2)}$, where the subscript 0 stands for the fields at $z \to 0$. Hence $(\epsilon_1/\kappa_1)\, E_{x0}^{(1)} = -(\epsilon_2/\kappa_2)\, E_{x0}^{(2)}$ and as $E_{x0}^{(1)} = E_{x0}^{(2)}$ due to the continuity of the tangential components of $\boldsymbol{E}$, we get:

$$\epsilon_2(\omega)/\kappa_2 = -\epsilon_1/\kappa_1. \tag{10}$$

Actually, eq. (10) already gives the $\omega(q_x)$ dependence. Using eqs. (7) for κ_1, we shall rewrite eq. (10) in the form:

$$q_x^2 = \frac{\omega^2}{c^2}\frac{\epsilon_1\epsilon_2(\omega)}{\epsilon_1+\epsilon_2(\omega)}. \tag{11}$$

It is the desired dispersion relation for a semi-infinite crystal with the dielectric function $\epsilon_2(\omega)$. Because κ_1, κ_2 and ϵ_1 are positive, it follows from eq. (10) that the surface waves considered can exist only in the region where the dielectric function of the crystal is negative: $\epsilon_2(\omega) < 0$.

It can be seen from eq. (11) that the surface-wave frequency for a given q value decreases with increasing ϵ_1. The dispersion dependence (11) is schematically shown in fig. 2. The frequency range where the surface modes exist is limited by the roots of the equations $\epsilon_2(\omega) = -\infty$ and $\epsilon_2(\omega) = -\epsilon_1$ (or $\epsilon_2(\omega) = -1$ in case of crystal–vacuum interface). The lower limit coincides with the bulk transverse phonon frequency ω_{TO}. The upper limit is reached at $q \gg \omega_{TO}/c$ and will be designated as ω_s. Putting $\epsilon(\omega)$ from eq. (1) equal to $-\epsilon_1$, we get for ω_s:

$$\omega_S = \left(\frac{\epsilon_0+\epsilon_1}{\epsilon_\infty+\epsilon_1}\right)^{1/2}\omega_{TO}. \tag{12}$$

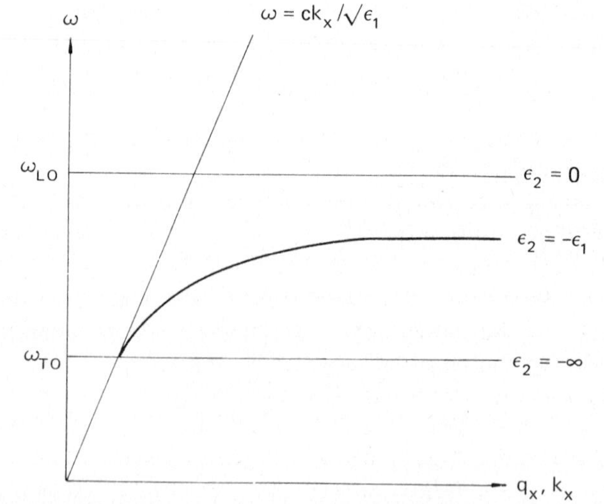

Fig. 2. The dispersion curve of the surface optical vibrations on a semi-infinite crystal bounded by a medium with dielectric constant ϵ_1. The positions of $q = 0$ frequencies of bulk optical vibrations and the corresponding values of $\epsilon_2(\omega)$ are shown as well as the "light line" $\omega = ck_x/\sqrt{\epsilon_1}$.

Equation (12) is similar to the well-known Lyddanne–Sachs–Teller relation: $\omega_{LO} = (\epsilon_0/\epsilon_\infty)^{1/2}\omega_{TO}$. With increasing ϵ_1, ω_s shifts to ω_{TO}.

The phase velocity of the surface wave is

$$v_{ph} \equiv \omega/q_x = c\frac{|\epsilon_2| - \epsilon_1}{|\epsilon_2| \times \epsilon_1}.$$

Over the whole range of surface waves existence $\omega_{TO} < \omega < \omega_s$ v_{ph} is smaller than the light velocity c and tends to it when $\omega \to \omega_{TO}(|\epsilon_2| \to \infty)$. Accordingly, in fig. 2 the dispersion curve of the surface modes runs to the right of the light line $\omega = ck_x/\sqrt{\epsilon_1}$ (which corresponds to the light propagating along the surface of the crystal) and does not cross the light line anywhere. This fact implies that the vibration modes considered are not radiative, i.e., they cannot be radiated as photons and cannot couple with light in conventional experiments on light absorption*. The presence of the crossing point is essential for the energy and momentum conservation laws in absorption and emission to be satisfied.

As was mentioned already, the electric vector of a surface wave lies in the xz-plane (the sagittal plane). In accordance with eqs. (8) and (9) there is a phase difference of $\pm\pi/2$ between E_x and E_z components of the electric field. Hence, the electric vector is rotating in the xz-plane. Inside the crystal it rotates from x to $-z$, i.e., in the clock-wise direction for the observer who sees the x-axis directed to the right. Besides, the ratio between E_z and E_x is q-dependent. When q_x is sufficiently large, E_x and E_z become equal.

The directions of $\boldsymbol{E}$, $\boldsymbol{H}$ and $\boldsymbol{S}$ (the Poynting vector), which describes the energy flow along the interface, are shown in fig. 3. The total energy flux decreases as q_x increases and becomes zero at $\epsilon_2(\omega) = -\epsilon_1$ (at the same time, the group velocity $\partial\omega/\partial q_x$, which describes the energy transfer velocity in the wave, also becomes zero).

The rate of the electromagnetic field decay normal to the interface is governed by the quantities κ_i and, in accordance with eq. (7), is different in both media and depends on q_x. For sufficiently short-wavelength vibrations $\kappa_i \approx q_x$ both in the crystal and in the surrounding medium, i.e., the region of localization is of the order of a wavelength. As $\omega \to \omega_{TO}$, $q_x \to \omega_{TO}/c$, $\epsilon(\omega) \to -\infty$ in harmonic approximation (i.e. when the damping in the crystal is neglected) and though the wavelengths in this region are comparatively long (small q), the surface polarization field in the crystal turns out to be very much "pressed" to the surface. On the contrary in the medium $z > 0$ $\kappa_1 \to 0$ as $\omega \to \omega_{TO}$, which implies an enlargement of the localization region.

*This prohibition is lifted in processes of higher orders, i.e., Raman scattering and excitation of surface waves by methods of nonlinear optics.

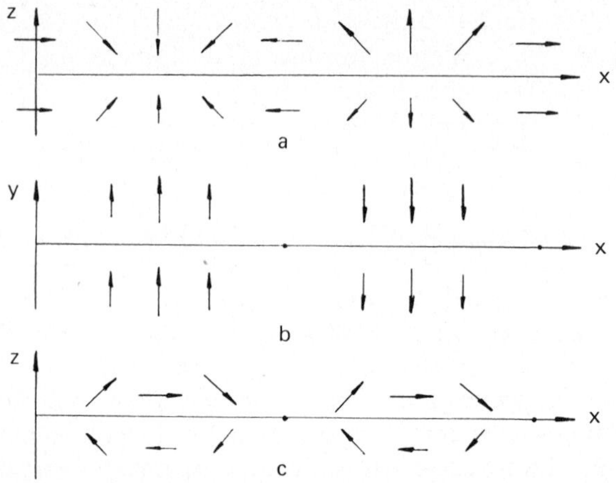

Fig. 3. (a) The electric field pattern of a surface polariton in the zx-plane. The field magnitudes are oscillatory in the x direction but decrease exponentially away from the boundary. (b) An analogous diagram of the magnetic field pattern in the xy-plane. (c) Poynting-vector pattern close to the boundary. The energy flow in medium $1(z>0)$ is greater than that in medium $2(z<0)$, giving a net flow of the energy to the right when q_x is positive (after Nkoma et al. 1974).

We have reproduced in detail the derivation of the dispersion relation and obtained the equations for the components of the electric vector in the case of a semi-infinite crystal. We consider next a slab of finite thickness, but we will confine ourselves to the analysis of the results omitting the details of the derivations. In the form presented, these results were obtained by Fuchs and Kliewer (1965) and Kliewer and Fuchs (1966a, b).

Let us direct the z-axis normal to the surface of the slab of thickness L and place the origin of the coordinate system in the centre of the slab (fig. 4). As long as the slab is sufficiently thick, the fields corresponding to the surface waves propagating along both boundaries do not interact: the two identical surface mode branches exist independently of one another. But when L becomes comparable or smaller than $1/\kappa_2$, the degeneracy is lifted producing two different branches of the surface modes. Their dispersion relations are:

for the high-frequency (ω_+) branch:

$$\epsilon_2(\omega)/\epsilon_1 = -(\kappa_2/\kappa_1)\tanh(\kappa_2 L/2), \tag{13}$$

for the low-frequency (ω_-) branch:

$$\epsilon_2(\omega)/\epsilon_1 = -(\kappa_2/\kappa_1)\coth(\kappa_2 L/2). \tag{14}$$

For $\kappa_2 L \gg 1$ eqs. (13) and (14) reduce to the already known eq. (10) describing the surface vibration of a semi-infinite crystal. Thus, the vibration frequencies in the slab of finite thickness depend not only on q but, at

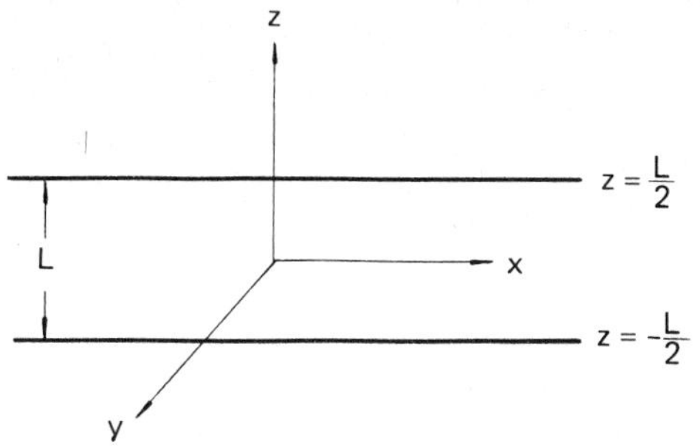

Fig. 4. Slab geometry and the coordinate system used.

a given q, on the value $\kappa_2 L$. The dispersion curves of both branches calculated from eqs. (13) and (14) for LiF slabs of different thicknesses are shown in fig. 5. The frequencies ω_+ and ω_- are plotted versus the dimensionless wave vector $q_x/q_T \equiv q_x c/\omega_{TO}$. The parameter in fig. 5 which governs the frequency at given q_x/q_T is $q_x L$. For $L > 10$ microns both branches merge into one. The surface mode branches calculated neglecting the retardation are also shown in fig. 5 (dashed lines). It can be seen that the high-frequency branch ω_+ is particularly strongly affected by retardation.

The coordinate dependence of the electric field (and the dipole moment as well) within the slab is for ω_+ branch:

$$E_x \sim \sinh(\kappa_2 z)\exp(iq_x x); \quad E_z \sim -\frac{iq_x}{\kappa_2}\cosh(\kappa_2 z)\exp(iq_x x), \tag{15}$$

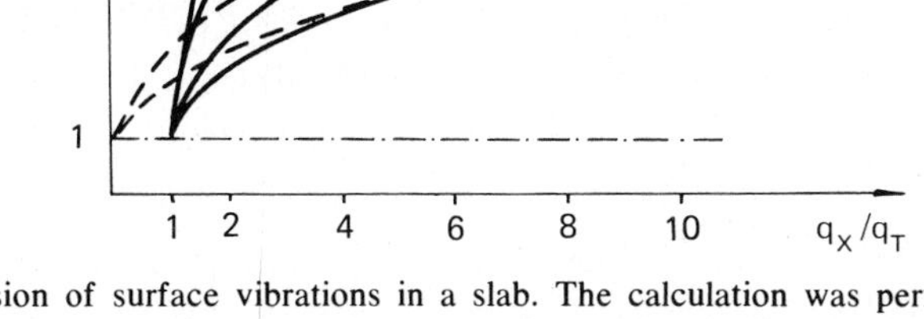

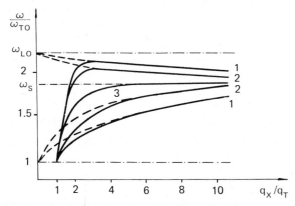

Fig. 5. Dispersion of surface vibrations in a slab. The calculation was performed for LiF crystal slabs of various thicknesses L bounded by a vacuum. Curves: (1) $q_T L = 0.1$ ($L \approx 0.5\ \mu$m); (2) $q_T L = 0.2$ ($L \approx 1\ \mu$m); (3) $q_T L > 2$ ($L > 10\ \mu$m). The solid curves represent the calculation with the account of retardation, the dashed curves that without retardation. Parameters used in the calculation were: $\omega_{TO} = 5.78 \times 10^{13}\ \text{cm}^{-1}$, $\epsilon_0 = 9.27$, $\epsilon_\infty = 1.92$ (after Bryksin et al. 1974).

for ω_- branch:

$$E_x \sim \cosh(\kappa_2 z)\exp(iq_x x); \quad E_z \sim \frac{-iq_x}{\kappa_2}\sinh(\kappa_2 z)\exp(iq_x x). \tag{16}$$

The field decreases exponentially outside the slab:

$$E_x \sim \exp(iq_x x \pm \kappa_1 z), \tag{17}$$

$$E_z \sim \frac{iq_x}{\kappa_1}\exp(iq_x x \pm \kappa_1 z), \tag{18}$$

where the signs "+" and "−" correspond to the regions $z < -L/2$ and $z > L/2$, respectively.

The electric vector of both branches lies in the xz-plane, as well as in the case of a semi-infinite crystal, and its components E_x and E_z have a $\pi/2$ phase difference. The vibrations in the high-frequency ω_+ branch, in accordance with eq. (15), are polarized mainly along z (for $\kappa_2 L \ll 1$ $E_x \approx 0$, $E_z \approx$ const). The vibrations in the ω_- branch, in accordance with eq. (16), are polarized mainly along x (for $\kappa_2 L \ll 1$ $E_x \approx$ const, $E_z \approx 0$)*.

Figure 6, taken from a paper by Kliever and Fuchs (1966a) shows the regions of the ω–q_x plane where different types of solutions for the slab exist. The nonradiative region L lies to the right of the line $\omega = cq_x/\sqrt{\epsilon_1}$. In this region $\kappa_1 = [q_x^2 - (\omega^2/c^2)\,\epsilon_1]$ is real, which implies that E is decaying outside the slab. The dispersion curves of the surface modes under investigation lie in the region L_2, in accordance with eqs. (11), (13) and (14). In the regions L_1 and L_1' lie the solutions related to the electromagnetic waves confined within the slab** (guided modes) rather than to the waves localized near the surface. Below, we will see the manifestation of these waves in the experiments on surface vibrations. The region N contains no solutions.

In the radiative region R(R_1 and R_2) $q_x < (\omega/c)\sqrt{\epsilon_1}$ and κ_1 is imaginary. This corresponds to the electromagnetic wave propagating from the slab. Such vibrations radiate the electromagnetic energy into the surrounding space and in this sense they are unstable, they were called virtual by Kliever and Fuchs (1966b). These vibrations are directly coupled to the light in conventional experiments on light absorption. The optical properties of the ionic crystal slab are completely described in terms of these vibrations (Fuchs et al. 1966).

The generalization of the results presented above can easily be made for

*This statement about polarization becomes invalid for q_x so large that $\kappa_2 L \approx q_x L$ cannot be $\ll 1$. It is also invalid for $q_x \to q_T$ when κ_2 increases due to increased ϵ_2 and $\kappa_2 L$ is again not small.

**This is due to the fact that their angle of incidence from within the crystal upon the boundary crystal–medium exceeds the angle of total internal reflection.

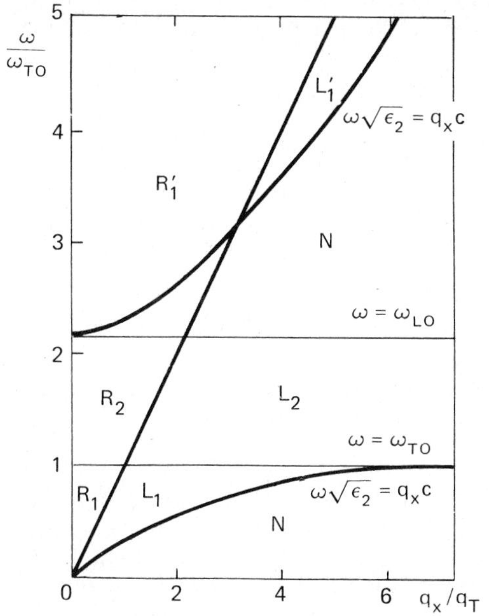

Fig. 6. ω–q_x diagram for an ionic crystal slab. The radiative modes lie in regions R, the nonradiative in regions L. Surface waves lie in region L_2. Regions N have no solutions (after Kliewer and Fuchs 1966a).

cubic crystals with more than two atoms in the unit cell. In this case the number of infrared-active vibrations can be two, three . . . and so on, and the number of poles in the equation for $\epsilon(\omega)$ increases respectively: ω_{TO1}, ω_{TO2}, ω_{TO3} . . .

$$\epsilon(\omega) = \epsilon_\infty + \sum_i \frac{S_i \omega_{TOi}^2}{\omega_{TOi}^2 - \omega^2}, \tag{19}$$

where S_i is the respective "oscillator strength". The number of surface polariton branches also increases. The maximum frequency for each branch of a semi-infinite crystal can be determined similarly to eq. (12) as the intersection of the $\epsilon_2(\omega)$ curve with the value $-\epsilon_1$ (or in the case of the crystal–vacuum boundary with the value -1). As an example one can mention the cubic crystal $SrTiO_3$ which has three infrared-active oscillators in the vibration spectrum and three surface polariton branches respectively.

So far we have not considered the phonon damping in the crystal and have consequently assumed $\epsilon_2(\omega)$ to be real. Actually the phonons have a finite lifetime determined mainly by anharmonic processes. Therefore, the dielectric function has an imaginary part. The specific form of $\epsilon(\omega)$ depends on the model used. It may include the frequency independent damping constant Γ, or take into account its frequency dependence $\Gamma(\omega)$, for instance, in the form:

$$\epsilon(\omega)=\epsilon_\infty+\frac{(\epsilon_0-\epsilon_\infty)\,\omega_{TO}^2}{\omega_{TO}^2-\omega^2-i\Gamma(\omega)\omega}. \tag{20}$$

The consistent account of anharmonics in the dynamics of the crystal lattice results in a complex value of damping in the expression for $\epsilon(\omega)$ (Maradudin and Wallis 1961 and 1962, Vinogradov 1962, Cowley 1963, Hisano et al. 1972, Placido and Hisano 1973). Two real functions introduced in so doing represent the anharmonic damping $\Gamma(\omega)$ and the anharmonic shift $\Delta(\omega)$ of the phonon frequency. Taking $\Gamma(\omega)$ and $\Delta(\omega)$ into account, one can write the dielectric function of a crystal with a single infrared-active phonon in the form:

$$\epsilon(\omega)=\epsilon_\infty+(\epsilon_0-\epsilon_\infty)\frac{\omega_0^2+2\Delta(\omega)\omega_0}{\omega_0^2-\omega^2+2[\Delta(\omega)+i\Gamma(\omega)]\omega_0}, \tag{21}$$

where ω_0 is the harmonic TO phonon frequency. These equations for $\epsilon(\omega)$ will be of use later for the analysis of the experimental results.

The surface, as well as the bulk polaritons are mixed photon-phonon excitations. Therefore, their damping depends on the partial magnitude of the phonon contribution $\Gamma(\omega)$ and in general differs from $\Gamma(\omega)$ (Nkoma et al. 1974). For instance, even for $\Gamma(\omega)=\text{const}\equiv\Gamma$ in eq. (20) the surface polariton damping is frequency dependent: it increases from the minimum value at $\omega\approx\omega_{TO}$,

$$\Gamma\frac{\epsilon_1}{\epsilon_1+\epsilon_0-\epsilon_\infty}<\Gamma$$

to the maximum value coinciding with the phonon damping Γ at $\omega\to\omega_s$ (Nkoma et al. 1974).

The dispersion curve for the surface polaritons, ranging from ω_{TO} to ω_s, crosses the bulk vibration spectrum of the crystal. This is shown schematically in fig. 7 which represents the dispersion curve of the surface polaritons $\omega(q_x)$ together with the continuous spectrum of the bulk modes. For these modes $q_x\approx 0$ but q_z changes over the entire Brillouin zone. Due to the crossing of the surface branch with the bulk spectrum the surface modes become in general nonstationary even in a harmonic approximation and acquire a finite lifetime. This explains the appearance of the term: quasi-surface modes. A similar situation is known in the dynamics of the lattice with impurities. The impurity levels within the lattice continuous spectrum are called quasi-local (in contrast to the local vibrations whose levels lie outside the continuous spectrum). The calculation made by Bryksin and Firsov (1972) (see also Bryksin et al. 1974) has shown that a noticeable harmonic broadening of the surface modes can be expected only

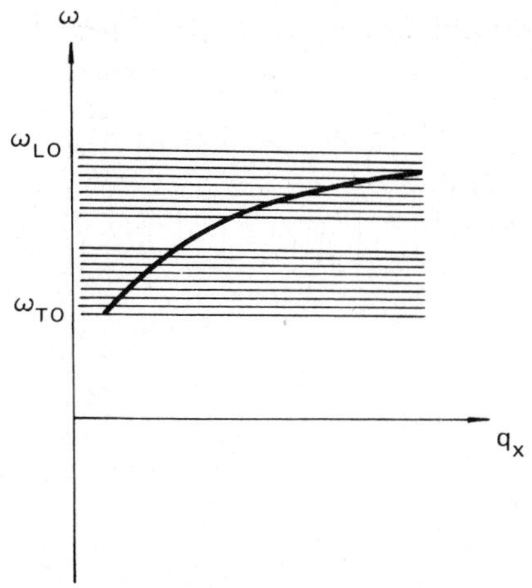

Fig. 7. The crossing of surface and bulk spectra. The heavy solid line is the dispersion curve of surface vibrations. The horizontal lines represent the energy levels of the bulk LO and TO phonons with $q_x \approx 0$ but various q_z. In the crossover area the surface mode changes into the "quasi-surface" one.

near the frequencies corresponding to the Van Hove singularities in the bulk spectrum. The broadening of the related surface modes has turned out to be of the order of $(q_x a)^{2/3}\omega_{TO}$ (where a is the lattice constant). In the infrared region, i.e. for $q_x \sim \omega_{TO}/c$, this amounts to a value of the order of $(10^{-2} + 10^{-3})\omega_{TO}$. For some crystals such broadening can be comparable to the anharmonic broadening*. The indicated effects may prove to be more significant at $q_x \gg \omega_{TO}/c$. However, the discussion of these questions connected with the regard for the spatial dispersion is beyond the scope of this chapter.

3. *Experimental Methods for the Excitation and Study of Surface Polaritons*

This section is a short review of the methods used for the excitation and experimental investigation of surface polaritons. The primary attention is focussed on the method of attenuated total reflection (ATR) by which the experimental results discussed in the following sections were obtained.

*It should be noted that special experiments in the infrared on KBr and MgO crystals have shown that the harmonic broadening in these crystals is, even at low temperatures, at least by one order of magnitude less than the anharmonic broadening (Gerbstein 1973). In this sense the phenomenological theory, which does not take into account the indicated effect, is a good approximation. It should be mentioned, however, that in KBr and MgO the anharmonic damping is comparatively large.

3.1. Methods of Electron Spectroscopy

Surface phonons were first observed experimentally more than a decade ago, when the energy losses of fast electrons traversing a LiF film were studied (Boersch et al. 1968). The film thickness was about 300 Å. For electrons with the energy of 25 keV the resolution of 0.012 eV was achieved. This technique permitted to detect in the spectrum of energy losses only a broad maximum which covered the entire q range. Later progress in monochromatizing low-energy electrons made it possible to use low-energy electron diffraction (LEED) for the investigation of surface phonons. By this method the energy spectrum of reflected electrons was studied and therefore not only thin films, but also massive crystals could be investigated. The first successful experiment of this kind was performed on ZnO crystal (Ibach 1970) and gave the energy spectrum of slow electrons (7.5 eV) reflected from different faces of the crystal. The experimental curves distinctly showed the zero-phonon reflection peak and the reflection peaks with the emission of one, two and three phonons and the absorption of one phonon (fig. 8). As was expected, the intensity of the energy-gain peak decreased with decreasing temperature. The measured frequencies of one-phonon peaks due to the reflection from different faces were 68.8 and 67.5 meV, thus a weak anisotropy could be observed. These frequencies within the experimental errors coincide with the calculated ω_s values. In his next work Ibach (1971) observed not only a surface vibration on an atomic-clean surface but also local vibrations due to the adsorption of impurity atoms.

However, the resolution of electron spectroscopy methods is comparatively low. For instance, it was reported by Ibach (1970) to be about 20 meV, i.e., approximately equal to the total dispersion range of the

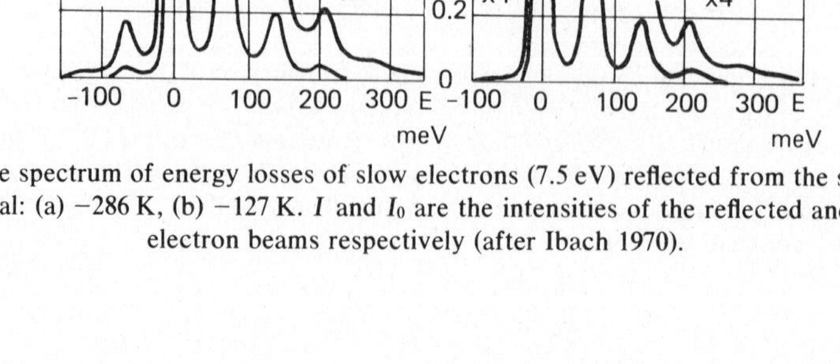

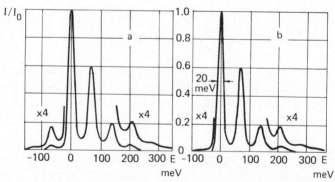

Fig. 8. The spectrum of energy losses of slow electrons (7.5 eV) reflected from the surface of ZnO crystal: (a) −286 K, (b) −127 K. I and I_0 are the intensities of the reflected and incident electron beams respectively (after Ibach 1970).

surface phonons. This restricts the possibilities of the technique in the measurement of dispersion, damping, etc. Though there was some progress in the LEED technique in later years, the basic success was achieved in the investigation of microscopic surface states due to the disruption of chemical bonds on the surface and to the adsorption of atoms and molecules (see, e.g. Ibach 1977, Ibach and Lehwald 1978).

3.2. Optical Methods

It was already pointed out that surface modes do not couple directly to the electromagnetic wave due to the fact that it is not possible in the frequency range of their dispersion to satisfy simultaneously the energy and momentum conservation laws

$$\hbar\omega = \hbar\Omega, \tag{22}$$

$$q_x = k_x, \tag{23}$$

where Ω is the frequency and k_x is the x-component of the wave vector of the incident photon. In fact, the frequency and the wave vector are related by the equation

$$\Omega = (c/\sqrt{\epsilon_1})(k_x^2 + k_z^2)^{1/2}, \tag{24}$$

where k_y is taken to be zero and ϵ_1 is the dielectric constant of the surrounding medium. Using eqs. (22) and (23) we can rewrite eq. (24) as

$$\omega = (c/\sqrt{\epsilon_1})(q_x^2 + k_z^2)^{1/2}. \tag{25}$$

But because the nonradiative surface modes obey the inequality (see fig. 2):

$$\omega < (c/\sqrt{\epsilon_1})\, q_x, \tag{26}$$

eqs. (25) and (26) cannot be valid simultaneously for $k_z^2 > 0$, i.e. for real k_z. Note that the light line in fig. 2 corresponds to the extreme case, when the electromagnetic wave is propagating along the interface ($k_z = 0$). For any other incidence angle of light upon the crystal ($k_z \neq 0$) the light line is more inclined to the ordinate axis. However, very essential for the future discussion is the fact that eqs. (25) and (26) can be satisfied for $k_z^2 < 0$. This inequality corresponds to imaginary k_z, i.e., to the electromagnetic wave decaying along the z-axis.

One of the comparatively old ways (see e.g. Beaglehole 1969) to excite surface plasmons in metals consists in inscribing on the surface a periodic line grating. Then the momentum conservation law in the coupling of light to the surface polaritons holds with an uncertainty $n(2\pi/t)$, where t is the groove spacing and n is an integer. In this case instead of eq. (23) we can write:

$$q_x = k_x + n2\pi/t. \tag{27}$$

As indicated above, k_x is the x-component of the incident photon momentum;

$$k_x = (\omega/c)\sqrt{\epsilon_1}\sin\varphi$$

where φ is the incidence angle of light upon the crystal.

For a given value of sin φ and t the excitation of the surface polariton of a certain frequency ω becomes possible. Accordingly, a series of dips appear in the frequency dependence of reflection coefficient corresponding to $n = 1, 2, 3\ldots$ The dispersion curve can be plotted in a certain range of q_x from the positions of dips for various φ and t values. In metals such a procedure gives only a limited section of the dispersion curve because t is usually large compared to the wavelengths of surface plasmons lying in the ultraviolet spectral region. In the case of surface plasmons in semiconductors it is possible to cover a relatively greater q_x range. By this procedure the dispersion curve of surface plasmons in n-InSb was obtained by Marschal et al. (1971), see also (Fischer et al. 1973). They used several inscribed line gratings with $t = 10$, 20 and 30 μm. However, the grating method has some disadvantages, which apparently were the cause of its limited application, in particular, for the investigation of surface phonons. The inscribing of a grating may destroy the surface layer and result in an uncontrolled change in the frequencies and damping of the surface polaritons. So, Fischer et al. (1973) ascribed the observed discrepancy between the surface plasmon frequencies obtained by the ATR method and by the grating method to the appearance of a depleted layer in the latter case. Another disadvantage of the grating technique is that, in general, the positions of the reflection minima do not coincide with the surface polariton frequencies and in order to evaluate the frequencies one has to know the exact profile of the groove. The ATR method is free from both disadvantages; the positions of the reflection minima practically coincide with the surface polariton frequencies, when certain conditions are fulfilled.

As was stated before, the energy and momentum conservation laws in the coupling of surface phonons to light, on the one hand, and the shape of the dispersion curve, on the other, do not contradict each other only if the light wave has an imaginary k_z, i.e., if it decays along the normal to the surface. Such a nonuniform wave arises on the interface of two media of different optical densities at the total internal reflection. This wave propagates in the plane of incidence along the interface and decays exponentially in the medium with lower density at a distance of the order of a wavelength. In the propagation direction along the interface the phase velocity of the transmitted wave is $v_{ph} = cn_{12}/\sin\varphi$ where φ is the angle of incidence upon the interface and $n_{12} = n_2/n_1 < 1$ is the relative refractive index (see, e.g. Born and Wolf 1964). On the interface of medium (1) with a higher density and vacuum (2) $n_{12} = 1/n_1$ and v_{ph} changes from c to c/n_1, as the angle of incidence changes from the critical value arcsin $1/n_1$ to $\pi/2$.

These considerations form the basis of the ATR method introduced by Otto (1968) to study surface plasmons in metals (see also Otto 1974a, b, 1975) and used subsequently for the investigation of surface phonons (Bryksin et al. 1971, 1974). Figure 9 gives the experimental setup of the ATR method. The reflection coefficient is measured of an electromagnetic wave impinging upon the interface of the prism base and the gap layer between the prism and the crystal at the angle of total internal reflection. The electromagnetic wave which propagates in the gap between the prism and the crystal has the "diminished" phase velocity $c/n \sin\varphi$. The dispersion curve of this wave is shown on the ω–q_x diagram in fig. 10 by the dashed line which crosses the dispersion curve of surface polaritons, and therefore ensures the fulfillment of the conservation laws. In this case the total internal reflection becomes frustrated, the absorption of light occurs and the reflection coefficient R becomes less than unity. The positions of the minima appearing in the reflection spectrum $R(\omega)$ correspond to the frequencies of surface polaritons $\omega(q_x)$ if certain conditions are satisfied (see below). Here $q_x = k_x$, where k_x is the projection of the incident photon momentum on the propagation direction x: $k_x = (\omega/c)n \sin\varphi$ (n is the refractive index of the prism).

The absorption of light in this situation is due to two mechanisms. The radiative broadening (the perturbation caused by the prism) makes possible the excitation of nonradiative modes by light*. The second broadening

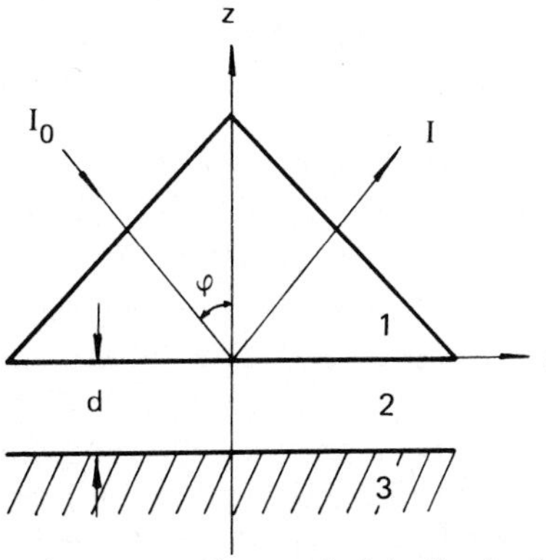

Fig. 9. Geometry of the experiment on surface mode detection by the ATR method: (1) a prism (e.g. silicon); (2) a gap (vacuum or a nonabsorbing dielectric); (3) a crystal under investigation. $R(\omega) = I(\omega)/I_0(\omega)$.

*The opposite process is made possible by the same mechanisms: the emission of nonradiative surface modes through the prism in the form of photons.

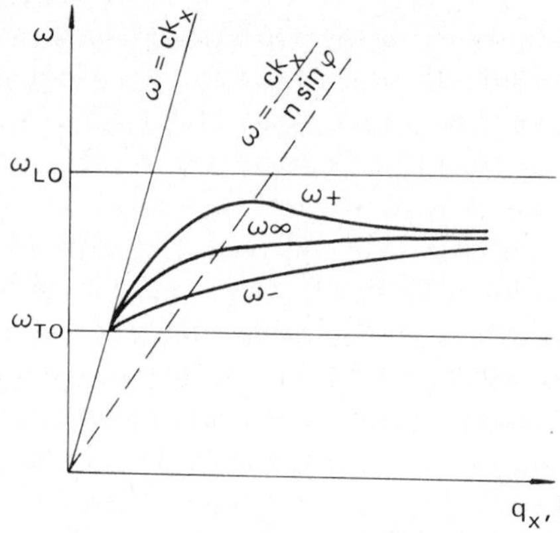

Fig. 10. Dispersion of surface vibrations in a thin slab (ω_+ and ω_-) and in a semi-infinite crystal (ω_∞). The dashed line corresponds to the light wave in the ATR method. Light is absorbed at frequencies where this line intersects the dispersion curves of the surface modes.

process – an anharmonic broadening is responsible for the dissipation of the absorbed energy. The value of the radiative broadening and the amplitude of the minimum in the reflection spectrum are proportional to $\exp(-\kappa_2 d)$, i.e., they decrease exponentially as the gap thickness d increases. For small gaps the coupling to the prism results in the frequency shift of the observed minimum downwards to ω_{TO}. For $d \to 0$ the range where the electromagnetic wave exponentially decaying in the z direction can exist completely disappears and the surface waves become unobservable. They are, as before, nonradiative modes, but on the interface with a medium of the prism having the dielectric constant ϵ_1.

The computations of ATR spectra of surface modes were performed and the results published in a number of papers (see e.g., Bryksin et al. 1972a, Otto 1974b). In these computations the dielectric function was assumed to be complex: $\epsilon(\omega) = \epsilon'(\omega) + i\epsilon''(\omega)$, i.e. the phonon damping mainly due to the anharmonic interaction was taken into account. The account of the damping is absolutely necessary because the absorption itself is conditioned by it and the depth of the minimum in the ATR spectrum is proportional to ϵ''. Damping is responsible also for some other effects: the anharmonic shift of the frequencies and the appearance of additional minima in ATR spectra. A more detailed discussion of anharmonic effects will be given in the sections that follow, where the experimental results for some crystals are presented.

It should be noted that the method for the study of surface waves which was considered here differs from the conventional ATR technique (see, for instance, Harrick 1967) only in the presence of a gap between the prism and the crystal. It was pointed out, above, that this gap is of absolute necessity in the investigation of nonradiative modes. The gap must be made sufficiently large to minimize the perturbation caused by the prism (in this case the position of the minimum is not affected by the further increase in d). However, for the reliable experimental investigation the minimum in the ATR spectrum must be still well pronounced. The optimal values of d are chosen experimentally and, as the calculation and the experiment show, are different for different φ values, i.e., for different q_x. Locating the positions of minima in ATR spectra for various q_x (using sufficiently large gaps), it is possible to find the dispersion curve $\omega_s(q_x)$. In the first work by Otto (1968), carried out in the visual spectral region, the change of q_x was achieved by changing φ for a fixed frequency. But later in the experiments making use of infrared spectrometers, another procedure appeared to be more suitable: a frequency scan at a fixed angle of incidence. It is this procedure that we will mean in further discussion, unless otherwise specified*.

Different crystals transparent in the infrared were used as the prism material (Si, CaF_2, etc). For a more accurate determination of the incidence angle (i.e. q_x) the provision was made in the experiment for the restriction of the angular divergence of the light beam. The divergence in the prism usually did not exceed 1°. Ideal in this sense would be the excitation of surface waves with lasers. The development of the technique of tunable lasers in infrared will greatly simplify the experimental procedure.

In the last few years other optical methods for the generation and investigation of surface phonons have come into use. Contrary to the ATR method, where the allowed q_x range has the value $(\omega_s/c)n$ as the upper limit, Raman scattering from surface phonons (Evans et al. 1973) is possible in a more extended range of wave vectors. Unfortunately, this technique is apparently limited to the investigation of thin films and therefore is not widely used. In processes of higher order than the conventional light absorption the conservation laws in the coupling of light to the surface modes are more easily satisfied. DeMartini and Shen (1976) proposed for the excitation of surface modes the methods of nonlinear optics. In the research by DeMartini et al. (1976) the dispersion curve of surface phonons in GaP has been obtained by difference frequency generation in optical mixing. Here we have restricted ourselves only to listing these methods, for there are special chapters in this book devoted to them.

*The dispersion curves obtained by this procedure resemble those shown in fig. 2. But if the ATR spectrum is angle-scanned at fixed frequency, then in the range of sufficiently large q_x even the shape of experimentally obtained dependence of ω_{min} on q_x (where ω_{min} is the position of the ATR minimum) is changed due to the anharmonisity. At greater length this problem will be discussed in the next section.

4. *Experimental Studies of Surface Polaritons in Cubic Crystals – Anharmonic Effects*

The first observation of surface phonons by means of the ATR method was made by Bryksin et al. (1971) in NaCl crystals. In the ATR geometry, described in the previous section, with a silicon prism and for $\varphi = 45°$ a well pronounced minimum was observed in the spectrum between ω_{TO} ($164\,cm^{-1}$) and ω_{LO} ($255\,cm^{-1}$). The frequency of this minimum was dependent on the gap thickness: the extreme high-frequency value with a vacuum gap was $214\,cm^{-1}$*. This value shifted to $197\,cm^{-1}$ in accordance with eq. (11) when the gap between the prism and the sample was filled up with paraffin ($\epsilon = 1.96$).

The absorption was observed only for p-polarized light, i.e. for the electric vector $\boldsymbol{E}$ lying in the plane of incidence in conformity with the polarization of surface polaritons (see sect. 2). The absorption disappeared for s-polarized light ($\boldsymbol{E}$ normal to the plane of incidence). When the crystal thickness was decreased to 3 μm the absorption minimum split into two. Apparently, this indicated the appearance of two branches of surface vibration: ω_+ and ω_- (see eqs. (13) and (14)).

The measured frequencies were lower than the calculated ones approximately by $20\,cm^{-1}$ (the cause of this discrepancy cleared up later). However, all of the obtained results stated unambiguously that the observed minima were connected with the absorption of infrared radiation by surface vibrations in the NaCl slab. A more thorough investigation of surface phonons in NaCl and other alkali halide crystals and in CaF_2 was reported by Bryksin et al. (1972a, b). The dispersion curves were plotted from the positions of minima in ATR spectra (fig. 11, curve 1). It was found that the above mentioned discrepancy between the experimental values of frequencies and the ones calculated from eq. (11) was reduced by two, when the temperature was decreased to 80 K. This indicated a noticeable influence of anharmonicity on the frequencies of surface vibrations. Bryksin et al. (1971, 1972a, b) have taken account of the anharmonic effects by introducing a second pole in $\epsilon(\omega)$ at $247\,cm^{-1}$. This resulted in a much better agreement between the experiment and the calculation (see fig. 11). A still better agreement between experimental results of Bryksin et al. (1972a) and the calculated values was achieved by Barker (1974), who used a three-pole approximation for $\epsilon(\omega)$ in NaCl.

The use of two- and three-pole approximations for $\epsilon(\omega)$ is of course purely a phenomenological approach. It was already noted in sect. 2 that a consistent account of anharmonicity in the lattice dynamics is manifested

*Although the units of ω in all equations are rad/s we shall express all frequencies in wave numbers (cm^{-1}).

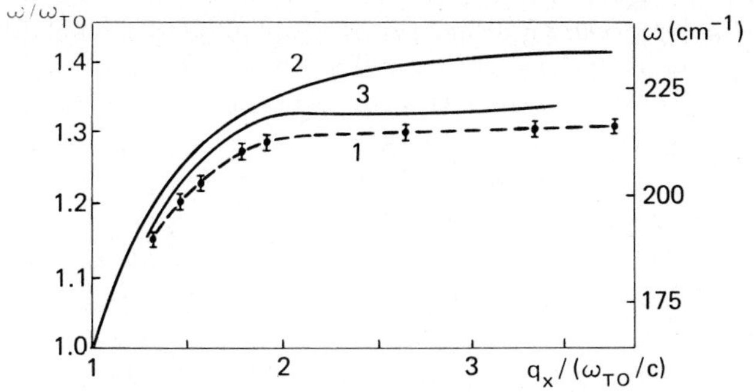

Fig. 11. Dispersion of surface vibrations on a NaCl crystal; (1) experiment, (2) calculation in harmonical approximation, (3) calculation with the account of anharmonism (after Bryksin et al. (1972a).

in the appearance of complex damping in $\epsilon(\omega)$. The real part of this damping corresponds to the anharmonic phonon damping $\Gamma(\omega)$ and the imaginary one to the anharmonic shift $\Delta(\omega)$. For a NaCl crystal these functions were evaluated by Hisano et al. (1972). Placido and Hisano (1973) used them to plot the dispersion curves of surface phonons. They compared these dispersion curves with the experimental results of Bryksin et al. (1972a, b) and found a good agreement (see fig. 12).

Bryksin et al. (1972c) also carried out a careful investigation of surface phonons in a thin polycrystalline NaCl film. The film evaporated on a polyethylene substrate was pressed to the base of an ATR prism that had been coated with paraffine. The paraffine layer played the role of the gap*.

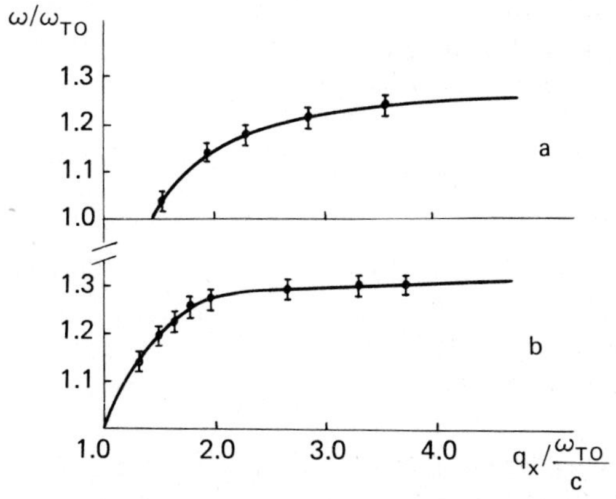

Fig. 12. Dispersion of surface vibrations on a NaCl crystal. Solid curves are the calculations by Placido and Hisano (1973), the points denote the experimental results of Bryksin et al. (1972a): (a) paraffin gap; (b) air gap.

*Polyethylene and paraffiin do not absorb in the region of optical lattice vibrations of NaCl.

In the ATR spectra shown in fig. 13 for three angles of incidence the most intense minima correspond to two surface mode branches: ω_+ eq. (13) and ω_- eq. (14). This follows, for instance, from the fact that the magnitude of the splitting, i.e., the frequency range between the minima, depended on the film thickness, increasing as it decreased. Moreover, the minima in ATR spectrum were observed only for p-polarized infrared radiation, in full agreement with the polarization of the surface modes. The dispersion curves plotted from the positions of minima in the ATR spectra at various angles of incidence are compared with the calculated dispersion curves in fig. 14. The discrepancy between the calculated and the experimental values lies within (5–10) cm^{-1}. This value is somewhat larger than in the case of a "semi-infinite" NaCl crystal when the same form of $\epsilon(\omega)$ was used in the calculation. Probably this increased discrepancy was associated with the poor quality of the polycrystalline film.

On the experimental curves in fig. 13 a feature (a dip) can be seen at a frequency close to ω_{TO}. This additional dip is due to the coupling of light to the bulk modes of polariton type lying in the nonradiative region (region L_1 on the ω–q_x diagram in fig. 6). The dip results from the capture of electromagnetic wave in the film: because of the total internal reflection on the boundaries of the NaCl slab with paraffin and polyethylene this wave

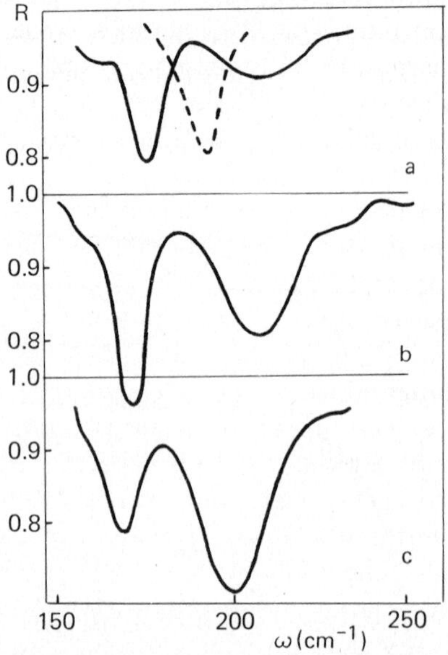

Fig. 13. ATR spectra of a NaCl film 2 μm thick at various angles of incidence: (a) 56°5; (b) 36°5; (c) 31°5. The dashed line shows the ATR spectrum for a thick "semi-infinite" crystal. (After Bryksin et al. 1972c.)

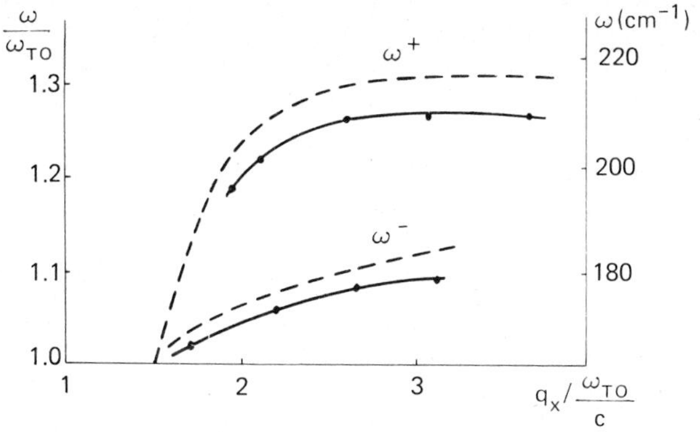

Fig. 14. Dispersion of surface vibrations in a NaCl film 2 μm thick. The solid curves are drawn through the experimental points, the dashed curves are calculated (after Bryksin et al. 1972c).

propagates along the slab, undergoing manyfold reflections from the slab boundaries. Such guided wave polaritons have recently been studied by Valdez et al. (1978) in thin GaP slabs by means of Raman scattering.

Surface phonons were observed in a number of simple cubic crystals. A good agreement between theory and experiment has been obtained for many crystals: GaP(Marshall and Fischer 1972), CaF_2, CdF_2 (Bryksin et al. 1972a, b). In the mentioned cases the surface mode frequencies calculated in the harmonic approximation are already close to the experimental values, i.e., the anharmonic corrections are small. On the contrary, for NaF, as noted by Fischer et al. (1973) in accordance with the results of Bryksin et al. (1972a), it was necessary to take into account the anharmonic effects.

A number of polyatomic cubic crystals with several infrared-active phonons have also been studied. It was already mentioned in sect. 2 that such crystals have several branches of surface vibrations. Yakovlev and Zhizhin (1974) measured the dispersion curves of three surface phonon branches in yttrium-iron-garnet crystals, $Y_3Fe_5O_{12}$ (see fig. 15). The frequencies of the longitudinal and transverse vibrations were found from the frequency dependence of the dielectric function $\epsilon(\omega)$ evaluated from the reflection spectra by Kramers–Kronig procedure. The surface polaritons in $SrTiO_3$ and $KTaO_3$ crystals were observed by Fischer et al. (1974). These crystals have three infrared-active vibrations, and, accordingly, three surface phonon branches. The dielectric function $\epsilon(\omega)$ of these crystals were also evaluated by Kramers–Kronig procedure (Fischer et al. 1974). For $SrTiO_3$ a sufficiently good agreement with the experiment was obtained when $\epsilon(\omega)$ in eq. (11) was replaced by its real part $\epsilon'(\omega)$. Such a procedure is the first approximation in taking account of anharmonic effects and it was often found wholly satisfactory. However, for $KTaO_3$ it was

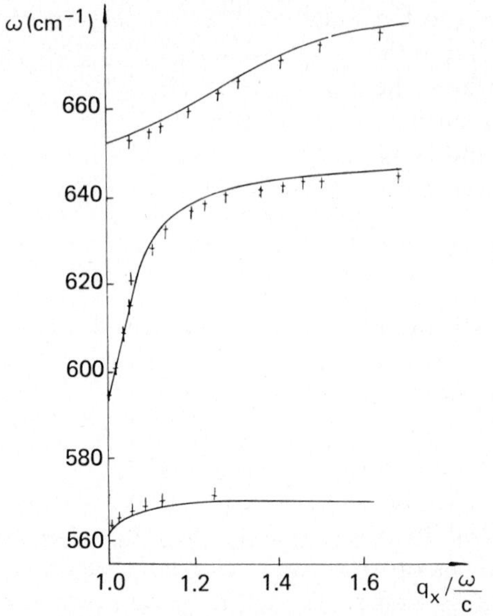

Fig. 15. Dispersion of surface phonons in yttrium-iron-garnet ($Y_3Fe_5O_{12}$) in the range 560–680 cm^{-1}. Solid curves are calculated without account of damping (after Yakovlev and Zhizhin 1974).

insufficient and the exact calculation of the ATR spectra proved to be necessary for the comparison with the experiment. In $KTaO_3$ the anharmonic effects are more pronounced and, as in the case of alkali-halide crystals, responsible for the additional structure in the spectra due to the combination frequencies. The additional poles introduced in $\epsilon(\omega)$ do reflect the account of these frequencies.

The surface vibrations in an organic crystal (the urotropin cubic crystal) were first observed by Zhizhin et al. (1975a). They obtained the dispersion curves of four surface phonon branches. Three of them fitted the curves calculated in harmonic approximation.

In conclusion of this incomplete review on surface phonon dispersion in cubic crystals it may be said that experimental and theoretical dispersion curves are in fairly good agreement if sufficiently full account of anharmonic effect has been taken.

The experimental results concerning the surface phonon damping are not so conclusive. Gammon and Palik (1974), analyzing the results of linewidth measurements by the ATR method for GaAs and GaP crystals, have shown that the experimental values can be adequately described when the bulk phonon damping constant is taken into account in the calculations. We will draw a similar conclusion from the analysis of the experimental results on MgO crystals (see below). It is felt that the additional line broadening observed sometimes in the ATR spectra, which does not correlate with the bulk damping, is due to the surface damage during the polishing procedure. One can recollect here the analogous line broadening in Raman spectra of bulk phonons in a surface layer of GaAs due to the surface damage during polishing (Evans and Ushioda 1972, 1974).

Equation (11), written for the crystal–vacuum boundary

$$q_x^2 = \frac{\omega^2}{c^2}\frac{\epsilon(\omega)}{\epsilon(\omega)+1}, \tag{28}$$

gives unambiguously the dispersion relation of surface vibration $q_x(\omega)$ or $\omega(q_x)$ as long as $\epsilon(\omega)$ is real. In general, when damping is taken into account and the imaginary part ϵ'' of the dielectric function $\epsilon(\omega) = \epsilon'(\omega) + i\epsilon''(\omega)$ is not zero, eq. (28) involves complex values of q and ω.

Here the question arises about the concept of the dispersion dependence in general. Above, it was actually presumed that the dispersion dependence is determined by the relationship between the positions of ATR minima and the angles of incidence (i.e. q_x) of the excitation light, provided the gap thickness is sufficiently large. However, when spectra in the presence of damping are measured (in IR absorption, Raman scattering, etc.) the results are always in a certain degree influenced by the way the measurements are performed*. So, it also appeared that the dispersion curves evaluated from the ATR spectra, even when measured with extremely large gaps, still depended on the way these spectra were obtained. ATR spectra presented above were frequency-scanned at a fixed angle of incidence (i.e. q_x). In other words, the frequency dependence of the reflectivity $R(\omega)$ was obtained. The results of such measurements can be adequately described, assuming q_x to be real. Then the real part of the complex frequency (obtained from eq. (28)) determines the position of the minimum and the imaginary part – its width (Gammon and Palik 1974). In another experimental arrangement ATR spectra were angle-scanned at fixed frequency (i.e. the dependence $R(\varphi)$ was obtained). It appeared that in this case the experimental points on the ω–q_x plane did not approach ω_s as q_x was increased (eq. (14)). Instead, a "back-bending" of the dispersion curve was observed. This effect, originally discovered by Arakawa et al. (1973) for surface plasmons, was also distinctly observed in the case of surface phonons (Schuller et al. 1975b, c, Zhizhin et al. 1976) (see fig. 16). The reason for such "anomaly" was explained by Kovener et al. (1976). Minima in the angle-scanned ATR spectra, occurring at frequencies higher than ω_s, are related to the singularities in the optical constants in the region $\epsilon(\omega) > -1$ where no "true" surface modes exist. The absorption arising in

*See, for instance, the discussion of this problem by Barker (1973, 1974). Considering these effects, he introduced the response function method for the analysis of ATR spectra.

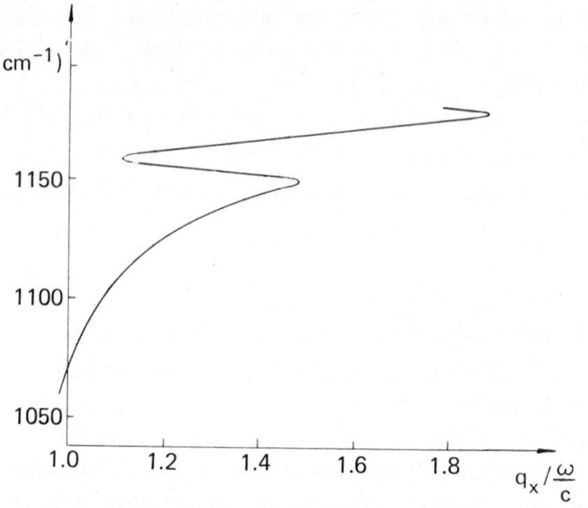

Fig. 16. Backbending of the SiO_2 dispersion curve when angle-scanned at fixed frequency (after Zhizhin et al. 1977a).

this case in the region $\omega_{LO} > \omega \geqslant \omega_s$ is described by Halevi (1978) in terms of the "evanescent waves" which exist only if $\epsilon'' \neq 0$. We will emphasize once more that such ambiguity in the interpretation of the dispersion curves is the result of phonon damping. However, the shape of the dispersion curve itself given by eq. (11) in the range of large q_x (the straight line parallel to the q_x-axis) makes it apparent that close to ω_s a well pronounced minimum in the ATR spectrum cannot be obtained by an angular scan at fixed frequency. Therefore, the angular scan is not very suitable to obtain the dispersion curves of purely surface excitations in this range. Evidently, for this purpose a frequency-scan is preferable. On the contrary, in the range of small q_x, where the dispersion curve is very steep, the angular scan at fixed frequency should be preferred.

5. *Evaluation of Anharmonic Functions from ATR Spectra*

The dielectric functions $\epsilon(\omega)$ used in calculations of the dispersion curves and ATR spectra of surface polaritons were usually evaluated from the transmission and reflection spectra. Meanwhile, an opposite problem is of considerable interest – evaluation of the optical constants of crystals from the spectra of surface phonons. In spite of the fact that the dielectric function of a crystal can be obtained by numerous methods –transmission spectra, external reflection, ellipsometry and so on– none of these methods can be regarded as universal and are efficient only in a limited range of optical constants. The advantage of evaluating optical constants from surface polariton spectra lies in a greater illustrativeness of this method: the frequency of the surface polariton is defined by the position of the minimum in the spectrum and the damping at this frequency by the width of this minimum. It proved to be possible to determine the frequency dependence of anharmonic functions $\Gamma(\omega)$ and $\Delta(\omega)$ (see eq. (21)) in the frequency range of the surface phonon dispersion by making use of the smooth dependence of the frequency upon the wave vector*.

Such a program for surface polaritons in MgO crystals was carried out by Reshina et al. (1976). The measurements of ATR spectra were restricted to the portion of the dispersion curve between 550 and 630 cm^{-1}**. The dielectric function was used in the most general form of eq. (21), i.e., involving frequency dependent anharmonic damping $\Gamma(\omega)$ and shift $\Delta(\omega)$. Within the frequency range of the ATR minimum $\Gamma(\omega)$ and $\Delta(\omega)$ were assumed to be frequency independent.

The unknown parameters Γ, Δ and also the gap values d were evaluated by the least-square method (further referred to as method I) from the fit of the experimental ATR spectra for a series of q_x values to the calculated ones. The gap should be included among the varying parameters because its value is usually known with an accuracy insufficient to calculate the ATR spectra. Reshina et al. (1976) also suggested an approximate method, further referred to as method II, which permitted to find the damping function $\Gamma(\omega)$ from the experimental line widths in the ATR spectra for different q_x values and consequently for different ω. This procedure was carried out taking into account the relation between $\Gamma(\omega)$ and the damping of the surface polariton (Nkoma et al. 1974).

Reshina et al. (1976) evaluated the frequency dependence of anharmonic functions $\Gamma(\omega)$ and $\Delta(\omega)$ of MgO crystal and also $\epsilon'(\omega)$ and $\epsilon''(\omega)$ (fig. 17). The obtained values were used to calculate the reflection coefficients $R(\omega)$, which appeared to be in satisfactory agreement with the experimental ones. In table 1 we compare the phonon damping for three values of ω evaluated by Reshina et al. (1976) by methods I and II with the corresponding values of "bulk" damping found from the graph presented by Genzel and Martin (1972). It can be seen that within the experimental accuracy the agreement is quite satisfactory. Thus, in this case too the damping of surface phonon polaritons is governed by the "bulk" damping.

It is likely that damping in some crystals has a weak frequency dependence at least in the reststrahlen region. Apparently to such crystals belong A_3B_5 semiconductors and, possibly, A_2B_6. Perry et al. (1973) used the ATR

*A similar problem was solved by Ushioda and McMullen (1972) by Raman scattering from bulk polaritons of the low-frequency branch (i.e. at $\omega < \omega_{TO}$).

**The frequencies ω_{TO} and ω_s in MgO at 300 K are 393 cm^{-1} and 640 cm^{-1}, respectively.

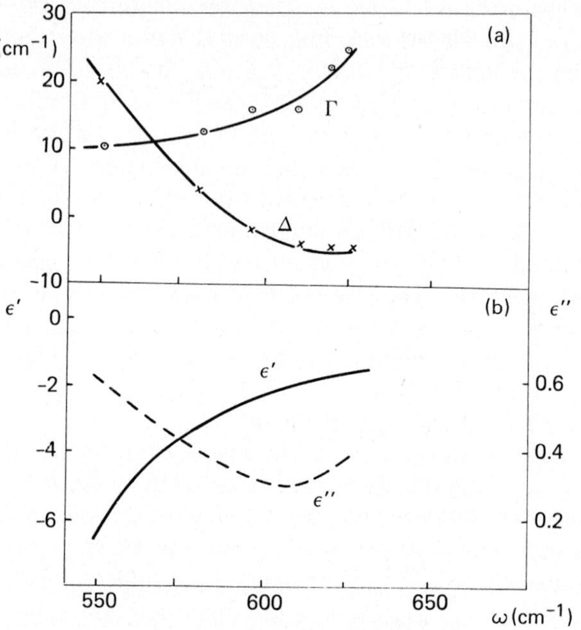

Fig. 17. The frequency dependence of anharmonic parameters Γ and Δ of a MgO crystal. $\epsilon'(\omega)$ and $\epsilon''(\omega)$ are also shown (after Reshina et al. 1976).

spectrum of surface polaritons in uniaxial (hexagonal) crystals of CdS to derive the optical vibration frequencies*. From the best fit with the experimental data these authors obtained the frequencies and damping of TO vibrations, $\omega_\parallel = 233\ \mathrm{cm}^{-1}$, $\omega_\perp = 240\ \mathrm{cm}^{-1}$, $\Gamma_\parallel = \Gamma_\perp = 4\ \mathrm{cm}^{-1}$. (The subscripts $\parallel$ and $\perp$ denote the polarization of the vibrations with respect to the optic axis.) They derived also the frequencies and the oscillator strengths of the subsidiary pole in $\epsilon(\omega)$. The dielectric functions $\epsilon'_{\parallel,\perp}(\omega)$ and $\epsilon''_{\parallel,\perp}(\omega)$ were generated from these data on the assumption that the damping was frequency independent. From the frequency dependence of dielectric functions they derived ω_s and also the frequency ω_F of the so-called Fröhlich mode. This latter corresponds to the maximum of infrared absorption spectrum by extremely small particles and satisfies the equation $\epsilon(\omega) = -2$. This equation gives for CdS particles $\omega_F = 283$–$286\ \mathrm{cm}^{-1}$, in good agreement with experimental results of Perry et al. (1973) on small particles of CdS ($285 \pm 5\ \mathrm{cm}^{-1}$).

*The peculiarities of surface vibrations in anisotropic crystals will be discussed in detail in sect. 6.

Table 1
Phonon damping in MgO crystals evaluated from ATR spectra of surface polaritons and from external reflection spectra.

Frequency (cm^{-1})	Damping $\Gamma(\omega)$ (cm^{-1})		
	ATR[a] method I	ATR[a] method II	$R^{(b)}$
550	10.2	11.0	11.0
610	15.4	17.7	19.5
625	24.0	25.0	23.5

[a]Reshina et al. (1976). [b]Genzel and Martin (1972).
$\Gamma(\omega)$ is the function which enters into eq. 20 for $\epsilon(\omega)$. In the paper by Reshina et al. (1976) it is designated $\gamma(\omega)$. The damping function $\gamma(\omega)$ introduced by Genzel and Martin (1972), fig. 4, is related to the phonon damping $\Gamma(\omega)$ by the equation $\Gamma(\omega)=\frac{1}{2}\gamma(\omega)(\omega/\omega_{TO})$. The values in the last column of the table are given after appropriate recalculation.

6. *Surface Vibrations in Anisotropic Crystals*

Surface polaritons in anisotropic crystals display several nontrivial features. Some of them will be discussed in this section.

The first theoretical study of surface vibration spectra in uniaxial crystals in electrostatic approximation was carried out by Agranovich and Dubovskii (1965), see also Agranovich (1968). A theoretical treatment, taking account of retardation, was performed by Dubovskii (1971), Lyubimov and Sannikov (1972), Bryksin et al. (1973), Hartstein et al. (1973a, b), Borstel (1973) and Agranovich (1975).

We will consider in detail the case of an uniaxial crystal bounded by a vacuum. It will be assumed that the z-axis is normal to the surface of the crystal and the dielectric tensor is diagonal in x, y, z. The propagation direction of the wave is, as before, choosen along the x-axis. The solution of the set of Maxwell equations results in the following dispersion relation for the surface TM waves:

$$\epsilon_x = -\frac{[(\epsilon_x/\epsilon_z)(\kappa_x^2-\epsilon_z)]^{1/2}}{(\kappa_x^2-1)^{1/2}}, \tag{29}$$

where $\kappa_x = q_x/(\omega/c)$ is the reduced dimensionless wave vector and it will be frequently used in this section and in the following. We will consider only the solutions of eq. (29) corresponding to $\kappa_x^2 > 1 (q_x > \omega/c)$, i.e., the non-radiative surface modes.

Let us denote the components of the $\epsilon_{ij}(\omega)$ tensor ($\epsilon_{ij}(\omega) = \delta_{ij}\epsilon_{ii}(\omega)$) as $\epsilon_\parallel$ and $\epsilon_\perp$, where the subscripts $\parallel$ and $\perp$ stand for the directions parallel and perpendicular to the optic axis. Let the optic axis C of the crystal be perpendicular both to the surface normal and to the propagation direction. This implies that both components, ϵ_x and ϵ_z in eq. (29), coincide: $\epsilon_x = \epsilon_z = \epsilon_\perp$. The dispersion relation (29) is then reduced to the expression

$$\kappa_x^2 = \epsilon_\perp(\omega)/(\epsilon_\perp(\omega) + 1). \tag{30}$$

Equation (30) is analogous to the dispersion relation (11) for an isotropic crystal. The dispersion of surface waves involves in this case a single dielectric function $\epsilon_\perp(\omega)$. In the limit $\kappa_x^2 \gg 1$ ($q_x \gg \omega/c$) the corresponding ω_s value is determined from the equation $\epsilon_\perp(\omega_s) = -1$ in analogy with eq. (12). The situation is more varied if both components of the dielectric tensor are simultaneously involved in the dispersion relation. This, for instance, is the case when the optic axis is parallel to the surface normal ($C \parallel z$). Then $\epsilon_z = \epsilon_\parallel$, $\epsilon_x = \epsilon_\perp$ and it follows from eq. (29):

$$\kappa_x^2 = \epsilon_\parallel(\omega)\frac{1 - \epsilon_\perp(\omega)}{1 - \epsilon_\perp(\omega)\epsilon_\parallel(\omega)}. \tag{31}$$

If the optic axis is parallel to the propagation direction ($C \parallel x$), then $\epsilon_x = \epsilon_\parallel$, $\epsilon_z = \epsilon_\perp$ and:

$$\kappa_x^2 = \epsilon_\perp(\omega)\frac{1 - \epsilon_\parallel(\omega)}{1 - \epsilon_\perp(\omega)\epsilon_\parallel(\omega)}. \tag{32}$$

Equations (31) and (32) can be transformed into each other when $\epsilon_\parallel$ is replaced by $\epsilon_\perp$. The limiting value ω_s at $\kappa_x^2 \gg 1$ in the last two cases is determined from the equation $\epsilon_\parallel(\omega)\epsilon_\perp(\omega) = 1$, where both $\epsilon_\parallel$ and $\epsilon_\perp$ are negative.

The dispersion relations for all three cases considered are, for clarity, presented in table 2.

It can be readily shown that the surface modes in anisotropic crystals exist only if $\epsilon_x(\omega) < 0$ (similarly, in isotropic crystals the condition of localization near the surface is leading to the requirement $\epsilon(\omega) < 0$). It follows from eq. (29) that for the real frequency values the inequality $\epsilon_x(\omega) < 0$ can be valid only in two cases:

$$\text{I} \quad \epsilon_x < 0, \quad \epsilon_z < 0 \tag{33a}$$

$$\text{II} \quad \epsilon_x < 0, \quad \epsilon_z > \kappa_x^2. \tag{33b}$$

The surface modes of type I exist at arbitrary values of $\kappa_x^2 > 1$. They can be

Table 2
The diagonal components of the dielectric tensor and the dispersion relations of surface waves in uniaxial crystals.

Configuration of the crystal	Dielectric tensor components ϵ_x	ϵ_y	ϵ_z	Dispersion relation
$C\parallel y$	$\epsilon_\perp$	$\epsilon_\parallel$	$\epsilon_\perp$	$\kappa_x^2 = \dfrac{\epsilon_\perp}{\epsilon_\perp + 1}$
$C\parallel z$	$\epsilon_\perp$	$\epsilon_\perp$	$\epsilon_\parallel$	$\kappa_x^2 = \epsilon_\parallel \dfrac{1-\epsilon_\perp}{1-\epsilon_\perp\epsilon_\parallel}$
$C\parallel x$	$\epsilon_\parallel$	$\epsilon_\perp$	$\epsilon_\perp$	$\kappa_x^2 = \epsilon_\perp \dfrac{1-\epsilon_\parallel}{1-\epsilon_\perp\epsilon_\parallel}$

derived theoretically neglecting the retardation. The account of retardation results only in "deformation" of their dispersion curves in the range of small q_x. In this sense the surface vibrations of type I, which were also called real surface polaritons by Hartstein et al. (1973a, b), are similar to the surface vibrations in isotropic crystals. One particular case of the type I surface modes was mentioned above: they are the ordinary surface vibrations, for which only one component of the dielectric tensor $\epsilon_x(\omega) \equiv \epsilon_z(\omega)$ enters the dispersion relation (30).

The surface modes of type II exist only at sufficiently small $q_x < \omega/(c/\sqrt{\epsilon_z})$ $(\kappa_x < \sqrt{\epsilon_z})$. Such branches have no analogy in the case of isotropic crystals. The vibrations in these branches are predominantly polariton in character (with large photon contribution). They are frequently referred to as virtual surface modes (Hartstein et al. 1973a, b). At large q_x the vibrations of these branches disappear. There exists a special point for these vibrations $q_x = \omega/(c/\sqrt{\epsilon_z})$ $(\kappa_x^2 = \epsilon_z)$, at which the phase velocity of the surface wave becomes equal to the phase velocity $c/\sqrt{\epsilon_z}$ of the bulk polariton polarized along z. Thus, the dispersion curve of the surface polariton terminates at this point of the dispersion curve of the bulk polariton with $\boldsymbol{q}\parallel x$ and $\boldsymbol{E}\parallel z$. Besides, at this point, in accordance with eq. (29), $\epsilon_x = 0$, i.e., the crossover point of the dispersion curves of the surface and bulk polaritons coincides also with the frequency of the longitudinal optical phonon polarized along x (see fig. 18). In a paper by Falge and Otto (1973) this point was called the "stop-point".

At $q_x > \omega/(c/\sqrt{\epsilon_z})$ the amplitude of the type II surface wave, which is proportional to $\exp(q_z z)$, does not decay into the crystal $(z < 0)$. It takes on the usual oscilating character of bulk vibrations because the quantity

$$q_z = (\omega/c)\ [(\epsilon_x/\epsilon_z)(\kappa_x^2 - \epsilon_z)]^{1/2}$$

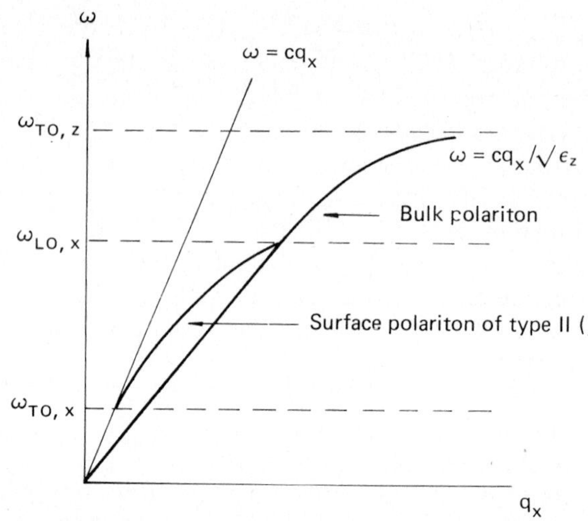

Fig. 18. The dispersion curve of a type II surface polariton enclosed between the "light-line" $\omega = cq_x$ and the dispersion curve of a bulk polariton $\omega = cq_x/\sqrt{\epsilon_z}$.

becomes imaginary at $\kappa_x^2 > \epsilon_z$. At the same time the frequency $\omega(q_x)$ found from eq. (29) becomes complex, i.e., a large broadening occurs in harmonic approximation already. This harmonic broadening is due to the above mentioned intersection of the surface polariton branch with the bulk polariton and the longitudinal phonon propagating along x. The energy of the surface polariton is transferred into the bulk one.

The type II surface modes can also occur when the regions of negative ϵ_x and ϵ_z are positioned differently from those in fig. 18. So, a situation may arise when $\epsilon_x < 0$ in a relatively broad frequency range, whereas ϵ_z changes its sign within the same range, being positive in one part of the range and negative in the other. Then in one part of this frequency range (where $\epsilon_z > 0$) the surface modes of type II can occur, whereas in the other part ($\epsilon_z < 0$) of type I. The measured dispersion curve in this case will be practically continuous, though the nature of the vibration changes at some point. We will come across such an example below, when describing the experimental data on the surface phonons in α-quartz crystals.

The frequency dependencies of the components of the dielectric tensor allow to predict the general character of the surface phonon spectrum. However, it should be kept in mind that eqs. (29)–(33), which determine the spectrum and the conditions of existence of the surface modes, were obtained in harmonic approximation. Analyzing the situation in real crystals with not very large damping, one can use instead of $\epsilon(\omega)$ its real part $\epsilon'(\omega)$.

Surface vibrations in anisotropic crystals were first studied experimentally by Bryksin et al. (1972d, 1973). These authors investigated the tetragonal crystals MgF_2 and TiO_2 which have the same structure (rutile structure) but different frequency dependence of $\epsilon_{\parallel}(\omega)$ and $\epsilon_{\perp}(\omega)$. The spectrum of bulk vibrations in these crystals is due to three oscillators polarized perpendicular to the tetragonal optic axis C_4 and one oscillator polarized parallel to C_4. The frequency dependencies $\epsilon'_{\parallel}(\omega)$ and $\epsilon'_{\perp}(\omega)$ for MgF_2 crystal are shown in fig. 19. It is seen that the lowest frequency range where $\epsilon'_{\perp}$ is negative ($\omega_{TO} = 247\ cm^{-1}$) overlaps the range where $\epsilon'_{\parallel}$ is positive. Recollecting the above discussion, one can see that if the z-axis is directed along C_4 (i.e. $x \perp C_4$), the conditions vital for the occurrence of the surface phonons of type II (virtual surface modes) are satisfied. The dispersion relation in this case is given by eq. (31). But if the crystal is cut so that $C_4 \| y$, then in the same frequency range $\epsilon_x = \epsilon_z = \epsilon_{\perp}$ and surface vibrations of type I should occur. In that case the dispersion relation is given by eq. (30). Both possibilities are illustrated in fig. 20, which shows schematically the configuration of the crystal in the experiment and presents the corresponding ATR spectra for two κ_x^2 values equal to 3.9 and 8.1, i.e., for values lower and higher than $\epsilon_z (\epsilon_z = 7.8$ and in this region has a weak frequency dependence). In $C_4 \| z$ configuration the ATR minimum at $\kappa_x^2 = 8.1$ disappears. This should be expected for the type II surface mode. On the contrary, in quasi-isotropic configuration, $C_4 \| y$, the observed ATR minima at both values of the wave vectors are of comparable intensity. The dispersion curves drawn from the positions of minima in the ATR spectra for these configurations are shown in fig. 21 and clearly reveal the

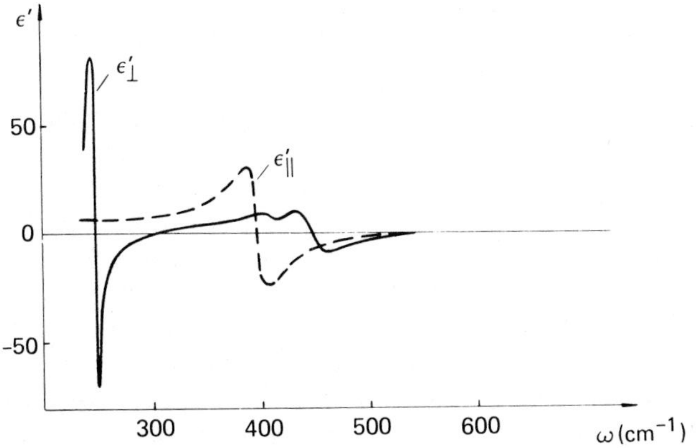

Fig. 19. The frequency dependence $\epsilon'(\omega)$ in a MgF_2 crystal for directions parallel ($\epsilon'_{\parallel}$) and perpendicular ($\epsilon'_{\perp}$) to the optic axis C_4 (after Bryksin et al. 1973). The values of resonance frequencies, damping constants and oscillator strength used in the calculation of dielectric functions were taken from Barker (1964).

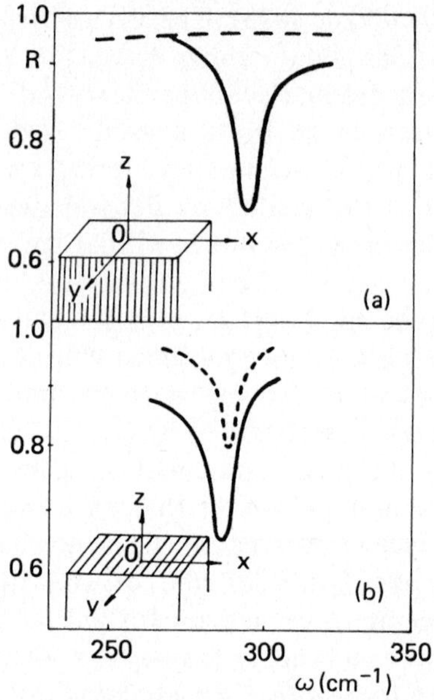

Fig. 20. Crystal orientation and ATR spectra in an uniaxial crystal MgF_2 (after Bryksin et al. 1973): (a) $C_4 \| z$; (b) $C_4 \| y$. The direction of C_4-axis is indicated by the dash-direction. For solid curves: $q_x = 1.98$; for dashed curves: $q_x = 2.85$.

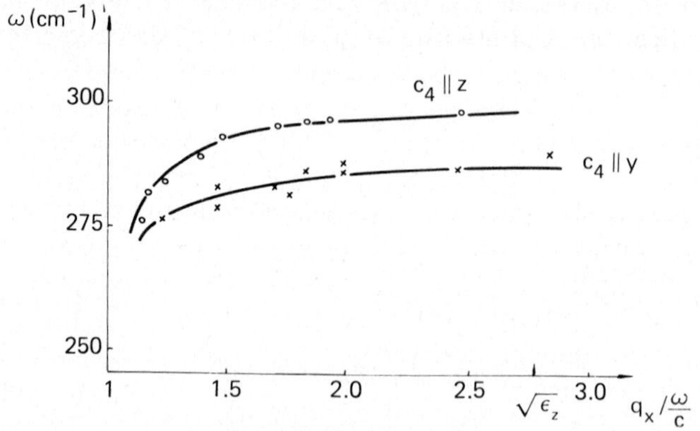

Fig. 21. The dispersion of surface modes in a MgF_2 crystal (after Bryksin et al. 1973). In the orientation $C_4 \| z$ the surface wave disappears at $\kappa_x = q_x/(\omega/c) = \sqrt{\epsilon_z}$. This value is marked on the abscissa axis.

anisotropy of the crystal. In the same frequency range but in the configuration $C_4 \| x$ $\epsilon_x = \epsilon_\| > 0$ and though $\epsilon_z = \epsilon_y = \epsilon_\perp < 0$ surface waves cannot exist and consequently no minima were observed in the ATR spectra. A minimum observed in the ATR spectrum near $500\,\mathrm{cm}^{-1}$ corresponded to the highest frequency resonance in the bulk spectrum. In accordance with the frequency dependence of $\epsilon'(\omega)$, fig. 19, this minimum was due to the type I surface phonon in all configurations when the optic axis was oriented along x, y, or z (in all three cases $\epsilon_x < 0$ and $\epsilon_z < 0$). No minima were observed due to the surface phonons in the region of the bulk resonance near $400\,\mathrm{cm}^{-1}$. In the author's opinion it can be attributed to the strong influence of the anharmonism: $\epsilon'_\perp$ in this region does not even become negative.

In a TiO_2 crystal all three ranges, where $\epsilon'_\perp$ is negative, are overlapped by a broad region of negative $\epsilon'_\parallel$. Accordingly, the three minima observed in the ATR spectra belong to the type I surface vibrations.

A very thorough investigation of surface phonons was carried out in α-quartz crystals. Probably, all theoretically predicted situations for a uniaxial crystal (including the above ones) can be realized in this case due to the complex character of the vibration spectrum. For simple configurations (i.e. $C \parallel x$, $C \parallel y$ and $C \parallel z$) the study of surface phonons in α-quartz by the ATR method was done by Falge and Otto (1973) and Reshina and Zolotarev (1973), see also a review article by Borstel et al. (1974). According to Lamprecht and Merten (1973), the spectrum of bulk vibrations in α-SiO_2 is composed of eight resonances of the ordinary ray ($\boldsymbol{E} \perp C$) and four resonances of the extraordinary ray ($\boldsymbol{E} \parallel C$). There are 22 branches of surface polaritons corresponding to the frequency dependence of $\epsilon_\parallel(\omega)$ and $\epsilon_\perp(\omega)$. Falge and Otto observed all predicted branches except the four lowest, which fell out of the frequency range of the spectrometers used. The dispersion curves obtained from these measurements agreed satisfactorily with the calculations based on formulae (30)–(32) derived in harmonic approximation. An elegant observation of the "stop point" of the type II surface polariton in α-SiO_2 made by Falge and Otto is illustrated in fig. 22 taken from their paper. The spectra are shown for two close values of q_x to the right and to the left of the "stop point". For the higher q_x value the minimum corresponding to the type II surface polariton ($550\,\mathrm{cm}^{-1}$) was absent even when very small gaps were used. At the same time the intensity of the minimum, corresponding to the type I surface vibration ($510\,\mathrm{cm}^{-1}$) did not practically change over the same range of q_x.

An interesting behaviour of the surface polariton due to the resonance $\omega_{TO\perp} = 450\,\mathrm{cm}^{-1}$ was indicated by Falge and Otto (1973) and Reshina and Zolotarev (1973). This surface polariton at great q_x values belongs to type I because $\epsilon'_z = \epsilon'_\parallel < 0$ (fig. 23). But as q_x decreases, the frequency of the surface phonon also decreases and near the resonance $\omega_{TO\parallel} = 495\,\mathrm{cm}^{-1}$ reaches a value where ϵ'_z changes its sign (see fig. 24). The dispersion curve does not practically have a discontinuity because $\epsilon'_z(\omega)$ is a very steep function in the region, where it changes its sign. Thus, the surface phonon

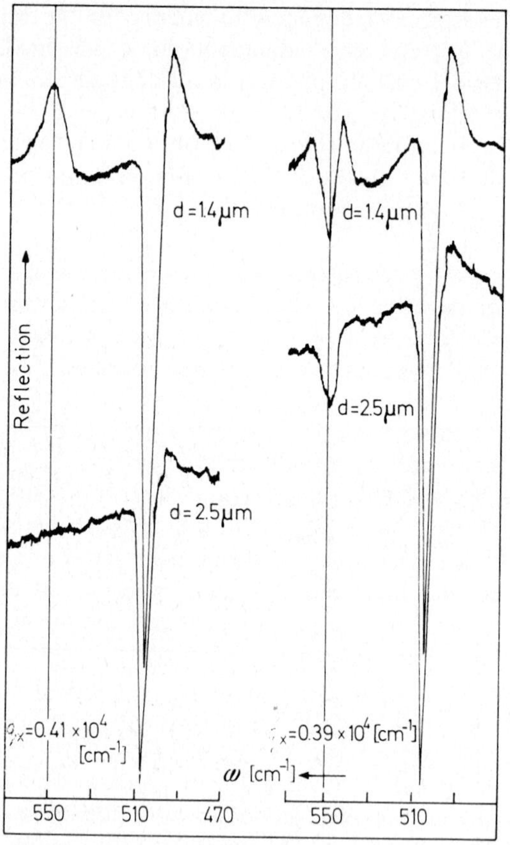

Fig. 22. The stop point of the type II surface polariton in α-SiO_2 at 550 cm^{-1}. The spectrum is shown for two gap values and two q_x values (after Falge and Otto 1973).

exists at all values of the wave vector. Rewriting the dispersion relation (31) for this case in the form

$$\kappa_x^2 = \frac{1-\epsilon_x}{(1/\epsilon_z)-\epsilon_x},$$

one can see that in a narrow region near the resonance where $1/\epsilon_z$ is very small the ϵ_z value has a very slight influence on the dispersion of surface polaritons.

In a general case of an arbitrary geometry the behaviour of surface TM waves in anisotropic crystals is very complicated. If the propagation direction of the surface wave is along one of the principal axes of the dielectric tensor, then the electromagnetic field decays exponentially into the crystal, $E \sim \exp(q_z z)$; $z<0$, $q_z>0$. In other cases q_z is complex and only Re $q_z>0$. Then the decay in the z-direction (normal to the surface) is

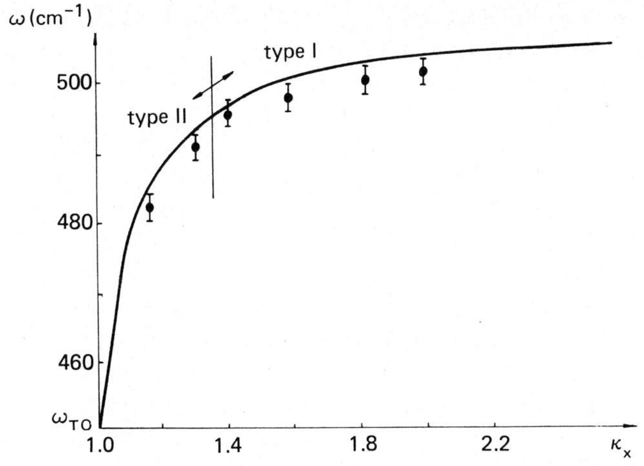

Fig. 23. Dispersion of one of the α-SiO_2 surface polariton branches. It remains continuous when its nature changes from the type II to type I. The calculated dispersion is shown by the solid curve. (after Reshina and Zolotarev 1973).

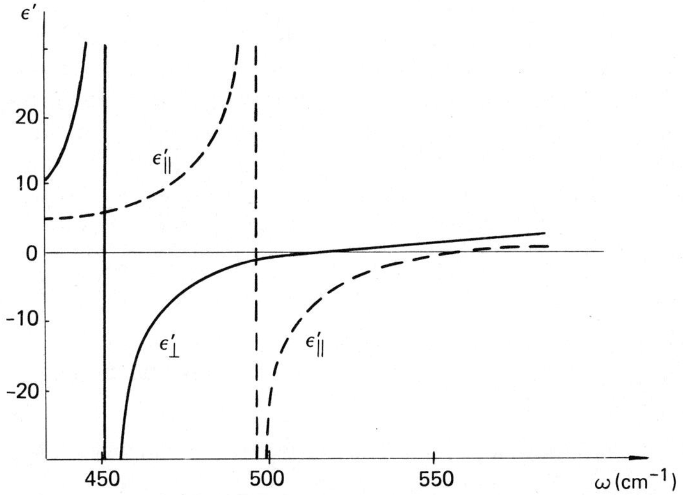

Fig. 24. Frequency dependence of $\epsilon'_{\parallel}$ (dashed curve) and $\epsilon'_{\perp}$ (solid curve) in α-SiO_2 in the range 440–580 cm^{-1}. $\epsilon'_{\perp}$ is negative in the range 450—510 cm^{-1}. Note, that $\epsilon'_{\parallel}$ changes its sign near the resonance at 495 cm^{-1} (after Reshina and Zolotarev 1973).

no longer monotonic: it is the product of the exponential and the oscillatory components. Such generalized surface waves were studied by Falge et al. (1974). The experimental dispersion for a 45° cut of an α-quartz crystal was obtained by the ATR method.

Surface waves in anisotropic crystals remain TM waves for special propagation directions only. For instance, in uniaxial crystals this is the case if the optic axis lies in the xz-plane at an arbitrary angle to the z-axis, i.e., to the surface normal (Agranovich 1975, Borstel et al. 1974). In a general case of an arbitrary geometry, surface waves are a superposition of two waves with different decay constants and polarizations and therefore they cannot be TM waves. The ATR spectra for such waves in α-SiO_2 were theoretically and experimentally studied by Schuller et al. (1975a). The 45°-cut of α-SiO_2 crystal was used with the optic axis lying in the xz-plane. However, surface polaritons in this case were excited not along the x-axis (they would be TM waves then) but along y, i.e. $|q| = q_y$.

From the theoretical point of view the situation is significantly simplified and the corresponding results can be obtained in an analytical form in the limiting case $q_x \gg \omega/c$. For this case, when retardation can be neglected, a simple expression was derived by Agranovich (1975), from which the limiting frequencies ω_s (in the above sense) could be evaluated for an arbitrary anisotropic crystal bounded by the medium with a dielectric constant ϵ_1. These frequencies fit the equation

$$\epsilon_1 = \sqrt{[\epsilon_{11}(\omega)\epsilon_{33}(\omega) - \epsilon_{13}(\omega)]}. \tag{34}$$

In the coordinate system for which eq. (34) is written* the components of the dielectric tensor $\epsilon_{ij}(\omega)$ depend on the orientations of the principal axes of the tensor. This results in a strong dependence of ω_s frequencies on the propagation direction of the surface wave. In a particular case when ϵ_{ij} tensor is diagonal in the xyz-axes $\epsilon_{13}(\omega) = 0$ and eq. (34) reduces to $[\epsilon_x(\omega)\epsilon_z(\omega)]^{1/2} = \epsilon_1$. This equation for ω_s for a vacuum boundary ($\epsilon_1 = 1$) was presented above.

7. *Coupled Plasmon–Phonon Surface Waves in Semiconductors*

7.1. *Dispersion and Damping of Plasmon–Phonon Surface Waves*

To solve the problem of coupled plasmon–phonon surface waves we must first of all, like in the cases considered previously, write the expression for the dielectric function. Taking into account the free-carrier contribution and in the same approximations as for eq. (1), i.e., in the long-wavelength limit and without damping, we have

$$\epsilon(\omega) = \epsilon_\infty + \frac{(\epsilon_0 - \epsilon_\infty)\omega_{\mathrm{TO}}^2}{\omega_{\mathrm{TO}}^2 - \omega^2} - \frac{\omega_p^2}{\omega^2}. \tag{35}$$

*As above, the z-axis is directed along the surface normal and $q \parallel x$.

The last term in eq. (35) is due to the free-carrier contribution in $\epsilon(\omega)$, $\omega_p = (4\pi ne^2/m^*)^{1/2}$ is the plasma frequency in a vacuum, n and m^* are the concentration and the effective mass of free carriers. For convenience, we will also introduce the designation $\tilde{\omega}_p$ for the plasma frequency in the crystal with a dielectric constant $\epsilon_\infty: \tilde{\omega}_p^2 = \omega_p^2/\epsilon_\infty$. The account of the free-carrier contribution in eq. (35) is taken in the long-wavelength limit in the sense that $qv/\omega \ll 1$, where v is the thermal velocity of electrons (or the Fermi velocity in a degenerate case). If such inequality is not valid, the effects of spatial dispersion (Zemski et al. 1975) must be taken into account for plasma vibrations. The condition $qv/\omega \ll 1$ is usually more stringent than the condition for the neglect of phonon spatial dispersion, $qa \ll 1$, where a is the lattice constant. Nevertheless, in experiments on ATR spectroscopy in the infrared region described below the condition $qv/\omega \ll 1$ holds well.

Putting $\epsilon(\omega)$ in eq. (35) equal to zero we obtain the equation for bulk coupled plasmon–phonon modes. As long as the concentration of free carriers is low ($\omega_p \ll \omega_{LO}$), the frequency of the longitudinal lattice vibration obtained from the equation $\epsilon(\omega) = 0$ remains unperturbed. But if ω_p and ω_{LO} are comparable, two mixed vibration branches occur. This phenomenon is well known and thoroughly studied by the methods of Raman and infrared spectroscopy (in the last case from reflection spectra). A similar situation exists for surface waves. It was discussed theoretically by Wallis and Brion (1971) and Chiu and Quinn (1971). The dispersion relation (12) for surface waves on the medium–vacuum interface can be rewritten in the form

$$\epsilon(\omega) = -\frac{q_x^2c^2}{\omega^2} \Big/ \left(\frac{q_x^2c^2}{\omega^2} - 1\right). \tag{36}$$

Equating eqs. (35) and (36) we get an equation for surface wave frequencies. This equation has two solutions describing two branches of plasmon–phonon modes. Let us label them ω_+ and ω_-, where $\omega_+ > \omega_-$ (they are not to be confused with two branches of surface vibrations in a thin slab!). These solutions, except the dispersion relations $\omega_+(q_x)$ and $\omega_-(q_x)$, give the concentration dependencies $\omega_+(n)$ and $\omega_-(n)$ at fixed q_x.

The measurements of surface plasmons dispersion in InSb were made by Marschall et al. (1971). Anderson et al. (1971) observed some manifestations of surface plasmon and phonon interactions. Measurements in both works were performed by using samples with line gratings mechanically produced on the surface. More informative were the investigations by Reshina et al. (1972), Bryksin et al. (1972e), Fischer et al. (1973) and Gammon and Palik (1974) who used the ATR technique. It should be noted at once that nearly all investigations on surface plasmon–phonon modes in semiconductors and the subsequent studies of these modes in an applied magnetic field were performed in n-InSb crystals. This is not accidentally so: the electron mobility in InSb is very high and even at room temperature amounts to 50 000 $cm^2V^{-1}s^{-1}$ for the concentration n about 10^{17} cm^{-3}*.

If damping is taken into account the free carrier contribution into $\epsilon(\omega)$ has the form

$$-\omega_p^2/\omega(\omega + i\gamma),$$

where $\gamma = 1/\tau$ and τ is the relaxation time that can be estimated from the Hall mobility (μ_H) data.

$$\tau = m^*\mu_H/e.$$

For $\mu_H = 5\times 10^4$ $cm^2V^{-1}s^{-1}$ and $m^* = 0.02\,m_0$, $\tau = 0.7\times 10^{-12}$ s, corresponding to the line width $\Delta\nu = 1/(2\pi c\tau) \approx 7.5$ cm^{-1}. The line width in the ATR spectra is in general governed by γ and by phonon damping. The closer the plasmon-phonon frequency to the phonon frequency, the more important is the role of phonon damping in the total width and, on the contrary, if the frequencies of the coupled modes are far removed from the phonon frequency, the line width is determined mainly by γ.

In other $A_{III}B_V$ semiconductors, e.g. GaAs, InP, etc. having a lower mobility, surface plasmon–phonon modes can be readily observed but the corresponding ATR lines in the plasma region are noticeably broader and, for this reason, less suitable for the investigation.

Figure 25 shows the dispersion curves of surface plasmon–phonon modes in InSb at $n = 2\times 10^{17}$ cm^{-3} (Bryksin et al. 1972e). The calculated dispersion curves of "intrinsic" uncoupled surface phonons and surface plasmons which have a crossover point are also shown. As was already mentioned the coupling results in a lifting of degeneracy and in the appearance of two coupled plasmon–phonon modes ω_+ and ω_-. One of these modes (ω_+) exists at $q_x > \omega_{TO}/c$, the other (ω_-) in the whole range of q_x, beginning from $q_x = 0$. At low frequencies it coincides with the light line. The experimental frequencies fit satisfactorily the theoretical curves.

The concentration dependence of the frequencies of both modes, $\omega_+(n)$ and $\omega_-(n)$, at fixed q_x is presented in fig. 26. The repulsion of the branches can be seen in the region of the strongest coupling, like in fig. 25. At small $\tilde{\omega}_p/\omega_{TO}$ the ω_- branch is predominantly plasmon-like and the upper ω_+ branch is phonon-like, at high concentrations vice versa. The picture is very similar to that obtained by Blum and Mooradian (1970) from Raman spectra for bulk plasmon–phonon modes in InSb.

The comparison of experimental and calculated ATR spectra of surface plasmon–phonon modes in InSb performed by Bryksin et al. (1972e),

*Just at such concentrations $\tilde{\omega}_p$ is close to phonon frequencies of InSb ($\omega_{LO} = 192$ cm^{-1}), and there is a strong coupling of plasma vibrations of free carriers to the longitudinal phonons.

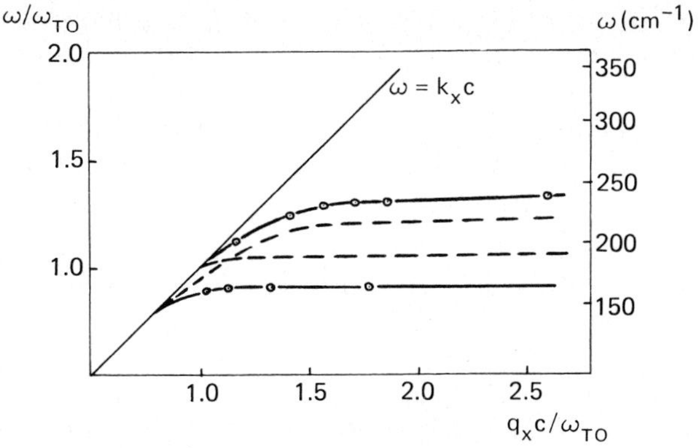

Fig. 25. Dispersion of surface plasmon–phonon modes in n-InSb ($n = 2 \times 10^{17}\,\text{cm}^{-3}$). The circles indicate the experimental data. The calculated dispersion of coupled plasmon–phonon modes is shown by solid lines and that of uncoupled modes by dashed lines (after Bryksin et al. 1972e).

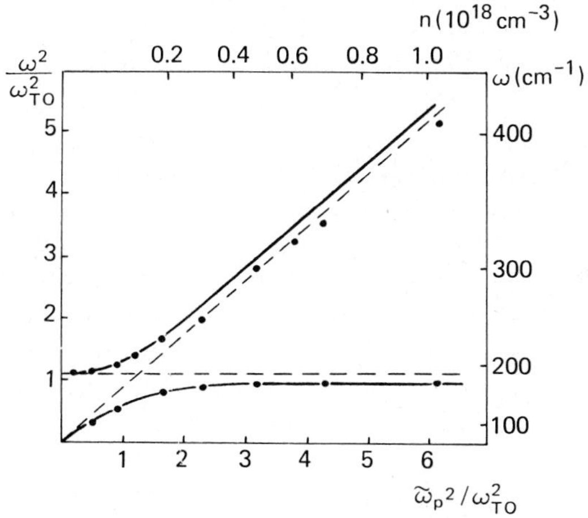

Fig. 26. The frequencies of surface plasmon–phonon modes in n-InSb as a function of free-carriers concentration (i.e. of plasma frequency $\tilde{\omega}_p$). The angle of incidence is 23°. The uncoupled plasmon and phonon branches are shown by dashed lines. The calculated curves for coupled plasmon–phonon modes are shown by solid lines. The experimental data are shown by circles (after Bryksin et al. 1972e).

Gammon and Palik (1974) and Palik et al. (1973) has shown that in the plasma region the experimental minima are 2–3 times broader than the calculated ones and amount to 20–30 cm^{-1}. γ values used in these calculations were estimated from the Hall mobility and plasma reflectivity. They are of the order of 7–10 cm^{-1} for $n \approx 10^{16}$–$10^{17}\,\text{cm}^{-3}$ (measurements of plasma reflection give slightly higher values of γ). Yet, in lightly doped crystals ($\tilde{\omega}_p \ll \omega_{TO}$) the ATR minimum width was about 3 cm^{-1}, this value being very close to the damping of bulk phonons. The calculated and experimental spectra are compared in fig. 27.

Various assumptions have been made to explain the additional broadening of the surface plasmon line in the ATR spectra of n-InSb. Most natural at first glance seems the assumption of the important part played by the surface electron scattering (Bryksin et al. 1972e, Gammon and Palik 1974), which, among other things, is known to cause the decrease of mobility in thin semiconducting layers. The penetration depth of light in the above ATR experiments was 2–3 μm, that is, several times less than the penetration depth in reflection (10–20 μm) in the region of the plasma minimum. Therefore in ATR experiments surface scattering might produce a more pronounced effect than in reflection experiments.

Another additional damping mechanism might be associated with the surface depletion layer usually present on the surface of n-InSb (Pinczuk and Burstein 1970). In such nonuniform plasma the energy can be transferred from the surface to bulk plasmons (Romanov 1964, Pakhomov and Stepanov 1967). The effect of a depletion (accumulation) layer on the dispersion dependence of surface plasmons in semiconductors was treated theoretically by Cunningham et al. (1974). Hartstein and Burstein (1974) observed the frequency change of ATR minima in InSb crystals after chemical etching. They ascribe it to the removal of the damaged layer with decreased carrier density and mobility developed during mechanical polishing (Hartstein et al., 1975). A shift (decrease) of ATR frequencies after mechanical polishing was also observed by Holm and Palik (1976) in GaAs crystals in quantitative correspondence with the changes observed after polishing in reflection spectra. An analysis of reflection and ATR spectra of an InSb sample (1.8 × 10^{17} cm^{-3}) was performed by Holm and Palik (1975) in

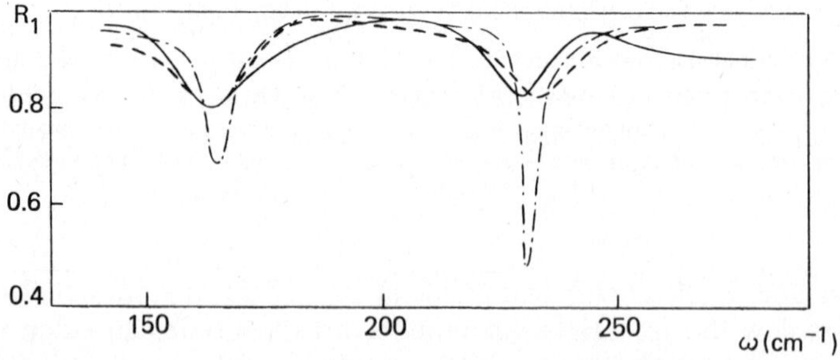

Fig. 27. Experimental and calculated ATR spectra for n-InSb sample. $n = 2 \times 10^{17}$ cm^{-3}. The experimental spectrum is shown by the solid line. The calculated spectrum for plasmon damping $\gamma = 10$ cm^{-1} is shown by the dash-dotted line and that for plasmon damping $\gamma = 30$ cm^{-1} by the dashed line (after Bryksin et al. 1972e).

the frame of a model taking into account the depletion layer. They found that minima in measured ATR spectra had lower frequencies than in the theoretical ones. They were not able to get a self-consistent fit to both the bulk reflectivity and the ATR spectra when using a uniform depletion layer model and assumed that a more realistic model (e.g. nonuniform depletion layer) could account for the discrepancy.

So far the discussion in this section has been concerned with the isotropic case. Surface plasmon–phonon modes in uniaxial semiconductors were discussed by Gurevich and Tarkhanian (1975) and Tarkhanian (1975). The main peculiarities of anisotropic semiconductors are as follows.

We have seen that in isotropic crystals two branches of surface plasmon–phonon modes occur. The lower branch ω_- exists in the whole q_x range, the upper ω_+ branch exists at $q_x > \omega_{TO}/c$. In anisotropic crystals the same is valid only in a quasi-isotropic configuration, when $C \parallel y$, i.e., the optic axis is perpendicular both to the propagation direction of the wave and to the surface normal (cp. fig. 20b). It is not valid in a general case. If the optic axis is normal to the surface ($C \parallel z$, cp. fig. 20a) or parallel to the propagation direction ($C \parallel x$) the number of possible branches increases and can amount to four. One of them is analogous to the ω_- branch and starts as always at $q_x = 0$. The number of branches existing at $q_x > \omega_\parallel/c$, $\omega_\perp/c$ depends on the concentration n. On the whole, it increases with n and can take on the values from one to three, depending on the position arrangement of the characteristic frequencies of optical vibrations and also on the orientation of the optic axis relative to the crystal surface. These modes exist either in the whole range of q_x, like in the isotropic crystal, or in the top limited q_x range. In the latter case the top q_x value also depends on the concentration n through the dielectric tensor. The penetration depth of the surface wave at the top q_x value becomes infinite. It means that the surface wave turns into the bulk one – the situation peculiar to the type II surface vibration (cp. sect. 6).

The effect of a finite thickness of an anisotropic crystal on the properties of plasmon–phonon modes, i.e. the case $\kappa_2 L \lesssim 1$ (cp. eq. (13)), was treated by Tarchanian (1975). To our knowledge, plasmon–phonon surface modes in anisotropic semiconducting crystals have not been studied experimentally.

7.2. Plasmon–Phonon Surface Waves in a Magnetic Field

The spectrum of plasma vibrations becomes quite complicated in an applied magnetic field. This is also true for mixed plasmon–phonon modes in semiconductors including also the surface modes. The effect of a magnetic field on the bulk plasma vibration of free carriers in semiconductors and, particularly, the optics of magnetoplasmons, i.e. plasma vibration in a magnetic field, was discussed, for instance, by Palik and Wright (1967). The effect of a magnetic field on the surface polaritons in semiconductors has been thoroughly studied lately both theoretically and experimentally. A short review of these works is given below.

We shall see that many properties of surface polaritons in semiconductors in a magnetic field are similar to those of surface polaritons in anisotropic crystals. This does not seem surprising, since the magnetic field produces anisotropy. However, the effect of a magnetic field is more complicated because it does not only change the diagonal components $\epsilon_{ii}(\omega)$ of the dielectric tensor* but results in the appearance of the off-diagonal ϵ_{ij} components. The measure of the plasma perturbation by a magnetic field is the ratio of the cyclotron frequency $\omega_c = eH/m^*c$ to the plasma frequency. Surface magnetoplasmons in semiconductors were treated theoretically by Brion et al. (1972) and Chiu and Quinn (1972a) for simple geometries, neglecting the phonon contribution to $\epsilon(\omega)$. In an earlier paper by Pakhomov and Stepanov (1967) the investigation of surface waves propagating along a flat plasma–vacuum interface in a magnetic field was performed for arbitrary directions of the magnetic field $\boldsymbol{H}$ and wave propagation q_x without account of retardation. In a weak magnetic field, as long as $\omega_c \ll \omega_p$ there are two branches for arbitrary q_x and arbitrary $\boldsymbol{H}$ direction relative to the interface. Their frequencies are close to $\omega_p/\sqrt{2}$, the frequency of the surface plasmon at $H = 0$, and differ from one another by the value of the order of ω_c. For $\omega_c > \omega_p$ the conditions for surface waves existence on a plasma interface are more rigid. So, if $\boldsymbol{H} \| xy$ (i.e. the interface plane) the possible propagation directions are concentrated around $q_x \perp \boldsymbol{H}$ (Voigt geometry), deviating from this direction not more than by the angle $\arcsin(\omega_p/\omega_c)$, which decreases with increasing H. Brion et al. (1972) have made a theoretical investigation of surface magnetoplasmons in semiconductors in the Voigt geometry for arbitrary ϵ_∞ values while taking retardation into account. They found that a gap exists in the surface magnetoplasmon dispersion curve if ϵ_∞ and the cyclotron frequency satisfy the inequality $\epsilon_\infty > (\omega_c^2 + \tilde{\omega}_p^2)^{1/2}/\omega_c$.

Surface plasmon–phonon modes in semiconductors in a magnetic field with the account of phonon contribution were discussed by Brion et al. (1973) and Chiu and Quinn (1972b, 1974). The dispersion of both magnetoplasmons and surface plasmon–phonon modes with damping were investigated by Martin et al. (1978). The calculated dispersion curves for the angle-scanned ATR spectra obtained by Martin et al. (1978) showed a

*If $\boldsymbol{H}$ is directed along one of the principal axes of the ϵ_{ij} tensor, e.g., along y, then only the ϵ_{yy} component remains unchanged. The ϵ_{xx} and ϵ_{zz} components change their magnitudes in the field, remaining equal to each other. Such a relationship between ϵ_{ii} components is characteristic for a uniaxial crystal.

backbending effect. Almost all calculations in the cited papers were performed for n-InSb*. Mainly three geometries were considered:
(1) $\boldsymbol{H}$ parallel to the surface normal, $\boldsymbol{H} \parallel z$;
(2) $\boldsymbol{H} \parallel \boldsymbol{q} (\boldsymbol{H} \parallel x)$ – Faraday geometry;
(3) $\boldsymbol{H} \perp \boldsymbol{q}$ and $\boldsymbol{H} \parallel y$ – Voigt geometry.

Surface waves remain TM waves only in the Voigt geometry. In this case $\boldsymbol{H}$ is perpendicular to the polarization plane xz of the surface wave, where its electric vector is rotating, and the cyclotron motion of electrons takes place in the same plane. For other $\boldsymbol{H}$ directions surface waves acquire both TM and TE components due to the cyclotron motion of electrons.

When ω_c becomes comparable with the frequencies of surface excitations the magnetic field strongly complicates the dispersion relations. As was already mentioned, this is due to the fact that in a magnetic field the diagonal components of the dielectric tensor become unequal and the off-diagonal ones appear. If $\boldsymbol{H} \parallel y$:

$$\epsilon = \begin{pmatrix} \epsilon_1(H) & 0 & i\epsilon_2(H) \\ 0 & \epsilon_3 & 0 \\ i\epsilon_2(H) & 0 & \epsilon_1(H) \end{pmatrix}, \tag{37}$$

where

$$\epsilon_1 = \epsilon_\infty + \frac{(\epsilon_0 - \epsilon_\infty)\omega_{TO}^2}{\omega_{TO}^2 - \omega^2} - \frac{\omega_p^2}{\omega^2 - \omega_c^2},$$
$$\epsilon_2 = \omega_p^2 \omega_c / \omega(\omega^2 - \omega_c^2),$$
$$\epsilon_3 = \epsilon_\infty + (\epsilon_0 - \epsilon_\infty)\omega_{TO}^2/(\omega_{TO}^2 - \omega^2) - \omega_p^2/\omega^2.$$

Only the component $\epsilon_3 = \epsilon_{yy}$ does not depend on the magnetic field directed along y. Other non zero components are dependent on H via ω_c. At $H = 0$ the tensor becomes diagonal, the anisotropy produced by the magnetic field disappears. Then the diagonal components have the form (35). We will not present here the dispersion relations following from eq. (37) for the Voigt geometry and the analogous rather complicated formulae for the Faraday geometry. Below for comparison with the experimental results we will restrict ourselves to the graphical presentation.

Let us shortly discuss the Faraday geometry. The dispersion relations describe waves having two damping constants κ_1 and κ_2 (Wallis et al. 1974, Palik et al. 1976).

The components of the electric field inside the crystal are:

*To the authors knowledge, all experimental investigations published so far were also made with this crystal, because in n-InSb it is simpler to obtain the necessary values of ω_c in accessible magnetic fields due to its small effective mass ($m^* \approx 0.02\, m_0$ at $n \sim 10^{17}\,\mathrm{cm}^{-3}$).

$$\left.\begin{aligned} E_x &= A\exp(\kappa_1 z) + B\exp(\kappa_2 z) \\ E_z &= C\exp(\kappa_1 z) + D\exp(\kappa_2 z \\ E_y &= F\exp(\kappa_1 z) + G\exp(\kappa_2 z) \end{aligned}\right\} \times \exp(\mathrm{i}(q_x x - \omega t))$$

where A, B, $C\ldots$ are rather complicated functions evaluated from the boundary conditions. In this geometry at $H \neq 0$ all three components of the electric field are nonzero, whereas at $H = 0$ only E_x and E_z are the nonzero components, $E_y = 0$. It turned out that the values of $\kappa_{1,2}$ can be real, imaginary or complex in different spectral regions. When $\kappa_{1,2}$ are both real, positive numbers, the wave is a standard surface wave. If $\kappa_{1,2}$ are complex and $\mathrm{Re}\,\kappa_{1,2} > 0$ the exponential decay in the wave perpendicular to the surface is modulated by an oscillatory component, i.e., the decay is not monotonic. These are the generalized surface waves. If one κ is real and positive but the other κ is imaginary, we have a superposition of the decaying component and the propagating component. Strictly speaking, such waves are not surface waves. They are called pseudosurface. Nevertheless, they also were detected by the ATR method (Palik et al. 1976). Another peculiarity of the ATR spectra in this case is that they are no longer completely p–p-polarized (p–p stand for p-incident and p-reflected) since $E_y \neq 0$. Therefore, except p–p spectra, p–s and s–p spectra may exist (p-incident, s-reflected and vice versa) and also s–s spectra.

The experimental investigations of surface plasmon–phonon modes in the Faraday geometry in the fields up to 110 kG were reported by Palik et al. (1973, 1974, 1976). In such a field the cyclotron frequency ω_c is approximately 2.5 times higher than the phonon frequencies of InSb ($\omega_{TO} = 180\ \mathrm{cm}^{-1}$, $\omega_{LO} = 192\ \mathrm{cm}^{-1}$). The concentration of carriers in the samples was in the range from $1.07 \times 10^{17}\ \mathrm{cm}^{-3}$ to $1.7 \times 10^{17}\ \mathrm{cm}^{-3}$, corresponding to $\tilde{\omega}_p$ in the range from 167 to $202\ \mathrm{cm}^{-1}$. Figure 28 shows H dependence of the main ATR minima. Two of the main minima only slightly change thier positions with the field and become merely a little broadened. The analysis shows that they may be related to the surface modes ω_+ and ω_- at $H = 0$ and, probably, to the pseudosurface modes at high fields. In the case of mixed polarizations p–s, s–p and also s–s the signal-to-noise ratio was poor and no distinct experimental minima could be detected.

Most striking was the appearance of the phonon-like mode between 180–190 cm^{-1} at $H > 43$ kG, when ω_c exceeds ω_{TO}. The appearance of this mode can be understood from the following considerations. In increasing fields the dispersion of the free-carriers is shifted to higher-frequencies along with ω_c. This results is the decoupling of plasmon–phonon modes and appearance of a pure phonon mode. Also the reststrahlen reflection spectra become just the same as in the case of low-carrier-density crystals.

Let us now direct our attention to the Voigt geometry. In this case ($H \parallel y$)

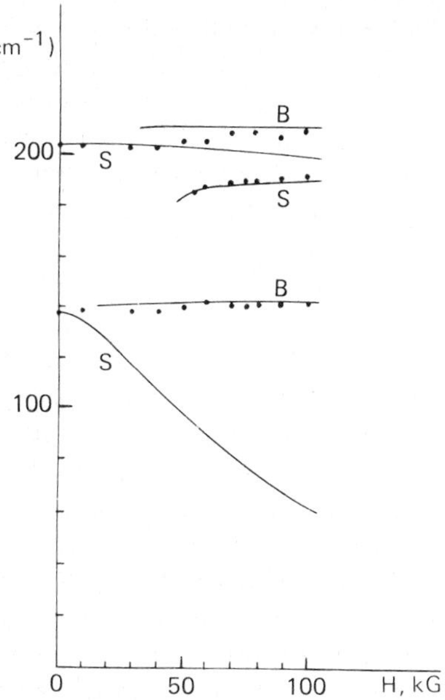

Fig. 28. The frequencies of ATR minima in *n*-InSb vs. magnetic field. Faraday geometry. Solid lines – calculation, points – experiment. Angle of incidence $\varphi = 35°$, $n = 1.07 \times 10^{17}\ cm^{-3}$. Surface modes are labeled S and the bulk ones B (after Palik et al. 1976).

the dielectric tensor has the form (37). The cyclotron motion, as was already noted, takes place in the *xz*-plane and surface waves remain the TM waves. It will be recalled that in the absence of a magnetic field the electric vector of the surface wave inside the crystal rotates clockwise in the *xz*-plane for the observer who sees the *x*-axis directed to the right (see fig. 3). The cyclotron motion, which occurs in the applied magnetic field, coincides with the intrinsic rotation of the electric vector of the surface wave for one direction of $\boldsymbol{H}$ but has a reverse rotation for the opposite $\boldsymbol{H}$ direction. For this reason, $+q_x$ and $-q_x$ directions are not equivalent and the frequencies corresponding to different field directions are different. The behaviour of the surface wave in a magnetic field is determined by the sign of $\boldsymbol{q} \times \boldsymbol{H}$. Therefore, either the $\boldsymbol{q}$ direction or the $\boldsymbol{H}$ direction can be changed in order to investigate the non-reciprocal nature of the surface polariton dispersion curves. In experiments it is more convenient to change the $\boldsymbol{H}$ direction, as was done by Hartstein and Burstein (1974) and Hartstein et al. (1974). The first of these works was concerned with a heavy-doped sample of InSb ($1.4 \times 10^{18}\ cm^{-3}$), when the plasmon–phonon coupling could be neglected. In the second work the plasmon–phonon modes were studied in a sample with $n = 2 \times 10^{17}\ cm^{-3}$ in fields up to 40 kG. The dispersion curves of this sample for two field directions are shown in fig. 29, which demonstrates clearly the non-reciprocal magnetic field dependence of the coupled surface plasmon–phonons. In a magnetic field "virtual" modes also arise which are analogous to the type II surface vibrations in anisotropic crystals. They are also shown in fig. 29. At $H = 0$ the spectrum of surface excitations reduces to the two branches similar to those in fig. 25.

8. *Interface Polaritons*

In the earlier sections we discussed surface polaritons on a crystal bounded by vacuum or a medium with the positive dielectric constant ϵ_1 frequency independent in the spectral region of interest. It is clear that such an approach is far from a general formulation of the problem, since ϵ_1 and ϵ_2 in the dispersion equation (11) are interchangeable. In this section we will specially discuss such situations when not only the frequencies but the pattern of the spectrum itself and the shape of the dispersion curves depend significantly on the properties of both contacting media. It is more correct to speak in this case about interface modes rather than surface ones. Situations are possible when both dielectric functions become negative in different spectral regions. We will denominate the medium with a negative dielectric function in a given spectral region as surface-active and

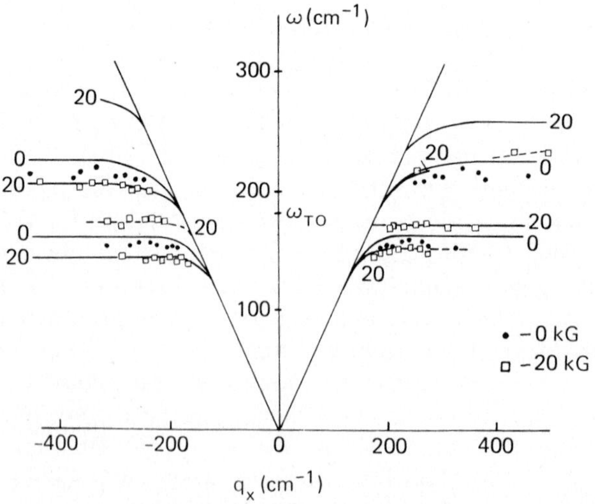

Fig. 29. Dispersion of plasmon-phonon modes in Voigt geometry. The data are for $H = 0$ and 20 kG, in the latter case for two field directions which is equivalent to the $+\boldsymbol{q}$ and $-\boldsymbol{q}$ data. The calculated curves for surface polaritons are shown by solid lines and for bulk excitations by dashed lines (after Hartstein et al. 1974).

the medium with a positive dielectric function as surface-inactive (Hartstein et al. 1973a). The region where surface waves exist is given by the inequalities:

$$\epsilon_2(\omega) \leqslant -\epsilon_1(\omega), \qquad \epsilon_1(\omega) > 0$$

(medium 2) is surface-active), or

$$\epsilon_1(\omega) \leqslant -\epsilon_2(\omega), \qquad \epsilon_2(\omega) > 0$$

(medium 1 is surface-active).

Let us consider the contact of two semi-infinite polar dielectrics 1 and 2. The relative position of their optical frequencies may be different. Various possibilities are illustrated in fig. 30 taken from the review article by Halevi (1978). In the case A medium 2 is surface-active in the region of higher frequencies, whereas medium 1 in the region of lower frequencies. In the case B the regions of negative dielectric functions partially overlap and in

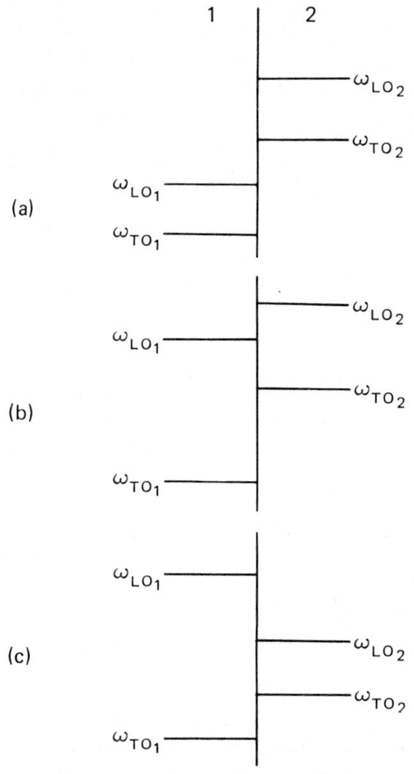

Fig. 30. Different possible positioning of LO and TO frequencies of two contacting polar dielectrics (after Halevi 1978).

the case C they entirely overlap. It is apparent that no interface modes exist in the region of overlap (in the case C in the region $\omega_{TO2} < \omega < \omega_{LO2}$). In the case B the interface modes can exist in the regions $\omega_{TO1} < \omega < \omega_{TO2}$ and $\omega_{LO1} < \omega < \omega_{LO2}$. A similar situation may occur for a semiconductor–dielectric interface. It is illustrated in fig. 31, where the plasma frequency of a semiconductor slightly exceeds the optical vibration frequencies of a dielectric*. In this case the surface plasmon branch of the semiconductor, has a discontinuity in the range $\omega_{TO2} < \omega < \omega_{LO2}$ because both $\epsilon_1(\omega)$ and $\epsilon_2(\omega)$ are negative in this range and the interface waves cannot propagate.

Interface modes between two active media were treated theoretically in a number of papers. Stern and Ferrel (1960) studied interface waves between two free-electron plasmas without account of damping. For the interface plasmon frequency ω_s an apparent result was obtained $\omega_s^2 = \frac{1}{2}(\omega_{p1}^2 + \omega_{p2}^2)$, where ω_{p1}, ω_{p2} are plasma frequencies of two adjacent semi-infinite media. Economou (1969) and Ngai and Economou (1971) studied plasmon interface modes in sandwich structures of different types (metal–dielectric–semiconductor, vacuum–metal–metal) and their possible manifestations in tunnel current characteristics, LEED and photoemission. It was predicted by Agranovich et al. (1972) that surface modes which exist on a dielectric–vacuum interface are destroyed on the interface between a dielectric and a semi-infinite metal (this effect of metal damping will be discussed below for a case of thin metallic films). Recently a phenomenological treatment of interface polaritons for systems of the type shown in fig. 30 has been made by Halevi (1976, 1977). He discussed also the propagation of plasmon polaritons on the interface of two metals (Halevi 1975). The review of these and other works was presented by Halevi (1978). Plasmon–phonon interface modes of the types shown in fig.

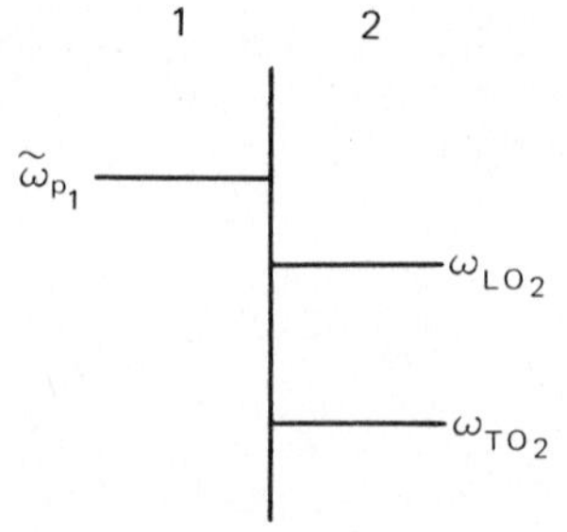

Fig. 31. The position of the plasma frequency of a nonpolar semiconductor (1), relative to LO and TO frequencies of polar dielectric (2).

*For simplicity, the semiconductor in fig. 31 is assumed to be nonpolar (e.g. Ge or Si) so that its optical vibration makes no contribution to the dielectric function.

31 were treated theoretically by Ikarashi (1976). He also considered the case of two semi-infinite media (an ionic crystal and a semiconductor) separated by a narrow gap (Ikarashi 1978). In this case subsidiary waves occur of TM and TE type which propagate along the gap. However, the calculations in the cited papers by Halevi and Ikarashi were performed for the interface of two semi-infinite media. This idealization is not always adequate for the experimental arrangements. In particular, in experiments on interface modes by ATR spectroscopy the thickness of one of the media must be of the order, or less, than the wavelength of the probing radiation, otherwise the incident light would not reach the interface. In a few experiments on interface modes conducted so far by the ATR method the geometry of the type shown in fig. 32 was applied: the film 2 deposited on the substrate 3 was separated from the ATR prism by the gap 1. Below, the dielectric functions of different layers of this "sandwich" are denoted by $\epsilon_i (i = 1, 2, 3)$. The three layer system 1–2–3 has two interfaces: film–gap (2–1) and film–substrate (2–3). Interface modes on both interfaces are observed in the ATR spectra. If the film thickness is sufficiently small, these modes are coupled. The materials for surface-active media, used by different authors in ATR experiments following the pattern of fig. 32, are listed in table 3.

The solution for nonradiative surface modes in a three-medium system 1–2–3 (two interfaces) can be obtained in the same manner as in sect. 2 for a two-medium system (one interface). The corresponding dispersion relation has the form (Mills and Maradudin 1973, Gerbstein and Mirlin 1974, Mirlin and Reshina 1974):

$$\left(\frac{\epsilon_1}{\kappa_1}+\frac{\epsilon_2}{\kappa_2}\right)\left(\frac{\epsilon_2}{\kappa_2}+\frac{\epsilon_3}{\kappa_3}\right)+\left(\frac{\epsilon_1}{\kappa_1}-\frac{\epsilon_2}{\kappa_2}\right)\left(\frac{\epsilon_2}{\kappa_2}-\frac{\epsilon_3}{\kappa_3}\right)\exp(-2\kappa_2 L)=0 \tag{38}$$

where L is the thickness of film 2 and $\kappa_i = (q_x^2 - (\omega^2/c^2)\epsilon_i)$. For a film sufficiently thick ($\kappa_2 L \gg 1$), neglecting the second term in the sum in eq. (38) we obtain for each of the interfaces 1–2 and 2–3 the dispersion equations of the type eq. (10):

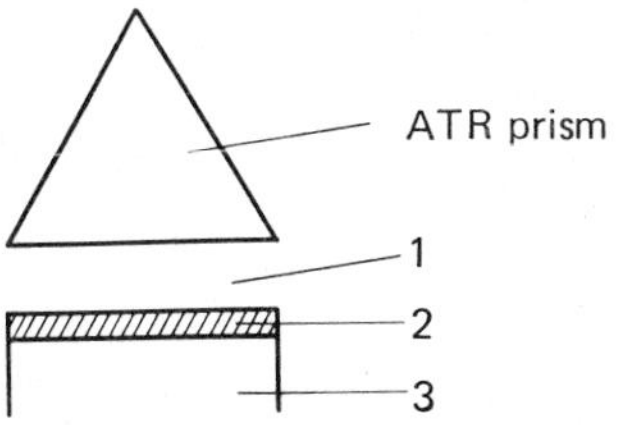

Fig. 32. Experimental set-up for studying interface modes by ATR method. The three-layer system under investigation includes: (1) air (vacuum) gap, (2) the film, (3) the substrate (semi-infinite crystal).

Table 3
Materials used as surface-active media in investigations of interface modes by the ATR method.

References	Film (2)	Substrate (3)
Gerbstein and Mirlin (1974)	ZnSe	InSb
Mirlin and Reshina (1974)	MgF_2	SiO_2
	Ag, Bi	SiO_2
	ZnSe	Ag
Zhizhin et al. (1974), (1975), (1977a)	Au	SiO_2
Yakovlev et al. (1975)	LiF	TiO_2, Al_2O_3
Zhizhin et al. (1977b)	LiF	TiO_2, Al_2O_3, $Y_3Fe_5O_{12}$
Holm and Palik (1978)	n^+-GaAs	n-GaAs

$$\text{(a)}\quad \epsilon_1/\kappa_1 = -\epsilon_2/\kappa_2, \qquad \text{(b)}\quad \epsilon_2/\kappa_2 = -\epsilon_3/\kappa_3. \tag{39}$$

This is illustrated in fig. 33 for a dielectric film with optical vibration frequencies ω_{TO2} and ω_{LO2} which is deposited on a nonpolar semiconducting substrate. Without the dielectric film the surface plasmon on the vacuum–semiconductor interface is characterized by dispersion curve III. After deposition of the film a gap appears in the energy spectrum: curve III splits into two dispersion branches labeled I and I′ in fig. 33. Furthermore,

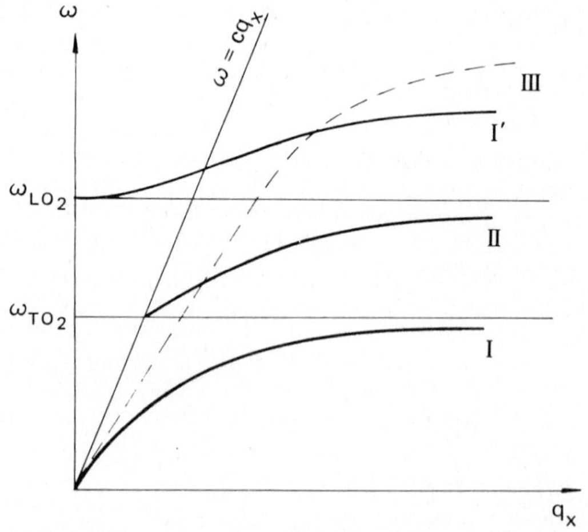

Fig. 33. The dispersion of surface polaritons in a three-layer system (vacuum gap–dielectric film (with $\kappa_2 L \gg 1$)–nonpolar semiconductor). I and I′ are interface vibrations on the semiconductor–dielectric interface, II on vacuum–dielectric interface. III is the surface plasmon on the semiconductor–vacuum interface in the absence of the dielectric film (after Gerbstein and Mirlin 1974).

surface phonon II appears on the dielectric–vacuum interface. As L decreases and $\kappa_2 L$ becomes of the order of unity or less, the frequencies of all branches change as a function of L and the attributing of the dispersion branches to a particular interface, like it was made before, becomes invalid. In this case only the coupled plasmon–phonon interface modes of the entire system remain meaningful. The calculated and experimental dispersion curves of the system vacuum-dielectric film of ZnSe–n-InSb substrate are shown in fig. 34 taken from a paper by Gerbstein and Mirlin (1974). The electron concentration in InSb was $5 \times 10^{17}\,\mathrm{cm}^{-3}$, implying $\tilde{\omega}_p = 325\,\mathrm{cm}^{-1}$. This value only slightly exceeded the phonon frequencies of ZnSe. The free-carrier concentration in ZnSe was sufficiently small, justifying the treatment of this semiconductor as a dielectric. In the frequency range covered in fig. 34 $\kappa_2 L$ was close to unity when L (the film thickness of ZnSe) was 2.5 μm. Therefore, the observed dispersion branches I, I′ and II are the coupled modes of the entire system, although one can trace the origin of branch II to the perturbed surface phonon on the ZnSe–vacuum interface and the origin of branches I and I′ to the perturbed surface plasmon on InSb substrate*.

The interface modes of two dielectrics were studied by Mirlin and Reshina (1974). The positions of the optical vibration frequencies in the investigated system: vacuum–MgF_2 film–SiO_2 crystal corresponded to the case C in fig. 30, i.e., to the entirely overlapping reststrahlen regions. In fact, one of the SiO_2 resonances ($\omega_{TO3} = 450\,\mathrm{cm}^{-1}$, $\omega_{LO3} = 495\,\mathrm{cm}^{-1}$) lies within the range of negative $\epsilon(\omega)$ in MgF_2 ($\omega_{TO2} - \omega_{LO2}$)**. The dispersion curves of interface modes for 1 μm MgF_2 film are shown in fig. 35 (curves I, II, III). The "surface phonon" of SiO_2 (curve II) is shifted to higher-frequencies and at large q_x "pushed out" into the region where SiO_2 dielectric function $\epsilon_3(\omega)$ is positive. The reason for this shift can be made clear from the following qualitative considerations. For a thick MgF_2 film, ($\kappa_2 L \gg 1$), eq. (39b) for the SiO_2 MgF_2 interface is $\epsilon_3(\omega) = -\epsilon_2(\omega)$. If $\epsilon_2 < 0$ the frequency of the interface mode is shifted into the range where $\epsilon_3 > 0$. If the film is not too thick, the inequality $\kappa_2 L \gg 1$ does not hold, $\kappa_2 L \lesssim 1$, but a shift to higher frequencies still remains and its magnitude depends on q_x and on the film thickness. We would like to draw attention to the fact that

*On the InSb–vacuum interface for $n = 5 \times 10^{17}\,\mathrm{cm}^{-3}$ apart from the plasmon-like mode ω_+, there exists a phonon-like mode near $180\,\mathrm{cm}^{-1}$. After deposition of a ZnSe film this mode also is the interface mode of the vacuum–ZnSe–InSb system, although it is only slightly perturbed by the ZnSe film, because of the small LO–TO splitting in InSb. The minimum corresponding to this branch in the ATR spectrum was erroneously attributed to the bulk excitation (Gerbstein and Mirlin 1974), as was noted by Holm and Palik (1978).

**The vibration spectrum of a polycrystalline MgF_2 film was approximated by one oscillator with $\omega_{TO2} = 400\,\mathrm{cm}^{-1}$, $\omega_{LO2} = 620\,\mathrm{cm}^{-1}$ (Mirlin and Reshina 1974).

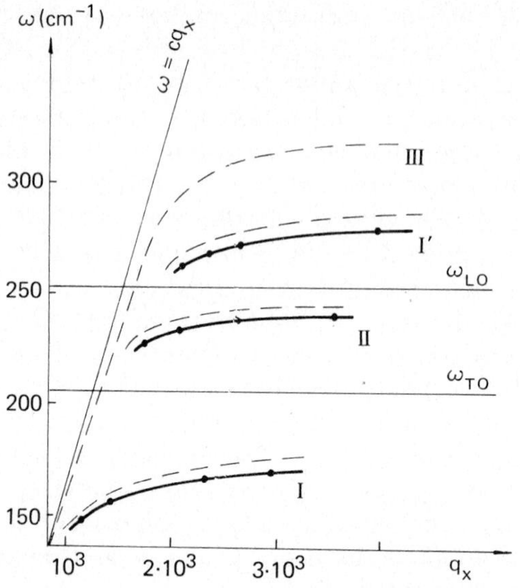

Fig. 34. Dispersion of surface polaritons in the vacuum–ZnSe–InSb system. The experimental curves are shown by solid lines, and the calculated ones by dashed lines. I, I′ and II are the coupled plasmon–phonon modes of the system. III is the surface plasmon of InSb substrate (in the absence of ZnSe). The LO and TO frequencies of ZnSe are shown (after Gerbstein and Mirlin 1974).

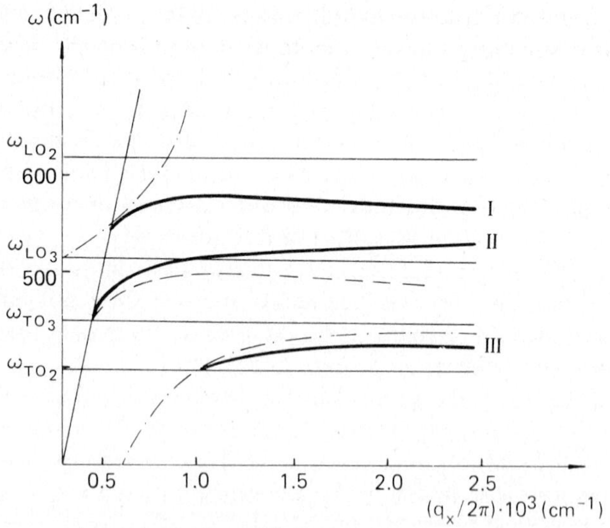

Fig. 35 Dispersion of nonradiative surface modes (I–II–III) in vacuum–MgF_2 ($L = 1\ \mu m$)–SiO_2 system. The surface phonon of SiO_2 substrate bounded by vacuum is shown by a dashed line. The polariton branches of SiO_2 are shown by the dash-dotted lines. The parameters used in the calculation are: $\omega_{TO2} = 399\ cm^{-1}$, $\epsilon_{02} = 4.6$, $\epsilon_{\infty 2} = 1.9$; $\omega_{TO3} = 450\ cm^{-1}$, $\epsilon_{03} = 4.03$, $\epsilon_{\infty 3} = 3.17$, (after Mirlin and Reshina 1974).

curves I and III in fig. 35 lie to the right of the SiO_2 bulk polariton dispersion curves $\omega^2 = c^2q_x^2/\epsilon_3(\omega)$, just as the nonradiative modes on the vacuum interface lie to the right of the line $\omega = cq_x$. Curves I and III can be considered as dispersion curves of a thin MgF_2 film bounded from one side by a vacuum and from the other by a SiO_2 substrate. The ATR minima corresponding to branches I and III were observed by Mirlin and Reshina (1974). The minimum corresponding to the "surface phonon of SiO_2" was noticeably broadened due to the dissipation of the electromagnetic energy in the MgF_2 film, which absorbs in this frequency range*.

The system studied by Holm and Palik (1978) was an epitaxial GaAs film with low carrier density deposited on a high density GaAs substrate. The peculiarity of such a system is the coincidence of the phonon frequencies, in particular, ω_{TO} and ω_{LO} of the film and the substrate. The spectrum of the interface modes was conditioned by the relation between plasma frequencies. In the experiment by Holm and Palik (1978) the plasma frequency of the film was $\tilde{\omega}_{p2} = 100\ cm^{-1}$ ($L = 1.2\ \mu m$) and that of the substrate $\tilde{\omega}_{p3} = 452\ cm^{-1}$. The computed ATR spectra of the system gap–film–substrate had four minima (by genesis two plasmon–phonons for each interface). In experiment two of these minima were observed for a series of q_x values.

The results of the discussed works indicate that the characteristics of surface and interface polaritons can be used for investigation of thin dielectric and semiconducting films, epitaxial films in particular. The same, to no lesser degree, is true for metallic films. A thin metallic film deposited on a dielectric substrate causes a metallic damping of the surface polaritons (Agranovich et al. 1972, Agranovich 1975). Metallic damping manifests itself as a broadening of the ATR minima due to the energy losses in the film where the energy of the surface wave is partly converted into Joule heat. This effect can be used for determination of the conductivity of metallic films in the infrared region, as was proposed by Agranovich (1975) and Agranovich and Leskova (1974).

Zhizhin et al. (1974) found that the ATR minima due to surface polaritons on SiO_2 disappeared when even a very thin Au film ($L \gtrsim 40$ Å) was vacuum-evaporated on the SiO_2 surface. A similar result was obtained by Mirlin and Reshina (1974), who observed the smearing of the ATR minimum when a thin ($L < 100$ Å) Ag film was vacuum-evaporated on the SiO_2 surface. For a semi-metallic Bi film with a noticeably smaller conductivity evaporated on SiO_2 Mirlin and Reshina (1974) observed a gradual evolution of the spectrum as a function of the film thickness. The ATR spectrum of SiO_2 with and without Bi in the range 450–550 cm^{-1} is shown in fig. 36 taken

*It will be noted once more that the dispersion curves of interface modes depend on the properties of all three media involved. The applied terms, such as "surface phonon on the 2–3 interface" and the like, are rather conventional and are used for descriptive purposes only.

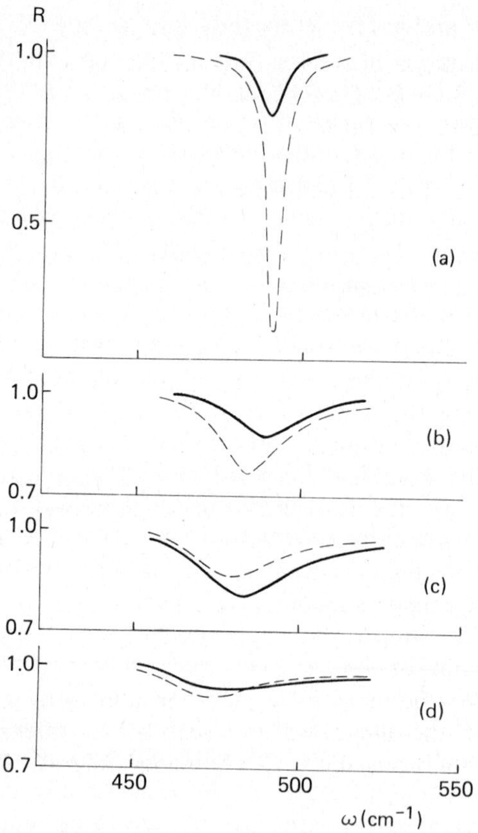

Fig. 36. ATR spectra of a surface polariton on α-SiO_2 (a) and on α-SiO_2 covered with Bi films of various thicknesses: (b) 100 Å, (c) 200 Å, (d) 350 Å. The experimental spectra are shown by solid lines, the calculated ones by dashed lines (after Mirlin and Reshina 1974).

from their paper. The increase in the film thickness up to 350 Å resulted in a gradual smearing of the ATR minimum accompanied by a shift to low frequencies. Qualitatively the same behaviour was observed in the calculated ATR spectra. The analysis carried out by the authors has shown that the dielectric function in Bi with account of the damping is positive in the frequency range of interest. This results in the shift to low frequencies just as in the case of a boundary with an inactive dielectric with $\epsilon'(\omega) > 0$. Thus, the contact with a metal, along with the shift of surface polaritons to higher frequencies (as may be inferred from the analogical case of two adjacent surface-active dielectrics discussed above) can also cause a shift in the opposite direction due to the damping. The sign and the magnitude of the resulting effect are determined by the complex dielectric function of the metallic film.

Zhizhin et al. (1974, 1975b, 1977a) and Yakovlev et al. (1978) carried out a thorough investigation of interface polaritons on α-SiO_2 crystals covered with a very thin Au film ($L < 100$ Å). The primary attention was focussed on one of the surface polaritons of α-SiO_2 with a wide dispersion range (1070–$1150\,cm^{-1}$). The analysis of the results presented some difficulties due to the absence of data concerning the optical properties of Au films in this range. Besides, these properties could be dependent on the film deposition technique. Therefore, the investigations of ATR spectra were supplemented by the measurements and analysis of the reflection spectra. It turned out that for very thin films ϵ'' is small and $\epsilon' > 0$, in accordance with the Maxwell Garnet model of island films (Heavens 1955). For such films the broadening of the surface polariton minimum in the ATR spectrum was small and the minimum was shifted to lower frequencies. This type of behaviour of the dispersion curves is illustrated in fig. 37 taken from the paper by Zhizhin et al. (1975b). When the film thickness was increased and the islands came into contact, ϵ'' increased rapidly and the ATR minimum was greatly broadened. The frequency of the ATR minimum depended both on ϵ' and on ϵ''. The width of the ATR minimum, as was shown by experiments and calculations, was governed by the product $\epsilon''L$, and increased when $\epsilon''L$ was increased (fig. 38). These works show that the investigation of the changed properties of surface polaritons of a substrate after deposition of a metallic film allows to find some physical characteristics of the film as a function of its thickness, evaporation conditions, etc. Such possibility is of special interest for films with the

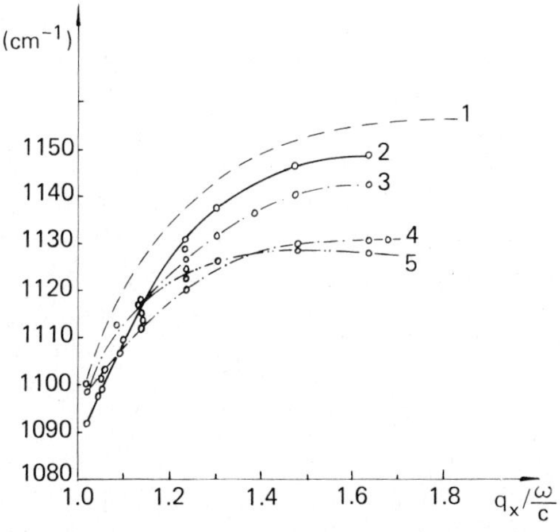

Fig. 37. Dispersion of a surface polariton on α-SiO_2 (1) and on α-SiO_2 covered with Au films of various thicknesses: (2) 38 Å, (3) 70 Å, (4) 90 Å, (5) 100 Å, after Zhizhin et al. (1975b).

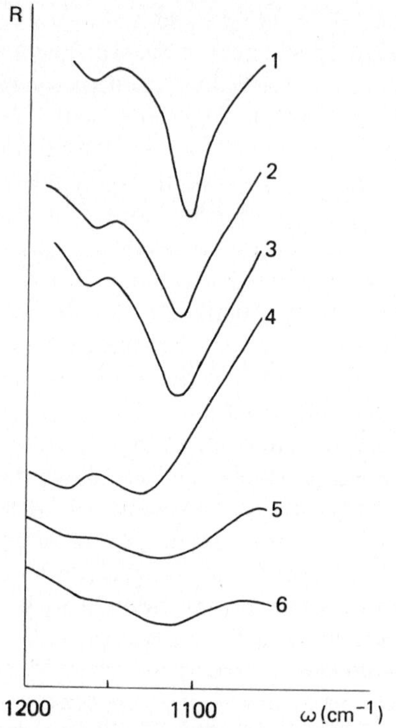

Fig. 38. The broadening of surface polariton minimum in the ATR spectrum of α-SiO_2 as a function of increased metallicity of a thin Au film deposited on the surface. From curve 1 to curve 6 the magnitude $\omega L\epsilon''$ increases from 0.18 to 2.5 (after Yakovlev et al. 1978).

thickness less than 100 Å when the film properties differ significantly from the properties of bulk metals.

Finally, let us consider a film of an active dielectric on a metallic substrate (vacuum–dielectric–metal). It is already known to the reader that an isolated dielectric film, so long as its thickness is small has two surface polariton branches: a high-frequency branch ω_+ and a low-frequency one ω_- (see sect. 2). In a sufficiently thin film ($q_xL \ll 1$) both branches are strongly polarized. So, for the ω_+ branch the E_x component of the electric field is zero and E_z is constant across the film: $E_x = 0$, $E_z = \text{const}$. For the ω_- branch vice versa: $E_x(z) = \text{const}$, $E_z = 0$. In a dielectric film deposited on a substrate with ideal conductivity the low-frequency branch ω_- must disappear so that the boundary condition $E_x = 0$ could be satisfied. This effect must also occur for a substrate having a finite but sufficiently high conductivity. Figure 39, taken from the paper by Mirlin and Reshina (1974), shows the ATR spectrum of a ZnSe film evaporated on a glass substrate precoated with a thick Ag film. In the reststrahlen region of ZnSe (ω_{TO}–ω_{LO})

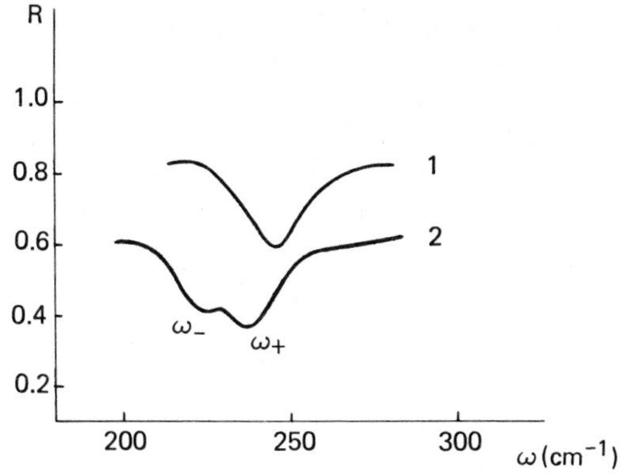

Fig. 39. ATR spectra of ZnSe film ($L = 1.1\ \mu m$) vacuum evaporated on Ag(1) and on TlBr–TlJ crystal (2). The contact with the metal results in the disappearance of ω_- minimum in spectrum 1. The high-frequency minimum ω_+ in spectrum 2 is shifted relative to spectrum 1 in accordance with the ϵ values of the substrates. (after Mirlin and Reshina 1974).

the spectrum has only one minimum, which was attributed to the high-frequency ω_+ branch. For reference sake, fig. 39 presents also an ATR spectrum of ZnSe film on an inactive substrate of TlBr–TlJ crystal (KRS-5). In this case two minima corresponding to both branches are distinctly seen. Thus, in fact, the contact with a metal results in a complete disappearance of one of the surface polariton branches (ω_-) of a thin active-dielectric film.

Acknowledgement

The author is very thankful to Dr. Irene Reshina for reading the manuscript and for her valuable comments.

References

Agranovich, V.M., 1968, Theory of Excitons (Nauka, Moscow).

Agranovich, V.M., 1975, Usp. Fiz. Nauk **115**, 199 (1975, Sov. Phys. Uspekhi, **18**, 99).

Agranovich, V.M. and O.A. Dubovskii, 1965, Fiz. Tverd. Tela, **7**, 2885. (1966, Sov. Phys.-Solid State, **7**, 2343.)

Agranovich, V.M. and T.A. Leskova, 1974, Fiz. Tverd. Tela, **16**, 1800. (1974, Sov. Phys.-Solid State, **16**, 1172.)

Agranovich, V.M., A.G. Mal'shukov and M.A. Mekhtiev, 1972, Fiz. Tverd. Tela, **14**, 849. (1972, Sov. Phys.-Solid State, **14**, 1725.)

Anderson, W.E., R.W. Alexander and R.E. Bell, 1971, Phys. Rev. Lett. **27**, 1057.

Arakawa E.T., M.W. Williams, R.N. Hamm and R.H. Ritchie, 1973, Phys. Rev. Lett. **31**, 1127.
Barker A.S., Jr., 1964, Phys. Rev. **136**, 1290.
Barker A.S., Jr., 1973, Surf. Sci. **34**, 62.
Barker A.S., Jr., 1974, Dispersion of Bulk and Surface Polariton Modes in Solids, in: Polaritons, eds. Burstein, E. and F. de Martini (Pergamon Press, New York) p. 127.
Beaglehole, D., 1969, Phys. Rev. Lett. **22**, 706.
Blum, F.A. and A. Mooradian, 1970, Light Scattering From Plasmons in InSb, in: Proc. Tenth Intern. Conf. Phys. Semicond., Cambridge, 1970, eds. Keller S.P., J.C. Hensel and F. Stern (USA Atomic Energy Commission, Oak Ridge) p. 755.
Boersch, H., J. Geiger and W. Stickel, 1968, Zs. Phys. **212**, 130.
Born, M. and K. Huang, 1954, Dynamical Theory of Crystal Lattices (Clarendon Press, Oxford) §8.
Born, M. and E. Wolf, 1964, Principles of Optics (Pergamon Press, London, New York) §1.5.
Borstel, G., 1973, Phys. Status Solidi B **60**, 427.
Borstel, G., H.J. Falge and A. Otto, 1974, Surface and Bulk Phonon Polaritons Observed by Attenuated Total Reflection, in: Springer Tracts in Modern Physics, Vol. 74, ed. Springer (Springer, Berlin,-Heidelberg, New York) p. 107.
Brion, J.J., R.F. Wallis, A. Hartstein and E. Burstein, 1972, Phys. Rev. Lett. **28**, 1455.
Brion, J.J., R.F. Wallis, A. Hartstein and E. Burstein, 1973, Surf. Sci. **34**, 73.
Bryksin, V.V. and Yu.A. Firsov, 1972, Fiz. Tverd. Tela, **14**, 1148, (1972, Sov. Phys.-Solid State, **14**, 981.)
Bryksin, V.V., Yu.M. Gerbstein and D.N. Mirlin, 1971, Fiz. Tverd. Tela, **13**, 2125 (1972, Sov. Phys.-Solid State, 13, 1779).
Bryksin, V.V., Yu.M. Gerbstein and D.N. Mirlin, 1972a, Phys. Status Solidi B **51**, 901.
Bryksin, V.V., Yu.M. Gerbstein and D.N. Mirlin, 1972b, Fiz. Tverd. Tela, **14**, 543. (1972, Sov. Phys.-Solid State, **14**, 453.)
Bryksin, V.V., Yu.M. Gerbstein and D.N. Mirlin 1972c, Fiz. Tverd. Tela, **14**, 3368. (1973, Sov. Phys.-Solid State, **14**, 2849).
Bryksin, V.V., D.N. Mirlin and I.I. Reshina, 1972d, Pis'ma JETF **16**, 445.
Bryksin, V.V., D.N. Mirlin and I.I. Reshina, 1972e, Solid State Commun. **11**, 695.
Bryksin, V.V., D.N. Mirlin and I.I. Reshina, 1973, Fiz. Tverd. Tela **15**, 1118. 1973, Sov. Phys.-Solid State **15**, 760.)
Bryksin, V.V., D.N. Mirlin and Yu.A. Firsov, 1974, Usp. Fiz. Nauk **113**, 29. (1974, Sov. Phys. Usp., **17**, 305.)
Burstein, E., A. Hartstein, J. Schoenwald, A.A. Maradudin, D.L. Mills and R.F. Wallis, 1974, Surface Polaritons–Electromagnetic waves at Interfaces, in: Polaritons, Proc. of the First Taormina Conf. on the Structure of Matter, Taormina, Italy, eds. Burstein, E. and F.De Martini (Pergamon, New York), p. 89.
Chiu, K.W. and J.J. Quinn, 1971, Phys. Lett. **35A**, 469.
Chiu, K.W. and J.J. Quinn, 1972a, Phys. Rev. **B5**, 4707.
Chiu, K.W. and J.J. Quinn, 1972b, Phys. Rev. Lett. **29**, 600.
Chiu, K.W. and J.J. Quinn, 1974, Magnetoplasma Surface Waves in Metals and Semiconductors, in: Polaritons, Proc. of the First Taormina Research Conf. on the Structure of Matter, Taormina, Italy, 1972, eds. E. Burstein and F. DeMartini (Pergamon, New York, Toronto, Oxford, Sydney) p. 259.
Cowley, R.A., 1963, Adv. Phys., **12**, 421.
Cunnigham, S.L., A.A. Maradudin and R.F. Wallis, 1974, Phys. Rev. **B10**, 3342.
DeMartini, F. and Y.R. Shen, 1976, Phys. Rev. Lett. **36**, 216.
DeMartini, F., G. Giuliani, P. Maratoni and E. Palange, 1976, Phys. Rev. Lett. **37**, 440.
Dubovskii, O.A., 1970, Fiz. Tverd. Tela **12**, 3054 (1971, Sov. Phys.-Solid State **12**, 2471.)
Economou, E.N., 1969, Phys. Rev. **182**, 539.

Evans, D.J. and S. Ushioda, 1972, Solid State Commun. **11**, 1043.
Evans, D.J. and S. Ushioda, 1974, Phys. Rev. **B9**, 1638.
Evans, D.J., S. Ushioda and J.D. McMullen, 1973, Phys. Rev. Lett. **31**, 369.
Falge, H.J. and A. Otto, 1973, Phys. Status Solidi B **56**, 523.
Falge, H.J., G. Borstel and A. Otto, 1974, Phys. Status Solidi B **65**, 1234.
Fischer, B., N. Marschall and H.J. Queisser, 1973, Surf. Sci. **34**, 50.
Fischer, B., B. Bäuerle and W.J. Buckel, 1974, Solid State Commun. **14**, 291.
Fuchs, R. and K.L. Kliewer, 1965, Phys. Rev. **140A**, 2076.
Fuchs, R., K.L. Kliewer and W.J. Pardee, 1966, Phys. Rev. **150**, 589.
Gammon, R.W. and E.D. Palik, 1974, J. Opt. Soc. Am. **64**, 350.
Genzel, L. and T.P. Martin, 1972, Phys. Status Solidi B **51**, 91.
Gerbstein, Yu.M., 1973, Experimental Investigation of Surface Polaritons, Thesis (Leningrad).
Gerbstein, Yu.M. and D.N. Mirlin, 1974, Fiz. Tverd. Tela, **16**, 2584. (1974, Sov. Phys. Solid State, **16**, 1680).
Gurevich, L.E. and R.G. Tarkhanian, 1975, Fiz. Tverd. Tela, **17**, 1944. (1975, Sov. Phys.-Solid State, **17**, 1273.)
Halevi, P., 1975, Phys. Rev. **12**, 4032.
Halevi, P., 1976, Bol. Inst. Tonantzintla **2**, 131.
Halevi, P., 1977, Opt. Commun. **20**, 167.
Halevi, P., 1978, Surf. Sci. **78**, 64.
Harrick, H.J., 1967, Internal Reflection Spectroscopy (Interscience Publ. New York, London, Sydney).
Hartstein, A. and E. Burstein, 1974, Solid State Commun. **14**, 1223.
Hartstein, A., E. Burstein, J.J. Brion and R.F. Wallis, 1973a, Surf. Sci. **34**, 81.
Hartstein, A., E. Burstein, J.J. Brion and R.F. Wallis, 1973b, Solid State Commun. **12**, 1083.
Hartstein, A., E. Burstein, E.D. Palik, R. Kaplan, R.W. Gammon and B.W. Henvis, 1974, Investigation of Magnetoplasmon-Phonon Type Surface Polariton on *n*-InSb, in: Proc. of the 12th Intern. Conf. on the Phys. of Semiconductors, Stuttgart, 1974, ed. M. Pilkuhn (Teubner, Stuttgart) p. 541.
Hartstein, A., E. Burstein, E.D. Palik, R.W. Gammon and B.W. Henvis, 1975, Phys. Rev. **B12**, 3186.
Heavens, O.S., 1955, Optical Properties of Thin Solid Films (Butterworths Scientific Publication, London) p. 176.
Hisano, K., F. Placido, A.D. Bruce and G.D. Holah, 1972, J. Phys. C (Solid State) **5**, 2511.
Holm, R.T. and E.D. Palik, 1975, CRC. Critical Reviews in Solid State Sci. **5**, 397.
Holm, R.T. and E.D. Palik, 1976, J. Vac. Sci. Technol. **13**, 889.
Holm, R.T. and E.D. Palik, 1978, Phys. Rev. **B17**, 2673.
Ibach, H., 1970, Phys. Rev. Lett. **24**, 1416.
Ibach, H., 1971, Phys. Rev. Lett. **27**, 256.
Ibach, H., 1977, Surf. Sci. **66**, 56.
Ibach, H. and S. Lehwald, 1978, Surf. Sci. **76**, 1.
Ikarashi, T., 1976, J. Phys. Soc. Japan, **41**, 1962.
Ikarashi, T., 1978, J. Phys. Soc. Japan, **44**, 551.
Kliewer, K.L. and R. Fuchs, 1966a, Phys. Rev. **144**, 495.
Kliewer, K.L. and R. Fuchs, 1966b, Phys. Rev. **150**, 573.
Kövener, G.S., R.W. Alexander, Jr. and R.J. Bell, 1976, Phys. Rev. **B14**, 1458.
Lamprecht, G. and L. Merten, 1973, Phys. Status Solidi B **55**, 33.
Landau, L.D. and E.M. Lifshitz, 1960, Electrodynamics of Continuous Media (Pergamon Press, New York), §68.
Lyubimov, V.N. and D.G. Sannikov, 1972, Fiz. Tverd. Tela, **14**, 675. (Sov. Phys.-Solid State, **14**, 575).

Maradudin, A.A. and R.F. Wallis, 1961, Phys. Rev. **123**, 777.
Maradudin, A.A. and R.F. Wallis, 1962, Phys. Rev. **125**, 1277.
Marshall, N. and B. Fisher, 1972, Phys. Rev. Lett. **28**, 811.
Marshall, N., B. Fisher and H. Queisser, 1971, Phys. Rev. Lett. **27**, 95.
Martin, B.G., A.A. Maradudin and R.F. Wallis, 1978, Surf. Sci. **77**, 416.
Mills, D.L. and A.A. Maradudin, 1973, Phys. Rev. Lett. **31**, 372.
Mirlin, D.N. and I.I. Reshina, 1974, Fiz. Tverd. Tela, **16**, 2241. (1974, Sov. Phys.-Solid State, **16**, 1463.)
Ngai, K.L. and E.N. Economou, 1971, Phys. Rev. **B4**, 2132.
Nkoma, J., R. Loudon and D.R. Tilley, 1974, J. Phys. C. **7**, 3547.
Otto, A., 1968, Zs. Phys. **216**, 398.
Otto, A., 1974a, Experimental Investigation of Surface Polaritons on Plane Interfaces, in: Festkörperprobleme XIV, Advances in Solid State Physics, eds. Madelung, O. and H.J. Queisser (Pergamon Press, London, New York) p. 1.
Otto, A., 1974b, The Surface Polariton Resonance in Attenuated Total Reflection, in: Polaritons, Proc. of the First Taormina Research Conf. on the Structure of Matter, Taormina, Italy, eds. Burstein, E. and F. de Martini (Pergamon Press, New York, Toronto, Oxford, Sydney) p. 117.
Otto, A., 1975, Spectroscopy of Surface Polaritons by Attenuated Total Reflection, in: Optical Properties of Solids, New Developments, ed. Seraphin, B.O. (North-Holland, Amsterdam) p. 677.
Pakhomov, V.I. and K.N. Stepanov, 1967, Zh. Tekh. Fiz. **37**, 1393 (1968, Sov. Phys.-Tech. Phys. **12**, 1011).
Palik, E.D. and G.B. Wright, 1967, Free-Carrier Magnetooptical Effects, in: Semiconductors and Semimetals, vol. 3, eds. Willardson R.K. and A.C. Beer (Academic Press, New York, London) p. 421.
Palik, E.D., R. Kaplan, R.W. Gammon, H. Kaplan, J.J. Quinn and R.F. Wallis, 1973, Phys. Lett. **45A**, 143.
Palik, E.D., R. Kaplan, R.W. Gammon, H. Kaplan, R.F. Wallis and J.J. Quinn, 1974, Magneto-Surface Polaritons on n-Type InSb for the Geometry $\boldsymbol{H} \| \boldsymbol{k}$, in: Proc. 12th Intern. Conf. Phys. Semicond., 1974, Stuttgart, ed. M. Pilkuhn (Teubner, Stuttgart) p. 546.
Palik, E.D., R. Kaplan, R.W. Gammon, H. Kaplan, R.F. Wallis and J.J. Quinn, 1976, Phys. Rev. **B13**, 2497.
Perry, C.H., B. Fischer and W. Buckel, 1973, Solid State Commun. **13**, 1261.
Pinczuk, A. and E. Burstein, 1970, Physics of Raman Scattering in Opaque Semiconductors, in: Proc. 10 Intern. Conf. Phys. Semicond., Cambridge, Massachusetts, 1970, eds. Keller, P., J.C. Hensen and F. Stern (USA Atomic Energy Commission) p. 727.
Placido, F. and K. Hisano, 1973, Phys. Status Solidi. B **55**, 113.
Reshina, I.I. and V.M. Zolotarev, 1973, Fiz. Tverd. Tela, **15**, 3020. (1974, Sov. Phys.-Solid State, **15**, 2012.)
Reshina, I.I., Yu.M. Gerbstein and D.N. Mirlin, 1972, Fiz. Tverd. Tela, **14**, 1280. (1972, Sov. Phys.-Solid State, **14**, 1104).
Reshina, I.I., D.N. Mirlin and A.G. Bantschikov, 1976, Fiz. Tverd. Tela, **18**, 506.
Romanov, Yu.A., 1964, Zh. Eksp. Theor. Fiz. **47**, 2119.
Schuller, E., G. Borstel and H.J. Falge, 1975a, Phys. Status Solidi B **69**, 467.
Schuller, E., H.J. Falge and G. Borstel, 1975b, Phys. Lett. **A54**, 317.
Schuller, E., H.J. Falge and G. Borstel, 1975c, Phys. Lett. A **55**, 109.
Stern, E.A. and R.A. Ferell, 1960, Phys. Rev. **120**, 130.
Tarkhanian, R.G., 1975, Phys. Status Solidi B **72**, 111.
Ushioda, S. and J.D. McMullen, 1972, Solid State Commun. **11**, 299.
Vinogradov, V.S., 1962, Fiz. Tverd. Tela, **4**, 712 (1962, Sov. Phys.-Solid State, **4**, 519).

Valdez, J.B., G. Mattei and S. Ushioda, 1978, Solid State Commun. **27**, 11.
Wallis, R.F. and J.J. Brion, 1971, Solid State Commun. **9**, 2099.
Wallis, R.F., J.J. Brion, E. Burstein and A. Hartstein, 1974, Phys. Rev. B **9**, 3424.
Yakovlev, V.A. and G.N. Zhizhin, 1974, Pis'ma JETF, **19**, 333.
Yakovlev, V.A., V.G. Nazin and G.N. Zhizhin, 1975, Opt. Commun. **15**, 293.
Yakovlev, V.A., G.N. Zhizhin, M.A. Moskalova and V.G. Nazin, 1978, in: Spectroscopy of molecules and crystals, (Naukova Dumka, Kiev) p. 82 (in Russian).
Zhizhin, G.N., M.A. Moskalova, V.G. Nazin and V.A. Yakovlev, 1974, Fiz. Tverd. Tela, **16**, 1402. (1974, Sov. Phys.-Solid State, **16**, 902.)
Zhizhin, G.N., M.A. Moskalova and V.A. Yakovlev, 1975a, Fiz. Tverd. Tela, **17**, 2217. (1976, Sov. Phys.-Solid State, **17**, 1468).
Zhizhin, G.N., O.I. Kapusta, M.A. Moskalova and V.A. Yakovlev, 1975b, Fiz. Tverd. Tela, **17**, 2008 (1976, Sov. Phys.-Solid State, **17**, 1313).
Zhizhin, G.N., M.A. Moskalova and V.A. Yakovlev, 1976, Fiz. Tverd. Tela, **18**, 252. (1976, Sov. Phys.-Solid State, **18**, 146).
Zhizhin, G.N., M.A. Moskalova, V.G. Nazin and V.A. Yakovlev, 1977a, Fiz. Tverd. Tela, **19**, 2309. (1978, Sov. Phys.-Solid State, **19**, 1352).
Zhizhin, G.N., M.A. Moskalova, V.G. Nazin and V.A. Yakovlev, 1977b, Zh. Eksp. Theor. Fiz. **72**, 687. (1978, Sov. Phys.-JETP, **45**, 360).
Zemski, V.I., E.L. Ivchenko, D.N. Mirlin and I.I. Reshina, 1975, Solid State Commun. **16**, 221.

CHAPTER 2

Surface Exciton Polaritons from an Experimental Viewpoint

J. LAGOIS and B. FISCHER

Max-Planck-Institut für Festkörperforschung
Heisenbergstr. 1, D-7000 Stuttgart 80
F.R. Germany

Surface Polaritons
Edited by
V.M. Agranovich and D.L. Mills

Contents

1. Introduction 71
   1.1. Wannier excitons and the polariton picture 71
   1.2. Spatial dispersion and additional boundary conditions 74
2. Surface polaritons in spatially dispersive media 76
   2.1. The dispersion relation 76
   2.2. The electromagnetic fields near the boundary 80
3. Experimental results 82
   3.1. Experimental observation of surface exciton polaritons 82
   3.2. Surface exciton polaritons in anisotropic crystals 85
   3.3. Modified exciton eigenenergies near the surface 88
4. Conclusions 90
References 91

1. Introduction

Surface exciton polaritons are the quanta of elementary excitations at the boundary of a crystal whose corresponding bulk excitation is a Wannier-type exciton. The description of excitonic polaritons requires inclusion of spatial dispersion, i.e., the wave-vector dependence of the dielectric function. This dependence leads to particular properties of excitonic surface polaritons which differ from those of surface polaritons belonging to other elementary excitations. The detailed study of surface exciton polaritons therefore gives information about the similarities of and the differences between various surface polaritons. Furthermore, the behavior of an exciton near a semiconductor surface is of great interest in connection with spatial dispersion. Surface exciton polaritons probe the decay of the excitonic polarization at the boundary of a crystal very sensitively. Experiments on surface exciton polaritons therefore offer the possibility to test macroscopic and microscopic models describing the behavior of excitons near a semiconductor surface.

In this introduction, we shall briefly review the concept of Wannier excitons and describe the effects of spatial dispersion on the optical properties of a crystal. Section 2 will provide the basics of the theoretical description of surface exciton polaritons, and in particular of their dispersion relation. The most important part of this article will be devoted to a review of experimental results (sect. 3). This section will include a description of the present knowledge of the behavior of exciton polaritons in the vicinity of a boundary. In sect. 4 we shall conclude with an outlook of further experimental and theoretical work.

1.1. Wannier Excitons and the Polariton Picture

Excitons are elementary excitations of solids. They consist of an electron and a hole bound together by the Coulomb attraction. Behavior and description of excitons are similar to those of hydrogen atoms (see, for instance, Knox 1963) where the excitonic hole is playing the part of the hydrogenic proton.

The properties of the polarizable medium in which the excitons are embedded determine the binding energy of electron and hole. Strong binding yields excitons with a hydrogen-like excitonic Bohr radius which is comparable to the lattice constant of the crystal. Those excitons are called Frenkel excitons. Excitons with small binding energy have a Bohr radius larger than the lattice spacing and are called Wannier excitons. They are delocalized in the crystal. The excitons formed by a hole from the valence band and an electron from the conduction band are such Wannier excitons in semiconductors. In this article we shall deal only with this type of exciton. (For Frenkel-type excitons see, for instance, Philpott and Swalen 1978).

The Coulomb attraction of electron and hole forming a Wannier exciton is effectively screened in a crystal with a large dielectric constant. Excitonic binding energies in semiconductors therefore are several milli-electron-volts, thus much smaller than the binding energy of an electron in a hydrogen atom. The screening also leads to excitonic Bohr radii of several nanometers which are much larger than the hydrogenic Bohr radius.

The energy of excitons is given by a hydrogen-like energy series with main quantum number n (Knox 1963):

$$E_n(k) = E_g - R_{exc}(1/n^2) + \hbar^2k^2/2M. \quad (1)$$

E_g is the energy of the band gap, R_{exc} is the Rydberg or binding energy of the exciton, k the exciton's wave vector, M the sum of electron and hole mass, and $\hbar$ is Planck's constant divided by 2π.

The energy of the pure exciton state versus the wave vector of the exciton is shown in fig. 1. The ground state of the crystal belongs to $E = 0$

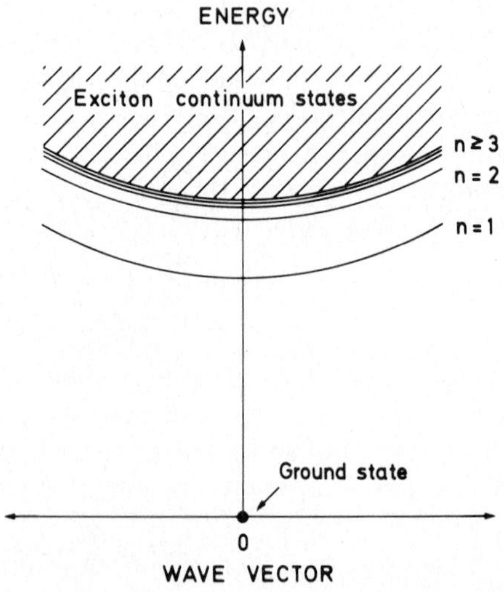

Fig. 1. Schematic energy versus wave-vector relation for hydrogen-like exciton series. n denotes the main quantum number of the energy series.

and $k = 0$ (no exciton). The exciton continuum states describe free electrons and holes in the conduction and valence band, respectively, with relative kinetic energy too large to produce a bound level.

The excitation of excitons is effectively accomplished by photons of appropriate energy. The electromagnetic radiation field in the crystal couples to the excitonic oscillators. Therefore, the excitons yield a contribution to the polarizability of the crystal. This contribution is described in macroscopic terms by the complex dielectric function $\epsilon(\omega, k)$ which depends in general on the frequency ω and the wave vector k.

It was shown by Huang (1951) for bulk waves that the dispersion relation of such a coupled state of elementary excitation and light has the form

$$c^2k^2/\omega^2 = \epsilon(\omega, k) \qquad (c \text{ is the vacuum velocity of light}). \tag{2}$$

The quanta of the eigenstates which obey this equation were called "polaritons". Hopfield (1958) used this name in connection with excitons. In the present literature, every coupled state consisting of an elementary excitation of a crystal and the electromagnetic radiation field is called a polariton. For a better classification, it has become common use to call them exciton polaritons, phonon polaritons, or plasmon polaritons depending whether the dielectric function ϵ describes excitons, phonons, or plasmons.

A possible elementary excitation forming a polariton is an optical phonon which we will deal with temporarily to introduce the polariton picture. A schematic energy versus wave-vector dispersion curve for phonon polaritons is shown in the upper part of fig. 2. The lower (1) and the upper (2) transverse bulk polariton branches are obtained from eq. (2) using the well-known dielectric function $\epsilon(\omega)$ for phonons (Huang 1951):

$$\epsilon(\omega) = \epsilon_\infty\left(1 + \frac{\omega_L^2 - \omega_T^2}{\omega_T^2 - \omega^2 - i\omega\Gamma}\right) \tag{3}$$

with ϵ_∞ being a frequency- and wave-vector-independent background dielectric constant ($\epsilon = \epsilon_\infty$ for $\omega \to \infty$). ω_T and ω_L are the transverse and the longitudinal resonance frequencies, and Γ is the empirical damping constant ($i = \sqrt{-1}$).

All bulk polaritons have their counterparts at the surface of the crystal. Those surface polaritons are solutions of Maxwell's equations and are bound to the surface. These modes are transverse magnetic waves with periodic electric fields along the surface and exponential decay of the fields perpendicular to the surface on both sides. It has been shown by Stern (1958) and by Ritchie and Eldridge (1962) that surface polaritons always exist in the frequency regime where the real part of the dielectric function ϵ of the crystal is negative and its absolute value exceeds the dielectric constant of the adjacent medium. This condition is fulfilled between a

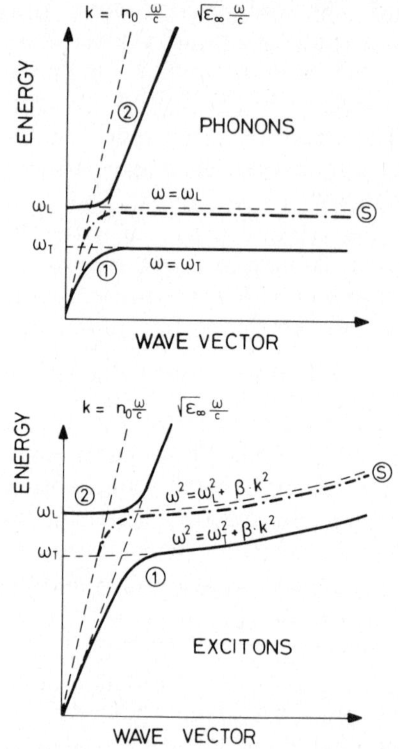

Fig. 2. Schematic energy versus real wave-vector dispersion relations for phonon polaritons without spatial dispersion (upper part) and exciton polaritons with spatial dispersion (lower part). The symbols (1) and (2) denote the lower and upper bulk polariton branches, (S) the surface-polariton dispersion curves. The refractive index n_0 of the adjacent medium was taken $n_0 = 1$ for all figures of this article.

transverse bulk resonance frequency and a limiting frequency close to the corresponding longitudinal bulk frequency. A dispersion curve typical for surface phonon polaritons is shown in the upper part of fig. 2 (branch (S)). Note that the phase velocity, ω/k, of the surface wave is always smaller than that of a wave in the adjacent medium with refractive index n_0. These surface polaritons are therefore called nonradiative since they cannot decay into photons radiating away from the surface, nor can they be excited by simply shining light onto the surface.

1.2. Spatial Dispersion and Additional Boundary Conditions

Let us now return to excitons as the elementary excitation under consideration. The description of excitonic polaritons requires one to include the kinetic energy of the exciton's center of mass motion besides the rest energy at $k = 0$. This inclusion leads to a dielectric function $\epsilon(\omega, k)$ for excitons (Hopfield and Thomas 1963)

$$\epsilon(\omega, k) = \epsilon_\infty\left(1 + \frac{\omega_L^2 - \omega_T^2}{\omega_T^2 - \omega^2 + \beta k^2 - i\omega\Gamma}\right). \tag{4}$$

The spatial dispersion arises from the term $\beta k^2 = (\hbar\omega_T/M)k^2$, in contrast to the dielectric function for phonons in eq. (3). Equation (4) may be used for calculating the optical properties of a crystal in the exciton energy region.

The dispersion curves of the bulk polaritons can be calculated from eqs. (2) and (4) which lead to an equation quadratic in k^2. There are two transverse modes 1 and 2 with dielectric functions ϵ_1 and ϵ_2, which propagate in the same direction of the crystal and with the same polarization for a given frequency ω. The longitudinal solution in the bulk is defined by

$$\epsilon_L(\omega, k) = 0, \tag{5}$$

which yields a mode with frequency $\omega^2 = \omega_L^2 + \beta k^2$ (neglecting damping, so $\Gamma = 0$).

The lower transverse and the longitudinal branches possess for excitons a k^2 dependence at large wave vectors as can be seen from the lower part of fig. 2. The coupling with the photon dispersion line forces the lower transverse branch to bend over into the light line at small wave vectors. This bending is typical for polaritons.

The coexistence of more than one polariton branch at a given energy is an immediate consequence of the wave-vector dependence of the exciton energy. Figure 2 shows two transverse and one longitudinal bulk modes at a given frequency above ω_L. Also below the longitudinal resonance frequency, one always has to deal with three modes. The wave vectors of two of them, however, are purely imaginary and therefore are not shown in fig. 2.

The coexistence of three bulk modes at a given energy complicates a "simple" reflection experiment and its theoretical fit by Fresnel's equations. The reason is that a transverse wave incident on a spatially dispersive crystal in general excites all three bulk modes in the crystal with certain amplitudes. The two Maxwell's boundary conditions for tangential electric and magnetic fields from which Fresnel's equations are derived do not suffice to determine the amplitude ratios of the three waves inside the crystal. One needs additional information expressed in the form of the so-called additional boundary condition, subsequently abbreviated ABC. The problems in connection with appropriate ABC are treated by V.M. Agranovich in ch. 5.

It was speculated for a long time that the dispersion curve of surface exciton polaritons extends somewhere in the gap between the transverse and the longitudinal bulk frequencies, in analogy to the case of optical phonons without spatial dispersion. Surface exciton polaritons are influenced by spatial dispersion in a similar way as the corresponding bulk polaritons. Since the transverse and longitudinal modes shift with the wave vector the surface mode may no longer be confined to an energy gap as is the case for phonons (Maradudin and Mills 1973). A possible dispersion curve is indicated in the lower part of fig. 2 (branch (S)).

The surface mode with such a dispersion curve coexists with all three bulk modes at a given energy. It was shown by Maradudin and Mills (1973) and by Mahan (1974) that there is an additional damping mechanism for the surface mode due to an admixture of bulk modes at the same energy. This type of damping is absent without spatial dispersion.

2. *Surface Polaritons in Spatially Dispersive Media*

2.1. *The Dispersion Relation*

The dielectric behavior of a crystal with an excitonic resonance (main quantum number $n = 1$) showing spatial dispersion is characterized by the dielectric function $\epsilon(\omega, k)$ of eq. (4). Let us consider the boundary between such a homogeneous insulating crystal of the described behavior and an adjacent medium of frequency-independent dielectric constant $\epsilon_0 = n_0^2$ (for vacuum: $n_0 = 1$) which both fill up infinite half spaces.

We search for the eigenstates of this boundary between the crystal with excitons having spatial dispersion and the adjacent medium. The eigenstates of the boundary are electromagnetic waves, one transverse mode outside the crystal (called mode "0") obeying the dispersion relation $k_0 = n_0\omega/c$, and three modes inside the crystal belonging to the two transverse polariton modes (called mode "1" and "2") and to the longitudinal one (called mode "L"). All these modes are matched together at the crystal surface by the Maxwell boundary conditions and the ABCs to form the eigenstate of the boundary. We call the quanta of this eigenstate at the boundary surface exciton polaritons if the mode "0" is localized at the boundary and is travelling nearly parallel to it. The behavior of the surface eigenstates is determined only by the dielectric function of the crystal bulk and by the refractive index of the adjacent medium.

Before we define the electric fields of the polariton modes we split the wave vectors into components parallel and perpendicular to the surface. (In this article, we follow the notation of Lagois and Fischer (1978a)). The component $k_\parallel$ parallel to the surface is the same for all waves outside and inside the crystal because of the phase matching condition. This common $k_\parallel$ together with the dielectric functions of all modes describes the eigenstate of the boundary. Therefore, $k_\parallel$ is usually plotted in energy versus wave-vector dispersion relations of surface polaritons.

The components perpendicular to the surface are then given by

$$k_{j\perp}^2 = k_j^2 - k_\parallel^2 = \epsilon_j k_v^2 - k_\parallel^2 \quad (j = 0, 1, 2, \mathrm{L}) \tag{6}$$

with the wave vector in vacuum being $k_v = \omega/c$. If $k_{j\perp}^2$ is negative and therefore $k_{j\perp}$ imaginary, the corresponding mode j is exponentially decaying perpendicular to the surface and periodically travelling parallel to the surface, thus being localized at the boundary. If $k_{j\perp}^2 > 0$, then mode j is not localized.

According to fig. 3, we define the electric field amplitudes $\boldsymbol{E}_0$ of the transverse mode "0" outside the crystal, the amplitudes $\boldsymbol{E}_1$ and $\boldsymbol{E}_2$ of the transverse modes and $\boldsymbol{E}_\mathrm{L}$ of the longitudinal mode in the crystal. The electric fields are all polarized parallel to the plane of incidence (the *xz*-plane in fig. 3) because only this geometry yields eigenstates of the boundary.

The conservation of the tangential component of the electric field amplitude may then be written:

$$E_0 k_{0\perp}/k_0 = E_1 k_{1\perp}/k_1 + E_2 k_{2\perp}/k_2 + E_\mathrm{L} k_\parallel / k_\mathrm{L}. \tag{7}$$

The magnetic field is already tangential. The conservation of the tangential magnetic field amplitude yields

$$E_0 \epsilon_0/k_0 = E_1 \epsilon_1/k_1 + E_2 \epsilon_2/k_2. \tag{8}$$

Equations (7) and (8) are the two Maxwell boundary conditions of the

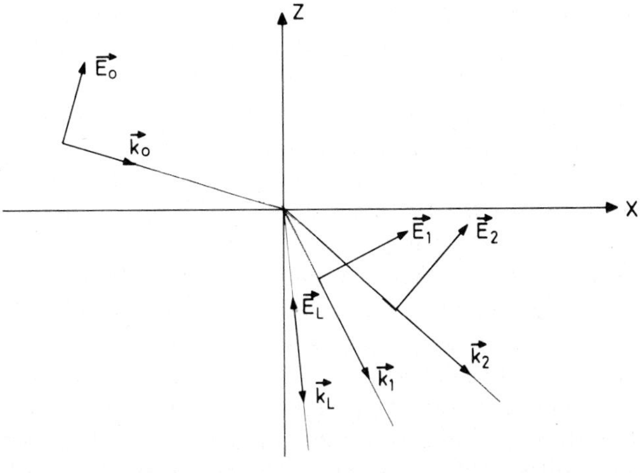

Fig. 3. The boundary between the crystal ($z < 0$) and the adjacent medium ($z > 0$). The wave vectors $\boldsymbol{k}$ and the electric field amplitudes $\boldsymbol{E}$ of one wave outside the crystal (mode "0") and three waves inside the crystal (modes "1", "2", and "L") are shown.

considered eigenstate. They connect the electric field amplitudes at the crystal surface of one wave outside and three waves inside the crystal which all have the same $k_\parallel$ but different $k_\perp$. The explicit behavior of the surface eigenstate depends on the ABC which connects the electric field amplitudes inside the crystal. It was proposed as ABC by Pekar (1957, 1958) and applied also by Hopfield and Thomas (1963) that the excitonic contribution to the macroscopic polarization should vanish at the surface. Splitting the polarization and thus the electric field vectors into components parallel and perpendicular to the surface yields two special ABC equations for the eigenstate of the boundary

$$E_1(\epsilon_1 - \epsilon_\infty)k_{1\perp}/k_1 + E_2(\epsilon_2 - \epsilon_\infty)k_{2\perp}/k_2 - E_\mathrm{L}\epsilon_\infty k_\parallel/k_\mathrm{L} = 0 \tag{9}$$

$$E_1(\epsilon_1 - \epsilon_\infty)k_\parallel/k_1 + E_2(\epsilon_2 - \epsilon_\infty)k_\parallel/k_2 + E_\mathrm{L}\epsilon_\infty k_{\mathrm{L}\perp}/k_\mathrm{L} = 0. \tag{10}$$

Equations (7) to (10) form a system of four linear equations for the four unknown electric field amplitudes E_0, E_1, E_2, and E_L. This system has a solution only if the determinant equals zero (Lagois and Fischer 1978a):

$$\begin{aligned}\epsilon_1(\epsilon_2 - \epsilon_\infty)(\epsilon_\infty k_{0\perp} - \epsilon_0 k_{1\perp})k_\parallel^2 - \epsilon_2(\epsilon_1 - \epsilon_\infty)(\epsilon_\infty k_{0\perp} - \epsilon_0 k_{2\perp})k_\parallel^2 \\ + \epsilon_\infty k_{\mathrm{L}\perp}[\epsilon_1(\epsilon_2 - \epsilon_\infty)k_{0\perp}k_{2\perp} - \epsilon_2(\epsilon_1 - \epsilon_\infty)k_{0\perp}k_{1\perp} + \epsilon_0(\epsilon_1 - \epsilon_2)k_{1\perp}k_{2\perp}] = 0.\end{aligned} \tag{11}$$

Equation (11) is the dispersion relation between the wave vector $k_\parallel$ and the frequency ω of surface exciton polaritons for the chosen ABC. This relation describes the behavior of the electromagnetic fields associated with surface exciton polaritons.

The dispersion relation contains in general complex quantities because the wave vectors $k_\perp$ perpendicular to the surface may have imaginary parts even without any empirical damping Γ. Thus, the eigenstate of the boundary must have complex solutions for the wave vector $k_\parallel$ or the frequency ω (Maradudin and Mills 1973, Rimbey 1977). Dispersion relations of surface exciton polaritons have been derived by several authors using various ABC. A comparison of their results is given by Fischer and Lagois (1979).

The dispersion relation, eq. (11), has to be solved numerically to obtain the energy versus wave-vector relation which describes the behavior of the surface exciton polaritons. Figure 4 shows a three-dimensional plot of the dispersion relation of surface exciton polaritons for real wave vector $k_\parallel$ and complex frequency ω (Lagois and Fischer 1978a). The dispersion relation is fulfilled along the full line ($\Gamma = 0$).

The dispersion curve of the surface exciton polariton lies at frequencies above the transverse resonance frequency ω_T and at wave vectors above that of wave vector k_0 in the adjacent medium because they are localized at the surface. At larger wave vectors the real part of the dispersion relation intersects the longitudinal mode with $\omega^2 = \omega_\mathrm{L}^2 + \beta k^2$. Figure 4 shows that

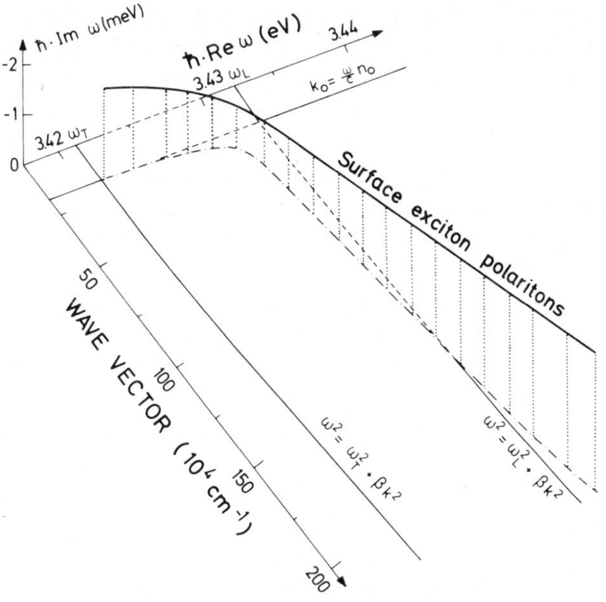

Fig. 4. Calculated dispersion curve of surface exciton polaritons for complex frequency ω and real wave vector $k_\parallel$. The calculation was done for the $C_{n=1}$ exciton in ZnO using the ABC by Pekar (1957, 1958) and Hopfield and Thomas (1963) (Lagois and Fischer, 1978a).

the energy of the surface eigenstate has an imaginary part which increases with increasing wave vector. In this region the projection of the dispersion relation into the real-ω real-k plane gives no longer a complete description and one always has to include the imaginary parts.

Dispersion relations of surface exciton polaritons obtained with different ABC deviate from each other especially at large wave vectors (see, for instance, Fischer and Lagois 1979). Experimental studies of surface exciton polaritons and of their dispersion relation especially for large wave vectors may offer a new possibility of obtaining information about the appropriate ABC for excitons. This information is not supplied by the experimental results which have been obtained up to now. All these experiments are as yet confined to the region at small $k_\parallel$ where the dispersion curves nearly coincide.

Surface exciton polaritons are attenuated by two different damping mechanisms. The first is described by the empirical damping constant Γ of the dielectric function in the bulk. This damping Γ is due to dissipative processes, for example by phonons or crystal inhomogeneities. We have no exact microscopic knowledge of these damping processes, and one always fits the empirical damping constant Γ to experimental results.

The second damping mechanism is very important for surface exciton polaritons and arises from the spatial dispersion of excitons (Maradudin and Mills 1973). A periodically travelling bulk mode always exists even in the energy region between the transverse and the longitudinal resonance frequencies. This lower polariton branch lies at wave vectors $k_1^2 = \epsilon_1 k_v^2$ greater than the wave vector $k_\parallel$ of the surface exciton polariton for a given frequency. Therefore, the component $k_{1\perp}^2$ of the mode "1" of the surface exciton polariton becomes mainly real and positive, as seen in eq. (6). Mode "1" is periodically travelling into the crystal bulk and transports intensity away from the surface. However, mode "1" builds up the surface exciton polariton together with all other modes, and their electric field amplitudes are matched together at the boundary. Therefore, all amplitudes have to decrease either along the surface for spatial damping or in time for temporal damping. This damping mechanism occurs for excitons even without any empirical Γ and is due only to spatial dispersion. The explicit strength of the damping via the periodical mode "1" depends on the frequency ω and the wave vector $k_\parallel$ and also on the additional condition. García-Moliner and Flores (1977) showed, for instance, that there is no damping of the surface exciton polariton by the coexisting bulk modes for the special ABC with vanishing derivative of the polarization at the surface.

2.2. The Electromagnetic Fields near the Boundary

In the previous section, we have considered the dispersion relation of surface exciton polaritons and their damping mechanism. This section presents a picture derived from the mathematical solution of the dispersion relation in order to give an impression of the behavior of the electromagnetic fields near the boundary.

Figure 5 shows a picture of the electromagnetic waves which build up the surface exciton polariton for spatial damping at fixed time (Lagois and Fischer 1978a). This picture represents the solution of the dispersion relation for a given real frequency ω and for the resulting complex wave vectors $k_\parallel$ and $k_{j\perp}$ without empirical damping (ABC of Pekar 1957, 1958). The upper part of fig. 5 represents the half space of the adjacent medium which is separated by the boundary from the crystal in the lower half space. The thicknesses of the wave fronts represent the electric field amplitudes.

The upper part of fig. 5 shows the mode "0" outside the crystal. This mode is nearly an evanescent wave propagating along and bound to the surface with constant amplitude along the propagation direction Re $\boldsymbol{k}_0$. The electric field amplitude is decreasing weakly along the surface due to the imaginary part of $k_\parallel$ and is decreasing strongly perpendicular to the surface due to the imaginary part of $k_{0\perp}$. As a consequence, the propagation

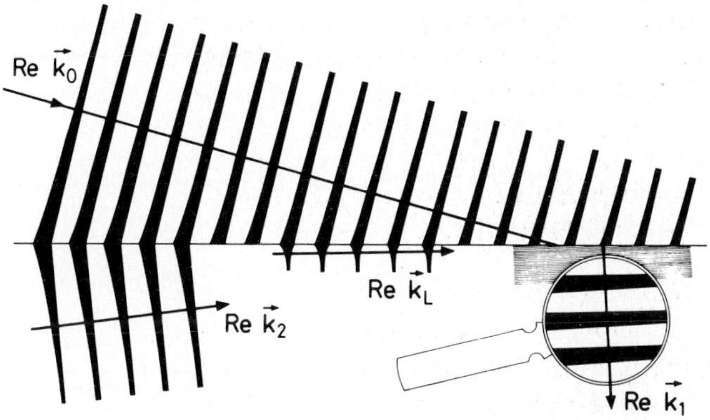

Fig. 5. Picture at fixed time of the electromagnetic waves which build up the surface exciton polariton for complex wave vector $k_\parallel$ and real frequency ω. Upper part: adjacent medium. Lower part: crystal with waves of the three modes which have to be extended throughout the whole crystal and have to be superimposed (Lagois and Fischer 1978a).

direction Re $\boldsymbol{k}_{0\perp}$ is not exactly parallel to the surface but slightly tilted with respect to the surface because the amplitude is constant along the propagation direction but the amplitude decreases parallel to the surface.

The lower part of fig. 5, which represents the crystal, is split for clarity into three internal sections. However, the waves of all three sections have to be extended throughout the half space of the crystal and have to be superimposed. The relative amplitudes of all waves are chosen equal at the boundary. In reality, they have to be calculated from the additional boundary conditions.

On the left-hand side and in the middle of fig. 5 the evanescent bound waves of mode "2" and mode "L" are shown. They are also slightly tilted with respect to the surface, and they have the same behavior as mode "0" outside the crystal.

Mode "1" is shown on the right-hand side of fig. 5. This mode gives to the surface eigenstate a contribution which is mainly periodic in the direction perpendicular to the surface because of Re $k_{1\perp} \gg$ Im $k_{1\perp}$. Parallel to the surface the amplitude is slightly decreasing as shown in the magnified inset. This periodical mode is responsible for the damping of the surface exciton polariton because it carries intensity from the surface into the crystal bulk. The propagation directions of the other modes bound to the surface are tilted to the surface because their amplitudes are coupled to that of mode "1" at the surface by the boundary conditions. The amplitudes have to decrease because of the intensity transport by mode "1".

Surface phonon polaritons which are damped by an empirical damping parameter show a similar behavior. The propagation directions inside and outside the crystal are also tilted to the surface, because the intensity dissipates in the crystal due to the damping along the propagation direction in the crystal. It should be emphasized that in case of excitons this tilting occurs even without empirical damping.

3. Experimental Results

3.1. Experimental Observation of Surface Exciton Polaritons

The dispersion relation of surface polaritons always lies at wave vectors $k_\parallel$ larger than those in the adjacent medium. The experimental excitation of nonradiative surface polaritons therefore can be achieved only by techniques which offer sufficiently large wave vectors parallel to the crystal surface. Different techniques fulfill this requirement: the use of periodic surface structures like gratings, the method of attenuated total reflection (ATR), nonlinear optical mixing processes, inelastic light scattering, and inelastic electron scattering.

Only attenuated total reflection and nonlinear optical mixing have been used experimentally to excite surface exciton polaritons. Data taken with electron energy loss spectroscopy have shown no structure related to surface exciton polaritons (Froitzheim and Ibach 1974). The problems of electronic excitations near the surface of a solid have been discussed by Mills (1977). Agranovich and Leskova (1979) proposed the observation of surface exciton polaritons in luminescence experiments. They assumed that either acoustic phonons or surface roughness enable the surface exciton polariton to become radiative. Experimental luminescence spectra show such structure (Brodin and Matsko 1980), however, it is not yet clear whether this structure is really coming from excitonic surface polaritons.

The excitation of surface polaritons in an ATR experiment occurs via an evanescent wave propagating along a prism base when light is totally reflected inside this prism (see fig. 6). The wave vector of this evanescent wave is given by

$$k_\parallel = n_p(\omega/c)\sin\alpha, \tag{12}$$

where n_p is the index of refraction of the prism, and α is the internal angle of incidence. It is obvious from eq. (12) that $k_\parallel$ can be varied between $k_\parallel = \omega/c$ and $k_\parallel = n_p\omega/c$ when the angle of incidence is varied within the range of total reflection ($\sin\alpha > 1/n_p$). The sample is brought at close distance to the prism for the coupling of intensity from the evanescent wave of the prism into a surface mode. The gap thickness d_{gap} has to be of the order of, or less than, the wavelength of the exciting light. This method of ATR was developed first by Otto (1968) for surface plasmon polaritons.

The scheme of an ATR experiment is shown in fig. 6. The upper part

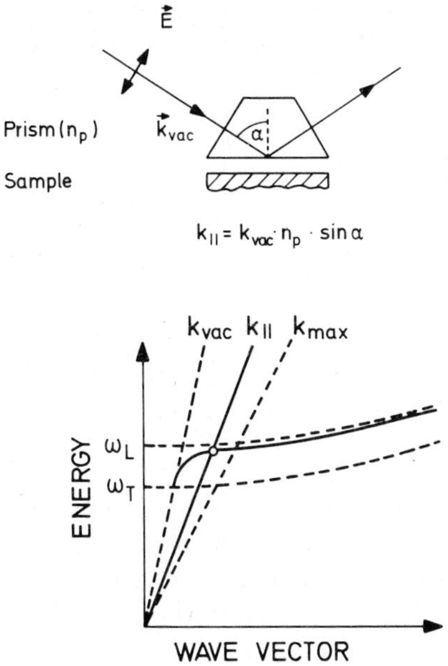

Fig. 6. Schematic drawing showing the principle of the attenuated total reflection method. Upper part: prism (refractive index n_p) and sample. Lower part: construction of the point of excitation on the surface polariton dispersion curve.

represents the experimental arrangement, the lower part gives a dispersion diagram. The electric field vector is polarized parallel to the plane of incidence. An excitation of surface polaritons occurs at the intersection of the line $k_{\|}(\omega)$ with the dispersion curve of a surface mode. An outcoupling of intensity from the totally reflected beam is observed as a dip in the ATR spectrum of internally reflected intensity versus energy. A variation of the angle of incidence in the ATR prism yields excitation at various points on the dispersion curve.

The application of this technique for the detection of surface exciton polaritons requires, for most materials to be investigated, cooling the whole prism arrangement down to liquid He temperatures. An additional complication is that the gap between prism and sample has to be controlled at these low temperatures.

An experimental ATR spectrum is shown in fig. 7 belonging to excitons in ZnO. Pronounced ATR minima occur between the transverse and longitudinal resonance frequencies of the $C_{n=1}$ and $C_{n=2}$ excitons of the hydrogen-like exciton energy series (n: main quantum number). The higher excited states with $n \geqslant 3$ are not resolved because the excitonic oscillator strength decreases with n^{-3} (Knox 1963). This hydrogen-like energy series occurs especially for excitonic surface polaritons and distinguishes them

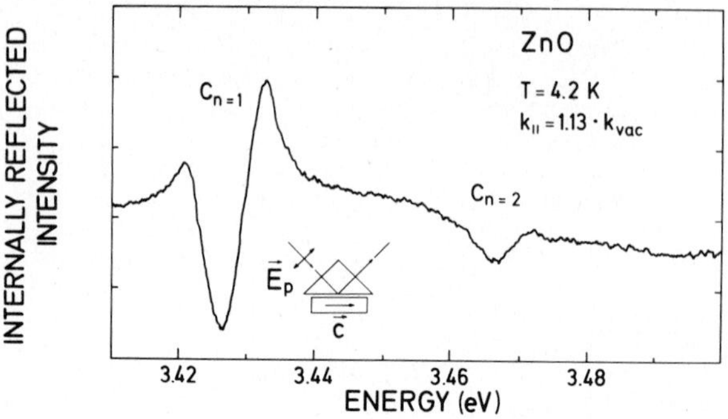

Fig. 7. Experimental ATR spectrum of the exciton series belonging to the C valence band in ZnO.

from other surface polaritons. The maximum in the ATR spectrum above the $C_{n=1}$ energy depends on the amount of coupling to the ATR prism. Further experiments on surface exciton polaritons on ZnO are reported by Lagois and Fischer (1976a, b, c, 1978b), Fischer and Lagois (1979), Lagois (1980, 1981a, b).

The ATR minima shift when the wave vector $k_\parallel$ is changed by variation of the angle of incidence at the prism base surface. Hirabayashi et al. (1976, 1977) excited surface exciton polaritons in CuBr and CuCl at different wave vectors. They evaporated thin crystalline films onto LiF as spacer material on the ATR prism. Experimental spectra of excitons in CuCl obtained for three different angles of incidence are shown in the left part of fig. 8. The incident light was polarized parallel (p) and perpendicular (s) to the plane of incidence. Coupling to surface modes occurs only for p-polarized light. The low energy tails of these spectra show interference structures of the CuCl films.

The corresponding calculated spectra in the right-hand part of fig. 8 were calculated with a multilayer reflection coefficient using a Lorentz oscillator model for the crystal without spatial dispersion [$\epsilon(\omega)$ of eq. (3)]. The calculations assumed a semi-infinite CuCl crystal. Therefore, the theoretical spectra do not reproduce the interferences at low energies.

Tokura et al. (1977) reported ATR spectra showing the excitation of surface exciton polaritons in cubic ZnSe single crystals. In their experimental setup Tokura et al. used a cryogenic adhesive with an index of refraction close to that of their glass prism. This arrangement improves the optical coupling between the prism and the film of MgF_2 evaporated as spacer layer on the ZnSe crystal. Surface exciton polaritons have also been observed in CdS using ATR techniques (Koda 1978).

A nonlinear excitation process for the generation of surface exciton polaritons has been applied by DeMartini et al. (1977). Their experimental

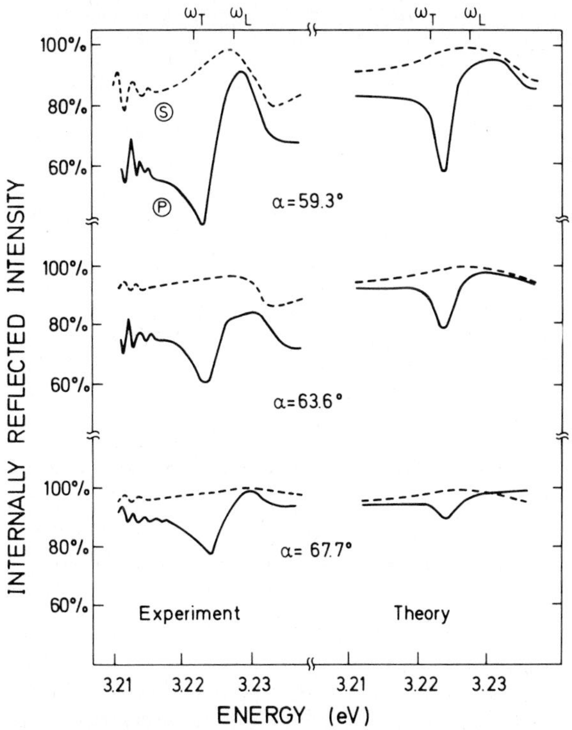

Fig. 8. Experimental and theoretical spectra of ATR for the Z_3 exciton in CuCl at 77 K with different angles α of incidence. The solid and dashed curves represent the curves for polarization parallel (p) and perpendicular (s) to the plane of incidence (after Hirabayashi et al. 1977).

dispersion relation for excitonic surface polaritons on ZnO is shown in fig. 9 together with the halfwidth of their experimental curves. They assumed that their experimental halfwidth equals the imaginary part of $k_{\parallel}$. The solid lines in fig. 9 are calculated without spatial dispersion [$\epsilon(\omega)$ of eq. (3)]. Spatial dispersion has been included by Fukui et al. (1979–1981) to describe the experiments by DeMartini et al. (1977). A detailed description of these experiments is given by Shen and DeMartini in ch. 14.

3.2. *Surface Exciton Polaritons in Anisotropic Crystals*

The dielectric function of solids with anisotropic crystal structure is a tensor which describes the anisotropic propagation of the elementary excitation in the crystal bulk. Thus, the optical properties of the crystal depend on the direction of the wave vector and on the polarization of the electric field. The anisotropy of the dielectric function influences the

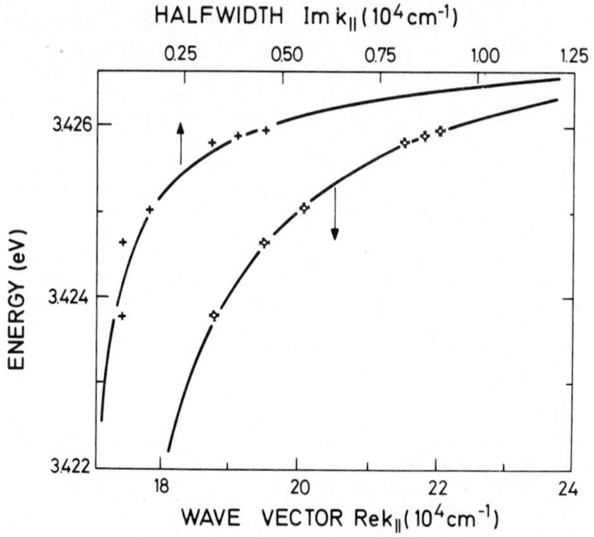

Fig. 9. Measured dispersion (⌖) and damping characteristics (+) of surface exciton polaritons belonging to the $C_{n=1}$ exciton in ZnO at 2 K. The solid curves are calculated without spatial dispersion (after DeMartini et al. 1977).

properties of surface exciton polaritons appreciably. Their dispersion relations lie at different energies depending on the geometry of the experimental arrangement, i.e., on the crystal orientation and the polarization of the light.

ZnO, for example, has a hexagonal crystal structure and therefore three valence bands called A, B, and C. They are split in energy by spin–orbit interaction and crystal field. There exist three different excitonic series belonging to the A, B, and C valence bands. The A- and B-excitons may be excited only if the electric field E of the light is polarized perpendicular to the hexagonal c-axis, whereas the light has to be polarized parallel to the c-axis to excite the C-excitons. It should be possible, therefore, to decide in an ATR experiment with light polarized parallel to the plane of incidence whether the electric field components parallel and/or perpendicular to the crystal surface excite the surface exciton polaritons. This distinction is not possible for isotropic media.

The upper two experimental spectra of fig. 10 show normal incidence reflection spectra for $E \perp c$ and $E \parallel c$. They reveal the reflectivity structure belonging to the A- and B-excitons and to the C-excitons, respectively. The lower part of fig. 10 shows experimental ATR spectra for three different orientations of ATR prism and c-axis for light polarized parallel (E_p) and perpendicular (E_s) to the plane of incidence (Lagois, 1981b). Coupling to surface exciton polaritons occurs only when the light is polarized parallel to the plane of incidence.

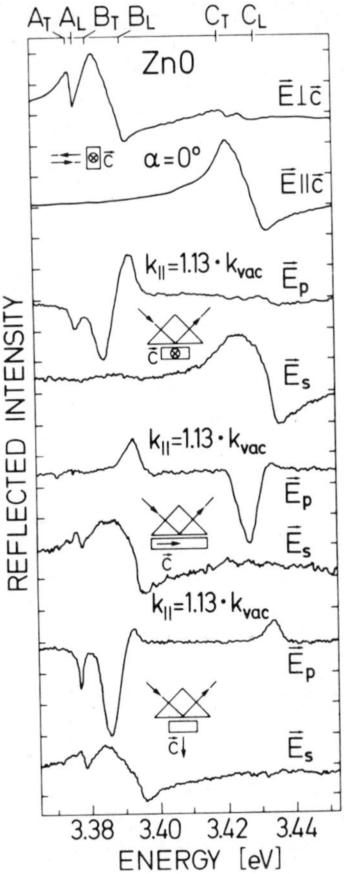

Fig. 10. Experimental normal-incidence reflection spectra and ATR spectra on ZnO at 4.2 K. The different orientations of the electric field $\boldsymbol{E}$ and the hexagonal $\boldsymbol{c}$-axis for light polarized parallel ($\boldsymbol{E}_p$) and perpendicular ($\boldsymbol{E}_s$) to the plane of incidence are shown. The transverse and longitudinal resonance energies of the $A_{n=1}$, $B_{n=1}$, and $C_{n=1}$ excitons are indicated on the upper scale (after Lagois 1981b).

ATR minima are seen for E_p at the energies of the A- and B- ($E \perp c$) or C-excitons ($E \| c$), respectively, depending on the orientation of the crystal c-axis with respect to the ATR prism. It is possible therefore to derive from the different geometries in fig. 10 for E_p that it is the electric field component E parallel to the crystal surface which is responsible for the coupling to surface exciton polaritons leading to ATR minima. The field component perpendicular to the surface leads to maxima at the different excitonic energies depending on the geometry. These features agree with calculations for surface exciton polaritons in anisotropic crystals including spatial dispersion (Lagois 1981b).

3.3. *Modified Exciton Eigenenergies near the Surface*

The properties of surface polaritons, and thus of the ATR spectra, are determined by the refractive index of the adjacent medium and by $\epsilon(\omega, k)$, the bulk excitation's dielectric function which determines also the reflection spectra. In the ideal case, the parameters entering the dielectric function $\epsilon(\omega, k)$ should be the same for fits to reflectivity or to ATR spectra taken at the same crystal surface.

The upper part of fig. 11 exhibits an experimental normal-incidence reflection spectrum of the $A_{n=1}$ and $B_{n=1}$ exciton polaritons in ZnO (Lagois 1980). The experimental spectrum is compared with a theoretical one obtained from a best fit to the experiment. The optical properties of the crystal are calculated using the dielectric function $\epsilon(\omega, k)$ of two excitonic transitions A and B and using an exciton-free surface layer (Lagois and Fischer, 1976b). The vanishing of the excitonic contribution to the

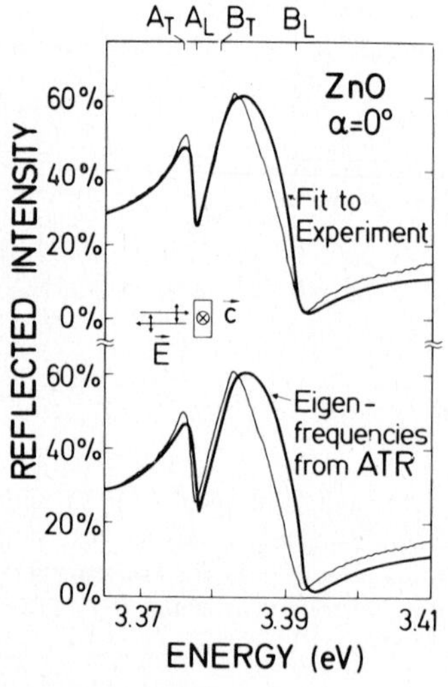

Fig. 11. Normal-incidence reflection spectra of the $A_{n=1}$ and $B_{n=1}$ excitonic polaritons in ZnO. Experiment (weak line) at 4.2 K. Upper part: theory (strong line) with parameters obtained from fit to the experimental reflection spectrum. Upper scale gives the transverse and longitudinal eigenenergies. Lower part: theory with eigenenergies obtained from fit to the experimental ATR spectrum (Lagois 1980).

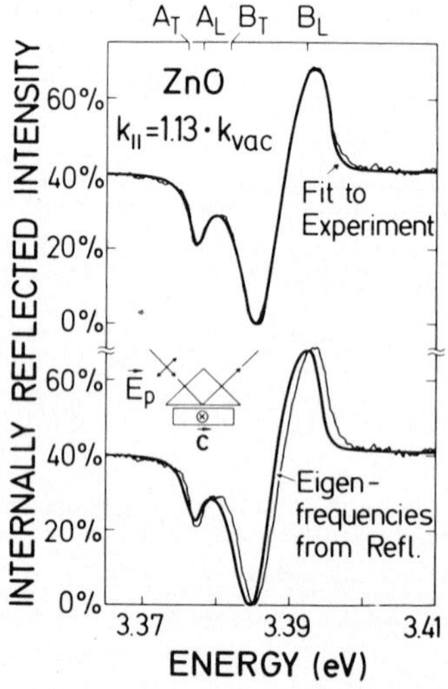

Fig. 12. Attenuated-total-reflection spectra of the $A_{n=1}$ and $B_{n=1}$ excitonic polaritons in ZnO. Experiment (weak line) at 4.2 K. Upper part: theory (strong line) with parameters obtained from fit to the experimental ATR spectrum. Upper scale gives the transverse and longitudinal eigenenergies. Lower part: theory with eigenenergies obtained from fit to the experimental reflection spectrum. (Lagois 1980).

macroscopic polarization at the internal interface is chosen as additional boundary condition (Pekar 1957, 1958).

The upper part of fig. 12 shows an experimental ATR spectrum taken on the same sample as the reflection spectrum of fig. 11. The experimental ATR spectrum is fitted with a theoretical one of the $A_{n=1}$ and $B_{n=1}$ excitonic surface polaritons in ZnO. The excitation of the longitudinal exciton branch is included.

A comparison of the parameters obtained from fits to reflection and to ATR spectra shows two striking results. First, the empirical damping constant Γ used in the dielectric function differs by a factor of about three. Second, the excitonic eigenenergies obtained by fitting ATR spectra are considerably higher than those obtained by fitting reflection spectra, although both theoretical spectra are determined by the same dielectric function containing the excitonic eigenenergies of the bulk crystal. This deviation amounts to about 10 percent of the splitting between the transverse and the longitudinal eigenenergies, and is conspicuous in the lower parts of figs. 11 and 12 where the reflection spectrum is calculated with eigenenergies obtained from the ATR fit, and the ATR spectrum with those obtained from the reflection fit. These theoretical spectra are compared with the same experimental spectra as in the upper parts of figs. 11 and 12, respectively.

The deviations of the parameters obtained for both experimental arrangements can be explained by considering the penetration depths of excitonic polaritons. The normal-incidence reflection spectra between the transverse and longitudinal excitonic eigenenergies are determined mainly by the lower polariton branch which is influenced by spatial dispersion (Hopfield and Thomas, 1963). These reflection spectra probe deeply into the crystal bulk. On the other hand, the structure of ATR spectra of excitonic surface polaritons is determined mainly by the upper polariton branch with a negative dielectric constant below the longitudinal eigenenergy. Therefore, the penetration depth of the electric field amplitude of the upper polariton branch is small in this energy region, (about 25 nm for ZnO), and ATR spectra probe regions near the surface (Lagois 1981a).

The comparison of spectra revealing excitation of bulk and surface exciton polaritons as shown in figs. 11 and 12 gives evidence for two conclusions. Primarily, the excitonic eigenenergies in near-surface crystal regions are increased considerably. Secondly, the damping of excitonic polaritons close to the surface is strongly enhanced. Model calculations including such a depth-dependence of excitonic energies and damping lead to an excellent agreement between experimental and theoretical reflection and ATR spectra as shown in fig. 13 (Lagois 1981a).

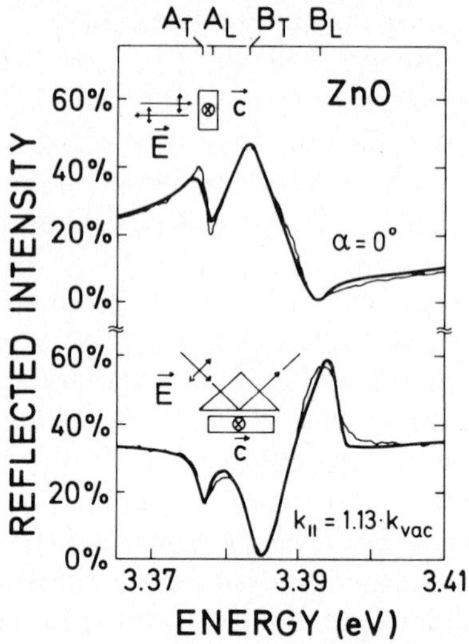

Fig. 13. Normal-incidence-reflection (upper part) and attenuated-total-reflection spectra (lower part) of the $A_{n=1}$ and $B_{n=1}$ excitonic polaritons in ZnO. The experimental spectra (weak lines) are the same as in figs. 11 and 12. The theoretical spectra (strong lines) are calculated with a depth-dependence of the excitonic eigenenergies and damping (Lagois 1981a).

4. *Conclusions*

We shall try to point out the status of knowledge and the current problems as a summary of the present article about surface exciton polaritons. We have seen that excitons as elementary excitation are dominated by spatial dispersion which arises from the kinetic energy of the exciton moving through the crystal. Spatial dispersion causes several coexisting polariton waves with different wave vectors but same energy and polarization. The problem of matching these bulk waves to the external radiation field at the crystal surface leads to the necessity of additional information about the behavior of exciton polaritons near the surface.

The information might emerge from microscopic theories which treat the problem of how the wave function decays when an exciton approaches the surface. Such theories are but now being developed (D'Andrea and Del Sole 1979, R. Zeyher 1980).

The experimental situation is more advanced. We showed rather detailed investigations of surface exciton polaritons, primarily studied with the attenuated-total-reflection technique. The optical spectra can be fitted rather well using a macroscopic additional boundary condition. Surface exciton polaritons are very sensitive for the parameters entering such a macroscopic model. A comparison of ATR and reflection spectra showed that the excitonic eigenfrequencies and damping change appreciably in regions close to the crystal surface. A microscopic theory which reproduces this experimental finding would be most desirable.

Future experiments should be devoted to the influence of external fields to clear up the origin of the modified exciton behavior near the surface. Also, the coupling of surface exciton polaritons with other surface modes, e.g. at interfaces between different surface active crystals, is still open to further studies.

Investigations of surface exciton polaritons are motivated by different interesting aspects. First of all, they are surface polaritons and show a general behavior like surface phonon or plasmon polaritons. In addition, however, they reveal the unique properties of excitons which are one of the elementary excitations in solids. Furthermore, surface exciton polaritons can be regarded as a sensitive probe for intrinsic and extrinsic contributions to band bending and the resulting field effects at the surface. They give, therefore, direct information on how an exciton behaves when approaching the boundary of a crystal.

Acknowledgment

The authors gratefully acknowledge many helpful discussions and a critical reading of the manuscript by H.J. Queisser and M. Altarelli. The authors also appreciate the permission to reprint figures from other publications. J. Lagois was supported by the Deutsche Forschungsgemeinschaft.

References

Agranovich, V.M. and T.A. Leskova, 1979, Pis'ma Zh. Eksp. Teor. Fiz. **29**, 151 [JEPT Lett. **29**, 135 (1979)].

D'Andrea, A. and R. Del Sole, 1979, Solid State Commun. **30**, 145.

Brodin, M.S. and M.G. Matsko, 1980, Solid State Commun, **35**, 375.

DeMartini, F., M. Colocci, S.E. Kohn and Y.R. Shen, 1977, Phys. Rev. Lett. **38**, 1223.

Fischer, B. and J. Lagois, 1979, Surface exciton polaritons, in: Excitons, Topics in Current Physics, Vol. 14, ed. K. Cho (Springer Verlag, Heidelberg), ch. 4, p. 183.

Froitzheim, H. and H. Ibach, 1974, Z. Phys. **269**, 17.

Fukui, M., V.C.Y. So and G.I. Stegeman, 1979, Solid State Commun. **30**, 683.

Fukui, M., V.C.Y. So and G.I. Stegeman, 1980, Phys. Rev. **B22**, 1010 (1980).

Fukui, M., A. Kamada and O. Tada, 1981, Solid State Commun. **37**, 455.

García-Moliner, F. and F. Flores, 1977, J. Physique **38**, 851.

Hirabayashi, I., T. Koda, Y. Tokura, J. Murata and Y. Kaneko, 1976, J. Phys. Soc. Japan **40**, 1215.

Hirabayashi, I., T. Koda, Y. Tokura, J. Murata and Y. Kaneko, 1977, J. Phys. Soc. Japan **43**, 173.

Hopfield, J.J., 1958, Phys. Rev. **112**, 1555.
Hopfield, J.J. and D.G. Thomas, 1963, Phys. Rev. **132**, 563.
Huang, K., 1951, Proc. Roy. Soc. (London) **A208**, 352.
Knox, R.S., 1963, Theory of excitons, in: Solid State Physics, Suppl. 5, eds. F. Seitz and D. Turnbull (Academic Press, New York).
Koda, T., 1978, as quoted in: Brillante, A., I. Pockrand, M.R. Philpott and J.D. Swalen, 1978, Chem. Phys. Lett. **57**, 395.
Lagois, J. and B. Fischer, 1976a, Phys. Rev. Lett. **36**, 680.
Lagois, J. and B. Fischer, 1976b, Solid State Commun. **18**, 1519.
Lagois, J. and B. Fischer, 1976c, Surface exciton polaritons, in: Proc. 13th Intern. Conf. Phys. Semicond., Rome, 1976, ed. F.G. Fumi (North-Holland, Amsterdam) p. 788.
Lagois, J. and B. Fischer, 1978a, Phys. Rev. **B17**, 3814.
Lagois, J. and B. Fischer, 1978b, Introduction to surface exciton polaritons, in: Festkörperprobleme – Advances in Solid State Physics, ed. J. Treusch (Vieweg, Braunschweig) Vol. XVIII, p. 197.
Lagois, J., 1980, Observation of modified exciton-polariton behavior near a semiconductor surface, in: Proc. 15th Intern. Conf. Phys. Semicond., Kyoto, 1980, ed. S. Tanaka J. Phys. Soc. Japan **49**, 397 (1980) Suppl. A.
Lagois, J., 1981a, Phys. Rev. **B23**, No. 10 (1981).
Lagois, J., 1981b, Solid State Commun. (in press).
Mahan, G.D., 1974, Electron interaction with surface modes, in: Elementary excitations in solids, molecules and atoms, Pt.B, eds. J. T. Devreese, A.B. Kunz, and T.C. Collins (Plenum Press, New York) p. 93.
Maradudin, A.A. and D.L. Mills, 1973, Phys. Rev. **B7**, 2787.
Mills, D.L., 1977, Progr. Surf. Sci. **8**, 143.
Otto, A., 1968, Z. Physik **216**, 398.
Pekar, S.I., 1957, Zh. Eksp. Teor. Fiz. **33**, 1022 [Sov. Phys. JETP **6**, 785 (1958)].
Pekar, S.I., 1958, J. Phys. Chem. Solids **5**, 11.
Philpott, M.R. and J.D. Swalen, 1978, J. Chem. Phys. **69**, 2912.
Rimbey, P.R., 1977, Phys. Rev. **B15**, 1215.
Ritchie, R.H. and H.B. Eldridge, 1962, Phys. Rev. **126**, 1935.
Stern, E.A., 1958, as quoted in: Ferrell, R.A., 1958, Phys. Rev. **111**, 1214.
Tokura, Y., I. Hirabayashi and T. Koda, 1977, J. Phys. Soc. Japan **42**, 1071.
Zeyher, R., 1980, Polaritons in a bounded spatially dispersive medium, in: Proc. 1980 Annual Conf. Cond. Matter Div. European Phys. Soc., Antwerpen 1980, ed. J.T. Devreese (Plenum Press, New York, 1981).

CHAPTER 3

Surface Electromagnetic Wave Propagation on Metal Surfaces

G. N. ZHIZHIN, M. A. MOSKALOVA, E. V. SHOMINA
and V. A. YAKOVLEV

Institute of Spectroscopy
USSR Academy of Sciences
142092, Troitzk, Moscow region
USSR

Surface Polaritons
Edited by
V.M. Agranovich and D.L. Mills

Contents

Introduction 95
1. Surface electromagnetic waves at the two media interface 96
2. Propagation of IR surface electromagnetic waves on metal and dielectric surfaces 99
3. Surface electromagnetic waves in layered systems 104
4. Methods of surface electromagnetic wave transformation 107
5. Set-up for SEW propagation experiments 109
6. Optimization of conditions of prism coupling of surface electromagnetic waves with volume EM radiation 111
7. Diffraction effects in propagation of surface electromagnetic waves on metal surfaces 117
8. Determination of optical constants for metals in the IR region by the propagation of surface electromagnetic waves 121
9. Effects of metal surface roughness on propagation length of surface electromagnetic waves 126
10. Selective absorption of surface electromagnetic waves by thin films on metal surfaces 129
11. Absorption of surface electromagnetic waves by natural metal oxides 134
12. Investigation of SiO_x films on metals 137
References 143

Introduction

Recently ever increasing interest is being shown in the studies of surface properties of condensed media, the effects of thin coatings on optical, electrical and mechanical properties of solid surfaces, physical properties of the transition layer at the interfaces. These studies are of great scientific and practical importance and are directly related to integrated optics, microelectronics, biological systems, laser resonator reliability, etc. The investigation of surface polaritons and plasmons plays an important role in surface studies. They are extremely sensitive probes for the surface properties as they have the maximal electromagnetic field intensity just on the surface. The properties of this surface have a major effect on the origin, life-time and spectra of its excited states.

Lately a considerable progress has been achieved in the systematic application of surface polaritons and plasmons as a diagnostic tool for obtaining the characteristics of the solid state surfaces. Physical properties of condensed media near the interface differ from bulk properties, and the presence of thin films, defects, coatings, results in further change of properties. The information on these properties can be obtained mainly by studying spectral characteristics of the vibrations related to the surface, i.e., of surface polaritons. Initially the studies of surface polaritons at optical frequencies were concentrated mainly at the determination of dispersion curves and transformation of bulk and surface electromagnetic waves whereas no attention was paid to the propagation distances of surface polaritons.

First experimental studies on the propagation of infrared (IR) surface excitations at the "metal–air" interface were reported in 1973 (Shoenwald et al. 1973). In this case surface polaritons have a very small portion of mechanical energy and are, in fact, surface electromagnetic waves (SEW) which can propagate over macroscopic distances ($\sim$cm) in the IR region (Schoenwald et al. 1973, Bell et al. 1973). The possibility of exciting a SEW at some surface point and of detecting it at another point, spaced at several centimeters, makes a surface plasmon polariton an extremely sensitive instrument for the study of metal surfaces and coatings on them (Bell et al. 1975, Agranovich 1975).

1. *Surface Electromagnetic Waves at the Two Media Interface*

The existence of SEWs is due to the interface. In connection with further analysis of the experimental data we consider –in a rather detailed way– here and in sect. 2 the conditions of SEW existence at the plane interface of two (1 and 2) semi-infinite isotropic media with complex dielectric functions $\epsilon_1 = \epsilon_1' + i\epsilon_1''$ and $\epsilon_2 = \epsilon_2' + i\epsilon_2''$. For simplicity we limit ourselves to media with the magnetic permeability $\mu_1 = \mu_2 \equiv 1$ and dielectric function independent of the wave vector in the absence of spatial dispersion effects. This is true for many materials.

The directions of coordinate axes are chosen os that the z-axis is perpendicular to the media interface and $z = 0$ corresponds to the interface; the x-axis is directed along SEW propagation, the y-axis lies in the interface plane (fig. 1). The SEW field intensity is maximal at the media interface and decays exponentially on both sides of the interface; therefore, the solutions of the Maxwell equations for the SEW field are sought in the form (Landau and Lifschitz 1948):

$$H = H_- \exp(ik_x x + \kappa_1 z) \qquad (z < 0),$$
$$H = H_+ \exp(ik_x x - \kappa_2 z) \qquad (z > 0), \tag{1}$$

where k_x is the wave vector component along SEW propagation, κ_1 and κ_2 are wave vectors in media 1 and 2, respectively. Such a field satisfies the Maxwell equations together with the boundary conditions for p-polarization (TM modes). The field with s-polarization (TE modes) at the two media interface with $\mu_1 = \mu_2 = 1$ cannot satisfy the boundary conditions at any field wave vector. Therefore, in the SEW field the magnetic field is perpendicular to the xz-plane $H = H_y$ and the electric field vector lies in the

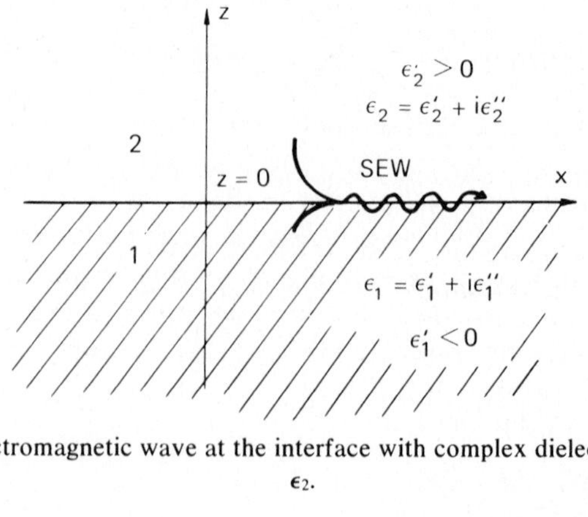

Fig. 1. Surface electromagnetic wave at the interface with complex dielectric functions ϵ_1 and ϵ_2.

xz-plane. The solution of the Maxwell equations with the given field and with the boundary conditions for the electric and magnetic fields at the transition from one medium to the other gives:

$$H_+ = H_-, \tag{2}$$

$$-\kappa_1/\epsilon_1 = \kappa_2/\epsilon_2. \tag{3}$$

The relationship (3) shows that a SEW exists only at the interface of media with dielectric functions having opposite signs. The solution of the Maxwell equations may also give the expression for the z-component of the SEW field wave vector:

$$\kappa_1 = (k_x^2 - \epsilon_1\omega^2/c^2)^{1/2},$$
$$\kappa_2 = (k_x^2 - \epsilon_2\omega^2/c^2)^{1/2}. \tag{4}$$

The sign is chosen in such a way that Re $\kappa_1 > 0$ and Re $\kappa_2 > 0$ since in the system under consideration the wave is decaying as it moves away from the surface.

From eqs. (3) and (4) the well-known dispersion equation for SEWs (the relationship between the wave vector and the frequency) is obtained:

$$k_x^2 = \frac{\omega^2}{c^2}\frac{\epsilon_1\epsilon_2}{\epsilon_1 + \epsilon_2}. \tag{5}$$

To simplify numerical calculations in the formulae given below we replace the circular frequency by the wavenumber (cm^{-1})$\omega/c = 2\pi\nu$. Then the dispersion equation becomes:

$$k_x = 2\pi\nu\left(\frac{\epsilon_1\epsilon_2}{\epsilon_1 + \epsilon_2}\right)^{1/2}. \tag{6}$$

The SEW field wave vector is complex: $k_x = k_x' + ik_x''$ and the imaginary part k_x'' characterizes the SEW field attenuation during its propagation along the media interface. The real part k_x' exceeds the wave vector of a bulk polariton in the surface-inactive ($\epsilon_2 > 0$) medium. Hence, bulk polaritons cannot transform linearly to SEW, and vice versa, SEW propagating along the smooth homogeneous surface cannot transform into bulk polaritons without breaking the translational symmetry. Thus, SEWs are non-radiative modes (Agranovich 1975).

The SEW electromagnetic field intensity decays exponentially with the distance away from the interface (fig. 1). The decay constants of the field (i.e. the distance at which the SEW field amplitude decreases by e times) are:

$$\begin{aligned} &\delta_1 = 1/\kappa_1' \quad \text{in the medium 1,} \\ &\delta_2 = 1/\kappa_2' \quad \text{in the medium 2.} \end{aligned} \tag{7}$$

z-Components of the wave vector (4) can be expressed in terms of the dielectric constants of the two contacting media (Ward et al. 1974):

$$\kappa_1' = -2\pi\left[\frac{(\epsilon_2'^2 - \epsilon_2''^2)(\epsilon_1' + \epsilon_2') + 2\epsilon_2'\epsilon_2''(\epsilon_1'' + \epsilon_2'') + (\epsilon_2'^2 + \epsilon_2''^2)\sqrt{D}}{2D}\right]^{1/2} \tag{8}$$

and

$$\kappa_1'' = 2\pi\left[\frac{-(\epsilon_2'^2 - \epsilon_2''^2)(\epsilon_1' + \epsilon_2') - 2\epsilon_2'\epsilon_2''(\epsilon_1'' + \epsilon_2'') + (\epsilon_2'^2 + \epsilon_2''^2)\sqrt{D}}{2D}\right]^{1/2},$$

where

$$D = (\epsilon_1' + \epsilon_2')^2 + (\epsilon_1'' + \epsilon_2'')^2,$$

and z-components of the wave vector in the medium 2 can be obtained from formulae (8) by reversing the positions of the components ϵ_1 and ϵ_2 and replacing κ_1' by $-\kappa_2'$ and κ_1'' by κ_2'', respectively.

Calculations by formulae (8) show that for metals in the IR region most of the energy of the SEW field lies above the metal surface, and inside the metal the SEW field decays rapidly.

To consider the structure of the SEW field we substitute the values of the wave vector components ($k_x = k_x' + ik_2''$, $\kappa_1 = \kappa_1' + i\kappa_2''$ and $\kappa_2 = \kappa_2' + i\kappa_2''$) into the expressions for the SEW field (1) and obtain

$$\begin{aligned} H_y &= H_0 \exp[i(k_x'x + \kappa_1''z) - (k_x''x - \kappa_1'z)] \quad \text{at} \quad z < 0, \\ H_y &= H_0 \exp[i(k_x'x - \kappa_2''z) - (k_x''x + \kappa_2'z)] \quad \text{at} \quad z > 0. \end{aligned} \tag{9}$$

It is easy to see that the plane of the SEW field constant amplitude intersects the interface of the two media at the angle $-\text{arc tg}(k_x''/\kappa_2')$ in the medium 2, and the constant phase plane intersects the interface at the angle $\text{arc tg}(k_x'/\kappa_2'')$; in this case the angle between the planes of the constant phase and the constant amplitude in the inactive medium is 90° since $k_x' \cdot k_x'' = \kappa_2' \cdot \kappa_2''$ (Chen 1977).

The graphical representation of the SEW field structure is shown in fig. 2a. Such waves in the case of one medium with absorption are called Zenneck modes. At the interface of two nonabsorbing media surface electromagnetic waves can also exist (Fano modes) providing $\epsilon_1 \cdot \epsilon_2 < 0$, but their structure is a little bit different (fig. 2b). The electric field vector lies in the incidence plane and the ratio of x and z components of the electric field can be determined from the Maxwell equations. In the inactive medium (medium 2):

$$\left(\frac{E_{2z}}{E_{2x}}\right)^2 = \left|\frac{k_x^2}{\kappa_2^2}\right| = |\epsilon_1|. \tag{10}$$

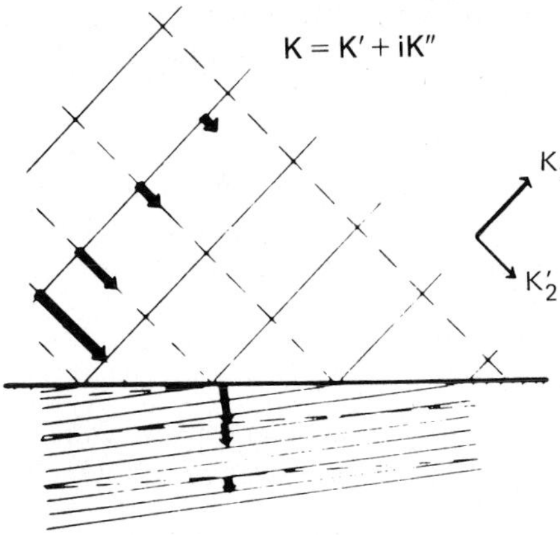

Fig. 2a. SEW field structure at the interface in the presence of absorption in one of the media (Zennek modes).

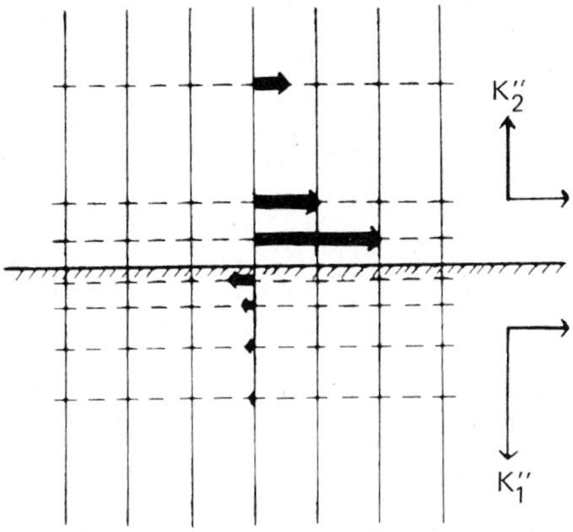

Fig. 2b. Structure of surface waves at the interface of nonabsorbing media with $\epsilon_1\epsilon_2 < 0$ (Fano modes). Dash lines – planes of constant amplitude, solid lines – planes of constant phase.

2. *Propagation of IR Surface Electromagnetic Waves on Metal and Dielectric Surfaces*

Crystal-optics methods are used to describe light propagation in bulk crystal materials. Agranovich (1975) developed the crystal optics of surface polaritons for the description of surface optical effects. The SEW refractive index is introduced in the same way as for volume waves

$$k_x \equiv 2\pi\nu\kappa_x. \tag{11}$$

Here κ_x is already dependent of dielectric functions of both contacting media

$$\kappa_x = \left(\frac{\epsilon_1\epsilon_2}{\epsilon_1+\epsilon_2}\right)^{1/2}, \tag{12}$$

while in traditional crystal optics the refractive index is defined as $n = \sqrt{\epsilon}$. With regard to damping the SEW index of refraction is complex $\kappa_x = \kappa_x' + i\kappa_x''$ and its imaginary part determines the SEW absorption coefficient α at the interface:

$$\alpha = 4\pi\nu\kappa_x''. \tag{13}$$

The value reciprocal to the absorption coefficient is called the SEW propagation length $L = 1/\alpha$ (the distance at which the SEW intensity decays to 1/e of its initial value).

On a smooth interface of two isotropic media the attenuation of SEWs is entirely due to the dielectric loss (for metals – to Joule losses) of the contacting media, and the SEW propagation length may be expressed from eq. (13) in terms of the media dielectric constants (Ward 1974):

$$\frac{1}{L} = 2k_x'' = \frac{2\pi^2\nu^2}{k_x'}\frac{B}{D}, \tag{14}$$

$$k_x' = 2\pi\nu\left\{\frac{1}{2D}[A + (A^2 + B^2)^{1/2}]\right\}^{1/2}, \tag{15}$$

where

$$A = \epsilon_2'(\epsilon_1'^2 + \epsilon_1''^2) + \epsilon_1'(\epsilon_2'^2 + \epsilon_2''^2),$$

$$B = \epsilon_2''(\epsilon_1'^2 + \epsilon_1''^2) + \epsilon_1''(\epsilon_2'^2 + \epsilon_2''^2)$$

and

$$D = (\epsilon_2' + \epsilon_1')^2 + (\epsilon_2'' + \epsilon_1'')^2.$$

At the interface of two metals or dielectrics SEW propagation is possible, if in a certain frequency region dielectric functions have opposite signs ($\epsilon_1 + \epsilon_2 < 0$) – which may be the case for metals with very different frequencies of plasma oscillations. Such systems were considered by Halevi (1978).

In the case when SEWs propagate over macroscopic distances ($L \geqslant 0.1$ cm) $k_x'' \ll k_x'$; and this is true, when the dielectric function of the surface-active medium ($\epsilon_1' < 0$) is more than that of the inactive medium and has the opposite sign: $-\epsilon_1' \gg \epsilon_2'$. Under these conditions approximate expressions for the wave vector and the SEW propagation length can easily be obtained:

$$\frac{1}{2k_x''} = L \simeq \frac{1}{2\pi\nu(\epsilon_2')^{3/2}}\frac{\epsilon_1'^2 + \epsilon_1''^2}{\epsilon_1''}, \tag{16}$$

$$k'_x \simeq 2\pi\nu(\epsilon'_2)^{1/2}. \tag{17}$$

The calculation precision with these formulae in the IR spectral region is not worse than 1% (Ward 1974).

For the calculation of the absorption coefficient or the SEW propagation length dielectric functions of the contacting media are required. First we consider SEWs on metal surfaces. The propagation of electromagnetic waves along smooth isotropic metal surfaces is described by the Maxwell equations with the complex dielectric function ϵ,

$$\epsilon = \epsilon' - \mathrm{i}\frac{4\pi\sigma}{\nu} \equiv \epsilon' + \mathrm{i}\epsilon'' \equiv (n - \mathrm{i}\kappa)^2, \tag{18}$$

where σ is the metal conductivity, n is the refractive index of the medium, κ is the absorption coefficient of the medium. In the case of a normal skin effect the field in a metal varies according to the exponential law; it occurs in those spectral regions where the local relation between the current and the field in the metal is valid. But when the average free path length of the conduction electron is comparable with the depth of the field penetration into the metal, the electron will pass through the areas with various field intensities which results in nonlocal relation between the current and the field. In this case the electric field in the metal is not exponential and the representation with the complex dielectric function is no more valid. This is the case of the anomalous skin effect.

For the normal skin effect the metal dielectric function ϵ_H is given (Ginzburg et al. 1955):

$$\begin{aligned} \epsilon_H &= \epsilon'_H + \mathrm{i}\epsilon''_H, \\ \epsilon'_H &= 1 - \frac{\nu_p^2}{\nu^2 + \nu_\tau^2}, \\ \epsilon''_H &= \frac{\nu_p^2 \nu_\tau}{\nu(\nu^2 + \nu_\tau^2)}, \end{aligned} \tag{19}$$

where $\nu_p^2 = Ne^2/\pi m^*$ is the plasma frequency (in cm^{-1}), N is the number of conduction electrons, e is the electron charge, m^* is the electron effective mass, ν_τ is the total effective frequency of electron collisions with phonons, electrons and impurities. A similar expression is used to describe the dielectric function of the free classical electron gas with the electron concentration N and the mass of the free electron m and, therefore, such an approximation in the case of metals is called the Drude approximation for optical constants of metals. The conditions when the skin effect is normal and the Drude approximation can be used are: $nv_F \ll c$ and $l_e \ll \delta$ (v_F is the electron velocity at the Fermi surface, l_e is the mean free path length of electrons, δ is the depth of the skin layer).

For metals in the crystalline state the case of the weakly anomalous skin effect in the near IR region (~ to 10 μm) is commonly found. The metal dielectric function here can be calculated with the correction for the anomalous skin effect ψ (Motulevich 1969)

$$\epsilon^{-1}(1-2\psi)=\epsilon_H^{-1},$$

$$2\psi \approx \frac{3}{8}\frac{v_F}{c}\frac{1+i(n/\kappa)}{(\nu_\tau/\nu)+i}\kappa. \tag{20}$$

The SEW propagation length on the metal surface in the case of the normal skin effect is described by

$$L \approx \nu_p^2/2\pi\nu_\tau\nu^2, \tag{21}$$

from which there follows a square dependence of the SEW propagation length on the light frequency. The SEW propagation length decreases with the increase of the frequency. For the visible spectral region the approximate formula (21) is not valid. In this case at the surface plasmon frequency (where $\epsilon_2=-\epsilon_1$) there is a resonance of the surface polariton refractive index and, therefore, the absorption coefficient for this frequency has its maximal value. In this region the SEW propagation length is negligibly small (several μm).

An equally strong SEW absorption can also be observed for phonon polaritons near the frequency of the surface phonon (ν_s). While moving away from the resonance, the absorption coefficient decreases. At the dielectric–air interface the surface polariton can propagate as a non-radiative mode within the frequency range $\nu_{TO}<\nu<\nu_s$ (McMullen 1975) (ν_{TO} is the frequency of the transverse phonon, and ν_s – see below). The SEW propagation length within this range is determined by eq. (16).

The dielectric function for one dipole-active oscillator is described by

$$\epsilon=\epsilon_\infty+\frac{(\epsilon_0-\epsilon_\infty)\nu_{TO}^2}{\nu_{TO}^2-\nu^2-i\gamma\nu}, \tag{22}$$

where ϵ_∞ is the high-frequency dielectric constant, ϵ_0 is the static dielectric constant, γ is the phonon damping in cm^{-1}. The real and imaginary parts of the complex dielectric function are:

$$\epsilon_1'=\epsilon_\infty+\frac{(\epsilon_0-\epsilon_\infty)\nu_{TO}^2(\nu_{TO}^2-\nu^2)}{(\nu_{TO}^2-\nu^2)^2+\gamma^2\nu^2},$$

$$\epsilon_1''=\frac{(\epsilon_0-\epsilon_\infty)\nu_{TO}\gamma\nu}{(\nu_{TO}^2-\nu^2)^2+\gamma^2\nu^2}. \tag{23}$$

The boundary frequency of the surface polariton ν_s is approximately determined from the condition $\epsilon_1'=-1$ and equals

$$\nu_s=\nu_{TO}\left(\frac{\epsilon_0+1}{\epsilon_\infty+1}\right)^{1/2}. \tag{24}$$

For the case when the second medium is air, i.e. $\epsilon_2' = 1$ and $\epsilon_2'' = 0$, the polariton propagation length near the transverse frequency may be expressed by an approximate formula (Ward 1974):

$$L_x \simeq \frac{1}{2\pi\nu}\left(\frac{\epsilon_1'^2 + \epsilon_1''^2}{\epsilon_1''}\right) \approx \frac{\epsilon_0 - \epsilon_\infty}{2\pi\gamma}. \tag{25}$$

Usually for dielectrics γ is of the order of $10\,\mathrm{cm}^{-1}$, and $\Delta\epsilon = \epsilon_0 - \epsilon_\infty$ is of the order of 1; then the appropriate propagation length of the surface polariton is about $100\,\mu\mathrm{m}$, i.e., well below the SEW propagation length on the metal (~ 1 cm). Propagation lengths of the surface polariton were measured experimentally for BeO (McMullen 1975) and α-$LiIO_3$ (Frolkov et al. 1979). The SEW propagation length on the semiconductor GaAs as a function of carrier concentrations was measured in the far IR region (Begley et al. 1979).

Now let us consider the interface of an isotropic medium with the dielectric function ϵ_1 and an anisotropic conductor with one principal axis of the dielectric tensor perpendicular to the interface whereas the second one is directed along the SEW propagation direction; tensor components along these directions are denoted by ϵ_z and ϵ_x, respectively. Then the SEW index of refraction becomes

$$\kappa_x = \left(\epsilon_1 + \epsilon_1^2 \frac{\epsilon_z - \epsilon_1}{\epsilon_1^2 - \epsilon_x\epsilon_z}\right)^{1/2}. \tag{26}$$

For SEW to exist the value ϵ_x must be negative; for ϵ_z two cases may be considered (Lubimov et al. 1972, Hartstein et al. 1973). The first one is that of weak anisotropy ($\epsilon_z < 0$) when SEW can propagate, if $\epsilon_x\epsilon_z > \epsilon_1^2$; the second one – with strong anisotropy ($\epsilon_z > 0$) when SEW exist at $\epsilon_z > \epsilon_1$.

For SEW propagation along considerable distances the condition $|\epsilon_x| \gg \epsilon_1$ has to be fulfilled and, if $|\epsilon_x\epsilon_z| \gg \epsilon_1^2$, an approximate formula is obtained:

$$\kappa_x \approx (\epsilon_1)^{1/2}\left[1 + \frac{\epsilon_1}{-2\epsilon_x}\left(1 - \frac{\epsilon_1}{\epsilon_z}\right)\right]. \tag{27}$$

With regard to damping $\epsilon_x = \epsilon_x' + i\epsilon_x''$ (similarly for ϵ_z) and the imaginary part of the refractive index determines the absorption coefficient α and the propagation length L:

$$\frac{1}{L} = \alpha \approx 2\pi\nu(\epsilon_1)^{3/2}\frac{\epsilon_x''}{|\epsilon_x|^2}\left[1 - \frac{\epsilon_1(\epsilon_z' - \epsilon_x'\epsilon_z''/\epsilon_x'')}{|\epsilon_z|^2}\right]. \tag{28}$$

From eqs. (27) and (28) it is seen that at $|\epsilon_z| \gg \epsilon_1$ SEW absorption is determined only by the value ϵ_x and is independent of the value ϵ_z; hence, an appreciable effect of anisotropy may only be expected, if $|\epsilon_z| \lesssim \epsilon_1$.

The ratio of the SEW propagation length with anisotropy to the SEW propagation length without anisotropy as a function of ϵ_z' is shown in fig. 3.

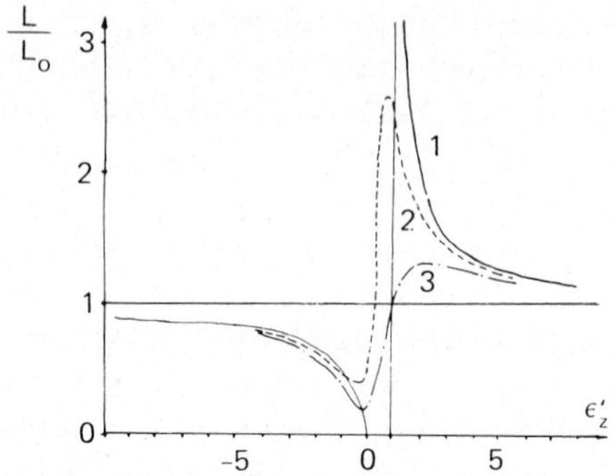

Fig. 3. Effect of the dielectric function value in the direction perpendicular to the surface on the SEW propagation length on "one-dimensional" metal: without damping (curve 1) and with damping ($\epsilon_z'' = 0.5$) at $\epsilon_x'/\epsilon_x'' = 0.5$ (curve 2) and $\epsilon_x'/\epsilon_x'' = 2$ (curve 3).

It is seen that at negative ϵ_z' the SEW propagation length decreases as $\epsilon_z' = 0$ is approached; in the region $0 < \epsilon_z' < 1$ SEWs do not exist at $\epsilon_z'' = 0$ (in the absence of absorption); at $\epsilon_z' > 1$ the SEW propagation length can become large and it decreases with the increase of ϵ_z'. At $\epsilon_z'' \neq 0$ the anisotropy effect decreases and its role depends also on the ratio ϵ_x'/ϵ_x'' (curves 2 and 3 in fig. 3).

Thus, if in the direction perpendicular to the surface there is no conductivity ($\epsilon_z'' = 0$), the SEW propagation length exceeds that in the isotropic case. Such a situation may be realized in quasi-one-dimensional and quasi-two-dimensional conductors with the high conductivity directed along the x-axis and the low conductivity perpendicular to the surface.

3. *Surface Electromagnetic Waves in Layered Systems*

A sharp interface of two media is rather rare to occur; more often various transition layers are present at this interface. In many cases the study of the spectra or optical properties of these transition layers is of the utmost interest. The properties of the transition layer often vary across the layer. Such a layer, inhomogeneous in thickness, is usually approximated by a set of homogeneous films with properties varying from one film to another. Then SEWs, propagating along such a layer, may be described by solving the Maxwell equations with usual boundary conditions: continuity of the tangential components of the SEW electric and magnetic fields at all interfaces. The fields in two semi-infinite media are of the same form as in the case of one interface (1); for the films the solutions are written as a sum of two waves:

$$H = [A_1 \exp(\kappa_i z) + A_2 \exp(-\kappa_i z)] \exp(-i\omega t + ik_x x). \tag{29}$$

Further the systems are discussed with no more than four media (two films), therefore, we consider the following system: medium 1 – film 2 of the thickness d – film 3 of the thickness l – medium 4. In such a system the dispersion equation for TM waves is

$$(\beta_1+\beta_2)A+(\beta_1-\beta_2)B\exp(-2\kappa_2 d)=0, \tag{30}$$

where

$$A=(\beta_2+\beta_3)(\beta_3+\beta_4)+(\beta_2-\beta_3)(\beta_3-\beta_4)\exp(-2\kappa_3 l),$$

$$B=(\beta_2-\beta_3)(\beta_3+\beta_4)+(\beta_2+\beta_3)(\beta_3-\beta_4)\exp(-2\kappa_3 l),$$

$$\beta_i=\epsilon_{ix}/\kappa_i, \qquad \kappa_i=2\pi\nu[\epsilon_{ix}(\kappa_x^2/\epsilon_{iz}-1)]^{1/2}, \quad i=1,2,3,4$$

ϵ_{ix} and ϵ_{iz} are x and z components of the dielectric tensor of the ith medium, respectively (the tensor is supposed to be diagonal and its principal axes coincide with the coordinate axes xyz).

Assuming that $d\to\infty$, we obtain for the three media:

$$A=0, \tag{31}$$

and for an interface (at $l\to\infty$)

$$\beta_1+\beta_2=0. \tag{32}$$

In contrast to the system with one interface of two media, for more complicated systems it is impossible to obtain a simple expression analogous to eq. (11) for the SEW refractive index which is sometimes also called "the reduced wave vector". Equations (30) and (31) are complicated and they can only be solved numerically by means of a computer.

The situation becomes easier in the case of thin films ($\kappa_2 d \ll 1$ and $\kappa_3 l \ll 1$) when approximate formulae can be obtained to calculate the SEW refractive index. Here the dispersion equation becomes

$$\frac{\kappa_1}{\epsilon_{1x}}+\frac{\kappa_4}{\epsilon_{4x}}+l\left(\frac{\kappa_3^2}{\epsilon_{3x}}+\frac{\epsilon_{3x}}{\beta_4^2}\right)+d\left(\frac{\kappa_2^2}{\epsilon_{2x}}+\frac{\epsilon_{2x}}{\beta_4^2}\right)\approx 0. \tag{33}$$

Let medium 4 be a metal and medium 1 a transparent dielectric. Since at $|\epsilon_4|\gg 1$, $\kappa_4\approx 2\pi\nu\sqrt{-\epsilon_{4x}}$, then

$$\kappa_x\approx\sqrt{\epsilon_{1z}}\left[1-\frac{\epsilon_{1x}}{2\epsilon_{4x}}+2\pi\nu l\frac{\epsilon_{1x}}{\sqrt{-\epsilon_{4x}}}\left(1-\frac{\epsilon_{1z}}{\epsilon_{3z}}-\frac{\epsilon_{3x}}{\epsilon_{4x}}\right)+\right.$$
$$\left.+2\pi\nu d\frac{\epsilon_{1x}}{\sqrt{-\epsilon_{4x}}}\left(1-\frac{\epsilon_{1z}}{\epsilon_{2z}}-\frac{\epsilon_{2x}}{\epsilon_{4x}}\right)\right]. \tag{34}$$

Thus, very thin films make an additive contribution to the SEW refractive index and, accordingly, the absorption by several films is also additive. The SEW absorption coefficient can be represented as

$$\alpha=\alpha_0+\Delta\alpha_1+\Delta\alpha_2, \tag{35}$$

where $\alpha_0 \approx 2\pi\nu \operatorname{Im}(\epsilon_{1x}\sqrt{\epsilon_{1z}})/2\epsilon_{4x}$ is the absorption at the interface without a film, $\Delta\alpha$ is the contribution of each of the films to absorption.

Formula (34) allows anisotropy to be taken into account. Experimental studies described in the present review were made on isotropic substrates, therefore, in most cases below we assume $\epsilon_{ix} = \epsilon_{iz} = \epsilon_i$, if no other special remarks are made.

It is seen from eq. (34) that even for transparent films SEW absorption increases when the film deposited has $\epsilon_3 > \epsilon_1$. This results from the redistribution of the SEW field in the presence of a film leading to the increase in the fraction of the SEW energy stored in the metal. For a metal substrate $1/\epsilon_4 \approx (\nu^2 - i\gamma\nu_\tau)/\nu_p^2$ at $\nu \gg \gamma$ and, if $\epsilon_1 = \epsilon_2 = 1$ and $\epsilon_3 = \epsilon_3'$, then

$$\Delta\alpha \approx \frac{2\pi^2\nu^2\nu_\tau l}{\nu_p}\frac{\epsilon_3 - 1}{\epsilon_3}, \tag{36}$$

i.e., a thin nonabsorbing film makes no change in the frequency dependence of the absorption coefficient, but only slightly decreases the SEW propagation length. If the film absorption is selective, its presence is easier to be found since in this case the frequency dependence of the absorption coefficient is no more a square one and the absorption maximum should exist. Bell et al. (1975) and Agranovich (1975) showed that at the vibration frequencies of the transition layer the resonances of the SEW refractive index appear. Then the absorption at the frequencies near the longitudinal vibrations of the layer is much stronger than that at the frequency of transverse vibrations.

A similar result can be obtained from the analysis of formula (34). The resonances of the refractive index correspond to the pole ϵ_{3x} and zero ϵ_{3z}, therefore, ȧbsorption maxima for SEW must be at the frequencies where $\operatorname{Im}(\epsilon_{3x})$ and $\operatorname{Im}(-1/\epsilon_{3z})$ are maximal with the second minimum much stronger, if $|\epsilon_{4z}| \gg 1$. For the isotropic metal surface in the spectral region $\nu_p \gg \nu \gg \nu_\tau$ the change of the absorption coefficient is

$$\Delta\alpha \approx 8\pi^2\nu^2\frac{\nu_\tau}{\nu_p} l \operatorname{Im}\left(-\frac{1}{\epsilon_{3z}}\right). \tag{37}$$

The resonance proposed in the works of Bell et al. (1975) and Agranovich (1975) can be detected not only by the SEW absorption maximum, but also by measuring the dispersion curve of the surface polariton (i.e. $\kappa_x'(\nu)$); however, in the case of a film on metals in the IR region very high precision of angular measurements is required (Chen 1977) since κ_x differs only slightly from unity (see formula (5)). For a film on a dielectric this resonance was observed by Zhizhin et al. (1977).

4. Methods of Surface Electromagnetic Wave Transformation

As noted above, the SEW field intensity is maximal at the interface and decays exponentially on both sides of the interface. For such a SEW field to exist one of the contacting media has to be surface-active [e.g. (3)], i.e., its dielectric function has to be negative (e.g. a metal in the IR wavelength region). The wave vector k_x of the SEW field, according to the dispersion equation (6), exceeds that of the bulk polariton at the same frequency in the surface inactive medium 2:

$$k_x > 2\pi\nu\sqrt{\epsilon_2}. \tag{38}$$

In other words, SEW phase velocity at the interface of two semi-infinite isotropic media is less than the respective phase velocities of volume EM waves in a surface-inactive medium ($\epsilon_2 > 0$) which results in the SEW nonradiativity. Wave vectors or phase velocities of volume EM waves and SEWs can be matched by various methods (since at the interface they correspond to TM modes, they interact with p-polarized polaritons, but not with s-polarized ones).

If there are some roughnesses on the surface or a grating is ruled on the surface, the coupling of surface and volume electromagnetic waves can be observed. In this case, the parallel component of the wave vector of the wave, falling onto the surface, increases by the wave vector connected with the Fourier components of the surface roughness or with the periodicity of the grating and some coupling occurs between volume EM waves and surface electromagnetic waves (Agranovich 1975). This type of coupling is called "grating coupling"; it is shown schematically in fig. 4. The radiation falling onto the grating excites SEWs if the angle of incidence

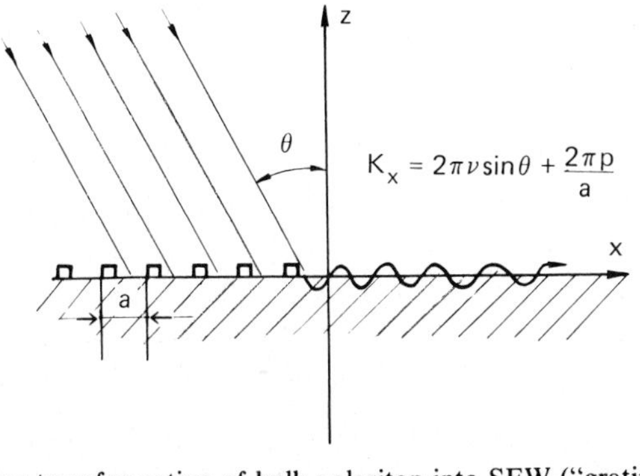

Fig. 4. Grating transformation of bulk polariton into SEW ("grating coupling").

onto the grating satisfies the condition,

$$k_x = 2\pi\nu\sqrt{\epsilon_2}\sin\theta + 2\pi p/a, \tag{39}$$

where a is the period of the grating, p is the integer.

The coupling of volume EM waves and SEW may also take place, if the third medium in the form of a prism (or a hemicylinder or a hemisphere) is placed close to the interface and its dielectric function exceeds that of the surface-inactive medium. The volume EM waves inside the prism which fall onto the interface at an angle θ bigger than the critical one of the total internal reflection, have the wave vector in excess of that of the volume EM waves in the surface-inactive medium. When the wave vectors of the volume EM wave inside the prism with the refractive index n coincide with the SEW wave vector, i.e., when

$$k_x = 2\pi\nu n\sin\theta, \tag{39a}$$

the p-polarized volume EM waves in the prism will excite the SEW on the surface. In this case the attenuation of the total internal reflection of volume EM waves takes place (fig. 5). This method is called "prism coupling" technique which uses the attenuated total reflection (ATR) effect (Otto 1968).

For the study of SEW propagation along the interface the two-prism coupling method is used (fig. 6) (Shoenwald et al. 1973, Bell et al. 1973). One prism couples SEW and the other decouples it.

Volume and surface waves can also be mutually transformed by means of nonlinear processes, e.g. Raman scattering or parametric mixing of volume EM waves and surface acoustic waves (Talaat et al. 1975). In the case of a high-conductivity metal (e.g. copper) and a prism with a low refractive index (KBr, $n = 1.5$) in the two-prism method it was found that SEW

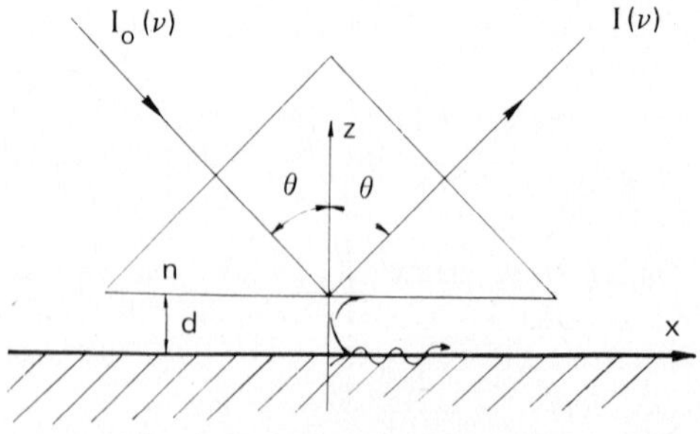

Fig. 5. Attenuated total internal reflection (ATR) prism. Radiation falling upon the prism base at the angle $\theta > \theta_{crit} = \arcsin(1/n)$ can excite SEW on metal surface spaced at a distance d from the prism base ("prism coupling").

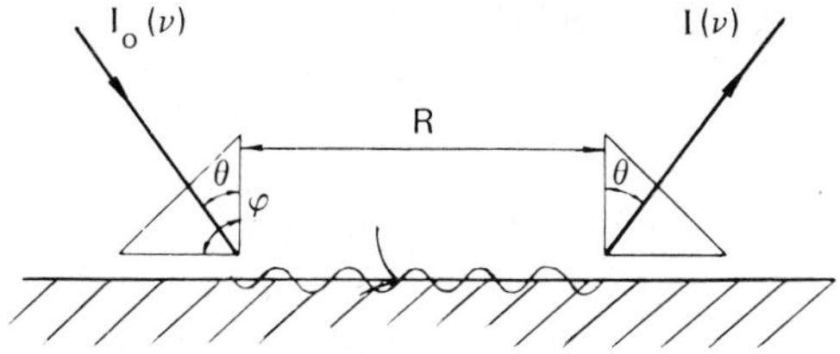

Fig. 6. Two-prism method of studying SEW propagation along the interface.

excitation is due to diffraction at the edge of the prism. Both experiments and calculations showed that in this case the angle of incidence θ is $\sim 0.1°$ less than the critical angle (Zhizhin et al. 1979). The field wave vector can also be changed by the violation of the translational symmetry at the edge of the obstacle in the following way: a gap should be provided over the metal surface by means of a blade edge. The light falls at the sliding angle onto this gap (fig. 7) and, passing through it, excites SEWs (Zhizhin et al. 1979).

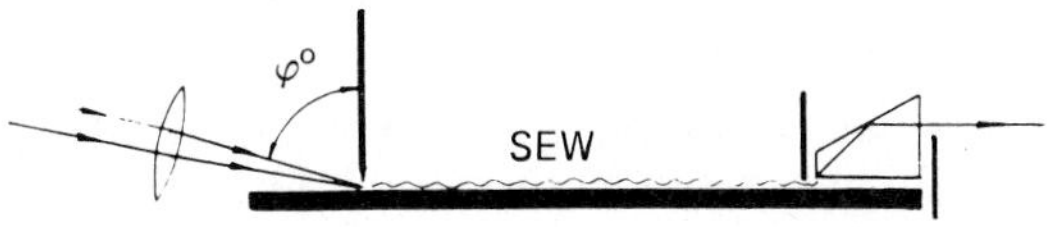

Fig. 7. Edge diffraction excitation of SEW.

5. *Set-Up for SEW Propagation Experiments*

Several experimental set-ups are known to study SEW absorption (Begley et al. 1976, Zhizhin et al. 1976, Zhizhin et al. 1978, Chabal and Sievers 1978). The basic scheme of the set-up is shown in fig. 8. Tunable IR lasers are used as radiation sources. Their modulated radiation falls onto the prism coupler of the optical unit. At the exit of the prism coupler the incident radiation is partially transformed into SEW which propagates along the sample surface. By means of the prism decoupler placed at a certain distance from the prism coupler the SEW is transformed into normal radiation which is recorded by a piroelectric detector. The laser power stability and the beam direction stability during measurements are controlled by the second radiation detector. The radiation is divided into two beams by a NaCl plate.

The main unit of the experimental set-up is the measuring device – the "optical unit" with a sample holder. As shown below, the SEW excitation and decay measurements require precise adjustment of the beam angle of

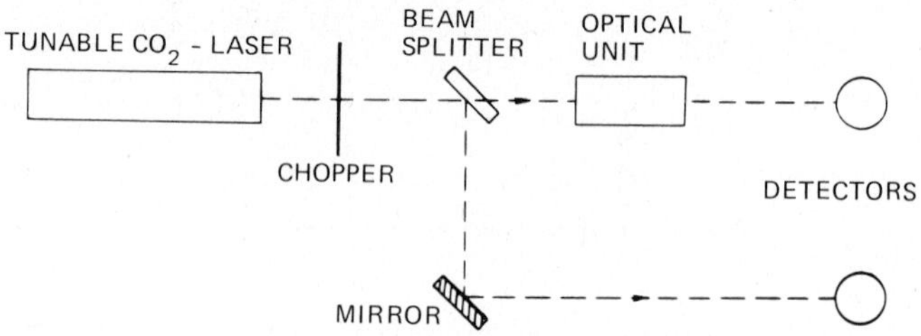

Fig. 8. Experimental set-up.

incidence into the prism coupler, of the air gaps between the prisms and the sample, and also of the controllable variation of the distance between the prisms. If prisms of the configuration shown in fig. 6 are used, the optical unit is mounted at the goniometer (Bryan et al. 1976) and the gaps between the prisms and the samples are adjusted by mylar film spacers.

When the form of the prism is as shown in fig. 9, the beams, entering the prism coupler and going out from the decoupler, are at the optical axis of the set-up. This allows independent variation of the gaps between the prisms and the sample by means of differential screws with the pitch 1 μm (fig. 10). Slight variation of the angle of incidence may be performed by varying the angle of inclination of the sample holder. In our optical unit the angle of incidence may be varied with the precision 0.1°.

The material for the prisms should be chosen so that the prism would almost not absorb laser radiation and had small dispersion in the range of the laser generation tuning. This allows us to avoid readjustment of the optical unit while tuning laser radiation over the whole frequency region. Table 1 gives the comparison of some materials for prisms to be used within the range of the tuned laser. Though silicon is one of the best materials for the prism couplers with its dispersion parameter of the refractive index, its application is limited due to the impurity absorption.

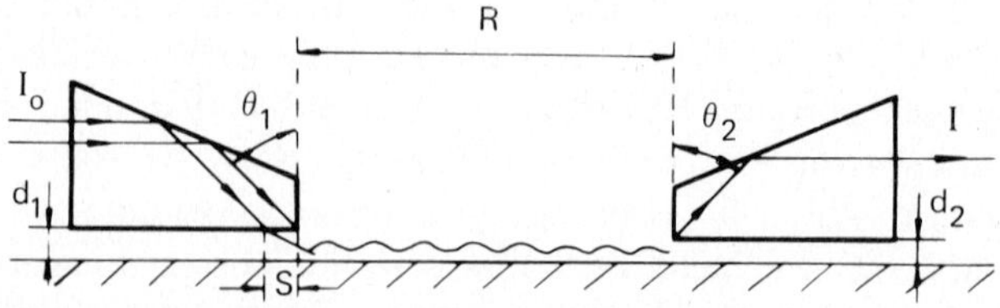

Fig. 9. Configuration of the coupling–decoupling prisms to provide coincidence of the direction of the incident beam with the direction of the beam coming out of the "optical unit". Parameters which affect the efficiency of coupling and decoupling of SEW: d_1, d_2 are the gap heights "prism–metal", θ_1 and θ_2 are the angles of beams in the prisms; S is the distance from the prism edge to the beam position; R is the prism separation distance.

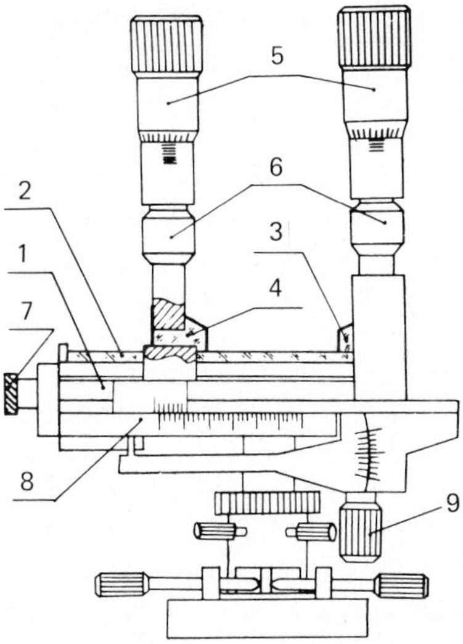

Fig. 10. "Optical unit": 1 – table; 2 – sample; 3,4 – prisms; 5 – micrometer screws (the pitch 1 μm); 6 – screw for rough gap height adjustment; 7 – adjustment of the prism separation distance; 8 – the scale of the prism position; 9 – adjustment of the table inclination angle.

Table 1
Comparison of the dispersion of prism materials suitable for SEW excitation by a tunable CO_2 laser

Prism material	Refractive index $n(\nu = 943\ \text{cm}^{-1})$	Dispersion	$\theta^0_{crit} = \arcsin 1/n$ $(\nu = 943\ \text{cm}^{-1})$	$\Delta\theta^0 = \theta_{crit}(\nu = 943\ \text{cm}^{-1}) - \theta_{crit}(\nu = 1080\ \text{cm}^{-1})$
Si	3.418	0.0003	17.012	0.002
Ge	4.002	0.0005	14.470	0.003
KBr	1.525	0.002	40.97	0.1
KCl	1.454	0.004	43.45	0.2
NaCl	1.490	0.006	42.15	0.3
BaF_2	1.392	0.013	45.92	0.8

6. *Optimization of Conditions of Prism Coupling of Surface Electromagnetic Waves with Volume EM Radiation*

The radiation falling upon the prism is not totally transformed into a surface EM wave. The coupling efficiency as depending on the coupled element was well studied in the radio-frequency range (Davarpanah et al. 1976). The maximal transformation coefficient (92%) was achieved when a rectangular waveguide was used as a coupling element. With the prism coupling the transformation coefficient was 60%.

In the IR range the prisms as a rule have been used as SEW excitation elements. To get the maximum coupling efficiency the effect of the following prism setting parameters must be studied: the angle of beam incidence onto the prism surface facing the metal θ_1, the height of the "prism–metal" gap (d_1, d_2), the place of the beam output at the prism S.

The dependence of the SEW input η_1 (the ratio of the SEW intensity to the intensity of incident radiation) on the angle of incidence was measured on a copper sample with the SEW propagation length $L = 1.8$ cm at the frequency 943 cm^{-1}. The output prism decoupler was separated by a distance of 4 cm from the prism coupler and this distance remained constant in the course of the experiment. The angle of incidence of the input beam in the prism coupler θ_1 was varied and the signal after the prism decoupler was recorded. The dependence obtained is shown in fig. 11a. The half-width of the SEW input efficiency equals the angular divergence of the CO_2-laser used (0.2°). The angle which provides the maximal input efficiency may be called optimal (θ_{opt}). In the case of a KBr prism θ_{opt} is less than the critical angle of the prism by $0.2 \pm 0.1°$. The angle was determined along the beam path under the prism without a sample and also from the calculation of the beam path geometry in the prism (all the angles of the prism as well as the angle between the laser beam and the sample were measured). A similar dependence of the SEW input efficiency on the angle of incidence was obtained by Begley et al. (1977), but since the absolute value of the angle of beam incidence θ_1 in the prism was not measured, the fact that a SEW is excited without total internal reflection remained unnoticed due to the smallness of the deviation from the critical angle. The design of the optical unit shown in fig. 10 allowed for observation of this.

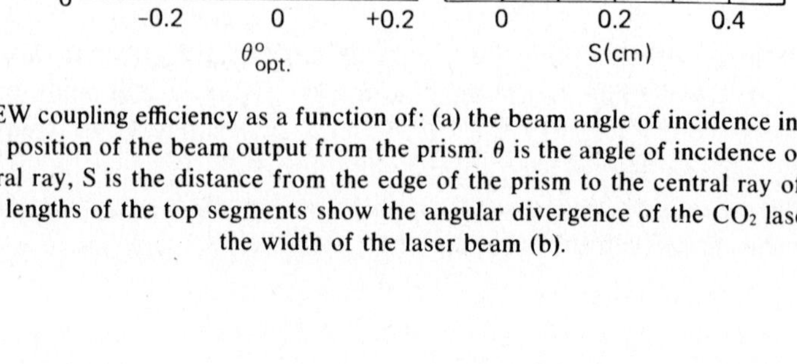

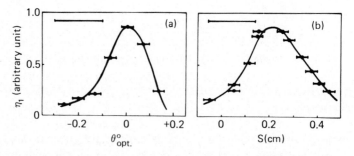

Fig. 11. SEW coupling efficiency as a function of: (a) the beam angle of incidence in the prism and (b) the position of the beam output from the prism. θ is the angle of incidence of the laser beam central ray, S is the distance from the edge of the prism to the central ray of the laser beam. The lengths of the top segments show the angular divergence of the CO_2 laser (α) and the width of the laser beam (b).

In order to explain the experimental result obtained, the prism influence on the metal SEW must be considered. The presence of a prism changes the surface polariton dispersion. Moreover, for a prism over a metal at the angles of incidence close to the critical angle (but smaller than the critical one) interference modes may exist in the gap between two surfaces with a large reflection coefficient (the metal and the prism at the angles of incidence close to 90°). These modes are analogous to those in a Fabry–Perot interferometer and they were reported by Gerbstein et al. (1975). These electromagnetic waves at not too large gaps ($\leq 100\ \mu m$) are not the surface waves as for them $k_x < 2\pi\nu$ and they do not decay exponentially with the distance away from the interface. It may be supposed, however, that such waves –due to diffraction– are partially transformed into surface waves at the edge of the prism. In the quasi-particle language the disturbance of the translational symmetry leads to the change of the wave vector. At such interfaces, in the same way as at two-metal interfaces (Schevchenko 1969), the wave transformation into a SEW may be accompanied by two additional processes: the reflection from the interface and emission of a volume electromagnetic wave. The problem of a wave outlet from under the prism over metal has not been solved as yet. Only the problem of the excitation of waveguide modes with an ATR prism, used in the integrated optics was intensely discussed in literature (Jogansen 1962, 1963, 1968; Tien 1971). In contrast to waveguide modes excited in a closed waveguide, the modes under consideration exist in an open resonator excited at the output from under the prism by their diffraction at the prism edge.

Normal modes in the "metal–air" gap can be calculated. Most metals have an oxide film on their surfaces, therefore, it is also included in calculations to clarify its effect on the SEW excitation. The system used for calculations was: a KBr prism (1) with $\epsilon_1 = 2.33$; an air gap ($\epsilon_2 = 1$) of the thickness $d(2)$; an oxide film on copper ($\epsilon_3 = 7.3$) of the thickness $l(3)$; copper (4); ϵ_4 was calculated by formula (19) with $\nu_p = 65\,100\ \mathrm{cm}^{-1}$ and $\nu_\tau = 340\ \mathrm{cm}^{-1}$. The dispersion was calculated by formula (30). The reflection coefficient in an ATR prism was also determined by

$$R = \frac{(\beta_1 - \beta_2)A + (\beta_1 + \beta_2)B \exp(-2\kappa_2 d)}{(\beta_1 + \beta_2)A + (\beta_1 - \beta_2)B \exp(-2\kappa_2 d)}. \tag{40}$$

The solutions of the complex eq. (30), corresponding to the minima in $R(\theta)$ at $k_x = (2\pi\nu\sqrt{\epsilon_4}) \sin\theta$, were found by the descent method in the gradient direction at various values of the gap d. The appropriate curves of the minima positions of reflection in the plane d–θ for $\nu = 943\ \mathrm{cm}^{-1}$ and $l = 0$ are shown in fig. 12. Note that all of them are below the line θ_{crit}, and only the left-hand curve intersects it at large d and approaches asymptotically the angle corresponding to SEW dispersion without a prism. It

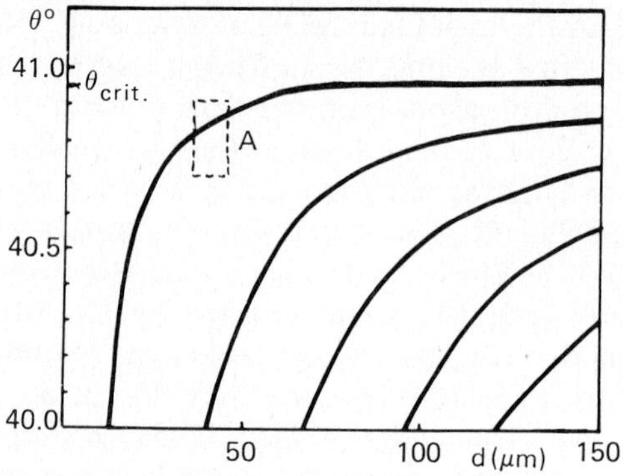

Fig. 12. Reflection minima positions as a function of the angle of incidence and the gap height "prism–copper".

means that only here the total internal reflection occurs while in other regions it is absent. Thus, the polariton dispersion curve for the system "prism–gap–metal" shifts, depending on the gap height, to the left of the photon straight line and coincides with the first Fabry–Perot mode, which, as noted above, is not a surface one. At the minimum of R the energy maximum of the incident beam falls within the "prism–metal" gap and partially transforms into a SEW. The conditions for SEW optimal excitation determined experimentally correspond to the section denoted by A in fig. 12.

The curves $(1 - R)$–d at various angles of incidence θ are shown in fig. 13.

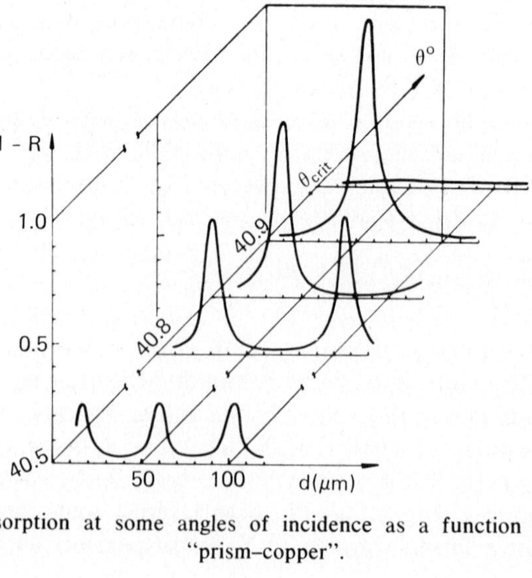

Fig. 13. Light absorption at some angles of incidence as a function of the gap height "prism–copper".

At $\theta > \theta_{crit}$ there is no absorption and at the angles smaller than the critical angle ($k \simeq 0.9993 \times 2\pi\nu$) the reflection inside the prism begins to decrease very fast. The incident wave therewith is almost entirely transformed into a wave under the prism. Then at the edge of the prism this wave is partially transformed into a SEW. It should be noted that the air-gap height in this case is close to that observed experimentally and the angle of incidence is less than the critical value – as it was also found in the experiment. If the divergence of the laser beam falling upon the prism is taken into account, and also the losses at the edge of the prism, then the efficiency should decrease appreciably.

The calculation has shown that the decrease of the metal conductivity by an order of magnitude does not change the positions of the curves shown in fig. 12, but the reflection has a minimum at lower values of the angles and gap heights. The decrease in the optimal gap heights with the decrease of the SEW propagation length was seen from the experimental data, but the variation of the optimal angles within the limits of the experimental errors was not observed.

The calculation of the oxide film effect has shown that a film of the natural oxide ~30 Å thick does not affect the efficiency of the SEW excitation; a remarkable influence of the film begins with the thickness ~100 Å, i.e., in spectral measurements of the films of such thicknesses the variation of the optimal coupling conditions have to be taken into account.

The SEW coupling efficiency coefficient varies with the angle φ of the prism coupler (fig. 6) (Begley et al. 1977). BaF_2 prisms were made. SEWs were excited on the surface of copper at various internal angles from 59.5 to 95.5° at 1048 cm^{-1}. The coupling efficiency was maximal (3.4%) at the angle $\varphi \doteqdot 66.8°$. In the case of the prism angle $\varphi = 90°$ the SEW coupling efficiency was 1%. Begley et al. (1977) did not measure the absolute value of the angle of incidence of the light beam.

The effect of the beam exit position from under the prism coupler on the coupling efficiency was studied with prism configurations shown in fig. 9. As would be expected, the optimal SEW excitation requires that the laser beam should be as close as possible to the prism edge (fig. 11b).

As it has already been shown by calculations, the SEW excitation efficiency is affected by both the angle of incidence in the prism and the gap height. The dependence of the SEW coupling efficiency on the size of both "prism–metal" gap heights was studied on a copper sample at the optimal angle of incidence in the prism coupler. The experimental values of coupling (1) and decoupling (2) efficiency as a function of the size of the gap between the prism and the copper surface are presented in fig. 14. The solid lines are drawn through the averaged experimental values obtained in two independent series of measurements. The gap dependence of the coupling efficiency is different for the coupler and decoupler prisms. In the

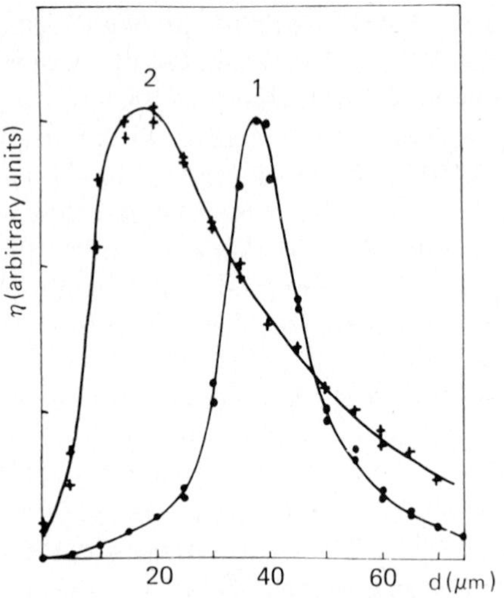

Fig. 14. SEW relative coupling (1) and decoupling (2) efficiency as a function of the gap height "prism–copper".

case of copper the curve of the SEW excitation efficiency has its maximum at the gap $d_1 = 38\ \mu m$ and a half-width $15\ \mu m$, while the decoupling efficiency is maximal at the gap $d_2 = 18\ \mu m$ and the curve is much broader (the half-width is $35\ \mu m$).

The existence of other maxima of SEW excitation efficiency at the same angle of incidence was shown by calculations (fig. 12). Besides the main maximum of the SEW excitation efficiency (at the gap $38\ \mu m$) an additional maximum with lower intensity at the gap $120\ \mu m$ was observed experimentally. If the angle of incidence is changed in the prism coupler, a SEW on the metal can also be excited, but already at a different gap height. The coupling coefficient in this case is less than in the optimal conditions.

The size of the gap "prism coupler–metal" at the optimal excitation depends on the metal and the conditions of its evaporation affecting its optical constants. Thus, for copper with the SEW propagation length $L = 1.8$ cm the optimal gap height is $38 \pm 5\ \mu m$, for copper with $L = 3.1$ cm the optimal gap height is $65 \pm 5\ \mu m$.

Comparison of the experimental results with calculations shows clearly that the optimal efficiency of SEW excitation in a system "prism–air–metal" is achieved at the beam incidence angles and air gaps which provide minimum reflection in the prism, i.e., far from the conditions of the total internal reflection. The excitation of the SEW on the metal surface at the output from under the prism is provided by the diffraction at the prism edge, due to which the agreement by the wave vector is achieved. A half-width of the SEW excitation efficiency curve vs the gap (fig. 14) may be explained by a series of reflection minima (fig. 13) over the range of the angles of incidence 0.2° due to the diffraction divergence of the laser beam.

The optimal gap under the prism coupler remains almost the same for various samples and various metals (Begley et al. 1977).

The effect of the "prism–metal" gap size on the angular distribution of radiation coming out of the prism was studied. This angular distribution was measured by scanning the detector height over the sample at various gap sizes (fig. 15). The highest intensity and the lowest divergence of the radiation at the output of the prism was obtained with the gap 15 μm.

The total efficiency η is equal to the product of the SEW excitation efficiency η_1 and the decoupling efficiency η_2 ($\eta = \eta_1 \cdot \eta_2$). It was determined by measuring the ratio of the output radiation intensity from the "optical unit" to the intensity of the incident radiation with optimized "prism–metal" gaps, angle of incidence and the beam position in the prism coupler. The studies were carried out on a copper sample with the SEW propagation length $L = 2.4$ cm. The prisms were spaced at a distance 6 cm with the optimal gaps (the prism coupler 45 μm, the prism decoupler 25 μm). In this case the total efficiency was $\eta = 3.5 \pm 0.3\%$.

7. *Diffraction Effects in Propagation of Surface Electromagnetic Waves on Metal Surfaces*

On the edge of a metal surface, where a SEW is propagating, surface electromagnetic waves can easily frustrate and transform into the bulk

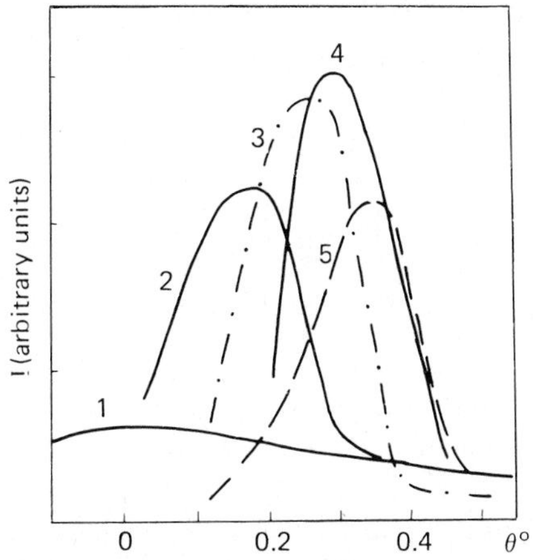

Fig. 15. Effect of the gap height between the prism decoupler (KBr) and metal (copper with $L = 1.8$ cm) on the angular distribution of radiation coming out of the prism. The gap heights are (d μm): (1) $d < 1$ (the prism is pressed to the metal); (2) 5; (3) 10; (4) 15; (5) 20.

radiation (this phenomenon is well known for the radio-frequency range (Agranovich 1975)). Figure 16 shows the radiation pattern (at $\nu = 943\ \text{cm}^{-1}$) for a SEW frustrating on the metal edge. This radiation pattern was measured by scanning the detector position over the sample in the direction perpendicular to its plane without a prism decoupler. To eliminate background (ghost) beams three diaphragms were placed over the sample at a distance 0.5 mm with the spacings ~1 cm. Several peaks were observed with the ratio of intensities 20:5:2:1 at the angles 0°; 2.4°; 3.8°; 5.1° (the angles were measured relative to the main peak).

An inverse process to the SEW frustration can be realized, i.e., SEW excitation at the sliding angle of incidence of a beam onto a sharp sample edge. This is known as the edge coupling technique (Chabal and Sievers 1978). The beam was focused onto the edge by a lens at the angle of incidence 88°. A SEW was actually excited on the metal surface edge which was further propagating along it (Zhizhin et al. 1979).

An artificial edge over the smooth metal surface can be provided by means of a razor blade held at a certain distance over the surface (fig. 7). The investigations have shown that the gap height for the optimal SEW excitation is about the same as in the case of a prism. With that way of excitation the efficiency of radiation transformation into a SEW obtained was equal to 1%.

If the metal is scratched (transverse metal stripes have been removed from the evaporated metal layer on the glass substrate), then a SEW approaching the scratch boundary, has to transform totally into bulk radiation, propagating almost parallel to the surface (see fig. 16). At the other scratch boundary an inverse transformation of a bulk radiation wave into a surface one takes place. SEW attenuation by scratches of various

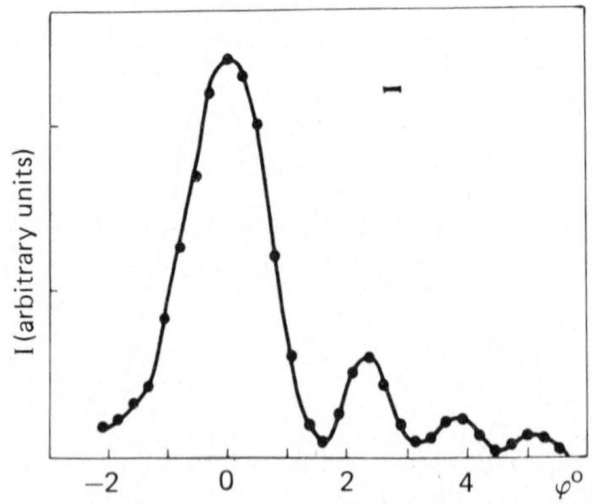

Fig. 16. Radiation pattern of a SEW frustrating on the metal edge. The segment length on the right shows the size of the IR radiation detector sensing area.

widths on the aluminium film was studied experimentally (Zhizhin et al. 1979a). SEWs were coupled and decoupled by prisms. Figure 17 shows in a half-logarithmic scale the intensity as a function of the displacement of the prism decoupler relative to the prism coupler. Two scratches were made on the metal: the first –of the width 20 μm–, the second –of the width 500 μm– (measured with a microscope). In the absence of scratches between the prisms the decoupled signal intensity decreases exponentially with the increase of the prism separation distance. The inclination of the straight line (1) in fig. 17 gives the SEW propagation length $L = 1.2$ cm at $\nu = 945\ \text{cm}^{-1}$. When the prism decoupler is over the scratch, the decoupled radiation which comes out at the angle, corresponding to the SEW output, decreases sharply; the radiation frustrates on the scratch edge and comes out of the prism decoupler at a different angle. When the prism decoupler is again over the metal and is moving away from the prism coupler, the recorded SEW decoupled intensity becomes smaller since the energy which has frustrated on the metal edge at one scratch boundary is not totally transformed into a SEW at the other boundary, but is partially lost as a radiated wave. A similar jump of SEW decoupled intensity is observed at the second scratch, only its depth is more due to the greater width of the scratch. The slope of the straight line in fig. 17 after the scratches does not change because the SEW is still propagating on the same metal, but the straight line shifts parallel to itself by the value of radiation losses.

The loss of the SEW decoupled signal intensity on the scratch depends on the scratch width. Figure 18 shows the dependence of the SEW decoupled signal intensity attenuation on the width of scratches on the aluminium sample. The inclination of the straight line allows us to evaluate the divergence of the radiation beam which has frustrated on the metal edge. It is assumed that a certain part of radiation at the second boundary

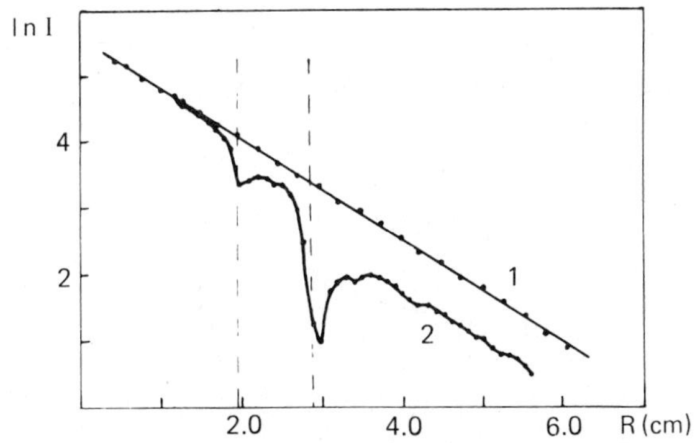

Fig. 17. SEW intensity as a function of the prism separation distance: (1) SEW propagation on metal without scratches; (2) with scratches.

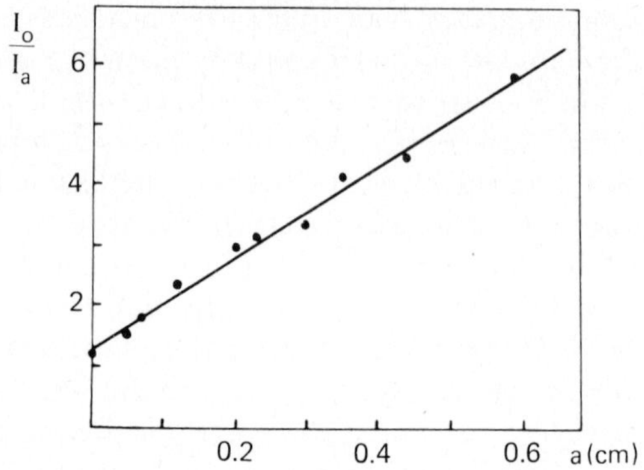

Fig. 18. SEW intensity decrease as a function of the scratch width a on the metal. I_0 is SEW intensity before scratching; I_a is SEW intensity after scratching.

of the scratch cannot transform to a SEW (since it is too far from the metal edge) due to the divergence. The divergence of the frustrated beam thus estimated is several degrees. Approximately the same divergence value was obtained by the direct measurement of the radiation pattern frustrated on the metal edge (fig. 16).

SEW attenuation by surface discontinuities of greater sizes was studied with the use of two identical glass plates covered with an evaporated copper layer. The distance between these plates on the same base was varied. In fig. 19 a dependence of the ratio of intensities of the incident and outcoming beams is shown as a function of the distance a between the glass plates at a fixed prism separation distance. It is seen from fig. 19 that at the distance $a = 2$ cm between the plates a half of the SEW intensity, frustrating on the edge, transforms to a SEW on the front edge of the other

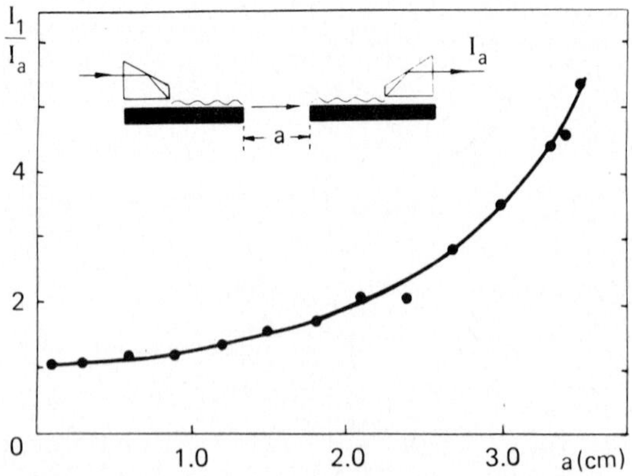

Fig. 19. SEW intensity decrease as a function of the distance a between the plates with metal surfaces. I_1 is the SEW intensity at the distance 0.1 cm; I_a – at the distance a cm.

sample. From the comparison of figs. 18 and 19 it can be noted that the curve slope at small values of surface discontinuities (fig. 19) is much less than the slope in fig. 18 which is probably due to the better quality of the edge of the evaporated metal layer as compared to the edge of the mechanical cutting of the metal used in the previous measurements.

8. *Determination of Optical Constants for Metals in the IR Region by the Propagation of Surface Electromagnetic Waves*

At a smooth homogeneous metal–air interface the SEW propagation length is determined basically by dielectric functions of the metal (eqs. 14, 21). Figure 20 illustrates the frequency dependence of SEW propagation lengths for various metals. Copper, silver, gold and aluminium samples were prepared in the form of films deposited by vacuum evaporation onto polished glass substrates under various evaporation conditions. A molybdenum sample used was a rolled foil 200 μm thick glued upon a polished glass plate. Experimentally measured SEW propagation lengths on various metals and the samples prepared in different ways are summarized in table 2. It is seen from fig. 20 and table 2 that the method of sample preparation

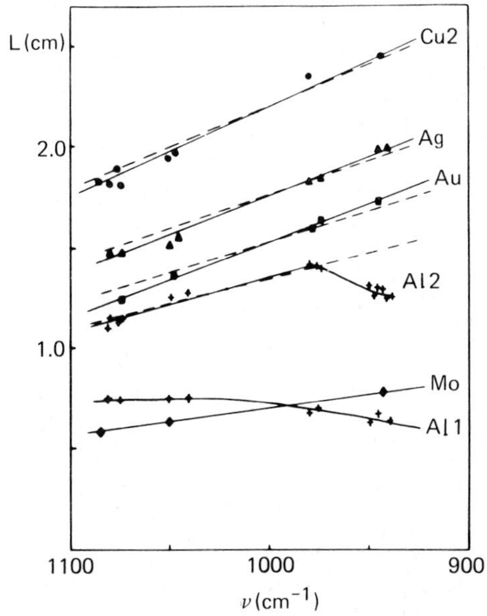

Fig. 20. Frequency dependence of SEW propagation length for various metals.

Table 2
Experimentally measured SEW propagation lengths on various metal surfaces

	SEW propagation length (cm)				
Metal	$\nu = 1080\ cm^{-1}$	$\nu = 1000\ cm^{-1}$	$\nu = 945\ cm^{-1}$	$\nu = 84.2\ cm^{-1}$	References
Ti		0.105			Bryan et al. (1976)
Ag			0.5		Shoenwald et al. (1974)
Cu		2.6		1.15	Begley et al. (1979)
Au			0.5	1.2	Mc Mullen 1975 Begley et al. (1979)
Pd		0.23		1.26	Begley et al. (1979)
W		0.41		1.25	Begley et al. (1979)
Ni		0.35		0.92	Begley et al. (1979)
Pt		0.37		1.07	Begley et al. (1979)
steel			0.16	1.52	Mc Mullen 1975 Begley et al. (1979)
Cu 1	0.92		1.25		Zhizhin et al. (1979b)
Cu 2	1.85		2.4		Zhizhin et al. (1979b)
Ag	1.5		2.0		Zhizhin et al. (1980a)
Au	1.25		1.72		Zhizhin et al. (1980a)
Al 1	0.75		0.65		Zhizhin et al. (1980a)
Al 2	1.1		1.25		Zhizhin et al. (1980a)
Mo	0.58		0.78		Zhizhin et al. (1980a)
In			0.24		Zhizhin et al. (1980a)
Fe			0.12		Zhizhin et al. (1980a)
Cr 1			0.1		Zhizhin et al. (1980a)
Cr 2			0.075		Zhizhin et al. (1980a)

has a considerable influence on the SEW propagation length as it is related to the optical constants of the given sample.

The measurement of the SEW propagation length on the metal surface does not allow simultaneous determination of both parameters of the metal (n and κ or ν_p and ν_τ), but only their combination (e.g. conductivity σ_0):

$$\sigma_0 = \pi L_0 \nu^2. \tag{41}$$

If a film with ϵ_2 and the thickness d is deposited on the surface, the SEW propagation length decreases and the dispersion equation becomes more complicated (eq. 30). If the film thickness and ϵ_2 are known, then from the measured SEW propagation lengths on the metal with a film and without a film it is possible to obtain –from eqs. (14) and (30)– the optical constants (or, using eq. (21), the plasma frequency and the collision frequency).

The calculation is simplified considerably, if the dielectric film is sufficiently thin, i.e. if the condition

$$2\pi\nu\epsilon_2 d \ll 1 \tag{42}$$

is fulfilled.

Then for a nonabsorbing film on the metal surface the variation of the SEW propagation length is described by an approximate formula

$$\frac{1}{L}-\frac{1}{L_0}\approx 8\pi^2\nu^2 d\,\frac{\epsilon_2-1}{\epsilon_2}\,\frac{n}{n^2+\kappa^2}, \tag{43}$$

where L is the SEW propagation length on the metal with a film. The use of the Drude model (eq. 21) for the optical constants of a metal gives

$$\frac{L_0-L}{L}=2\pi d\,\frac{\epsilon_2-1}{\epsilon_2}\,\nu_p, \tag{44}$$

$$\nu_p=\frac{L_0/L-1}{2\pi d(1-1/\epsilon_2)}, \tag{45}$$

and

$$\nu_\tau=\nu_p^2/2\pi L_0\nu^2. \tag{46}$$

The optical constants of a metal are calculated by formulae

$$n=\frac{4AB^2}{A^4+4B^2};\qquad \kappa=\frac{2A^3B}{A^4+4B^2}, \tag{47}$$

where

$$A=\frac{8\pi\nu^2 d(1-1/\epsilon_2)}{1/L-1/L_0};\qquad B=\frac{2\pi\nu}{1/L_0}.$$

Formulae (47) may also be used in the case of a metal with an anomalous skin effect.

Thus, if a film is thin (condition (42)), the optical constants of the metal may be obtained by formula (47); for thick films, when approximate formulae (44)–(46) cannot be used, the exact solution of eqs. (30) is required. In this case it is possible, for example, using a graphical method, to determine the optical constants (n, κ) or microcharacteristics (ν_p, ν_τ) of a metal.

In order to determine the optical constants of a metal by the SEW propagation length (Zhizhin et al. 1979b, 1980) dielectric films of the thickness of several hundredths of an Å were deposited along their full length on a half of the metal sample. The requirements placed upon the dielectric film were as follows: (1) it should not absorb in the spectral range under investigation (900–1100 cm^{-1}) and (2) it should have a sufficiently high refractive index so that the error in its determination would only slightly affect the determination precision of the optical constants of the metal. The most appropriate material is germanium.

The thickness of the deposited germanium film was measured by the Tolanskii interferometric technique (Glang 1970) and was equal to 900 ±

30 Å on the sample Cu 1 and 750 ± 30 Å on the sample Cu 2. The value ϵ_2 of the deposited germanium film was measured over the wavelength range around 5 μm by the positions of maxima in the interference pattern in the reflectivity spectrum (a germanium film was simultaneously deposited on quartz), it was close to $\epsilon_2 = 16$.

Copper samples were prepared in the following way: an oxygen-free copper (the purity 99.99%) was thermally evaporated onto a polished plane-parallel glass plate. The copper samples Cu 1 and Cu 2 under investigation had different evaporation conditions. The film Cu 1 was prepared in vacuum $\sim 1 \times 10^{-5}$ Torr with the evaporation rate ~20 Å/s, and the film Cu 2 in vacuum $\sim 5 \times 10^{-6}$ Torr with the evaporation rate ~120 Å/s. The film thickness ~0.5 μm is sufficient to consider the metal as a semi-infinite medium.

For the determination of the optical constants of copper films prepared by vacuum thermal evaporation at various conditions the frequency dependence of the SEW propagation length on copper and on germanium coated copper was measured. The results obtained for the sample Cu 1 are given in fig. 21. The presence of a germanium film causes a vast decrease in the SEW propagation length, but it does not alter the square dependence of the propagation length on the frequency. The thickness of the films under study was too large to limit oneself to approximate formulae (the condition

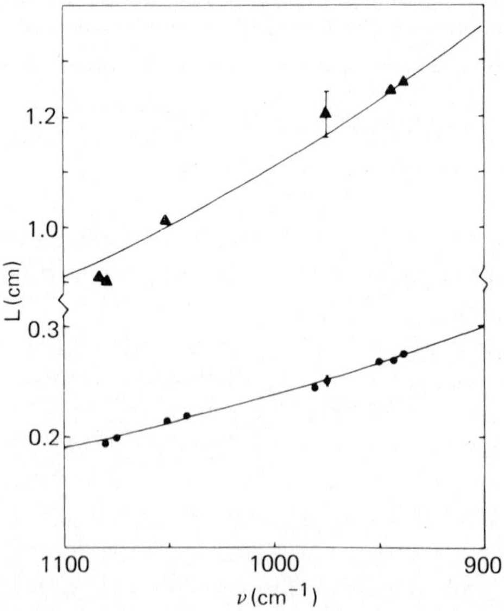

Fig. 21. Frequency dependence of SEW propagation length on copper surface (▲) and on copper with a germanium layer (●) for a copper sample 1. The lines are SEW calculated propagation lengths with the parameters given in table 3.

(42) is not satisfied), but they can be used as the first approximation for further calculations by the accurate formulae. The calculation was made in the following sequence. First the ratio ν_p^2/ν_τ [eq. (21)] was determined by the measured L_0 value. Figure 21 shows the frequency dependence (the upper curve) of the SEW propagation length on the copper surface calculated by this ratio, together with the experimental values. The next step was to compute ν_p and ν_τ by the approximate formulae [eqs. (45, 46)] with the measured L, L_0, d and ϵ_2. Using the obtained approximate values ν_p and ν_τ and solving the dispersion eq. (30) by means of a computer, a family of curves L_p was found at various thicknesses d (the ratio ν_p^2/ν_τ remained constant). Such a family of curves for $\nu = 1000\ \text{cm}^{-1}$ is shown in fig. 22. The plasma frequency was determined from the abscissa of the intercept point of the curve for $d = 900$ Å with the horizontal line, corresponding to the SEW propagation length on a metal with the film at the given frequency. The error caused by the inaccuracy in the determination of the film thickness may also be determined from these curves. The values ν_p and ν_τ as well as the values ϵ_4', ϵ_4'', n and κ calculated with their use are presented in table 3. In fig. 21 the lower curve gives the frequency dependence of the SEW propagation length for a metal with a film calculated for the values ν_p and ν_τ given in table 3. The agreement with the experiment is quite satisfactory.

The similar calculations have also been made for the copper sample Cu 2. The calculated optical constants are given in table 3. SEW propagation lengths on copper Cu 2 is well in excess of those on copper Cu 1, but the plasma frequencies almost do not vary, while the collision frequency of conduction electrons is half as much. The invariance of the plasma frequency is an indication of the invariability of the ratio of free

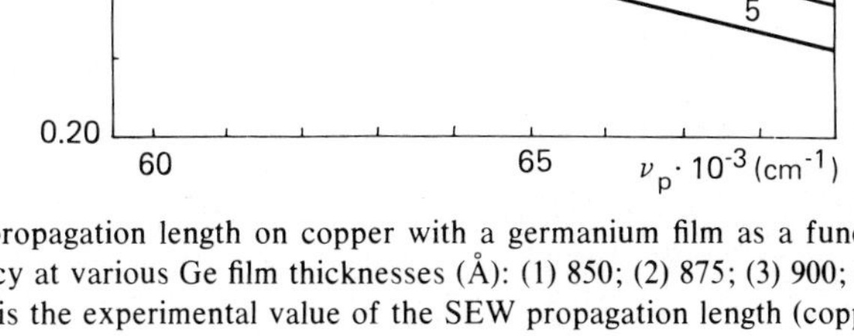

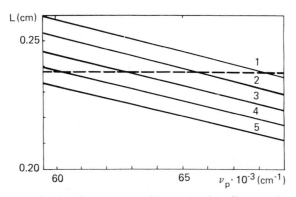

Fig. 22. SEW propagation length on copper with a germanium film as a function of copper plasma frequency at various Ge film thicknesses (Å): (1) 850; (2) 875; (3) 900; (4) 925; (5) 950. A dash curve is the experimental value of the SEW propagation length (copper sample 1).

Table 3
Optical constants of copper samples at $\nu = 1000\ cm^{-1}$

	Cu 1	Cu 2
L_0 (cm)	1.11 ± 0.03	2.20 ± 0.05
$\nu_p \times 10^{-3} (cm^{-1})$	63 ± 3	63.5 ± 2.5
$\nu_\tau\ (cm^{-1})$	570 ± 54	293 ± 23
ϵ'_H	-3000 ± 140	-3710 ± 240
ϵ''_H	1710 ± 240	1090 ± 160
n	15.0 ± 1.7	8.8 ± 1.0
κ	56.8 ± 1.7	61.6 ± 2.1

electron densities to the effective masses in copper samples Cu 1 and Cu 2 and the variation of the collisional frequency of free electrons indicates that the samples have different surface structures, i.e. granularity, etc. (Schkl'arevskij et al. 1966). The layer of natural copper oxide which has no resonance absorption in the spectral region under investigation was taken into consideration (formula 30); its influence on the results obtained is insignificant. If a metal oxide has the resonance absorption in the spectral region under study, for example aluminium oxide (fig. 20), the effect of the oxide layer should be taken into account in spite of its small (20–30 Å) thickness.

The method proposed may also be used to determine optical constants of other metals as well as the quality of the deposited metal layers.

9. *Effects of Metal Surface Roughness on Propagation Length of Surface Electromagnetic Waves*

On a perfectly smooth metal surface an incident light wave, inducing surface currents, spends its energy only on Joule losses. But on a rough metal surface the distribution of currents is modulated in accordance with the distribution of surface defects and their correlation along the surface which results in light scattering. Moreover, the roughnesses increase light absorption due to the decrease of the electron free paths and this effect predominates over light scattering in the IR region, when the metal optical properties are determined by the number of conduction electrons in the metal and their average propagation length. SEW propagation length on a smooth metal surface also depends on the value of Joule losses of the free conduction electrons [eq. (21)]. The presence of defects and scratches leads to the decrease in SEW propagation length. This decrease is due to two mechanisms of losses: SEW scattering by the edges of the defects (Talaat et al. 1975, Mills 1975) and the increase of the collisional frequency of conduction electrons. It was studied on real metal mirrors with different roughnesses of which the mechanisms predominates in the case of SEW propagation length decrease.

The study was carried out on four mirror samples of oxygen-free copper of 1 cm thickness and a diameter of the polished surface of 5 cm. The mirrors were polished by free lapping with the use of foamed resins and diamond powders. The statistical analysis of the defects and surface roughnesses was performed by means of an instrument "Quantimet-720" used to obtain histograms of surface defect distributions, in particular, cavities, as functions of their diameters. It turns out that for the four samples studied the distribution of defects by diameters within the range from 0.5 to 3 μm can be described by the right-hand branch of the Gaussian curve with maxima at the diameters below 0.5 μm and the average diameter 1.16 μm. Defects of the diameter in excess of 3 μm accounted for 6–9% of the total number of defects, and their distribution differed from that of the Gaussian type. The whole area occupied by the defects varied from 3 to 7.7% of the total area of the polished surface. The average height of roughnesses h obtained by the processing of profilograms of sample surfaces was 370–580 Å.

As a reference sample with the surface assumed to be without defects a plane-parallel optically polished glass plate covered with a copper layer (~5000 Å) with a chromium sublayer (~1000 Å) was used. The thickness of the plate was 1 cm and the diameter 5 cm. The copper layer was thermally evaporated in vacuum (10^{-6} Torr) with the evaporation rate 120 Å/s. No surface defects were recorded on this sample by "Quantimet-720".

Measurements of SEW propagation lengths at two wavelengths were made on metal mirror surfaces. The results of these measurements are summarized in table 4; the samples are numbered in order of decreasing SEW propagation length. The greatest SEW propagation length, as one

Table 4
Comparison of optical characteristics of samples and their roughness parameters

Sample No	SEW propagation length (mm)		S (%)	$S(Q > 3\ \mu m)$ % of S	$(1-R)$ % calculation $\nu = 1000\ cm^{-1}$
	$\nu = 941\ (cm^{-1})$	$\nu = 1041\ (cm^{-1})$			
1	24	19.7	0	0	0.94
2	15.2	12.8	3.29	7.5	1.5
3	14	10	4.39	6	1.7
4	12.4	9.3	4.67	9	1.9
5	12	8.6	7.07	7	2

would expect, was found on the reference sample. Table 4 gives also the values of the total area of defects S and the area of defects with the diameter in excess of 3 μm (S($\theta > 3\ \mu$m)). From the data given in table 4 it is seen that the SEW propagation length correlates with the total area of surface defects. Figure 23 shows the SEW absorption coefficient $\alpha = 1/L$ as a function of the total area of defects, SEW absorption increases monotonously with the area of defects and this increase is practically linear with the increase of the total area of defects. The presence of the defects on the area of 7% on the sample 5 reduces the SEW propagation length by a half relative to the standard.

As noted before, such decrease in the SEW propagation length can be attributed either to the increase in the number of the collision frequency of conduction electrons in the subsurface layer (~200 Å) comparable to the roughness depths, or to SEW scattering by roughnesses, resulting in the SEW transformation to bulk radiation. The latter mechanism of scattering on the surface with the real and negative dielectric function was considered theoretically by Mills (1975). It was shown that SEW losses at the scattering by roughnesses with the average spacings between the defects Q and the average roughness height h result in the decrease of the SEW propagation length L:

$$1/L = 1/L_0 + 1/l, \tag{48}$$

where L_0 is the SEW propagation length on a smooth surface with $|\epsilon'| \gg |\epsilon''|$ and l is given by

$$l \approx 3|\epsilon|^{1/2}/2\pi^6\nu^5h^2Q^2, \tag{49}$$

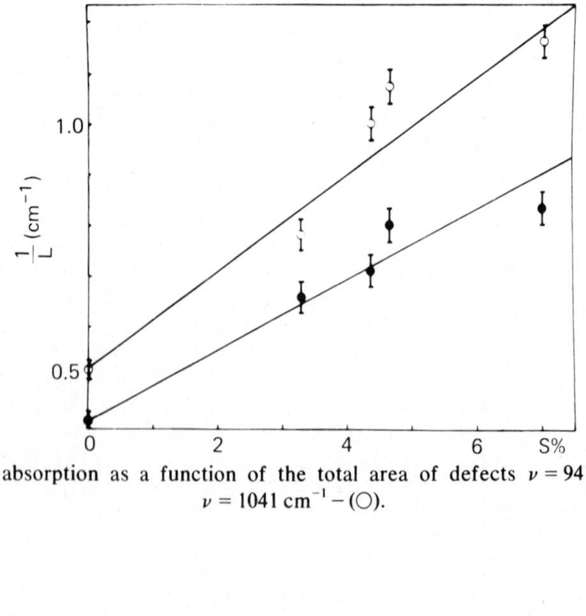

Fig. 23. SEW absorption as a function of the total area of defects $\nu = 941\ \text{cm}^{-1}$ (●) and $\nu = 1041\ \text{cm}^{-1}$ – (○).

in the case of the Gaussian distribution of roughnesses. Numerical estimations by the formula (49) for copper with the average height of roughnesses $h = 500$ Å and $Q = 1\ \mu$m corresponding to copper mirrors under study, give 1.6 cm at $\nu = 1000\ \text{cm}^{-1}$. But experimentally a considerably greater decrease of the SEW propagation length was observed. This means that light scattering by roughnesses does not play a dominant role. This was also confirmed by the frequency dependence of the SEW propagation length observed in the present experiment. It is close to the square dependence on the wavelength, whereas light scattering by roughnesses should give a much stronger frequency dependence, i.e. $l \sim 1/\nu^6$ (with regard to $|\epsilon|^{1/2} \sim 1/\nu$). The estimations made allow us to suppose that the main contribution to losses is made by the variation of the collision frequency of conduction electrons due to the electron scattering by microdefects, i.e., the source of losses is the same as in the case of light reflection. The number of microdefects increases with the deterioration of the surface quality quantitatively characterized by the area of defects. The collision frequency of conduction electrons in the subsurface layer calculated with this assumption [eq. (21)] varies from $300\ \text{cm}^{-1}$ for copper on the reference sample to $650\ \text{cm}^{-1}$ on the sample 5 of a metal copper mirror.

In the case of light beam reflection the increase in the collisional frequency of conduction electrons ν_τ leads to the increase of absorption and to additional heating of the polished metal mirror. The reflection coefficient in the approximation $\nu_\tau^2 \ll \nu^2$ (valid for many metals in the IR wavelength range) is determined by

$$1 - R = 2\nu_\tau/\nu_p. \tag{50}$$

The absorption coefficient $1 - R$ given in table 4 is calculated by the formula:

$$1 - R = \nu_p/\pi\nu^2 L. \tag{51}$$

Plasma frequency was assumed to be constant for all samples: $\nu_p = 63\,500\ \text{cm}^{-1}$. Good correlation between SEW attenuation, reflection losses and the area of defects should be noted (see table 4).

10. *Selective Absorption of Surface Electromagnetic Waves by Thin Films on Metal Surfaces*

As it has already been noted in sect. 3, a thin dielectric film deposited on a metal increases SEW attenuation which becomes stronger as eigenfrequencies of the film vibrations are approached. This fact allows to develop a new effective spectroscopic method – the spectroscopy of surface electromagnetic waves.

Bell et al. (1975) and Agranovich (1975) showed that the presence of a thin film transition layer at the interface leads to resonances of the SEW refractive index at the frequencies of TO and LO modes of the layer, i.e., to the increase of the SEW absorption coefficient at these frequencies. It can be shown therewith that for a film on a metal the SEW absorption at the frequency of a TO phonon is much weaker than at the frequency of a LO phonon. It was confirmed by an experiment (Zhizhin et al. 1976) carried out with a 200 Å silicon monoxide film on copper and a 200 Å apatite film on silver. The frequency dependence of SEW propagation length on silver with apatite films is shown in fig. 24. The SEW propagation length has its minimum close to the apatite longitudinal frequency ($\nu_{LO} = 1080\ cm^{-1}$) while the transverse frequency ($\nu_{TO} = 1040\ cm^{-1}$) is noticable only at large film thicknesses (300 Å according to calculations). Experimental values are given for an apatite film of the thickness 200 Å. It is seen that L decreases considerably as the resonance (ν_{LO}) is approached. Not a very good agreement between calculations and the experiment is probably due to the fact that dielectric functions used for calculations were taken for the bulk sample and not for the film.

Possible application of SEW absorption to obtain the spectra of films on metals was illustrated by cellulose acetate films (Bhasin et al. 1976) 15 Å and 75 Å thick on copper. At the thickness 75 Å the absorption peak

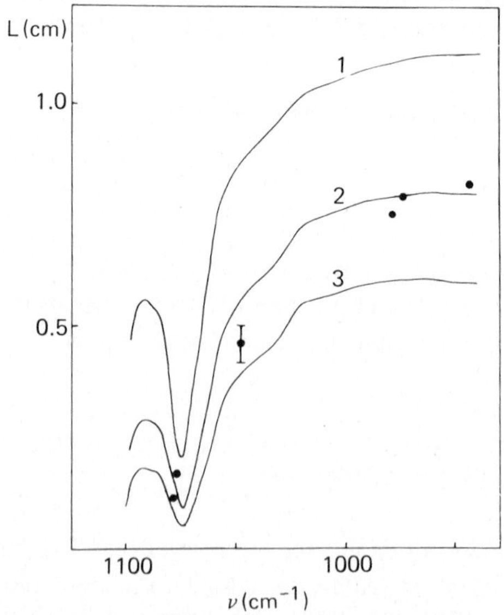

Fig. 24. Frequency dependence of SEW propagation length on silver with apatite films. Curves – calculations with the film thicknesses (Å): (1) 100; (2) 200; (3) 300. Points – experimental results.

(1050 cm^{-1}) is well defined in the spectra while at the film thickness 15 Å it is hardly noticeable and lies within the limits of the measurement errors. In the case of thin films the absorption spectrum can be more conveniently obtained not by direct measurement of the SEW propagation length, but with a "differential" method. In the "differential" method the ratio of intensities is measured of a SEW which has passed on the copper surface with a film (I) and without a film (I_0) at various distances R between the prisms. The values of gaps between the prisms and the samples should in both cases be kept at optimal values (in the experiment the intensities have to tend to unity when the distance between the prisms approaches zero). The SEW absorption is determined by

$$\frac{1}{L} - \frac{1}{L_0} = \Delta\alpha = \ln\frac{I_0(\nu)}{I(\nu)} R^{-1}. \tag{52}$$

In fig. 25 a SEW absorption spectrum is given for one monomolecular layer and three monolayers of a Langmuir film of sym-di(γ-carboxydecyl)tetramethylsiloxane (compound 1) on copper (Zhizhin et al. 1980b). This compound has an absorption band with the maximum at 1075 cm^{-1}, corresponding to the stretching Si–O–Si vibration (the absorption spectrum has been obtained for the compound 1 pressed into a KBr pellet). The films of compound 1 were deposited on a half-width of the substrate which was a plane-parallel polished glass plate (13 × 3 × 0.6 cm) with a thick (~1 μm) vacuum evaporated copper layer. The surface pressure in a monolayer during coating by the Langmuir–Blodgett technique was 25 dyn/cm^{-1}.

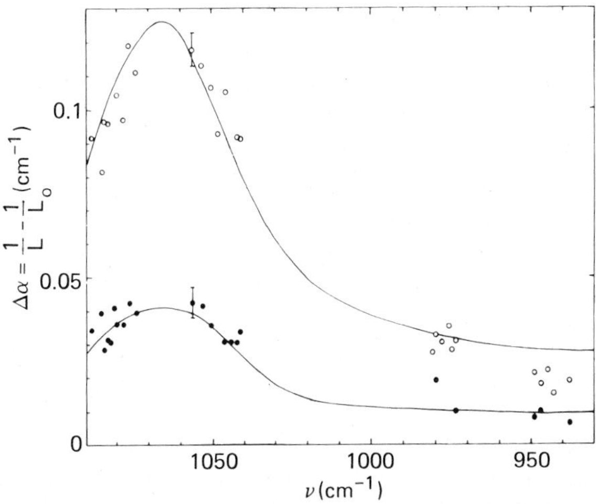

Fig. 25. SEW absorption spectrum for sym-di (γ-carboxydecyl)tetramethylsiloxane on copper: (●) one monolayer; (○) three monolayers. Solid curves correspond to calculated data.

Before the coating with a Langmuir film the substrates with a copper layer were held for three days to stabilize the copper oxide film of a thickness of about 30 Å. For measuring SEW absorption the prisms were separated at the maximum possible distance $R = 11$ cm (the distance was limited by the sample size) and at each frequency of the tunable CO_2 laser the data points were obtained after averaging 5–10 measurements. For better reliability the SEW absorption by a monolayer was measured independently on two identical samples. The absorption difference measurement accuracy was $\pm 0.005\ cm^{-1}$.

For the computation of the SEW absorption spectra shown in fig. 25 the values of the compound 1 film optical constants were obtained from the multiple ATR spectra. In the SEW absorption spectra, the same as in multiple ATR spectra, the maximum of the stretching Si–O–Si vibration of the compound 1 film occurs at the frequency $1063\ cm^{-1}$.

It should be noted that the high frequency dielectric constant ϵ_∞ of the films differs considerably from that of the initial substance. The calculations showed that the absorption in the region $940–980\ cm^{-1}$ (far from resonance) was considerably less than that expected for $\epsilon_\infty > 2$ (which is characteristic for most of the organic substances). To obtain the agreement between the calculated and the experimental data (the solid lines in fig. 25) the value $\epsilon_\infty = 1.2$ was used. Such a small value of the dielectric constant can be explained in terms of a skeleton film structure, i.e., by incomplete packing of molecules in the film plane. The computation performed for SEW absorption spectra with regard to a 30 Å copper oxide layer between copper and the substance film has shown that the difference between the computed curves with and without taking the oxide into consideration is small.

The thickness dependence of SEW selective absorption could be followed by the absorption character according to the number of monolayers deposited. Figure 26 shows the comparison of the measured and computed SEW absorption spectra of a monolayer film with the absorption spectra of three- and eleven-monolayered films of compound 1 reduced to one monolayer. From the coincidence of the curves it is seen that SEW absorption is proportional to the number of monolayers. This fact together with the absence of the absorption maximum shift and the broadening of the absorption band, while going from one to eleven monolayers, indicates that the stretching Si–O–Si vibration does not interact appreciably with copper and other monolayers, and the incompact packing of compound 1 remains constant in all layers.

The above examples of the observation of vibrational IR spectra of very thin films of organic and inorganic substances on metal surfaces (when the film thicknesses reached limiting small values down to monolayers (~20 Å) are a reliable indication of the very high sensitivity of the SEW spec-

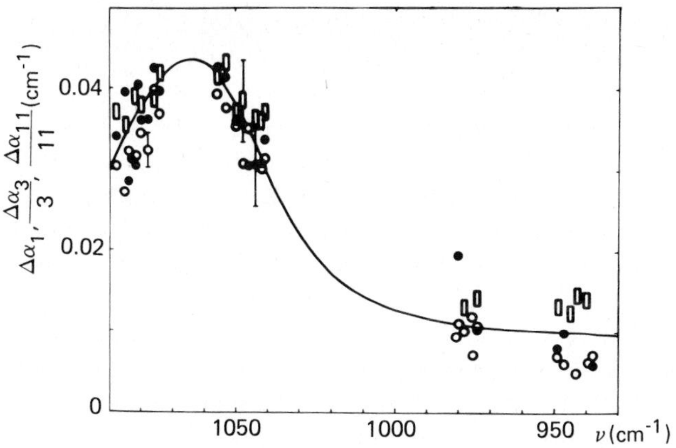

Fig. 26. Frequency dependence of reduced to one monolayer SEW absorption by eleven monolayers $\Delta\alpha_{11}/11$ (□); three monolayers of $\Delta\alpha_3/3$(○) and one monolayer $\Delta\alpha_1$(●). The solid curve corresponds to calculation with the following parameters: $\epsilon_\infty = 1.2$, $\nu_{TO} = 1060\ cm^{-1}$, $\nu_{LO} = 1063\ cm^{-1}$, $\gamma = 55\ cm^{-1}$.

troscopy method under discussion. Such high sensitivity is due to two circumstances: firstly, the maximum of the SEW electric field amplitude is right on the metal surface in contrast to the methods of ordinary reflection when it is small on the metal surface; secondly, the energy loss is accumulated along the sample length but not across the film thickness as in conventional spectroscopic methods of reflection and transmission, though the thickness dependence is present in this case too.

At the next step of this application method it would be appropriate to go from the reliable detection of the film on the metal surface to the detection of the indications of a thin layer interaction with a metal which shows up in vibrational spectra. Such works are likely to be represented only by Bhasin et al. (1976) where the "differential" method was used to obtain the absorption spectra of phys-adsorbed benzene on copper at 190 K. From the three spectra at the benzene film thicknesses 5 Å, 10 Å and 25 Å it is seen that the benzene molecule, while being adsorbed, becomes asymmetric; the film spectra differ from that of the solid benzene. The appearance of plane and non-plane C–H vibrations in the SEW spectrum is an indication of the copper roughness influence in the atomic scale on the disorder of benzene orientation at phys-adsorption. Chemisorption of hydrogen on tungsten was studied by Chabal et al. (1980).

11. *Absorption of Surface Electromagnetic Waves by Natural Metal Oxides*

The surface of most of the metals is covered –in the air– with films of natural oxides of the thickness of several Ängströms. IR absorption or reflection spectra of the films allow to make conclusions about the growth of oxide films, their composition and structural changes in the film. But the obtaining of the spectra of such thin natural oxide films presents great difficulties. Taking into account the above advantages of the SEW spectroscopy they can reasonably be studied by means of this method.

The dependence of the SEW propagation length on copper on the thickness of the oxide film obtained by heating was studied by Bryan et al. (1976). The copper oxide film has no selective absorption in the tuning region of a tunable CO_2 laser, therefore, the formation of an oxide film should result in a practically uniform decrease of the SEW propagation length at various frequencies. Thus, the copper oxide film of the thickness 20 Å, formed in natural conditions, causes the change in SEW absorption $\Delta\alpha = 0.031\ \text{cm}^{-1}$ and, accordingly, the SEW propagation length decreases from $L = 2.2$ cm ($\nu = 1000\ \text{cm}^{-1}$) on pure copper to $L = 2.06$ cm on copper with an oxide layer of the thickness 20 Å (the optical constants used were the same as those observed by Ward et al. (1975)). The situation observed experimentally was: the SEW propagation length right after the sample preparation was $L = 2.45$ cm ($\nu = 943\ \text{cm}^{-1}$) and in a month it decreased to $L = 1.75$ cm. This increase in the SEW attenuation constant cannot be explained only by the formation of an oxide film since the thickness of the natural oxide on copper is 20–30 Å while the observed attenuation could be caused only by an oxide film of the thickness ~110 Å. To explain such an increase in SEW absorption it may be supposed that the "ageing" of the copper film itself occurs with time and, as a result, the structure of the polycrystalline film changes and not only the upper copper layer is oxidized by oxygen from the air, but this oxygen diffuses along the film crystallites which leads to the variation of the copper optical constants in the subsurface layer.

In contrast to copper oxide an aluminium oxide film has selective absorption in the spectral range of the tunable CO_2 laser in use. Due to this absorption the frequency dependence of the SEW propagation length on aluminium differs from the dependence calculated with the Drude approximation (fig. 20). The absorption maximum of the aluminium oxide film is near $940\ \text{cm}^{-1}$, therefore, SEW propagation lengths were measured by the tunable $^{12}CO_2 + {}^{13}CO_2$ laser. An aluminium layer (~2000 Å) was deposited by thermal evaporation in vacuum $\sim 5 \times 10^{-6}$ Torr onto a polished glass plate with a chromium sublayer (~1000 Å). The purity of the aluminium was 99.99%. The sample was taken out of the vacuum chamber in the air after its cooling down to room temperature. The frequency dependence of the SEW propagation length was measured on this sample. The results of these measurements are given in fig. 27. At the frequency 945 cm^{-1} the SEW propagation length minimum was observed. The position of the absorption band is closed to the absorption maximum in the aluminium spectrum of multiple reflection (Mertens 1978) (8 reflections at the angle of incidence 72°) and in the measurements with the use of the photoacoustic method (Nardal et al. 1978) since these methods are sensitive also to the radiation absorption at the frequency of the film LO mode.

Calculations were made with the use of a single-oscillator model [eq. (22)]. With regard to data obtained from the SEW absorption spectrum ($\nu = 940\ cm^{-1}$ and $\gamma = 50\ cm^{-1}$) the value according to Harris (1955), was assumed to equal 3 and the frequency of the TO mode 650 cm^{-1}, according to measurements of Mainland et al. (1974) made for thicker anodized films. For aluminium the plasma frequency was assumed to be 83 000 cm^{-1}, the collision frequency of conduction electrons –by estimation with the use of approximated formulae– was about 650 cm^{-1} and varied in calculations.

The SEW absorption is proportional to the product of the film thickness by the function of losses [eq. (37)]. The absorption is maximal at the frequency where $-1/\epsilon_3$ has a pole, i.e., at the frequency of the film LO mode; the second absorption maximum at the frequency of the film TO mode (the pole ϵ_3) is much less and is observed experimentally only for thick (>300 Å) films. Thus, the function of losses is proportional to the oscillator strength of the LO mode: $\alpha \sim l(\nu_{LO}^2 - \nu_{TO}^2)$. The relation between the thickness and the frequency of the TO mode does not allow simultaneous determination of these two values from the SEW spectrum. For the films of natural aluminium oxide the thickness values available in literature vary from 15 Å for the dry air to 30–40 Å for the humid air. The dispersion analysis (DA) of the frequency dependence of SEW propagation

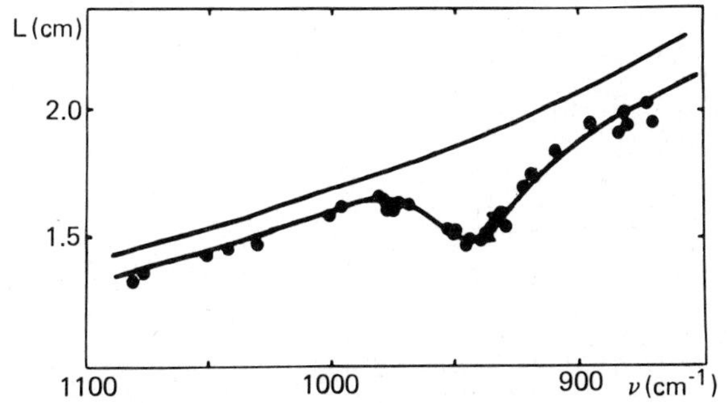

Fig. 27. Frequency dependence of SEW propagation length on aluminium with a natural oxide film.

length has resulted rather unexpectedly. If the film thickness is assumed to be 15 Å, the frequency of the TO mode should be strongly varied: the DA gives 880 cm^{-1} instead of 650 cm^{-1}. From the assumption about the coincidence of the frequency of TO mode with the literature data for thick oxide films it follows that the calculated film thickness has to be 5 Å (which is too small for a natural oxide film).

To explain the results obtained it can be supposed that either the homogeneous oxide layer of the thickness 15 Å has the TO phonon frequency which differs sharply from the frequency in thicker films of anodized aluminium oxide at approximately the same LO frequency, or the oxide layer is inhomogeneous – then various methods of measurement will yield different thicknesses. Its effective thickness determined by the absorption is ~5 Å. This would mean that pure aluminium oxide is only at the surface (the monomolecular layer is of the thickness ~3.7 Å), and below it there is a transition layer where the oxide is mixed with aluminium and the concentration of components and the optical constants of the layer vary with the depth. If in a transition layer free carriers screen the Coulomb interaction, thus leading to the decrease in the oscillator strength, the transition layer does not produce any selective absorption in SEW spectra. In the absence of a correct layer model the effective dielectric functions of a natural oxide on aluminium have been calculated by means of a homogeneous layer (15 Å) model and the values obtained are given in fig. 28.

In contrast to copper and aluminium molybdenum is stable to the air oxygen at room temperature. The oxidation begins only with the temperature increase up to 250°C in humid atmosphere or up to 400°C in dry atmosphere (Lebedinskij 1966). The growth of the oxide film on molybdenum at 397°C and above was observed by Gratton et al. (1978) by IR emission spectra. Using the high sensitivity SEW spectroscopy, we in-

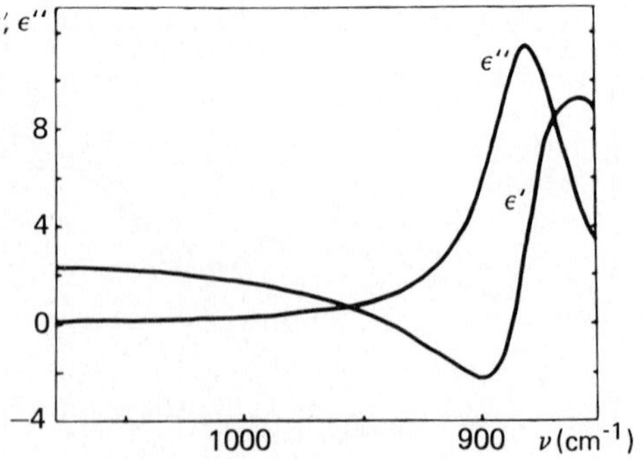

Fig. 28. Effective values of real (ε′) and imaginary (ε″) parts of the dielectric function of the natural aluminium oxide in the IR.

vestigated the early stages of the oxide growth on molybdenum. The samples were prepared by annealing molybdenum foil in the oven. Already at 220°C within 10 min an oxide layer is formed which is noticeable by SEW absorption. The frequency dependence of SEW propagation length for this sample is shown in fig. 29 (curve 2). The upper curve 1 is the SEW spectrum of the molybdenum foil before annealing. The SEW spectrum of the molybdenum oxide formed within 35 min of oxidation at 310°C is shown in fig. 29 (curve 3). An absorption band of the molybdenum oxide with the absorption maximum at 940 cm^{-1} is well seen to appear on heating. In calculations the plasma frequency of molybdenum was assumed to be 40 000 cm^{-1} (according to Siddiqui et al. 1977) and the collision frequency 390 cm^{-1} (the SEW propagation length therewith is in good agreement with the experimental value); for the molybdenum oxide film the TO frequency was supposed to equal 800 cm^{-1} as it is the case with most molybdates. The oxide film thicknesses obtained by dispersion analysis of SEW spectrum were: 35 Å (for the curve 2) and 210 Å (for the curve 3).

12. Investigation of SiO_x Films on Metals

Silicon monoxide is widely used in laser technology as a dielectric coating for mirrors. Thermally deposited in vacuum silicon monoxide films usually consist of various silicon oxides SiO_x. The film composition depends on the conditions of its deposition onto the substrate and the film optical properties vary with the film composition. As it has already been exemplified by

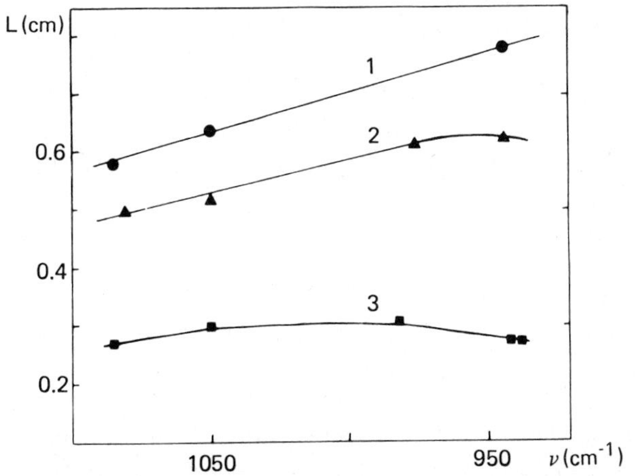

Fig. 29. Frequency dependence of SEW propagation length on molybdenum before annealing (curve 1) and after annealing (curves 2 and 3).

the film of aluminium oxide, SEW absorption spectra –at a known film thickness– allow to determine the film optical constants.

In 1976 we (Zhizhin et al. 1976) showed the SEW absorption spectrum for a silicon monoxide layer ~200 Å thick on copper and the calculation was made with the use of the silicon monoxide dielectric function obtained by Cox et al. (1975). The agreement between the experiment and the calculation was poor. This indicated the necessity of the use of optical constants of the mixtures SiO_x. In this case the film may be supposed to consist of a mixture of silicon dioxide with the dielectric constant ϵ_A (concentration $q = x - 1$) and silicon monoxide (ϵ_B, concentration $1 - q = 2 - x$). The dielectric function of the mixture with the approximation of its refraction additivity is calculated by

$$\frac{\epsilon - 1}{\epsilon + 2} = q\frac{\epsilon_A - 1}{\epsilon_A + 2} + (1 - q)\frac{\epsilon_B - 1}{\epsilon_B + 2}. \tag{53}$$

To check this supposition three samples were prepared. A copper layer (the thickness ~3000 Å) was deposited simultaneously by vacuum thermal evaporation onto polished glass substrates. After holding copper in the air and measuring the SEW propagation length, the whole length of the plate and a half area of the substrate were covered with a film of silicon monoxide thermally evaporated from a molybdenum evaporator in vacuum 10^{-5}–10^{-6} Torr at a rate ~5 Å/s. The silicon monoxide was first deposited onto only one substrate (sample 3) and the other two were shut with a screen; then the screen was partly opened and the films were now being deposited onto two substrates (samples 3 and 2) and, finally, the screen was taken away and simultaneous deposition was going on onto all the three substrates. Such a method of deposition allowed to hope that the films obtained would not have such great differences in their compositions and structures as the films obtained by different deposition cycles. Figure 30 shows the measurement results for SEW propagation lengths on copper and on copper with SiO_x films on these samples. For all the three samples the SEW propagation lengths on copper measured before the deposition of silicon oxide coincides within the limits of the measurement error ($L = 1.75 \pm 0.05$ cm at $\nu = 948\ \mathrm{cm}^{-1}$ and $L = 1.35 \pm 0.05$ cm at $\nu = 1080\ \mathrm{cm}^{-1}$).

The thicknesses of SiO_x films were measured ellipsometrically by means of a laser photoelectric ellipsometer LEF-2 and their values were: 120 Å (sample 1), 180 Å (sample 2) and 235 Å (sample 3). To check the thickness determination of the thickest film (sample 3) its value was also measured by the interferometric Tolanskii technique. The band shift of the multi-beam interference was measured at the step formed by the film edge. Used as mirrors in the interferometer were the aluminium film, covering the step, and a semi-transparent silver mirror. The band shift at the step was small, therefore, the interference pattern was photographed and the shift was

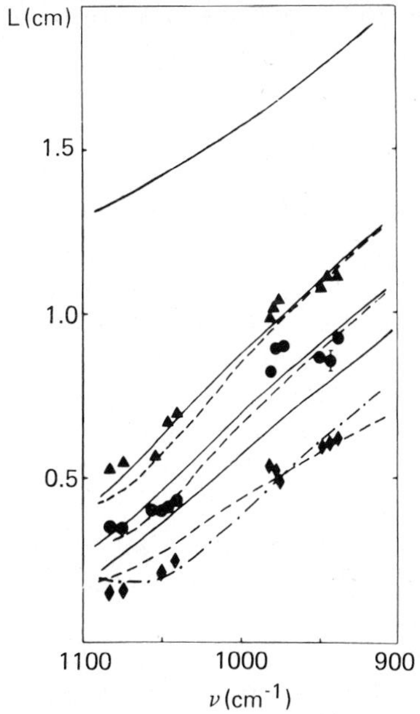

Fig. 30. Frequency dependence of SEW propagation length on copper surface with silicon oxide films for three samples and calculation with the use of various models shown in table 5: (▲) sample 1; (●) sample 2; (◆) sample 3; upper curve, Cu.

measured by means of a scanning microphotometer. The thickness 235 ± 10 Å thus obtained coincides with that measured ellipsometrically.

The IR reflection spectra (at the angles ~70°) allowed for an estimation of $x \approx 1.2$, i.e., the mixture consists of 20% SiO_2 and 80% SiO. The calculation of SEW attenuation and reflection spectra (dielectric functions of silicon monoxide were taken from Cox et al. (1975) and Zolotarev (1972)) with this value of x does not provide a satisfactory agreement with the experiment which may result from the inapplicability of the refraction additivity approximation to this case. Thus, the calculated spectra can still be used as the first approximation for the more precise calculations of the optical constants of SiO_x films.

The best agreement between the experimental and the calculated data for SEW propagation length on copper surface was achieved with the use of the following parameters for copper: $\nu_p = 65\,100\ cm^{-1}$, $\nu_\tau = 430\ cm^{-1}$, if the oxide film is neglected. The consideration of the copper oxide film 30 Å thick with $\epsilon_3 = 7.3$ decreases the collision frequency to 390 cm^{-1} at the same plasma frequency. At the calculations of the SEW propagation length and reflection spectra for the samples 1–3, the oxide layer under the SiO_x film was taken into account.

The dielectric function of the film (SiO_x) on a metal was chosen in the so called "factorized" form (Gerwais et al. 1974):

$$\epsilon_2 = \epsilon_\infty \prod_j^2 \frac{\nu_{LO_j}^2 - \nu^2 - i\gamma_j\nu}{\nu_{TO_j}^2 - \nu^2 - i\gamma_j\nu} \tag{54}$$

with two oscillators. This formula is more convenient for the dispersion analysis (DA) than the formula commonly used with the summation over the oscillators (Spitzer et al. 1961). In the calculations the damping γ_j was supposed to be the same for the longitudinal and transverse modes.

The dispersion analysis is performed according to the following scheme: initial values of oscillator parameters are taken and the frequency dependence of SEW propagation length (or reflection) is calculated for a SiO_x film on copper along with the total deviation from the experimental values. Then all parameters are alternatively varied, and the direction of the total deviation gradient is determined where afterwards a descent is made, until the minimum difference between the calculation and the experiment is achieved. The programs for DA were compiled in the FORTRAN IV language. Since the precision of the SEW spectra measurements was better than that for reflection spectra, the DA of the SEW spectrum played a leading role, while the reflection spectra were mainly used to eliminate the errors resulting from the narrow spectral interval, where SEW spectra had been measured.

The DA results are presented in table 5; different variants of model calculations are compared with the experiment in fig. 30. Solid lines (fig. 30) have been obtained with one and the same model (N 1 in table 5) for all the three films. For the films 1 and 2 the agreement between the calculations and the experiment is fairly good, but for the film of the greatest thickness the calculation does not agree with the experiment. The agreement can be improved, if we assume that the thicknesses of the last sample is 330 Å instead of 235 Å as it has been determined by direct measurements. But such a large difference in the thickness is unreasonable, therefore, it should be supposed that the optical constants of either the film or the metal varied. The SEW absorption coefficient should depend linearly on the film thick-

Table 5
Oscillator parameters for models of SiO_x films on copper surface

Model No	ν_T (cm^{-1})	ν_{TO_1} (cm^{-1})	ν_{LO_1} (cm^{-1})	γ_1 (cm^{-1})	ν_{TO_2} (cm^{-1})	ν_{LO_2} (cm^{-1})	γ_2 (cm^{-1})	ϵ_∞
1	390	995	1115	100	1135	1160	60	3.3
2	390	980	1100	100	1132	1162	67	3.3
3	540	980	1100	100	1132	1162	67	3.3
4	390	915	1075	100	1125	1156	60	3.3

ness [eq. (36)], but the experimental data give the linear dependence only for three thicknesses 0, 120 and 180 Å. For the thickness 235 Å SEW absorption exceeds considerably the value predicted by calculations.

In order to diminish the deviation of the calculations from the experiment, it may be supposed that the optical constants of the metal under the film have changed as compared to those before the film deposition. It could result, for example, from the greater heating of this substrate during the SiO film deposition, or from impurities present in the silicon monoxide (technical grade) being evaporated, since the film 3 was the first to be deposited and its deposition was proceeding for a longer time and it was placed closer to the evaporator than the other samples. The variation of the copper optical constants can be allowed for by increasing the electron collision frequency. The copper optical constants in the visible region therewith must not change considerably, since the collision frequency is much less than that, where the measurements were made, and the contribution of free electrons to the real part of the dielectric function is well described by the Drude formula eq. (19) with no regard for damping, while the imaginary part of the dielectric function is far below the real one and, besides free electrons, interzone transitions make a considerable contribution to the imaginary part.

The DA performed with the assumption of the coincidence of the film dielectric functions and the difference only in the collision frequency of electrons in copper, allows to choose model N 2 for the films 1 and 2 and model N 3 for the film 3 (see table 5). The calculation by means of these models is shown in fig. 30 by dash lines. The agreement for the film 3 has improved, but for 1 and 2 it has become worse. It shows that by varying only the optical properties of the substrate, it is impossible to obtain good agreement. Evidently, the optical properties of the films should be considered as being different.

If we suppose that the deposition of silicon oxide caused no changes in copper under the film 3, then the optical properties of this film should differ sharply from the properties of two other films. The DA in this case gives the oscillator parameters of the model N 4 (table 5). The agreement between the experimental and calculated (dash-and-dotted line in fig. 30) SEW spectra in this case is good, but the frequency of the transverse oscillation must be 915 cm^{-1} instead of 980 cm^{-1}. The shift is very big as compared to the silicon monoxide and the model with the refraction additivity.

The choice between models N 3 and N 4 for the film 3 can be made by means of the reflection spectra. For the film 2 the calculation by model N 1 gives the spectrum close to the experimental one. The calculation for the film 3 made with the use of model N 3 is also in good agreement with the

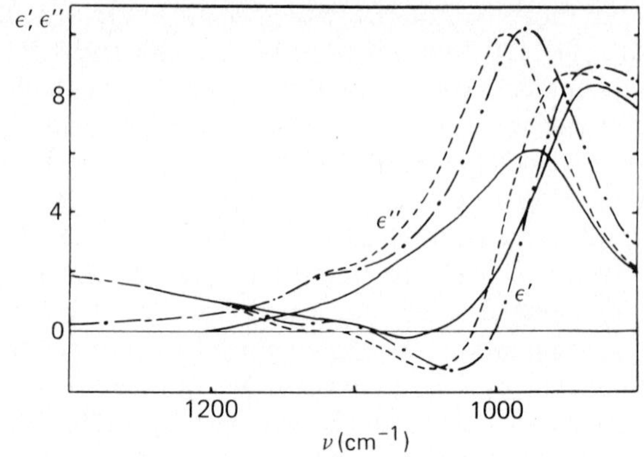

Fig. 31. Real (ϵ') and imaginary (ϵ'') parts of the dielectric function of SiO_x films on copper. Solid curves – silicon monoxide (Cox et al. 1975), dash lines – SiO_x films of thicknesses 120 and 180 Å, dash-and-dot lines – a film of the thickness 235 Å.

experiment, whereas for model N 4 the discrepancy with the experiment is very large.

Dielectric functions obtained for silicon oxide films are shown in fig. 31. The calculation was made by formula (54). For the films 1 and 2 model N 1 was used (dash lines in fig. 31), for the film 3 – model N 3 (dash-and-dot lines). For comparison the dielectric function for silicon monoxide is also given (solid lines) obtained by Cox et al. (1975) for appreciably thicker films. The difference between dielectric functions of the film 3 and the films 1 and 2 is not large; the main contribution to the difference in SEW spectra of these films is made by the variation of the electron collision frequency in copper.

In conclusion it should be noted that SEW spectroscopy is a new method of spectral studies of metal surfaces and superthin (monomolecular) films on them (natural oxides or those artificially applied by evaporation, Langmuir–Blodgett method, deposition from solutions, etc.). This method is based on the measurements of the frequency dependence of the propagation length of SEW excited by laser radiation. The SEW spectroscopy technique promises a wide application. Here its applicability is demonstrated on three types of problems:

(1) Measurement of metal optical constants on the IR region based on the difference in SEW propagation on a metal with a film for which the thickness and the refractive index are known and on a metal without a film.

(2) Detection of IR absorption spectra of super thin films on metals, revealing of the interaction effects of elementary excitations of a dielectric film with a metal, determination of the film optical constants at known optical characteristics of the metal.

(3) The control of the quality of metal surface preparation, in particular, laser mirrors, by the SEW attenuation on metal surface roughnesses.

References

Agranovich, V.M., 1975, Usp. Fiz. Nauk **115**, 199 (Sov. Phys. Usp. **18**, 99).

Begley, D.L., D.A. Bryan, R.W. Alexander, R.J. Bell and C.A. Goben, 1976, Surf. Sci. **60**, 99.

Begley, D.L., D.A. Bryan, R.W. Alexander and R.J. Bell, 1977, Appl. Opt. **16**, 1549.

Begley, D.L., R.W. Alexander, C.A. Ward and R.J. Bell, 1979, Surf. Sci. **81**, 238.

Bell, R.J., R.W. Alexander, W.F. Parks and G.S. Kovener, 1973, Opt. Commun. **8**, 147.

Bell, R.J., R.W. Alexander and C.A. Ward, 1975, Surf. Sci. **48**, 253.

Bhasin, K., D. Bryan, R.W. Alexander and R.J. Bell, 1976, J. Chem. Phys. **64**, 5019.

Bryan, D.A., D.L. Begley, K. Bhasin, R.W. Alexander, R.J. Bell and R. Gerson, 1976, Surf. Sci. **57**, 53.

Chabal, Y.J. and A.J. Sievers, 1978, Appl. Phys. Lett. **32**, 90.

Chabal, Y.J. and A.J. Sievers, 1980, Phys. Rev. Lett. **44**, 944.

Chen, W., 1977, Surface EM waves and spectroscopy in ATR configurations, Dissertation, University of Pennsylvania, Philadelphia.

Cox, F.T., G. Hass and W.R. Hunter, 1975, Appl. Opt. **14**, 1247.

Davarpanah, M., C.A. Goben, D.L. Begley and S.L. Griffith, 1976, Appl. Opt. **15**, 3066.

Frolkov, J.A., G.A. Puchkovskaya and V.M. Stoljarov, 1979, Ukr. Fiz. Zhurnal, **24**, 706.

Gerbstein, J.M. and M.A. Merkulov, 1975, Fiz. Tverd. Tela **17**, 1501 (Sov. Phys. Solid State **17**, 977).

Gervais, F. and B. Piriou, 1974, J. Phys. **C7**, 2374.

Ginzburg, V.L. and G.P. Motulevich, 1955, Usp. Fiz. Nauk **55**, 469.

Glang, R., 1970, Vacuum evaporation, in: Handbook of thin film technology, eds. L.I. Maisell and R. Glang (McGraw Hill Book Company, New York) ch. 1.

Gratton, L.M., S. Paglia, F. Scattaglia and M. Cavallini, 1978, Appl. Spectr. **32**, 310.

Halevi, P., 1978, Surf. Sci. **76**, 64.

Harris, L., 1955, J. Opt. Soc. Am. **45**, 27.

Harstein, A., E. Burstein, J.J. Brion and R.F. Wallis, 1973, Surf. Sci. **34**, 81; Solid State Commun. **12**, 1083.

Jogansen, L.V., 1962, Zh. Techn. Fiz. **32**, 406.

Jogansen, L.V., 1963, Zh. Techn. Fiz. **33**, 1323.

Jogansen, L.V., 1968, Zh. Techn. Fiz. **36**, 2056.

Landau, L.D. and E.M. Lifshitz, 1948, Elektrodinamika sploshnich sred (Electrodynamics of the continuum media) (Moscow) §68.

Lebedinskij, M.A., 1966, Electrovacuum materials (Energia, Moscow-Leningrad) p. 100 (in Russian).

Lubimov, J.N. and D.G. Sannikov, 1972, Fiz. Tverd. Tela **14**, 675.

Maeland, A.J., R. Rittenhouse, W. Lahar and P.V. Romano, 1974, Thin Solid Films, **21**, 67.

McMullen, J.D., 1975, Solid State Commun. **17**, 331.

Mertens, F.P., 1978, Surf. Sci. **71**, 161.

Mills, D.L., 1975, Phys. Rev. **B12**, 4036.

Motulevich, G.P., 1969, Usp. Fiz. Nauk, **97**, 211.

Nordal, P.E. and S.O. Kanstad, 1978, Opt. Commun. **24**, 95.

Otto, A., 1968, Z. Physik, **216**, 398.

Schevchenko, V.V., 1969, Smooth transitions in open waveguides (Nauka, Moscow) §8 (in Russian).

Schkl'arevskij, I.H., R.G. Jarovaya, V.P. Kostyuk and L.G. Lelyuk, 1966, Opt. i Spectr. **20**, 1075.

Shoenwald, J., E. Burstein and M. Elson, 1973, Solid State Commun. **12**, 185.

Siddiqui, A.S. and D.M. Treherne, 1977, Infrared Physics, **17**, 33.

Spitzer, W.A. and D.W. Kleinman, 1961, Phys. Rev. **121**, 1324.

Talaat, H., W.P. Chen, E. Burstein and J. Schoenwald, 1975, Scattering of volume and Surface electromagnetic waves by surface acoustic waves, in: Ultrasonics Symposium Proceedings, p. 441.

Tien, P.K., 1971, Appl. Opt. **10**, 2397.

Ward, C.A., R.J. Bell, R.W. Alexander and G.S. Kovener, 1974, Appl. Opt. **13**, 2378.

Ward, C.A., K. Bhasin, R.J. Bell, R.W. Alexander and J. Tyler, 1975, J. Chem. Phys. **62**, 1674.

Zhizhin, G.N., M.A. Moskalova, E.V. Shomina and V.A. Yakovlev, 1976, Pis'ma Zh. Eksp. Teor. Fiz. **24**, 221 (JETP Lett. **24**, 196).

Zhizhin, G.N., M.A. Moskalova, V.V. Nazin and V.A. Yakovlev, 1977, Zh. Eksp. Teor. Fiz. **72**, 687 (Sov. Phys. JETP **45**, 360).

Zhizhin, G.N., M.A. Moskalova, E.V. Shomina and V.A. Yakovlev, 1978, Study of films on the metal surfaces using surface electromagnetic wave attenuation, in: Spectroscopy of molecules and crystals (Naukova Dumka, Kiev) p. 92 (in Russian).

Zhizhin, G.N., M.A. Moskalova, E.V. Shomina and V.A. Yakovlev, 1979a, Pis'ma Zh. Eksp. Teor. Fiz. **29**, 533 (JETP Lett. **29**, 486).

Zhizhin, G.N., M.A. Moskalova, E.V. Shomina and V.A. Yakovlev, 1979b, Fiz. Tverd. Tela **21**, 2828 (Sov. Phys. Solid State, **21**, 1630).

Zhizhin, G.N., M.A. Moskalova, E.V. Shomina and V.A. Yakovlev, 1980a, Fiz. Met. i Met. **50**, N 3.

Zhizhin, G.N., H.H. Morozov, M.A. Moskalova, A.A. Sigarov, E.V. Shomina, V.A. Yakovlev and V.I. Grigos, 1980b, Opt. i Spectr. **48**, 181; Thin Solid Films, **70**, 163.

Zolotarev, V.M., 1972, Zh. Prikl. Spectr. **17**, 1052.

CHAPTER 4

Thermally Stimulated Emission of Surface Polaritons

E. A. VINOGRADOV, G. N. ZHIZHIN and V. I. YUDSON

Institute for Spectroscopy
USSR Academy of Sciences
Moscow
U.S.S.R

Surface Polaritons
Edited by
V.M. Agranovich and D.L. Mills

Contents

1. Introduction 147
2. Vibrational states in ionic crystal films 149
   2.1. s-Polarized modes 150
   2.2. p-Polarized vibrations, surface modes 150
   2.3. p-Polarized vibrations, longitudinal modes 153
   2.4. p-Polarized vibrations, transverse modes 154
3. Interaction with the transverse field (consideration of retardation) – IR absorption. 154
   3.1. s-Polarized modes, $\omega = \omega_{TO}$ 156
   3.2. p-Polarized vibrations, surface modes 157
   3.3. p-Polarized vibrations, longitudinal modes, $\omega = \omega_{LO}$ 157
   3.4. p-Polarized vibrations, transverse modes, $\omega = \omega_{TO}$ 157
4. Experiment 158
   4.1. Sample preparation 158
   4.2. Quantitative investigation of thermal emission spectra of the samples 159
5. Emissivity of surface polaritons of ZnSe films in the region $q < \omega/c$ 161
6. Emission of surface polaritons by films in the region $q > \omega/c$ 168
7. Emissivity of surface polaritons of ZnSe single crystals 174
8. High-frequency phonon polaritons in a $Gd_2(MoO_4)_3$ single crystal 177
9. Conclusions 181
10. Appendix 182
References 183

1. Introduction

Thermal vibrations of the crystal lattice ions give rise to the appearance of the alternating component of dipole moments at the characteristic frequencies of the substance (the frequencies of optical phonons). The second time derivative of such dipole moments differs from zero and, therefore, such ion vibrations are attended with the formation of an electromagnetic field. This electromagnetic field created by thermal vibrations of atoms can "leave" the crystal under certain conditions and be recorded as thermal emission of a nonblack body. Spectral measurements of the thermal emissivity of various samples find extensive application in the studies of their optical properties (Stepanov 1961, Stierwalt and Potter 1965, Huong 1978). However, until recently this method has been used only to study bulk vibrations in single crystals and vibrational states of liquids and gases.

The presence of the crystal boundary leads to specific surface vibrations of the crystal atoms. These vibrations are localized near the boundary and are characterized by the wave vector $\boldsymbol{q}$ parallel to the crystal surface. Thermal excitation of surface vibrations is also accompanied by the occurrence of alternating dipole moments and, therefore, of an electromagnetic field. If the field is produced by surface vibrations with the wave vector less than ω/c (ω is the vibration frequency, c is the light velocity in vacuum) it satisfies the emission conditions and is capable of "leaving" the crystal. It is recorded as thermal emission of surface states. But, if the wave vector $\boldsymbol{q}$ of surface vibrations of the crystal atoms at the interface "crystal–vacuum" exceeds ω/c, i.e., it exceeds the light wave vector in vacuum, the electromagnetic field of such modes is "tied" to the surface and cannot be emitted (the emission would break the condition of conservation of the wave vector component parallel to the interface*. The interaction of these nonradiative modes with the transverse electromagnetic field results in the renormalization of the dispersion law – the states thus formed are surface polaritons (Agranovich 1975, Ruppin and Englman 1970, Bryksin et al. 1974, Fuchs and Kliewer 1966a, b).

*In the present paper we do not consider the case of fine-dispersed media (single crystal powder) and cylindrical crystals of the radius $r \ll \lambda$ for which there exist radiative surface states (Ruppin and Englman 1970).

The study of surface polaritons by means of optical spectroscopy has become possible with the use of prisms with the attenuated total internal reflection (ATR) which equalize the wave vector of the surface polariton with the wave vector of the electromagnetic field (light) wave in the prism (Otto 1968, Bryksin et al. 1971, 1972, Marschall and Fischer 1972, Fischer et al. 1973). The method is used by Teng and Stern (1967) of equalizing wave vectors of the surface polariton and the external electromagnetic wave by means of periodic rulings (grating) on the crystal surface which partially break the conservation law for the wave vector parallel to the interface. In this case the wave vector of the surface polariton is conserved with the precision up to $m(2\pi/a)$ ($m = 0, 1, 2, 3, \ldots$) a is the period of grating which allows the external electromagnetic wave to excite surface states of the crystals. The emission of surface polaritons may also be caused by surface roughnesses (defects). Thus, an ATR prism over the crystal and periodic rulings on the crystal surface turn nonradiative vibrational surface states of the "crystal–vacuum" interface with $q > \omega/c$ to light absorbing states.

According to the Kirchhoff law, a system which can absorb light may emit it, i.e., the dipole field may become radiative.

Thermal emission of surface vibrational states for single crystals with $q > \omega/c$ was observed by Vinogradov and Zhizhin (1976) and for the films with $q > \omega/c$ by Vinogradov et al. (1977a, b). Moreover, as it has already been noted above, in the films there exist also radiative surface states with $q < \omega/c$ capable of being created (or annihilated) at light absorption (emission) under ordinary conditions (without the prism) (Vinogradov et al. 1980). These radiative surface states were first observed by Berreman (1963) in the transmission of thin films at inclined angle of incidence, but he referred to them as to the longitudinal optical phonons. In all the studies which followed (Berreman 1963, i.e., for example, Ruppin and Englman 1970, Proix and Balkanski 1969, Fuchs et al. 1966, Hisano et al. 1970, Vodopianov and Vinogradov 1974) the light absorption band in thin films near the frequency of the single crystal longitudinal optical phonon and actually due to the surface radiative states of the films (Gerbstein and Mirlin, 1974) was erroneously classified as longitudinal optical phonons of the crystalline slab.

Experimental studies of the thermal spectral emission in single crystals and films accompanied by the detailed theoretical analysis yield more reliable and complete information on the vibrational states of the samples.

In this connection the present paper deals with the experimental and theoretical investigations of surface vibrational states of thin films and single crystals. In sect. 2 a detailed theoretical analysis of Coulomb normal vibrational modes of a thin dielectric film on a metal substrate is performed. Polarization fields accompanying each normal mode of the system under consideration and their dispersion laws have been found. In sect. 3 the interaction of these normal modes with the transverse electromagnetic field is considered i.e., the retardation effects are allowed for, and the absorption of infrared (IR) emission by each mode is calculated. Section 4 describes the experimental procedure and sects. 5 and 6 present the experimental results of the study of the film emissivity at surface polariton frequencies in radiative and nonradiative regions, respectively, and experimental results are correlated quantitatively with the theoretical calculation. The results for thermally stimulated emission of surface polaritons in a ZnSe single crystal are given in sect. 7; sect. 8 presents the investigation of surface polaritons in a $Gd_2(MoO_4)_3$ single crystal in the vicinity of the phase transition temperature; the main results of the studies performed are briefly summarized in the conclusions.

2. *Vibrational States in Ionic Crystal Films*

Optical properties of a system "vacuum–semiconductor film–metal substrate" have been studied experimentally. Light absorption and emission in such a system may be described theoretically with the use of the Fresnel formulae and the Kirchhoff law (see sect. 5). However, this approach fails to describe in detail what is the absorbed energy transformed to and which vibrations are responsible for this or that peak in the spectrum. Therefore, another approach is also used: at first Coulomb normal modes of the system are found, i.e., the states calculated upon the neglect of the interaction with the transverse electromagnetic field, and afterwards (in sect. 3) this interaction is taken into account. It is known (Agranovich 1968) that the equations for the Coulomb states are obtained from the solutions of the Maxwell equations when the light velocity formally tends to infinity:

$$\operatorname{div} \boldsymbol{D} = 0, \tag{1}$$

$$\operatorname{curl} \boldsymbol{E} = 0. \tag{2}$$

When spatial dispersion effects are neglected, the material relation between the vectors of induction $\boldsymbol{D}$ and the electric field $\boldsymbol{E}$ is local:

$$\boldsymbol{D}(\omega, \boldsymbol{r}) = \epsilon(\omega)\boldsymbol{E}(\omega, \boldsymbol{r}). \tag{3}$$

This locality allows to substitute into this relationship the value $\epsilon(\omega)$ known for the bulk single crystal:

$$\epsilon(\omega) = \epsilon_\infty + \frac{(\epsilon_0 - \epsilon_\infty)\omega_{TO}^2}{\omega_{TO}^2 - \omega^2 + i\gamma\omega}. \tag{4}$$

The validity of this substitution is confirmed by the coincidence of the calculated and experimental IR spectra of the system (Vinogradov et al. 1977b, 1978b).

The solutions of eqs. (1) and (2) are sought in the form $\boldsymbol{E} = \boldsymbol{E}_q(z)\exp(iqx)$ (the x-axis is chosen along the two-dimensional wave vector $\boldsymbol{q}$ in the film plane, the z-axis is perpendicular to the film). The system of eqs. (1, 2) falls into equations relating E_x and E_z field components (p-polarization) and the equations for the component E_y (s-polarization).

2.1. s-Polarized Modes

For s-polarization we have from eq. (2): $E_y = 0$. The specific polarization value in the medium $\boldsymbol{P} = (\boldsymbol{D} - \boldsymbol{E})/4\pi$, however, can still be different from zero. The function $P_y(z)$ is indefinite since any dependence $P_y = P_y(z)$ at $-d < z < 0$ agrees at $\omega = \omega_{\mathrm{TO}}$ both with the condition $E_y = 0$ and with the boundary conditions (the normal component of the induction vector $\boldsymbol{D}$ and the tangential component of the field $\boldsymbol{E}$ for the vibrations under consideration and everywhere equal to zero). The arbitrary function $P_y(z)$ in the interval $-d < z < 0$ can be represented as a linear combination of the following linearly independent functions:

$$P_y^{(s)}(z) = A_n \sin\left[\frac{\pi n}{d}(z + d)\right], \quad n = 1, 2 \ldots \tag{5a}$$

$$P_y^{(c)}(z) = B_n \cos\left[\frac{\pi n}{d}(z + d)\right], \quad n = 0, 1, 2 \ldots \tag{5b}$$

Functions (5a) and (5b) correspond to the sought set of s-polarized Coulomb eigenmodes of the film. These modes are highly degenerated – the frequencies of all modes, as shown above, are equal: $\omega_g^{(s)} = \omega_g^{(c)} = \omega_{\mathrm{TO}}$. Similar to the transverse vibrations in a single crystal, the film vibrations under discussion are not accompanied by the occurrence of a bulk charge:

$$\rho = -\operatorname{div} \boldsymbol{P} = 0.$$

Surface charges do not appear either. For brevity these modes are called below the transverse modes of the film, but it should be kept in mind that strict unambiguous classification of vibrations into longitudinal and transverse ones is possible only for homogeneous systems. It is shown in the next subsection that the interaction with the transverse electromagnetic field of free photons makes the states with $q = \omega_{\mathrm{TO}}/c$ radiative (damping).

2.2. p-Polarized Vibrations, Surface Modes

In this section only those modes are considered which have $\epsilon(\omega) \neq 0$ and the electric field does not turn identically to zero. Here the dielectric function for the metal substrate ϵ_{M} is introduced.

Cancelling $\epsilon(\omega)$ from eq. (1), we have in each medium:

$$\mathrm{d}E_z/\mathrm{d}z + \mathrm{i}qE_x = 0. \tag{6}$$

Equation (2) for the p-polarized field has the form:

$$\mathrm{d}E_x/\mathrm{d}z - \mathrm{i}qE_z = 0. \tag{7}$$

Taking into account that at $|z| \to \infty$ the field falls to zero, and using usual boundary conditions at the vacuum-film and film-substrate interfaces, the field inside the film $(-d < z < 0)$ is determined:

$$E_x(z) = E^\circ\left[\cosh qz - \frac{1}{\epsilon(\omega)}\sinh qz\right],$$

$$E_z(z) = \mathrm{i}E^\circ\left[\frac{1}{\epsilon(\omega)}\cosh qz - \sinh qz\right], \tag{8}$$

and also the dispersion equation for the mode frequency:

$$\epsilon^2(\omega)\tanh qd + \epsilon(\omega)(1+\epsilon_{\mathrm{M}}) + \epsilon_{\mathrm{M}}\tanh qd = 0. \tag{9}$$

The latter equation falls into two relationships for $\epsilon(\omega)$ which at $|\epsilon_{\mathrm{M}}| \gg 1$ are:

$$\epsilon(\omega) = -\tanh qd, \tag{10a}$$

$$\epsilon(\omega) = -\epsilon_{\mathrm{M}}\coth qd. \tag{10b}$$

The appropriate modes are described below.

(a) Substituting eq. (10a) into eq. (8), we obtain the field in the film:

$$E_x(z) = E\sinh[q(z+d)],$$

$$E_z(z) = -\mathrm{i}E\cosh[q(z+d)]. \tag{11}$$

It is seen from eq. (11) that at large film thicknesses, i.e. at $qd \gg 1$, the field in this mode is concentrated near the vacuum–film interface and decreases exponentially with the depth into the film. At $qd \to \infty$ the dispersion [eq. (10a)] turns to a known (Agranovich 1975, Ruppin and Englman 1970, Bryksin et al. 1974) equation (obtained upon the neglect of retardation) for a surface phonon at the "semi-infinite medium–vacuum" interface:

$$\epsilon(\omega) = -1. \tag{12}$$

From eq. (12) for the dielectric function in the form (4) the frequency of this surface phonon ω_s at $d \to \infty$ is obtained:

$$\omega_{\mathrm{s}} = \omega_{\mathrm{TO}}\left(\frac{\epsilon_0 + 1}{\epsilon_\infty + 1}\right)^{1/2}.$$

For arbitrary d eq. (10a) with due regard to eq. (4) is solved as follows:

$$\omega_{\mathrm{s}}(q, d) = \omega_{\mathrm{TO}}\left(\frac{\epsilon_0 + \tanh qd}{\epsilon_\infty + \tanh qd}\right)^{1/2}. \tag{13}$$

The type of vibrations under discussion is not connected with the bulk charge (div $\boldsymbol{E} = 0$ at $-d < z < 0$), but it is accompanied by the appearance of a surface charge at the film boundaries:

$$\rho_s = \mathrm{i}E\,\frac{\epsilon(\omega)-1}{4\pi}\cosh qd, \quad \text{at } z=0,$$
$$\rho_s = -\mathrm{i}E\,\frac{\epsilon(\omega)-1}{4\pi}, \quad \text{at } z=-d. \tag{14}$$

This vibration will be called "a surface mode". As it is shown in the next section, the interaction with the transverse electromagnetic field makes these states with $q < \omega/c$ nonstationary (radiative). Note the fact which is rather important for further interpretation of the experimental data, namely, that at $qd \to 0$ the value $\epsilon(\omega)$, according to eq. (10a), tends to zero, while the frequency $\omega_s(qd) \to \omega_{LO} = \omega_{TO}(\epsilon_0/\epsilon_\infty)^{1/2}$, where ω_{LO} is the frequency of longitudinal phonons in a bulk single crystal. It should also be pointed out that in a thin film ($qd \ll 1$) the electric field of the mode under consideration is directed almost perpendicularly to the film surface and is practically homogeneous along its thickness.

(b) Consider a mode which corresponds to the dispersion relationship (10b). This equation includes the dielectric function of the metal substrate. An ideally conductive substrate corresponds to the equality $|\epsilon_M| = \infty$ and in this case from eqs. (10b) and (4) it follows that $\epsilon(\omega) = \infty$ and $\omega_s(qd) = \omega_{TO}$.

We are also interested in the field distribution for this mode. But from the finiteness of the induction vector $\boldsymbol{D} = \epsilon(\omega)\boldsymbol{E}$ in the film it follows that at $|\epsilon_M| \to \infty$ the value of the electric field in the film tends to zero. However, in this case the vector of the medium specific polarization $\boldsymbol{P} = (\epsilon(\omega)-1)\boldsymbol{E}/4\pi$ remains finite. Determination of this value $\boldsymbol{P}$ by eqs. (8) at the finite value $|\epsilon_M|$ with subsequent transition to $|\epsilon_M| \to \infty$ gives for $\boldsymbol{P}$ in this mode:

$$P_x(z) = P\cosh qz,$$
$$P_z(z) = -\mathrm{i}P\sinh qz. \tag{15}$$

It can be seen from eq. (15) that at large film thickness ($qd \gg 1$) the vibration being considered is concentrated in the region of maximum $|z|$ values possible in the film, i.e., at the interface with metal ($z = -d$) and decreases exponentially with the distance from this interface. Thus, we are dealing here with the second surface mode of the system "vacuum–film–metal". Note that in a thin film ($qd \ll 1$) the vector $\boldsymbol{P}$ of the medium specific polarization in this mode is directed almost parallel to the film surface and its value is practically constant across the film thickness.

It should be pointed out that the finiteness of the metal substrate conductivity results in that the frequency of the surface mode at hand is somewhat different from ω_{TO} and is only slightly dependent on q and d.

From eq. (10b) we find (at $|\epsilon_M| \gg 1$):

$$\omega_s(q, d) = \omega_{TO} + \frac{\omega_{TO}(\epsilon_0 - \epsilon_\infty)}{2\epsilon_M} \tanh qd. \tag{16}$$

In a thin ($qd \ll 1$) film the frequency shift (proportional to $\mathrm{Re}(1/\epsilon_M)$) and the damping of this mode (proportional to $\mathrm{Im}\,(1/\epsilon_m)$) turn out to depend linearly on d.

2.3. p-Polarized Vibrations, Longitudinal Modes

Now we consider the p-polarized modes which satisfy the relationship:

$$\epsilon(\omega) = 0. \tag{17}$$

From eq. (2) there follows eq. (7) which holds both in vacuum and in the film. But eq. (1) with the condition (17) is fulfilled identically inside the film. From the boundary condition $E_z(0+\delta) = \epsilon(\omega)E_z(0-\delta)$ and eqs. (6) and (7) it follows that the field in vacuum is identically equal to zero and, accordingly, the boundary condition for the field in the film gives $E_x(0-\delta) = 0$. The latter condition together with the relationship $E_x(-d) = 0$ determines the form of eigenmodes in the film:

$$E_x^{(n)}(z) = C_n \sin\left(\frac{\pi n}{d}(z+d)\right), \tag{18}$$

$$E_z^{(n)}(z) = -\mathrm{i}C_n \frac{\pi n}{d} \cos\left(\frac{\pi n}{d}(z+d)\right), \tag{19}$$

where $n = 1, 2, 3 \ldots$. from eqs. (17) and (4) the frequency of these modes is found:

$$\omega = \omega_{TO}(\epsilon_0/\epsilon_\infty)^{1/2} \equiv \omega_{LO}. \tag{20}$$

The vibrations under consideration are accompanied by the creation of both a bulk charge

$$\rho(q) = -\frac{\mathrm{i}C_n}{4\pi}\left[q + \left(\frac{\pi n}{d}\right)^2 \frac{1}{q}\right] \sin\left(\frac{\pi n}{d}(z+d)\right), \tag{21}$$

and a surface one:

$$\rho_s(q) = \frac{\mathrm{i}Cn}{4\pi} \frac{\pi n}{qd} \begin{cases} (-1)^n, & z = 0 \\ 1, & z = -d. \end{cases} \tag{22}$$

The properties of these modes (they are called longitudinal modes of the film or, for shortness, longitudinal modes) are close to those of single crystal longitudinal modes and their frequencies coincide.

2.4. p-Polarized Vibrations, Transverse Modes

Consider, finally, the p-polarized modes of the film for which the electric field $\boldsymbol{E}$ is identically equal to zero (it should be recalled that we are considering the Coulomb modes – with no regard for retardation). The medium polarization $\boldsymbol{P}$ and the induction vector $\boldsymbol{D}$ can still differ from zero. From the condition $\boldsymbol{D} = \epsilon(\omega)\boldsymbol{E}$ at $\boldsymbol{E} = 0$ it follows that $\epsilon(\omega) = \infty$ and the frequency of the modes under study coincides with the transverse phonon frequencies in a single crystal: $\omega = \omega_{\mathrm{TO}}$. From eq. (1) we obtain for the polarization $\boldsymbol{P}$:

$$iqP_x(z) + \mathrm{d}P_z(z)/\mathrm{d}z = 0 \tag{23}$$

Equation (2) for the modes at hand is, evidently, an identity. From the continuity condition for the tangential component of the electric field $\boldsymbol{E}$ at the "vacuum–film" interface it follows that the value E_x and, consequently, E_z in vacuum equal zero. The boundary condition $P_z(0) = 0$ results from the value E_z being equal to zero in vacuum and from the continuity condition for the normal component of the induction vector $\boldsymbol{D}$. At the finite value of the dielectric function ϵ_{M} for the metal substrate it is possible to obtain in a similar way also the second boundary condition for P_z: $P_z(-d) = 0$, which retains its form also at the transition to the case of the ideally conducting substrate ($|\epsilon_{\mathrm{m}}| \to \infty$). The form of the function $\boldsymbol{P}(z)$ is determined by eq. (23) and both boundary conditions:

$$P_x(z) = \sum_n C_{n,q} \cos\left[\frac{\pi n}{d}(z + d)\right],$$

$$P_z(z) = -\frac{iqd}{\pi} \sum_n \frac{1}{n} C_{n,q} \sin\left[\frac{\pi n}{d}(z + d)\right]. \tag{24}$$

The function $\boldsymbol{P}(z)$ can be seen from (24) to be a linear superposition of independent modes. The frequency of each of these modes is shown above to equal ω_{TO}. These vibrations are not accompanied by the appearance of either a surface charge (since $P_z(0) = P_z(-d) = 0$) or a bulk one (due to eq. (23)). These film modes may naturally be called transverse. Similar to the s-polarized transverse modes of the film, the p-polarized transverse modes just considered can interact with the electromagnetic field to result in the finite radiative width of the modes with $q < \omega/c$.

3. Interaction with the Transverse Field (Consideration of Retardation) – IR Absorption

The above Coulomb modes of the films can interact with the transverse electromagnetic field $\boldsymbol{E}^{\perp}$ of free photons. The Hamiltonian of this interaction is:

$$H_{\text{int}} = -\int \hat{\boldsymbol{P}}\hat{\boldsymbol{E}}^{\perp}\, \mathrm{d}V, \tag{25}$$

where $\hat{\boldsymbol{P}}$ is the polarization, corresponding to the film Coulomb modes. Generally speaking, the interaction (25) leads to a radical change of the film eigenmodes due to their mixing with the transverse field (polariton effect). In the problem on the absorption of an incident transverse wave the consideration of multiple reabsorption processes described by the Hamiltonian (25) is also important: it leads, in particular, to the renormalization of the incident wave field inside the film in accordance with the Fresnel formulae. However, with the decrease of the film thickness the role of the processes of the higher order by the interaction (25) becomes less important. In the film with the small optical thickness as compared to the wavelength of the transverse field, the consideration can be limited to the lower order interaction (25).

In this section we consider light absorption by a thin film, elucidate the relative contribution of each of the modes found in the previous section to the light absorption and, furthermore, obtain the dependence of the absorption of the film thickness and the angle of the light incidence. Naturally, light absorption and radiation by a film on the substrate can be found in a similar way – by means of the Kirchhoff law and the Fresnel formulae (Ruppin and Englman 1970, Vinogradov et al. 1977b, 1978a; see also this chapter sect. 5), omitting preliminary determination of the system Coulomb modes. However, the formulae thus obtained which describe the light absorption by a system do not allow to describe in detail what the absorbed energy is transformed into and which vibrations in the system are responsible for this or that peak in the spectrum.

Consider the interaction of the Coulomb modes in a thin film with the electromagnetic field of free photons described by the Hamiltonian (25). At photon absorption accompanied by the Coulomb mode excitation the energy and the wave vector parallel to the film plane should be retained:

$$\omega(q) = \omega, \tag{26}$$

$$q = (\omega/c)\sin\theta, \tag{27}$$

where ω and $\omega(q)$ are the frequencies of the incident light and Coulomb mode, respectively, θ is the angle of light incidence. The relationships (26) (27) give the inequality which limits the number of modes capable of absorbing (or emitting) light in one-quantum processes:

$$q < \omega(q)/c. \tag{28}$$

Thus, the modes with $q > \omega(q)/c$ do not participate in the linear absorption or emission of light. Their interaction with the electromagnetic field results only in the renormalization of the dispersion law – the formation of surface polaritons.

When the conditions (26) and (27) are fulfilled, the probability of phonon absorption (integrated over the appropriate band) with the precision up to the normalization factor independent of the film thickness d and the angle of incidence θ is –according to the Fermi "golden" rule– proportional to the square of the matrix element of the interaction Hamiltonian (25):

$$A \sim \frac{\left| \int \boldsymbol{P}_q^*(z) \mathscr{E}_q(z)\, \mathrm{d}z \right|^2}{\int |\boldsymbol{P}_q(z)|^2\, \mathrm{d}z}. \tag{29}$$

Functions $\boldsymbol{P}_q(z)$, corresponding to the Coulomb modes under investigation, were obtained in the previous section. Functions $\mathscr{E}_q(z)$ describe the transverse electromagnetic field with no regard for the interaction with the film, but with the allowance made for the reflection from the metal:

$$\mathscr{E}_q(z) \sim (0, \sin[k_z(z+d)], 0) \tag{30}$$

for s-polarization, and

$$\mathscr{E}_q(z) \sim (-\mathrm{i} \cos\theta \sin[k_z(z+d)], 0, \sin\theta \cos[k_z(z+d)]) \tag{31}$$

for p-polarization, where $k_z = (\omega/c)\cos\theta$.

Light absorption by various Coulomb modes is considered below in the same sequence as these modes were obtained in the previous section.

3.1. s-Polarized Modes, $\omega = \omega_{\mathrm{TO}}$

These modes interact, evidently, only with the s-polarized light. Substituting the expression (30) for the amplitude of the incident wave and formulae (5a) and (5b) for the polarization of sinusoidal and cosinusoidal modes into eq. (29), we have at $k_0 d \ll 1$ ($k_0 = \omega/c$) for the sinusoidal modes (5a):

$$A \sim [(k_0 d)^3/n^2] \cos^2\theta, \quad n = 1, 2, \ldots \tag{32}$$

and for cosinusoidal modes (5b):

$$A \sim \begin{cases} (k_0 d)^3 \cos^2\theta, & n = 0 \\ [(k_0 d)^7/n^4] \cos^6\theta, & n = 2, 4, 6, \ldots \\ (k_0 d)^3/n^4, & n = 1, 3, 5, \ldots. \end{cases} \tag{33}$$

Light absorption by all s-polarized modes gives a peak in the absorption spectrum of a thin film at the frequency ω_{TO} (note that all the modes considered here have the frequency ω_{TO}). As it can be seen from eqs. (32) and (33), the main contribution to this peak at small $k_0 d$ is made by the sinusoidal modes (5a) and cosinusoidal modes with $n = 0$ and $n = 1$, 3,

5 . . . . The summation of the contributions of these modes gives the integral intensity A of the absorption peak of the s-polarized light:

$$A \sim (k_0 d)^3 \cos^2 \theta. \tag{34}$$

3.2. p-Polarized Vibrations, Surface Modes

The interaction of the p-polarized light with the first surface mode (described in sect 2.2(a) results in an absorption peak near the frequency ω_{LO} (according to formula (13) at $qd \ll 1$). Using formulae (11), (29) and (31), we obtain the integral absorption value:

$$A(\omega_s^a \approx \omega_{\mathrm{LO}}) \sim k_0 d \sin^2 \theta. \tag{35}$$

The interaction with the second surface mode described in sect. 2.2 gives the absorption peak near the frequency ω_{TO}. Its integral intensity is found by means of (15):

$$A(\omega_s^b \approx \omega_{\mathrm{TO}}) \sim (k_0 d)^3. \tag{36}$$

3.3. p-Polarized Vibrations, Longitudinal Modes, $\omega = \omega_{\mathrm{LO}}$

Using formulae (15), (16) and (31) and integrating (29), we obtain that the matrix element $|\int P^* E \, \mathrm{d}z|^2$ equals zero and, accordingly, we have

$$A(\omega_{\mathrm{LO}}) \equiv 0. \tag{37}$$

Thus, the modes with the frequency ω_{LO} considered in sect. 2.3 do not interact at all with the free electromagnetic field in the same way as the longitudinal modes of the bulk single crystal. This justifies the name "longitudinal" given to the modes considered.

3.4. p-Polarized Vibrations, Transverse Modes, $\omega = \omega_{\mathrm{TO}}$

Using eq. (24), we obtain for the absorption intensity of each mode:

$$A_n(\omega_{\mathrm{TO}}) \sim \begin{cases} (k_0 d)^3 / n^4, & n = 1, 3, 5, 7 \\ [(k_0 d)^7 / n^4] \cos^4 \theta, & n = 2, 4. \end{cases} \tag{38}$$

Therefore, the total contribution of all these transverse modes to the absorption is:

$$A(\omega_{\mathrm{TO}}) \sim (k_0 d)^3. \tag{39}$$

Summarizing this section, we see that the light interaction with the Coulomb modes of the system leads to the appearance of absorption peaks at the frequencies of these modes. In the s-polarized light only one peak [eq. (34)] at the frequency ω_{TO} appears and it results from light absorption by s-polarized modes. In p-polarized light there are, generally speaking, three absorption peaks. One of them eq. (35) is near the frequency ω_{LO} and it is due to light absorption by a surface vibration of the (a) type. It is a rather intense peak: it is $(k_0d)^{-2}$ times as large as the other peaks. Note that the absorption peak near the frequency ω_{LO} observed experimentally by Berreman (1963), Proix and Balkanski (1969), Hisano et al. (1970), Vodopianov and Vinogradov (1974) is connected with just the surface mode [eq. (11)] and not with light absorption by longitudinal modes as it was supposed by Berreman (1963) and many subsequent experimental and theoretical works (Ruppin and Englman 1970, Bryksin et al. 1974, Proix and Balkanski 1969, Vodopianov and Vinogradov 1974)*. According to the expression (37), the longitudinal modes of the film do not interact at all with the transverse electromagnetic field.

In the p-polarized light absorption there is also a peak eq. (39) at the frequency ω_{TO}; this peak results from the light absorption by the p-polarized transverse modes. Moreover, in the vicinity of the frequency ω_{TO} there is a peak [eq. (36)] which results from light absorption by the second surface mode [eq. (15)]. The frequency of this peak is rather close to ω_{TO} (the difference between these frequencies is due to the imperfection of metals and is proportional –according to eq. (26)– to a small value k_0d) and the dependence of the intensity of this peak on the angle and the film thickness d is the same as for the ω_{TO} peak of absorption by the transverse modes. Therefore, both these peaks merge into one slightly asymmetric peak at the frequency ω_{TO}. The integral intensity of this peak –according to eqs. (36) and (39)– is:

$$A(\omega_{TO}) \sim (k_0d)^3. \tag{40}$$

The relationship derived in this section will be shown in sect. 5 to agree well with the experiment.

4. Experiment

4.1. Sample Preparation

The research was performed on single crystals and films of a broad-gap semiconducting compound ZnSe. It is one of the most studied simplest compounds with the zinc blende structure (the space symmetry group T_d^2) with two atoms (ions) in the primitive cell. The procedure of a perfect single crystal growth and the crystalline film deposition on various substrates has been well developed. Single crystals were grown by the

*Surface modes in radiation region were first observed by Gerbstein and Mirlin (1974) in ATR spectra.

Bridgeman method. According to the crystal growth regimes, it was possible to obtain either single crystals with the zinc blende structure or crystals with a lot of alternating layers, the same structure separated by plane-parallel twinning boundaries perpendicular to the growth axis. RS spectra and IR reflection spectra of these crystals give us only the optical phonon frequencies of the centre of the Brillouin zone for a cubic ZnSe crystal.

The alternating layers of the twinning zinc blende can be seen well in crossed polarizers of the microscope. The thicknesses of the twinning slabs for some crystals turn to be rather similar and are of the order of several dozens of microns. Thus, such a crystal with the sides parallel to the growth axis is a sufficiently periodic structure analogous to the diffraction grating. The white light, when passing through such an optically polished plate undergoes spectral decomposition.

The films of ZnSe were prepared by vacuum evaporation onto various metal substrates of Al, Cr, Ti. In the course of evaporation the temperature of the substrates was maintained at 150°C; as a rule, the evaporation provides amorphous films of homogeneous thickness. Long recrystallizing annealing in an argon atmosphere at 300°C yields solid crystalline films of the mosaic structure with grain (crystal) diameters in excess of 10 μm at the film thickness 0.3–2 μm. The crystallinity degree of the films was checked by IR absorption spectra at the optical phonon frequency. In this work the films investigated had the crystallinity degree in excess of 95%.

4.2. Quantitative Investigation of Thermal Emission Spectra of the Samples

The method of thermally stimulated emission of vibrational states in single crystals and films was used in our experiments for two reasons. Firstly, this method permits easy measurement of the angular dependence of the emissivity since the sample in the thermostat replaced a standard instrumental IR source in the spectrometer and the axis of the sample rotation in the thermostat coincides with the sample plane. Secondly, the sensitivity of this method for the samples with low optical density is several dozens of times higher as compared to that of the conventional methods used to measure the absorption. The absorption by surface polaritons at large gaps between the ATR prism and the sample is a low contrast. The same low contrast is characteristic for the absorption spectra of thin films.

The investigation of thermal emission spectra at the sample temperature differing only slightly from the temperature of the spectral instrument is a rather complicated problem. It is due to the fact that in the IR spectral region the intensity of the wall emission in conventional spectral instruments, choppers of the light flow, is comparable by its intensity with the emission of samples heated up (or cooled down) by 100–150°C and having low optical density. Furthermore, spectral intensity of thermal emission of individual elements of the spectrometer are not known with sufficient accuracy, and they enter (additive to the sample emission) the expression for the intensity of the signal obtained from the IR emission detector (Stepanov 1961, Stierwald and Potter 1962, 1965, Huong 1978, Hisano et al. 1970, Vinogradov et al. 1976):

$$I_1 \simeq kT_\lambda\{W_s(\lambda, t_s^0) - W_{ch}(\lambda, t_{ch}^0) + \beta W_w(\lambda, t_w^0) + W_{th}(\lambda, t_{th}^0)\}, \tag{41}$$

where k is the coefficient of the light flow transformation into the output signal of the spectrometer, T_λ is the spectral transmission of the spectrometer at the wavelength λ; W_s, W_{ch}, W_w, W_{th} are the thermal emission intensities of the sample, chopper, instrument walls and thermostat, respectively; $\beta \leq 1$ is the coefficient determined by the geometry of the spectral instrument. In the region of large wavelengths $\lambda > 10\,\mu$m the right-hand part of eq. (41) is proportional to the absolute value of the temperature difference between the sample and the instrument elements. Therefore, in contrast with the thermal emission we call our recorded sample emission "thermally stimulated".

If the sample is replaced by a metal mirror and the signal I_2 is measured (a good metal mirror is characterized by very little absorption and, therefore, its emission is small; in this case it tends to zero) and then the "black" body emission is measured at the same temperature, I_3, the ratio of the signal difference is:

$$\frac{I_1 - I_2}{I_3 - I_2} = \frac{W_s - W_M}{W_{bb} - W_M} \simeq \frac{W_s}{W_{bb}} = E_s, \tag{42}$$

where $E_s \simeq (1 - \exp(-\alpha d))$, αd is the optical density of the sample. The relation (42) is true in the absence of interference of the emitted electromagnetic waves in a plane-parallel sample, i.e. in the frequency region, where the optical thickness of the sample satisfies the relationships $\tilde{d} \gg \lambda$ and $\tilde{d} \ll \lambda$.

As black body in the experiments we used a cylindrical cavity of the copper thermostat with the known (from separate measurements) coefficient of the specular and diffuse reflection, i.e., with the known "greyness" coefficient. While measuring the emissivity of single crystals an aluminium mirror was placed at their back sides between the sample and the thermostat wall. In this way the emissivity of bulk vibrational states of single crystals and films was measured along with radiative surface states of the films and "layered" crystal ZnSe. Nonradiative surface states (with $q > \omega/c$) of single crystals and films become radiative if we put close to them a half-cylinder of single-crystalline silicon or KRS-6 at emission angles greater than the critical one, $\theta_c > \text{arc tg}\, n$, where n is the refractive index of the half-cylinder material. The sample, together with the half-cylinder, was fastened in the thermostat holder and the gap between the crystal and the half-cylinder was set with the aid of a frame-like mylar film spacer. The emissivity spectra of the systems "sample/gap/half-cylinder I_1", "mirror/gap/half-cylinder I_2", "black body I_3 and the mirror I_4" were measured successively. Then the ratio

$$(I_1 - I_2)/(I_3 - I_4) \simeq E_s/E_{bb} \tag{43}$$

was calculated. The thermostat temperature in all four measurements was maintained constant with the precision ±0.05°.

The measurements were made by means of a modified single-beam far IR spectrometer FIS-21 (Hitachi). A pneumatic Golay detector was used as an emission detector.

5. Emissivity of Surface Polaritons of ZnSe Films in the Region $q < \omega/c$

In sects. 2 and 3 the Coulomb modes of the film were obtained and their interaction with the transverse electromagnetic field was investigated. This approach is sufficiently simple and permits not only the correct description of the dependence of the intensity of absorption peaks on the film thickness and the angle of light incidence, but it also allows to explain which Coulomb modes are responsible for each peak. However, the simplicity and fruitfulness of this approach were conditioned by the fact that it was used mainly to consider a thin film ($qd \ll 1$), i.e., in the case when the consideration may be limited to the interaction (25) in the lower order of the perturbation theory. In thicker films (for example, already in those which have the optical thickness comparable with the light wavelength) it is necessary to take into account the interaction (25) in all orders. In principle, the squareness of the total Hamiltonian of the system allows to make all calculations, but the method becomes rather cumbersome.

In the following, the emissivity of the system "film–metal substrate" with the arbitrary thickness of the film is calculated without resorting to the above procedure. The Kirchhoff law is used instead. The Kirchhoff law which relates the system emissivity, reflectivity and transmissivity at the fixed frequency and the angle of observation holds true only at small light scattering. The smallness of scattering in the IR region chosen is provided by rapid decrease of scattering intensity with the increasing wavelength ($I_s \sim \lambda^{-4}$).

It follows from the Kirchhoff law that the emission intensity of the film on the metal substrate (the system transmissivity is zero) is:

$$E(\omega, \theta, T) = [1 - R(\omega, \theta)]\, E_{b.b}(\omega, T), \tag{44}$$

where $E_{b.b}(\omega, T)$ is the spectral intensity of the radiation emitted from a black body at the temperature T; $R(\omega, \theta)$ is the reflectivity of the system for a plane wave incident at an angle θ.

The reflectivity $R(\omega, \theta)$ of the system can easily be found by solving the Maxwell equations with suitable boundary conditions for an insulator (semiconductor) film on a metal substrate (see sect. 2). Here the p-polarized emission is of the utmost interest. For the p-polarized light the reflectivity $R(\omega, \theta)$ is given by the expression (Vinogradov et al. 1978a):

$$R(\omega, \theta) = \left| \frac{\epsilon(\omega) u \cos\theta - (k_z/k_0) V}{\epsilon(\omega) u \cos\theta + (k_z/k_0) V} \right|^2, \tag{45}$$

where

$$u = e^{-ik_z d} + Q\, e^{ik_z d}; \qquad V = e^{-ik_z d} - Q\, e^{ik_z d};$$

$$Q = \left(i\frac{\epsilon_M(\omega)}{\kappa} + \frac{\epsilon(\omega)}{k_z} \right)\left(i\frac{\epsilon_M(\omega)}{\kappa} - \frac{\epsilon(\omega)}{k_z} \right)^{-1};$$

$$k_z = k_0(\epsilon(\omega) - \sin^2\theta)^{1/2}; \quad \kappa = k_0(\sin^2\theta - \epsilon_M(\omega))^{1/2}; \quad k_0 = \omega/c,$$

and $\epsilon(\omega)$ and $\epsilon_M(\omega)$ are the dielectric functions of the film and the metal substrate, respectively.

In the region with the surface state of the film near the frequency ω_{LO} (sect. 2.2a) $\epsilon(\omega)$ is small, consequently, $k_z d \ll 1$. In this case eq. (45) can be written in the form:

$$R(\omega, \theta) \simeq 1 - \frac{2\pi\gamma(d/\cos\theta)\sin^2\theta(1/\epsilon_\infty - 1/\epsilon_0)\,\omega_s^2}{(\omega_s - \omega)^2 + \gamma^2} + \frac{1}{\cos\theta}\frac{1}{\mathrm{Im}\,(-\epsilon_M)^{1/2}}.$$

If the dielectric function of the metal is expressed in the form

$$\epsilon_M(\omega) = i\,4\pi\sigma(\omega)/\omega, \tag{47}$$

where $\sigma(\omega)$ is the conductivity of the metal substrate, we obtain from eqs. (46) and (47) with due regard for eq. (44):

$$E(\omega, \theta) = \frac{2\pi\gamma}{\cos\theta}\frac{d\sin^2\theta\,(1/\epsilon_\infty - 1/\epsilon_0)\,\omega_s^2}{(\omega_s - \omega)^2 + \gamma^2} + \frac{1}{\cos\theta}\left(\frac{2\omega}{\pi\sigma}\right)^{1/2}. \tag{48}$$

The last term in eq. (48) corresponds to the emission (absorption) of electromagnetic waves by the metal substrate, and it represents the well-known Hagen–Rubens correction (Ziman 1964). And the first term in eq. (48) represents the emission band of a surface polariton (surface states eq. (11)). This band, as it follows from eq. (48), has the Lorentzian form with the maximum at the frequency $\omega_s \leqslant \omega_{LO}$. Its intensity at the maximum is proportional to $d\sin^2\theta$, which is in full agreement with the results obtained in sect 3.2 (the expression (35)).

Figures 1 and 2 show the experimental data on the angular dependence of the p-polarized emissivity of ZnSe film on the aluminium substrate. As it follows from the formulae given in sect. 3, two emission bands are observed in these spectra: one of them increases rapidly with the film thickness (the band $\omega_1 \approx \omega_{TO} = 200\,cm^{-1}$) and is practically only slightly dependent on the angle of emission, and the other one ($\omega_2 \approx \omega_{LO} =$

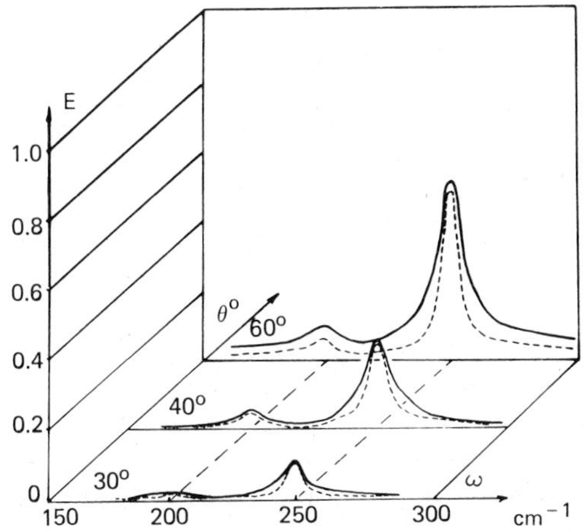

Fig. 1. Emissivity of ZnSe film 0.3 μm thick evaporated on an aluminium mirror (p-polarization, spectral resolution $1\,cm^{-1}$). The continuous curves are experimental and the dashed curves are calculated by formula (47).

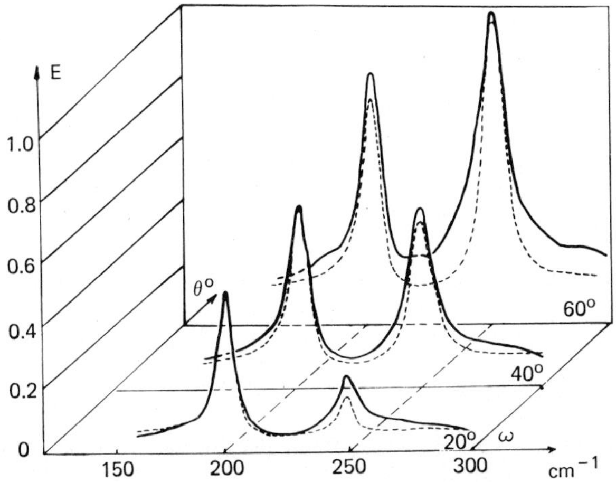

Fig. 2. Emissivity of ZnSe films in p-polarization (the film thickness 1 μm on an aluminium mirror). The continuous curves are experimental, the dashed curves are calculated.

250 cm^{-1}) increases rapidly with the angle of emission, but not so rapidly as ω_l increases with the film thickness.

It should be noted that in the s-polarized emission of the system ZnSe–Al only one emission band ω_l is observed at the frequency $\omega_l \approx \omega_{TO} = 200\ cm^{-1}$ and there is no emission band ω_2. This also agrees well with the results of the calculation from sects. 2.1 and 3.1.

Figures 1 and 2 show –besides the experimental curves– also the calculated ones (dashed) for the emissivity of the system ZnSe–Al obtained by means of the formula (45). The values $\epsilon_0 = 8.9$; $\epsilon_\infty = 5.8$; $\omega_{TO} = 200\ cm^{-1}$ and $\gamma = 6\ cm^{-1}$ known from measurements on single crystals and the substrate conductivity known for a freshly evaporated thick aluminium layer ($\sigma_{Al} = 2 \times 10^4\ ohm^{-1}cm^{-1}$) were substituted into this formula.

Considering that the perfection degree of the crystalline structures of films and single crystals is different (it affects primarily half-widths of the absorption bands) the agreement between the experimental and the calculated emissivities of the sandwiches "semiconductor–metal" should be regarded as good.

Figure 3 shows the experimental and theoretical dependences of the intensities of emission bands on film thicknesses at various emission angles θ. Theoretical dependences were obtained by numerical calculation using the formula (45). The analysis of the dependences presented in fig. 3a shows that the intensity of the band ω_2 at small d and all emission angles is proportional to the film thickness, and from fig. 3b it follows that the intensity of the band ω_l at the maximum ($\omega = \omega_l$) is proportional to d^3. This dependence is most clearly observed at small angles of radiation.

Figure 4 shows the experimental and theoretical dependences of the

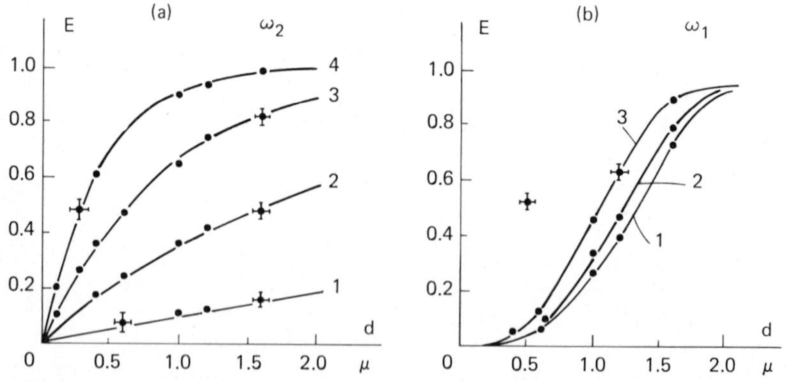

Fig. 3. Emissivity of ZnSe films on Al vs. ZnSe film thickness: (a) for the emission band $\omega_2 = 250\ cm^{-1}$; (b) for the emission band $\omega_l = 200\ cm^{-1}$. (1) $\theta = 15°$, (2) $\theta = 30°$, (3) $\theta = 45°$, (4) $\theta = 60°$: (●) experimental intensities of the bands ω_l and ω_2; the continuous curves are calculated.

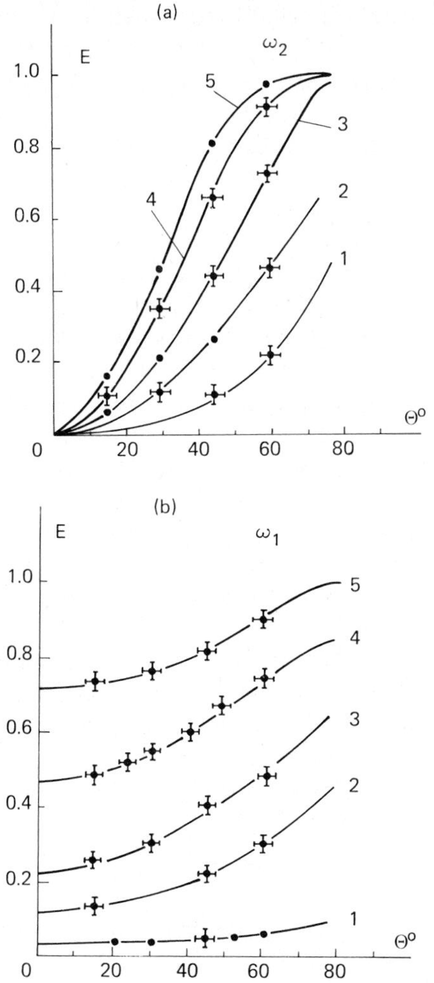

Fig. 4. Emissivity of ZnSe films on Al vs. the emission angle: (a) for the band ω_2; (b) for the band ω_1: (1) $d = 0.3$ μm; (2) $d = 0.5$ μm; (3) $d = 1$ μm; (4) $d = 2$ μm (●) experimental data, the continuous curves are calculated.

intensities of emission bands on the emission angles for films of various thicknesses. Theoretical dependences were also obtained by numerical calculation using the formula (45). The analysis of the dependence presented in fig. 4 shows that the intensity of the band ω_2 is proportional to $\sin^2 \theta$ (at small d it is strictly followed) and the intensity of the band ω_1 increases very slightly with the emission angle, i.e., at small d it remains practically constant.

The totality of the experimental results obtained allows us to state that the band $\omega_1 = 200\,\text{cm}^{-1}$ can really be conditioned by the emission of vibrational states described by surface (15) and transverse (24) modes which have the same thickness and angular dependence of the intensity (40). The band $\omega_2 = 250\,\text{cm}^{-1}$, however, is conditioned by the surface states (11), for which the experimental dependence of the absorption (emission) intensity on the angle and film thickness coincides well with that described by the expression (35).

The investigations carried out show also that the polarization field of vibrations with the frequency ω_1 is practically parallel to the film plane whereas the polarization field of vibrations with the frequency ω_1 is perpendicular to the film plane. These conclusions obtained from the polarization and angular measurements of the emissivity are in good agreement with the calculations made in sects. 2 and 3 and with the results of studying the effect of the metal substrate conductivity on optical properties of ZnSe films (Vinogradov et al., 1978a). The substrate conductivity, as stated above, enters, in its explicit form, the dispersion law for the low-frequency surface mode with $\omega \sim \omega_{TO}$ (see sect. 3.2, expression (10b)).

Figure 5 presents the emissivity spectra (p-polarization) of the system "ZnSe film–metal". As metal substrates thick deposited films of Al, Cr, Ti were used. All three samples of "ZnSe–metal" were prepared in the same technological cycle and all of them were ZnSe evaporated at the same time; this was followed by recrystallizing thermal treatment in the same quartz ampoule. Therefore, the degree of crystallinity of ZnSe films may be assumed to be the same. This was indicated by the identical half-widths of emission bands $\omega_2 = 250\,\text{cm}^{-1}$ of all three samples*. The emissivity of these

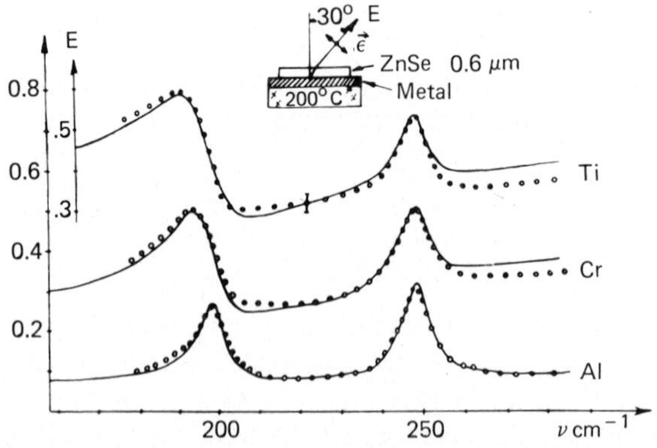

Fig. 5. Emissivity of ZnSe films on aluminium, chromium and titanium (p-polarization). The points are the experimental data, the continuous curves are calculated.

*The dispersion law for the surface vibration of the frequency $\omega_2 \simeq \omega_{LO}$ is independent of the substrate eq. (10a).

samples (continuous curves in fig. 5) was calculated by substituting the experimental frequencies ω_1, ω_2 and the half-width γ for a film on an aluminium mirror in formula (45); the value of ϵ_∞ was taken to be the same as for a single crystal. The only variable was the substrate conductivity. The best agreement between the calculated and experimental values of the emissivity (fig. 5) was obtained for $\sigma_{Al} = 8 \times 10^3\ \text{ohm}^{-1}\text{cm}^{-1}$ ($R_{Al} = 94\%$), $\sigma_{cr} = 5 \times 10^2\ \text{ohm}^{-1}\text{cm}^{-1}$ ($R_{Cr} = 76\%$) and $\sigma_{Ti} = 3 \times 10^2\ \text{ohm}^{-1}\text{cm}^{-1}$ ($R_{Ti} = 67\%$). The small and insignificant difference between the experimental and calculated emissivity curves in the frequency range $\omega > 260\ \text{cm}^{-1}$ observed for the ZnSe films on chromium and titanium is probably due to the fact that the substrate conductivity is no longer described satisfactorily by the Hagen–Rubens formula. The values of the chromium and titanium substrate conductivity obtained in this way are in good agreement (the error is below 20%) with the results of independent measurements of the conductivity of pure chromium and titanium films (in the absence of a ZnSe film).

The shift observed in fig. 5 and the ω_1 band broadening with the decrease of the metal substrate conductivity agrees qualitatively well with the expression (16). Expression (16) also gives a qualitatively good description of the decrease of the ω_1 band frequency with the increase of ZnSe film thickness. Figure 6 shows the experimental and theoretical (calculated by formula (45)) dependences of the emission band frequencies of ZnSe films on the aluminium substrate on the film thickness d. The data available do not permit us to say whether the band ω_1 is conditioned only by the low-frequency surface mode (15) since the frequency of s-polarized ω_{TO} transverse modes of the film (analogous to the p-polarized ω_{TO} modes) also decreases with the increase of the film thickness (Vodopianov and Vinogradov 1974). The band ω_1 is most likely to be conditioned by both surface

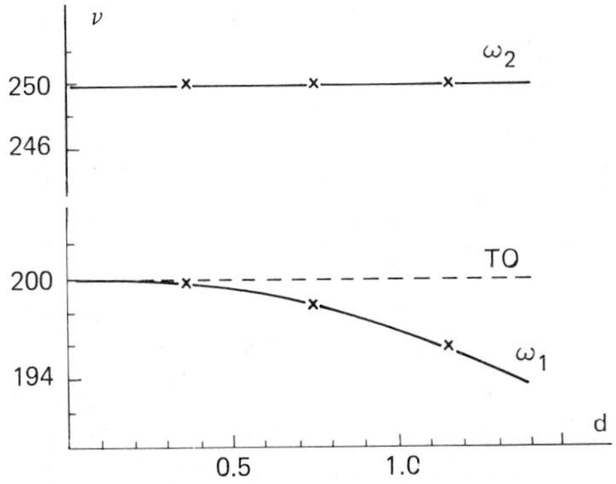

Fig. 6. Band frequencies of p-polarized emission vs. the thickness of ZnSe film on Al. The points are the experimental data, the continuous curves are calculated.

modes (15) and by the p-polarized ω_{TO} transverse mode (Vinogradov et al. 1978a).

6. *Emission of Surface Polaritons by Films in the Region* $q > \omega/c$

In sect. 2 the Coulomb modes of a semiconductor film on a metal substrate were investigated. Among these modes there were also considered two surface modes (formulae (10a) and (10b)). As noted above, the modes with $q > \omega/c$ under normal conditions do not participate in single-quantum processes of light emission or absorption. However, the interaction with the transverse electromagnetic field results in the renormalization of the dispersion laws for these modes known as "surface polaritons" (Agranovich 1975, Ruppin and Englman 1970, Bryksin et al. 1974, Fuchs and Kliewer 1966a, b). Since the electromagnetic field of these modes (with $q > \omega/c$) is damping exponentially with the distance from the film boundaries, it is evident that it does not satisfy the emission conditions, i.e., surface polaritons with $q > \omega/c$ are nonradiative in the system under consideration.

If an ATR prism is placed close to the film from the vacuum side, then, as it is known, the light wave vector in the ATR prism can be fitted to the wave vector q of the surface polariton. In case this condition is fulfilled, the surface polariton may turn into light. Thermally stimulated emission of surface polaritons in the regime of the inverted ATR was first observed for single crystals by Vinogradov and Zhizhin (1976) and for films by Vinogradov et al. (1977a, b). In the case of a film on a substrate –depending on the substrate conductivity– there can be observed either one surface polariton ω_+ (connected mainly with the film–vacuum interface) or, in addition to it, the second surface polariton ω_- (localized near the film–metal interface). It will be seen from a further analysis that the possibility of observing the emission of the second surface polariton ω_- is connected with the finite conductivity of the substrate. For an ideally conducting substrate the frequency of the surface polariton is equal to the frequency ω_{TO} of the film transverse mode and an additional peak, which is due to the surface polariton emission, impossible to observe on the background of the emission of this mode.

In order to obtain the dispersion equation for surface polaritons with regard to retardation we use –the same as in the previous section– a system of the Maxwell equations supplemented in each of the three media with a material relation $\boldsymbol{D}(\omega) = \epsilon(\omega)\boldsymbol{E}(\omega)$ with the appropriate local value of the dielectric function. The solution for the p-polarized field is sought in the form $\boldsymbol{E}(\boldsymbol{r}) = \boldsymbol{E}(z)\exp(\mathrm{i}qx)$, and at $|z| \to \infty$ the field is required to tend to zero. Using the boundary conditions for the fields at the media interfaces ($z = 0$ and $z = -d$), we find the dispersion equation for surface polaritons:

$$\epsilon(\omega) = \frac{\kappa}{2}\coth(\kappa d)\left[\frac{1}{\kappa_0} + \frac{\epsilon_M}{\kappa_M}\right]\left\{-1 \pm \left[1 - \frac{4\epsilon_M \tanh^2 \kappa d}{\kappa_0\kappa_M(1/\kappa_0 + \epsilon_M/\kappa_M)^2}\right]^{1/2}\right\}, \tag{49}$$

where $\epsilon(\omega)$ and ϵ_M are dielectric functions at the frequency for the film and the metal substrate, respectively

$$\begin{array}{lll} \kappa = [q^2 - k_0^2\epsilon(\omega)]^{1/2}; & \kappa_0 = (q^2 - k_0^2)^{1/2}; & \kappa_M = [q^2 - k_0^2\epsilon_M]^{1/2} \\ \mathrm{Re}\,\kappa > 0 & & \mathrm{Re}\,\kappa_M > 0. \end{array} \tag{50}$$

The field of the surface polariton in the film is

$$E_x \sim \frac{i\kappa}{q}\left[\frac{1}{\epsilon(\omega)}\sinh \kappa z - \frac{\kappa_0}{\kappa}\cosh \kappa z\right],$$

$$E_z \sim \frac{1}{\epsilon(\omega)}\cosh \kappa z - \frac{\kappa_0}{\kappa}\sinh \kappa z. \tag{51}$$

Instead of $\epsilon(\omega)$ an appropriate solution of eq. (49) should be substituted into eqs. (51).

In the IR region the value $|\epsilon_m|$ for the medium with the metal conductivity is large: $|\epsilon_M| \gg |\kappa_M| \gg 1$ and in the square brackets in eq. (49) it is small as compared to unity. Therefore, expanding the root in eq. (49), we obtain for the frequencies ω_+ and ω_- of the surface polaritons the following equations:

$$\epsilon(\omega_+) \simeq -\frac{\kappa}{\kappa_0}\tanh(\kappa d)\left[1 - \frac{\kappa_M}{\kappa_M\kappa_0}(1 + \tfrac{1}{4}\tanh^2 \kappa d)\right], \tag{52}$$

$$\epsilon(\omega_-) \simeq -\frac{\epsilon_M}{\kappa_M}\kappa \coth \kappa d. \tag{53}$$

Consider eq. (52). In its derivation the small term allowing for the final conductivity of the substrate is retained in the first order (the second term in the square brackets in eq. (52)). It follows from eq. (52) that the value $\epsilon(\omega_+)$ is small ($\epsilon(\omega_+) \sim qd$) and the frequency $\omega_+(q)$ is at $q \gtrsim \omega/c$ close to the frequency ω_{LO}. The neglect of the second term in the square brackets of eq. (52) and the consideration of only such wavelengths of surface polaritons, for which $qd \ll 1$, give a simplified dispersion equation for ω_+:

$$\epsilon(\omega_+) = -\frac{q^2 d}{(q^2 - k_0^2)^{1/2} - k_0^2 d}. \tag{54}$$

Substituting the value $\epsilon(\omega_+)$ from eq. (52) into eq. (51), we find the field in the film which corresponds to the given mode:

$$E_x(z) \sim i(\kappa_0/q)\sinh[\kappa(z + d)],$$

$$E_z(z) \sim -(\kappa_0/\kappa)\cosh[\kappa(z + d)]. \tag{55}$$

As it is seen from eq. (55), the field in this mode has the highest magnitude at the vacuum–film interface and decreases with the distance into the film. In a thin film $qd \ll 1$ this inhomogeneity of the field as a function of the thickness is rather small. The field in a thin film is almost perpendicular to the film: $|E_z| \gg |E_x|$.

Since dissipative processes in a metal are taken into account by the imaginary part ϵ_M, the term with κ_M/ϵ_M in eq. (52) allows us to determine metal quenching of the surface polariton. The appropriate width of the line γ is of the order:

$$\gamma \sim |\omega_+ - \omega_{LO}| \, \mathrm{Im}(\kappa_M/\epsilon_M). \tag{56}$$

Equation (53), in contrast to eq. (52), yields a solution $\omega_-(q)$; its dispersion law is strongly affected by the properties of the metal substrate. For an ideally conducting substrate it follows from eq. (53) that $\omega_-(q) = \omega_{TO}$. The difference $\omega_-(q) - \omega_{TO}$ increases as the "metal" properties of the substrate deteriorate. The distribution of the field (connected with this mode) in the film is obtained by substituting into eq. (55) the dielectric function $\epsilon(\omega_-)$ described by eq. (53). At $|\epsilon_M| \gg 1$ we have:

$$\boldsymbol{E} = (\cosh \kappa z, 0, -(\mathrm{i}q/\kappa) \sinh \kappa z). \tag{57}$$

As it is seen from eq. (57), the field in the film at this vibration has its maximum at the film–substrate interface (at $z = -d$).

The experimental results have been obtained by the method of thermally stimulated emission of the system "ATR prism–gap–ZnSe film on a metal substrate". The temperature of the system maintained during the experiment was 150 ± 0.5°C. The experimental emissivity was calculated by formula (42). It should be noted at once that the intensity of the surface polariton emission is strongly dependent on the value of the gap between the ATR prism (a half-cylinder made of single-crystalline silicon) and the ZnSe film. As the emission intensity increases with the decrease of the gap size, there will be shifts of the surface polaritons to the region of lower frequencies. Analogous relationships are also observed in the spectra obtained by the conventional ATR method (Bryksin et al. 1972, Barker 1973).

Figure 7 shows the emission spectra of the system at the emission angle 20° (the critical angle $\theta_{cr} = 17°$) at various values of the gap between the prism and ZnSe film on a thick (the thickness of the aluminium substrate is much more than the thickness of the skin-layer in the studied frequency region) aluminium mirror. The thickness of the ZnSe film and the thickness of the aluminium film were 1 μm each. The size of the gap was set by a spacer in the form of a frame of a mylar film. The numerical calculation of the emissivity of the system with a prism gives the relationships identical (qualitatively) with the experimental ones; however, the theoretical emis-

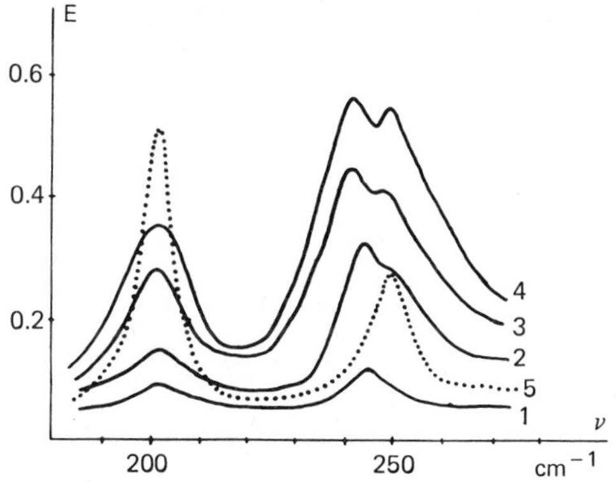

Fig. 7. Emissivity of the system "prism Si–gap–ZnSe film on Al" in p-polarization at the emission angle 20° and various values of the gap: (1) $l = 20\ \mu m$; (2) $l = 12\ \mu m$; (3) $l = 6\ \mu m$; (4) $l = 0$ (mechanical contact); (5) the film without a prism.

sivity turns to be much in excess of the experimental ones. It occurs due to some experimental inaccuracies: (1) The material of the ATR prism has its own nonselective emissivity (the emission intensity of the system "prism–gap–mirror" makes up for 80% of the emission of the system "prism–gap–film"). The light waves emitted by the film surface modes are partially absorbed in the prism material. (2) The optical scheme for the illumination of the monochromator entrance slit in the spectrometer FIS-21 (Hitachi) used in the experiment has large aberrations (coma and astigmatism); it is due to these aberrations that the nonparallelism of the light flux inside the ATR half-cylinder reaches the value $\Delta\theta \pm 2°$. The error in the value θ leads to the error in the determination of the wave vector $q = 2\pi\nu n_{Si} \sin\theta$ and the latter one is responsible for the emission band broadening.

Figure 8 shows the experimental and calculated dependences of the emissivity of surface polaritons for a ZnSe film 1 μm thick on an aluminium substrate at two emission angles*. The gap size l was used as an adjustable parameter. Under the condition of the best agreement between the experimental and calculated spectra the deviation of l from the value given in the experiment is 50%.

The influence of the metal substrate conductivity on the dispersion law of surface polaritons was checked in two ways: either by varying the material of the substrate or by decreasing its thickness to the depth of the skin layer. Figure 9 presents the spectra of the thermally stimulated emission for a system "half-cylinder Si–ZnSe film on chromium" in the

*Figures 7 and 8 show the emissivity of the system for p-polarization. In the s-polarized emission only one band near the frequency ω_{TO} of the single crystal is observed.

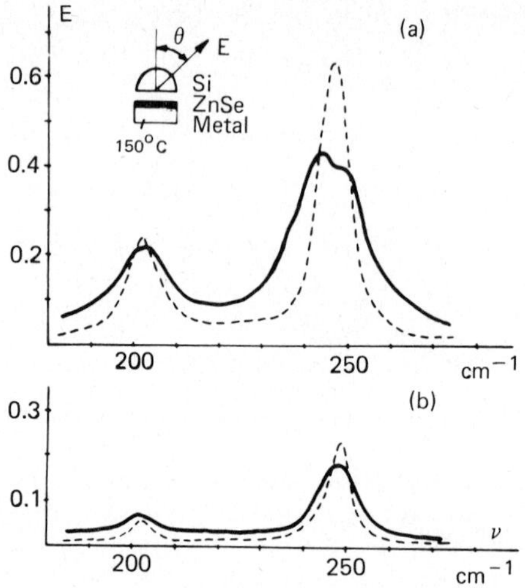

Fig. 8. Emissivity of surface polaritons in a ZnSe film on Al with $q > \omega/c$. The continuous curves are experimental, the dashed curves are calculated: (a) $\theta = 18°$; (b) $\theta = 30°$.

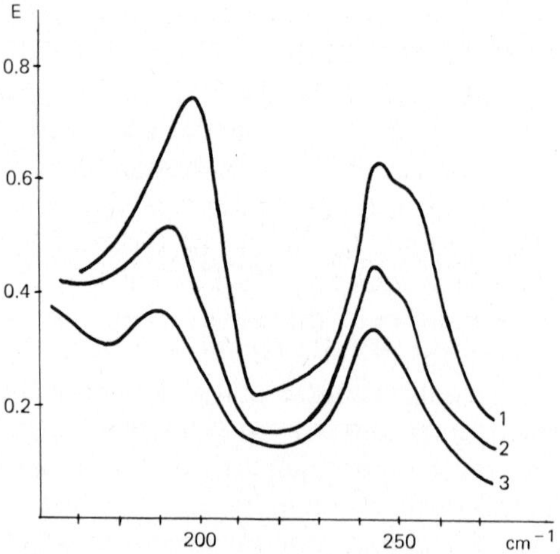

Fig. 9. Emission spectra of the system: "Si-prism–ZnSe on Cr": (1) $\theta = 20°$; (2) $\theta = 30°$; (3) $\theta = 45°$.

p-polarization at various emission angles. In this system an emission band ω_- is observed split from ω_{TO}-band. At $q > \omega/c$ the splitting value $\omega_{TO} - \omega_-$ is in good agreement with the expressions (16) and (53) and is determined mainly by the substrate conductivity. The frequency position of the bands ω_+ is practically independent of the substrate conductivity which agrees well with the results of calculation.

Figure 10 shows the dispersion curves for surface polaritons constructed from the experimental data along with the theoretical dispersion laws calculated by the formulae (52) and (53). The data used to calculate ϵ_0, ϵ_∞, ω_{TO}, γ and σ of the substrate are from the experimental results of sect. 5, fig. 5.

In fig. 11 the emissivity spectra are shown for the system "prism Si–gap–ZnSe film (1 μm thick) on an aluminium mirror 0.1 μm thick on glass" in the p-polarization at various emission angles. In these spectra, the same as in those shown in fig. 9, a band of the surface polariton ω_- is observed. The splitting value $\omega_- - \omega_{TO}$ enables the evaluation of the value of the equivalent conductivity in a thin aluminium film (Vinogradov et al. 1977a, b): it turns out to be about 200 times less than the conductivity of a

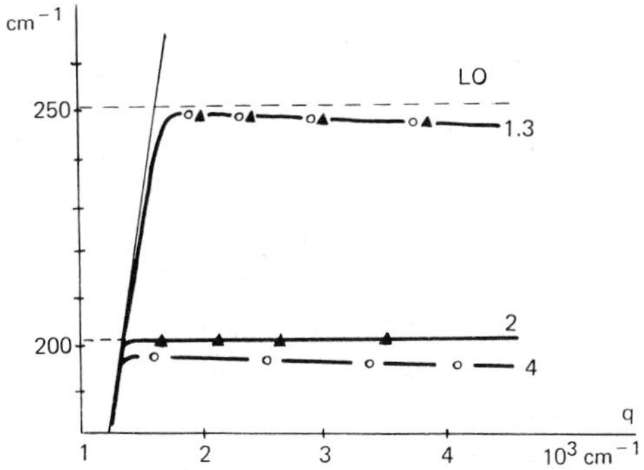

Fig. 10. Dispersion of surface polaritons with $q > \omega/c$ for ZnSe on metal: (1) and (2) ω_+ and ω_- respectively for ZnSe on Al, (3) and (4) ω_+ and ω_- for ZnSe on Cr.

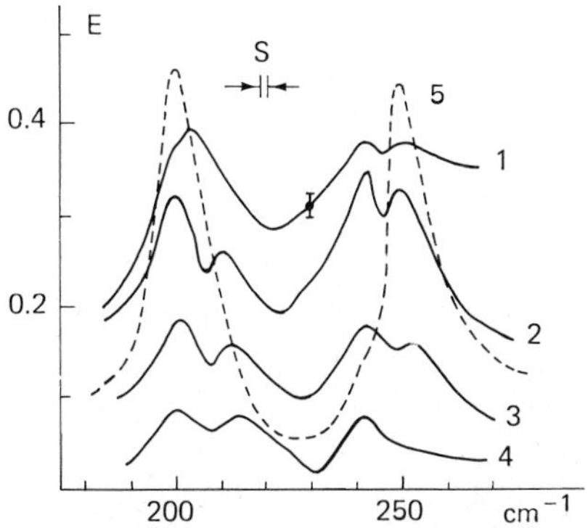

Fig. 11. Emissivity of surface polaritons with $q > \omega/c$ of ZnSe film (1 μm thick) on a thin Al mirror (0.1 μm): (1) $\theta = 17°$; (2) $\theta = 19°$; (3) $\theta = 24°$; (4) $\theta = 40°$: (5) radiation of the film without a prism ($\theta = 40°$).

thick film of freshly evaporated aluminium and the real part of its dielectric function is positive, i.e., in a thin film it has changed its sign.

7. *Emissivity of Surface Polaritons of ZnSe Single Crystals*

In the presence of periodic inhomogeneities on the surface of a bulk sample the tangential component of the wave vector may take on the values (Fischer et al. 1973, Teng and Stern 1967):

$$q_m = k_0 \sin\theta + m(2\pi/a), \quad m = 0, \pm 1; \pm 2 \ldots \tag{58}$$

where θ is the angle between a normal to the sample and the observation direction for a reflected (emitted) electromagnetic wave with the wave vector $k_0 = \omega/c$. Thus, surface polaritons with $q > \omega/c$ become accessible in conventional optical experiments.

In fig. 12 IR reflection spectra are shown of a ZnSe single crystal with periodically repeated twinning planes in the p-polarized light and with the plane of the light incidence (reflection) parallel to the axis of the crystal growth. A reflection spectrum of the same single crystal is also shown, but turned on 90° so that its growth axis was perpendicular to the incidence plane. In this case the wave vector $q = (\omega/c)\sin\theta$ of the light wave is parallel to the twinning planes, in contrast to the previous case, when it was perpendicular to the twinning boundaries. As it is seen from this figure, a minimum is observed in the band of the reststrahlen region ($200 < \nu <$

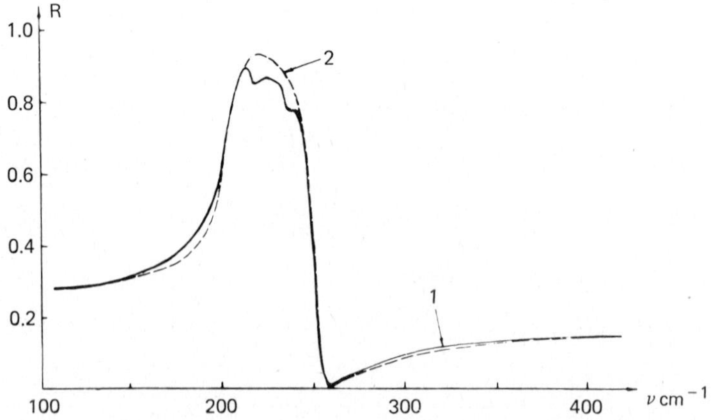

Fig. 12. Reflectivity spectrum of a "layered" ZnSe crystal in the p-polarized light ($\theta = 16 \pm 12°$): (1) $q = 2\pi\nu \sin\theta$ is perpendicular to twinning boundaries; (2) q is parallel to twinning boundaries.

250 cm^{-1}) which is due to the absorption of their emission by a surface polariton.

Figure 13 shows the spectra of thermally stimulated emission of the same ZnSe "layered" crystal in the p-polarized light at the thermostat temperature 150 ± 5°C. The spectra of fig. 13 differ from one another by emission angle and the mutual positions of the twinning boundaries and the emission plane. The emission of surface polaritons is observed only in the case when twinning planes are perpendicular to the emission plane.

The dispersion law for the surface polaritons in a ZnSe single crystal obtained from these measurements is shown in fig. 14. The possibility of

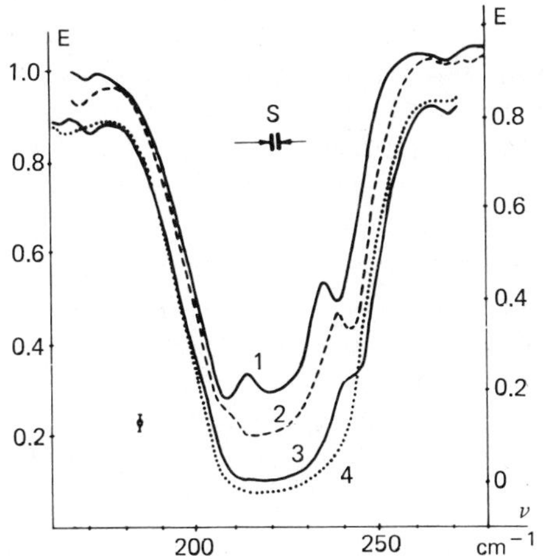

Fig. 13. Emission spectra of a "layered" ZnSe crystal at various emission angles (q is perpendicular to twinning boundaries): (1) $\theta = 20°$; (2) $\theta = 40°$; (3) $\theta = 60°$; (4) $\theta = 40°$ but q is parallel to the twinning boundaries.

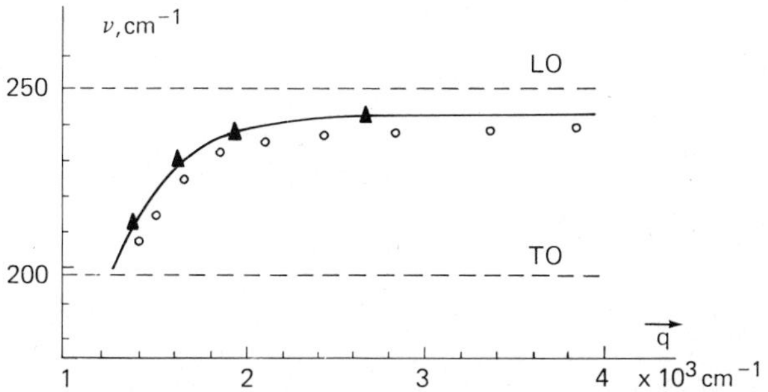

Fig. 14. Dispersion law for a surface polariton in a ZnSe single crystal. The continuous curve is calculated and the points are experimental: (○) layered ZnSe; ▲ emission with the prism.

smooth variation of the emission angle from 0° to 60° enables us to follow the dispersion of the polariton branch since the crystal had a suitable twinning period $a \approx 40\ \mu$m. This figure shows also the theoretical dependence $\omega(q)$ for a surface polariton in a ZnSe single crystal. The calculation was performed according to the formulae

$$q = \frac{\omega}{c}\left(\frac{\epsilon(\omega)}{1+\epsilon(\omega)}\right)^{1/2} \tag{59}$$

with $\epsilon(\omega)$ in the form (4) and parameters $\epsilon_\infty = 5.8$, $\epsilon_0 - \epsilon_\infty = 3$, $\omega_{TO} = 198\ \mathrm{cm}^{-1}$ (Vinogradov et al. 1977a, 1976).

ZnSe single crystals without the twinning planes have no emission (absorption) bands conditioned by surface vibrations of atoms. However, surface polaritons with $q > \omega/c$ of such a single crystal emit electromagnetic waves, if a half-cylinder of silicon is placed above it (Vinogradov and Zhizhin 1976, Vinogradov et al. 1977b) (inverted ATR).

Figure 15 shows emission spectra obtained for a ZnSe single crystal with a prism heated up to 150 ± 0.5°C. The spectra were measured in two ways. In fig. 15a the emission spectra are presented measured at fixed emission angles (the ordinary spectral recording), and fig. 15b shows the emissivity of the system with scanning of the emission angle at fixed frequencies. The positions of maxima of emission bands correspond to a point in the $(\omega - q)$ space on the dispersion curve of the surface polariton.

The dispersion law for a surface polariton in a ZnSe single crystal obtained from measurements of the emissivity of the system with a prism is presented in fig. 14. Good agreement with the theoretical curve is observed.

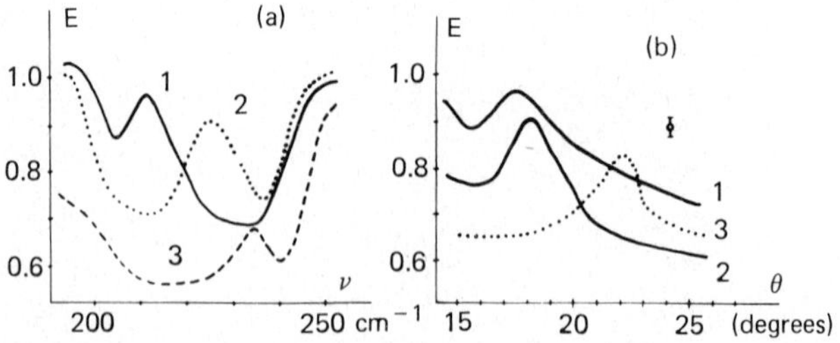

Fig. 15. Emissivity of the system "prism Si–gap–ZnSe single crystal–mirror": (a) frequency scanning at fixed emission angles: (1) $\theta = 17°$; (2) $\theta = 19°$; (3) $\theta = 25°$. (b) scanning of the emission angle at fixed frequencies: (1) $208\ \mathrm{cm}^{-1}$; (2) $220\ \mathrm{cm}^{-1}$; (3) $234\ \mathrm{cm}^{-1}$.

8. High-frequency Phonon Polaritons in a $Gd_2(MoO_4)_3$ Single Crystal*

Optical properties of crystals having the structural phase transitions were investigated in a great number of works. The changes in the bulk vibrational states in the vicinity of phase transition temperature were studied there. In such studies in the IR spectral region there is, as a rule, a danger of obtaining wrong information due to the errors in the spectrometer photometric scale resulting from the recording of the sample intrinsic heat emission. The method of studying the spectra of thermally stimulated emission is free from this and some other shortcomings. Moreover, it enables easy investigation of angular dependences of emissivities of both bulk and surface vibrational states.

In the study of the mechanism of phase transitions the effect of the surface may be important. However, to our knowledge, surface vibrational states of single crystals in the vicinity of the phase transition temperature have not been studied as yet.

Single crystals of $Gd_2(MoO_4)_3$ (GMO) are extrinsic segnetoelastics with the Curie temperature $T_c = 159°C$. In the high-temperature phase they belong to tetragonal syngony with the space group symmetry D_{2d}. The unit cell in this phase contains two formula units. Below T_c the GMO has an orthorhombic (C_{2v}) unit cell with four formula units (Aizu 1969).

The GMO surface polaritons were studied by means of a two-beam IR spectrophotometer IKS-16 with an ATR-1 attachment. A KRS-6 half-cylinder was used as an ATR element. A special holder for the sample and the ATR cylinder made it possible to vary the temperature from room temperature to 250°C and to maintain it with a precision up to ±0.5°.

Spectra of surface polaritons of GMO single crystals were obtained by both the ATR method commonly used for this purpose and by measuring the thermally stimulated emission of the system "crystal–air gap–ATR element". It should be noted that ATR spectra at the sample temperature $T > 100°C$ become a low contrast due to the fact that ordinary spectrophotometers without special schemes for subtracting the intrinsic heat emission record this emission. At the same time, the spectra of thermally stimulated emission of surface polaritons with $q > \omega/c$ are much more contrasting. In fig. 16 the ATR spectra of surface polaritons are shown which were obtained at various sample temperatures, and in fig. 17 the spectra of their thermally stimulated emission under the same conditions. In accordance with the Kirchhoff law the minima of the spectra in fig. 16 correspond to the maxima of the spectra in fig. 17. In recording the

*The results presented in this section were obtained with the active participation of I.I. Khammadov and V.A. Yakovlev.

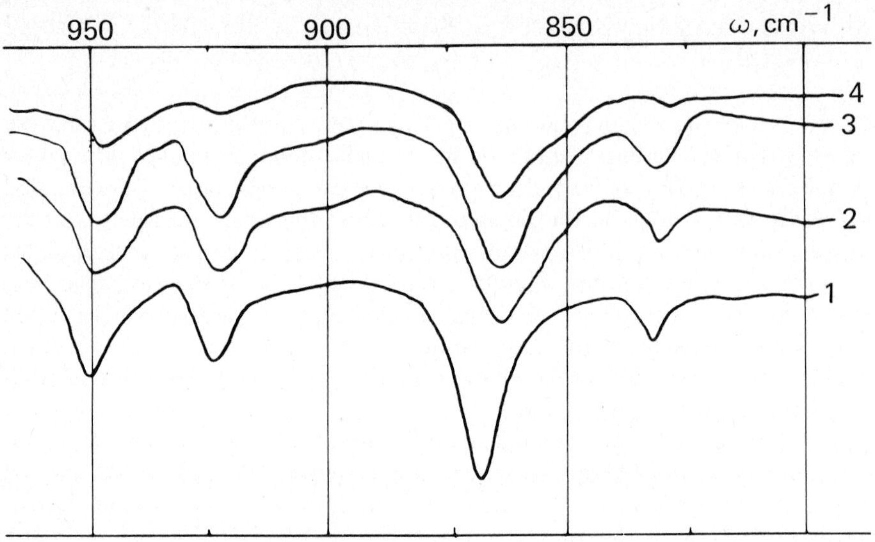

Fig. 16. ATR spectra of surface polaritons in a $Gd_2(MoO_4)_3$ single crystal (p-polarization, $\theta = 35°$): (1) $T = 20°C$; (2) $T = 105°C$; (3) $T = 120°C$; (4) $T = 155°C$.

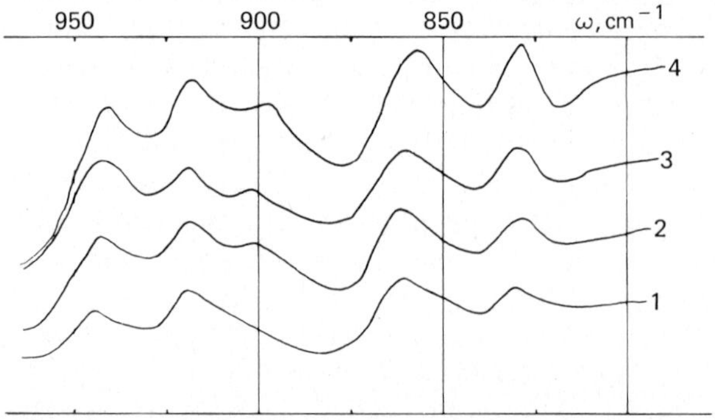

Fig. 17. Spectra of thermally stimulated p-polarized emission of GMO surface polaritons ($\theta = 35°$): (1) $T = 120°C$; (2) $T = 160°C$; (3) $T = 190°C$; (4) $T = 210°C$.

emission spectra of surface polaritons (fig. 17), in contrast to the absorption spectra (fig. 16), the light flux of the IR emission source in the spectrophotometer is cut off in the sample channel by a nontransparent shield at the inlet to the cell compartment. In this case we are interested not in the absolute value of the spectral emissivity of surface polaritons in the system "crystal–air gap–ATR prism", but in its variation with the temperature. Therefore, we neglect here the intrinsic thermal emission of the ATR prism, and the spectrum normalization is performed in the instrument automatically by the light flux of the conventional IR emission source attenuated in the spectrophotometer reference channel by the standard neutral attenuator commonly used for equalizing light fluxes in both channels.

Figure 18 shows the dispersion branches of surface polaritons in GMO single crystals for $c \parallel y$ obtained from ATR spectra at room temperature. Continuous curves in this figure were obtained by calculating the dispersion curves with the use of the frequency dependence of the crystal complex dielectric function calculated from the GMO reflection spectra at room temperature in the linearly polarized light. In the appendix to this chapter the GMO reflection spectrum and the spectra of the real and imaginary parts of $\epsilon(\omega)$ are presented.

Figure 18 shows also the dispersion curves of surface polaritons obtained from the emission spectra of the system "crystal–gap–prism". Some discrepancy between the dispersion curves of surface polaritons obtained by various methods can easily be explained by the increase of

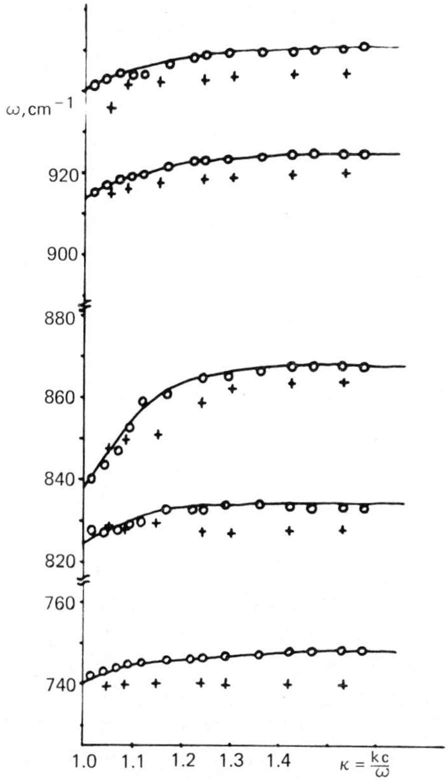

Fig. 18. Dispersion of GMO surface polaritons: (O) from ATR spectra; (+) from emission spectra; the continuous curves are calculated.

phonon damping with the increasing temperature (the emission spectra were obtained at the temperature of the sample and the prism $T = 140°C$ while the ATR spectra were recorded at room temperature).

The variation of surface polariton frequencies ($\theta = 35°$, $q = 1.25\ k_0$) with the increase of the sample temperature from 20°C up to 215°C is shown in fig. 19. These temperature dependences show clearly sufficiently narrow minima at $T = T_c = 159°C$. The depth of these frequency minima reaches $10\ cm^{-1}$. As the temperature increases, the frequencies of surface polaritons ($T < T_c$) at first decrease slowly and then at $T = T_c$ an abrupt minimum is reached with the subsequent increase of the surface polariton frequencies at $T > T_c$. At $T = 160°C$ all emission bands of surface polaritons increase their frequency up to the values occurring at room temperature. At $T > T_c$ the frequencies of surface polaritons again decrease smoothly. Moreover, near $T = T_c$ a new band with $\omega = 905\ cm^{-1}$ appears in the spectra of the thermally stimulated emission of surface polaritons which was not present there at $T < T_c$. The temperature behaviour of this band is similar to that of the other bands (fig. 19) with the only difference that its frequency minimum is $\approx 10°$ higher. It is difficult to explain the appearance

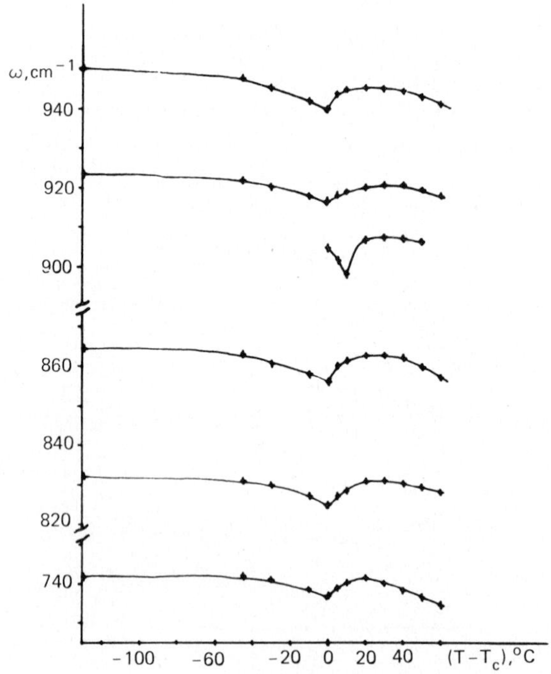

Fig. 19. Temperature dependence of surface polariton frequencies in GMO ($\theta = 35°$).

of this new band, taking into account the fact that at $T > T_c$ the number of formula units in the GMO elementary cell has decreased from four to two. A possible methodical error is excluded by the experimental results obtained from the spectra of thermally stimulated emission of the same GMO sample recorded by means of a single-beam IR spectrometer IKS-31. This instrument provides accumulation and averaging of the spectra up to the preset signal-to-noise ratio with the aid of a minicomputer (Vinogradov et al. 1978b). Figure 20 shows the emission spectra of a GMO single crystal in the linearly polarized light at two values of temperature: $T = 100°C$, $T = 170°C$ ($T_c = 159°C$). In the second spectrum a weak new band with $\omega \approx 905\ cm^{-1}$ is seen.

The appearance of minima at the temperature dependence of the surface polariton frequencies (fig. 19) is most likely connected with the movement of the domain boundaries near the crystal surface. At $T = T_c = 159°C$ a ferroelastic phase transition takes place in a GMO crystal which is accompanied by the rearrangement of domains, and the formation of the embryo of the new phase is most likely to occur near the crystal boundaries.

9. *Conclusions*

The investigation of the spectra of thermally stimulated emission of films and single crystals enables the following conclusions to be made on the vibrational states of the samples:

(1) Longitudinal optical modes of the film do not interact with the external electromagnetic field in single-quantum processes under any conditions including the inclined incidence of light onto a thin film.

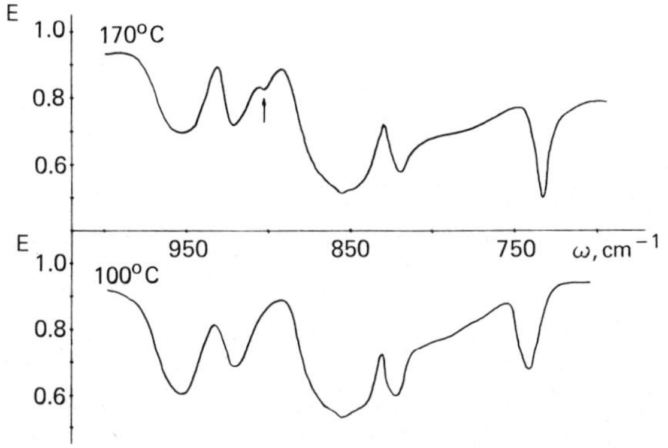

Fig. 20. Emission spectra of a GMO single crystal at $T = 100°$ (curve 1) and $T = 170°$ (curve2).

(2) Radiative surface vibrational states exist in thin films which participate in the processes of light absorption and emission (in thin films, however, such states are not localized near their boundaries, but propagate throughout the whole thickness). The frequencies of the surface states depend on the values of the wave vector, film thickness and substrate conductivity. In a thin film on a substrate of high conductivity the frequencies of surface radiative vibrational states lie near the frequencies ω_{LO} and ω_{TO} of the single crystal. They emit the p-polarized electromagnetic (light) waves.

(3) Nonradiative surface vibrational states of single crystals and films become radiative when a prism (or the ATR type) is placed close to the samples being investigated. For their experimental observation a temperature difference between the sample and the spectral instrument is required. The emission of these states is observed in the p-polarized light at the emission angles $\theta > \text{arc tg}\, n$, n is the refractivity index of the prism.

(4) The use of the technique of measuring the thermally stimulated emission of vibrational states of various samples makes it possible to obtain a more complete and reliable information on their optical properties as compared to the spectroscopic method commonly used for these purposes since the sample is not radiated by the emission from an external source. This method is mostly promising for the study of vibrational states in objects with low optical density and for the investigation of spectral peculiarities in the vicinity of structural phase transitions in the samples when the temperature of the phase transition differs from room temperature.

The authors regard it as their pleasant duty to thank V.P. Grachev for the preparation of thin-film samples and Prof. V.M. Agranovich for the discussion of the results obtained.

10. Appendix

The external reflection spectra of $Gd_2(MoO_4)_3$ single crystals oriented along the crystallographic axes were recorded in the linearly polarized (s-polarization) light by a spectrophotometer developed by us on the basis of a single-beam diffraction IR spectrometer IKS-31 and a special electronic-digital computer "Vikhr". The spectrum was recorded by the points with the use of a "cell-in–cell-out" technique in the regime of accumulation and averaging at each point of reflection coefficients up to the preset signal-to-noise ratio (Vinogradov et al. 1978). In this regime the instrument enables photometric reproducibility with an error of ±0.1% independent of the value of the reflection coefficient and spectral resolution. The photometric error in the measurement did not exceed ±0.5%, the spectral resolution

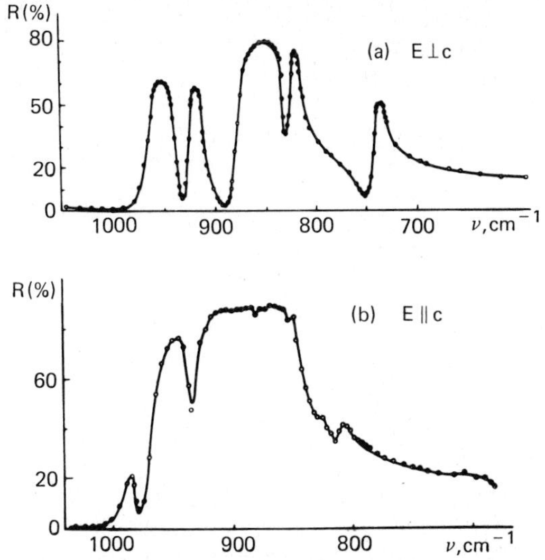

Fig. A1. External reflection spectrum of GMO: (○) experimental; the continuous curve resulted from the dispersion analysis.

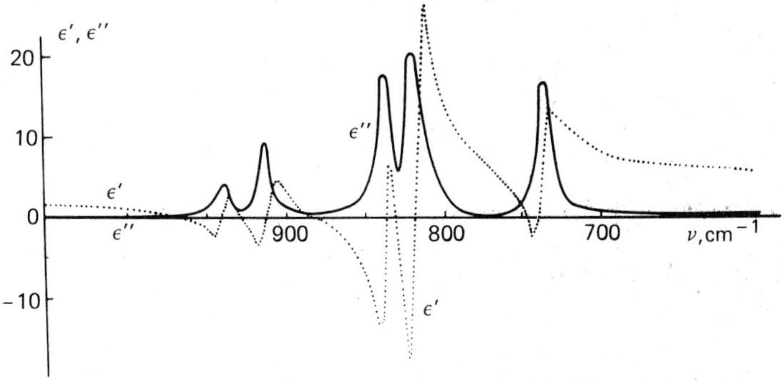

Fig. A2. Real and imaginary parts of the GMO dielectric function.

was $\simeq 0.5\ \text{cm}^{-1}$ over the whole range of interest to us. Figure A1 shows the reflection spectra of GMO single crystals obtained by the above technique. The same figure shows also the calculated results (the continuous curve) for the reflectivity obtained by the method of dispersion oscillators (the Lorentz analysis) and fig. A2 shows frequency dependences for the real and the imaginary parts of the GMO dielectric function.

References

Agranovich, V.M., 1968, The theory of excitons (Nauka, Moscow) in Russian.
Agranovich, V.M., 1975, Sov. Phys. Usp. **18**, 99 (in Russian 1975, **115**, 199).

Aizu, K., 1969, J. Phys. Soc. Japan, **27**, 387.
Barker, A.S., 1973, Surf. Sci., **34**, 62.
Berreman, D.W., 1963, Phys. Rev. **130**, 2193.
Bryksin, V.V., Yu.M. Gerbstein and D.V. Mirlin, 1971, Fiz. Tverd Tela, **13**, 2125 (in Russian).
Bryskin, V.V., Yu.M. Gerbstein and D.V. Mirlin, 1972, Phys. Status Solidi (b) **51**, 901.
Bryksin, V.V., D.N. Mirlin and Yu.A. Firsov, 1974, Sov. Phys. Usp. **17**, 305 (in Russian **113**, 29).
Fischer, B., N. Marschall and H.J. Qucisser, 1973, Surf. Sci. **34**, 50.
Fuchs, R., K.L. Kliewer, 1966a, Phys. Rev. **144**, 425.
Fuchs, R., K.L. Kliewer, 1966b, Phys. Rev. **150**, 573.
Fuchs, R., K.L. Kliewer and W.J. Pardee, 1966, Phys. Rev. **150**, 589.
Gerbstein, Yu.M. and D.N. Mirlin, 1974, Fiz. Tverd. Tela, **16**, 2584 (in Russian).
Hisano, K., Y. Okamoto and O. Matumura, 1970, J. Phys. Soc. Japan, **28**, 425.
Huong, P.V., 1978, in: Advances in Infrared and Raman Spectroscopy, V4, ed. R.S-H. Clark and R.E. Chester (Heyden, London) p. 85.
Marschall, N. and B. Fischer, 1972, Phys. Rev. Lett. **28**, 811.
Otto, A., 1968, Z. Phyzik, **216**, 399.
Proix, F. and M. Balkanski, 1969, Phys. Status Solidi **32**, 119.
Ruppin, R. and R. Englman, 1970, Rep. Progr. Phys. **33**, 149.
Stepanov, B.I., 1961, The principles of the spectroscopy of negative light fluxes (Minsk) in Russian.
Stierwald, D. and P. Potter, 1962, J. Phys. Chem. Solids **23**, 99.
Stierwald, D. and P. Potter, 1965, Phys. Rev. **137A**, 1007.
Teng, Y.Y. and E.A. Stern, 1967, Phys. Rev. Lett. **19**, 511.
Vinogradov, E.A. and G.N. Zhizhin, 1976, Pis'ma ZhETF, **24**, 71 (in Russian).
Vinogradov, E.A., G.N. Zhizhin, N.N. Melnik and O.K. Filipov, 1976, Fiz. Tverd. Tela, **18**, 2847 (in Russian).
Vinogradov, E.A., G.N. Zhizhin and A.G. Mal'shukov, 1977a, ZhETF, **73**, 1480 (in Russian).
Vinogradov, E.A., G.N. Zhizhin, A.G. Mal'shukov and V.I. Yudson, 1977b, Solid State Commun. **23**, 915.
Vinogradov, E.A., V.L. Grachev, G.V. Grushevoi, G.N. Zhizhin and V.I. Yudson, 1978a, ZhETF, **75**, 1919 (in Russian).
Vinogradov, E.A., G.N. Zhizhin, I.A. Ivanov and O.A. Lubeznikov, 1978b, in: Proc. IV Intern. Conf. Computers in Chemical Research and Educttion, Novosibirsk, 1978, Abstract N 1–18.
Vinogradov, E.A., G.N. Zhizhin, T.A. Leskova, N.N. Mel'nik and V.I. Yudson, 1980, ZhETF, **78**, 1030 (in Russian).
Vodopianov, L.K. and E.A. Vinogradov, 1974, Fiz. Tverd. Tela, **16**, 1432 (in Russian).
Ziman, J.M., 1964, Principles of the theory of solids (University Press, Cambridge).

PART II

Surface Polaritons as a Probe of Surface and Interface Properties

F. ABELES
V. M. AGRANOVICH
D. M. KOLB
V. E. KRAVTSOV
T. A. LESKOVA
T. LOPEZ-RIOS
A. A. MARADUDIN
H. RAETHER
V. A. YAKOVLEV
G. N. ZHIZHIN

[illegible]

[illegible]

[illegible] electronic [illegible]

[illegible] surface [illegible]

[illegible] wave [illegible]

$k_s = n_1(\omega/c)$ [illegible]

where k_s is the [illegible] of incidence [illegible] table [illegible] distance [illegible] wave [illegible] the [illegible] thickness [illegible]

The schematic [illegible] is shown in Fig. [illegible] The [illegible]

CHAPTER 5

Effects of the Transition Layer and Spatial Dispersion in the Spectra of Surface Polaritons

V.M. AGRANOVICH

Institute of Spectroscopy
USSR Academy of Sciences
Troitsk, Moscow Oblast 142092
USSR

Surface Polaritons
Edited by
V.M. Agranovich and D.L. Mills

Contents

1. Introduction: Spectroscopy of surface polaritons and properties of the surface . . . 189
2. Dispersion of surface polaritons with macroscopic transition layers taken into account . . . 191
   2.1. Effective boundary conditions for fields in the presence of a macroscopic homogeneous and isotropic transition layer . . . 191
   2.2. Effective boundary conditions for fields in the presence of an anisotropic homogeneous transition layer . . . 194
   2.3. Surface polariton dispersion in the case of a macroscopic isotropic transition layer – additional surface polaritons and the problem of additional boundary conditions . . . 196
   2.4. Effects of transition layer anisotropy on the surface polariton spectra – splitting anisotropy upon resonance with oscillations in the transition layer . . . 208
3. Effects of thin semiconductor films on surface polariton spectra . . . 211
4. Surface wave diffraction by an impedance step when an additional surface polariton is taken into account . . . 220
5. Surface polaritons in exciton luminescence spectra . . . 225
   Surface polariton "radiation" width and transition layer effects in molecular crystals and semiconductors . . . 225
6. Is self-focusing of surface polaritons feasible? . . . 233
   Nonlinear properties of surface polaritons due to the presence of a transition layer . . . 233
References . . . 236

1. Introduction

Spectroscopy of Surface Polaritons and Properties of the Surface

In deriving the dispersion laws for surface polaritons (SP) in chs. 1 and 2, the boundary conditions applied referred to a sharp boundary, and surface currents and charges were not taken into consideration. In this approximation, the properties of the SPs were completely determined by the values of the dielectric tensors of the contacting media. Hence, these properties, like those of bulk polaritons, represent only the features of the spectrum obtained in bulk excitations of the medium.

Actually, however, the dielectric properties of the subsurface region of a crystal can, in general, differ substantially from the dielectric properties of the medium in the parts that are sufficiently distant from the surfaces or interfaces of media. This leads to the formation, near the interface, of the so-called transition layer within which Maxwell's material equation, i.e., the relation between the induction vector and the electric vector $\boldsymbol{E}$, turns out to be different than in the bulk of the medium.

If L is the thickness of the transition layer (usually $L \approx 10$ to $100\ \text{Å} \ll \lambda_0 \equiv 2\pi c/\omega$), then at $L \gg a$, where a is the lattice constant, the transition layer is macroscopic. Only in this case is it valid to apply the relationship between vectors $\boldsymbol{D}$ and $\boldsymbol{E}$, mentioned above, inside the transition layer, i.e. at $0 \leq z \leq L$. For fields with the frequency ω, this relation within the transition layer can be written in the form

$$D_i(\omega, x, y, z) = \tilde{\epsilon}_{ij}(\omega, z) E_j(\omega, x, y, z), \tag{1}$$

where $\tilde{\epsilon}_{ij}$ is the dielectric constant of the transition layer. Equation (1) relates the values of the fields at the same point in space, i.e., it is a local relation and, consequently, neglects spatial dispersion within the transition layer*. Nevertheless, as will be shown later on, the fact itself that the

*Transition layers can also be produced artificially by applying thin films on various kinds of substrates. In the cases when the film is a semiconductor, and its thickness L is less than, or of the order of the effective Bohr radius a_0 of an exciton in a sufficiently thick semiconductor, the dielectric constant $\tilde{\epsilon}_{ij}$ of the film becomes a function of L. Under these conditions, it may prove essential to take into account, for the region of exciton resonances, the dependence of $\tilde{\epsilon}_{ij}$, not only on the frequency ω, but on the wave vector $\boldsymbol{k}$ of the surface wave as well. Section 3 discussed these effects.

transition layer is taken into account leads (see Agranovich 1975 and 1979a, b) to many interesting effects and, in particular, to the effects of spatial dispersion with respect to parameter kL, where $\boldsymbol{k} = (k_1, k_2, 0)$ is a two-dimensional wave vector. Since $L \gg a$ in the case of macroscopic transition layers, the parameter $kL \gg ka$. In most cases, this circumstance enables the "bulk" spatial dispersion to be neglected, i.e., the dependence of the dielectric constant of contacting media and of the transition layer on the wave vector to be ignored. An exception is the region of exciton resonances in one or both of the contacting media where, when spatial dispersion is taken into consideration, additional light waves appear in the bulk of the contacting media. This significant and special case of the influence of spatial dispersion will not be discussed here because it is specifically dealt with in ch. 2. in this book.

If the transition layer is of microscopic thickness, i.e. if $L \approx a$, it is no longer possible to speak of the values of fields $\boldsymbol{D}$ and $\boldsymbol{E}$ within the layer in the scope of macroscopic electrodynamics, and relation (1) becomes meaningless. In this case, as has been shown by Sivukhin (1948, 1952 and 1956) for a number of simple models, the presence of a transition layer can be taken into consideration by introducing certain corrections into the boundary conditions for the fields. The deriving of more accurate boundary conditions for this case (i.e. when $L \approx a$) requires reasoning that is beyond the scope of Maxwell's phenomenological equations.

Even when the transition layer is not taken into account and the properties of surface polaritons (SP) are completely determined by the bulk values of the dielectric constants, SP spectroscopic techniques open up new ways of investigating the properties of media. For example, the investigation of wave propagation along the surfaces of metals (Shoenwald et al. 1973) enables the electrical conductivity of metals at SP frequencies to be studied. New possibilities are offered for studying phase transitions in nontransparent media by means of the scattering of surface waves by order parameter fluctuations (Agranovich 1976, Agranovich and Leskova 1977), etc.

But surface physics, proper in this field of spectroscopy, begins where its methods are capable of providing information on the structure of the excitation spectrum in the transition layer, i.e., in the region of the crystal whose properties are perturbed by the presence of an interface. These, precisely, are the situations that offer definite promise in using surface polaritons to investigate various kinds of two-dimensional and quasi-two-dimensional structural and electronic phase transitions which may occur on the surface of condensed media and which provoke such great interest today.

In passing over to a consecutive discussion in the following sections of various kinds of optical effects stemming from the presence of a transition layer, we point out that the problem of a transition layer in the optics of crystals and liquids, is, in itself, quite an old one (see Drude 1912) and has been the subject of a great many investigations (for a review of them, see Kizel 1973). In these works, however, the attention of the investigators was concentrated on the problem of the effect the transition layer has on the reflection of light. Only in recent years, owing to the developments in the spectroscopy of surfaces and to the dramatic advances in the spectroscopy of surface polaritons in particular, the problem of taking the transition layer into account has again become an especially timely one (Agranovich 1975).

In the following discussion, as in most of the above-mentioned investigations, a phenomenological description will be used for the properties of both macroscopic and microscopic transition layers. Such an approach seems quite natural at the present time because it makes it possible to introduce certain phenomenological parameters, characterizing the transition layer and to investigate in a sufficiently general form the most significant effects due to the presence of a transition layer. The phenomenological parameters of the transition layer mentioned above appear in the boundary conditions for the fields. The next section will show how to obtain, for various models of the transition layer, effective boundary conditions of a kind more general than those employed in chs. 1 and 2. Further on, these boundary conditions will be made use of to analyse the dispersion of SPs (sect. 2 and 3). Section 4, in taking additional SPs into account, deals with the diffraction of SPs by the impedance step, and sect. 5 discusses the possible role of SPs in the spectra of exciton luminescence. The last subsection of sect. 6 is devoted to a discussion of the possibility of self-focusing and self-trapping SPs under conditions when the nonlinear properties of SPs are due to the presence of a transition layer.

2. *Dispersion of Surface Polaritons with Macroscopic Transition Layers Taken into Account*

2.1. *Effective Boundary Conditions for Fields in the Presence of a Macroscopic Homogeneous and Isotropic Transition Layer*

We shall begin with the simplest situation in which the anisotropy of the dielectric properties in the transition layer can be neglected. We assume that the transition layer corresponds to region III and that this layer achieves contact between media I and II (see fig. 1).

For monochromatic fields $\boldsymbol{E}, \boldsymbol{D}$ and $\boldsymbol{H}$, varying in space according to the law

$$\mathbf{A}(r, t) = \mathbf{A}(z) \exp \mathrm{i}(k_1 x + k_2 y - \omega t), \tag{2}$$

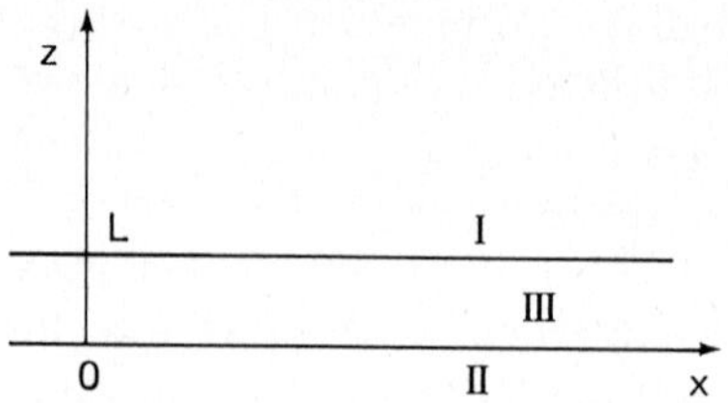

Fig. 1. Transition layer (III) between media I and II.

the boundary conditions, relating the values of the fields in media I (at $z = 0$) and II (at $z = L$), can be obtained directly from Maxwell's equations. As a matter of fact, from the equation

$$\operatorname{div} \boldsymbol{D} = 0, \tag{3}$$

taken at $0 < z < L$, it follows that

$$\mathrm{i}k_1 D_1^{\mathrm{III}}(z) + \mathrm{i}k_2 D_2^{\mathrm{III}}(z) + \frac{\mathrm{d}D_3^{\mathrm{III}}(z)}{\mathrm{d}z} = 0.$$

Integrating this relation with respect to z between the limits $z = 0$ and $z = L$ we obtain

$$D_3^{\mathrm{III}}(L) - D_3^{\mathrm{III}}(0) = -\mathrm{i}k_1 \int_0^L D_1^{\mathrm{III}}(z)\,\mathrm{d}z - \mathrm{i}k_2 \int_0^L D_2^{\mathrm{III}}(z)\,\mathrm{d}z. \tag{4}$$

Since we assume sharp boundaries between media I and III, and II and III,

$$D_3^{\mathrm{I}}(L) = D_3^{\mathrm{III}}(L), \quad D_3^{\mathrm{II}}(0) = D_3^{\mathrm{III}}(0) \tag{4a}$$

and, consequently, it follows from eq. (4) that

$$D_3^{\mathrm{I}}(L) - D_3^{\mathrm{II}}(0) = -\mathrm{i}k_1\tilde{\epsilon}(\omega) \int_0^L E_1^{\mathrm{III}}(z)\,\mathrm{d}z - \mathrm{i}k_2\tilde{\epsilon}(\omega) \int_0^L E_2^{\mathrm{III}}(z)\,\mathrm{d}z. \tag{5}$$

If in the theory of the transition layer we take into consideration only small terms of the order of $L/\lambda = L(\omega/2\pi c) \ll 1$ and neglect terms of a higher order of smallness, it is necessary to neglect the dependence of the fields on z under the integral on the right-hand side of eq. (5). In this approximation, taking into account the boundary conditions

$$\boldsymbol{E}_t^{\mathrm{II}}(0) = E_t^{\mathrm{III}}(0), \qquad \boldsymbol{E}_t^{\mathrm{I}}(L) = \boldsymbol{E}_t^{\mathrm{III}}(L),$$

relation (5) can be rewritten to obtain

$$D_3^{\mathrm{I}}(L) - D_3^{\mathrm{II}}(0) = -\mathrm{i}\tilde{\epsilon}L\boldsymbol{k}\boldsymbol{E}_t^{\mathrm{I}}(0). \tag{5a}$$

The quantity $D_3^{\mathrm{I}}(L)$ in eq. (5a) can be formally represented as the expansion

$$D_3^{\rm I}(L) \approx D_3^{\rm I}(0) + L\left(\frac{{\rm d}D_3^{\rm I}}{{\rm d}z}\right)_{z=0} + \cdots,$$

hence, in place of eq. (5a), we obtain

$$D_3^{\rm I}(0) - D_3^{\rm II}(0) = -{\rm i}\tilde{\epsilon}L\boldsymbol{k}\boldsymbol{E}_t^{\rm I}(0) - L\left(\frac{{\rm d}D_3^{\rm I}}{{\rm d}z}\right)_{z=0}.$$

It follows from condition (3) in medium I that ${\rm d}D_3^{\rm I}/{\rm d}z = -{\rm i}\boldsymbol{k}\boldsymbol{D}^{\rm I} = -{\rm i}\epsilon_1\boldsymbol{k}\boldsymbol{E}_t^{\rm I}$, hence

$$D_3^{\rm I}(0) - D_3^{\rm II}(0) = -{\rm i}\gamma\boldsymbol{k}\boldsymbol{E}_t^{\rm I}(0), \tag{6}$$

where

$$\gamma \equiv \gamma(\omega) = L(\tilde{\epsilon} - \epsilon_1). \tag{6a}$$

As could be expected, the right side of relation (6) vanishes at $\tilde{\epsilon} = \epsilon_1$, where ϵ_1 is the dielectric constant of medium I. In a similar manner it can be shown that in the first order with respect to kL, the quantity $H_3(z)$ does not have a discontinuity, i.e.,

$$H_3^{\rm I}(0) - H_3^{\rm II}(0) = 0. \tag{7}$$

This result is analogous to eq. (6) and is a consequence of the fact that we assume the transition layer and media I and II to be nonmagnetic, having the magnetic permeability $\mu = \mu_1 = \mu_2 = 1$.

To obtain the boundary conditions for the tangential components of the electric field strength $\boldsymbol{E}$, we shall make use, within the transition layer, of the equation

$$\operatorname{curl} \boldsymbol{E} = {\rm i}\frac{\omega}{c}\boldsymbol{H}. \tag{8}$$

For the xth component and fields of the type of eq. (2), we obtain for the values of z within the transition layer:

$${\rm i}k_2E_3^{\rm III} - \frac{\partial E_2^{\rm III}}{\partial z} = {\rm i}\frac{\omega}{c}H_1^{\rm III},$$

which, after integrating with respect to z between the limits 0 and L, becomes

$$E_2^{\rm III}(L) - E_2^{\rm III}(0) = {\rm i}k_2\int_0^L E_3^{\rm III}(z)\,{\rm d}z - {\rm i}k_0\int_0^L H_1^{\rm III}(z)\,{\rm d}z, \tag{8a}$$

where $k_0 = \omega/c$.

Reasoning along the same lines as in deriving relation (6), we find that

$$E_2^{\rm I}(0) - E_2^{\rm II}(0) = {\rm i}k_2\mu D_3^{\rm I}(0), \tag{8b}$$

where $\mu = L\left(\frac{1}{\tilde{\epsilon}} - \frac{1}{\epsilon_1}\right)$.

An analogous expression is obtained for E_1. Hence, in a more general form,

$$\boldsymbol{E}_t^{\mathrm{I}}(0) - \boldsymbol{E}_t^{\mathrm{II}}(0) = \mathrm{i}\boldsymbol{k}\mu D_3^{\mathrm{I}}(0). \tag{9}$$

The application of the equation

$$\operatorname{curl} \boldsymbol{H} = -\mathrm{i}\frac{\omega}{c}\boldsymbol{D} \tag{10}$$

enables the fourth boundary condition to be obtained:

$$\boldsymbol{H}_t^{\mathrm{I}}(0) - \boldsymbol{H}_t^{\mathrm{II}}(0) = \mathrm{i}k_0\gamma[\boldsymbol{n}\boldsymbol{E}_t^{\mathrm{I}}(0)], \tag{11}$$

where $\boldsymbol{n}$ is a unit vector directed along the z-axis.

When the boundary conditions (6), (7) (9) and (11) are applied, the effect of the transition layer, as has been previously underlined, is taken into account with an accuracy to within first-order terms with respect to the small parameter L/λ_0, where $\lambda_0 = 2\pi c/\omega$. No difficulties, in principle, are encountered in refining the theory.

In dealing with an isotropic transition layer and isotropic media I and II, the preceding remark is sufficiently obvious. Any analysis of the situation for anisotropic media turns out to be more cumbersome. In deriving the boundary conditions for this case in the following subsection, we shall, as before, take into consideration only terms of the order of L/λ because the present volume of experimental research conducted in this field makes the development of a more precise theory inexpedient as yet.

It should be noted that in the presence of a microscopic transition layer, the form of the boundary conditions for the fields, i.e. relations (6), (7), (9) and (11), is retained. However, the values of quantities μ and γ appearing in these equations cannot be found in the general form because these values depend essentially on the features of the transition layer model being considered. The quantities γ and μ for dielectrics were expressed, in particular, by Sivukhin (1948, 1952, 1956) in terms of the polarizability of the boundary molecules. The same works presented the corrections to the boundary conditions with respect to higher orders of L/λ (see also sect. 3).

2.2. *Effective Boundary Conditions for Fields in the Presence of an Anisotropic Homogeneous Transition Layer*

It proves sufficient to generalize only the boundary conditions that contain characteristics of the transition layer.

For this purpose, recalling (4a), we find it convenient to write relation (4) in the form

$$D_3^{\mathrm{I}}(L) - D_3^{\mathrm{II}}(0) = -\mathrm{i}L\boldsymbol{k}\boldsymbol{D}_t^{\mathrm{III}}(0), \tag{12}$$

where

$$D_1^{\mathrm{III}} = \tilde{\epsilon}_{1j} E_j^{\mathrm{III}}, \qquad D_2^{\mathrm{III}} = \tilde{\epsilon}_{2j} E_j^{\mathrm{III}}. \tag{13}$$

Unlike D_3, the quantity E_3 has discontinuities at the interfaces. Therefore, making use of the relation $D_3^{\mathrm{III}} = \tilde{\epsilon}_{3j} E_j^{\mathrm{III}}$, it proves convenient to cancel out quantity E_3^{III} from eq. (13), and to express D_1^{III} and D_2^{III} only in terms of E_1^{III}, E_2^{III} and D_3^{III}, which are continuous at the interfaces.

Since

$$E_3^{\mathrm{III}} = \frac{1}{\tilde{\epsilon}_{33}}(D_3^{\mathrm{III}} - \tilde{\epsilon}_{31} E_1^{\mathrm{III}} - \tilde{\epsilon}_{32} E_2^{\mathrm{III}}), \tag{13a}$$

we make use of eq. (13) and obtain

$$D_1^{\mathrm{III}} = \left(\tilde{\epsilon}_{11} - \frac{\tilde{\epsilon}_{13}\tilde{\epsilon}_{31}}{\tilde{\epsilon}_{33}}\right) E_1^{\mathrm{III}} + \left(\tilde{\epsilon}_{12} - \frac{\tilde{\epsilon}_{13}\tilde{\epsilon}_{32}}{\tilde{\epsilon}_{33}}\right) E_2^{\mathrm{III}} + \frac{\tilde{\epsilon}_{13}}{\tilde{\epsilon}_{33}} D_3^{\mathrm{III}}, \tag{13b}$$

$$D_2^{\mathrm{III}} = \left(\tilde{\epsilon}_{21} - \frac{\tilde{\epsilon}_{23}\tilde{\epsilon}_{31}}{\tilde{\epsilon}_{33}}\right) E_1^{\mathrm{III}} + \left(\tilde{\epsilon}_{22} - \frac{\tilde{\epsilon}_{23}\tilde{\epsilon}_{32}}{\tilde{\epsilon}_{33}}\right) E_2^{\mathrm{III}} + \frac{\tilde{\epsilon}_{23}}{\tilde{\epsilon}_{33}} D_3^{\mathrm{III}},$$

so that the boundary condition analogous to eq. (6) for the case being discussed is

$$\begin{aligned} D_3^{\mathrm{I}}(0) - D_3^{\mathrm{II}}(0) = -\mathrm{i}\Big[&(k_1\gamma_{11} + k_2\gamma_{21}) E_1^{\mathrm{I}}(0) \\ &+ (k_1\gamma_{12} + k_2\gamma_{22}) E_2^{\mathrm{I}}(0) + \frac{k_1\tilde{\epsilon}_{13} + k_2\tilde{\epsilon}_{23}}{\tilde{\epsilon}_{33}} D_3^{\mathrm{I}}(0) L\Big], \end{aligned} \tag{14}$$

where

$$\gamma_{11} = \left(\tilde{\epsilon}_{11} - \epsilon_1 - \frac{\tilde{\epsilon}_{13}\tilde{\epsilon}_{31}}{\tilde{\epsilon}_{33}}\right) L, \qquad \gamma_{12} = \left(\tilde{\epsilon}_{12} - \frac{\tilde{\epsilon}_{13}\tilde{\epsilon}_{32}}{\tilde{\epsilon}_{33}}\right) L,$$

$$\gamma_{21} = \left(\tilde{\epsilon}_{21} - \frac{\tilde{\epsilon}_{23}\tilde{\epsilon}_{31}}{\tilde{\epsilon}_{33}}\right) L, \qquad \gamma_{22} = \left(\tilde{\epsilon}_{22} - \epsilon_1 - \frac{\tilde{\epsilon}_{23}\tilde{\epsilon}_{32}}{\tilde{\epsilon}_{33}}\right) L. \tag{14a}$$

Applying eq. (13a), we readily see that in the case being discussed eq. (8a) leads to the relation

$$E_2^{\mathrm{I}}(0) - E_2^{\mathrm{II}}(0) = \mathrm{i}k_2 L\left[\left(\frac{1}{\tilde{\epsilon}_{33}} - \frac{1}{\epsilon_1}\right) D_3^{\mathrm{I}}(0) - \frac{\tilde{\epsilon}_{31}}{\tilde{\epsilon}_{33}} E_1^{\mathrm{I}}(0) - \frac{\tilde{\epsilon}_{32}}{\tilde{\epsilon}_{33}} E_2^{\mathrm{I}}(0)\right],$$

or, in a more general form,

$$\mathbf{E}_t^{\mathrm{I}}(0) - \mathbf{E}_t^{\mathrm{II}}(0) = \mathrm{i}\mathbf{k}L\left[\frac{\mu}{L} D_3^{\mathrm{I}}(0) - \frac{\tilde{\epsilon}_{31}}{\tilde{\epsilon}_{33}} E_1^{\mathrm{I}}(0) - \frac{\tilde{\epsilon}_{32}}{\tilde{\epsilon}_{33}} E_2^{\mathrm{I}}(0)\right], \tag{15}$$

where

$$\mu = \left(\frac{1}{\tilde{\epsilon}_{33}} - \frac{1}{\epsilon_1}\right) L.$$

In an analogous manner, but making use of Maxwell's equation (10) and taking relation (13b) into consideration, we find that

$$H_1^{\mathrm{I}}(0) - H_1^{\mathrm{II}}(0) = -\mathrm{i}k_0\left[\gamma_{21}E_1^{\mathrm{I}}(0) + \gamma_{22}E_2^{\mathrm{I}}(0) + L\frac{\tilde{\epsilon}_{23}}{\tilde{\epsilon}_{33}}D_3^{\mathrm{I}}(0)\right],$$

$$H_2^{\mathrm{I}}(0) - H_2^{\mathrm{II}}(0) = \mathrm{i}k_0\left[\gamma_{11}E_1^{\mathrm{I}}(0) + \gamma_{12}E_2^{\mathrm{I}}(0) + L\frac{\tilde{\epsilon}_{13}}{\tilde{\epsilon}_{33}}D_3^{\mathrm{I}}(0)\right]. \qquad (16)$$

We emphasize that in deriving boundary conditions (16), (15) and (14), the form of tensor $\tilde{\epsilon}_{ij}$ was not specified, nor was the symmetry of medium II assumed to be given. In the following, however, medium II, like medium I, is to be assumed isotropic.

2.3. *Surface Polariton Dispersion in the Case of a Macroscopic Isotropic Transition Layer – Additional Surface Polaritons and the Problem of Additional Boundary Conditions*

Function $\mathbf{A}(z)$, appearing in eq. (2), will be sought for in the following form

$$\mathbf{A}(z) = \begin{cases} \mathbf{A}^{I}\mathrm{e}^{-\kappa_1 z}, & z>0, \quad \mathrm{Re}\,\kappa_1>0, \\ \mathbf{A}^{\mathrm{II}}\mathrm{e}^{\kappa_2 z}, & z<0, \quad \mathrm{Re}\,\kappa_2>0. \end{cases} \qquad (17)$$

In these relations $\mathbf{A}^{\mathrm{I}}$ and $\mathbf{A}^{\mathrm{II}}$ are field amplitudes independent of the coordinates. It can readily be shown (see chs. 1 and 2) that fields of the form (2) and (17) satisfy the Maxwell equations in both media, I and II, if

$$\kappa_1 = \left(k_1^2 + k_2^2 - \frac{\omega^2}{c^2}\epsilon_1\right)^{1/2},$$

$$\kappa_2 = \left(k_1^2 + k_2^2 - \frac{\omega^2}{c^2}\epsilon_2\right)^{1/2}. \qquad (18)$$

The x-axis is directed along vector $\boldsymbol{k}(k_1, k_2)$. This means that in the chosen coordinate system $k_2 = 0$ and, consequently, $k = k_1$. At $z>0$ and $z<0$, it follows from the equation div $\boldsymbol{H} = 0$ that

$$kH_1^{\mathrm{I}}(0) + \mathrm{i}\kappa_1 H_3^{\mathrm{I}}(0) = 0,$$
$$kH_1^{\mathrm{II}}(0) - \mathrm{i}\kappa_2 H_3^{\mathrm{II}}(0) = 0, \qquad (19a)$$

or, taking eq. (7) into consideration,

$$k[H_1^{\mathrm{I}}(0) - H_1^{\mathrm{II}}(0)] + \mathrm{i}H_3^{\mathrm{I}}(0)(\kappa_1 + \kappa_2) = 0. \qquad (19b)$$

Since the transition layer is assumed to be isotropic, it follows from eq. (11) that

$$H_1^{\mathrm{I}}(0) - H_1^{\mathrm{II}}(0) = -\mathrm{i}k_0\gamma E_2^{\mathrm{I}}(0). \qquad (11a)$$

Substituting eq. (11a) into eq. (19b), we find

$$-ik_0k\gamma E_2^{\rm I}(0) + iH_3^{\rm I}(0)(\kappa_1 + \kappa_2) = 0. \qquad (19c)$$

But it follows from eq. (8) that

$$ik_0 H_3^{\rm I}(0) = ikE_2^{\rm I}(0),$$

so that relation (19c) can be written in the form

$$H_3^{\rm I}(0)(\kappa_1 + \kappa_2 - k_0^2\gamma) = 0. \qquad (20)$$

This indicates that irrespective of whether or not the inequality

$$\kappa_1 + \kappa_2 - k_0^2\gamma \neq 0 \qquad (20a)$$

is satisfied, we can assume that in the surface waves being considered (see also eqs. (19a) and (11a))

$$H_3^{\rm I}(0) = H_3^{\rm II}(0) = H_1^{\rm I}(0) = H_1^{\rm II}(0) = E_2^{\rm I}(0) = E_2^{\rm II}(0) = 0. \qquad (20b)$$

Thus, the structure of the surface waves remains unchanged, even in the presence of an isotropic transition layer. Namely, these waves remain H-waves (TM waves) with the only nonzero components of fields E_1, E_3 and H_2. To derive the required dispersion law, we make use of the boundary condition (11) for H_2:

$$H_2^{\rm I}(0) - H_2^{\rm II}(0) = ik_0\gamma E_1^{\rm I}(0), \qquad (21a)$$

as well as the boundary condition (9) for E_1:

$$E_1^{\rm I}(0) - E_1^{\rm II}(0) = ik\mu\epsilon_1 E_3^{\rm I}(0). \qquad (21b)$$

Employing eq. (10), the quantities E_1 and E_3 in both media, I and II, are expressed in terms of H_2. Namely:

$$-ik_0\epsilon_1 E_1^{\rm I}(0) = \kappa_1 H_2^{\rm I}(0),$$

$$-ik_0\epsilon_2 E_1^{\rm II}(0) = -\kappa_2 H_2^{\rm II}(0),$$

$$-k_0\epsilon_1 E_3^{\rm I}(0) = kH_2^{\rm I}(0). \qquad (21c)$$

Eliminating the quantities E_1 and E_2 from eqs. (21) by means of the preceding relations, we obtain the following system of the linear equations for $H_2^{\rm I}(0)$ and $H_2^{\rm II}(0)$:

$$H_2^{\rm I}(0)\left(1 + \frac{\kappa_1}{\epsilon_1}\gamma\right) - H_2^{\rm II}(0) = 0,$$

$$H_2^{\rm I}(0)\left(\frac{\kappa_1}{\epsilon_1} + k^2\mu\right) + \frac{\kappa_2}{\epsilon_2} H_2^{\rm II}(0) = 0.$$

The solvability condition for this system,

$$\frac{\kappa_1}{\epsilon_1}+\frac{\kappa_2}{\epsilon_2}+k^2\mu+\frac{\kappa_1\kappa_2}{\epsilon_1\epsilon_2}\gamma=0, \tag{22}$$

is precisely what determines the dispersion law for surface waves. As $L\to 0$, the quantities μ and γ vanish and eq. (22) becomes the dispersion law at a sharp boundary. Within the limits of accuracy used in this discussion, relation (22) can also be written in the form

$$\frac{\epsilon_1}{\kappa_1}+\frac{\epsilon_2}{\kappa_2}+\gamma-k^2\mu\frac{\epsilon_1^2}{\kappa_1^2}=0, \tag{22a}$$

or else

$$\frac{\kappa_1}{\epsilon_1}+\frac{\kappa_2}{\epsilon_2}+k^2\mu-\frac{\kappa_1^2}{\epsilon_1^2}=0. \tag{22b}$$

In order to find the dependence of the frequency of surface polaritons on the wave vector $\omega=\omega(k)$ by means of relations (22), it is necessary to employ the explicit expressions for the functions $\epsilon_1(\omega)$, $\epsilon_2(\omega)$, $\gamma(\omega)$ and $\mu(\omega)$. Such an analysis is impossible in the general form. We shall therefore discuss only certain features of the dispersion laws of SPs, occurring in response to the transition layer under conditions when a certain frequency $\tilde{\omega}_0$ of dipole oscillations in the transition layer falls within the range of the frequency change region of zeroth-order approximation SP.

Zeroth-order approximation SPs are here understood to be SPs that could occur at the boundary between media I and II if there were no transition layer (i.e. at $L=0$). The frequencies of these SPs satisfy the condition

$$\frac{\epsilon_1}{\kappa_1}+\frac{\epsilon_2}{\kappa_2}=0. \tag{23}$$

If $\omega=\omega_s^\circ(k)$ complies with the solution of the preceding equation, then at $\omega\approx\omega_s^\circ(k)$ the left-hand side of eq. (23) can be written, for a fixed value of k, in the form

$$\frac{\epsilon_1}{\kappa_1}+\frac{\epsilon_2}{\kappa_2}=\frac{\omega-\omega_s^\circ(k)}{C(k)}, \tag{23a}$$

where, in virtue of the fact that $d\epsilon_1/d\omega>0$ and $d\epsilon_2/d\omega>0$,

$$C^{-1}(k)=\frac{\partial}{\partial\omega}\left(\frac{\epsilon_1}{\kappa_1}+\frac{\epsilon_2}{\kappa_2}\right)_{\omega=\omega_s^\circ(k)}>0.$$

In accordance with eqs. (6a) and (8a), resonance γ corresponds to condition $\tilde{\epsilon}(\tilde{\omega}_\perp)=\infty$, and resonance μ to condition $\tilde{\epsilon}(\tilde{\omega}_\parallel)=0$. We must underline, however, that the effect of resonances in the frequency range $\tilde{\omega}_\perp$ on the dispersion of SPs always turns out to be relatively weaker because of the fulfilment of the inequality

$$\frac{\kappa_1^2}{k^2\epsilon_1} = \frac{k^2 - (\omega^2/c^2)\epsilon_1}{k^2\epsilon_1} < 1. \tag{24}$$

In the region of resonances being discussed ($\tilde{\omega}_0 = \tilde{\omega}_\perp$ or $\tilde{\omega}_0 = \tilde{\omega}_\parallel$),

$$\gamma - k^2\mu \frac{\epsilon_1}{\kappa_1^2} = \frac{A\tilde{\omega}_0}{\tilde{\omega}_0 - \omega},$$

where $A(k) > 0$. Taking eq. (23a) into account, we find that eq. (22a) can be rewritten in the form

$$[\omega - \omega_s^\circ(k)](\omega - \tilde{\omega}_0) = A(k)\tilde{\omega}_0 C(k).$$

Solving this equation for ω, we find the SP dispersion law in the region being discussed to be

$$\omega_{1,2}(k) = \tfrac{1}{2}[\tilde{\omega}_0 + \omega_s^\circ(k)] \pm \tfrac{1}{2}\{[\omega_s^\circ(k) - \tilde{\omega}_0]^2 + 4A\tilde{\omega}_0 C(k)\}^{1/2}. \tag{25}$$

If frequency $\tilde{\omega}_0$ is within the range of the frequency change region of zeroth-order approximation SP, i.e., if at a certain $k = k_0$, the frequency $\tilde{\omega}_0 = \omega_s^\circ(k)$, a gap of magnitude $\Delta = 2[A\tilde{\omega}_0 C(k_0)]^{1/2}$ is formed in the SP spectrum due to the effect of the transition layer. Since $A \sim L$, the gap magnitude Δ is found to be of the order of $\sqrt{(L/\tilde{\lambda}_0)}$ where $\lambda_0 = (2\pi/\omega_0)c$. Subsequently, we shall return to a more detailed evaluation of Δ. Here we note only that the presence of a transition layer usually leads to effects of the order of $L/\tilde{\lambda}$. The occurrence of a root dependence in the case being discussed is due to the existence of resonance and to the renormalization of the spectrum of the surface waves. This, precisely, is the circumstance that makes Δ large in many cases in comparison with the width of the SP lines, thereby enabling the gap to be experimentally observed.

The splitting of the SP dispersion curve and the root dependence of Δ on L were first observed by Yakovlev et al. (1975, 1977) for the infrared region of the spectrum in studying SP propagating along the surface of sapphire coated with a film of LiF (at $L \approx 100$ Å, $\Delta \approx 20\ \mathrm{cm}^{-1}$). The width of the gap increases substantially for the visual range of the spectrum (Lopez-Rios et al. 1978). In this last work, the splitting effect was observed for SP propagating along the surface of aluminium coated with a silver film ($L \approx 20$ to 60 Å). In this case, the gap magnitude Δ at $L = 26$ Å was found to be approximately 0.4 eV, which agrees with theory (see also chs. 6 and 7).

Evidently, resonance of the oscillations in the transition layer with SPs is a quite common phenomenon. In particular, its possibility must also be taken into account in analyzing the spectra of attenuated total reflection of light from the surfaces of molecular crystals (for example, anthracene, see Sherman and Philpott 1978, Turlet and Philpott 1975a, b, 1976, Syassen and Philpott 1978, Glockner and Wolf 1943, Turlet et al. 1978, Sugakov 1970, Philpott 1974) as well as in studying Fermi resonance with SPs (Agranovich and Lalov 1976).

In connection with the aforesaid, a further, more detailed analysis of SP dispersion in the case of resonance with the oscillations in the transition layer becomes timely and, in particular, so does an analysis of possible effects due to the taking of spatial dispersion into account. This analysis has been carried out by the author (Agranovich 1975) for the nonresonance situation. In this work (Agranovich 1975) it was shown, among other matters, that in the vicinity of the Coulomb frequency ω_s of a surface polariton at the boundary with vacuum (frequency ω_s satisfies the condition $\epsilon(\omega_s) = -1$, where $\epsilon(\omega)$ is the dielectric constant of the substrate), the influence of the transition layer leads to the linear relation $\omega_s(\boldsymbol{k})$ with respect to $\boldsymbol{k}$, where $\boldsymbol{k}$ is the wave vector of the SP. This results in the appearance of an additional electromagnetic surface wave. Damping is very great, however, in the frequency region $\omega \approx \omega_s$, and this should particularly prevent the propagation of the additional surface wave.

It should be noted, in this connection, that damping is sufficiently strong for SPs propagated along the surfaces of dielectrics, not only at $\omega \approx \omega_s$, but at $\omega < \omega_s$ as well, i.e., throughout the whole region of the SP spectrum. But for SPs propagated along the surfaces of metals this is found, in general, to be a different matter. For waves with frequency $\omega \approx \omega_s \approx \omega_p/\sqrt{2}$ (where ω_p is the frequency of a bulk plasmon), the field of the surface polariton penetrates considerably into the metal and the SP in this spectral region is strongly damped. In the frequency region $\omega \ll \omega_p/\sqrt{2}$, however, the field of the surface wave penetrates only negligibly into the metal, the damping of the surface polariton is weak and its length of propagation turns out to be macroscopically large (of the order of several centimeters, see Shoenwald et al. 1973). The relatively small damping is also retained in many cases in the frequency region of resonance of the oscillations in the transition layer with those of SPs, only if the frequency of these oscillations is $\tilde{\omega}_0 \ll \omega_p/\sqrt{2}$, and if the transition layer is sufficiently thin. Consequently, additional surface waves will evidently prove most accessible to observation and study when they propagate along the surfaces of metals.

In what follows, for the sake of definiteness, we shall deal with the dispersion of surface waves under conditions that comply with experiments (Lopez-Rios et al. 1978). Specifically, we shall assume that medium II is formed by a metal with a plasma frequency $\omega_p \gg \tilde{\omega}_0$, that the transition layer is macroscopic and obtained by applying a film of another metal with the plasma frequency $\omega_0 \equiv \tilde{\omega}_0$ on the surface $z = 0$, and that medium I is a vacuum.

Since for electrons on a Fermi surface $k_F \approx 10^8\,\mathrm{cm}^{-1}$, such a film can be regarded as macroscopic when inequality $k_F L \gg 1$ is satisfied, as will be assumed. Assuming, in addition, that we have a case with normal skin effect and neglecting the possible contribution of interband transitions, we shall presume that $\epsilon_1 = 1$,

$$\epsilon_2 \equiv \epsilon(\omega) = 1 - \frac{\omega_p^2}{\omega(\omega + \mathrm{i}\Gamma)}, \quad \tilde{\epsilon}(\omega) = 1 - \frac{\omega_0^2}{\omega(\omega + \mathrm{i}\tilde{\Gamma})}, \tag{26}$$

where Γ and $\tilde{\Gamma}$ are the frequencies of collisions of electrons in the metal II and in the transition layer. Substitution of expressions (26) into (22) results in a relation enabling the dispersion of surface waves to be determined for the case when damping occurs. In accordance with eqs. (26), the quantity $\tilde{\epsilon}(\omega)$ has no poles at $\omega \neq 0$. It is therefore sufficient in eq. (22) to retain the term containing the quantity μ.

Of interest to us here is only the situation in which the frequency of the surface wave is a real quantity, specified by the pumping source of the surface waves. Under these conditions, $k = k' + \mathrm{i}k''$ is a complex quantity and the length of propagation of the surface wave is $l = (2k'')^{-1}$. If the damping of the surface wave is sufficiently weak (i.e. if $k' \gg k''$) then damping can be entirely neglected as a first approximation in deriving the dispersion law. In this case, the dispersion law for a surface wave, i.e., the relationship $\omega_s(k)$ is determined by the equation

$$F(\omega) = -cLk^2\omega^2(\omega^2 - \omega_0^2)^{-1}, \tag{27}$$

where

$$F(\omega) = \frac{\omega^2(k^2c^2 + \omega_p^2 - \omega^2)^{1/2}}{\omega^2 - \omega_p^2} + (k^2c^2 - \omega^2)^{1/2}. \tag{27a}$$

If $\omega_{0s}(k)$ is the frequency of a surface polariton at a sharp boundary (i.e. under conditions when the presence of a transition layer is not taken into account), then $F(\omega_{0s}^2) = 0$ and, in the frequency region $\omega \approx \omega_{0s}(k)$,

$$F = (\mathrm{d}F/\mathrm{d}\omega^2)_0(\omega^2 - \omega_{0s}^2).$$

It follows from eq. (27a) that at $\omega^2 \ll \omega_p^2$ and $k^2c^2 \ll \omega_p^2$,

$$F(\omega^2) = -\omega^2/\omega_p + (k^2c^2 - \omega^2)^{1/2}.$$

Hence

$$\omega_{0s}^2(k) \approx k^2c^2 - k^4c^4/\omega_p^2 + \cdots,$$
$$(\mathrm{d}F/\mathrm{d}\omega^2)_0 \approx -\omega_p/2\omega_{0s}^2$$

and, consequently,

$$F(\omega^2) \approx -\omega_p[\omega^2 - \omega_{0s}^2(k)]/2\omega_{0s}^2(k).$$

Thus, for the frequency region $\omega \approx \omega_0 \approx \omega_{0s}(q_0)$ being discussed, eq. (27) can be written as follows:

$$(\omega^2 - \omega_0^2)[\omega^2 - \omega_{0s}^2(k)] = 2cLk^2\omega^2\omega_{0s}^2(k)/\omega_p.$$

Solving this equation with respect to ω^2, we find two solutions $\omega_{1,2}^2(k)$:

$$\omega_{1,2}^2(k) = \tfrac{1}{2}[\omega_0^2 + \omega_{0s}^2(k) + 2cLk^2\omega_{0s}^2(k)/\omega_p]$$
$$\pm \tfrac{1}{2}\{[\omega_0^2 - \omega_{0s}^2(k)]^2 + 4cLk^2\omega_{0s}^2(k)[\omega_0^2 + \omega_{0s}^2(k)]\omega_p^{-1}\}^{1/2}. \tag{28}$$

Relatively small terms, proportional to k^2L^2, have been omitted under the radical sign in eq. (28).

At $k = q_0$, where the frequency $\omega_{0s}(q_0) = \omega_0$, branch splitting occurs, so that for the values of frequencies $\omega_{1,2}(q_0)$ we have

$$\omega_{1,2}^2(q_0) \approx \omega_0^2[1 \pm (2cLq_0^2/\omega_p)^{1/2}].$$

The magnitude of the gap $\Delta \equiv \omega_1(q_0) - \omega_2(q_0)$ is determined by the relation

$$\Delta \approx \omega_0(2L\omega_0^2/c\omega_p)^{1/2}, \tag{29}$$

or, in terms of wavelengths

$$\mathrm{d}\lambda/\lambda_0 = 2(\pi L/\lambda_p)^{1/2}(\lambda_p/\lambda_0), \tag{29a}$$

where

$$\lambda_p = 2\pi c/\omega_p \quad \text{and} \quad \lambda_0 = 2\pi c/\omega_0.$$

Along with the splitting $\Delta = \omega_1(q_0) - \omega_2(q_0)$, we can also introduce the quantity $\Delta_1 = \omega_1(\min) - \omega_2(\max)$. In this equation, $\omega_1(\min)$ is the minimum value of the frequency on the upper branch, corresponding to the value of $k_{\min} = \omega_{\min}/c$ where the upper branch abuts against straight line $\omega = ck$. From eq. (27) it follows that at $\omega_0 \ll \omega_p$,

$$\omega_1(\min) - \omega_0 \approx (L/2c)\omega_0\omega_p,$$

i.e. it depends linearly on L, so that the principal value of Δ is determined by the lowering of the lower branch. Note also that $\Delta_1 < \Delta$ (see fig. 2).

The magnitude of splitting decreases with an increase in ω_p: $\Delta \sim \omega_p^{-1/2}$. This is due to the fact that with an increase in ω_p, the electric field strength at $z \approx 0$ decreases in the surface wave and, correspondingly, the interaction of the wave with the transition layer is reduced. If, on the contrary, the quantity $\omega_p \to 0$ (this case evidently corresponds to a metal film in vacuum), eq. (27) takes the form

$$2(k^2c^2 - \omega^2)^{1/2} = cLk^2\omega^2(\omega_0^2 - \omega^2)^{-1}.$$

It follows from this equation that the nonradiational surface waves discussed here occur only for values of ω and k, for which $kc > \omega$ and $\omega < \omega_0$. No splitting of the surface wave spectrum occurs. As to relation $\omega(k)$, it coincides in this case with the relation for two-dimensional systems (see Agranovich and Dubovski 1966).

But let us return to a consideration of the dispersion of surface waves for the case of $\omega_p \gg \omega_0$. Note, first of all, that at $k \gg q_0$, when $kc \gg \omega_p$, i.e. in the nonrelativistic limit, eq. (27) is simplified and takes the form

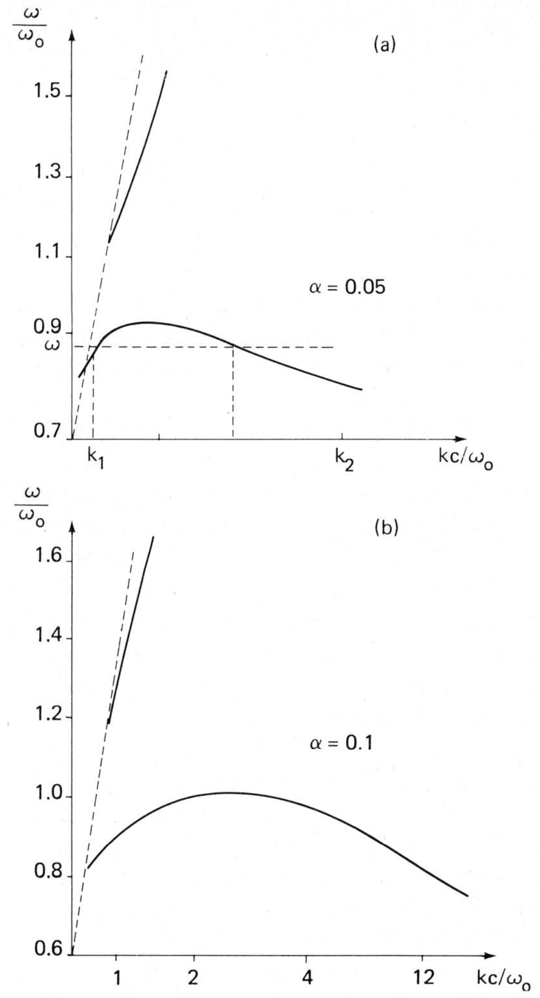

Fig. 2. SP dispersion law with the transition layer taken into account: (a) $y_p = 15$ and $\alpha = 0.05$; (b) $y_p = 15$ and $\alpha = 0.1$.

$$\frac{\omega_p^2 - 2\omega^2}{\omega^2 - \omega_p^2} = \frac{\omega^2 kL}{\omega^2 - \omega_0^2}. \tag{30}$$

It should be borne in mind that this nonrelativistic equation is valid only when the inequality $\omega_p L/c \ll 1$ is satisfied. It is only in this case that the passage to the nonrelativistic limit in eq. (27) does not contradict the inequality $kL \ll 1$ that was used in writing boundary conditions (21).

It follows from eq. (30) that for the upper branch of frequencies at large k,

$$\omega_1(k) \approx 2^{-1/2}\omega_p(1 + kL/4), \tag{31a}$$

while for the lower branch,

$$\omega_2(k) \approx \omega_0(1 - kL/2). \tag{31b}$$

The fact that the lower branch of frequencies of SPs obeys the linear dispersion law with a negative slope, leads, as is evident in fig. 2, to the occurrence of an additional surface wave in the frequency region $\omega \lesssim \omega_0$. There are two SPs rather than one in this frequency region. They have one and the same frequency and different values of the wave vector. Naturally, this circumstance can, in general, prove to be of importance in studying various optical phenomena that develop with the participation of SPs (see sect. 4).

The polariton dispersion law for the case being discussed (a metal film on a massive metal substrate) without taking damping into account is represented in fig. 2 for various values of the parameters $\alpha \equiv \omega_0 L/c$ and $y_p = \omega_p^2/\omega_0^2$. Note that for aluminium coated with a film of silver, the values are $y_p \approx 15.2$ and $\omega_0/c = 2 \times 10^5\,\text{cm}^{-1}$. For a film of LiF applied to a substrate of silver, $y_p = 45$ and $\omega_0/c = 4.2 \times 10^3\,\text{cm}^{-1}$. In the work (Lopez-Rios et al. 1978) mentioned above, polariton spectrum splitting for the pair Ag/Al was studied for films corresponding to the values of 5×10^{-2}, 8×10^{-2} and 12×10^{-2} of the parameter α.

As has already been underlined, additional surface waves can be observed if their damping is not too strong. In order to evaluate this damping, we note that when the dissipation of energy in the substrate and film is taken into account, the quantities ϵ and μ, appearing in eq. (22) when $\gamma = 0$, are complex (see eq. (26)) even for real ω: $\epsilon = \epsilon' + i\epsilon''$ and $\mu = \mu' + i\mu''$. In this case, relation (22) enables the real and the imaginary part of $k = k' + ik''$ to be determined as functions of ω. Since the frequencies of electron collisions in the metal are Γ and $\tilde{\Gamma} \approx 10^{14}\,\text{s}^{-1}$ (see Abeles 1972), for the frequency band $\omega \sim \omega_0 \sim 10^{15}\,\text{s}^{-1}$ being discussed, $\Gamma/\omega \ll 1$ and $\tilde{\Gamma}/\omega \ll 1$, so that $|\epsilon'| \gg \epsilon''$ and $|\tilde{\epsilon}'| \gg \tilde{\epsilon}''$. This means that, in the case being considered, a first approximation with respect to ϵ'' and $\tilde{\epsilon}''$ can be made use of in determining the value of k''. Taking the aforesaid into account, we find from relation (22) (at $\gamma \equiv 0$ and $\epsilon_1 = 1$) that

$$k''(\omega) = -\frac{k^2\mu'' - \epsilon''(\kappa/\epsilon^2 + \omega^2/2c^2\epsilon\kappa)}{k(1/\epsilon\kappa + 1/\kappa_1 + 2\mu)}. \tag{32}$$

In accordance with the approximation mentioned above, the value of $k \equiv k(\omega)$ on the right-hand side of this equation is determined by the dispersion law (27), while the quantities ϵ and μ should be regarded as being real, i.e., with $\Gamma = \tilde{\Gamma} = 0$ (see eqs. (26) and (8b)).

At $\mu = 0$, i.e., under conditions when the transition layer is not taken into consideration

$$k_0'' = \frac{\epsilon''(\kappa/\epsilon^2 + \omega^2/2\epsilon\kappa c^2)}{k(1/\epsilon\kappa + 1/\kappa_1)},$$

and for ω, where $\omega_p \gg \omega \gg \Gamma$, we have

$$k_0''(\omega) = \frac{\Gamma}{2c}\left(\frac{\omega}{\omega_p}\right)^2.$$

When the transition layer is taken into account in the frequency region $\omega \leqslant \omega_0$, each value of ω, as previously mentioned (see fig. 2b), corresponds to two SPs (ordinary and additional), and relation (32) enables their corresponding attentuation lengths l to be compared.

The evaluation of the propagation lengths of SPs, carried out by Agranovich (1979a, b), indicates that, as in the case of bulk polaritons, the observation of the additional wave requires considerable effort and can be accomplished only by a suitable choice of substrates and films. It should be pointed out in this connection, that for the surfaces of metals thin dielectric films only slightly reduce the path length of SPs even within the resonance region in the film (see Zhizhin et al. 1976). Evidently, it is precisely under these conditions that additional surface waves are to be sought.

It has previously been noted (Agranovich 1975) that optical effects due to the additional surface wave can be observed by making use of the excitation of SPs originating in the diffraction of light (for example, the light of a laser) by wedges. Recently (see Zhizhin et al. 1979) such a method of excitation was accomplished for the infrared region by a metallic wedge. No less interesting was the investigation of the interference of surface waves and the bulk wave that occurs in the diffraction of SPs by an impedance step (Schlesinger and Sievers 1979). A further development of experiments of the type described by Zhizhin et al. (1979) and Schlesinger and Sievers (1979) will evidently enable effects due to the interference of additional and ordinary surface waves of the same given frequency to be investigated as well (in this connection, see sect. 4).

When additional surface waves are taken into account ordinary boundary conditions prove to be insufficient to determine the amplitudes of the fields. Here, as in bulk crystal optics, the problem of additional boundary conditions (ABC) arises.

These ABCs become necessary in solving problems involving the diffraction of bulk and surface waves by a wedge or by a step, i.e., under conditions in which the dielectric properties of the surface are found to differ at its different regions. Since the diffraction of surface waves by an impedance step is taken up in sect. 4, here we shall discuss the possible form of ABC for this specific case. Namely, we shall assume that an isotropic film of constant thickness L covers the boundary of the substrate

$z=0$ only at values of $x>0$ (see fig. 3, $\epsilon_1=1$). The aforesaid signifies that in boundary conditions (6), (9) and (11), the quantities $\mu=\gamma=0$ at $x<0$. Hence, the situation being considered is a special case of the more general case $\mu=\mu(x)$, $\gamma=\gamma(x)$.

The form of the ABC for fields at $x=0$ and $z=0$ should depend, in general, on the type of film and the nature of the dipole oscillations in this film that lead to resonance with SPs. Hence, we shall first consider the type of ABC for the case that corresponds to dispersion relation (22) at $\gamma=0$.

For the sake of simplicity we shall assume that the wave vector of the surface waves is parallel to the x-axis ($k_2=0$, $k_1=k$). Taking into account the relation $\mu=\mu(x)$, the boundary conditions for the fields at $z=0$ can be written as follows (see eqs. (6) through (11)):

$$D_3^{\rm I}(0)-D_3^{\rm II}(0)=0, \qquad \boldsymbol{H}^{\rm I}(0)-\boldsymbol{H}^{\rm II}(0)=0,$$
$$E_1^{\rm I}(x,0)-E_1^{\rm II}(x,0)=\frac{\partial}{\partial x}[\mu(x)E_3^{\rm I}(x,0)]. \tag{33}$$

In the case of a sharp impedance step $[\mathrm{d}\mu/\mathrm{d}x=\mu_0\delta(x)]$,

$$\frac{\partial}{\partial x}[\mu(x)E_3^{\rm I}(x,z=0)]=\mu_0\delta(x)E_3^{\rm I}(x,0)+\mu(x)\frac{\partial E_3^{\rm I}(x,0)}{\partial x},$$

and the discontinuity in the component $E_1(x,z)$ of the electric field at $x=0$ and $z=0$ is found to be infinite. Of essential importance is the fact that an infinite discontinuity for field E_1 at $x=0$ and $z=0$ in the impedance approximation also occurs when there is no additional surface wave (see Vainshtein 1966, para. 60). In this case, however, the value itself of field $E_1^{\rm I}$

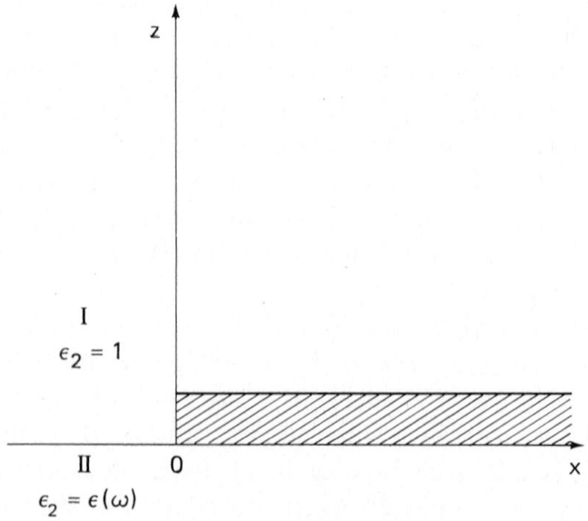

Fig. 3. Impedance step.

turns out to be limited. Since, in taking the additional surface wave into account, its intensity is usually small (see sect. 4), we can employ as ABC, needed in this case to find all the amplitudes of the fields, the requirement of limitedness of field E_1^{I} at the step:

$$E_1^{\mathrm{I}}(x, z=0) \approx C, \qquad x \approx 0, \tag{34}$$

where C is a certain constant.

If, however, "bulk" spatial dispersion is taken into consideration in the film and the number of additional SPs can equal two (see sect. 3), still another, i.e. a second, ABC is required. Since the need for this second ABC is due to spatial dispersion being taken into consideration, its derivation requires a model of the film. In this connection, we note, first of all, that the correction to boundary condition (8b) (or (34)) is due to the polarization of the film along a normal to it. Therefore, in seeking a second ABC for the case when the film is dielectric, we can assume that the dipole moment of the transition in the film is directed along the z-axis. Since the variation of the field along the film thickness is neglected in the approximation linear with respect to L/λ, the film can be assumed, in general, to be two-dimensional. If such a two-dimensional system (two-dimensional crystal) is bounded along the axis $x=0$ and if the deformation of the outer-edge molecules is not considered, then, similar to the three-dimensional case (see Pekar 1957) the boundary condition at $x=0$ for the polarization of the film is the condition

$$p_z(x, y) = 0, \qquad x \approx 0,$$

where p_z is the polarization of the film per unit of its area.

The fulfilment of this condition means that at $x=0$, the film cannot bring about the discontinuity in E_1 of the form (8a) since at $x \approx 0$

$$E_1^{\mathrm{II}}(x,0) - E_1^{\mathrm{I}}(x,0) = -\frac{\partial}{\partial x}\int_0^L E_3^{\mathrm{III}}(z', x)\, \mathrm{d}z', \tag{35}$$

where $E_3^{\mathrm{III}}(z', x)$ is the normal component of the field $\boldsymbol{E}$ in the film. Since at $x=0$ in the film $p_z \approx 0$, we come to the conclusion that inside the film $E_3^{\mathrm{III}}(z, x=0) \approx D_3^{\mathrm{III}}(z, x=0)$ and, consequently,

$$\int_0^L E_3^{\mathrm{III}}(z, x=0)\, \mathrm{d}z = \int_0^L D_3^{\mathrm{III}}(z, x=0)\, \mathrm{d}z$$
$$= LD_n^{\mathrm{I}}(0, x=0) = LE_3^{\mathrm{I}}(0, x=0).$$

Thus, at $x \approx 0$, no term of the order of $\mu = L/\tilde{\epsilon}$ that resonates at $\tilde{\epsilon} = 0$ appears on the right-hand side of eq. (35). Since in investigating waves in the resonance region, nonresonance terms of the order of L were omitted, then, at $x \approx 0$ it is to be presumed that

$$E_t^{\text{I}}(0, x = 0) - E_t^{\text{II}}(0, x = 0) = 0. \tag{35a}$$

Comparing eqs. (35a) and (8b), we arrive at the conclusion that the condition $p_x(x = 0) = 0$ for polarization in the film leads to the sought ABC in the form

$$\frac{\partial}{\partial x} E_n^{\text{I}}(0, x = 0) = 0. \tag{36}$$

Since $E_n(\text{I}) = D_n(\text{II})$, condition (36) can be replaced by the equivalent condition

$$\frac{\partial}{\partial x} D_n^{\text{II}}(0, x = 0) = 0. \tag{36a}$$

ABC having the form $\partial E_t^{\text{I}}(0, x = 0)/\partial x = 0$ can also be found in a similar manner for dielectric films in the vicinity of resonance $\tilde{\epsilon}(\omega)$. If, however, this resonance is due to a two- or three-dimensional Wannier–Mott exciton, it may prove important to introduce a dead layer of the thickness $l_0 \sim r_B$, where r_B is the Bohr radius of the exciton, analogous to the way in which this is done in the theory of ABC for bulk waves (Hopfield and Thomas 1963).

The problem of the form of the ABC for metallic films requires special examination.

We shall demonstrate in sect. 4 how the ABCs "operate" in solving problems of the diffraction of surface waves.

2.4. *Effects of Transition Layer Anisotropy on the Surface Polariton Spectra – Splitting anisotropy upon Resonance with Oscillations in the Transition Layer*

We begin by seeking a solution for surface polaritons, assuming that these polaritons are H(TM)-waves. In this case, according to eq. (20a), relation (19a) is automatically complied with, and boundary conditions (15) and (16) for the amplitudes of the field, which must be satisfied, take the following form:

$$\begin{aligned}
&E_1^{\text{I}}(0) - E_1^{\text{II}}(0) = ik\left[\mu D_3^{\text{I}}(0) - \frac{\tilde{\epsilon}_{31}}{\tilde{\epsilon}_{33}} L E_1^{\text{I}}(0)\right], \\
&H_2^{\text{I}}(0 = -H_2^{\text{II}}(0) = ik_0\left[\gamma_{11} E_1^{\text{I}}(0) + \frac{\tilde{\epsilon}_{13}}{\tilde{\epsilon}_{33}} L D_3^{\text{I}}(0)\right], \\
&0 = -ik_0\left[\gamma_{21} E_1^{\text{I}}(0) + \frac{\tilde{\epsilon}_{23}}{\tilde{\epsilon}_{33}} L D_3^{\text{I}}(0)\right].
\end{aligned} \tag{37}$$

In the case of an isotropic transition layer, $\gamma_{21} = \tilde{\epsilon}_{23} = 0$ and the third of the preceding relations becomes an identity. If and only if this is the case, the system of eqs. (37) can be reduced to a system of two linear equations for determining H_2^{I} and H_2^{II}. This leads us to the dispersion laws (22). If, on the other hand, the transition layer is anisotropic, eqs. (37) are reduced to three equations for the same two unknowns. This circumstance indicates that with an anisotropic transition layer, the surface waves, in general, are no longer H-waves: in them $E_2 \neq 0$ and, consequently, $H_1 \neq 0$ and $H_3 \neq 0$.

The derivation of a new dispersion relation for surface waves in the general form and the determination of all the components of the field is quite a cumbersome matter. Fortunately, however, there is no special necessity for such a derivation if, as previously, only terms of the first order with respect to kL are retained in the boundary conditions.

Since the field components H_1, H_3 and E_2 are nonzero in the surface polariton only when the anisotropy of the transition layer is taken into account, their ratios to the field components E_1, E_3 and H_2 are found to be of the order of kL. Therefore, to find the dispersion law in the zeroth-order approximation, we can put $H_1 = H_3 = E_2 = 0$ in eqs. (15) and (16). Then for the dispersion law thus found we can determine the quantities E_2, H_1 and H_3, making use of ordinary perturbation theory.

In view of the aforesaid and making use of eqs. (21c), the first two relations of system (37) can be written as follows:

$$H_2^{\mathrm{I}}(0)\left(\frac{\kappa_1}{\epsilon_1} + k^2\mu + ikL\frac{\kappa_1}{\epsilon_1}\frac{\tilde{\epsilon}_{31}}{\tilde{\epsilon}_{33}}\right) + \frac{\kappa_2}{\epsilon_2} H_2^{\mathrm{II}}(0) = 0,$$
$$H_2^{\mathrm{I}}(0)\left(1 + \gamma_{11}\frac{\kappa_1}{\epsilon_1} + ikL\frac{\tilde{\epsilon}_{13}}{\tilde{\epsilon}_{33}}\right) - H_2^{\mathrm{II}}(0) = 0. \tag{38}$$

Since $\tilde{\epsilon}_{13} = \tilde{\epsilon}_{31}$ when there is no damping in the transition layer, the condition for equating the determinant of the system of eqs. (38) to zero can be written in the following way:

$$\frac{\kappa_1}{\epsilon_1} + \frac{\kappa_2}{\epsilon_2} + k^2\mu + \gamma_{11}\frac{\kappa_1\kappa_2}{\epsilon_1\epsilon_2} + ikL\frac{\tilde{\epsilon}_{13}}{\tilde{\epsilon}_{33}}\left(\frac{\kappa_1}{\epsilon_1} + \frac{\kappa_2}{\epsilon_2}\right) = 0.$$

Thus, in the approximation being discussed, i.e., taking into account only terms linear with respect to L, the equation that determines the dispersion of surface polaritons in the presence of an anisotropic transition layer is of the form

$$\frac{\kappa_1}{\epsilon_1} + \frac{\kappa_2}{\epsilon_2} + k^2\mu + \gamma_{11}\frac{\kappa_1\kappa_2}{\epsilon_1\epsilon_2} = 0 \tag{39}$$

or

$$\frac{\epsilon_1}{\kappa_1} + \frac{\epsilon_2}{\kappa_2} = k^2\mu\frac{\kappa_1^2}{\epsilon_1^2} - \gamma_{11}, \tag{39a}$$

where (see also eqs. (14a) and (15))

$$\mu = \left(\frac{1}{\tilde{\epsilon}_{33}} - \frac{1}{\epsilon_1}\right)L, \qquad \gamma_{11} = \left(\tilde{\epsilon}_{11} - \epsilon_1 - \frac{\tilde{\epsilon}_{13}\tilde{\epsilon}_{31}}{\tilde{\epsilon}_{33}}\right)L. \tag{39b}$$

Equation (39) is a natural generalization of dispersion relation (22) for an isotropic transition layer and contains this relation as its limiting case.

In what follows we shall show (see also Agranovich et al. 1980a) how the application of eq. (25) enables us to examine the dependence of the dispersion of surface waves on the direction of their propagation. Here we note that a no less important effect of transition layer anisotropy is the occurrence in the surface polariton of the nonzero field component E_2. The existence of this field component enables, for example in applying the attenuated total reflection (ATR) method, surface polaritons to be excited not only in p-polarization, but in s-polarization as well. This, like the study of splitting anisotropy (see in the following), opens the way for investigation the orientation of the molecules of the transition layer.

We assume that the transition layer is uniaxial, i.e., that its dielectric constant in the system of principal axes has the form

$$\tilde{\epsilon}_{ij} \sim \delta_{ij}, \quad \text{with} \quad \tilde{\epsilon}_{11} = \tilde{\epsilon}_{22} = \tilde{\epsilon}^{\perp} \quad \text{and} \quad \tilde{\epsilon}_{33} = \tilde{\epsilon}^{\parallel}.$$

Next, let θ be the angle which the optical axis in the transition layer makes with a normal to the interface (i.e. with the z-axis), and φ be the angle which the projection of the optical axis on the plane $z = 0$ makes with the x-axis (it will be recalled that by assumption and without loss of generality, the two-dimensional wave vector of the surface polariton is directed precisely along this axis). In this case the components of tensor $\tilde{\epsilon}_{ij}$ contained in eq. (39a) are determined, in the coordinate system being used, by the relations

$$\begin{aligned}
\tilde{\epsilon}_{11} &= \tilde{\epsilon}^{\perp}(1 - \sin^2\theta\cos^2\varphi) + \tilde{\epsilon}^{\parallel}\sin^2\theta\cos^2\varphi, \\
\tilde{\epsilon}_{13} &= (\tilde{\epsilon}^{\parallel} - \tilde{\epsilon}^{\perp})\cos\varphi\sin\theta\cos\theta, \\
\tilde{\epsilon}_{33} &= \tilde{\epsilon}^{\perp}\sin^2\theta + \tilde{\epsilon}^{\parallel}\cos^2\theta.
\end{aligned} \tag{40}$$

Applying relations (40), we find that the right-hand side of eq. (39a) can be written in the following manner:

$$\begin{aligned}
\Phi \equiv \frac{L}{\tilde{\epsilon}_{33}}\Big\{k^2\frac{\epsilon_1^2}{\kappa_1^2} - \tilde{\epsilon}^{\perp}[\,\tilde{\epsilon}^{\perp}\sin^2\varphi\sin^2\theta \\
+ \tilde{\epsilon}^{\parallel}(1 - \sin^2\varphi\sin^2\theta)\,]\Big\} - \frac{Lk^2\kappa_1^2}{\epsilon_1^3} + L\epsilon_1.
\end{aligned} \tag{41}$$

It follows from this expression that at $\theta \neq \pi/2$, the effect of the transition layer can be especially substantial in the frequency region where either

$\tilde{\epsilon}_{33}(\omega) = 0$ or $\tilde{\epsilon}^{\perp}(\omega) = 0$. If, however, $\theta \approx \pi/2$ and $\tilde{\epsilon}_{33} \approx \tilde{\epsilon}^{\perp}$ (see eq. (40)), the effect of the layer is significant in the resonance region $\tilde{\epsilon}^{\|}(\omega)$ as well.

At $\theta \neq \pi/2$ and $\theta \neq 0$ in the resonance region $\tilde{\epsilon}^{\perp}(\omega)$, expression (41) is equal to the quantity $-L\tilde{\epsilon}^{\perp}(\omega)\sin^2\varphi$. This means that in this frequency region the size of the gap magnitude depends on the azimuth angle φ in the following manner:

$$\Delta = \Delta_0^{\perp}|\sin\varphi|,$$

where $\Delta_0^{\perp}$ is the gap magnitude that would exist for an isotropic transition layer having thickness L and dielectric constant $\tilde{\epsilon} \equiv \tilde{\epsilon}^{\perp}(\omega)$. At $\theta = 0$, expression (41) is simplified and the splitting anisotropy disappears. Splitting anisotropy in the frequency region where $\tilde{\epsilon}_{33}(\omega) = 0$ can be dealt with in an analogous manner. The condition $\tilde{\epsilon}_{33}(\omega) = 0$ determines the frequencies of the longitudinal Coulomb excitations, which, in the medium that the transition layer is formed of, propagate at the angle θ to the optical axis. Since the quantity $(k^2/\kappa_1^2)\epsilon_1^2$ is always greater than unity, it is the resonance with these oscillations that may prove to be most strongly pronounced, as is also the case for isotropic transition layers. At $\theta = 0$, these resonances correspond to frequencies ω for which $\tilde{\epsilon}^{\|}(\omega) = 0$, and at $\theta = \pi/2$, to frequencies for which $\tilde{\epsilon}^{\perp}(\omega) = 0$. It is of interest that at $\theta = \pi/2$ and $\tilde{\epsilon}^{\perp}(\omega) \approx 0$

$$\Phi = \frac{Lk^2}{\tilde{\epsilon}^{\perp}(\omega)} \frac{\epsilon_1^2}{\kappa_1^2},$$

and thus, as at $\theta = 0$, there is no splitting anisotropy.

An analysis of the more general situations is presented by Agranovich et al. (1980a). No experimental investigations have yet been conducted on the dispersion of surface waves in the presence of anisotropic transition layers.

3. *Effects of Thin Semiconductor Films on Surface Polariton Spectra*

In examining the effects on the SP spectra of semiconductor films in the spectral region of their exciton states, a number of interesting features are to be found at a film thickness $L \lesssim a_0$, where a_0 is the effective Bohr radius of an exciton in an unbounded semiconductor. Since large values of the static dielectric constant, $\epsilon \sim 10$ to 100, are typical of semiconductors, Coulomb interaction in them is greatly weakened, the hydrogen-like bound states of the electron and hole, i.e., Wannier–Mott excitons, have macroscopically large effective radii

$$a_0 = \frac{\epsilon\hbar^2}{me^2} \gtrsim 10^{-6}\ \text{cm},$$

where m is the effective mass of the exciton, and the inequality $L \leqslant a_0$ is complied with, even at $L \leqslant 100$ Å. In such thin films, the interaction between the electron and hole in the exciton depends essentially on the dielectric properties of the substrate because it is determined by the magnitude of the potential set up by the charge in the film in taking the image forces into account.

If the radius of the electron plus hole state $a_0 \gtrsim L$, the field, set up by these charges in the medium surrounding the film, begins to play an appreciable role. In the case when the substrate is metallic (this was discussed by Agranovich and Lozovik 1973, Agranovich et al. 1975, Lozovik and Nishanov 1978), this field is strongly screened by the electrons of the metal, while the interaction (attraction) between the electron and hole is substantially weakened. On the contrary, in the case when the static dielectric constant of the medium surrounding the film is small in comparison with the static dielectric constant of the film (this was discussed by Keldysh 1979a, b) the electron–hole interaction becomes stronger. Clearly, the effect of the image forces in the first case should increase the radius of the bound state of the electron plus hole in comparison with a_0, while in the second case it should reduce this radius. Naturally, in both cases mentioned above, the binding energy of the electron and hole (effective Rydberg) should diminish or increase respectively. This effect, however, is not the only possible one. Under the conditions being discussed (i.e. at $L \ll a_0$), due to the effect of a metallic substrate on semiconductor films with a narrow gap even a certain instability becomes possible with respect to the spontaneous creation of electron–hole pairs and the dielectric–semimetal transition. Without going into details of the theory (Agranovich and Lozovik 1973, Agranovich et al. 1975), we point out that in the creation of an electron and hole, located within the film at points $(\boldsymbol{\rho}, z)$ and $(0, z')$ ($\boldsymbol{\rho} = (x, y)$ being the coordinate in the plane of the film), the corresponding potential energy at $\rho \gg L$ (where L is the film thickness) is

$$V(\boldsymbol{\rho}, z; z') = \varphi(\boldsymbol{\rho}, z; z') + V_0(z) + V_0(z'),$$

where φ $(\boldsymbol{\rho}, z; z')$ is the electron–hole interaction (attraction) in the film screened by metallic coating, while $V_0(z)$ is the energy of attraction of the quasi-particle to its image. This energy is always negative and, for instance, at $z < L$, is determined by the well-known formula $V_0(z) = -e^2/4\epsilon z$. The formulas for φ and V_0 appear especially simple in the case when the thickness L of the semiconductor film is small in comparison with the width d of the vacuum (dielectric) clearance between the film and the metal. In this case and at $\rho \gg L$, the quantity V depends neither on z nor on z'. Thus

$$V(\rho) = \pm\varphi(\rho) - 2V_0(d),$$

where

$$\varphi(\rho) = -\frac{e^2}{\epsilon\rho} + \frac{e^2}{\epsilon(\rho^2 + 4d^2)^{1/2}},$$

$$2V_0(d) = \frac{e^2}{2\epsilon d}.$$

The upper sign in the equation for $V(\rho)$ refers to quasi-particles with opposite charges; the lower sign to those with the same charge. The second term in the equation for $\varphi(\rho)$ takes into account the interaction of a quasi-particle with the image of the other quasi-particle; the quantity $-2V_0 = -e^2/2\epsilon d$ is equal to the sign-independent energy of attraction of the quasi-particles to their own images. The latter is precisely what can lead to a cardinal rearrangement of the spectrum of a narrow-zone semiconductor near the metal. As a matter of fact, the minimum energy that must be expended for the creation of a particle and hole near the boundary is equal, not to E_G, the width of the forbidden zone in a massive specimen, but to $E_G^{eff} = E_G - 2V_0$. The interaction of quasi-particles that are not bound to one another is unimportant here because, on an average, they are at infinitely great distances from one another. At $2V_0 \geqslant E_G$, the ground state of a thin film of intrinsic semiconductor becomes unstable with respect to the formation of electron–hole pairs until the increasing kinetic energy of the quasi-particles compensates for the value of $2V_0 - E_G$. Then the stable state of the semiconductor corresponds to a semimetal, i.e., by the action of the electrostatic image forces a transition of the film of semiconductor into a semimetal takes place near the boundary. Image forces can initiate a Mott semiconductor–metal transition even in quite thick films of an extrinsic semiconductor near the boundary with the metal. This occurs due to the substantial weakening near the boundary of the attraction $\varphi(\rho)$ of an electron to an impurity (at $\rho \gg L$, $\varphi(\rho) \sim -2L^2e^2/\rho^3$). This last leads to a considerable reduction in the ionization energy of the impurity and to an increase in the radius a of the isolated impurity state (Agranovich and Lozovik 1973, Agranovich et al. 1975, Lozovik and Nishanov 1978). This facilitates the fulfilment of the Mott transition criterion $a \gtrsim d$, where d is the distance between the impurities. A similar weakening of the attraction between electrons and holes near the boundary results in an E_G-dependent reduction of the binding energy of the Wannier–Mott exciton, an increase in its radius and the disappearance of the excited states (Lozovik and Nishanov 1978). Though the feasibility of the above-discussed semiconductor–semimetal transition is sometimes taken into consideration in analyzing the properties of layer systems (see Dubovsky and Kagan 1977) such a transition could be observed, for instance, in investigating the damping of surface waves as a function of the distance from the film to the surface of the metal. In virtue of the above-mentioned, the gap in the semiconductor spectrum should decrease with this distance and this fact could be detected by the increased damping of surface waves in the infrared region of the spectrum.

Let us now discuss the effect of thin semiconductor films on the spectra of surface polaritons of the substrate in the previously mentioned cases when the static dielectric constants ϵ_1 and ϵ_2 of the media surrounding the film are small in comparison with the static dielectric constant ϵ of the film. To analyze this problem we can make use of the results of works (Keldysh 1979a, b) in which the exciton states in thin films were investigated and corrections to the boundary conditions (see sects. 1 and 2), due to the presence of a film, were found within the scope of the microscopic theory. Keldysh (1979a, b) showed that at $L < a_0 \equiv \epsilon \hbar^2(me^2)^{-1}$ (where e is the charge of the electron, and m is the reduced mass of the electron and hole) the binding energy $R(L)$ of an exciton in the film increases in comparison with that of a bulk specimen and becomes equal to

$$R(L) = \frac{e^2}{\epsilon d}\left\{\ln\left[\left(\frac{2\epsilon}{\epsilon_1 + \epsilon_2}\right)^2 \frac{L}{a_0} - 0.8\right]\right\}, \tag{42}$$

and the effective radius a of the exciton is reduced to

$$a = \tfrac{1}{2}\sqrt{a_0 L}. \tag{43}$$

In such a situation excitons are practically two-dimensional. For dipole-active excitons, the intensity of the optical transition is determined by the quantity $|\Phi(0)|^2$, where $\Phi(\boldsymbol{r})$ is a wave function of the relative motion of the electron and hole. Since the quantity $|\Phi(0)|^2$ is inversely proportional to the "volume" of the exciton, by virtue of the two-dimensional nature of the exciton, $|\Phi(0)|^2 \sim L^{-1}a^{-2} \sim a_0^{-1}L^{-2}$. Consequently, as the thickness of the film is reduced, along with an increase in the binding energy of the exciton, its corresponding oscillator strength* should also increase. On account of the aforesaid, the polarizability of the film does not depend upon its thickness according to a linear law but in a more complex manner.

It was shown (Keldysh 1979b) that the contribution of an exciton transition in the film to its polarization at point z (per unit area), produced by the electric field $\boldsymbol{E}$ with frequency ω and with the projection $\boldsymbol{k}$ of the wave vector on the plane of the film, is determined by the nonlocal relations $P(z)$ and $E(z)$:

$$P_i(k, \omega, z) = \Lambda_{ij}\chi(k, \omega)\bar{E}_j \cdot 2\sin^2(\pi z/L), \tag{44}$$

*In comparison with an exciton in thick crystals, the increase in the oscillator strength is $\sim(a_0^3/a^2L) \sim (a_0/L)^2$.

where

$$\chi(k,\omega)=\frac{1}{\epsilon}\left[\frac{e^2}{\mathrm{d}\mathscr{E}(k)}\right]^2 m|V_{cv}|^2\frac{2\mathscr{E}(k)}{\mathscr{E}^2(k)-(\hbar\omega)^2},$$

$$\bar{\boldsymbol{E}}=2d^{-1}\int_0^L \boldsymbol{E}_{k\omega}(z')\sin^2(\pi z'/L)\,\mathrm{d}z'.$$

In the relation for χ, the quantity V_{cv} is the matrix element of the electron velocity operator corresponding to the electron's transition from the valence band to the conduction band; function $(2/L)^{1/2}\sin(\pi z/L)$ describes the dimensionally quantized motion, transverse with respect to the film, of electrons and holes: and $\mathscr{E}(k)$ is resonance energy of the exciton transition. If the materials of the film and substrate are isotropic, tensor Λ_{ij} contains only two components: $\Lambda\equiv\Lambda_{zz}$ and $\Lambda_{\|}$. In the Kane model the quantity $\chi(k,\omega)$ is reduced to

$$\chi(k\omega)=\tfrac{2}{3}(e^2/L)^2[\mathscr{E}^2(k)-(\hbar\omega)^2]^{-1}. \tag{45}$$

The existence of nonlocal relation (44) requires a certain generalization of relations (6), (7), (9) and (11). In this connection, using eq. (44), we can express the induction $\boldsymbol{D}(k,\omega,z)$ in the form

$$D_i(k,\omega,z)=\epsilon E_i(k,\omega,z)+a_{ii}(z)\int_0^L b(z')E_i(z')\,\mathrm{d}z', \tag{46}$$

where

$$a_{ii}(z)=4\pi\Lambda_{ii}\chi(k,\omega)\cdot 2\sin^2(\pi z/L),\qquad i=1,2,3, \tag{46a}$$

$$b(z)=(2/d)\sin^2(\pi z/L). \tag{46b}$$

The existence of relation (46) leads to relations (6) and (11), in which

$$\tilde{\epsilon}=\tilde{\epsilon}_{\|}(k,\omega)=\epsilon+4\pi\Lambda_{\|}\chi(k,\omega). \tag{47}$$

To obtain a relation analogous to eq. (9), consideration must be given to the fact that, according to eq. (46),

$$D_3^{\mathrm{III}}(z)=\epsilon E_3^{\mathrm{III}}(z)+a_{\perp}(z)\int_0^L b(z')E_3^{\mathrm{III}}(z')\,\mathrm{d}z'.$$

Hence the quantity appearing in eq. (8a) is

$$\int_0^L E_3(z)\,\mathrm{d}z=\frac{LD_3^{\mathrm{III}}}{\epsilon}\frac{\epsilon+2\pi\Lambda_{\|}\chi}{\epsilon+6\pi\Lambda_{\|}\chi}.$$

Therefore, parameter μ for the model being discussed is determined by the relations

$$\mu=L\left(\frac{1}{\epsilon}\frac{\epsilon+2\pi\Lambda_{\perp}\chi}{\epsilon+6\pi\Lambda_{\perp}\chi}-\frac{1}{\epsilon_1}\right). \tag{47a}$$

A knowledge of expressions (47) and (47a) enables us to examine the effects of semiconductor films on surface wave spectra. In this connection we first carry out certain evaluations. Note, first of all, that for the frequency region in which $\omega > (1/\hbar)\mathscr{E}(k)$, yet $\omega < (1/\hbar)E_G$, where E_G is the width of the forbidden zone (this, precisely, is the frequency region corresponding to relation (46)), the quantity ϵ_∞ appearing in eq. (46) can be assumed approximately equal to ϵ. This is due to the fact that even in thin films of narrow-gap semiconductors, the contribution of exciton resonance to the static dielectric constant turns out to be small in comparison with the contribution of zone–zone $v \to c$ transitions.

Assuming, for instance, that $\epsilon = 30$ and $m = 0.1m_0$, where m_0 is the mass of the free electron, we find that $a_0 \approx 150$ Å. Hence, for this kind of semiconductor films of a thickness $L = 20$ Å $\ll a_0$ we find that in the region of exciton resonance at the indicated values of the parameters and at $\mathscr{E} \approx \frac{1}{2}$ eV,

$$4\pi\chi(k, \omega) \approx \frac{A\mathscr{E}(0)}{\mathscr{E}(k) - \hbar\omega}, \tag{48}$$

where

$$A = \frac{4\pi}{\epsilon}\left(\frac{e^2}{L}\right)^2 \cdot \frac{1}{\mathscr{E}^2(0)} \approx 1. \tag{48a}$$

Since the components of the dimensionless tensor $\Lambda_{ij} \sim 1$, the result of the evaluation enables us to arrive at the conclusion that in sufficiently thin films of narrow-gap semiconductors, the oscillator strength of the exciton transition and, correspondingly, the integrated absorption associated with this transition, are found to have an order of magnitude that is the same as for excitons in molecular crystals (for example, for the first transition in anthracene $A \approx 0.2$; see Agranovich and Ginzburg 1979, para. 7). Owing, however, to the relatively large values of $\epsilon_\infty \approx \epsilon(0)$, the frequency regions where resonances of quantities $\tilde{\epsilon}_\parallel$ or $\tilde{\epsilon}_\perp$ are manifested are found to be relatively narrower, though still sufficiently large to observe the effects of the influence of semiconductor films on surface polariton spectra. In fact, as is evident from eqs. (46) and (48), the difference in frequencies $\omega_2 - \omega_1$, where, for instance, the quantity $\tilde{\epsilon}_\parallel$ has the resonance ($\omega = \omega_1 = \mathscr{E}/\hbar$), or becomes zero ($\omega = \omega_2$) (we are dealing here, actually, with an analog of longitudinal–transverse splitting in massive specimens) is determined from the relation

$$\omega_2 - \omega_1 \approx (A/\epsilon\hbar)\mathscr{E}.$$

For the parameters chosen above ($\epsilon = 30$ and $\mathscr{E} = \frac{1}{2}$ eV), the difference

$$\omega_2 - \omega_1 \approx \tfrac{1}{60}\ \text{eV} \approx 150\ \text{cm}^{-1},$$

which substantially exceeds the width of the exciton lines, even at not very low temperatures.

If in one of the media adjacent to the semiconductor film (for example, medium I) the dielectric constant $\epsilon_1(\omega)$ in the frequency region $\omega \sim (\omega_1$ to $\omega_2)$ has a negative value (i.e., $\epsilon_1(\omega) < 0$; this condition is not, of course, inconsistent with the previous assumption that the static quantity $\epsilon_1(0)$ is positive and much less than ϵ) resonance may possibly be attained between a two-dimensional exciton in the film and a surface polariton at the boundary with medium I. Since the oscillator strength of the exciton transition being discussed is $\sim L^{-2}$ (see eqs. (48) and (48a)), the dependence of the gap magnitude Δ on the film thickness, which is due to splitting (see sect. 2), is determined for this case by the relation

$$\Delta \sim \frac{\mathscr{E}(0)}{\hbar}\left(\frac{AL}{\lambda_0}\right)^{1/2} \sim \frac{1}{\hbar}\frac{e^2}{\sqrt{(\lambda_0 L)}}\left(\frac{4\pi}{\epsilon}\right)^{1/2}, \tag{49}$$

where

$$\lambda_0 = 2\pi c\hbar/\mathscr{E}(0).$$

Thus, at $\mathscr{E}(0) \approx \frac{1}{2}\,\mathrm{eV}$, $L = 20$ Å and $\epsilon = 30$, the value of $\Delta \approx 100\,\mathrm{cm}^{-1}$. In the preceding expression for the magnitude of the gap Δ (at $\omega = \omega_1$ and $\tilde{\epsilon}_{\|}(\omega_1) = \infty$), we have not written out the factor of eq. (24). As has been mentioned previously, this factor leads to a result in which the magnitude of the gap in the resonance region $\tilde{\epsilon}_{\|}$ is found to be, under ordinary conditions (see sect. 2), somewhat smaller than the magnitude of the gap for the region of zeros $\tilde{\epsilon}_{\perp}$. In the given case of large ϵ values ($\epsilon \gg 1$), the situation changes since, if we take eqs. (47a) and (48) into account,

$$\mu = \left[\frac{1}{\epsilon} - \frac{A\Lambda_{\perp}\mathscr{E}(0)/\epsilon^2}{\mathscr{E}(k) - \hbar\omega + \frac{3}{2}A\Lambda_{\perp}\mathscr{E}(0)/\epsilon} - \frac{1}{\epsilon_1}\right]L.$$

Consequently, compared to eq. (49), the gap magnitude at $\omega = \omega_2$, where $\omega_2 = [\mathscr{E}(0)/\hbar](1 + 1.5A\Lambda_{\perp}/\epsilon)$, is reduced by a factor of $\epsilon/\sqrt{\Lambda_{\perp}}$. This makes the observation of the gap practically impossible in all cases when the parameters of the semiconductor differ only slightly from those specified above.

As to the observation of the gap in the spectrum of a polariton at ω_1, it can be feasible only in the temperature range in which the width of the SP lines is $\delta < \Delta$.

Finally, it should be noted that when spatial dispersion is taken into account in eq. (48), i.e., in using for the energy $\mathscr{E}(k)$ of the two-dimensional polariton, for example, the expression

$$\mathscr{E}(\boldsymbol{k}) = \mathscr{E}(0) + \hbar^2 k^2/2m_2,$$

where $\boldsymbol{k}$ is the two-dimensional wave vector and m_2 is the effective mass of the two-dimensional exciton in the film, the occurrence of resonance in eq. (47) at $\hbar\omega = \mathscr{E}(k)$ and, therefore, in the expression for γ (see eq. (6a); in the approximation being discussed, $\gamma \equiv \gamma(\omega, k)$) leads to the appearance of still another additional surface polariton. This conclusion follows from eq. (22a) which, for frequencies $\omega < \mathscr{E}(0)/\hbar$, leads to three values of the wave vector of SPs. The occurrence of two additional SPs is due, as usual (see Agranovich and Ginzburg 1979), to the distinctive features of wave dispersion in the nonrelativistic limit. In fact, in this limit and if only the terms of eq. (22a) that are linear with respect to L are taken into account, it follows that the dispersion of SPs in the case being discussed and at $\omega \approx \mathscr{E}(0)/\hbar$ is of the form

$$\omega(k) = \mathscr{E}(0)/\hbar - AkL + \hbar k^2/2m_2, \quad A > 0, \tag{50}$$

It is evident from this equation that the relation $\omega(k)$ has a minimum at $k_0 = ALm_2/\hbar$, leads to two additional SPs in the frequency region $\omega \gtrsim \omega(k_0)$, and to a single additional SP for frequencies in the region of the gap or higher.

No experimental investigation of the distinctive features of surface waves appearing due to the effects of thin semiconductor films has yet been carried out. In this connection, it should be pointed out that the results of such research could be utilized not only to check the theory developed by Keldysh (1979a, b), but also for an experimental study of possible electron phase transitions in semiconductor films occurring due to the effects of doping and external actions, or at high levels of excitation. In the last case, it should also prove of interest to observe polariton luminescene with reflecting prisms and without them.

A few remarks should be made in conclusion concerning the theory of surface polaritons in crystals with a nonhomogeneous transition layer. The case of a nonhomogeneous transition layer discussed in this section is instructive, in this sense, because relation (46) was used in analyzing surface waves for the region inside the layer. This relation (46) corresponds to a quite general nonlocal relationship between the induction $\boldsymbol{D}$ and the electric field strength $\boldsymbol{E}$. We know of no other work on SP theory in which the relation between quantities $\boldsymbol{D}$ and $\boldsymbol{E}$ in the transition layer was also assumed to be nonlocal. Within the scope of the local relation (1), the distinctive features of SP spectra were investigated in a number of works (see Cunningham et al. 1974, Guidotti et al. 1974, Conwell 1975, Conwell and Kao 1976, Kao and Conwell 1976, as well as Blank and Berezinsky 1978). The work of Cunningham et al. (1974) was stimulated by experimental research of SP dispersion in crystals n-InSb (Marschall et al. 1971, Hartstein and Burstein 1974, Anderson et al. 1971, Marschall et al. 1973). It was shown in these investigations that SP dispersion essentially depends on the dimensions of the region occupied by the spatial surface charge. In this connection, when the SP dispersion law was derived by Cunningham et al. (1974), the dependence of the dielectric constant $\epsilon(z)$ on z in the transition layer was presented as a piecewise-continuous function with a linear dependence of ϵ on z within each of several layers into which the transition region was divided. At the same time, in the work of Guidotti et al. (1974), the relation $\epsilon(\omega, z)$ was assumed to be of the form

$$\epsilon(\omega, z) = \epsilon_b(\omega) + \Delta\epsilon(\omega)\, e^{-z/L}, \qquad z > 0, \tag{51}$$

where L is a parameter of an order of magnitude equal to the thickness of the region occupied by the spatial surface charge. Guidotti et al. (1974) were interested in the effects of SPs, propagated along the machined surfaces of metals, on the dispersion law. These effects lead to the formation of a transition layer either depleted of or enriched with charge carriers. The existence of such effects was indicated by experiments performed by this group.

In all the works mentioned above, it was assumed that the appearance of a z-dependent dielectric constant in eq. (51) is due to the dependence of the plasma frequency on z:

$$\omega_p^2(z) = 4\pi N(z) e^2/\epsilon_\infty m^*(z),$$

where $N(z)$ is the carrier concentration, and e and $m^*(z)$ are their charge and effective mass. As to the dielectric constant in the region where $\tau \gg 1/\omega$, it was supposed to be of the form (see Conwell 1975, Conwell and Kao 1976, Kao and Conwell 1976):

$$\epsilon(\omega, z) = \epsilon_\infty\left[1 - \frac{\omega_p^2(z)}{\omega^2}\right] + i\epsilon_\infty \frac{\omega_p^2(z)}{\omega^2}\frac{1}{\omega\tau}. \tag{52}$$

The results of the research conducted by Cunningham et al. (1974), Guidotti et al. (1974), Conwell (1975), Conwell and Kao (1976), Kao and Conwell (1976) are essentially based on relation (51). They are cumbersome and we shall not repeat them here. We only emphasize their most important qualitative result which, incidentally, is directly concerned with the contents of the preceding sections. Thus, it was shown in the cited works (Guidotti et al. 1974) that in taking into account the carrier-depleted or -enriched transition layer, a SP spectrum of a complex nature may appear, in particular one with an additional (new) branch. Through the conclusion of the appearance of a new branch or new branches in SP spectra was reached in these works (Cunningham et al. 1974, Guidotti et al. 1974, Conwell 1975, Conwell and Kao 1976, and Kao and Conwell 1976) on the basis of very special models of transition layer structure, it seems to be understandable in view of the investigations to which we are led when taking transition layers into account and which were discussed in sect. 1. Thus, for example, if a layer with a lowered charge-carrier concentration were present at the surface, the plasma frequency within the layer would be small compared to the plasma frequency in the bulk. In this case, the oscillations in the transition layer would have to resonate with the "ideal" SP and change its spectrum, leading to the formation of a gap and a new branch. The nonhomogeneity of the transition layer is, however, roughly speaking, equivalent to the presence in it of a continuous spectrum of plasmons. Thus their resonance with the SP may, in general, lead to more complex patterns than splitting in their spectra. This is indicated by numerical computations (Cunningham et al. 1974, Guidotti et al. 1974, Conwell 1975, Conwell and Kao 1976, Kao and Conwell 1976).

If the surface layer is enriched by carriers, there is no resonance with SPs, but an additional wave may appear (see also Agranovich 1975) in the region of bulk plasma frequency owing to the linear term in the dispersion law at $k \gg \omega_p/c$:

$$\omega_s(k) = \frac{\omega_p(\infty)}{\sqrt{2}}[1 + A|\boldsymbol{k}| + B|\boldsymbol{k}|^2].$$

For the model (52) but at $\tau = \infty$, Blank and Berezinsky (1978) showed that $A = A' + \mathrm{i}A''$, and

$$A' = \frac{\omega_p(\infty)}{4\sqrt{2}}\left\{\int_{-\infty}^{\infty}\left[\frac{1}{\epsilon(z)} - \epsilon(z)\right]\mathrm{d}z - \frac{4\pi a|r|\sin\varphi_0}{R_0}\right\},$$

$$A'' = \frac{\pi a \omega_p(\infty)}{2R_0\sqrt{2}}(1 - |r|^2),$$

where $R = r\exp \mathrm{i}\varphi_0$ is the reflection factor of the plasmons from the surface; $R_0 = (1 + |r|\cos\varphi_0)^2 + |r|^2\sin^2\varphi_0$, and a is a quantity having the dimensionality of length and the order of magnitude equal to the thickness of the transition layer (see Blank and Berezinsky 1978). Thus, the imaginary addition to the plasmon spectrum was found by Blank and Berezinsky (1978) to be related to the factor of plasmon reflection from the boundary. Taking electron scattering into consideration ($\tau \neq \infty$), as has been noted above, leads to additional damping of SPs and both of these effects should be kept in mind in interpreting the results of experiments with thin films (see sect. 1).

4. *Surface Wave Diffraction by an Impedance Step when an Additional Surface Polariton is Taken into Account*

In sect. 2.3, when discussing the distinctive features of SP dispersion in the region of resonance with the oscillations of the transition layer, it has been underlined that taking an additional SP (or additional surface polaritons, see sect. 3) into account may prove essential in investigating various kinds of optical processes accompanied by excitation of surface waves. Evidently, however, the role of additional waves should be most distinctly manifested in studying SP diffraction. If the frequency of the radiation that excites the diffracting waves is within the existence region of several types of SPs, then all of these SPs should be taken into consideration. In this sense, the situation that arises here is analogous, so to speak, to the one in bulk crystal optics (see Agranovich and Ginzburg 1979) for the region of exciton resonances, though its analysis requires the application of much more complex mathematical means – those of diffraction theory.

To clear up this matter we shall discuss in this section a relatively simple case of SP diffraction by an impedance step. The pertinent theory, neglecting additional waves, is discussed, for example, by Vainshtein (1966).

Thus, we assume that along its plane $z = 0$ medium I (vacuum with $\epsilon_1 = 1$) borders on medium II (dielectric constant $\epsilon_2 \equiv \epsilon(\omega) < 0$). In this case, eq. (23), determining SP dispersion, takes the form

$$\kappa_1 + \kappa_2/\epsilon = 0,$$

where $\kappa_1 = \sqrt{(k^2 - k_0^2)}$, $\kappa_2 = \sqrt{(k^2 - k_0^2\epsilon)}$, $k_0 = \omega/c$ and, for the values of $k \ll k_0|\epsilon|^{1/2}$, can be written in the so-called impedance approximation

$$\kappa_1 - k_0/\sqrt{|\epsilon|} = 0. \tag{53}$$

The impedance approximation used below is valid if $|\epsilon| \gg 1$. In solving electrodynamic problems with the impedance approximation, there is no need to determine the structure of the field in medium II, which has larger, in absolute value, dielectric constant. Appearing in the theory, instead, is the boundary condition at $z = 0$:

$$\boldsymbol{E}_t(\mathrm{II}) = \zeta[\boldsymbol{n}\boldsymbol{H}(\mathrm{II})], \tag{54}$$

where ζ is the impedance of the surface, and $\boldsymbol{n}$ is the vector of the normal to this surface, directed along the z-axis. In the case being discussed $\zeta = -\mathrm{i}/\sqrt{|\epsilon|}$. Hence, it follows from eq. (54) that at $z = 0$,

$$E_x(\mathrm{II}) = (\mathrm{i}/\sqrt{|\epsilon|})H_y(\mathrm{II}). \tag{55}$$

Applying this boundary condition, as well as boundary conditions (21a) and (21b) with $\gamma = \mu = 0$, and taking into account the relation

$$E_x(\mathrm{I}) = \frac{1}{\mathrm{i}k_0}\frac{\partial H_y(\mathrm{I})}{\partial z} = -\frac{\kappa_1}{\mathrm{i}k_0}H_y(\mathrm{I}),$$

following from Maxwell's equations, we arrive precisely at approximation (53).

In the following we shall concern ourselves with the region of resonance $\mu(\omega)$ when we take the transition layer and additional wave into consideration. In this case quantity γ, appearing in eq. (21a) can be put equal to zero. Hence SP dispersion in the impedance approximation is determined by the relation (see eq. (22b):

$$\kappa_1 - k_0/\sqrt{|\epsilon|} + k^2\mu = 0. \tag{53a}$$

For the region of frequencies where $\mu < 0$, this equation leads to an additional SP.

It should be pointed out that in the presence of a transition layer with $\mu \neq 0$ and $\gamma = 0$, boundary conditions (21a), (21b) and (55) can be reduced to the following boundary condition for the function $H_y^{\mathrm{I}}(x, z)$ at $z = 0$:

$$\frac{\partial}{\partial z} H_y^{\mathrm{I}} + \frac{k_0}{\sqrt{|\epsilon|}} H_y^{\mathrm{I}} + \frac{\partial}{\partial x} \mu \frac{\partial}{\partial x} H_y^{\mathrm{I}} = 0. \tag{56}$$

We shall require this boundary condition in the following to find the fields set upon SP diffraction. If in eq. (56) we substitute the function $H_y^{\mathrm{I}}(x, z)$ in the form of eqs. (2) and (17), we obtain dispersion relation (53a), which, of course, is in no way a surprise.

Let us assume now that a surface wave $H_0(r) = H_0 \exp(ikx - \kappa_0 z)$ is incident on the impedance step from the metal side free from the film. For this wave, $\kappa_0 = k_0/\sqrt{|\epsilon|}$, and $k = (\kappa_0^2 + k_0^2)^{1/2}$. In accordance with the general method (see Vainshtein 1966) we shall seek the field $H_y^{\mathrm{I}}(x, z) = H(r)$ in the region $z > 0$ in the following form:

$$H(r) = H_0(r) + \int_{-\infty}^{+\infty} \frac{\mathrm{d}w}{2\pi} \exp(-iwx - vz) \frac{F(w)}{v - \kappa_0}, \tag{57}$$

where $v = \sqrt{w^2 - k_0^2}$, $F(w)$ is an unknown function. We can readily see that the field (57) satisfies the Maxwell equations. To determine $F(w)$ the field (57) must comply with the boundary condition (56).

Substituting eq. (57) into eq. (56) and making the inverse Fourier transformation, we obtain the following integral equation for the function $\Psi(w) = -(k/w)F(w)$,

$$\Psi(w) + \int_{-\infty}^{+\infty} \frac{\mathrm{d}w'}{2\pi} \Psi(w') \frac{(w')^2 \mu(w - w')}{v' - \kappa_0} = -H_0 k^2 \mu(w + k). \tag{58}$$

Assuming that $\mu(x)$ is a step (impedance step, see fig. 3)

$$\mu(x) = u\theta(x),$$

we rewrite eq. (58) in the form

$$\Psi(w)+\frac{i\mu k^2 H_0}{w+k}=i\mu\int_{-\infty}^{+\infty}\frac{dw'}{2\pi}\Psi(w')\frac{(w')^2}{(v'-\kappa_0)(w'-w-i\delta)}. \quad (59)$$

It is easy to check that the function

$$\Psi(w)=-\frac{i\mu k^2 H_0}{\Phi_-(-k)\Phi_+(w)(w+k)} \quad (60)$$

is the solution of eq. (59) if the representation of the function $\Phi(w)=(v-\kappa_0+uw^2)/(v-\kappa_0)$ is known in the form of the product

$$\Phi(w)=\Phi_+(w)\Phi_-(w),$$

where the functions $\Phi_+(w)$ and $\Phi_-(w)$ are holomorphic and have no zeros, respectively, in the upper and the lower half-planes of the complex variable w.

Thus the required field $H(r)$ is completely determined:

$$H(r)=H_0(r)-\frac{i\mu k H_0}{\Phi_-(-k)}\int_{-\infty}^{+\infty}\frac{dw}{2\pi}\exp(-iwx-vz),$$

$$\times\frac{w}{(w+k)(v-\kappa_0)\Phi_+(w)}. \quad (62)$$

We can readily see that the whole field (62) is determined by superposition of surface and bulk waves.

In particular, at $x<0$ (it is the free surface of the metal) the pole contribution to the integral (62) describes the reflected surface wave. For the region of space $x>0$ there are three pole contributions: one of them quenches an incident surface wave and two other describe two transmitted surface waves.

The integral along the cuts gives the expression for the bulk radiation sliding along the surface and it decreases as $|x|^{-3/2}\exp(ik|x|)$.

If z is very large, we can readily obtain an asymptotic expression for the field of the bulk (cylindrical) waves.

It should be pointed out that for an additional wave (see fig. 2) the group and phase velocities for the same value of k have opposite directions. In this wave, according to the aforesaid, the energy flux in vacuum is directed opposite to that in normal waves. It is just because of this that the main part of the energy flux in the additional wave is concentrated in the film, and the directions of the energy fluxes in the film and vacuum are opposite at $\epsilon_1<0$.

The method of finding functions $\Phi_+(w)$ and $\Phi_-(w)$ is given by Vainshtein

(1966). For the case being discussed their expressions are given by Agranovich et al. (1981a). We shall not go into the details of these calculations, but shall only give certain consequences that follow from the results obtained. At first we give the expressions for the energy flux of all transformed waves. The energy flux in the reflected surface wave is equal to

$$\frac{W}{W_0}=\frac{\kappa_0^2k^2(k+k_2)^2}{(k+k_1)^2(k_2-k)^4}. \tag{63}$$

There, k_1 and k_2 are the wave vectors of a normal and an additional surface wave (see eq. (53a)). The energy flux in the normal wave is,

$$\frac{W_1}{W_0}=\frac{4\kappa_0\kappa_1k_1^2k^2(k_2+k)(k_2-k_1)}{(k_1^2-k^2)^2(k_2-k)^2(k_2+k_1)^2}, \tag{64}$$

and in the additional wave,

$$\frac{W_2}{W_0}=\frac{4\kappa_0k^2(k_2-k_1)}{\kappa_2(k_2-k)^2(k_1+k)}. \tag{65}$$

The expression for the angular distribution of the intensity of the bulk emission is:

$$\frac{W^{(b)}}{W_0}=\frac{2\kappa_0k^2(k_2+k)}{(k_1+k)(k_2-k)^2}\,\frac{k_x^2k_z^2(k_2-k_x)}{(k_1-k_x)(k+k_x)(k_2+k_x)^2(k-k_x)^2}, \tag{66}$$

where $k_x=k_0\sin\theta$, $k_z=k_0\cos\theta$.

There we assumed that there is no damping either in the metal or in the film.

It should be pointed out that frequency dependence of the intensity of the bulk radiation is very strong near the frequency of the oscillation in the film. Since in this case the reflected and transmitted waves are very weak (at $|\epsilon|\gg 1$) the main part of the transformed radiation is of the bulk-wave type, and the wave in the very narrow spectral region will be present in the radiation spectrum. Namely, the width of band radiation is equal to the true width of a gap in the spectrum of a surface polariton $g\sim\pi\omega_0(d/\lambda)|\epsilon|^{1/2}$.

Very interesting phenomena appear also when the bulk wave is incident on the impedance step. This case was considered by Agranovich et al. (1981b). A narrow dip was shown to appear in the region of frequency vibrations in a film in the spectra of mirror reflection.

This can evidently be effectively utilized to study the resonances of the dielectric constant of thin films by the surface-wave diffraction.

5. Surface Polaritons in Exciton Luminescence Spectra

Surface Polariton "Radiation" Width and Transition Layer Effects in Molecular Crystals and Semiconductors

On an ideal plane surface and with fluctuations of the dielectric constant being absent or not taken into consideration, SPs are not capable of being transformed into bulk radiation. Valid in this sense is the statement that the radiation width of SPs equals zero. As a matter of fact, however, there is always roughness on the surfaces of a crystal and the dielectric constant fluctuates due to the influence of thermal motion. That is why SPs are capable of undergoing scattering. When a bulk polariton propagates in an unbounded medium, only a bulk polariton can be formed in the final state after scattering. Scattering in the case of SPs can proceed along two channels. In the first of these the SP retains its surface wave nature after scattering. In the second channel the SP is transformed, after scattering, into a bulk light wave which propagates in the one of the contacting media having a positive dielectric constant. The possibility of the transformation of an SP into a bulk wave when one or another kind of perturbing factors (roughness, phonons, etc.) are taken into account signifies that, in general, SPs are capable of contributing to the spectra of exciton fluorescence. Later on we shall return to a discussion of the features of exciton fluorescence spectra which can be related to the SP scattering processes. For the present, we shall turn to a rather old problem concerning the radiation width of Coulomb excitons in two-dimensional systems (Agranovich and Dubovsky 1966, Agranovich 1968; see also Philpott and Sherman 1975 and Orrit et al. 1980) and trace its evolution in going over to polaritons within the scope of an approach that consistently takes retardation interaction into account.

It is known that in three-dimensional boundless crystals, a consistent account of the retardation interaction between charge (Hopfield 1958, Agranovich 1959) leads, for the exciton region of the spectrum, to the formation of polaritons, quasi-particles of a new type. These quasi-particles, photons in a medium, become attenuated only as a result of exciton–phonon interaction, defects, etc. But the radiation width of the polaritons is equal to zero in a three-dimensional crystal. Hence, the situation in this case essentially differs from that characteristic of the excited states of molecules in a gas or solution.

For excitons in one- and two-dimensional crystals, however, the radiation width is, in a certain sense, restored (Agranovich and Dubovsky 1966, Agranovich 1968). In this case, as in three-dimensional crystals, the influence of the retardation interaction also leads to renormalization of the energy of the quasi-particles, resulting in the formation of several branches of their spectrum. For some of them (i.e. for the so-called nonradiational states) decay into photons is found to be forbidden; for others the radiation width is nonzero and gigantic: many orders of magnitude greater than the radiation width of a molecule in vacuum. We shall describe in more detail the situation for two-dimensional excitons, i.e., ones whose state is characterized by a two-dimensional wave vector. In this sense, two-dimensional excitons are excitons in two-dimensional crystals, as well as surface excitons in three-dimensional crystals.

We shall assume that the dispersion law is known for the two-dimensional exciton $E(\boldsymbol{k})$, where $\boldsymbol{k}$ is the two-dimensional wave vector found when only Coulomb interaction between charges is fully taken into account. As shown by Agranovich and Dubovsky (1966) and Agranovich (1968), taking the retarded nature of the interaction into consideration leads to the mixing of two-dimensional excitons and transverse photons, thereby changing their spectrum. The dispersion law for the Coulomb exciton is shown by dash lines in fig. 4; that for polaritons is shown by a solid line. The lower branch of the spectrum for $\mathscr{E}_1(k)$ is of zero radiation width: $\Gamma_1 = 0$. The upper branch corresponds to two-dimensional excitations having a large radiation width Γ_2. This radiation width increases drastically with k, leading to the end point of the spectrum at a certain value of $k = k^*$. At $k > k^*$ the radiation width $\Gamma_2(k)$ is not small in comparison with the excitation energy $\mathscr{E}_2(k)$ and it is meaningless in this region of the spectrum to employ the concept of elementary excitations of the type being considered. But if $k \ll k^*$ the lifetime of a two-dimensional excitation with respect to the process of its transformation into bulk photons $T = 2\pi\hbar/\Gamma_2(k)$ is found to be much greater than the period $2\pi\hbar/\mathscr{E}_2(k)$. Under these conditions, i.e., at $\Gamma_2(k) \ll \mathscr{E}_2(\boldsymbol{k})$, a two-dimensional excitation is found to be sufficiently stable. Let us point out, however, as shown by Agranovich and Dubovsky (1966) and Agranovich (1968), that the value of Γ_2 turns out to be of the order of $\Gamma_2 \approx \Gamma_0(\lambda_0/2\pi a)^2$, where Γ_0 is the radiation width in an isolated molecule, $\lambda_0 = 2\pi\hbar c/\mathscr{E}_2$ and a is the lattice constant. Since $\lambda_0/a \approx 10^3$ for the visible spectral range, the quantity $\Gamma_2 \approx 10^4\Gamma_0$, and at $\Gamma_0 \approx 10^{-3}\ \mathrm{cm}^{-1}$, the radiation width $\Gamma_2 \approx 10\ \mathrm{cm}^{-1}$. The quantity Γ_0 is related in a known manner with the transition oscillator strength and, for instance, in the case of semiconductors with Wannier-Mott excitons, Γ_0 can be much smaller than the value indicated above. In particular, at $\Gamma_0 \approx 10^{-4}$ to

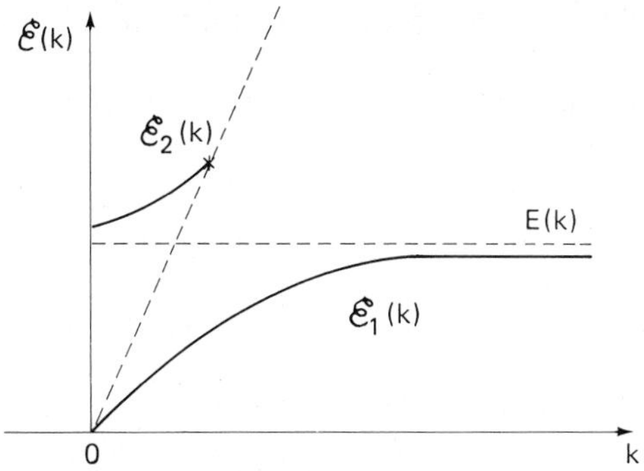

Fig. 4. Radiation and nonradiation branches of SPs.

10^{-5} cm^{-1}, the radiation width Γ_2 of the surface exciton of the upper branch may be of the order of ≈ 1 to 0.1 cm^{-1}.

For the sake of definiteness, we shall, in the following, concern ourselves with the states that arise in the hybridization of surface Coulomb excitons and transverse vacuum photons. Here the nonradiation states (the lower branch of $\mathscr{E}_1(k)$) are, in essence, the same SPs that were discussed above within the scope of macroscopic electrodynamics in utilizing one or another model expression for the dielectric constant. But the states of the upper branch are usually not examined within the scope of the macroscopic approach, because these states do not correspond to any normal solutions of Maxwell's equations. These states are decaying even when the processes of energy dissipation of the electromagnetic field are not taken into account. In a certain sense the states are analogous to the quasi-local or quasi-surface states in the dynamic theory of crystal lattices. The existence of the radiation width indicated above means that the state of the surface exciton of the upper branch with $k \ll k^*$, created at some instant of time, subsequently decays, changing into bulk photons with the same energy and with the value of the tangentional projection of the wave vector being equal to $\boldsymbol{k}$.

In connection with the aforesaid, it becomes clear that in spectra of exciton luminescence, the states being discussed can make a contribution in the energy region $\hbar\omega \approx \mathscr{E}_2(0)$ of an extent of the order of $\Gamma_2 \approx \Gamma_0(\lambda_0/2\pi a)^2$. In cases when Γ_2 is relatively small, radiation decay of the SP states of the upper branch can lead, in general, even to the appearance of more or less distinct maxima that are to be called type-A maxima in the following.

At the same time, in the reflection spectra, the SP states of the upper branch can appear only (see fig. 7a) as an effect of the exciton–phonon interaction or of the scattering of SPs by irregularities (roughness). Otherwise, light energy is first transformed into the energy of the upper branch SPs and subsequently, after the radiation lifetime T, is transformed into the energy of mirror-reflected light with the same frequency values and tangential component of the wave vector. Hence, "true" absorption of light does not occur in a crystal. Such delay (during the time T) of the appearance of a photon in the energy region $E = \hbar\omega \approx \mathscr{E}_2(0)$, reflected from the surface, could probably be observed experimentally in investigating the reflection of sufficiently short light pulses with a duration of $t_p \lesssim T$. Since pulses of a length as short as $t_p \approx 10^{-12}$ s can be feasibly produced today, we have in mind, primarily, experiments for the region of exciton transitions in semiconductors where, as mentioned above, the quantity Γ_2 can be sufficiently small.

Let us turn our attention (see also Agranovich and Leskova 1979) to a discussion of the features of exciton luminescence spectra which may be related to the lower branch SP. As has been pointed out, the transformation of these SPs into photons of fluorescence is impossible on an ideal plane boundary and if scattering processes are ignored. But when roughness of the boundary and scattering by phonons are taken into account, this forbiddenness is violated, leading to the appearance of the radiation width of SPs.

At not particularly low temperatures, both excitons and SPs with the energy $E \gtrsim \hbar\omega_\perp$ can be considered to comply with the Boltzmann distribution $C(E) = \rho(E)\exp(-E/T)$. At the boundary with vacuum and with the transition layer not taken into account, the dispersion law for lower branch SPs (see fig. 5) is determined by the relation

$$k^2 = \frac{\omega^2}{c^2}\frac{\epsilon(\omega)}{\epsilon(\omega)+1}, \tag{67}$$

where for the region of exciton resonance $\epsilon(\omega) = \epsilon_\infty + s\omega_\perp^2/(\omega_\perp^2 - \omega^2)$, $s = \epsilon_0 - \epsilon_\infty$. It follows from eq. (67) that the density of the SP states per unit area is

$$\rho(E) = \frac{\pi}{\hbar}\frac{(\omega_s^2 - \omega^2)^2\epsilon_\infty(1+\epsilon_\infty) + s\omega_s^2\omega_\perp^2}{(\epsilon_\infty + 1)^2(\omega_s^2 - \omega^2)^2},$$

and has a sharp maximum at the frequency $\omega = \omega_s$, $\epsilon(\omega_s) = -1$. SPs in the region $\omega \approx \omega_s$ have a large value of k (see eq. (67)) and are practically elementary Coulomb excitations because the contribution of the transverse field to their energy is small. Precisely for this region SPs in the spectrum

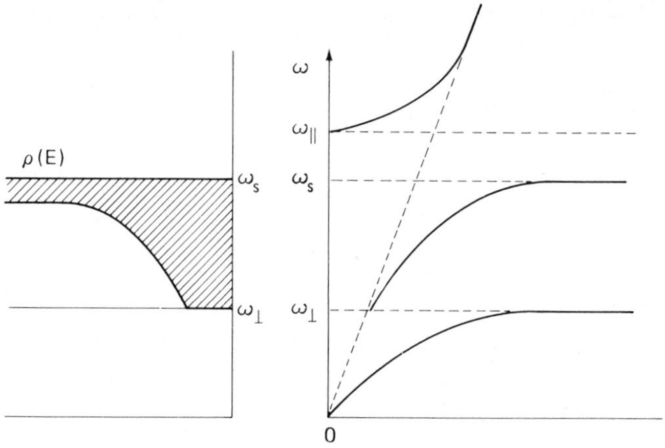

Fig. 5. SP dispersion law and density of states in a dielectric.

region being discussed (i.e. at $\omega \approx \omega_s$) interact comparatively strongly with phonons. This may result, in the exciton fluorescence spectra, in the occurrence of a maximum at the frequency ω_s in the region of longitudinal-transverse splitting (called a B-type maximum in the following), as well as in oscillatory iteration at the frequencies $\omega_s - \Omega$ and $\omega_s - 2\Omega, \ldots$, where Ω is the frequency of the optical (high-frequency) phonon.

Let us next consider SPs with the frequencies $\omega \gtrsim \omega_\perp$. Here the lower branch states have a small value of $k \approx \omega/c$, weakly interact with phonons, but are scattered by the roughness on the boundary. For this spectrum region $|\epsilon| \gg 1$, and scattering by roughness, as shown by Mills (1975), is generally accompanied by breaking away of the SPs and their transformation into bulk radiation photons. Thus, SP scattering by irregularities (roughness) may lead to the appearance of fluorescence at frequencies $\omega \gtrsim \omega_\perp$ (C-type emission band).

With a transition layer, whose frequency of electric oscillations falls within the region of SP frequency change, the structure of the fluorescence spectrum becomes more complex in the SP region.

To explain what has been said and for definiteness let us consider a case of molecular crystals of the anthracene type. In such crystals the level of molecule excitation in the bulk is lowered in response to Van der Waals interaction with the surroundings. But the molecules located in the surface monolayer are deprived of part of their nearest neighbors. Hence, the excitation energy of these molecules differs from the corresponding energy of molecules in the bulk and can, in certain cases, fall within the region of SP energy change. In this case a transition monolayer is formed in which the frequency ω_l of oscillation resonates with that of SP. Such a situation has already been discussed. As was shown in sect. 2, in this case a gap is formed in the spectrum of nonradiation SPs in the resonance region. Since the spectrum in the frequency region $\omega \approx \omega_l$ being discussed can be assumed to be formed as the result of the interaction of the two-dimensional exciton in the monolayer with the field of transverse photons, it is necessary to take also into account, along with the above-mentioned spectrum of nonradiation SPs, SPs having a radiation width and a spectrum end point (see fig. 6). Therefore, in the region $\omega \approx \omega_l$ of fluorescence spectrum of molecular crystals, A-type bands may, in general, appear (due to radiation SP decay where exciton-phonon interaction is unimportant), and B-type bands as well (due to nonradiation SP decay as a result of SP scattering by phonons).

The frequency of oscillation in the next monolayer $\omega = \omega_2$ may also fall within the region of frequency of SPs. In this case, the features of the fluorescence spectrum mentioned above can also exist in the frequency region $\omega \approx \omega_2$, etc.

In connection with the aforesaid, we point out that in the fluorescence spectra of anthracene at low temperatures (see, for instance, Turlet and Philpott 1975a, b, 1976, Syassen and Philpott 1978), as well as the reflection spectra of this crystal (Glockner and Wolf 1943, Turlet et al. 1978) a number of distinctive features are observed that can be associated with the excited states of the surface region in the crystal. The width of the corresponding spectral features in both types of spectra is of a magnitude of the order of several reciprocal centimeters. The origin of these features

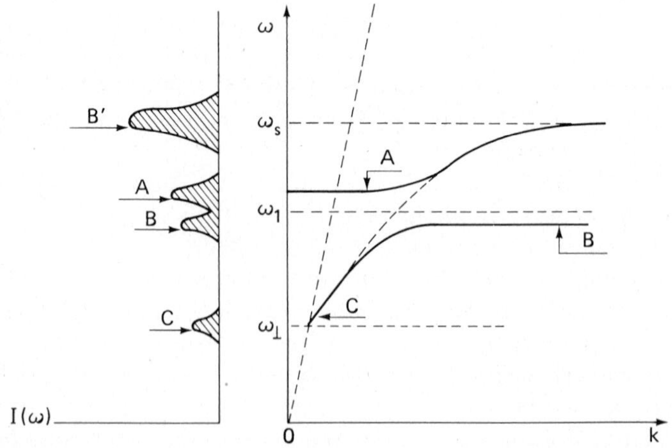

Fig. 6. SP dispersion law upon resonance with vibrations in the transition layer, and peculiarities of fluorescence spectra.

is still not quite clear and subsequent analysis should, in particular, answer the question: are the observed distinctive features associated with A-, B- or C-type fluorescence (see above)?

In a molecule of anthracene the radiation width is $\Gamma_0 \approx 10^{-3}\,\mathrm{cm}^{-1}$ and, consequently, $\Gamma_2 \approx \Gamma_0(\lambda_0/2\pi a)^2 \approx 10\,\mathrm{cm}^{-1}$. In its order of magnitude, such a width Γ_2 is in agreement with the widths of the observed lines of fluorescence which are associated (Glockner and Wolf 1943, Sugakov 1970, Turlet et al. 1978), in fact, with A-type luminescence, i.e., with luminescence originating upon radiation decay of the upper branch SP states.

It has been pointed out above that these same states, but only when scattering by phonons is taken into account, should be displayed in reflection spectra as well. This fact can be understood, not only on the basis of the quantum-mechanical reasoning used above, but within the scope of ordinary classical crystal optics as well, if only we bear in mind the presence of a transition layer on the surface of the crystal. Such a transition layer, formed of "spoiled" molecules, leads to the occurrence of surface currents and charges. If $\sigma(\omega)$ is the conductivity in the plane of the transition layer, the surface current is $\boldsymbol{j} = \sigma \boldsymbol{E}_t$. Hence, the following relations can be utilized as boundary conditions for fields at normal incidence:

$$\boldsymbol{H}_t^{(1)} - \boldsymbol{H}_t^{(2)} = (4\pi/c)\sigma \boldsymbol{E}_t^{(1)},$$

$$\boldsymbol{E}_t^{(1)} - \boldsymbol{E}_t^{(2)} = 0.$$

Making use of these relations, it can be shown that the reflection factor of light normally incident on a crystal from vacuum is

$$R = \left| \frac{1 - n_v - (4\pi/c)\sigma}{1 + n_v + (4\pi/c)} \right|^2,$$

where $n_v(\omega)$ is the refractive index of light at the frequency ω in the bulk of the crystal. In the region of longitudinal–transverse splitting, and neglecting "bulk" damping, $n_v = \mathrm{i}|n_v|$. In this case, and at a purely imaginary value of $\sigma(\omega)$, total reflection of light occurs because here the quantity $R = 1$. If the weak damping of the surface states is taken into account, then $\sigma = \sigma' + \mathrm{i}\sigma''$ and a deep and narrow minimum appears for the value of R at the resonance frequency $\sigma(\omega)$. Actually, for a dielectric transition layer in the region of an isolated resonance at the frequency ω_1, and conductivity is

$$\sigma(\omega) = -\mathrm{i}\omega \frac{\sigma_0}{\omega_1 - \omega - \mathrm{i}\gamma},$$

where γ is the attenuation constant, and σ_0 is a positive quantity proportional to the oscillator strength of vibrations in the transition layer and the layer thickness. In the region of longitudinal–transverse splitting, the reflection factor is

$$R = \left|\frac{1 - \mathrm{i}a - b}{1 + \mathrm{i}a + b}\right|^2,$$

where

$$a = |n_v| - (4\pi/c)\sigma'',$$

and

$$b = \frac{4\pi}{c}\sigma' = \frac{4\pi\omega\sigma_0\gamma}{c[(\omega_1 - \omega)^2 + \gamma^2]}.$$

Ordinarily $b < 1$, even at $\omega = \omega_1$. In this case, only the terms linear with respect to b need be taken into account in the expression for R. Thus

$$R \approx 1 - \frac{4b}{1 + |n_v|^2}.$$

Hence, at $\omega \approx \omega_1$, the quantity $R(\omega)$ has (see fig. 7b) a minimum of width γ (see also Philpott 1974). In a similar way it can be shown that the reflection factor R has a small peak (spike) at the frequency ω_1 if this frequency lies outside the region of longitudinal-transverse splitting.

If a transition layer of thickness $L \ll \lambda_0$ can be regarded as macroscopic, the expression for R assumes the form (see also Sugakov 1970, Philpott 1974):

$$R = \left|\frac{1 - n_v + \mathrm{i}(\omega/c)L(n_s^2 - n_v^2)}{1 + n_v - \mathrm{i}(\omega/c)L(n_s^2 - n_v^2)}\right|^2,$$

where $n_s = \sqrt{\tilde{\epsilon}_s(\omega)}$, and $\tilde{\epsilon}_s$ is the dielectric constant in the transition layer. This relation is similar to that discussed above and also leads to distinctive features in the relation $R(\omega)$ in the region of longitudinal–transverse splitting, provided that resonances of magnitude $n_s(\omega)$ occur in this region and damping is taken into account.

The first observations of the distinctive features of spectra of exciton fluorescence in semiconductors, due evidently to SP decay, were reported by Brodin and Matsko (1979). In this work, the fluorescence of a ZnTe crystal was investigated at liquid-helium temperatures. In the region of longitudinal–transverse splitting of the exciton, two fairly sharp maxima (with a halfwidth $\gamma \approx 2\,\mathrm{cm}^{-1}$) were found in the fluorescence spectrum. Only subsequent investigation of this effect can indicate to what extent its interpretation, given by Brodin and Matsko (1979), is substantiated. In this work, using the work of Agranovich and Leskova (1979), the new maxima in the fluorescence spectrum are associated with the processes of SP radiation decay, occurring when SPs are scattered by phonons and irregularities (roughness) of the boundary.

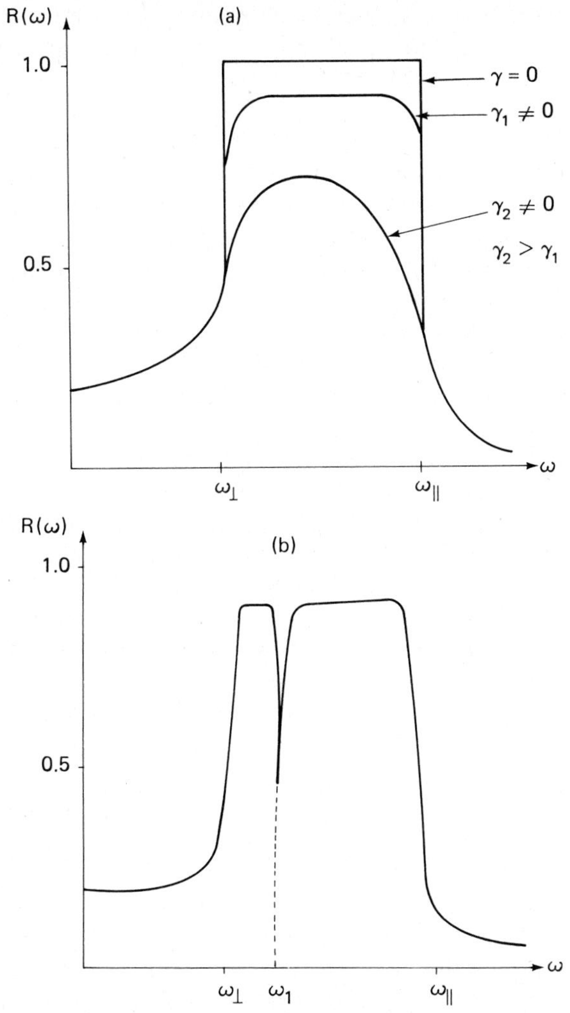

Fig. 7. Dependence of the reflection factor of light on the frequency upon normal incidence in the region of longitudinal–transverse splitting: (a) according to the magnitude of bulk damping γ: (b) in the presence of a surface current.

6. *Is Self-Focusing of Surface Polaritons Feasible?*

Nonlinear Properties of Surface Polaritons due to the Presence of a Transition Layer

A number of quite effective methods of SP pumping are used at the present time. Under these conditions, attempts to investigate their nonlinear properties, in conjunction with the corresponding theoretical analysis, are timely and even urgent. Special chapters of this volume are devoted to the problem of nonlinear excitation of SPs and the generation of the second harmonic. In the present section we shall touch upon only one of the possible questions of nonlinear SP optics, namely, the question of the effect of the transition layer on the self-focusing of SPs in the plane of their propagation. The dependence of the refractive index on the strength of the electric field in SPs is to be found and the length of self-focusing is to be assessed for SPs of the infrared range, propagating along the surface of a metal coated by a film of a transparent nonlinear material.

Before beginning the discussion of these questions, let us assume that the plane of SP propagation (plane $z = 0$) is divided by the axis $x = 0$ into two half-planes in which the refractive indices of SPs differ. Then upon reflection of SPs from the boundary line $x = 0$, we have the usual relation $n_2 \sin\theta = n_1 \sin\varphi$, where φ is the angle of incidence, θ is the refraction angle, and n_1 and n_2 are the indices of refraction of SPs in the half-planes $x < 0$ and $x > 0$, respectively. If $n_1 > n_2$, then at $\varphi > \varphi_0$ (where $\sin\varphi_0 = n_2/n_1$) there is no refracted wave and total internal reflection of the SPs occurs. Though under these same conditions a part of the energy of the SPs is expended in forming a spatial cylindrical wave, the question included in the title of this section is posed because the refractive index of SPs, like that of bulk light waves, may depend upon the SP intensity. Self-focusing of a SP in the plane of propagation could become feasible if its refractive index $n(|\boldsymbol{E}|^2)$ is increased with the strength of the electric field. If, on the contrary, a strong field of the SP reduces its refractive index, the reverse effect –defocusing– may occur rather than self-focusing.

In seeking the dependence $n(|\boldsymbol{E}|^2)$, we shall consider here SPs of the infrared range propagating along the surfaces of metals. Owing to the poor penetration of the field into the metal, the SPs are weakly damped and are capable of travelling macroscopic distances $l \approx 2$ to 5 cm. It is clear that only under these conditions can we rely upon inequality $l > l_f$ being complied with, where l_f is the self-focusing length. Since length l_f decreases with an increase in the field strength in SPs (see below), the preceding inequality leads to the lower power threshold for SPs, corresponding to self-focusing.

We next assume that a transition layer (film) of thickness L is on the surface of the metal and that nonlinearity only in this layer is essential. Taking into account the transition layer at the boundary with vacuum, the dispersion law for SPs (see eq. (22a)) is determined by the relation

$$\frac{\epsilon}{\kappa} + \frac{1}{\kappa_1} = \frac{L(1-\tilde{\epsilon})}{\tilde{\epsilon}\kappa_1^2}\left[k^2(1+\tilde{\epsilon}) - \frac{\omega^2}{c^2}\tilde{\epsilon}\right], \tag{68}$$

where ϵ and $\tilde{\epsilon}$ are the dielectric constants of the metal and the transition layer, k is the wave vector of the SP, $\kappa = \sqrt{[k^2 - (\omega^2/\psi^2)\epsilon]}$ and $\kappa_1 = \sqrt{[k^2 - \omega^2/c^2]}$. Since relation (68) corresponds to the taking of only terms linear with respect to L into account, the quantity k on the right-hand side of the equation should be determined from the condition $\epsilon/\kappa + 1/\kappa_1 = 0$. In this approximation (i.e., ignoring the transition layer)

$$k^2 = \frac{\omega^2}{c^2}\frac{\epsilon}{\epsilon+1} \equiv \frac{\omega^2}{c^2} n_0^2(\omega).$$

At $\omega \ll \omega_p$, where ω_p is the plasma frequency of the metal, $|\epsilon| \gg 1$ and hence $k^2 \approx (\omega^2/c^2)(1 + 1/|\epsilon|)$ and $n_0 \approx 1$. Assuming that frequency ω is far from resonances $\tilde{\epsilon}$, instead of equation (68) we have

$$\frac{\epsilon}{\kappa} + \frac{1}{\kappa_1} \approx L \frac{1-\tilde{\epsilon}}{\tilde{\epsilon}} |\epsilon|. \tag{69}$$

It follows from relation (69) that in the case being discussed the refractive index of the surface wave is

$$n^2 = n_0^2 - 2\frac{\omega}{c}\frac{L}{\sqrt{|\epsilon|}}\frac{1}{\tilde{\epsilon}}. \tag{70}$$

Since the change of the field inside the transition layer was neglected in obtaining eq. (68), we can assume that the dielectric constant of the transition layer is

$$\tilde{\epsilon} = \tilde{\epsilon}_0 + \tilde{\epsilon}_2 |\boldsymbol{E}(x, y)|^2, \tag{71}$$

where $\boldsymbol{E}(x, y)$ is the electric field strength in the SP at the surface of the metal at point (x, y). Making use of eqs. (70) and (71) we obtain

$$n^2(\omega, |\boldsymbol{E}|^2) = n_0^2(\omega) + A|\boldsymbol{E}(x, y)|^2_{z=0}, \tag{72}$$

where

$$A = 2\frac{\omega}{c} L \frac{\tilde{\epsilon}_2}{\tilde{\epsilon}_0^2 \sqrt{|\epsilon|}}. \tag{73}$$

It follows from this equation that the quantity A increases in the frequency region ω_0, for which $\tilde{\epsilon}(\omega_0) \approx 0$. At $|\epsilon| = 4000$, $(\omega/c)L = 0.15$, $\tilde{\epsilon}_0 = 0.1$ and $\tilde{\epsilon}_2 = 2 \times 10^{-11}$ cgse, $A = 10^{-11}$ cgse.

Relation (72) corresponds to the two-dimensional wave equation

$$\frac{\partial^2 E}{\partial x^2} + \frac{\partial^2 E}{\partial y^2} - \frac{n_0^2}{c^2}\frac{\partial^2 E}{\partial t^2} - \frac{A}{c^2}\frac{\partial^2}{\partial t^2}(|E|^2 E) = 0.$$

In investigating the problem of self-focusing, an analysis of such an equation was carried out in many works. According to Kelley (1965) and Akhmanov et al. (1966), the focusing length that corresponds to this equation is

$$l_f \approx \frac{a}{2E_m}\left(\frac{n_0^2}{A}\right)^{1/2},$$

where a is the initial transverse size of the beam, and E_m is the peak value of the field. At $a = 0.1$ cm, $E_m = 2 \times 10^3$ cgse and $n_0 = 1$, we obtain $l_f \approx$ 10 cm, which is of the order of magnitude of the length of propagation of SPs of the infrared range along the surfaces of metals. At such substantial or even somewhat less peak values of the field, the absorption of the energy of the SPs by the metal leads to its overheating and to the destruction of its surfaces. Therefore, it may be promising to employ only pulsed lasers and seek transition layers having sufficiently large non-linearity. Recently Zeldovich et al. (1980) showed that in certain liquid crystals the quantity ϵ_2 can be anomalously large (even up to $\epsilon_2 \approx 10^{-4}$ cgse). If these liquid crystals are used to form a film on metal, the length of self-focusing should become much less than the length of propagation of SPs on metal even if lasers of comparatively low power are employed. This instils confidence that the observation of SP self-focusing can, in general, be possible with a proper selection of materials. Also of interest are the investigation of the nonlinear properties of SPs when transition layers are absent or ignored, an analysis of questions analogous to those that arise in the theory of self-trapping of intensive laser radiation in the bulk, as well as a discussion of the feasibility of SP self-focusing in the centimeter range, in which the SP lengths of propagation are especially large.

References

Abeles, F., 1972, Optical Properties of Solids, ed. F. Abeles (North-Holland, Amsterdam) p. 93.

Agranovich, V.M., 1959, Zh. Eksp. Teor. Fiz. **37**, 340. (1960, Sov. Phys. – JETP **10**, 307.)

Agranovich, V.M., 1968, Teoriya Eksitonov (Theory of Excitons) (Nauka Publishers, Moscow).

Agranovich, V.M., 1975, Usp. Fiz. Nauk **115**, 199.

Agranovich, V.M., 1976, Zh. Eksp. Teor. Fiz. Pisma **24**, 602.

Agranovich, V.M., 1979a, Zh. Eksp. Teor. Fiz. **77**, 1125.

Agranovich, V.M., 1979b, Proc. 2nd USA–USSR Conf. Light Scattering in Solids, May 1979, New York, eds. J.L. Birman, H.Z. Cummins, K.K. Rebane (Plenum Press, New York) p. 113.

Agranovich, V.M., 1981, Proc. VII Intern. Conf. Raman Scattering (North-Holland) in press.

Agranovich, V.M. and O.A. Dubovsky, 1966, JETP Lett. **3**, 223.

Agranovich, V.M. and V.L. Ginzburg, 1979, Kristallooptika pri uchote prostranstvennoi dispersii i teoriya ekcitonov (Spatial Dispersion in Crystal Optics and the Theory of Excitons) (Nauka Publishers, Moscow) (Springer Verlag, 1982).

Agranovich, V.M. and I.I. Lalov, 1976, Opt. Commun. **16**, 239.

Agranovich, V.M. and T.A. Leskova, 1977, Solid State Commun. **21**, 1065.

Agranovich, V.M. and T.A. Leskova, 1979, Zh. Eksp. Teor. Fiz. Pisma **29**, 151.

Agranovich, V.M. and Yu.E. Lozovik, 1973, Zh. Eksp. Teor. Fiz. – Pisma **17**, 209.

Agranovich, V.M., Yu.E. Lozovik and A.G. Malshukov, 1975, Usp. Fiz. Nauk. **117**, 570.
Agranovich, V.M., S.A. Darmanyan and A.G. Malshukov, 1980a, Opt. Commun. **33**, 234.
Agranovich, V.M., V.E. Kravtsov and T.A. Leskova, 1981a, Zh. Eksp. Teor. Fiz. (in press).
Agranovich, V.M., V.E. Kravtsov and T.A. Leskova, 1981b, Solid State Commun. (in press).
Akhmanov, S.A., A.P. Sukhorukov and R.B. Khokhlov, 1966, Zh. Eksp. Teor. Fiz. **50**, 1535.
Anderson, W.E., R.W. Alexander and R.J. Bell, 1971, Phys. Rev. Lett. **27**, 1057.
Blank, A.Ya. and V.L. Berezinsky, 1978, Zh. Eksp. Teor. Fiz. **75**, 2317.
Brodin, M.S. and M.G. Matsko, 1979, Zh. Eksp. Teor. Fiz. Pisma **30**, 571.
Conwell, E.M., 1975, Phys. Rev. **B11**, 1508.
Conwell, E.M. and C.C. Kao, 1976, Solid State Commun. **18**, 1123.
Cunningham, S.L., A.A. Maradudin and R.F. Wallis, 1974, Phys. Rev. **B10**, 3342.
Drude, P., 1912, Lehrbuch der Optik (Verlag von S. Hirzel, Leipzig).
Dubovsky, E.B. and Yu.M. Kagan, 1977, Zh. Eksp. Teor. Fiz. **27**, 335.
Glockner, E. and H.C. Wolf, 1943, Z. Naturforsch. **24a**, 943.
Guidotti, D., S.A. Rice and H.L. Lemberg, 1974, Solid State Commun. **15**, 113.
Hartstein, A. and E. Burstein, 1974, Solid State Commun. **14**, 1223.
Hopfield, J.J., 1958, Phys. Rev. **112**, 1555.
Hopfield, J.J. and D.G. Thomas, 1963, Phys. Rev. **132**, 561.
Kao, C.C. and E.M. Conwell, 1976, Phys. Rev. **B14**, 2464.
Keldysh, L.V., 1979a, Zh. Eksp. Teor. Fiz. Pisma **29**, 716.
Keldysh, L.V., 1979b, Zh. Eksp. Teor. Fiz. Pisma **30**, 244.
Kelley, P.L., 1965, Phys. Rev. Lett. **15**, 1005.
Kizel, V.A., 1973, Otrazheniye Sveta (Reflection of Light) (Nauka Publishers, Moscow).
Lopez-Rios, T., F. Abeles and G. Vuye, 1978, J. de Phys. **39**, 645.
Lozovik, Yu.E. and V.N. Nishanov, 1978, Fiz. Tverd. Tela **20**, 3654.
Marschall, N., B. Fischer and H.J. Queisser, 1971, Phys. Rev. Lett. **27**, 95.
Marschall, N., B. Fischer and H.J. Queisser, 1973, Surf. Sci. **34**, 50.
Mills, D.L., 1975, Phys. Rev. **B12**, 4036.
Mittra, R. and S.W. Lee, 1971, Analytical Techniques in the Theory of Guided Waves (The Macmillan Co., New York; Collier-Macmillan Limited, London).
Orrit, M., C. Aslangul and P. Kottis, 1980, Phys. Rev. (in press).
Pekar, S.I., 1957, Zh. Eksp. Teor. Fiz. **33**, 1022.
Philpott, M.R., 1974, J. Chem. Phys. **60**, 1410.
Philpott, M.R. and P.G. Sherman, 1975, Phys. Rev. **B12**, 5381.
Schlesinger, Z. and A.J. Sievers, 1979, Infrared Surface Wave Interferometry, Report No. 4177, Materials Science Center, Cornell University, Ithaca, N.Y.
Sherman, P.G. and M.R. Philpott, 1978, J. Chem. Phys. **68**, 1729.
Shoenwald, J., E. Burstein and J.M. Elson, 1973, Solid State Commun. **12**, 185.
Sivukhin, D.V., 1948, Zh. Eksp. Teor. Fiz. **18**, 976.
Sivukhin, D.V., 1952, Zh. Eksp. Teor. Fiz. **21**, 367.
Sivukhin, D.V., 1956, Zh. Eksp. Teor. Fiz. **30**, 374.
Sugakov, V.I., 1970, Ukr. Fiz. Zh. (Russ. ed.) **15**, 2060.
Syassen, K. and M.R. Philpott, 1978, J. Chem. Phys. **68**, 4870.
Turlet, J.M. and M.R. Philpott, 1975a, J. Chem. Phys. **62**, 2777, 4260.
Turlet, J.M. and M.R. Philpott, 1975b, Chem. Phys. Lett. **35**, 92.
Turlet, J.M. and M.R. Philpott, 1976, J. Chem. Phys. **64**, 3852.
Turlet, J.M., J. Bernard and P. Kottis, 1978, Chem. Phys. Lett. **59**, 506.
Vainshtein, L.A., 1966, Teoriya- diffraktsii i metod faktorizatsii (Theory of Diffraction and the Factorization Method) (Soviet Radio Publishers, Moscow).
Yakovlev, V.A., V.G. Nazin and G.N. Zhizhin, 1975, Opt. Commun. **15**, 293.

Yakovlev, V.A., V.G. Nazin and G.N. Zhizhin, 1977, Zh. Eksp. Teor. Fiz. **72**, 687.
Zeldovich, B.Ya., N.F. Pilipetsky, A.V. Sukhov and N.V. Tabiryan, 1980, Zh. Eksp. Teor. Fiz. Pisma, **31**, 287.
Zhizhin, G.N., M.A. Moskalyova, E.V. Shomina and V.A. Yakovlev, 1976, Zh. Eksp. Teor. Fiz. Pisma, **24**, 221.
Zhizhin, G.N., M.A. Moskalyova, E.V. Shomina and V.A. Yakovlev, 1979, Zh. Eksp. Teor. Fiz. Pisma, **29**, 533.

CHAPTER 6

Surface Polaritons at Metal Surfaces and Interfaces

F. ABELES and T. LOPEZ-RIOS

Laboratoire d'Optique des Solides†
Université P. et M. Curie
4 place Jussieu, 75230 Paris Cédex 05
France

†Equipe de Recherche Associée au CNRS no. 462

Surface Polaritons
Edited by
V.M. Agranovich and D.L. Mills

Contents

1. Introduction 241
2. General properties of SPWs, their excitation and detection 243
   2.1. General properties of surface plasma waves (SPW) 243
   2.2. Optical excitation of surface plasma waves (SPW) 245
   2.3. Detection of surface plasma waves (SPW) 249
3. Modification of SPWs by a surface (transition) layer: general case 251
4. Modification of SPWs by a surface layer in the vicinity of its plasma frequency; . . nonlocal effects 256
5. Investigations of surface modifications in ultra-high vacuum 263
6. Use of SPWs for the investigation of solid–liquid and solid–solid interfaces . . . 268
7. Conclusion 271

References 272

1. *Introduction*

We consider in this chapter the use of surface plasma waves (SPW) for the investigation of superficial layers, surfaces and interfaces of metals. To begin with, a few comments are in order, which should, hopefully, clarify the present situation.

Historically, since the beginning of the quantum theory of solids, it became obvious that their optical properties are closely related to their electronic structure, especially for metals and semiconductors (Mott and Jones 1936, Wilson 1936). Unfortunately, it soon appeared that optical studies of metals lead to erratic results, and these were related to the difficulty of preparing good surfaces. The measurements were obscured by the uncontrollable quality of the surface, that is to say by the unavoidable presence of transition layers, surface roughness, etc. These difficulties were later overcome either by using vacuum evaporated and well annealed films, or by a sophisticated preparation of clean surfaces. This led to the conclusion that light can indeed be used as a standard tool for the investigation of the electronic structure of metals and alloys (Abelès 1966) and for condensed matter in general (Abelès 1972, Seraphin 1976).

This point having been established, it was natural to ask whether optical techniques could be used for nondestructive studies of surfaces, interfaces and very thin superficial layers. The penetration depth of an electromagnetic wave in condensed matter is at least of the order of 50 atomic layers, which means that the ratio of surface to bulk response function should be very small. However, optical methods sensitive to surface phenomena have been worked out. Ellipsometry is a very sensitive technique, mainly because it measures both amplitudes and phases and because it gives only ratios or differences for the two main polarization directions (perpendicular and parallel to the plane of incidence). If a change of $0.001° \simeq 10^{-5}$ rad can be detected by the ellipsometer, it can easily be shown that, assuming a very thin homogeneous layer (which is a pure fiction), the minimum layer thickness that can be detected is of the order of (Aspnes 1976) $10^{-5}\ \lambda/10$ which, for a wavelength $\lambda = 500$ nm, leads to ~ 0.005 Å. This indicates that ellipsometry should be sensitive to surface distributions of atoms of the order of a hundredth of a monolayer. Thus, in ultra-high vacuum, ellipsometry can be even more sensitive than Auger spectroscopy or LEED

(Low Energy Electron Difraction) and, moreover, it can be used in much more complicated atmospheres (ambient pressure, various gases, etc.).

In semiconductors, optical absorption studies using multiple reflections were able to provide evidence for surface states (Chiarotti et al. 1971).

Another highly sensitive technique is differential reflectivity which was introduced mainly for the study of dilute alloys (Hummel et al. 1970, Beaglehole et al. 1972) and then used for the investigation of surface modifications of electrodes in solution. Using this method, Kolb and McIntyre (1971) followed the anodic oxidation of gold, which corresponds to one or two monolayers of oxide. Differential-reflectivity methods were extended to the vacuum–ultraviolet too and measurements by Cunningham et al. (1979, 1980) showed that rare gases adsorbed on metals have the ionic characteristics of the excited configurations and only weak coupling to the host electron gas.

Following this, modulated reflectance measurements were performed in ultrahigh vacuum at the solid–gas interface by Rubloff et al. (1973, 1974) and Anderson et al. (1973, 1974), showing that small coverages of H_2, CO and O_2 on W(100) surfaces can be measured with well stabilized spectrophotometers. When using polarized surface-reflectance-spectroscopy on the H/W(110) system, Blanchet et al. (1980) were able to deduce the hydrogen surface configuration.

This brief historical view definitely shows that technical advances were able to overcome the low sensitivity of light to surface and interface effects. The main difficulty resides now in the theoretical analysis of the data. The problems related to electromagnetic effects at metal surfaces and more specifically to nonlocal effects have been recently summarized by Kliewer (1980).

This chapter is devoted to an optical technique based on the optical excitation of surface plasma waves (SPW) or surface plasmons (SP). This technique, being essentially based on resonance effects, is much more sensitive to surface and interface effects than those which were briefly reported above. For rough surfaces, this is already known since forty years, when Fano (1941) explained Wood anomalies in diffraction gratings with the help of SPWs. More recently, normal incidence reflectivity measurements were used to study corrosion films on metals from the shift of the surface plasmon resonance (Stanford 1970). Eventually, it was shown by Otto (1968) and Kretschmann (1971) that SPWs can be excited at a smooth surface. The configurations suggested by both make use of the method known as attenuated total reflection (ATR) and some review papers discussing them are now available (Otto 1976, Kliewer and Fuchs 1974, Raether 1977, Burstein et al. 1974). Here, we shall review some of the investigations of metal surfaces and interfaces which were performed with the help of SPWs.

2. *General Properties of SPWs, their Excitation and Detection*

2.1. *General Properties of Surface Plasma Waves (SPW)*

We will first give a brief discussion of the conditions of existence of surface plasmons at the plane interface between two media, together with their principal characteristics, and then the methods of excitation and detection of these waves.

Consider two isotropic media with dielectric constants ϵ_0 and ϵ_2 separated by a plane surface as indicated in fig. 1; $\boldsymbol{s}$ and $\boldsymbol{n}$ are unit vectors in the y (in the surface plane) and z (normal to the surface) directions respectively. The electromagnetic waves propagating in an infinite medium are transverse and there are no polarization charges: $\nabla \boldsymbol{P} = 0$. The presence of a surface separating two different media gives rise to a discontinuity in the normal component of the electric field E_z and to surface charges which are proportional to this discontinuity. We examine here the electromagnetic field associated with the oscillations of such surface charges in the absence of any exciting field. This field is generated at the surface and its amplitude must decay on both sides of the surface where the charges are localized. As there must necessarily be a component of the electric field normal to the surface, we consider the TH polarization only: the electric field is in the sagittal plane defined by $\boldsymbol{n}$ and the unit vector in the x direction. We assume that both media have a dielectric response for the electric field only ($\mu = 1$) and that this response is both linear and local: $\boldsymbol{D}(\omega) = \epsilon(\omega) \cdot \boldsymbol{E}(\omega)$.

The system being invariant under a translation in the x direction, the TH (p-polarized) field in each medium can be written:

$$E(x, z, t) = (E_x(z), 0, E_z(z))\, e^{iK_x x}\, e^{-i\omega t},$$
$$H(x, z, t) = (0, H_y(z), 0)\, e^{iK_x x}\, e^{-i\omega t}. \tag{2.1}$$

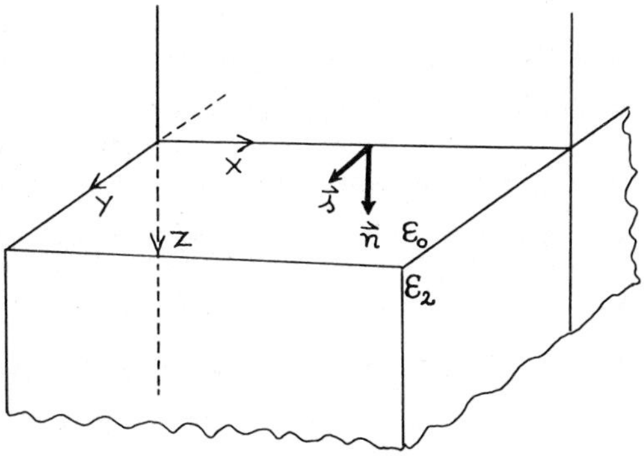

Fig. 1. Geometry of the surface region between the ϵ_0 and ϵ_2 media.

Equations (2.1) must verify Maxwell's equations, be bounded in both media and correspond to plane waves. They can be written:

$$E_i(r, t) = (s \wedge K_i)\, E_i\, e^{i(k_i \cdot r - \omega t)},$$

$$H_i(r, t) = s \cdot \epsilon_i E_i\, e^{i(k_i \cdot r - \omega t)},$$

with $K_i = (c/\omega)k_i = (c/\omega)(k_x, 0, k_{zi})$, $i = 0$ for the ϵ_0 medium and $i = 2$ for the ϵ_2 medium. K_i is the reduced wave vector and $k_{z0} = -\mathrm{i}[k_x^2 - \epsilon_0(\omega/c)^2]^{1/2}$, $k_{z2} = \mathrm{i}[k_x^2 - \epsilon_2(\omega/c)^2]^{1/2}$. We shall use the notation $S = K_x$ for the component of the reduced wave vector parallel to the surface.

The continuity conditions for the components parallel to the surface of the electric and magnetic fields lead to the continuity of their ratios, that is to say, to the continuity of the surface admittance:

$$Z_0 + Z_2 = 0, \tag{2.2}$$

where $Z_i = \epsilon_i/(S^2 - \epsilon_i)^{1/2}$. This is the existence condition for the surface plasmons, which leads to the well-known relation:

$$S^2 = (S^r + \mathrm{i}s^i)^2 = \frac{\epsilon_0 \epsilon_2}{\epsilon_0 + \epsilon_2}. \tag{2.3}$$

For very large wave vectors ($S \to \infty$), the surface wave frequency is given by the equation:

$$\epsilon_0(\omega) + \epsilon_2(\omega) = 0. \tag{2.4}$$

In order to have electromagnetic waves "bound" to the surface, one of the two adjacent media must have negative dielectric constant (active medium) so that electromagnetic waves cannot propagate in this medium. This is a general conclusion when looking for solutions which are related to the presence of a surface. For instance, surface elastic waves (Rayleigh waves), which are also polarized in the sagittal plane, exist only for frequencies for which no elastic wave can propagate in the infinite crystal. Analogously, surface electronic states in semiconductors can be found in the forbidden band of energies only, that is to say, for energies for which there are no electronic states in the infinite crystal; these surface states are described by evanescent wavefunctions (purely imaginary K) on each side of the surface.

Let ϵ_2 be the medium with negative dielectric constant. We assume that $|\epsilon_2^r| \gg \epsilon_2^i$, while ϵ_0 is real and positive. Equation (2.3) together with the inequality $|\epsilon_2^r| \gg \epsilon_2^i$ leads to $S^r \gg S^i$. The normal components of the reduced wave vector become now:

$$K_{z0} = \epsilon_0/(\epsilon_0 + \epsilon_2)^{1/2} \equiv (K_0)_n, \qquad K_{z2} = -\epsilon_2/(\epsilon_0 + \epsilon_2)^{1/2} \equiv (K_2)_n,$$

and we find that:

$$\mathrm{Im}[(K_0)_n] \gg \mathrm{Re}[(K_0)_n], \qquad \mathrm{Im}[(K_2)_n] \gg \mathrm{Re}[(K_2)_n].$$

Thus, only the component of the wave vector along the surface (the same in both media) has an important real part, indicating wave propagation along the surface. Surface electromagnetic waves can be described as being inhomogeneous waves, one in each medium, for which the isophase planes $\mathrm{Re}(\boldsymbol{K}_0 \cdot \boldsymbol{r}) = \mathrm{Re}(\boldsymbol{K}_2 \cdot \boldsymbol{r}) = \text{Const}$ are perpendicular to the surface and the constant amplitude planes, $\mathrm{Im}(\boldsymbol{K}_0 \cdot \boldsymbol{r}) = \text{Const}$ and $\mathrm{Im}(\boldsymbol{K}_2 \cdot \boldsymbol{r}) = \text{Const}$ in the ϵ_0 and ϵ_2 media respectively, are parallel to the surface. The phase velocity is the same in both media and equal to $v_{ph} = c/S^r$. For large S values, the phase velocity becomes very small compared to the velocity of light. The waves are then mainly longitudinal and we have a quasi-electrostatic situation. For small S values, on the other hand, the surface waves are similar to light waves.

A significant quantity is the penetration depth in each medium, defined as the distance δ in the z direction from the surface for which the electromagnetic energy is reduced by 1/e. One finds that:

$$\delta_0 = (2\,\mathrm{Im}[(K_0)_n])^{-1},$$

$$\delta_2 = (2\,\mathrm{Im}[(K_2)_n])^{-1},$$

leading to:

$$\frac{\delta_0}{\delta_2} \simeq \frac{\epsilon_2^r}{\epsilon_0}. \tag{2.5}$$

The penetration depth is much smaller in the active medium than in the other one. Consider, for instance, a metal which behaves like a free-electron gas in contact with vacuum ($\epsilon_0 = 1$). For wavelengths λ larger than the plasma wavelength λ_p, we find $\delta_2 \simeq \lambda_p/4\pi$, showing that the penetration depth in the metal is wavelength independent and depends on the free electron density only; for Al ($\lambda_p \simeq 800$ Å), one finds $\delta_2 \simeq 60$ Å. On the other hand, $\delta_0 \simeq \lambda^2/4\pi\lambda_p$, indicating that the penetration depth in vacuum is highly wavelength dependent. Figure 2 shows δ_0 and δ_2 vs. wavelength between 0.25 and 1 μm for Al. These values were computed using the dispersion relation with $\epsilon_0 = 1$ and the experimental dielectric constant for Al, which, in this spectral region, is not exactly that of a free-electron gas.

Another interesting quantity is the distance L along the surface for which the intensity of the surface electromagnetic wave is reduced by 1/e. We find $L = \lambda/4\pi S^i$. We have indicated in fig. 2 the L values corresponding to Al.

2.2. *Optical Excitation of Surface Plasma Waves* (*SPW*)

SPWs are nonradiative waves, which cannot be directly excited by light waves propagating in one of the two media. Indeed, we have seen that their

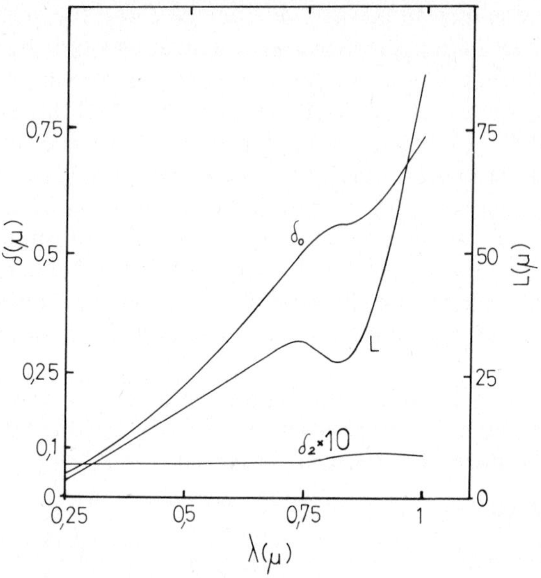

Fig. 2. Penetration depths in vacuum (δ_0) and in the metal (δ_2) together with the propagation distance (L) vs. wavelength (λ) for SPW propagating at the Al/vacuum surface. The measured optical constants are used for Al, and this explains the anomalies in the 0.8 μm region.

phase velocity is $v_{ph} = c/S^r$, from which, using eq. (2.3), we find:

$$v_{ph} = \frac{c}{\sqrt{\epsilon_0}} \operatorname{Re} \sqrt{\frac{\epsilon_0 + \epsilon_2}{\epsilon_2}},$$

v_{ph} is thus smaller than the phase velocity of light in the ϵ_0 medium, namely $c/\sqrt{\epsilon_0}$. In other terms, for a given frequency ω, the component parallel to the surface of the light wave vector will always be smaller than the parallel component K of the SPW wave vector, and the conditions for the energy (ω) and wave vector ($\boldsymbol{k}$) conservation cannot be simultaneously satisfied. Figure 3 gives a schematic representation of the dispersion relation $\omega(\boldsymbol{k})$ for SPW for the simple situation where both media in contact are nonabsorbing, that is to say $\epsilon_2^i = 0$, together with the dispersion relation for light in the ϵ_0 medium: $\omega/\boldsymbol{k} = c/\sqrt{\epsilon_0}$. When $\epsilon_2^i \neq 0$, the dispersion relations for fixed ω and variable $\boldsymbol{k}$ display a backbending in the vicinity of $\omega = \omega_p/(1+\epsilon_0)^{1/2}$, wherefrom there appears the possibility of intersection of the $\omega(\boldsymbol{k})$ curve with the "light line" (Arakawa et al. 1973, Alexander et al. 1974). This extreme case, where one cannot really speak of surface waves any more, does not concern us here. At the bottom of the same fig. 3 we also give a schematic representation of the electric field distribution at the surface for various values of the wave vector $\boldsymbol{k}_x$. The SPW

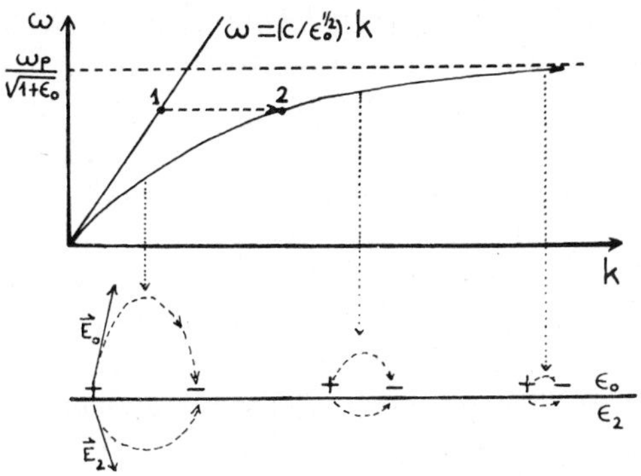

Fig. 3. Schematic representation of the dispersion relation $\omega(k)$ for SPW propagation at a vacuum–nonabsorbing plasma surface. The straight line $\omega = (c/\epsilon_0^{1/2})k$ is the dispersion curve for light. At the bottom of the figure are represented the electric field distributions for various k values.

excitation by light waves would correspond to the $1 \to 2$ process indicated on this figure.

SPW excitation by light waves can nevertheless happen in some particular cases:

(a) the possibility of nonlinear coupling between transverse and longitudinal electromagnetic fields through intraband electronic transitions (simultaneous excitation of a surface plasmon and an electron-hole pair) has been considered, but the absorption probability by this second-order process is weak (Ritchie 1965);

(b) if the surface is sinusoidally modulated with period d (a grating with rulings parallel to the y direction), Fano (1941) has shown that, if the amplitude of modulation is small with respect to the wavelength of light, the effect of the modulation is to modify the component of the wave vector parallel to the surface $k_0 = (\omega/c)\epsilon_0^{1/2}\sin\theta$ (where θ is the angle of incidence) into:

$$k_n = k_0 + 2\pi n/d, \tag{2.6}$$

with $n = 0, \pm 1, \pm 2$, etc. It might be possible then to find a value of n such that the condition $k_n = (\omega/c)S^r$ is verified and SPWs can be excited. This explains, for instance, Wood anomalies in gratings, which were experimentally known since the beginning of the century.

It is thus possible to excite SPWs via a periodic modulation of the surface by varying, for instance, the angle of incidence in order to satisfy eq. (2.3); one should observe then a minimum in the reflectivity. Teng and Stern (1967) could in this way determine the dispersion relation for SPWs in Al, and Ritchie et al. (1968) in Au and Al. Inversely, SPWs can have their wave vector modified by a surface modulation and become radiative.

Teng and Stern have also experimentally demonstrated the radiative de-excitation of SPWs which were excited in an Al grating by 7 to 12 keV electrons.

Fourier analysis shows that a random rough surface can be described by a set of wave vectors k_α (and amplitudes A_α), each of them being able to give rise to a relation analogous to (2.6): $k_{n\alpha} = k_0 + nk_\alpha$; some SPW excitation will thus always be possible. Jasperson and Schnatterly (1969) have indeed given, by normal incidence reflectivity measurements on rough Ag films, an experimental proof of SPW excitation via surface roughness.

(c) In order to be able to excite SPWs on a plane smooth surface, the light waves in the ϵ_0 medium should have, for a given frequency, the same wave vector as the SPW and, therefore, a phase velocity smaller than $c/(\epsilon_0)^{1/2}$. Otto (1968), by suggesting a method to create "slow photons", has opened the way to an important development in the optical studies of SPW. The basic idea of this method, called attenuated total reflection (ATR), is shown schematically in fig. 4. Before reaching the ϵ_0 medium, the incident light wave first enters a prism with dielectric constant $\epsilon_p > \epsilon_0$. According to Snell's law, the tangential component of the wave vector is conserved at the passage through a surface of discontinuity and, for angles of incidence θ greater than $\sin^{-1}(\epsilon_p/\epsilon_0)^{1/2}$ (critical angle for total reflection), we shall have $k_x > (\omega/c)\epsilon_0^{1/2}$. But, under these conditions, the normal component $(K_0)_n$ of the wave vector becomes purely imaginary and the wave is rapidly attenuated in the ϵ_0 medium in the z direction over a distance of the order of the wavelength. It is therefore necessary that the prism should be at a distance d from the active medium ϵ_2, of the order of the wavelength λ. This method of using attenuated total reflection for SPW excitation is known as Otto configuration.

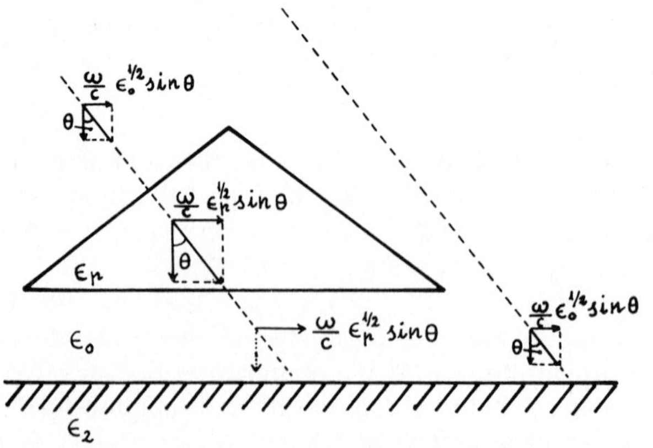

Fig. 4. SPW excitation by ATR in the Otto (1968) configuration.

Both Snell's law and eq. (2.3) being symmetrical with respect to ϵ_0 and ϵ_2, one can interchange the positions of these two media with respect to the prism. We have indicated in fig. 5 that SPWs can be excited at the ϵ_0/ϵ_2 interface by ATR when the active medium is a thin film deposited on the prism. Due to the absorption in the active medium, the film thickness must be sufficiently small in order to have enough energy reach the ϵ_0/ϵ_2 interface. This method is known as the Kretschmann (1971) configuration.

2.3. *Detection of Surface Plasma Waves (SPW)*

We have seen how SPWs can be excited by a light wave. Their observation can proceed in different ways:

(a) The most direct method is the detection of the reflected light in an ATR experiment using either of the above indicated configurations.

One measures the reflectivity R_p for p-polarized light as a function of the angle of incidence or wavelength. For perfect coupling, $R_p = 0$ and all the energy carried by the light wave is transferred to SPWs. In the absence of any radiative de-excitation (due, for instance, to roughness) the transmittivity ought to be null and the optical absorption is then $A = 1 - R_p$. $A(S)$ has a Lorentzian shape. One can use a more sophisticated detection method, measuring not only the amplitude of the reflected wave but its phase change at reflection δ_p too. This can be performed by ellipsometry, which gives the two parameters $\psi = \tan^{-1}(R_p/R_s)^{1/2}$ and $\Delta = \delta_p - \delta_s$ (where the index s refers to s-polarized light). R_s and δ_s being practically constant, all the information on the SPW contained in ψ and Δ comes in fact from R_p and δ_p.

(b) Another method uses the detection of light reemitted because of surface roughness, the SPW being excited, for instance, in an ATR

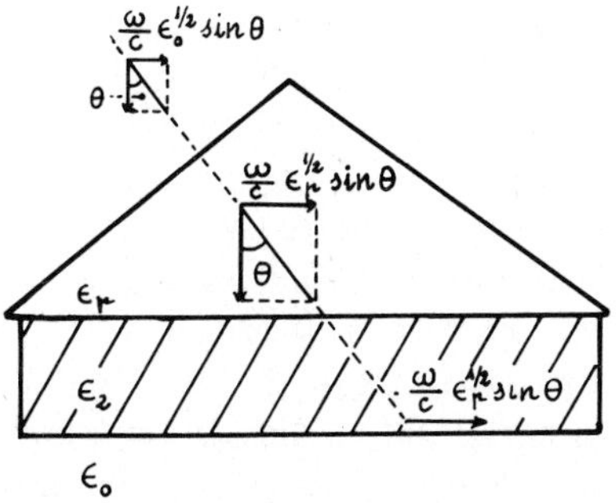

Fig. 5. SPW excitation by ATR in the Kretschmann (1971) configuration.

configuration. We have already indicated that SPWs not only can be excited by light via surface roughness, but, conversely, they can be scattered by such roughness. Two processes can take place: (i) a directional change of the SPW wave vector with modulus conservation, leading to absorption only; (ii) a modification of the wave vector modulus, which can bring it in the radiative region (where $\boldsymbol{k} < \omega/c$). In the latter case, there will be light emission, which can be measured. SPW scattering can also be due to volume inhomogeneity of the sample, or dielectric constant fluctuations. All real surfaces have some superficial or volume roughness and therefore scatter more or less the SPW. The emitted intensity can be written $I(\theta) = |A(\theta)|^2\xi$, where A is the intensity of SPW electromagnetic field at the surface and ξ a usually rather complicated function describing the roughness. In the vicinity of the excitation angle, $A \propto (S - S_{\mathrm{m}})^{-1}$, where S_{m} is the value of S given by (2.3) and $|A(\theta)|^2$ is a Lorentzian. The angular position of the emission maximum, θ_{m}, gives the real part of the wave vector via $\epsilon_{\mathrm{p}}^{1/2} \sin \theta_{\mathrm{m}} = \mathrm{Re}(S_{\mathrm{m}})$, whereas the average angular width of the emission curve, $\Delta\theta$, gives the imaginary part of the same wave vector via $\epsilon_{\mathrm{p}}^{1/2} \cos \theta_{\mathrm{m}} \Delta\theta = \mathrm{Im}(S_{\mathrm{m}})$ (Weber et al. 1975). The value I of the intensity itself provides information on the characteristic roughness dimensions. Important results have already been obtained in this field (see the review articles by Otto (1976) and Raether (1977)).

(c) Another means of detecting SPWs is the measurement of the surface photoelectric yield (Bösenberg 1971, Macek et al. 1972, Hincelin and Septier 1977, Chabrier 1978). For clean surfaces this method is limited to energies larger than the work function. Then, the SPWs deexcited into emitted photoelectrons are detected, and a maximum is observed for the yield Y when the excitation conditions are satisfied. The shape of the resonance curve $Y(\theta)$ is very similar to that of the $I(\theta)$ curve mentioned above. It is related to the value of the electromagnetic field of the SPW in a region corresponding to the escape depth of the photoelectrons, that is to say 5 to 100 Å thick according to the energy.

(d) Finally, we shall mention a method for the SPW observation based on their propagation along the surface properties, the propagation length reaching large (macroscopic) values in the infrared. Schoenwald et al. (1973) have shown that SPWs could be excited by ATR with a half-prism and detected, with a certain attenuation, by the same method at a certain distance from the point where they were generated. The same technique can be used in the microwave region ($\lambda = 3.55$ cm) (Bell et al. 1975), the excitation and detection being accomplished via a grating. The experimental values for the attenuation length L along the surface for wavelengths between 9.3 and 10.6 μm are in relatively good agreement with the theoretical values for some metals: Ni, Pd and W (Begley et al. 1976). In the far infrared ($\lambda = 118\,\mu$m) on the other hand, there is a large discrepancy between theory and experiment for many metals: Cu, Au, Pd, W, Ni, Pt and steel (Begley et al. 1979b), whereas there is a good agreement for n-type GaAs (Begley et al. 1979a).

3. *Modification of SPWs by a Surface (Transition) Layer: General Case*

SPWs are guided waves, propagating along the surface, and the electromagnetic field associated with them has its maximum amplitude at the surface. It is easy to understand that they should be very sensitive to any change in the dielectric constants of the media located in the immediate vicinity of the surface. Moreover, the optical excitation of SPWs is a resonant phenomenon: it happens only when both the frequency ω and the wave vector K of the exciting light waves verify the SPW dispersion relation. When the dispersion relation is modified due to a change in the superficial conditions, the excitation does not occur any more for these ω and K values, leading to an important variation of the quantities which are measured in such experiments. For these two complementary reasons, the sensitivity of these "bound photons" or surface polaritons to surface phenomena is considerably increased with respect to that of "ordinary" photons.

It has been theoretically shown (Chen et al. 1976) that SPW excitation by ATR at a silver surface increases by a factor of about 100 the intensity of the Raman effect in a surface layer displaying Raman activity. The large amplification of the surface electromagnetic field is very useful in order to observe nonlinear effects too. Second harmonic generation in Ag films could thus be obtained (Simon et al. 1974). More recently, a two photon photoemission process in Ag films was observed using a laser (Rudolf et al. 1977). The most important limitation for the use of ATR in such experiments is related to the fact that, for experimental reasons, one works in the Kretschmann configuration, that is, on thin films which can be heated and even deteriorated by the powerful lasers which are used.

As has been indicated above, any modification at the metal surface will lead to a modification of the surface plasmon dispersion relation. We shall not discuss here roughness effects, which have been extensively investigated, and will limit our discussion to the examination of a transition layer only.

We assume now that between the ϵ_0 and ϵ_2 media there is a transition layer extending over a distance $l(l \ll \lambda)$. Assuming a tensorial dielectric constant $\bar{\bar{\epsilon}}$ for this layer, the relation between displacement $\boldsymbol{D}$ and electric

$\boldsymbol{E}$ fields is given by:

$$\boldsymbol{D}(z, \omega) = \int \bar{\bar{\epsilon}}(z, z', \omega), \boldsymbol{E}(z', \omega)\, \mathrm{d}z'.$$

The dispersion relation can be written now:

$$Z_0 + Z_2 = \mathrm{i}(2\pi/\lambda)[\alpha - Z_0 Z_2(S^2\gamma - l)], \tag{3.1}$$

with

$$\alpha = \int_0^l \frac{D_x(z)\,\mathrm{d}z}{E_x(l)}, \qquad \gamma = \int_0^l \frac{E_z(z)\,\mathrm{d}z}{D_z(l)}.$$

The small term in the right-hand side of eq. (3.1) gives a general description of the effect of a transition layer on the dispersion relation. The α and γ integrals are, in general, difficult to compute. The point to be noticed is the contribution of both the parallel and the normal components of the electric field.

When the dielectric constant can be treated in the local approximation, and has only diagonal components, then $D_x(z) = \epsilon_{xx}(z) \cdot E_x(z)$ and $D_z(z) = \epsilon_{zz}(z) \cdot E_z(z)$ and

$$\alpha = \int_0^l \epsilon_{xx}(z)\,\mathrm{d}z, \qquad \gamma = \int_0^l \epsilon_{zz}^{-1}(z)\,\mathrm{d}z.$$

If the transition layer is a homogeneous and isotropic film with thickness d_f and dielectric constant $\epsilon_\mathrm{f} = \epsilon_\mathrm{f}^\mathrm{r} + \mathrm{i}\epsilon_\mathrm{f}^\mathrm{i}$, then $\alpha = \epsilon_\mathrm{f} d_\mathrm{f}$, $\gamma = d_\mathrm{f}/\epsilon_\mathrm{f}$ and eq. (3.1) becomes:

$$Z_0 + Z_2 = -\mathrm{i}(\omega/c) d_\mathrm{f}[Z_0 Z_2(1 - S^2/\epsilon_\mathrm{f}) + \epsilon_\mathrm{f}].$$

The right-hand term of this equation being very small, we can use in it the Z_0 and Z_2 values corresponding to the dispersion relation in the absence of the transition layer, which verify eq. (2.2). We can keep either ω or S fixed. In the first case, one finds that the modifications of the wave vector corresponding to surface plasmon excitation is given (in reduced units) by:

$$\delta S = \delta S^\mathrm{r} + \mathrm{i}\delta S^\mathrm{i} = \frac{(-\epsilon_0\epsilon_2)^{1/2}}{-\epsilon_2 + \epsilon_0} S^2 \frac{(\epsilon_\mathrm{f} - \epsilon_2)(\epsilon_0 - \epsilon_\mathrm{f})}{\epsilon_\mathrm{f}(\epsilon_0 + \epsilon_2)} \frac{2\pi d_\mathrm{f}}{\lambda}. \tag{3.2}$$

It can be noticed that δS is proportional to the thickness d_f of the transition layer and that $\delta S = 0$ for $\epsilon_\mathrm{f} = \epsilon_2$ and $\epsilon_\mathrm{f} = \epsilon_0$. For a highly reflecting substrate, δS^i is proportional to $\epsilon_\mathrm{f}^\mathrm{i}$, indicating that the imaginary part of δS is strongly related to the absorption in the surface layer.

For fixed S and frequency independent ϵ_0, the change in the resonance frequency $\delta\omega$ is given by:

$$\delta\omega = \delta\omega^\mathrm{r} - \mathrm{i}\delta\omega^\mathrm{i} = \frac{2\epsilon_2^2(\epsilon_0 + \epsilon_2 - \epsilon_0\epsilon_2/\epsilon_\mathrm{f} - \epsilon_\mathrm{f})}{[-(\epsilon_0 + \epsilon_2)]^{1/2}(\epsilon_2 - \epsilon_0)(\partial\epsilon_2/\partial\omega)} \frac{2\pi d_\mathrm{f}}{\lambda}. \tag{3.3}$$

Again $\delta\omega$ is proportional to d_f and $\delta\omega = 0$ when $\epsilon_f = \epsilon_0$ or ϵ_2. The imaginary part $\delta\omega^i$ measures the modification of the lifetime of the SPW.

In an ATR experiment, the presence of a transition layer leads to both a shift and a broadening of the resonance as was observed without this layer. The shift is given by the real part δS^r of the wave vector, whereas the broadening is related to its imaginary part δS^i (Abelès 1977). The modification of the depth of the resonance is also related to the absorption in the surface layer. Figure 6, after Gordon and Swalen (1977) gives results of reflectivity measurements in the ATR Kretschmann configuration at fixed incidence performed on bare Au and on the same film covered with various numbers of monolayers of cadmium arachidate. Notice that the R_p/R_s curves (R_s is nearly constant) have almost the same value at the minima, indicating that the organic layers have a real dielectric constant. Thus, one can immediately detect the presence of absorption in the surface layer from the inspection of the modification of the resonance curve. Figure 7 is characteristic of the opposite situation, where absorption in the layer is large. It corresponds to an experiment performed in our laboratory using again ATR and shows the resonance curves for a clean Ag surface and for the same surface covered by several Pd monolayers. Here we notice both a shift and a broadening of the resonance together with a large decrease of its intensity.

The shift δS^r can be either positive or negative, depending essentially on the relative values of the dielectric constant of the substrate and the surface layer, as can be seen from eq. (3.2). Both possibilities are illustrated in fig. 8 which gives the dispersion curves for Ag covered by Au

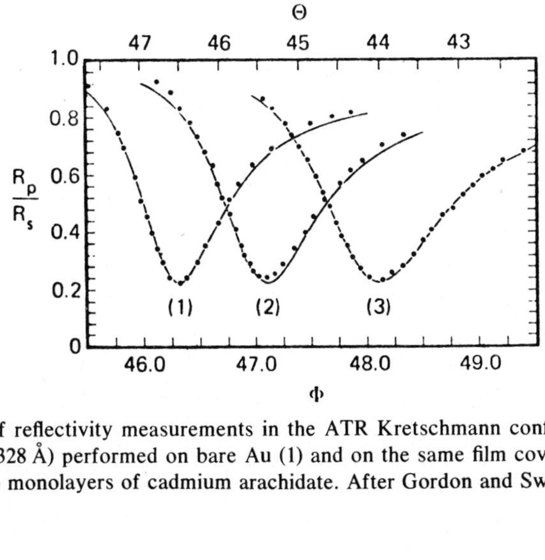

Fig. 6. Results of reflectivity measurements in the ATR Kretschmann configuration at fixed frequency ($\lambda = 6328$ Å) performed on bare Au (1) and on the same film covered with two (2) and four (3) monolayers of cadmium arachidate. After Gordon and Swalen (1977).

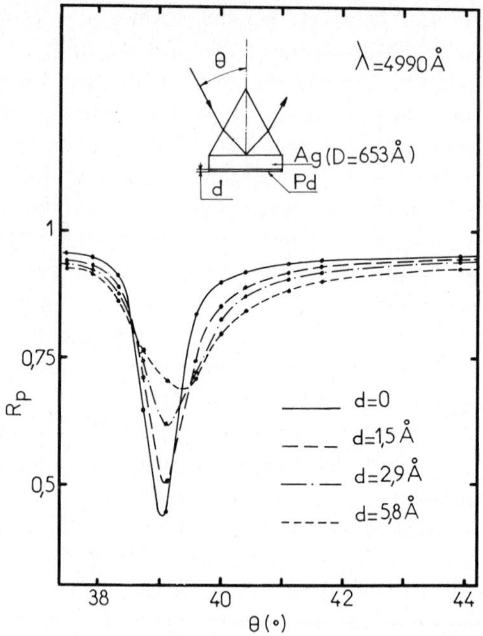

Fig. 7. Reflectivity versus angle of incidence for ATR measurements in the Kretschmann configuration performed on Ag covered with Pd layers. After Vuye (1980).

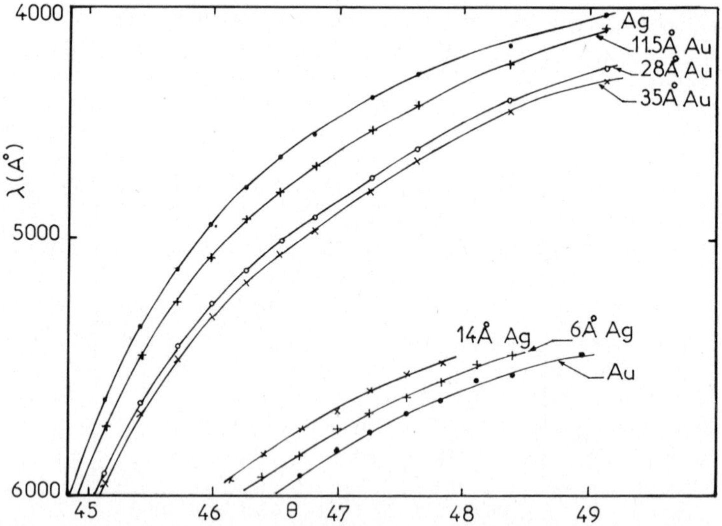

Fig. 8. Experimental dispersion relations λ (in Å) versus θ (degrees) for a Ag film (655 Å thick) covered with Au layers and for a Au film (930 Å thick) covered with Ag layers. The data correspond to the positions of the minima of $R_p(\theta)$ curves measured at fixed wavelength in the Kretschmann configuration. After Lopez-Rios (1980).

layers and for Au covered by Ag layers (Lopez-Rios 1980). As $|\epsilon_{Ag}| > |\epsilon_{Au}|$, deposition of very thin Au layers on Ag leads to a $\delta S^r > 0$, which means that, for measurements at a fixed frequency, the shift of the angular position of the resonance, given by $\delta S^r = \delta(n_p \sin\theta) = (n_p \cos\theta)\delta\theta$ with n_p = refractive index of the prism, is positive. The opposite occurs when very thin Ag layers are deposited on Au.

A superficial layer modifies the propagation length of SPWs too. If L is the value corresponding to the ϵ_0/ϵ_2 interface, the presence of a very thin layer between the two media leads to a propagation length $L' = L[1 + \delta S^i/S^i]^{-1}$, where S^i and δS^i are given by the eqs. (2.3) and (3.2) respectively.

A point which is important to keep in mind when performing spectroscopic investigations of surface layers using SPWs is that all the information provided from such experiments is the corresponding value of δS. That is to say, for a given frequency, we know only two quantities (δS^r and δS^i) related to the surface layer. If we assume that the latter is homogeneous and isotropic and can, therefore, be characterized by a (complex) dielectric constant $\epsilon_f(\omega)$ and its thickness d_f, it is, in principle, possible to deduce ϵ_f when d_f is known. We shall briefly comment on this possibility. The accuracy of the determination of ϵ_f is related to the value of $C = |\partial\alpha/\partial\epsilon_f|$, where $\alpha = \delta S/(2\pi d_f/\lambda)$. It is clear that $C = 0$ corresponds to an impossibility for the determination of ϵ_f, the latter becoming more and more accurate with increasing C. Equation (3.2) leads to:

$$C = |m|\,|\epsilon_0\epsilon_2/\epsilon_f^2 - 1|,$$

with $m = \epsilon_0\epsilon_2[(\epsilon_2 - \epsilon_0)\{-(\epsilon_2 + \epsilon_0)\}^{3/2}]^{-1}$, a quantity independent of the superficial layer characteristics. The relation $C = 0$ is verified when $\epsilon_f = (\epsilon_0\epsilon_2)^{1/2}$ and in this case the determination of ϵ_f is impossible. Figure 9 gives the iso-values of C in the $(\epsilon_f^r, \epsilon_f^i)$ plane when $\epsilon_0 = 1$ and $\epsilon_2 = -10$, which approximately corresponds to silver in the visible region. The point of zero precision ($C = 0$) is $\epsilon_f = (-10)^{1/2} = \mathrm{i}\,3.16$. The C = Constant curves are symmetrical with respect to the ϵ_f^i axis. When ϵ_2 is complex, these curves are symmetrical with respect to an axis joining the origin to the point corresponding to $(\epsilon_0\epsilon_2)^{1/2}$.

For a given value of δS, we have two values of ϵ_f which are solutions of eq. (3.2), one of them being unphysical. In the case corresponding to fig. 9, one solution corresponds to $\epsilon_f^r > 0$ and the other to $\epsilon_f^r < 0$.

Lopez-Rios and Vuye (1979) pointed out that, although it is impossible to have a simultaneous determination of both the (complex) dielectric constant ϵ_f and thickness d_f of a surface layer, it is nevertheless possible in some cases to have rather accurate values for these quantities by taking a parametric representation for $\epsilon_f(\omega)$. The method consists in fitting all the experimental data (reflectivity as a function of both angle of incidence and

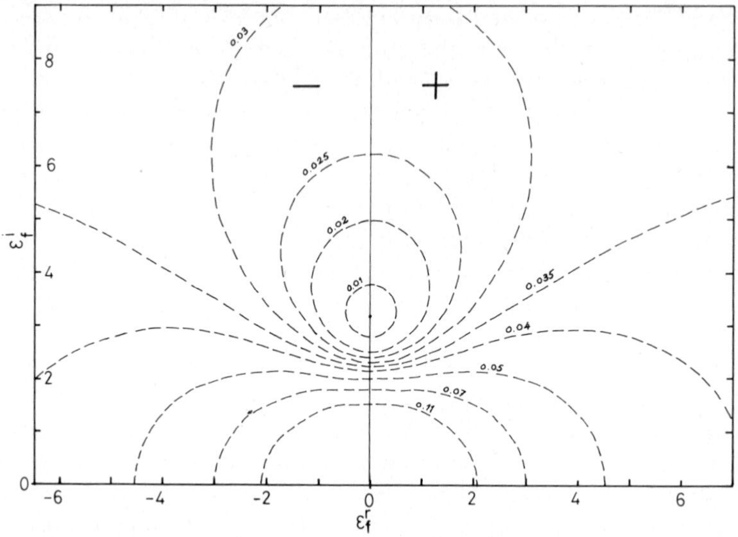

Fig. 9. Iso-precision curves $C = \text{constant}$ in the $(\epsilon_f^r, \epsilon_f^i)$ plane for the interface between vacuum and a medium with dielectric constant $\epsilon_2 = -10$. Notice that $C = 0$ for $\epsilon_f = (-10)^{1/2} = i\,3.16$. The curves are symmetrical with respect to the ϵ_f^i axis (given by $\epsilon_f^r = 0$). After Lopez-Rios (1980).

frequency) to their theoretical expression, which allows the determination of the parameters entering the $\epsilon_f(\omega)$ expression and the thickness. The latter being determined, it is then possible to obtain ϵ_f for each frequency. This has been tested on Au layers deposited on Ag, the "mass" thickness of which was also measured with a quartz microbalance. The results are given below:

Mass thickness (in Å)	13	28.4	40.6
Calculated thickness (in Å)	14.4	28.7	40.2

$\epsilon_f(\omega)$ was approximated in the considered spectral range (4000–6000 Å) by the simplest possible expression, namely $P_1 + P_2\omega + i(P_3 + P_4\omega)$, $P_1, \ldots, P_4$ being four unknown parameters. The five unknown parameters $P_1, \ldots, P_4$ and d_f were obtained by a fit to 150 to 180 experimental points.

4. *Modification of SPWs by a Surface Layer in the Vicinity of its Plasma Frequency; Nonlocal Effects*

We have already indicated that a superficial layer, even of very small thickness, can lead to an important modification in the characteristics of SPW propagation. This change is, of course, a function of the dielectric constant of the layer. According to eqs. (3.2) or (3.3) a singularity arises when the dielectric constant of the superficial layer becomes either infinite (or very large) or small. These two cases can occur for different frequencies in ionic crystals with dielectric constant well represented by an expression of the type: $\epsilon_f = \epsilon_\infty(\omega_{LO}^2 - \omega^2)/(\omega_{TO}^2 - \omega^2)$, ω_{LO} and ω_{TO} being the longitudinal and transverse vibrational frequencies respectively. The reflectivity has singularities for $\omega = \omega_{TO}$ (ϵ_f term) and for $\omega = \omega_{LO}$ (ϵ_f^{-1} term). This latter condition ($\epsilon_f = 0$) also occurs in a free-electron metal for the plasma frequency ω_{pf}.

Agranovich and Malshukov (1974) have shown that a superficial layer of an ionic crystal leads to a splitting of the surface polaritons of the substrate for the frequencies ω_{TO} and ω_{LO} of the superficial layer proportional to $d_f^{1/2}$, whereas the shift of the SPW resonance due to a thin superficial layer is usually proportional to d_f (see eq. (3.3)). Ngai and Economou (1971) had also found two branches in the dispersion relations for thin layers of an undamped plasma on another plasma in various configurations.

We examine here metals which can display an optical longitudinal mode for the plasma frequency ω_{pf} for which $\epsilon_f^r = 0$ and $\epsilon_f^i < 1$. Metallic plasma frequencies are located in the ultraviolet and the corresponding wavelengths are approximately two orders of magnitude smaller than in ionic crystals.

We have seen (in sect. 3) that a superficial layer leads, at fixed frequency, to a modification δK of the SPW wave vector, δK^r being either positive or negative according to the relative values of the dielectric constants of the substrate and superficial layer. If one works at fixed angle of incidence, there is a frequency shift $\delta\omega$ given by eq. (3.3). $\delta\omega$ is usually a complex quantity $\delta\omega = \delta\omega^r - i\delta\omega^i$, the imaginary part of which indicates a reduction ($\delta\omega^i > 0$) or increase ($\delta\omega^i < 0$) of the SPW life-time. When $\epsilon_f = 0$, the approximation leading to eq. (3.3) is no more valid. Nevertheless, we can investigate the situation occurring in the vicinity of the frequency for which $\epsilon_f = 0$ by setting $\epsilon_f \simeq (\partial\epsilon_f/\partial\omega)\delta\omega$. Neglecting then terms which are small with respect to ϵ_f^{-1} in eq. (3.3), we get:

$$(\delta\omega)^2 = \frac{2\epsilon_0\epsilon_2^3}{[-(\epsilon_0+\epsilon_2)]^{1/2}(\epsilon_0-\epsilon_2)(\partial\epsilon_2/\partial\omega)(\partial\epsilon_f/\partial\omega)}\,\frac{2\pi d_f}{\lambda}. \tag{4.1}$$

If the substrate is a free electron gas with dielectric constant $\epsilon_2 = 1-(\lambda/\lambda_p)^2$ and if the superficial layer is also a free electron gas with dielectric constant $\epsilon_f = 1-(\lambda/\lambda_{pf})^2$, and if $\lambda_{pf} \gg \lambda_p$, then eq. (4.1) leads, for $\epsilon_0 = 1$, to:

$$\delta\lambda/\lambda_{pf} = \pm\sqrt{\pi d_f/\lambda_p}. \tag{4.2}$$

The two roots of eqs. (4.1) and (4.2) correspond to the two branches of the dispersion relation, that is to say, to the two possible modes of the system.

Equation (4.2) shows that the two branches are at equal frequency distance from ω_{pf}.

It should be noted that, for a damped system, the splitting in the SPW dispersion relation occurs only if real values of the wave vector K are considered. In ATR experiments, this means working at fixed incidence. The situation is very different when the frequency is purely real, i.e., when working at fixed frequency in ATR experiments. In this case, a backbending happens at $\omega = \omega_{pf}$ and the splitting of the dispersion relation does not exist any more. Therefore, two minima in the reflectivity curves can only be observed in ATR experiments conducted at fixed angle of incidence as a function of frequency. Lopez-Rios (1976) has computed the dispersion curves in both cases for a K layer on an Al surface and has discussed the effects of damping. A schematic representation of both situations is shown in fig. 10.

An important problem arises when considering SPW dispersion relations (and the multiplicity of branches) for an inhomogeneous surface layer represented by a local dielectric constant $\epsilon_f(z, \omega)$ which vanishes for a given value of $z = z_0$. Cunningham et al. (1974) have investigated the SPW dispersion relations at semiconductor surfaces in the presence of a depletion, accumulation or inversion layer. They found a supplementary branch for the inversion layer only. This result is in disagreement with that of Guidotti et al. (1973), who find a supplementary branch in the SPW dispersion relation at a Hg surface covered with an exponential transition layer in all the situations which they examined and for any thickness of the transition layer. The existence of this supplementary branch has been the object of a controversy between the Chicago group (Guidotti and Rice 1976, Conwell 1976).

Lopez-Rios et al. (1978) observed a splitting in the SPW dispersion

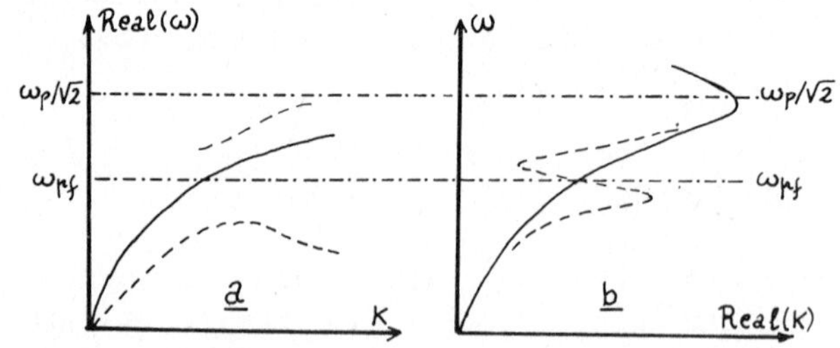

Fig. 10. Schematic representation of the SPW dispersion relation at a metallic surface (continuous line) and in the presence of a superficial plasma layer with plasma frequency ω_{pf} (dashed line). The plasma frequency of the bulk active medium is ω_p and $\omega_{pf} < \omega_p/\sqrt{2}$: (a) corresponds to measurements at fixed angle of incidence (splitting of the resonance); (b) corresponds to measurements at fixed frequency (double backbending of the dispersion curve).

curves of Al covered by Ag layers, 26, 39 and 58 Å thick, around the Ag plasma frequency. Figure 11 shows the dispersion curves ω versus reduced wave vector S, computed with the bulk optical constants of Al and Ag. The experimental points are also reported in this figure. The arrows indicate the frequencies for which the reflectivity measured from the external side (vacuum side) of the sample displays a minimum, corresponding to the Ag plasma frequency. One can see that the splitting indeed occurs around this frequency.

For films thinner than 26 Å, the splitting was not observed in these experiments. In order to explain this result and other experimental discrepancies with respect to theory, Lopez-Rios et al. (1979) suggested consideration of the following two effects:

(i) nonlocal effects associated with longitudinal plasma wave generation at a surface by electromagnetic waves. These effects are important when

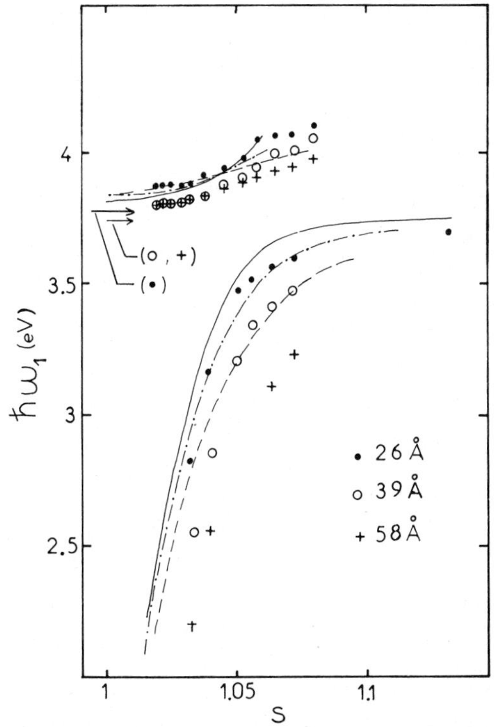

Fig. 11. Experimental and computed dispersion curves for Ag layers on a thick (214 Å) Al film. The experimental data correspond to the minima of the ATR reflectivity curves measured at fixed incidence and variable frequency. The computed curves (bulk optical constants) are represented by a continuous line (26 Å), dot-dashed line (39 Å) and dashed line (58 Å). After Lopez-Rios et al. (1978).

the layer thickness is of the order of the wavelength of longitudinal waves λ_L, which is much smaller than the wavelength of transverse electromagnetic waves: $\lambda/\lambda_L \sim c/v_F$, v_F being the Fermi velocity of conduction electrons in the superficial layer;

(ii) the nonspecular character of the collisions of free electrons with surfaces.

We now consider these two points successively. Nonlocal effects were taken into account in a very simple way, following the ideas put forwards by Forstmann (1967) and Melnyk and Harrison (1970). They consider, besides the electromagnetic (transverse) waves propagating in the metal with dispersion $k_T^2 = (\omega/c)^2\epsilon_T$, longitudinal polarization waves verifying the dispersion relation: $\epsilon_L(\omega, k_L) = 0$. The dielectric constants ϵ_T and ϵ_L of the Ag superficial layer were both taken as a sum of the contributions of free-electron interband transitions ϵ^c and d-electron interband transitions ϵ^b: $\epsilon = \epsilon^b + \epsilon^c$. ϵ^b was deduced from the results of optical measurements on thin Ag films (Dujardin and Thèye 1971). The free-electron contribution was given by a Drude expression, $\epsilon_T^c = 1 - \omega_{pf}^2/\omega(\omega + i\tau^{-1})$ for the transverse dielectric constant, and by the equation:

$$\epsilon_L^c = 1 - \frac{\omega_{pf}^2}{\omega(\omega + i\tau^{-1})}\left[1 + \frac{3}{5}\, v_F^2 \frac{K_L^2}{(\omega + i\tau^{-1})^2}\right]$$

valid for not too large values of K_L, for the longitudinal dielectric constant. The reflectivity and the SPW dispersion relation were obtained after taking suitable additional boundary conditions (Abelès et al. 1980).

Figure 12a shows the results of ATR experiments performed as a function of wavelength for fixed angle of incidence ($\theta = 37.22°$), under ultra-high vacuum, on an Al film, 114 Å thick. The spectra correspond first to the clean Al surface, then to this surface covered by Ag superficial layers, with thickness from 8 to 51 Å. Figure 12b shows the corresponding ATR spectra computed with the usual Fresnel equations, whereas fig. 12c shows the curves obtained when taking into account the longitudinal polarization waves as explained above. Notice that consideration of longitudinal waves leads to no splitting for very thin layers, which is in agreement with the experimental results.

The curves shown in figs. 13a, b, c are analogous to those of figs. 12a, b, c but correspond to ATR experiments carried out as a function of the angle of incidence at the Ag plasma frequency $\lambda = 3273$ Å.

We now consider the second point. For very thin surface layers, with thicknesses much smaller than the free-electron mean free path, strong absorption ought to be present due to the collisions of free electrons with the surfaces of the layer. Figure 14 shows the real and imaginary parts of the dielectric constant of Ag layers determined with the usual Fresnel equations from ATR experimental results similar to those represented in

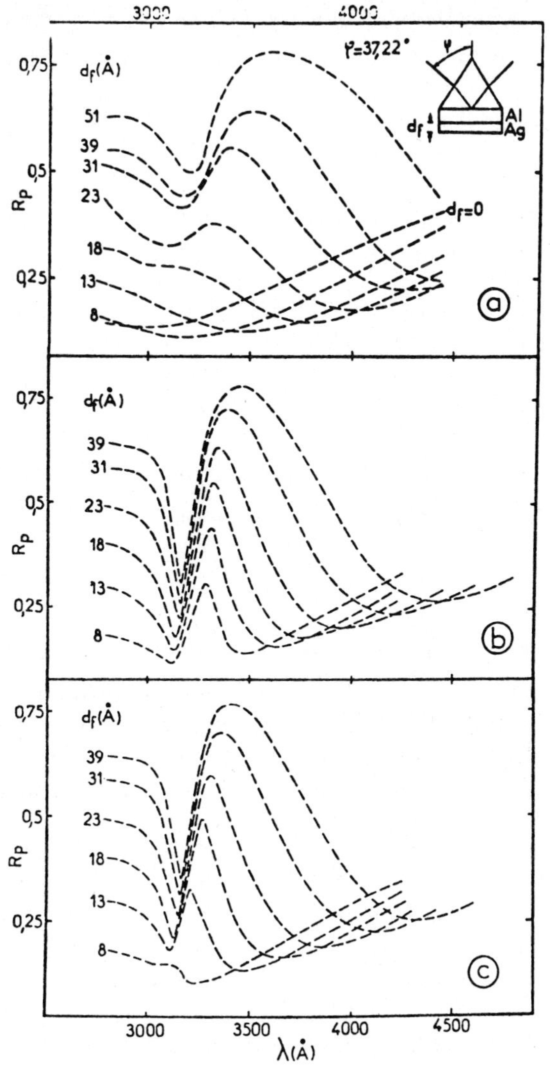

Fig. 12. ATR reflectivity R_p versus wavelength λ for fixed incidence 37.22° corresponding to an Al film 114 Å thick, bare and covered with Ag layers of various thicknesses d_f: (a) experimental curves; (b) computed curves using the usual Fresnel equations; (c) computed curves with longitudinal polarization waves taken into account. After Lopez-Rios (1980).

figs. 12 and 13. The imaginary part of ϵ_f displays such an extra-absorption, which increases with decreasing film thickness. The real part of ϵ_f also shows a variation with thickness, due to the inadequacy of the employed model. Figure 15 shows the real and imaginary parts of the dielectric

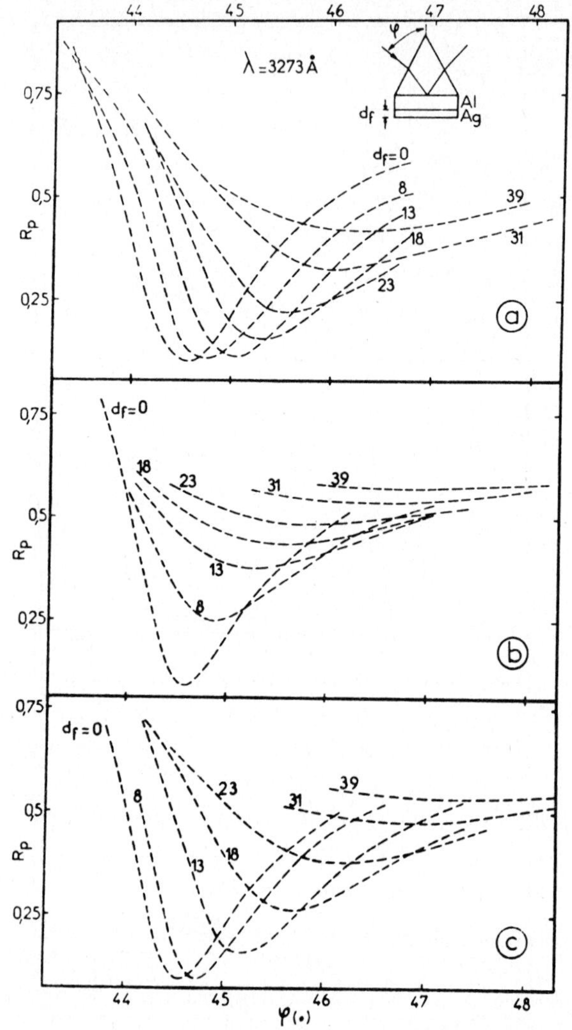

Fig. 13. ATR reflectivity R_p versus angle of incidence φ for fixed wavelength $\lambda = 3273$ Å, corresponding to the same layers as in fig. 12: (a) experimental curves; (b) computed curves using the usual Fresnel equations; (c) computed curves with longitudinal polarization waves taken into account. After Lopez-Rios (1980).

constant of the same Ag layers obtained when including longitudinal polarization waves, assumed to be unperturbed by the layer surfaces. In this case, there is still an important absorption, varying with layer thickness, but now the real part of ϵ_f is for all layers very close to the bulk Ag values. This example illustrates the difficulty of interpreting the optical properties of very thin surface layers, the usual simple models being no more valid.

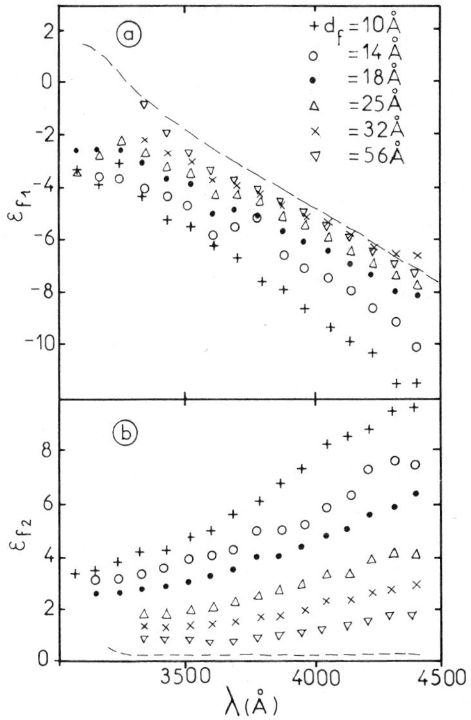

Fig. 14. Real part ϵ_{f1} (a) and imaginary part ϵ_{f2} (b) of the dielectric constant of Ag layers vs. λ in Å computed with the usual Fresnel equations from ATR measurements for various angles of incidence and frequencies. The Ag layers are deposited on a 220 Å thick Al film. After Lopez-Rios et al. (1979).

5. *Investigations of Surface Modifications in Ultra-high Vacuum*

Most of the work on surface polaritons was first performed in air, the aim being the investigation of the physical properties of the SPs themselves. Nevertheless, the sensitivity of surface polaritons to surface phenomena was rapidly acknowledged and ATR experiments were developed to study surface properties. Holst and Raether (1970) investigated Ag surfaces covered with C and LiF deposits, using the scattered light for SPW detection. Similar systems were later considered by Pockrand (1978). Abelès and Lopez-Rios (1974a) studied the tarnishing of Ag at ambient atmosphere with ellipsometric detection. The same problem was more recently examined by Kovacs (1978) with photometric detection. In the infrared, the reduction of the SPW propagation length along the surface

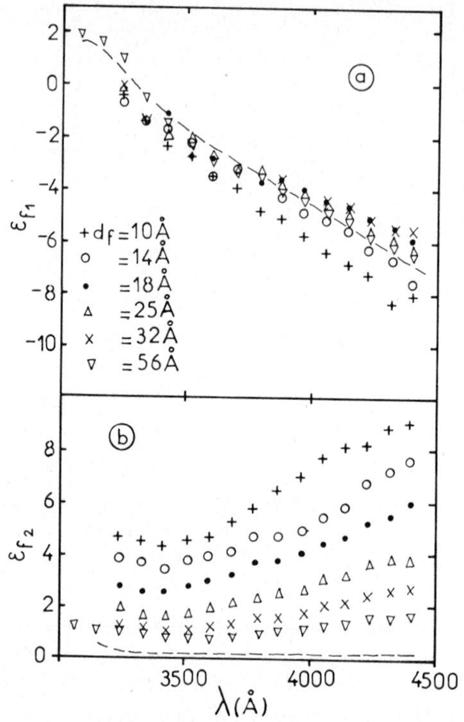

Fig. 15. Real part ϵ_{f1} (a) and imaginary part ϵ_{f2} (b) of the dielectric constant of Ag layers vs. λ in Å computed with the inclusion of longitudinal polarization waves from ATR measurements for various angles of incidence and frequencies. The Ag layers are deposited on a 220 Å thick Al film. After Lopez-Rios et al. (1979).

allowed the investigation of physisorbed molecules and oxide layers. Bhasin et al. (1976) used a tunable CO_2 laser to study physisorbed benzene films, 5 to 25 Å thick, whereas Bryan et al. (1976) characterized Cu oxide layers, 20 to 2000 Å thick, grown on Cu, in the 9.2–10.8 μm spectral range.

Only a few investigations have been performed under ultra-high vacuum. We shall now review some of them and describe a few experimental set-ups which have been used. Bösenberg (1971) detected surface plasmons generated by ATR by the increase of the photoelectric yield of Ag surfaces covered by Cs submonolayers; Cs was deposited on the Ag surface in order to lower the workfunction of Ag from about 4.5 eV to below 2 eV. His experimental set-up is schematically shown in fig. 16. The incident and reflected beams are antiparallel for any angle of incidence. The incoming beam enters a 90° prism and is successively reflected on its two faces. One of them is covered with a thick reflecting coating, the other one with a thin metal film (~ 500 Å thick). SPWs are excited at the film–vacuum surface by

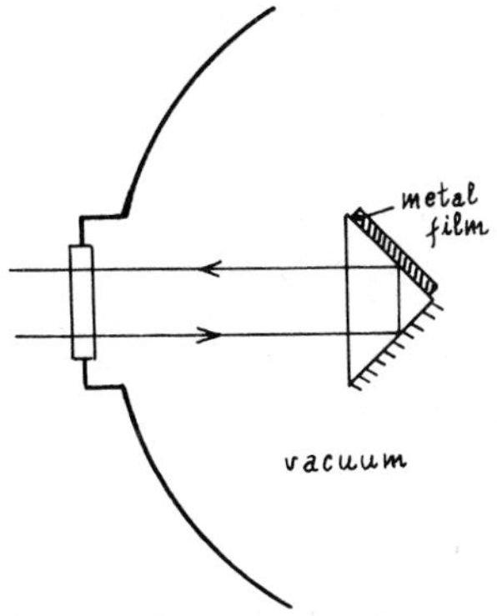

Fig. 16. Schematic description of the experimental set-up used by Bösenberg (1971) for ATR excitation of surface plasmons in the Kretschmann configuration.

the Kretschmann method. The incoming and outgoing beams pass through the same window. The outgoing beam is focussed on the photomultiplier in order to avoid spurious effects due to the beam shift accompanying the variations of the angle of incidence. With this same set-up geometry, Hincelin and Septier (1977) have studied photoemission yields of Ag covered with Cs and Cs oxide, then of cesiated Al (Hincelin and Septier 1980). The same technique was used by Chen and Chen (1980) who also investigated Ag covered with submonolayers of Cs and Cs oxide. These authors found that in the 0.45–1.1 μm spectral range, the Cs surface layer is strongly absorbing whereas the Cs-oxide layer is transparent.

Another set-up geometry, employed by Weber (1977) and Eagen and Weber (1979), is represented in fig. 17. Here the prism, located outside the vacuum chamber, is coupled to a window by means of an index-matching oil. A metal film is deposited on the vacuum side of the window and SPWs are generated at the metal–vacuum surface, again by the Kretschmann method. The advantage of this system is that all the mechanical parts are in air. Weber (1977) and Eagen and Weber (1979) detected the SPW by measuring the scattered light. As indicated in ch. 2, the intensity I of the scattered light as a function of angle of incidence has a Lorentzian shape: $I = A|S - S_m|^{-2}$ where $S = n_p \sin\theta$ and S_m is given by eq. (2.3), A is a function of the dimensions and shape of the surface roughness θ being related to the angle of incidence on the entrance face of the prism by the relation: $n_p \sin(\theta - \gamma) = \sin\alpha$, where n_p and γ are the refractive index and the interior angle of the prism respectively. α was the recorded angle in Weber's experiment. Modulation of this angle allowed the determination of the first and second angular derivatives of $I(\alpha)$. At the angle of resonance α_m, $dI/d\alpha = 0$, $I \propto A(\Delta\alpha)^{-2}$ and $d^2I/d\alpha^2 \propto A(\Delta\alpha)^{-4}$, where $\Delta\alpha$ is the half-width of the resonance directly related to $\mathrm{Im}(S_m)$. A servo-mechanical

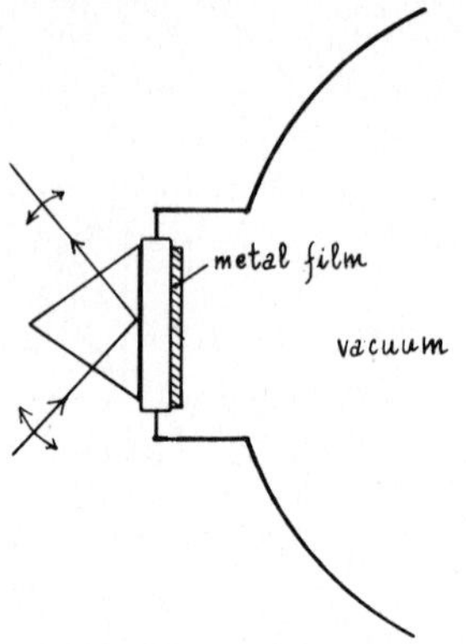

Fig. 17. Schematic description of the experimental set-up used by Weber (1977) for ATR excitation of surface plasmons at a metal–vacuum surface.

system was used to keep $\mathrm{d}I/\mathrm{d}\alpha = 0$ and allowed a recording of $I(\alpha_\mathrm{m})$ and $(\mathrm{d}^2I/\mathrm{d}\alpha^2)_{\alpha=0}$. For a Cu surface exposed to oxygen, Weber (1977) found that the quantity $I^2(\alpha_\mathrm{m})[\mathrm{d}^2I(\alpha_\mathrm{m})/\mathrm{d}\alpha^2]^{-1}$, which is proportional to surface roughness scattering, remained unchanged during oxygen exposure. He therefore assigned all the observed changes to the modifications of the SPW wave vector. Figure 18 taken from this work shows a plot of the half-width of the resonance $\Delta\alpha$ versus the angular position of the resonance α_m for various oxygen coverages. Two straight lines with different slopes are clearly apparent, the transition between the two occurring at 1–2×

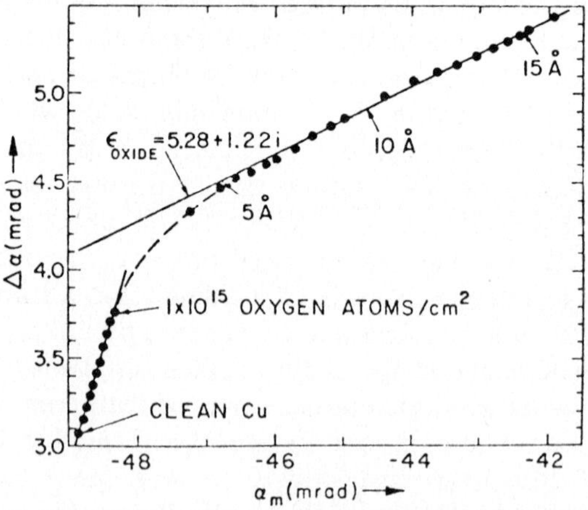

Fig. 18. Plot of the halfwidth of the SP resonance $\Delta\alpha$ vs. the angular position of the resonance α_m as O_2 adsorbs on Cu. The lower line corresponds to low coverages and its slope is proportional to the number of adsorbed atoms per cm^2. The upper line is drawn using eq. (3.2) with $\epsilon_\mathrm{oxide} = 5.28 + \mathrm{i}\,1.22$ and $\epsilon_\mathrm{Cu} = -15.06 + \mathrm{i}\,0.85$ for $\lambda = 6328$ Å. After Weber (1977).

10^{15} atoms/cm^2. The change in slope is interpreted as a transition between the presence of an oxygen chemisorbed state and the oxide formation. The upper straight line was computed with eq. (3.2) using the values of the optical constants of the oxide taken from the literature. Weber (1977) claimed that O_2 coverages as small as 0.2% of a monolayer could be detected.

A different experimental set-up is used in our laboratory. The underlying idea is illustrated in fig. 19. In this case a goniometer is located inside the ultra-high vacuum chamber, its angular precision being about 2×10^{-2} degrees. The outgoing beam follows the same path as the incoming beam after reflection on a mobile mirror, which means that there are two reflections on the basis of the prism. One therefore measures the square of the reflection coefficient. Figure 20 shows schematically the optical arrangement. A beam splitter gives a beam entering the vacuum chamber and incident on the sample, and a reference beam. These two beams are modulated by a chopper at two different frequencies, which allows to obtain the ratio of their intensities with a lock-in technique, as indicated in fig. 20. An advantage of this set-up is that it enables to perform, not only ATR measurements, but also reflectivity measurements on the basis of the prism from the vacuum side, just by turning the prism. Auger spectroscopy and film DC resistance measurements can be performed *in situ* simultaneously. The changes in the film resistance during deposition of surface ad-atoms give information about the superficial layer growth (Pariset and Chauvineau 1978).

Very recently, Chabal and Sievers (1980) reported high-resolution measurements of infrared absorption due to the ν_1 vibrational mode of H

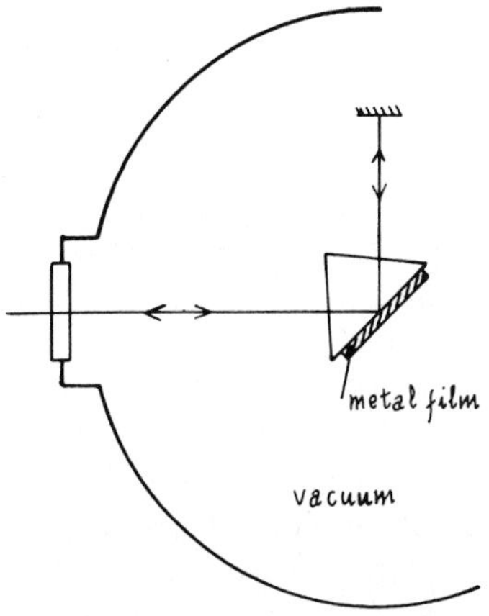

Fig. 19. Schematic description of the experimental set-up used by Lopez-Rios (1980) for ATR excitation of surface plasmons in the Kretschmann configuration.

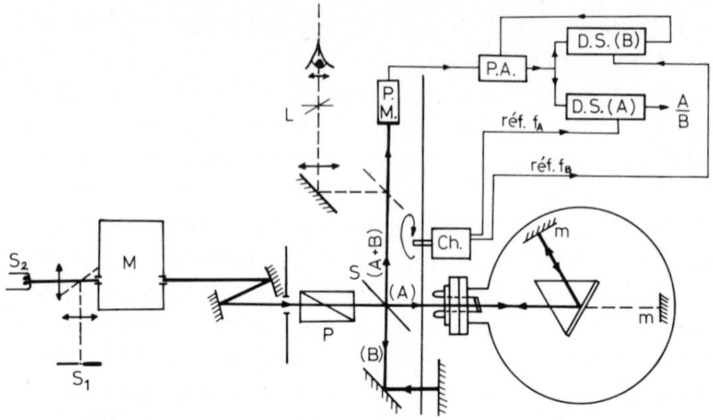

Fig. 20. A more detailed description of the experimental set-up of fig. 19. The beam-splitter S gives rise to a reference beam (B) and a measuring beam (A). S_1 and S_2 are two sources which can be used for different spectral regions M: monochromator, P: polarizer, Ch: chopper, P.M.: photomultiplier, P.A.: preamplifier, D.S.: synchronous detector, m: rotating mirror.

chemisorbed on W(100). Their technique makes use of surface plasmons propagating (over 1.3 cm) along the W surface and gives the transmission I of the guided wave as well as its modification ΔI induced by H adsorption. Chabal and Sievers find a good agreement between these results and those of high-resolution electron energy-loss spectroscopy as far as frequencies are concerned. But there is a strong disagreement when comparing the linewidths, which the authors explain by invoking dynamical coupling between hydrogen and the tungsten surface atoms.

6. *Use of SPWs for the Investigation of Solid–Liquid and Solid–Solid Interfaces*

Methods based on light spectroscopy are very well adapted to interface studies, which is not the case for methods using electron spectroscopy. In particular, optical methods can be employed for the investigation of metal/electrolyte interfaces in the region of transparency of the solution ($\sim$1 eV to $\sim$6 eV). SPW excitation was first used by Abelès et al. (1975) to study the Au/electrolyte interface in the Kretschmann configuration. This geometry is indeed very convenient in electrochemical work because light enters the electrolyte as an evanescent tail only. Besides, the electromagnetic field associated with SPW is highly concentrated at the interface and sensitive to charge modifications occurring in a region a few Å thick only in the metal. By this method, Kötz et al. (1977) determined the zero charge potential of Ag and Au/0.5M $NaClO_4$ using ATR excitation of SPWs at variable frequency. This technique is now employed by several groups for investigations of metal/electrolyte interfaces. A general review is given in ch. 8 of this book by Kolb.

Another interesting aspect of SPWs is that they offer the possibility to investigate solid–solid interfaces. If a free-electron metal is in contact with a transparent medium with dielectric constant ϵ_{diel}, SPWs can be excited at their interface by the ATR method provided one uses a prism with refractive index n_p such that $n_p^2 > \epsilon_{diel}$. Light scans both media but the electromagnetic field extends mostly in the transparent medium, as indicated by eq. (2.5). Moreover, for layered systems it is possible to examine the two surfaces of a layer independently by generating SPWs at each of them. A schematic representation of the dispersion curves for a semi-transparent metal film bounded by two transparent media with dielectric constants ϵ_0 and $\epsilon_2(\epsilon_0 > \epsilon_2)$ is shown in fig. 21. If the metal film is thick enough, the SPW modes at the two surfaces are decoupled and SPW can be excited at each surface by varying the angle of incidence at fixed frequency or the frequency at a given angle of incidence. Two distinct resonances are thus obtained. The insert to fig. 21 represents a possible geometry for the ATR excitation of both modes. Figure 22, taken from Abelès and Lopez-Rios (1974b), shows the values of the ellipsometric parameter tan ψ versus the reduced wave vector (and angle of incidence) for a Au film, 511 Å thick, deposited on a MgF_2 film, 1732 Å thick. The full curve corresponds to the experimental results, the dotted curve to values computed with $\epsilon_p^{1/2} = 1.886$, $\epsilon_0^{1/2} = 1.384$, $d_1 = 1732$ Å, $\epsilon = -9.75 - i1.20$, $\epsilon_2^{1/2} = 1$, $\lambda = 6093$ Å. As already pointed out, $\tan\psi = \sqrt{(R_p/R_s)}$ essentially describes the R_p behavior, R_s being nearly constant in this case. These curves display two SPW resonances corresponding to the two surfaces of the Au film. The minimum at about 34° corresponds to the Au/air surface, that at about 56° to the Au/MgF_2

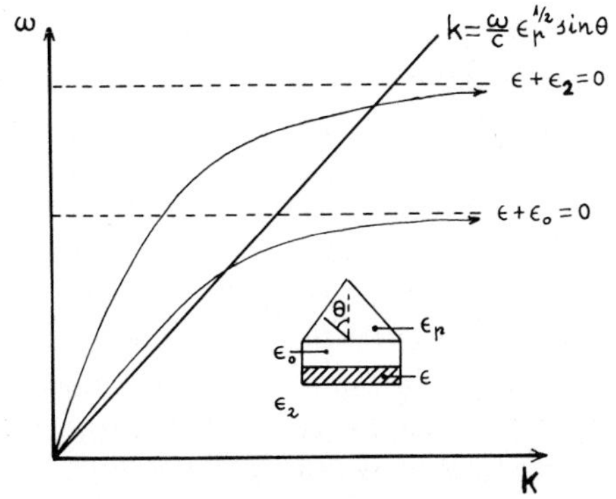

Fig. 21. SPW dispersion curves corresponding to the two surfaces of a metal film (ϵ) bounded by two dielectric media (ϵ_0 and ϵ_2).

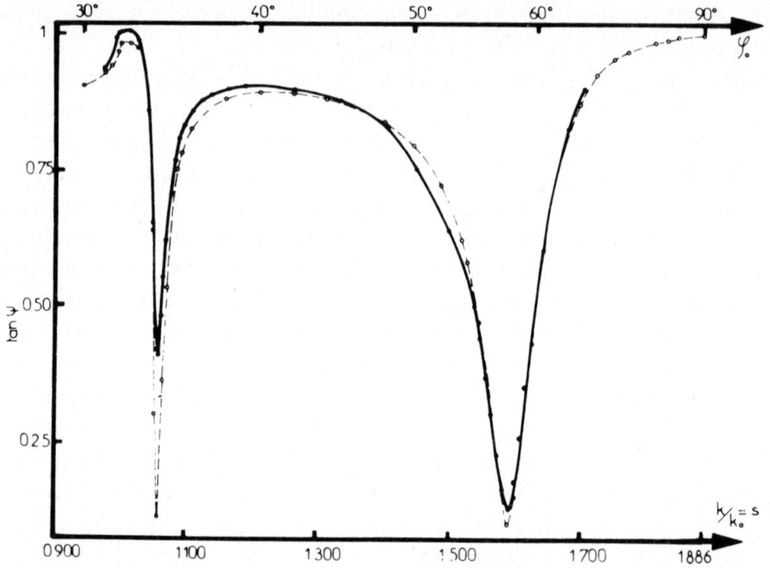

Fig. 22. tan ψ vs. reduced wave vector (and angle of incidence) for a 511 Å thick gold film deposited on a MgF_2 film, 1732 Å thick ($\epsilon_0^{1/2} = 1.384$, $\epsilon_2 = 1$). ATR experiments are performed at fixed wavelength $\lambda = 6093$ Å. Full curve: experimental results; dashed curve: computed with $\epsilon = -9.75 + i1.20$ for a prism with $\epsilon_p^{1/2} = 1.886$. After Abelès and Lopez-Rios (1974b).

interface. A possible application of this result is to the investigation of diffusion phenomena through metal films. Loisel and Arakawa (1980) studied the diffusion of Al into Au by using SPW excitation at the Au–dielectric interface. More complicated experimental situations corresponding to layer systems of alternate metal and dielectric films can also be employed, as illustrated by Kovacs and Scott (1977). These authors have computed the intensities of the electromagnetic fields and of the Poynting vector throughout the different layers.

An interesting situation occurs at the interface between two metals in the spectral region between the plasma frequencies of these metals. It was thoroughly examined by Halevi (1975, 1978) who computed the dispersion curves for SPWs at the Al/Mg interface, as well as the reflectivity curves corresponding to ATR experiments in the Otto and Kretschmann configurations. The basic idea is that, if ω_{p1} and ω_{p2} are the plasma frequencies of the two metals with dielectric constants ϵ_1 and ϵ_2 in contact ($\omega_{p1} > \omega_{p2}$), then for $\omega_{p2} < \omega < \omega_{p1}$, one has simultaneously $\epsilon_2 > 0$ and $\epsilon_1 < 0$. Thus the two media have dielectric constants of opposite sign, and this is the condition for the existence of SPWs at the interface. On the other hand, as pointed out by Halevi, $\epsilon_2 < 1$ means that the ATR excitation can occur without prism. For systems with several different films, SPWs can be excited at the various interfaces in complete analogy to the metal–dielectric case discussed above. Nevertheless, here the different interfaces can be scanned by varying the wavelength only. Figure 23 (Lopez-Rios 1980) shows the results of ATR measurements performed on a system consisting of an Al film, 220 Å thick, and a Ag film, 240 Å thick, as represented in the insert, together with a curve computed with the bulk optical constants for both metals. The same figure also shows the effects on the reflectivity of a thin Al superficial layer, 9 Å thick, on top of the Ag film, before and after oxidation of this layer. Two resonance minima are clearly apparent on these spectra. The high-energy (low-wavelength) one is related to SPW excitation at the Ag/Al interface, whereas the low-energy (high-wavelength) one is due to SPW excitation at the Ag/vacuum interface. Notice that only this low-energy resonance is affected by deposition of the Al superficial layer.

7. Conclusion

Our contribution was devoted to surface plasmons at metal surfaces. We have tried to summarize the work already done on this subject as well as on surface layers and adsorbed molecules when examined by using SPWs. Two questions are still to be answered. The first concerns the complete

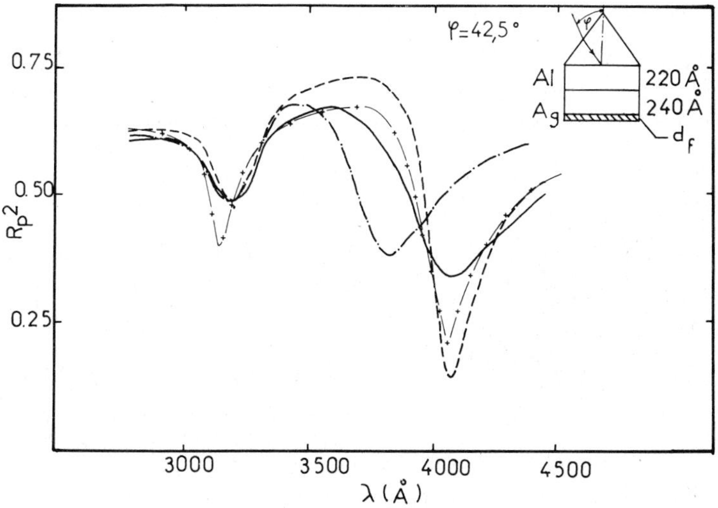

Fig. 23. Experimental curves $R_p^2(\lambda)$ for ATR measurements on the system Al(220 Å)–Ag(240 Å) deposited on a fused silica 60° prism. Dashed curve (---): bare Ag; dash-dotted curve (-.-.-.): the Ag film is covered by a 9 Å Al layer; full line (—): the Al layer has been oxidized. The crosses are computed values for the bare Ag surface. After Lopez-Rios (1980).

theoretical description of the interaction of electromagnetic waves with a metal surface in general and of SPWs at the surface. Some work has already been done on this subject, but we are still lacking a full solution of the problem, which definitely leads to nonlocal effects in many cases. It is very likely that the theoretical effort will be pursued and lead to a satisfactory result.

The second question calls for a more cautious answer. It concerns the future use of SPW to the investigation of metal surfaces and interfaces. Our guess is that, provided we have a good answer to the first question, the experimentalists will endeavor to still improve this technique, which will become one of the most powerful at hand, especially when extended both to the ultraviolet and to the infrared.

Acknowledgements

We thank Dr. M.L. Thèye for her critical reading of the manuscript.

References

Abelès, F., ed., 1966, Optical properties and electronic structure of metals and alloys (North-Holland, Amsterdam).
Abelès, F., ed., 1972, ed., Optical properties of solids (North-Holland, Amsterdam).
Abelès, F., 1977, J. Physique **C-5**, 67.
Abelès, F. and T. Lopez-Rios, 1974a, in: Proc. Taormina Conf. on Structure of Matter. eds., E. Burstein and F. DeMartini (Pergamon, New York) 241; 1974b, Opt. Comm. **11**, 89.
Abelès, F., T. Lopez-Rios and A. Tadjeddine, 1975, Solid State Commun. **16**, 843.
Abelès, F., Y. Borensztein, M. De Crescenzi and T. Lopez-Rios, 1980, Surf. Sci. **101**, 123.
Agranovich, V.M. and A.G. Malshukov, 1974, Opt. Commun. **11**, 169.
Alexander, R.W., G.S. Kovener and R.J. Bell, 1974, Phys. Rev. Lett. **32**, 154.
Anderson, J., G.W. Rubloff and P.J. Stiles, 1973, Solid State Commun. **12**, 825.
Anderson, J., G.W. Rubloff, M.A. Passler and P.J. Stiles, 1974, Phys. Rev. **B10** 2401.
Arakawa, E.T., M.W. Williams, R.N. Hamm and R.M. Ritchie, 1973, Phys. Rev. Lett. **31**, 1127.
Aspnes, D., 1976, in: Optical Properties of Solids: New Developments, ed., B. Seraphin (North-Holland, Amsterdam), p. 819.
Beaglehole, D. and E. Erlbach, 1972, Phys. Rev. **B6**, 1209.
Begley, D.L., D.A. Bryan, R.W. Alexander, R.J. Bell and C.A. Goben, 1976, Surf. Sci. **60**, 99.
Begley, D.L., R.W. Alexander, C.A. Ward and R.J. Bell, 1979a, Surf. Sci. **81**, 238.
Begley, D.L., R.W. Alexander, C.A. Ward, R. Miller and R.J. Bell, 1979b, Surf. Sci. **81**, 245.
Bell, R.J., M. Davarpanam, C.A. Goben, D.L. Begley, K. Bhasin and R.W. Alexander, 1975, Appl. Opt. **14**, 1579.
Bhasin, K., D. Bryan, R.W. Alexander and R.J. Bell, 1976, J. Chem. Phys. **64**, 5019.
Blanchet, G.B., P.J. Estrup and P.J. Stiles, 1980, Phys. Rev. Lett. **44**, 171.
Bösenberg, J., 1971, Phys. Lett. **37A**, 439.
Bryan, D.A., D.L. Begley, K. Bhasin, R.W. Alexander, R.J. Bell and R. Gerson, 1976, Surf. Sci. **57**, 53.

Burstein, E., W.P. Chen, Y.J. Chen and A. Hartstein, 1974, J. Vac. Sci. Technol. **11**, 1004.
Chabal, Y.L. and A.J. Sievers, 1980, Phys. Rev. Lett. **44**, 944.
Chabrier, G., 1978, Thesis, Dijon (France).
Chen, W.P. and J.M. Chen, 1980, Surf. Sci. **91**, 601.
Chen, Y.J., W.P. Chen and E. Burstein, 1976, Phys. Rev. Lett. **36**, 1207.
Chiarotti, G., S. Nannarone, R. Pastore and P. Chiaradia, 1971, Phys. Rev. **B4**, 3398.
Conwell, E.M., 1976, Phys. Rev. **B14**, 5515.
Cunningham, J.A., D.K. Greenlaw and C.P. Flynn, 1980, Phys. Rev. **B22**, 717.
Cunningham, J.A., D.K. Greenlaw, C.P. Flynn and J.L. Erskine, 1979, Phys. Rev. Lett. **42**, 328.
Cunningham, S.L., A.A. Maradudin and R.F. Wallis, 1974, Phys. Rev. **B10**, 3342.
Dujardin, M.M. and M.-L. Thèye, 1971, J. Phys. Chem. Solids, **32**, 2033.
Eagen, C.F. and W.H. Weber, 1979, Phys. Rev. **B19**, 5068.
Fano, U., 1941, J. Opt. Soc. Am. **31**, 213.
Forstmann, F., 1967, Z. Physik **203**, 495.
Gordon, J.G. and J.D. Swalen, 1977, Opt. Commun. **22**, 374.
Guidotti, D. and S.A. Rice, 1976, Phys. Rev. **B14**, 5518.
Guidotti, D., S.A. Rice and M.L. Lemberg, 1973, Solid State Commun. **15**, 113.
Halevi, P., 1975, Phys. Rev. **B12**, 4032.
Halevi, P., 1978, Surf. Sci. **76**, 64.
Hincelin, G. and A. Septier, 1977, Proc. 7th Int. Vac. Congr. and 3rd Int. Conf. on Solid Surfaces, Vienna, eds., R. Dobrozemsky, F. Rüdenauer, F.P. Viehböck and A. Breth (F. Berger and Söhne, Austria), vol. II, 1269.
Hincelin, G. and A. Septier, 1980, J. Physique Lettres, **41**, L127.
Holst, K. and H. Raether, 1970, Opt. Commun. **2**, 312.
Hummel, R.E., D.B. Dove and J. Alfaro Holbrook, 1970, Phys. Rev. Lett. **25**, 290.
Jasperson, S.N. and S. Schnatterly, 1969, Phys. Rev. **188**, 759.
Kliewer, K.L., 1980, Surf. Sci. **101**, 57.
Kliewer, K.L. and R. Fuchs, 1974, in: Adv. Chem. Phys., eds., I. Prigogine and S.A. Rice (John Wiley, New York) **27**, 355.
Kolb, D.M. and J.D.E. McIntyre, 1971, Surf. Sci. **28**, 321.
Kötz, R., D.M. Kolb and J.K. Sass, 1977, Surf. Sci. **69**, 359.
Kovacs, G.J., 1978, Surf. Sci. **78**, L245.
Kovacs, G.J. and G.D. Scott, 1977, Phys. Rev. **B16**, 1297.
Kretschmann, E., 1971, Z. Physik **241**, 313.
Loisel, B. and E.T. Arakawa, 1980, Appl. Opt. **19**, 1959.
Lopez-Rios, T., 1976, Opt. Commun. **17**, 342.
Lopez-Rios, T., 1980, Thesis, University P. and M. Curie, Paris.
Lopez-Rios, T. and G. Vuye, 1979, Surf. Sci. **81**, 529.
Lopez-Rios, T., F. Abelès and G. Vuye, 1978, J. Physique, **39**, 645.
Lopez-Rios, T., F. Abelès and G. Vuye, 1979, J. Physique Lettres, **40**, L-343.
Macek, C.H., A. Otto and W. Steinmann, 1972, Phys. Status Solidi. **B51**, K59.
Melnyk, A.R. and M.J. Harrison, 1970, Phys. Rev. **B2**, 835 and 851.
Mott, N.F. and H. Jones, 1936, The Theory of the Properties of Metals and Alloys (Oxford University Press, London).
Ngai, K.L. and E.N. Economou, 1971, Phys. Rev. **B4**, 2132.
Otto, A., 1968, Z. Physik, **216**, 398.
Otto, A., 1976, in: Optical Properties of Solids: New Developments, ed., B. Seraphin (North-Holland, Amsterdam) p. 677.
Pariset, C. and J.P. Chauvineau, 1978, Surf. Sci. **78**, 478.
Pockrand, I., 1978, Surf. Sci. **72**, 577.
Raether, H., 1977, in: Physics of Thin Films, eds., G. Hass, M.H. Francombe and R.W. Hoffman (Academic Press, New York) p. 145.

Ritchie, R.H., 1965, Surf. Sci. **4**, 497.
Ritchie, R.H., E.T. Arakawa, J.J. Cowan and R.N. Hamm, 1968, Phys. Rev. Lett. **21**, 1530.
Rubloff, G.W., J. Anderson and P.J. Stiles, 1973, Surf. Sci. **37**, 75.
Rubloff, G.W., J. Anderson, M.A. Passler and P.J. Stiles, 1974, Phys. Rev. Lett. **32**, 667.
Rudolf, H.W. and W. Steinmann, 1977, in Proc. 7th Int. Congr. and 3rd Int. Confer. Solid Surfaces, Vienna, eds., R. Dobrozemski, R. Rüdenauer, F.P. Viehböck and A. Breth, p. 1265.
Schoenwald, J., E. Burstein and J.M. Elson, 1973, Solid State Commun. **12**, 185.
Seraphin, B.O., ed., 1976, Optical properties of solids. New Developments (North-Holland, Amsterdam).
Simon, H.J., D.E. Mitchell and J.G. Watson, 1974, Phys. Rev. Lett. **33**, 1531.
Stanford, J.L., 1970, J. Opt. Soc. Am. **60**, 49.
Teng, Y.Y. and E.A. Stern, 1967, Phys. Rev. Lett. **19**, 511.
Vuye, G., 1980, Thesis, University P. and M. Curie, Paris.
Weber, W.H., 1977, Phys. Rev. Lett. **39**, 153.
Weber, W.H. and S.L. McCarthy, 1975, Phys. Rev. **B12**, 5643.
Wilson, A.H., 1936, The Theory of Metals (Cambridge University Press, Cambridge).

CHAPTER 7

Resonance of Transition Layer Excitations with Surface Polaritons

G. N. ZHIZHIN and V. A. YAKOVLEV

Institute of Spectroscopy
USSR Academy of Sciences
142093 Troitzk, Moscow region
USSR

Surface Polaritons
Edited by
V.M. Agranovich and D.L. Mills

Contents

1. Introduction 277
2. The theory of resonance 280
3. Resonance of the film phonon with the substrate surface phonon polaritons . . . 284
4. Other cases of resonance with the transition layer 293
References 296

1. Introduction

In many cases thin transition layers exist on solid surfaces and their structure differs from the bulk ones. Certainly, the other properties of such layers are different from those of bulk materials. Thin films on solid surfaces (adsorbed molecules, optical coatings, etc.) can also be considered as transition layers. Their properties are one of the subjects of surface polariton spectroscopy (Agranovich 1975).

Surface polaritons are localized in the vicinity of the interface. Their electromagnetic field intensity is the highest at the interface and decreases exponentially with the distance from the interface. This means that such excitations must be sensitive to the surface state and thin transition layers and thus they can be used as a good tool for surface studies.

The spectrum of the crystal surface excitations can be changed by a thin film. This change yields information on the thickness and optical properties of the transition layer.

Two cases should be considered here. In the first one, studied experimentally by Zhizhin et al. (1974, 1977b) and Mirlin and Reshina (1974) there is no selective absorption of the film; then the surface polariton line is shifting and broadening. For the films of high-conductivity metals this broadening is large and can be used for the determination of conductivity as it was shown by Agranovich (1975) and Agranovich and Leskova (1974). The second case corresponds to the coincidence of the film excitation frequencies with the frequency range of the substrate surface polaritons. The theoretical consideration (Agranovich 1975, Agranovich and Mal'shukov 1974) shows the splitting of the substrate surface polariton peak at the frequency of the film excitation given schematically in fig. 1.

Transition layers should be subdivided by their thicknesses into microscopic (if the thickness is of the order of the interatomic distance) and macroscopic (when the thickness is well in excess of the lattice constant). In the first case the transition layer is characterized by molecular polarizabilities and its effect is taken into account by the variation of boundary conditions. The macroscopic transition layer can be described by the dielectric function ϵ. Note that the effective dielectric function is introduced also for microscopic transition layers and the results in many cases coincide with those obtained with a more rigorous microtheory.

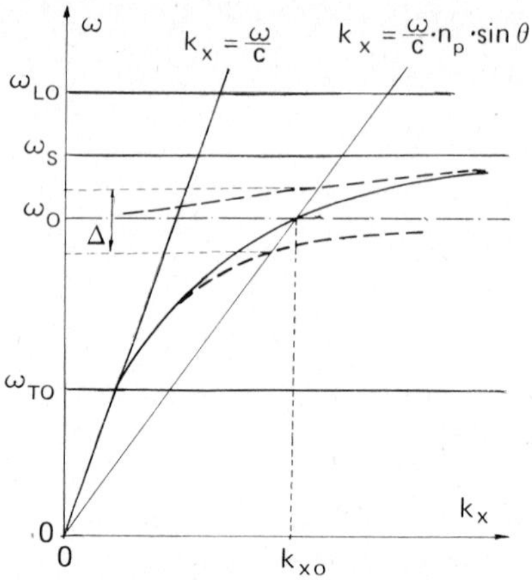

Fig. 1. A gap on the dispersion curve of the surface polariton at the resonance with the vibration of the transition layer.

Theoretical treatment of the splitting was performed for both microscopic and macroscopic isotropic transition layers.

The most interesting feature of the splitting predicted is the square root dependence on the thickness whereas the majority of the optical effects are proportional to the film thickness. This square root dependence increases the sensitivity of the spectroscopic method in the case of very thin films.

In the attenuated total reflection (ATR) experiment this effect should be observed in the splitting of a single surface polariton line (peak) into two lines. The damping limits the possibilities of the method: the splitting value less than the line width cannot be recorded. The other effect of damping is that the dispersion curve obtained by ATR at the angular scanning at various fixed frequencies differs from the dispersion curve obtained from the ATR spectra at fixed angles of incidence (frequency scan).

If an ATR spectrum is recorded at a fixed angle of incidence θ in a prism with an index of refraction n_p, then the reduced wave vector is given by

$$\kappa_x = n_p \sin\theta = k_x c/\omega \tag{1}$$

where k_x is the wave vector, $\omega = 2\pi\nu c = 2\pi c/\lambda$ is the frequency of the surface polariton, λ is the wavelength. If damping is not taken into account, then both the frequency and the wave vector are real and the dispersion is defined as the dependence of the surface polariton frequency on the wave vector (or on the reduced wave vector what is more convenient for the comparison with the experiment). If damping is taken into account, the dispersion is no more defined unambiguously and becomes dependent on the specific experimental situation. In the case of ATR spectra (frequency scan) measured at fixed reduced wave vectors, the damping determines the line width in the spectrum and the dispersion (κ_x dependence of the minimum frequency in the spectrum) differs only slightly from that obtained with no regard for damping. In this case the frequency increases with the wave vector and tends asymptotically to a limiting value as it occurs in the absence of damping. At the angular scanning the frequency is fixed and the dispersion can be defined as the frequency dependence of the angular position of the ATR minimum which appears due to the excitation of the surface polariton. Away from the limiting frequency both curves coincide; as this frequency is approached, a discrepancy appears since frequency and angular dependences of the reflectivity R are different sections of the surface $R(\omega, \theta)$. In the vicinity of the limiting frequency the angle at which the surface polariton is excited is maximum and the value of this maximum depends on the magnitude of damping. At higher frequencies the excitation angle decreases while in the absence of damping there is no surface polaritons in this frequency range.

Such behaviour of dispersion curves may be explained, if one remembers that at the given frequency the reduced wave vector of a surface polariton is equal to the refractive index of a surface wave introduced by Agranovich (1975):

$$\kappa_x = \left(\frac{\epsilon(\omega)}{\epsilon(\omega)+1}\right)^{1/2} \tag{1a}$$

for a crystal with the dielectric function $\epsilon(\omega)$ bordering upon vacuum. In the region lying between the limiting frequency ω_s, ($\epsilon(\omega_s) = -1$) and the longitudinal frequency ω_{LO}, ($\epsilon(\omega_{LO}) = 0$) the refractive index eq. (1a) is an imaginary value (in the absence of damping) and surface waves do not exist. When damping is taken into account, the refractive index becomes complex and, as in the case of crystal optics, in the region with no waves (in the absence of damping) a region of anomalous dispersion appears in the real part of the refractive index where the absorption is very high. This anomalous dispersion results in the appearance of back bending on the dispersion curves at angular scan ATR – which was observed experimentally for a metal by Arakawa et al. (1973), see also Alexander et al. (1973), and for dielectrics by Schuller et al. (1975a, b), Zhizhin et al. (1976a, 1977b).

The region where the surface polariton splitting is observed should correspond to the region of anomalous dispersion of the surface polariton refractive index and, therefore, to the back bending obtained at the angular scan ATR at fixed frequencies. In the resonance region the absorption of surface waves increases, which leads to the broadening of the line at the angular scan ATR and to the decrease of the wave path length. Thus, the resonance of transition layer excitations with the substrate surface polariton can be detected experimentally from the splitting of the surface polariton line, from the appearance of the back bending, from the decrease of the propagation length or from the broadening of the minimum in the angular dependence of the reflectivity. In this case the value of splitting follows the square root dependence and the other effects depend linearly on the film thickness.

2. *The Theory of Resonance*

If the condition of the film thickness smallness as compared to the wavelength is not fulfilled, the exact equation of surface polariton dispersion in the layered system must be used. For a film of the thickness l with the dielectric tensor ϵ_2 between semi-infinite media ϵ_1 and ϵ_3 (dielectric tensors are considered to be diagonal in the Cartesian coordinates with the axis x along the propagation direction of the surface polariton and the axis z perpendicular to the film) the equation for surface polariton dispersion can be written in the form:

$$(\beta_1+\beta_2)(\beta_2+\beta_3)+(\beta_1-\beta_2)(\beta_2-\beta_3)\exp(-2\kappa_2 l] = 0, \tag{2}$$

where

$$\beta_i = \epsilon_{ix}/\kappa_i; \quad \kappa_i = \frac{\omega}{c}\left(\frac{\epsilon_{ix}}{\epsilon_{iz}}\kappa_x^2 - \epsilon_{ix}\right)^{1/2}.$$

When damping is taken into account, eq. (2) becomes complex and relates two, generally speaking, complex values: the frequency and the wave vector. Usually one of these values, namely that given by the experimental conditions, is assumed to be real. Thus, the solution of eq. (2) with the real frequency corresponds to the experiment with "the angular scan ATR"; the real part of the complex reduced wave vector corresponds to the angular position of the minimum, and the imaginary part – to its width. The solution with the real reduced wave vector corresponds to the spectra obtained at fixed angles of incidence. The real part of the complex frequency gives the line frequency in the spectrum, and the imaginary part gives a half of the line width.

Equation (2) can be solved only numerically with the exception of two cases. One of them is the above mentioned case of very thin films. The other extreme case (very thick films) leads to the splitting of the equation into two equations for two film boundaries

$$\beta_1+\beta_2=0, \tag{3a}$$

$$\beta_2+\beta_3=0. \tag{3b}$$

Such "boundary" surface polaritons were discussed by Halevi (1977,

1978) for the case of a contact of two metals and two dielectrics. They are not considered in detail here as we are interested in the thin transition layer. We only point out that, if the region with the negative dielectric function of the film lies in the region of the substrate surface polariton, then three surface polaritons can be observed. One surface polariton is localized at the interface between the film and the outer medium and lives in the frequency region, where the dielectric function of the film is negative, i.e., between the frequencies of transverse and longitudinal vibrations. In this region no polaritons are available at the film–substrate interface: they are present in two other regions – one at the frequencies below the transverse frequency of the film, the other – above the longitudinal frequency. In the vicinity of the transverse and longitudinal frequency of the film there exist regions where surface polaritons do not exist.

With the decrease of thickness when surface polaritons at two film boundaries cannot be considered independent any more and eq. (2) must be solved, there remain three surface polaritons which are now already mixed up. It was shown experimentally (Gerbstein and Mirlin 1974) for the ZnSe film on the InSb substrate with high concentration of free carriers.

The frequencies of transverse and longitudinal film vibrations fall in the range where there exists a surface plasmon of the substrate. Three branches of mixed plasmon–phonon surface polaritons were observed; more details on this experiment are given in ch. 1 by Mirlin.

For very thin films when the exponent index in eq. (2) is small, it is shown by Agranovich (1975) and Agranovich and Mal'shukov (1974) that the splitting value is equal to

$$\Delta = 2\sqrt{(\omega_0 A C(\kappa_{x0}))} \tag{4}$$

where ω_0 is the resonance frequency, κ_{x0} is the reduced wave vector at which the substrate surface polariton frequency (without a film) is equal to ω_0,

$$C(\kappa_{x0}) = \frac{\partial}{\partial\omega}\left(\frac{\epsilon_1}{\kappa_1} + \frac{\epsilon_3}{\kappa_3}\right)_{\omega=\omega_0,\ \kappa_x=\kappa_{x0}}. \tag{5}$$

For the macroscopic transition layer on the assumption that

$$\epsilon_2(\omega) = \epsilon_{\infty 2}\frac{\omega_{LO2}^2 - \omega^2}{\omega_{TO2}^2 - \omega^2} = \epsilon_{\infty 2} + \frac{(\epsilon_{02} - \epsilon_{\infty 2})\omega_{TO2}^2}{\omega_{TO2}^2 - \omega^2} \tag{6}$$

the following formulae can be obtained (Agranovich 1975):

$$A_{TO} = l\epsilon_{\infty 2}\frac{\omega_{LO2}^2 - \omega_{TO2}^2}{2\omega_{TO2}^2} = l\frac{\epsilon_{02} - \epsilon_{\infty 2}}{2}, \tag{7a}$$

$$A_{LO} = \frac{l}{\epsilon_{\infty 2}}\frac{\omega_{LO2}^2 - \omega_{TO2}^2}{2\omega_{LO2}^2}\frac{k_{x0}^2\epsilon_3^2(\omega_0)}{\kappa_3^2(\kappa_{x0})} = l\frac{\epsilon_{02} - \epsilon_{\infty 2}}{2\epsilon_{02}\epsilon_{\infty 2}}\epsilon_1|\epsilon_3(\omega_0)|, \tag{7b}$$

i.e., both transverse and longitudinal vibrations of the transition layer can resonate with the surface polariton, but the gap width for them is different. Near the frequency of transverse vibrations for the dielectric substrate or –for a metal– far away from the plasma frequency ($|\epsilon_3| \gg 1$) the longitudinal phonon of the film should also give a wider gap. In the region where $\epsilon_3 \sim -1$ a stronger resonance is realized for the transverse phonon of the film. In studying surface electromagnetic waves on metals with $|\epsilon_3| \gtrsim 1000$ in the IR spectral region the absorption by films can only be observed at logitudinal oscillation frequencies (see ch. 3 by Zhizhin et al.).

Coefficient $C(\kappa_{x0})$ in eq. (4) is

$$C(\kappa_{x0}) = \frac{1}{c} \frac{\epsilon_3^2}{(\epsilon_1 - \epsilon_3)(|\epsilon_1 + \epsilon_3|)^{1/2}} \frac{(\omega_{\mathrm{TO3}}^2 - \omega_0^2)^2}{\epsilon_{\infty 3}(\omega_{\mathrm{LO3}}^2 - \omega_{\mathrm{TO3}}^2)}$$
$$= \frac{1}{c} \frac{\epsilon_3^2(\omega_{\mathrm{TO3}}^2 - \omega_0^2)^2}{(\epsilon_1 - \epsilon_3)(|\epsilon_1 + \epsilon_3|)^{1/2}\, \omega_{\mathrm{TO3}}^2(\epsilon_{03} - \epsilon_{\infty 3})} \tag{8}$$

where for $\epsilon_3(\omega)$ a dependence analogous to ϵ_2 eq. (6) is assumed. For $|\epsilon_3| \gg \epsilon_1$ the formula is simplified as follows

$$C(\kappa_{x0}) = \frac{1}{c} \frac{\sqrt{|\epsilon_3|}(\omega_{\mathrm{TO3}}^2 - \omega_0^2)^2}{\omega_{\mathrm{TO3}}^2(\epsilon_{03} - \epsilon_{\infty 3})}. \tag{9}$$

The difference between eq. (9) here and the similar formula used by Zhizhin et al. (1977a) is due to the fact that the differentiation of eq. (5) was made at fixed κ_x, but not k_x as in the cited paper because ATR spectra were measured at the given reduced wave vector (see above).

Let us consider the frequency dependence of the splitting value shown in fig. 2. For the substrate the following model was used: $\epsilon_{\infty 3} = 3$, $\epsilon_{03} = 10$,

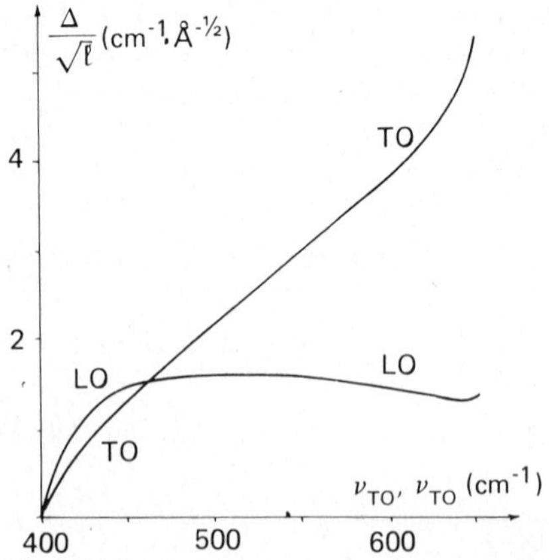

Fig. 2. Splitting as a function of the resonance frequency for the film transverse oscillation (TO curve) and longitudinal oscillation (LO curve) for the film thickness 1 Å. Approximate calculation by formulae (4–9).

$\nu_{TO3} = \omega_{TO3}/2\pi c = 400\,cm^{-1}$. Formulae (4–9) were used to calculate the splitting values for the film of the thickness 1 Å as a function of the resonance frequency for the transverse and longitudinal phonons of the film with $\epsilon_{\infty 2} = 2$, $\epsilon_{02} = 9$. The TO curve corresponds to the resonance at various frequencies of the film transverse phonon which falls in the region of the substrate surface polariton; the LO curve corresponds to the longitudinal phonon falling within this region. At the resonance frequency close to the substrate transverse frequency the splitting is small, but for the resonance with the longitudinal frequency of the film it is much greater. The splitting in the case of TO frequency increases with the resonance frequency, while the splitting for the LO frequency in the wide frequency interval is scarcely variable, since $|\epsilon_3|$ (see eq. (7b)) decreases with the increase of frequency.

If we assume that splitting can be recorded at its value compared with the line width ($\sim$10–20 cm^{-1}), then the films of the thickness $\sim$100 Å can be detected by the method described. To increase the sensitivity of the method it is desirable that the damping should not be too high both in substrates (to obtain a narrow line to be splitted) and in films (for the attenuation of metal quenching which occurs due to the complex dielectric function of the film).

In this section the frequency scan ATR has been considered up to now. For the angular scan ATR damping must be taken into account. An approximate formula for the refractive index can be obtained from eq. (2). Such formula was obtained by Pockrand and Swalen (1978) and Pockrand et al. (1979). With the designations used above for changes in the surface electromagnetic wave (SEW) refractive index due to the coating with an anisotropic film it can be rewritten as follows:

$$\kappa_x - \kappa_{x0} \approx 2\pi\nu l \left(\frac{\epsilon_1\epsilon_3'}{\epsilon_1+\epsilon_3'}\right)^2 \frac{[(\epsilon_{2x}' + \epsilon_{2z}'\epsilon_3'/|\epsilon_{2z}|^2 - \epsilon_3' - 1) + i(\epsilon_{2x}'' - \epsilon_{2z}''\epsilon_3'/|\epsilon_{2z}|^2)]}{\sqrt{-\epsilon_1\epsilon_3'}\,(\epsilon_1 - \epsilon_3')}. \tag{10}$$

Here $\epsilon_{2x} = \epsilon_{2x}' + i\epsilon_{2x}''$ is the dielectric function of the film along the surface, $\epsilon_{2z} = \epsilon_{2z}' + i\epsilon_{2z}''$ is perpendicular to the surface. Resonance of the refractive index occurs at the frequency of transverse oscillations polarized along the film and at the frequency of longitudinal oscillations in the direction perpendicular to the film. Resonance of the refractive index of surface polariton vibrational states in the anisotropic film is discussed in more detail by Pockrand and Swalen (1978) and Pockrand et al. (1979) for the resonance of film excitons with the surface plasmon polariton in a metal.

Figure 2 shows the results of the calculations made for phonon excitations in the film and in the substrate; however, formula (6) may also be used for excitonic excitations and, if we assume $\omega_{TO} = 0$, the same formula describes also plasma excitations in metals and semiconductors. This means that a similar calculation with formulae (4–9) can also be made for the above excitations. The comparison of the theoretically predicted square root dependence of splitting on the transition layer thickness with the experimental one has been performed up to now only for the film phonon resonance with the substrate surface phonon polaritons. Accordingly, in the next section just this type of resonance will be considered in detail.

3. *Resonance of the Film Phonon with the Substrate Surface Phonon Polaritons*

Experimental proof of the effects predicted theoretically by Agranovich and Mal'shukov (1974) and Agranovich (1975) was obtained with the use of resonance between surface phonon polaritons in crystals and a longitudinal phonon of lithium fluoride (LiF) film by Yakovlev et al. (1975a) and Zhizhin et al. (1975, 1977a). The following crystals were used as substrates: rutile (TiO_2), sapphire (α-Al_2O_3), magnesium oxide (MgO) and yttrium iron garnet ($Y_3Fe_5O_{12}$) which will further be abbreviated to YIG. The last two crystals are cubic, and sapphire and rutile are uniaxial, therefore, for these two crystals the spectrum and the dispersion of surface polaritons depend on the direction of the optical axis. The MgO crystal has one dipole-active phonon; due to anharmonicity its dielectric function cannot be described by a single Lorentz oscillator, but Kachare et al. (1972) approximated one by two oscillators. A surface polariton exists in the region 400–630 cm^{-1} (Reshina et al. 1976). For YIG surface polaritons we are interested in two high-frequency ones with the frequencies within the regions 590–647 and 652–680 cm^{-1} (Yakovlev et al. 1974, 1978, Yakovlev 1976). For sapphire the high-frequency surface polariton for various crystal orientations was observed by Yakovlev and Zhizhin (1974), Yakovlev et al. (1975b) in the region 635–810 cm^{-1}; the anisotropy in this region is rather small. The rutile surface polariton which is of interest to us has the dispersion region 500–750 cm^{-1} (Bryksin et al. 1973).

The LiF crystal, as the MgO one, has only one IR active phonon. The dispersion analysis made by Kachare et al. (1972) for the reflection spectrum gives $\nu_{TO} = 307.5\ cm^{-1}$, $\epsilon_\infty = 1.9$, the oscillator strength $\Delta\epsilon = 6.67$ and damping $\gamma/\nu_{TO} = 0.057$; in addition, due to anharmonicity, the second oscillator has to be introduced with $\nu_{TO} = 501.4\ cm^{-1}$, $\Delta\epsilon = 0.116$ and damping 0.173. Dielectric function of the crystal equals the sum of increments of oscillators:

$$\epsilon(\nu) = \epsilon_\infty + \sum_j \frac{\Delta\epsilon_j \nu_{TOj}^2}{\nu_{TOj}^2 - \nu^2 - i\nu\gamma_j} \tag{11}$$

The frequency of the longitudinal phonon of the LiF crystal is about $670\,cm^{-1}$.

LiF films were deposited by thermal evaporation in vacuum (10^{-5}–10^{-6} Torr) onto crystal substrates at room temperature. The film thickness was determined by means of a quartz resonator technique; in addition, for greater thicknesses an interferometric Tolanski method was applied (Glang 1970).

ATR spectra were measured with IR spectrophotometers IKS-16 and Hitachi-225 with an ATR unit NPVO-1 (LOMO, Leningrad). All measurements were made in Otto (1968) configuration; the sample was placed at a small (several μm) distance from the ATR element – the KRS-5 or KRS-6 semi-cylinder and the gap was filled with either air or nujol and the size of this gap was adjusted by means of mylar spacers on the periphery of the sample and was chosen big enough to exclude radiation distortion of the spectrum caused by a prism placed close to the sample. The ATR spectra were measured at various fixed angles of incidence (frequency scan ATR) and in some cases measurements were also made of the angular dependences of reflectivity at fixed frequencies (angular scan ATR). Since with large gaps the prism effect is not considerable, the minimum position frequencies in the spectra and the positions of the minima in the angular dependences allow to build dispersion curves at given reduced wave vector and at given frequency, respectively, and to observe the splitting in the first case and the back bending in the second one.

In ATR spectra of sapphire and rutile crystals the splitting could be observed for the film of the thickness just about 100 Å. Figure 3 shows the variation of ATR spectra (at two angles of incidence) of the rutile crystal with thin films of LiF of 160 Å and 270 Å thickness. The optical axis of the rutile crystal is perpendicular to the surface. It can be seen that for the film of thickness 160 Å the splitting of the minimum is only beginning to appear and the lines are almost overlapped, but for the film of thickness 270 Å the splitting is sufficiently strong and well observable. Figure 4 gives ATR spectra of the sapphire crystal at three angles of incidence without a film and with the LiF film. In this case the gap between the ATR element and the sample is filled with nujol and surface polaritons appear at $\kappa_x > 1.4$ (the refractive index of nujol). One of the two minima in the ATR spectra of a clear sapphire crystal splits into two minima in the presence of the film. The intensities of both minima resulting from the splitting are clearly seen to be the same only near the resonance. Farther away from the resonance the spectrum approaches that in the absence of a film – one line is seen of almost the same frequency as in the case of the pure substrate. The intensity of the line adjacent to the gap decreases rapidly while moving away from the resonance. Therefore, it is desirable to work in the vicinity

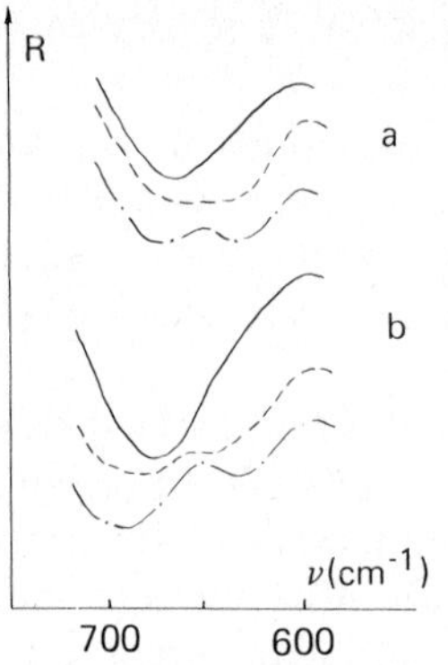

Fig. 3. Splitting of the surface polariton line in the ATR spectrum of the rutile crystal with lithium fluoride films 160 Å thick (dash lines) and 270 Å thick (dot-and-dash lines) for two angles of incidence, (a) $\kappa_x = 1.2$; (b) $\kappa_x = 1.26$; without a film (solid curves).

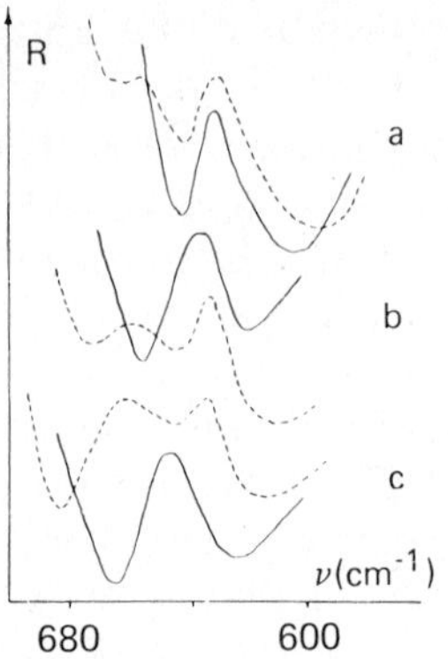

Fig. 4. Splitting of the surface polariton line in the ATR spectrum of the sapphire crystal with a lithium fluoride film 160 Å thick (dash lines) (a) $\kappa_x = 1.5$; (b) $\kappa_x = 1.56$; (c) $\kappa_x = 1.62$. The gap between the film and the sample was filled with nujol; without a film (solid lines). a film (solid lines).

of the resonance though in this case the splitting is minimum, but both components of splitting are clearly seen. Formulae (4–9) were obtained just for the case of the resonance.

Dispersion curves calculated with the use of the data obtained for the dielectric functions of LiF by Kachare et al. (1972) and sapphire by Barker (1963) and shown in fig. 5 disagree with the experimental dispersion: the resonance frequencies do not coincide. It means that the frequency of a longitudinal phonon in a thin film differs from that in a LiF single crystal. To imporve the agreement between the calculated and the experimental

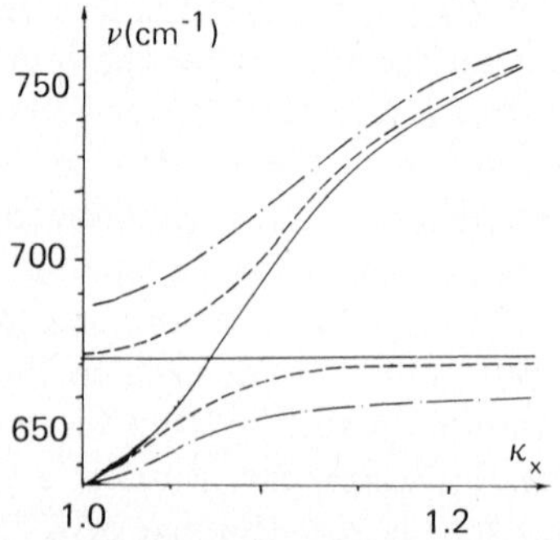

Fig. 5. Dispersion of high-frequency branch of sapphire surface polariton (with the optical axis c ∥ y) without the film (solid line) and its splitting in the presence of LiF films of the thickness 100 Å (dash line) and 500 Å (dot-and-dash line). Calculated with the use of literature data on dielectric function.

data the oscillator parameters were slightly changed. The dielectric function of the film near the longitudinal frequency was approximated by only one oscillator with the same frequency and damping as for the bulk crystal (see above) with the oscillator strength equal to 6.6. No errors should have resulted from the neglect of the band 500 cm^{-1} (due to anharmonicity) since this band is far away from the spectral region (600–700 cm^{-1}) of interest. The frequency of the longitudinal phonon in the film of such dielectric function is 650 cm^{-1}.

Calculated and experimental dispersion curves are shown in figs. 6 and 7 for three orientations of the optical axis in a uniaxial sapphire crystal – perpendicular to the plane of incidence ($c \parallel y$), perpendicular to the crystal plane ($c \parallel z$) and parallel to the propagation direction of the surface polariton ($c \parallel x$). Figure 6 corresponds to the films bordering upon the air gap, while fig. 7 to the system sapphire–lithium fluoroide–nujol.

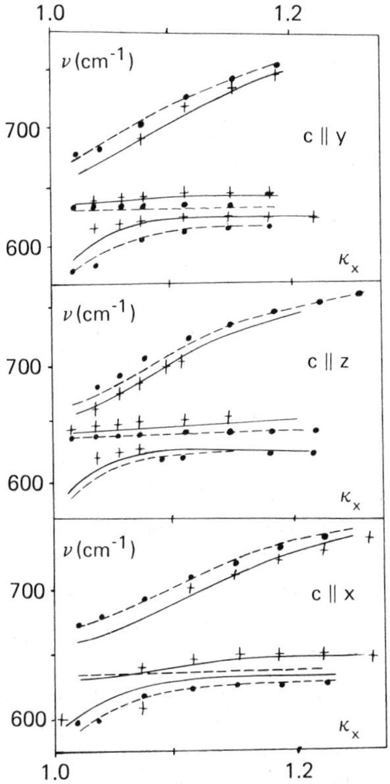

Fig. 6. Dispersion of sapphire surface polaritons with LiF films. Solid lines and crosses for $c \parallel y$ and $c \parallel x$ correspond to the calculation and the experiment for 200 Å; for $c \parallel z$ – 100 Å. Dash lines and circles for $c \parallel y$ and $c \parallel x$ correspond to the thickness 800 Å, for $c \parallel z$ – 400 Å (calculated with the changed frequency of LiF longitudinal oscillation).

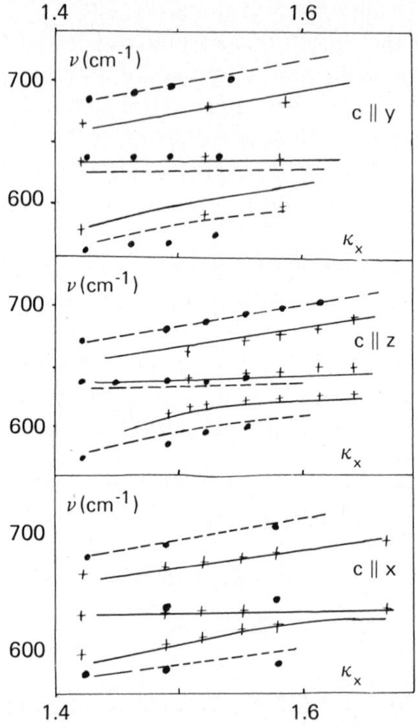

Fig. 7. The same as in fig. 6, but the film borders not onto air, but onto nujol.

Comparison of the calculation with the experiment shows that the gaps at the dispersion curves of surface polaritons for various orientations of the optical axis in the presence of a thin LiF film appear at the frequency $650\,cm^{-1}$. For this frequency it makes no difference whether the system crystal–film borders upon air or upon nujol. The presence of a gap in the spectrum of sapphire surface polaritons caused by the vibration $635\,cm^{-1}$ leads to a small dispersion of the low-frequency component of splitting (the middle branch in figs 6 and 7) and at large thicknesses its frequency is practically constant. The low-frequency branch (below $635\,cm^{-1}$) exists also in the absence of films (Yakovlev et al. 1975b) and, when films are deposited, its frequency decreases.

The decrease in the frequency of the longitudinal phonon explains also the absence of splitting in the case when a film obtained in similar conditions as those described above is deposited onto a YIG crystal. The frequency $650\,cm^{-1}$ falls in the gap on the dispersion curve due to the YIG own normal vibrations with the same frequency (Yakovlev and Zhizhin 1974). In this case only the increase of the existing gap is observed. In fig. 8 the dispersion of YIG surface polaritons is shown without the film and with

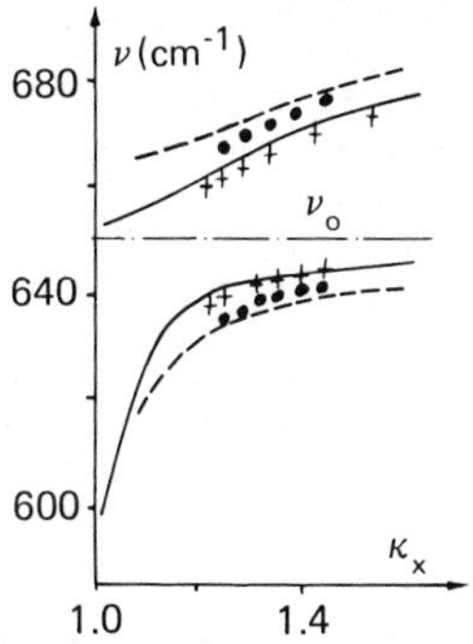

Fig. 8. Increase in the gap of the dispersion curve for surface polaritons of yttrium iron garnet (crosses and solid curves) after the deposition of a LiF film 500 Å thick (circles and dash lines).

a LiF film 500 Å thick (calculation and experiment). When the film is deposited, the upper branch goes up and the lower one goes down, thus increasing the splitting.

The above experimental results were obtained by the frequency scan ATR. At angular scan ATR near the frequency 650 cm^{-1} the back bending was observed. Figure 9 shows the calculated and experimental values of the dispersion curves at the angular ATR for the LiF film 800 Å thick at the sapphire crystal with the optical axis perpendicular to the plane of incidence. Worse (as compared to fig. 6) agreement between the calculated and the experimental results is attributed to the fact that dispersion curves of this type are characterized by strong dependence on the damping value while the effect of damping is usually small in the case of dispersion

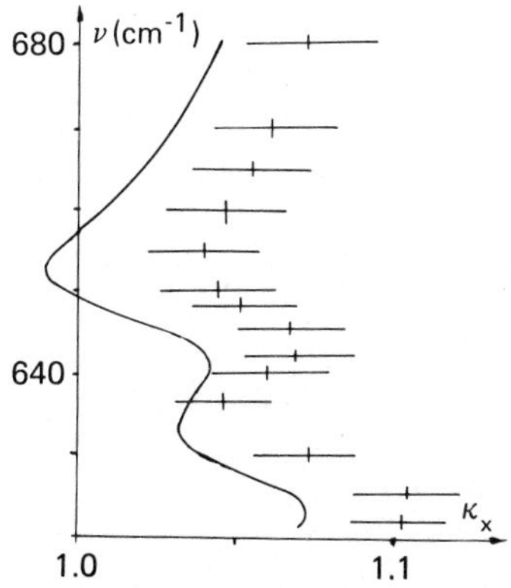

Fig. 9. Appearance of the back bending near 650 cm^{-1} on the dispersion curve of sapphire surface polaritons at angular scanning in the presence of the LiF film of thickness 300 Å (calculation and experiment).

obtained from frequency scan ATR spectra. Therefore, if damping is actually stronger than that taken for the calculation from literature, the dispersion calculated at the real reduced wave vector is in good agreement with the experiment while the dispersion at the real frequency, more sensitive to damping, does not coincide with the experimental one. The assumption that the damping increases in the case of a film (relative to the bulk material) is also confirmed by measurements on a series of samples of greater thicknesses. In this case Zhizhin et al. (1980a) obtained that line widths of thick films in transmission spectra must exceed those with damping value for the bulk LiF.

The decrease of the frequency of longitudinal vibrations and the increase of damping for a thin film as compared to a LiF crystal is, perhaps, due to the small size of crystallites in the film. The shift may also be attributed to the incompact packing of crystallites. The structure and packing of crystallites depend on the condition of the film preparation, therefore, by varying the conditions of evaporation it is possible to change also the film properties. All the films described above have been obtained in the same conditions and, accordingly, their frequencies of longitudinal vibrations are the same. For thicker films higher rates of evaporation were used so that the frequency of longitudinal vibrations decreased still more and for some samples it became 630–610 cm^{-1}.

Such frequency of longitudinal vibration falls not in the gap, but in the region with an YIG surface polariton (fig. 8); therefore, measurements on ATR spectra of YIG with a LiF film give the splitting of the surface polariton line at the frequency 620 cm^{-1} shown in fig. 10. The frequency 630 cm^{-1} falls also in the region of the surface polariton dispersion of the MgO crystal. The splitting of the MgO surface polariton with the increase of the LiF film thickness is shown in fig. 11

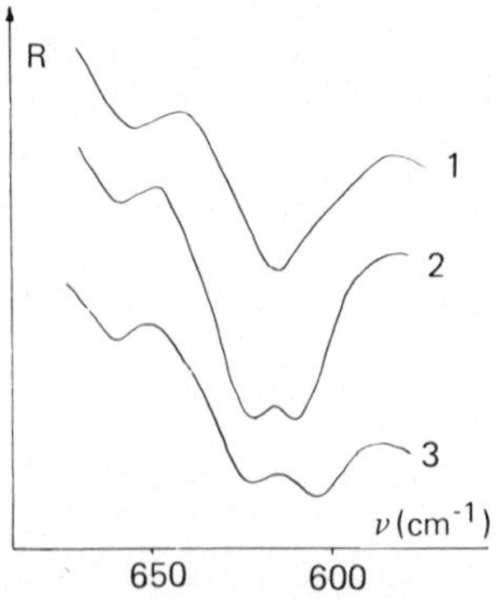

Fig. 10. ATR spectra of a YIG crystal without a film (1) and with films 500 Å (2) and 1000 Å (3) thick.

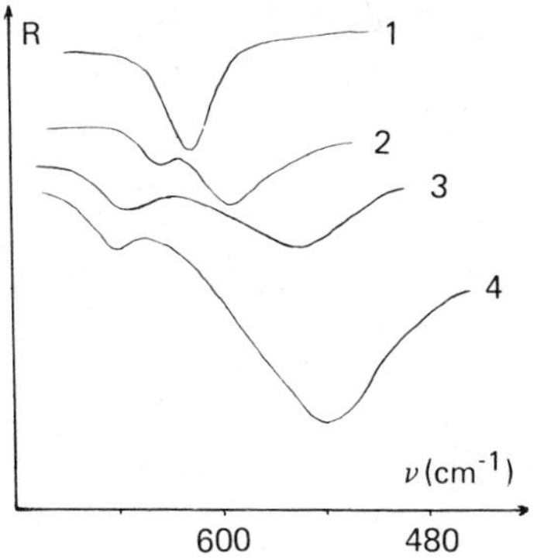

Fig. 11. Splitting of surface polaritons in MgO (1) covered with thin LiF films of the thicknesses (Å): (2) – 600, (3) – 3200, (4) – 5900; the angle of incidence is 40°.

The thickness dependence of splitting predicted theoretically was checked by means of two series of samples: LiF films of the thicknesses 100, 200, 400 and 800 Å with the frequency of longitudinal oscillations 650 cm^{-1} on sapphire and films of the thicknesses 600–5900 Å with the frequency of longitudinal oscillations 630 cm^{-1} on the MgO crystal. The samples of the first series were applied simultaneously in the same conditions, the samples on the MgO crystal were evaporated one at a time and we tried to provide more or less uniform conditions of evaporation for different samples.

For sapphire the experimental points and the thickness dependence calculated by formula (2) are in good agreement (fig. 12) with the square root dependence predicted theoretically. In this case the calculation by the formulae (4)–(9) is rather difficult since the sapphire dielectric function is not described by a simple one-oscillator model used to derive formulae (8)

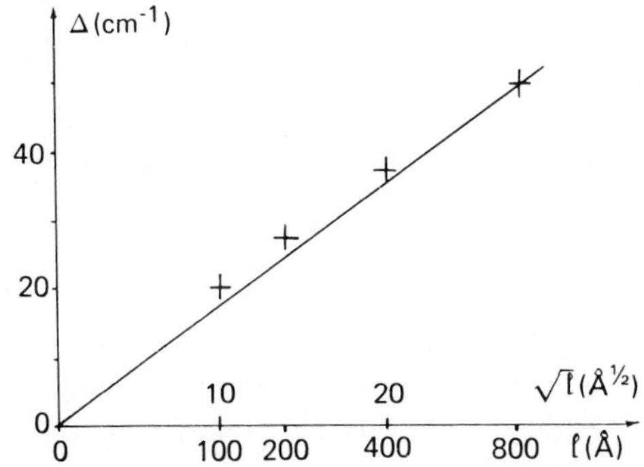

Fig. 12. Dependence of splitting on the film thickness in the case of a LiF film on sapphire. The straight line is a calculation by formula (2).

and (9). For MgO the additional oscillator introduced artificially to take into account the anharmonicity (Kachare et al. 1972) has the frequency 645 cm^{-1}. Far away from this frequency the effect of the additional oscillator weakens, therefore, for sufficiently thick films, when the splitting is large, the one oscillator model can be used. If the parameters of only the basic oscillator are used, we have for the splitting $\Delta \approx 1.85 \sqrt{(l)}$, where l is the film thickness in Å, Δ is the splitting in cm^{-1}. There is rather good agreement between the experimental points and this dependence (fig. 13) and a certain point scattering may be attributed to the differences in various films as deposition was not performed simultaneously and the films had different structures.

The above experiments have been made for transition layers obtained by applying a film of another substance. It would be of interest to study in the same way transition layers obtained by some action on the crystal surface. Such action may be exemplified by an ionic implantation. The first experiments (Zhizhin et al. 1979) on studying surface polaritons of crystalline quartz irradiated with nitrogen ions have not revealed resonance as yet. This may result either from a small shift of frequencies in the transition layer which causes no resonance or from big damping in the transition layers which smears the resonance. We have observed such smearing of resonance for silicon monoxide films deposited onto crystalline quartz. In this case the frequency of the longitudinal oscillation of SiO 1120 cm^{-1} obtained from the data given by Cox et al. (1975) falls within the region where quartz surface polaritons (Zhizhin et al. 1974) exist; however, the damping for SiO is an order higher than that for quartz and, accordingly, neither the experiment nor the calculation cause splitting, but only the shift of the dispersion curve. Such broadening is analogous to metal quenching of surface polaritons.

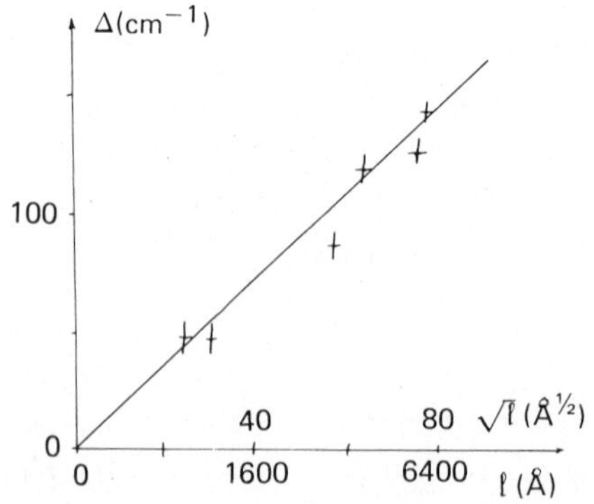

Fig. 13. Dependence of splitting on the film thickness in the case of a LiF film on MgO. The straight line is the approximate calculation by formulae (4–8).

Another method to provide an artificial transition layer is deformation of the crystal. The effect of uniaxial compression and extension in the direction of the x-axis on surface polariton spectra of crystals of MgO, $SrTiO_3$, SiO_2 was investigated experimentally by Novak et al. (1975) and Banshchikov et al. (1977). A transition layer was found to exist where the shift of frequencies at elastic deformation is several times as much as that in the bulk of the crystal. At pressures up to 10 kg/mm^2 this shift does not yield resonance effects.

Note that uniaxial deformation may lead to anisotropy in cubic crystals while the shift of frequencies caused by pressure may be different in the bulk and in the transition layer. The effect of anisotropy in the transition layer for the type of resonance under discussion have not been studied as yet. To study the role of anisotropy the hetero-epitaxial layers of the substance with high anisotropy may be used. In this case, depending on the orientation of the optical axis, the resonance frequency may vary. According to formula (10), the resonance should be expected at the frequency of transverse oscillations along the film and at the frequency of longitudinal oscillations across the film.

4. *Other Cases of Resonance with the Transition Layer*

As shown above, the resonance of the film phonon with the surface phonon polariton of the substrate is a sensitive method of studying thin surface films. A considerable drawback of this method is that surface polaritons of substrates exist in limited relatively small frequency regions and, therefore, it is always necessary to choose the pair "substrate–film". The dispersion region is much greater for surface plasmon polaritons; for doped semiconductors it extends from 0 cm^{-1} up to the frequency of the surface plasmon in the IR range (except a small region near the frequency of lattice oscillations of the semiconductor). Such substrate was used in the experiment already mentioned in sect. 2 of this chapter (the ZnSe film on InSb studied by Gerbstein and Mirlin 1974). Still greater is the dispersion region of the surface plasmon polariton in metals (up to the visible or to the UV spectral region). However, in the IR region the dispersion for a metal practically coincides with the "light line" ($k_x \approx \omega/c$ since $|\epsilon_3| \gg 1$) and, therefore, the splitting of the surface polariton line, the same as of the line itself, cannot be observed in practice because of the angular divergence of beam in the ATR element. In this case there are several possible methods to investigate the films. Firstly, the dispersion can be studied by the angular scan ATR and – at sufficiently high precision of angular measurements and small divergence of the beam – the back bending can be obtained as well as the increase in the angular width of the minimum at resonance (Chen 1977).

The second method is the investigation of the external reflection at grazing incidence (Greenler 1966, 1969). As shown by Gerbstein and Mirlin (1974), the minimum observed in this case at the longitudinal vibrational frequency of the film is on the extension of the upper branch of the dispersion curve into the radiational region. Finally, the spatial damping of the metal surface polariton can be studied at the propagation along the surface (the spectroscopy of surface electromagnetic waves described in ch. 3 and in a set of works of Bell et al. (1975), Zhizhin et al. (1976a, 1980b), Bhasin et al. (1976)). For the films of the thickness ~ 1 μm the ATR spectra may also be studied far away from resonance. In ch. 4 on thermostimulated emission the results are presented for ZnSe films on metals with different conductivities (Vinogradov et al. 1977).

The above methods of studying IR spectra of thin films on metal surfaces along with the resonance studies described in the previous section, are characterized by high sensitivity. They allow to obtain spectra up to monomolecular layers on metals.

In the visible and UV spectral regions a surface plasmon polariton has already an appreciable dispersion and, therefore, the resonance with excitations, having their frequencies within this range, can be studied by means of both angular and frequency scan ATR.

The resonance with a bulk plasmon with the frequency below the frequency of the substrate surface plasmon is interesting in that it may allow to investigate the transition layer with reduced concentration of carriers (a depletion layer). The numerical calculation made by Lopez-Rios (1976) showed that for potassium films on aluminium (the film 5 Å thick) a clearly marked resonance may be expected. Such resonance was studied experimentally by Lopez-Rios et al. (1978). Thin silver films of the thicknesses 2; 6.5; 26; 39; 58 Å were applied under vacuum, 10^{-10} Torr, onto an aluminium film of thickness 214 Å evaporated on the ATR prism in Kretschmann's configuration (Kretschmann and Raether 1968). For the two thinnest films the splitting was not observed which possibly is due to the change of the optical properties of such thin films. For those thicker samples the splitting is clearly seen. Dispersion curves were obtained by the frequency and angular scan ATR.

For the films with strong damping the angular scanning should preferably be used, as the back bending can usually be noticed, even if the splitting is not observed due to the large line width. Gold films of the thicknesses of 10 and 40 Å at (111) silver face were studied (Lopez-Rios and Vuye 1979) (films of the thicknesses 655 and 685 Å). At the angular scanning the method of the least squares was used to obtain the optical constants for gold films which differ slightly from the constants of the bulk metal. The above experiments are described in more detail in ch. 6.

Resonance of the surface plasmon polariton with excitonic excitations of the film allows to study the spectra of monomolecular films of dyes. Such films are characterized not only by their small thickness so that a microtheory may be required, but also by their usual anisotropy due to the orientation of the molecules. The calculation with the use of the microtheory for Otto's and Kretschmann's configurations gives the splitting at the frequency of the excitonic excitation at the frequency scan ATR and the back bending at the angular scan ATR (Sherman and Philpott 1978). For an anisotropic film the resonance occurs at the transverse frequency for the transition oriented in the film plane and at the longitudinal frequency for the perpendicular oriented transition. The absence of one of the resonances is an indication of the film anisotropy, but it should be taken into account that for isotropic films with strong damping two resonances are smeared and one heavily broadened resonance is left so that the effects of damping and anisotropy cannot be separated as Pockrand et al. (1978) showed.

Resonance was studied by Pockrand and Swalen (1978), Pockrand et al. (1978) experimentally for two dyes N,N′-di(methyloctadecyl)squarillium (subsequently referred to as squarillium) and N,N′-dioctadecyloxacarbocyanine (subsequently referged to as cyanine). The Kretschmann's configuration was used. A silver film (~ 600 Å) was evaporated in vacuum, 10^{-5} Torr, onto glass and this film was afterwards partially covered with monolayers: first of the dye and then of cadmium arachidate using the Langmuir–Blodgett technique described, for example, by Srivastava (1973). Cadmium arachidate served to protect the dye monolayer against oxidation.

The dye molecules are characterized by strong absorption in the visible region while the protective layers are transparent. Angular dependences of ATR for various laser frequencies have been measured. The presence of a dye leads to the shift and broadening of the minimum and the variation of the reflectivity at the minimum. In the vicinity of the resonance the back bending and the maximum of the line width are observed. Optical constants and dielectric functions for the film have been calculated both on the assumption of the film isotropy and with regard to anisotropy. Both assumptions do not contradict the experimental angular dependences, but judging by the molecule structure, the transition should be expected to be polarized in the plane of the layer and there is no absorption perpendicular to the layer. The values of dielectric functions are quite reasonable while at the same time for cyanine the supposition that the film thickness equals the length of the molecule (28, 1A) leads to $\epsilon_\infty \approx 1$. The studies on mixed cyanine-cadmium arachidate monolayers allow to obtain the real thickness of pure cyanine – below 10Å. After the thickness has been determined optical constants and dielectric functions were calculated from angular measurements of ATR with regard and with no regard to anisotropy. For squarillium with regard to anisotropy the following parameters have been obtained for the oscillator of the band falling within the interval under investigation $\nu_{TO} = 18900\ cm^{-1}$ (530 nm), $\Delta\epsilon = 0.85$, $\epsilon_\infty = 1.4$, $\gamma = 1600\ cm^{-1}$. For cyanine the bands overlap strongly, therefore, only those wavelengths have been determined which correspond to absorption minima: for anisotropic model ~ 500 nm and 464 nm.

Note that for Langmuir films of carbonic acid syloxane the decrease in the effective film thickness was also observed (Zhizhin et al. 1980b) similarly to that obtained for cyanine; in this case the calculated absorption spectra of surface electromagnetic waves in the IR range coincide with the measured ones on the assumption that $\epsilon_\infty = 1.2$ (at the film thickness 20 Å equal to the length of the molecule). Such a decrease in the dielectric permeability (or the effective film thickness) may be attributed to the film skeletization, i.e., incompact packing of the molecules on the metal.

In conclusion it may be noted that the splitting of the surface polariton line has not been observed as yet at the resonance with the film exciton because of large line widths. Sherman and Philpott (1978) showed that for J-bands of dyes with smaller line widths the splitting may be revealed. Small line width (10–100 cm^{-1}) is characteristic of excitons in the films of inorganic substances, especially at low temperatures. In this case the splitting of the line of a surface plasmon polariton may be expected at very small thicknesses of films – up to 10 Å; however, in this case it should be taken into account that at such small thicknesses microscopic effects must be allowed for and also the possibility of the existence of an excitonless layer near the film boundary (Hopfield and Thomas 1963).

The experiments described show that in the case of thin films the sensitivity of the surface polariton spectroscopy increases considerably with the transition to the visible and UV spectral regions which allows for reliable detection of monomolecular layers on metal surfaces. Such increase in the sensitivity is due to the decrease in wavelengths and the increase in the ratio of the layer thickness to the wavelength. In this case the square root dependence of splitting on this ratio provides high sensitivity of the method.

References

Agranovich, V.M., 1975, Sov. Phys.-Usp. **18**, 99.

Agranovich, V.M. and T.A. Leskova, 1974, Sov. Phys.-Solid State **16**, 1172.

Agranovich, V.M. and A.G. Mal'shukov, 1974, Opt. Commun. **11**, 169.

Alexander, R.W., G.S. Kovener and R.J. Bell, 1973, Phys. Rev. Lett. **32**, 154.

Arakawa, E.T., M.W. Williams, R.N. Hamm and R.H. Ritchie, 1973, Phys. Rev. Lett. **31**, 1127.

Banshchikov, A.G., V.E. Korsukov and I.I. Novak, 1977, Piesospectroscopic effect for the surface polaritons in crystals, in: Theoretical spectroscopy, (Moscow) (in Russian).

Barker Jr., A.S., 1963, Phys. Rev. **132**, 1474.
Bell R.J., R.W. Alexander and C.A. Ward, 1975, Surf. Sci. **48**, 253.
Bhasin, K., D. Bryan, R.W. Alexander and R.J. Bell, 1976, J. Chem. Phys. **64**, 5019.
Bryksin, V.V., D.N. Mirlin and I.I. Reshina, 1973, Sov. Phys.-Solid State **15**, 760.
Chen, W., 1977, Surface EM waves and spectroscopy in ATR configurations, Thesis (Philadelphia).
Cox, F.T., G. Hass and W.R. Hunter, 1975, Appl. Opt. **14**, 1247.
Gerbstein, Yu.M. and D.N. Mirlin, 1974, Sov. Phys.-Solid State **16**, 1680.
Glang, R., 1970, Vacuum evaporation, in: Handbook of Thin Film Technology, eds. L.I. Maissel and R. Glang (McGraw Hill Book Company, N.Y.) ch. 1.
Greenler, R.G., 1966, J. Chem. Phys. **44**, 310.
Greenler, R.G., 1969, J. Chem. Phys. **50**, 1963.
Halevi, P., 1977, Opt. Commun. **20**, 167.
Halevi, P., 1978, Surf. Sci. **76**, 64.
Hopfield, J.J. and D.G. Thomas, 1963, Phys. Rev. **132**, 563.
Kachare, A., G. Anderman and L.R. Brantley, 1972, J. Phys. Chem. Solids **33**, 467.
Kretschmann, E. and H. Raether, 1968, Z. Naturforsch. **A23**, 2135.
Lopez-Rios, T., 1976, Opt. Commun. **17**, 342.
Lopez-Rios, T., F. Abeles and G. Vuye, 1978, J. de Phys. **39**, 645.
Lopez-Rios, T. and G. Vuye, 1979, Surf. Sci. **81**, 529.
Mirlin, D.N. and I.I. Reshina, 1974, Sov. Phys. Solid State **16**, 1463.
Novak, I.I., V.E. Korsukov and A.G. Banshchikov, 1975, Dokl. Akad. Nauk SSSR **224**, 1297; Sov. Phys.-Dokl. **20**, 695.
Otto, A., 1968, Z. Phys. **216**, 398.
Pockrand, I. and J.D. Swalen, 1978, JOSA **68**, 1147.
Pockrand, I., J.D. Swalen, R. Santo, A. Brilliante and M.R. Philpott, 1978, J. Chem. Phys. **69**, 4001.
Pockrand, I., J.D. Swalen, J.G. Gordon II and M.R. Philpot, 1979, J. Chem. Phys. **70**, 3401.
Reshina, I.I., D.N. Mirlin and A.G. Banshchikov, 1976, Fiz. tverd. tela **18**, 506; Sov. Phys.-Solid State **18**, 292.
Scherman, G. and M.R. Philpott, 1978, J. Chem. Phys. **68**, 1729.
Schuller, E., H.J. Falge and G. Borstel, 1975a, Phys. Lett. **A54**, 317.
Schuller, E., H.J. Falge and G. Borstel, 1975b, Phys. Lett. **A55**, 109.
Srivastava, V.K., 1973, Langmuir molecular films and their applications, in: Physics of thin films, vol. 7, eds. G. Hass, M.H. Francombe and R.W. Hoffman, ch. 5.
Vinogradov, E.A., G.N. Zhizhin, V.A. Yudson and A.G. Mal'shukov, 1977, Solid State Commun. **23**, 915.
Yakovlev, V.A., 1976, Surface polaritons and effects of thin metal and dielectric coatings, Thesis (Moscow) (in Russian).
Yakovlev, V.A. and G.N. Zhizhin, 1974, Sov. Phys.-JETP Lett. **19**, 189.
Yakovlev, V.A., V.G. Nazin and G.N. Zhizhin, 1975a, Opt. Commun. **15**, 293.
Yakovlev, V.A., G.N. Zhizhin, M.I. Musatov and N.M. Rubinina, 1975b, Sov. Phys.-Solid State **17**, 1996.
Yakovlev V.A., G.N. Zhizhin, M.A. Moskalova and V.G. Nazin, 1978, Surface polariton spectroscopy and properties of thin metal and dielectric films on the crystal, in: Spectroscopy of molecules and crystals, ed. T.M. Shpak (Naukova Dumka, Kiev) (in Russian).
Zhizhin, G.N., M.A. Moskalova, V.G. Nazin and V.A. Yakovlev, 1974, Sov. Phys.-Solid State **16**, 902.
Zhizhin, G.N., O.I. Kapusta, M.A. Moskalova, V.G. Nazin and V.A. Yakovlev, 1975, Uspekhi fiz. nauk **117**, 573; Sov. Phys.-Usp. **18**, 927.

Zhizhin, G.N., M.A. Moskalova, E.V. Shomina and V.A. Yakovlev, 1976a, Sov. Phys.-JETP Lett. **24**, 196.
Zhizhin, G.N., M.A. Moskalova and V.A. Yakovlev, 1976b, Sov. Phys.-Solid State **18**, 146.
Zhizhin, G.N., M.A. Moskalova, V.G. Nazin and V.A. Yakovlev, 1977a, Sov. Phys.-JETP **45**, 360.
Zhizhin, G.N., M.A. Moskalova, V.G. Nazin and V.A. Yakovlev, 1977b, Sov. Phys.-Solid State **19**, 1352.
Zhizhin, G.N., V.A. Yakovlev and G. Schirmer, 1979, Sov. Phys.-JETP Lett. **29**, 315.
Zhizhin. G.N., E.I. Firsov and V.A. Yakovlev, 1980a, Fiz. tverd. tela **22**, 2106.
Zhizhin, G.N., N.N. Morosov, M.A. Moskalova, A.A. Sigarov, E.V. Shomina, Y.A. Yakovlev and V.I. Grigos, 1980b, Thin solid films **70**, 163.

CHAPTER 8

The Study of Solid–Liquid Interfaces by Surface Plasmon Polariton Excitation

DIETER M. KOLB

Fritz-Haber-Institut der Max-Planck-Gesellschaft,
Faradayweg 4–6, D-1000 Berlin 33,
West Germany

Surface Polaritons
Edited by
V.M. Agranovich and D.L. Mills

Contents

1. Introduction 301
2. The electrochemical interface – some basic properties 302
3. Experimental 304
4. The dispersion of surface plasmon polaritons 307
5. The optical properties of single crystal surfaces as studied by SPP excitation . . 309
6. SPP excitation studies at the electrode–electrolyte interface 312
7. SPP excitation on adsorbate covered electrode surfaces 315
8. The use of field enhancements in SPP excitation 318
9. SPP excitation by external reflection 322
   9.1. SPP excitation on rough surfaces 322
   9.2. SPP excitation on stepped surfaces 324
   9.3. Adsorbate induced SPP excitation 325
10. Conclusion and outlook 327
References 327

1. *Introduction*

The use of surface plasmon polaritons (SPP) for the investigation of electrode surfaces has a twofold bearing. For the electrochemist SPP excitation –as other optical techniques– is a valuable new tool for characterization of the electrochemical interface. It can be used in-situ, that is, in the electrochemical cell without interfering with electrode processes, and it can yield information on structural and electronic properties of the interface, whereas classical electrochemical methods usually yield thermodynamic data only. On the other hand it is increasingly recognized that the special and unique properties of the electrochemical interface can be used to obtain new surface physical and chemical information, not easily accessible under UHV conditions (see e.g. Gerischer et al. 1978). A particularly intriguing aspect of the electrode–electrolyte interface is the presence of large surface charges and high electric fields, the sign and magnitude of which are easily controlled by the electrode potential. Another important point is the following. At the electrochemical interface electric forces on all electrolyte phase species play a dominant role beside chemical forces. As a consequence there is a strong dependence of adsorption on the electrode potential, especially for polar or charged species. The superimposed chemical interaction between the electrode surface and the adsorbed species can be assumed –to a first approximation– to be independent of the electric double layer. Hence, such systems are well suited for the study of adsorption isotherms at equilibrium conditions since the free enthalpy of the adsorbents can be varied to an enormous extent in the electrolyte by chemical or electric means. A good example are metal adsorbates in the submonolayer or monolayer range on metallic substrates (Kolb 1978). These adsorbates undergo a strong interaction with the substrate and they are most conveniently studied at the electrode–electrolyte interface where the chemical potential of the adsorbate and hence the coverage can be varied reversibly over a wide range by a simple potential variation. Such conditions are not very often reached at the solid–vacuum interface, definitely not for the case of metal adsorbates.

2. The Electrochemical Interface – Some Basic Properties

In the absence of any electrochemical reaction the electrochemical interface behaves like a capacitor of molecular dimensions. One side of this condensor is made up by the electrode surface, the other side by the solvated ions which approach the electrode surface to a distance of approximately 3 Å, the radius of the solvation shell. These ions are held in position merely by the electrostatic forces. The capacity of such a condensor, with water as the dielectric, should be about 240 $\mu F\,cm^{-2}$ if the dielectric constant were that of bulk water, namely 81. This, however, is one order of magnitude larger than what is observed, indicating that a more realistic value for the H_2O in the electrochemical interface is between 5 and 10. Obviously, the H_2O molecules in the Helmholtz layer, as this interface is commonly called, have lost several degrees of freedom (mostly dipolar relaxation and some rotational degrees). The capacity of the Helmholtz layer is still tremendous and hence, by simple charging of the electrode, high electric fields (some $10^7\,V\,cm^{-1}$) and large surface charges (up to about 20–40 $\mu C\,cm^{-2}$ corresponding to about 0.1–0.2 electrons per surface atom) can easily be obtained, a fact, which makes this interface also attractive for surface physical studies as will be demonstrated later.

If our simple model of the rigid plate condensor were correct, the capacity C_{dl} of the Helmholtz layer should be independent of the electrode potential. This, however, is not true. When we determine the differential capacity of the interface, $C_{dl} = \Delta q/\Delta U$, we find a characteristic potential dependence of the capacity as is shown in fig. 1a (Parsons 1961). Besides structural details in the curve, it becomes evident that the double layer capacity is much larger when the electrode is charged positively than in case of a negative surface charge. The reason for the increase at the more anodic potentials is found in the much weaker solvation of anions, which in this region build up the ionic counter charge of the double layer. Here the attractive forces become so strong that the anions partly loose their

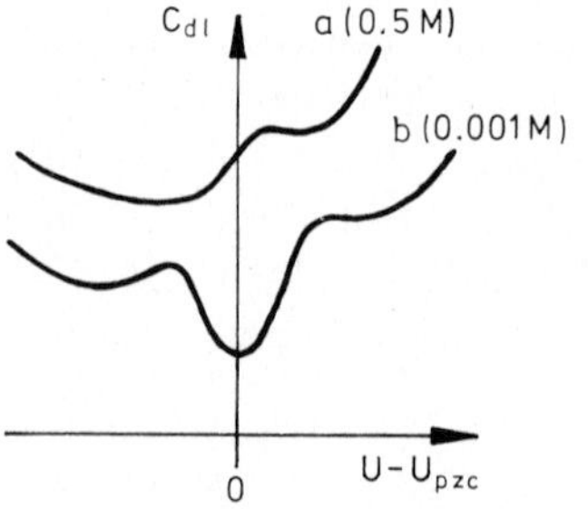

Fig. 1. Differential double layer capacity, C_{dl}, for the metal-electrolyte interface as a function of electrode potential. (Two different electrolyte concentrations.)

solvation shell and move closer to the electrode surface. This does not occur at the most cathodic potentials since the cations usually have a strongly bound hydration shell. In this potential range as well as in the range around the potential of zero charge (pzc), where no excess free charge is on the surface, the variations of the double layer capacity with potential are usually explained by structural changes in the water layer on the surface. To rationalize these changes, a two stage model is frequently employed where the H_2O molecules have their dipoles oriented towards or away from the electrode surface (see e.g. Yeager 1977). By measuring the temperature dependence of the inner double layer capacity it has been shown that the solvent surface excess entropy passes a maximum at potentials slightly negative of the pzc (Harrison et al. 1973). This is understood as a chemical interaction being present besides the electrostatic one which causes the negative end of the H_2O dipole (oxygen!) to be bound to the metal surface more strongly than the positive end and hence the net dipole orientation does not change exactly at the pzc. A text book model of the Helmholtz layer is shown in fig. 2. The main features include, as already mentioned above: a sheet of adsorbed water molecules, solvated ions at closest distance from the metal surface (their position defining the so-called outer Helmholtz plane, the location of the electrostatically bound ions) and ions which have partly lost their solvation shell and have become directly attached to the surface by chemical bonds. The latter ions are called specifically adsorbed and their location defines the inner Helmholtz plane. In the present context we will not discuss the double layer in greater detail and the reader is referred to text books on electrochemistry for a more thorough introduction (Bockris and Reddy 1970, Vetter 1967, Sparnaay 1972).

As indicated in the potential diagram in fig. 2b almost all of the potential drop in the electrochemical interface occurs across the compact double layer, that is between the electrode surface and the outer Helmholtz plane, when the electrolyte is sufficiently concentrated. At low electrolyte concentrations, the accumulation of charge at the outer Helmholtz plane leads to such an increase in the ion concentration with respect to the bulk electrolyte that the corresponding gradient in the chemical potential causes an extension of the double layer into the solution. This part is called the diffuse layer, the potential drop across it is φ_2 (see fig. 2b). The complete double layer, compact and diffuse part, can be described by two plate condensors in series, the first one with a fixed, the second one with a variable plate distance which depends on the electrolyte concentration (Debye length). At low concentrations the measured double layer capacity C_{dl} is mostly determined by the diffuse layer capacity and it has a minimum for $\varphi_2 = 0$, which (in the absence of specific adsorption) is found when the excess charge in the whole double layer is zero (Parsons 1961). This (in

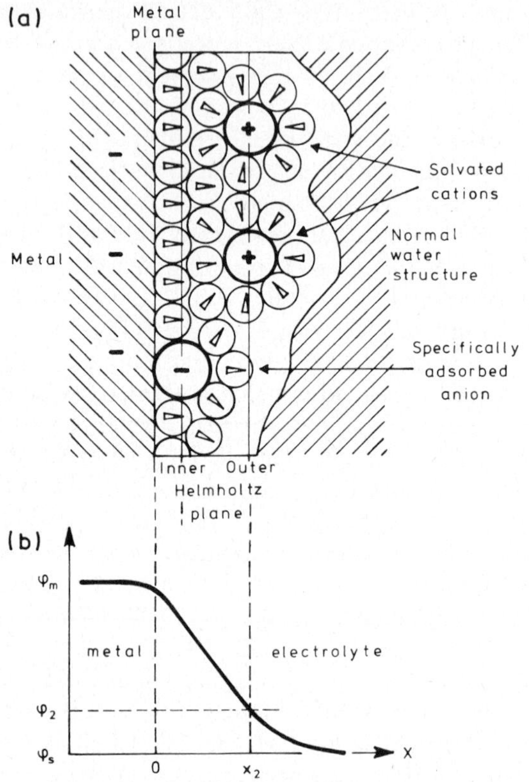

Fig. 2. (a) Model of the ionic double layer, after Bockris et al. (1963). (b) Potential distribution across the electrochemical interface.

dilute solutions very pronounced) minimum in C_{dl} can be used (see fig. 1b) to determine the potential of zero charge (pzc), a quantity which corresponds to the work function of the electrode in solution (Trasatti 1977). When changing the electrode potential away from the pzc, the electrochemical potential of the electrons in the electrode and hence the work for emitting electrons into the electrolyte is changed accordingly (Gerischer et al. 1978). A schematic drawing of the potential energy of the electron at the interfaces electrode–electrolyte and metal–vacuum is given in fig. 3.

3. Experimental

SPP excitation is usually studied by attenuated total reflection (ATR). The two commonly employed experimental arrangements are those put forward by Otto (1968) and Kretschmann (1971). In the Otto configuration the solid

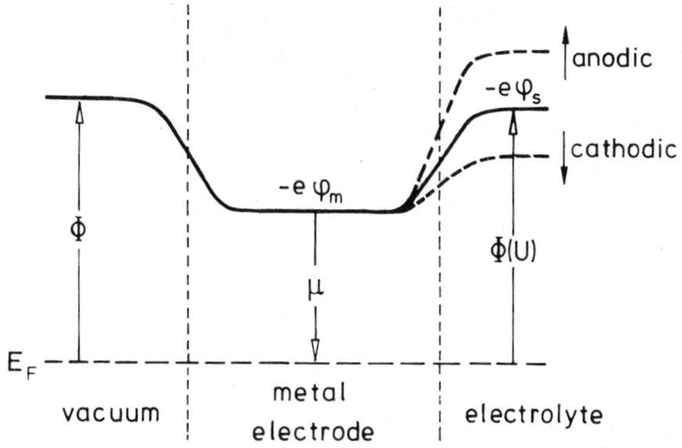

Fig. 3. Electron potential for the metal–vacuum and metal–electrolyte interfaces. Φ: work function, μ: chemical potential, U: electrode potential.

under investigation is separated by a thin air or electrolyte gap from the ATR element (in our case always a hemicylinder rather than a prism), while in the Kretschmann configuration a thin film is evaporated directly onto the ATR element and SPP excitation at the film–ambient interface is achieved through the film. Because of easy preparation of thin metal films of well-defined thicknesses on quartz or glass, the Kretschmann configuration has been used in recent years nearly exclusively for SPP studies. These metal films of course were polycrystalline. Although the exact adjustment of the air or ambient gap thickness, which ranges between 500 and 1000 nm, is a difficult experimental task, the Otto configuration has the distinct ability to study single crystal surfaces.

Various experimental set-ups for measuring SPP excitation under ATR or external reflection conditions have been described in the literature. The two arrangements used in the author's laboratory are shown in fig. 4. The external reflection studies were mostly carried out with a dual-beam rapid scan spectrometer (RSSB, Harrick Inc., Ossining, N.Y.), which allowed fast and accurate determination of relative reflectance changes (fig. 4a). This set-up was especially designed for detection of small reflectance changes of electrode surfaces due to potential variation or electrochemical reactions. The measuring procedure has been described elsewhere in detail (Kolb and Kötz 1977a). ATR measurements in the Kretschmann configuration at constant angle of incidence $[R_{\varphi_1} = f(\lambda)]$ were also done with this equipment.

ATR measurements at constant wavelength $[R_\lambda = f(\varphi_1)]$ were carried out at focussed ATR conditions, as proposed by Kretschmann (1978) (see fig. 4b). Instead of using parallel light, the beam is *focussed* onto the flat side of the hemicylinder (ATR element) thus supplying a wide range of momenta. SPP excitation is then seen in the reflected light cone by the

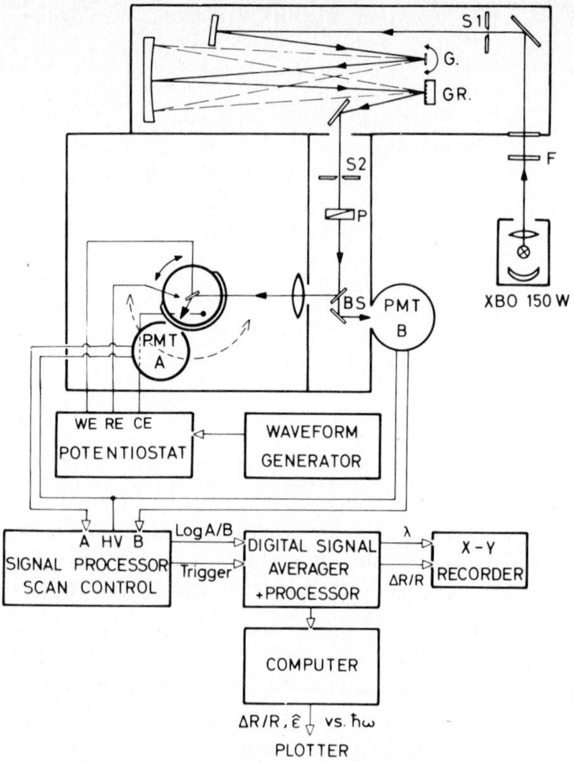

Fig. 4a. Rapid scan spectrometer for determining minute reflectance changes. PMT A, B: photomultiplier tubes for sample and reference beam; BS: beam splitter; F: filter; G: galvanometer mirror; GR: grating; P: polarizing prism; S1, S2: entrance and exit slits; WE, RE, CE: working, reference and counter electrode. After Kolb and Kötz (1977a).

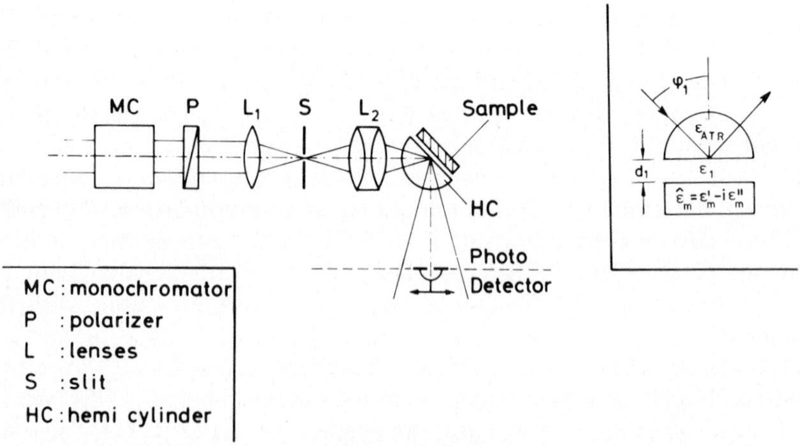

Fig. 4b. Schematic diagram for determining SPP excitation by focussed attenuated total reflection (FATR).

appearance of a dark narrow line. By either moving a photo detector across the light cone or imaging the cone onto an optical multichannel analyser, the reflectance as a function of angle of incidence can be recorded.

In our electrochemical cell for SPP studies in the Otto configuration the gap thickness was adjusted by a micrometer screw, which moved the single crystal electrode towards or away from the hemicylinder. The optimum gap thickness was found empirically by observing the dark line in the reflected light cone. When measuring the dispersion curve for SPP over a wide wave-length range it turned out that a gap readjustment had to be done for the various wavelength regions.

4. The Dispersion of Surface Plasmon Polaritons

Since details on SPPs are given in this book by more competent authors, I will restrict the discussion of SPP dispersion relations to a few points which are of direct relevance to the experimental results shown in the following sections.

SPPs are collective propagating charge density waves, which are confined to the surface or interfacial region of a solid (Ritchie 1973, Otto 1976, Raether 1977). The charge perturbation is strongly decaying in the direction normal to the surface and is coupled to the electromagnetic fields inside and outside the solid. With the outer normal as $+z$-direction, this charge density is given by (Forstmann and Stenschke 1978):

$$\rho(x, z, t) = \rho_0 \exp[\mathrm{i}(\boldsymbol{k}_x \boldsymbol{x} - \omega t)] \exp(\boldsymbol{q} z) \quad (1)$$

with

$$q = [k_x^2 - \alpha(\omega^2 - \omega_P^2)]^{1/2}. \quad (1a)$$

The coupled transverse electric fields inside $(-)$ and outside $(+)$ the solid are described by:

$$\boldsymbol{E}^{\pm}(x, z, t) = \boldsymbol{E}_0^{\pm} \exp[\mathrm{i}(\boldsymbol{k}_x \boldsymbol{x} - \omega t)] \exp(\boldsymbol{k}^{\pm} z) \quad (2)$$

with

$$k^{+} = -[k_x^2 - (\omega/c)^2 \epsilon_1]^{1/2} \quad (2a)$$

$$k^{-} = [k_x^2 - (\omega/c)^2 \hat{\epsilon}_m]^{1/2}, \quad (2b)$$

where $\hat{\epsilon}_m$ stands for the complex dielectric function of the solid (in our case always a metal), ϵ_1 is the ambient dielectric constant, ω_p is the plasma frequency and $\alpha = (\frac{3}{5} v_F^2)^{-1}$. The electric field of the SPP represents a harmonic wave along the surface (x, y-plane) with frequency ω and wave vector $\boldsymbol{k}_x$, parallel to the surface, while perpendicular to the surface (z-direction) the fields are exponentially decaying. Then, $|k^+|$ is a measure of the penetration depth of the SPPs into the adjacent medium. Since $\alpha^{-1/2}$ is of the order of the Fermi velocity v_F, we find $q \gg k^{\pm}$ (about a factor of c/v_F) for $\omega \lesssim \omega_p/\sqrt{2}$, while k_x is of the order of (ω/c). Hence, the charge density perturbation (eq. (1)) decays much faster than the transverse fields (eq. (2)) and we therefore can approximate the charge perturbation at the surface by a δ function as done in standard optics. This allows us to treat SPPs by classical optics and we can derive their dispersion relation from the usual boundary conditions:

$$\epsilon_1 k^- - \hat{\epsilon}_m k^+ = 0. \tag{3}$$

Solving eq. (3) leads to the more explicit form of the SPP dispersion relation (for $\omega < \omega_p$) (e.g. Otto 1976):

$$k_x = \left(\frac{\omega}{c}\right)\left(\frac{\epsilon_1 \hat{\epsilon}_m}{\epsilon_1 + \hat{\epsilon}_m}\right)^{1/2}. \tag{4}$$

Such a dispersion curve is schematically shown in fig. 5.

Excitation of SPPs is usually achieved either by p-polarized light, where the normal component of the electromagnetic field induces a surface charge, or by electrons (Krane and Raether 1976, Raether 1980). Direct excitation with p-polarized light, however, is only possible in the ATR configuration, where the light line can be inclined towards larger k values so that it intersects with the dispersion curve (Otto 1968). There are two somewhat different techniques to map out the $R(\omega, k)$ surface for p-polarized light. In the one case R is measured at constant angle of incidence, φ_1, as a function of wavelength, in the second case R is

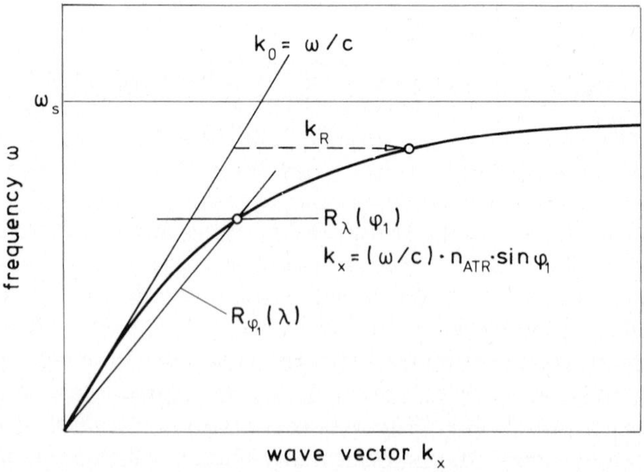

Fig. 5. Surface plasmon dispersion curve. Plasmon excitation is indicated by circles. k_R: wave vector supplied by roughness.

determined at constant wavelength as a function of φ_1 (hence at constant energy as a function of $k_x = (\omega/c)n_{ATR} \sin \varphi_1$). These two cuts are schematically shown in fig. 5. While for low energies $R_\lambda(\varphi_1)$ yields a more well-defined minimum in reflectance than $R_{\varphi_1}(\lambda)$, the reverse is true for higher energies. However, even with ATR elements made with high refractive index glass, only the low wave vector range (typically up to some 10^{-3} Å^{-1}) can be studied. Direct excitation of SPPs by light at large k values is not possible because of lack of momentum. Here, additional momentum sources are required, such as surface roughness or surface gratings, to facilitate excitation (see k_R in fig. 5).

The use of SPPs for interfacial studies on electrodes is twofold. First, since the dispersion relation depends on the dielectric functions of the media sampled by the SPPs, SPP excitation provides a sensitive tool for determining surface and film optical properties. Secondly, since excitation at large k values requires additional momentum which may be provided by the surface, the study of SPPs can yield structural information on surfaces as they act as momentum sources. This is especially valuable for the electrochemical interface, where such information is sparse.

5. *The Optical Properties of Single Crystal Surfaces as Studied by SPP Excitation*

The use of SPPs for the study of single crystal metal surfaces is still at the beginning because of experimental difficulties, encountered with obtaining an accurate gap adjustment. In the author's laboratory experiments on silver single crystals were started as these crystals are frequently used as electrodes in spectroelectrochemical experiments. The question was whether SPPs can reveal the differences in the silver surface optical properties due to the differences in surface package and symmetry. The experimental results for studies in air are summarized in fig. 6. Not only do we find clear differences in the dispersion curves for the various low index faces, we also see marked changes with crystallographic directions at low symmetry surfaces such as the Ag(110). For the latter surface SPP excitation in $[00\bar{1}]$ and in $[1\bar{1}0]$ direction occurs at different points in the ω–k-plane, while no such directional dependencies were found for Ag(111) and Ag(100). The various dispersion curves in fig. 6 show different SPP frequencies at larger k values, indicating that the most densely packed (111) surface has the highest effective electron density (hence the highest value for the plasmon frequency ω_p), while the more open (110) surface in $[00\bar{1}]$ direction has the lowest effective density (Tadjeddine et al. 1980).

Another interesting feature is seen in the curves for Ag(110). The dispersion curve for excitation in $[00\bar{1}]$ direction, that is across the sub-

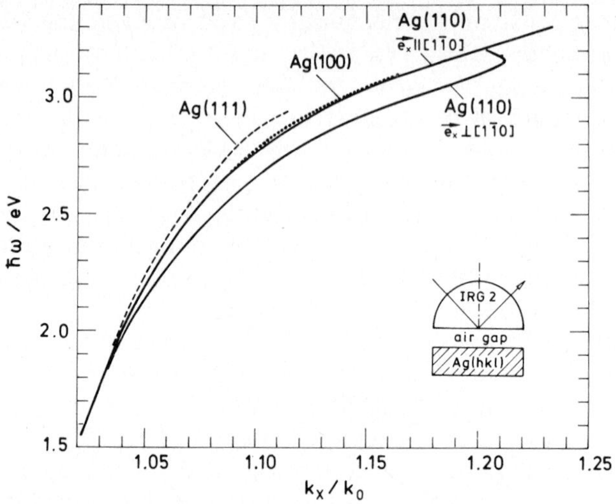

Fig. 6. Surface plasmon dispersion curves for Ag single crystal surfaces. After Tadjeddine and Kolb (1980a).

strate atom rows, shows a back bending at the highest obtainable k values, while for the other direction such an effect is not observed. Such back bendings have been observed for rough Ag surfaces, the detailed form depending somewhat on the degree of roughness (Orlowski and Raether 1976, Orlowski et al. 1979). This has been interpreted as a change in the 3-dimensional $R(\omega, k)$ surface, causing the position of the reflectance minimum to shift back. However, the position of the minimum under these circumstances is no longer believed to represent an eigenvalue of the SPP (Raether 1977). In this connection it is interesting to note that in the case of rough polycrystalline Ag the back bending was observed only for $R = R_\lambda(\varphi_1)$ and not for $R = R_{\varphi_1}(\lambda)$, which demonstrates that the position of the reflectance minimum depends now on the cut in the $R(\omega, k)$ plane. Unfortunately, for the single crystal studies only $R = R_\lambda(\varphi_1)$ was determined.

From the dispersion curve, eq.(4), the real part of the complex dielectric function of the metal can be calculated. With $k_0 = \omega/c$ and for $\epsilon_1 = 1$ and Im $\hat{\epsilon}_m \ll 1$ we obtain:

$$\epsilon_m' = \frac{(k_x/k_0)^2}{1-(k_x/k_0)^2}. \tag{4a}$$

Some values for ϵ' of Ag(hkl) are summarized in table 1*. There is a small

*Since these measurements were done in air the formation of a tarnishing film on Ag cannot be avoided. This imposes some doubt onto the accuracy of the absolute numbers. To obtain data from bare Ag surfaces experiments should be performed in UHV or in an electrochemical cell at potentials where any tarnishing film is reduced.

Table 1
Real part of the complex dielectric constant of various Ag surfaces, after Tadjeddine and Kolb (1981).

$\hbar\omega$/eV	$\epsilon'_{Ag(111)}$	$\epsilon'_{Ag(100)}$	$\epsilon'_{Ag(110)}$ $e_x \parallel [1\bar{1}0]$	$\epsilon'_{Ag(110)}$ $e_x \perp [1\bar{1}0]$	$\epsilon'_{Ag(poly)}$
1.7	−19.3	−19.3	−19.0	−19.0	—
1.8	−17.4	−17.0	−16.6	−16.5	—
1.9	−15.3	−15.0	−14.8	−14.5	—
2.0	−13.8	−13.3	−13.4	−12.8	−13.6
2.1	−12.5	−11.8	−12.0	−11.3	−12.0
2.2	−11.2	−10.6	−10.7	−9.9	−10.5
2.3	−10.1	−9.5	−9.6	−8.8	−9.3
2.4	−9.1	−8.6	−8.6	−7.8	−8.2
2.5	−8.3	−7.8	−7.8	−7.0	−7.4
2.6	−7.5	−7.0	−7.0	−6.2	−6.7
2.7	−6.8	−6.3	−6.2	−5.5	−6.0
2.8	−6.2	−5.6	−5.5	−4.9	−5.4
2.9	−5.5	−5.0	−4.9	−4.3	−4.8
3.0	—	−4.4	−4.3	−3.7	−4.2
3.1	—	−3.8	−3.8	−3.3	−3.7
3.2	—	—	−3.3	—	−3.2

but distinct difference in ϵ' for the two main crystallographic directions of Ag(110), which is easily determined to a high accuracy in SPP excitation by measuring the position of the reflectance minimum. This result, which corresponds to a difference in reflectances of less than one per cent, would be much more difficult to obtain with normal incidence reflectance spectroscopy by measuring the absolute reflectance values.

Since the dielectric function $\hat{\epsilon}$ can be expressed by the optical conductivity σ:

$$\hat{\epsilon} = \hat{\epsilon}_b - 4\pi \mathrm{i}\, \sigma/\omega \tag{5}$$

($\hat{\epsilon}_b$: bound electron contribution to the dielectric function), the result from fig. 6 may be explained as follows. The surface conductivity for a (110) surface is smaller perpendicular to the atomic rows where the electrons see larger variations of the surface potential than in $[1\bar{1}0]$ direction where a nearly constant density profile parallel to the surface is expected. The more densely packed the surface is in the direction of the electric field, the higher is σ and hence the higher is the SPP excitation energy (Tadjeddine et al. 1980).

6. SPP Excitation Studies at the Electrode–Electrolyte Interface

SPP dispersion curves for the metal–electrolyte interface have been measured and the curves were found to be shifted with respect to the metal–air interface according to the different dielectric constant of the ambient (Kötz et al. 1977). The more interesting question is whether the influence of the electrode potential on the SPP excitation can be seen. This is indeed the case as is demonstrated in fig. 7, where the reflectance minimum for $R_{\varphi_1}(\lambda)$ due to SPP excitation at the Ag–electrolyte interface for two different electrode potentials is shown. The shift in the SPP excitation energy to lower values with more anodic electrode potentials is clearly evident. The results of a detailed investigation of this influence as studied for poly-crystalline Ag and Au films in the Kretschmann configuration is shown in fig. 8. We find a rather asymmetric behavior of the SPP excitation energy, $\hbar\omega_{sp}$, on electrode charging (note that the potentials of zero charge (pzc) are different for Ag and Au as indicated by the arrows in fig 8). When the electrode is charged positively, $\hbar\omega_{sp}$ decreases linearly with electrode potential, the slopes being -25 and -13 meV/V for Ag and Au, respectively (Kötz et al. 1977). This follows the expected trend, since the electron density at the electrode surface is lowered at positive potentials. However, when the electrode is charged negatively the shift in $\hbar\omega_{sp}$ ceases and $\hbar\omega_{sp}$ becomes independent of the potential. Model calculations have shown that a change in the molar refractivity of the ionic (Helmholtz) layer with electrode potential cannot account for the observed shift at potentials

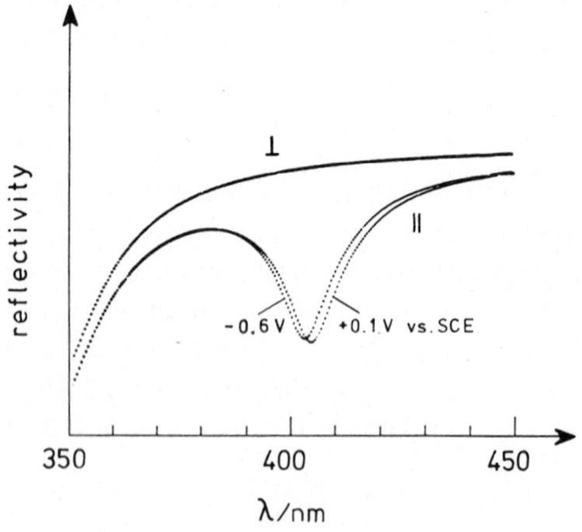

Fig. 7. ATR spectra of an Ag film on a glass hemicylinder in 0.5 M $NaClO_4$ at two different electrode potentials, $\varphi_1 = 60°$. The curves for s- and p-polarization are not normalized. After Kötz et al. (1977).

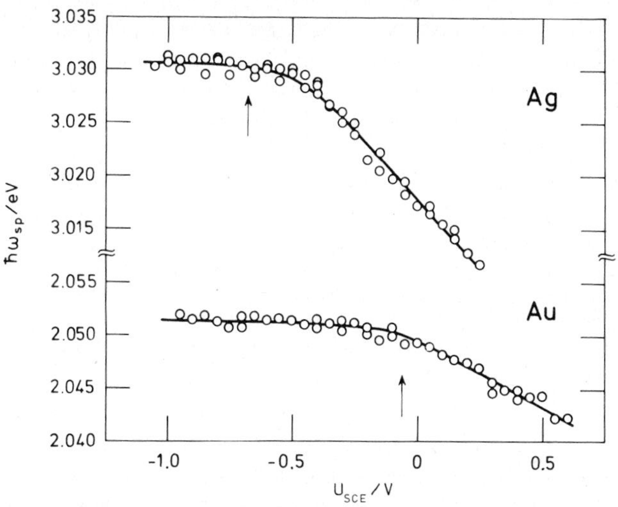

Fig. 8. Surface plasmon excitation energy, $\hbar\omega_{sp}$, for polycrystalline Ag and Au electrodes in 0.5 M $NaClO_4$ as a function of electrode potential U. $\varphi_1 = 60°$ (51°) for Ag (Au). IRG 2. The respective potentials of zero charge are marked by arrows. After Kötz et al. (1977).

anodic of the pzc (Kötz 1978). Measurements in electrolytes with ions of widely varying polarizability yielded nearly identical results, supporting the idea that the effect originates from changes in the metal surface optical properties (Kötz 1978).

Since detailed knowledge of the electron distribution at the metal–electrolyte interface and its dependence on electrode potential is completely missing, a model analogous to that used for the metal–vacuum interface (Lang 1971) may be helpful. As sketched in fig. 9, the electron density is assumed to decay in an evanescent tail from bulk density inside the metal to zero well outside. Electrolyte phase species in the inner and outer Helmholtz layer will feel this tail to various degrees depending on

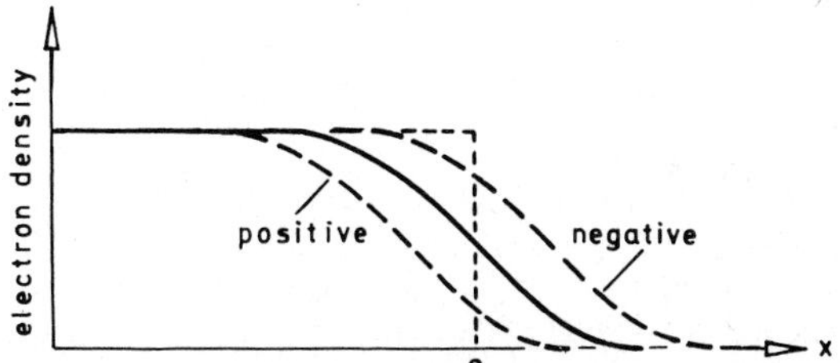

Fig. 9. Model for the electron distribution at the electrode–electrolyte interface and its change with electrode potential.

their distance from the surface. A change in the electrode potential is believed to shift the electron tail back and forth, with respect to the positive background which also should mark roughly the spatial extension of the bound electrons. This is quite important since the SPPs in Ag and Au at these energies require the presence of bound electrons (mainly d electrons), which screen the charge density waves in the conduction electron system to such an extent that the plasma frequency is shifted considerably to lower values. From such a simple picture one gets the impression that the electron density at the surface is indeed reduced at positive potentials by pushing the evanescent tail into the metal, whereas the electron density is not really increased at negative potentials as now by pushing the tail outward electrons are accumulated more in front of the surface. In the latter case the number of electrons is increased, but not the density. This would at least tentatively explain the observed potential dependence of $\hbar\omega_{sp}$, although other and better explanations could be found. The role of the pzc seems also reflected in the potential dependence of $\hbar\omega_{sp}$ for single crystal surfaces. The pzc of a surface depends on the crystallographic orientation in much the same way as the work function does (Trasatti 1977). E.g. the pzc is most positive for Ag(111), less for Ag(100) and least for Ag(110) (Valette and Hamelin 1973). When we look at the results for the potential dependence of $\hbar\omega_{sp}$ at the Ag(hkl) – electrolyte interface (see fig. 10) we not only find the same asymmetric behavior again, but also a shift in the bending point which seems to follow –at least qualitatively– the pzc. This is certainly true when comparing the curves for Ag(111) and Ag(poly). The curve for Ag(100) (and also that for Ag(110),

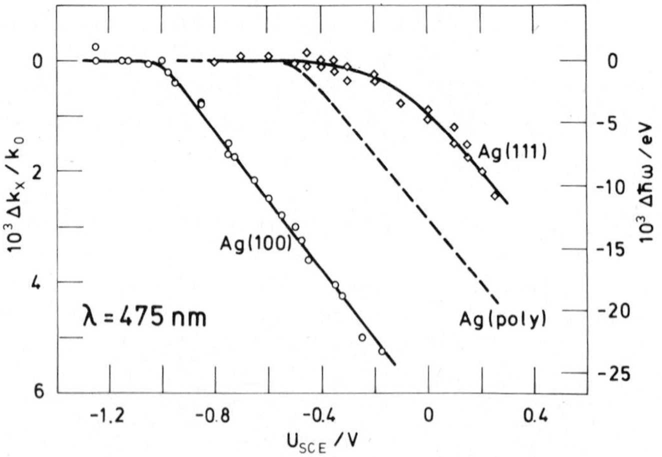

Fig. 10. Potential dependence of the surface plasmon excitation for various Ag surfaces: IRG 2, 0.5 M $NaClO_4$. After Tadjeddine and Kolb (1980a).

which is not shown here) is too far negative to be solely explained by differences in the pzc. However, one should keep in mind, that these potential dependences were determined for somewhat different ω–k areas (e.g., see dispersion curves in fig. 6) and then shifted relative to each other to match at the cathodic side. Here we assume that the bend always corresponds to the pzc. If, however, for the more open surfaces the electron density could indeed be increased somewhat before now at potentials negative of the pzc the limiting value is reached, then the plateau in $\hbar\omega_{sp} = f(U)$ should be drawn above the arbitrary zero line shown in fig. 10. This, however, is at present speculation and we must await more experimental data to reach a conclusive explanation.

7. *SPP Excitation on Adsorbate Covered Electrode Surfaces*

The properties of the SPPs, such as their spatial extension, sensitively depend on the dielectric constants of the adjacent media. Hence, it is not surprising that the SPP dispersion curve for a certain interface is changed when the interface optical properties are altered, e.g., by surface reactions. One of the most studied reactions at the metal–air interface is the formation of a thin overlayer, such as a tarnishing film due to oxide formation (see e.g. Kloos 1968), metal (see e.g. Lopez-Rios et al. 1978) or dye (see e.g. Pockrand et al. 1978) deposition. When a thin film of thickness $d \ll \lambda$ and with a dielectric constant $\hat{\epsilon}_2$ is brought on top of a metal film ($\hat{\epsilon}_m$) in contact with an ambient (ϵ_1), then the change in the SPP wave vector, Δk_x, parallel to the surface, is given by (Kretschmann 1971):

$$\Delta k_x = \frac{(-\hat{\epsilon}_m \epsilon_1)^{1/2}}{-\hat{\epsilon}_m + \epsilon_1} k_x^2 \left[1 - \left(\frac{k_x}{k_0}\right)^2 \left(\frac{1}{\hat{\epsilon}_2} + \frac{\hat{\epsilon}_2}{\epsilon_1 \hat{\epsilon}_m}\right)\right] d, \tag{6}$$

where k_x is the wave vector for the bare metal surface as given by eq. (4). We see that in this approximation for very thin films ($d \ll \lambda$) the change in wave vector is directly proportional to the film thickness, assuming the film optical properties do not change with d. Hence SPP excitation can be used for direct coverage monitoring. An example is given in fig. 11 for the gold–electrolyte interface, where the change in the energetic position of the reflectance minimum for $R_{\varphi_1} = f(\lambda)$ due to electrochemical reactions is monitored. The measurements were done in the Kretschmann configuration at constant angle of incidence. One reaction, monitored by SPP, is the formation of a surface oxide on gold by anodic decomposition of water. The oxide starts to form at +0.9 V, as is indicated by the onset of an anodic current at that potential. On the reversed potential scan the oxide layer is reduced near +0.3 V, seen in the stripping peak in the cyclic voltammogram (see insert in fig. 11). Oxide formation and reduction are

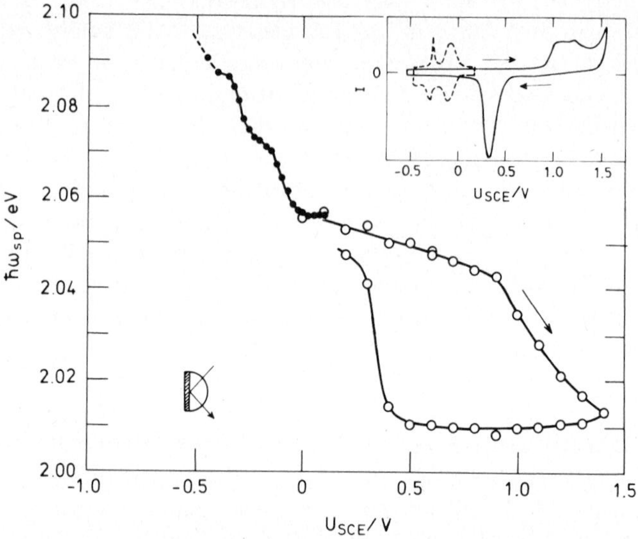

Fig. 11. Shift in the surface plasmon excitation energy $\hbar\omega_{sp}$ for Au in 0.5 M $NaClO_4$ due to oxide formation (-○-○-○-) and, after adding 2×10^{-4} M $Pb(NO_3)_2$ to the solution, due to Pb monolayer formation (-●-●-●-): $\varphi_1 = 51°$; IRG 2. After Kolb (1977).

clearly seen in the shift of $\hbar\omega_{sp}$ with potential, and the overall shape closely resembles that for the charge (coverage) – potential curve (Kolb 1977, Chao et al. 1977a). As a second example, lead ions were added to the electrolyte, and one monolayer of lead deposited in the potential range between 0 and −0.5 V. As is seen in the current–potential curve by the appearance of two adsorption and stripping peaks, monolayer formation occurs in two distinctly different steps, indicating that the adsorption energies for formation of the first half of the monolayer and for completion of the monolayer differ markedly (Kolb et al. 1974). Again, the coverage dependence of the Pb monolayer formation is clearly reflected in the shift of $\hbar\omega_{sp}$. Another example which has been studied is halide adsorption on silver electrodes (Kolb 1977, Gordon and Ernst 1980).

More information, beside just monitoring the coverage dependence at constant angle or wavelength, is gained by a complete analysis of the reflectance minimum, which yields real and imaginary part of the film dielectric function (Pockrand et al. 1978). This, however, requires the use of a model, such as a stratified multilayer model with sharp boundaries, which bears problems of its own. However, it has been demonstrated that measuring the frequency dependence of Δk_x can reveal already interesting information on the absorptive properties of the surface layer without the use of further model calculations. This was shown by Pockrand and Swalen (1978) for the system reproduced in fig. 12. The interesting feature

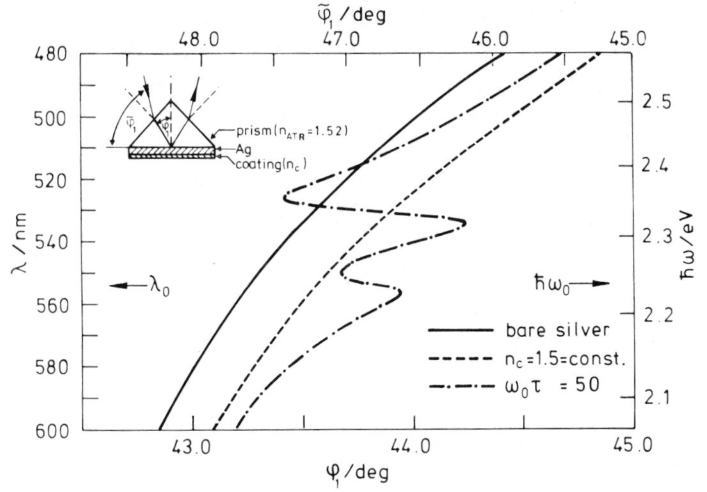

Fig. 12. Calculated SPP resonance angle as a function of wavelength for two different coatings ($d = 2.5$ nm) on Ag: (– – –) dielectric coating; (–·–·–) coating with an absorption band at $\lambda_0(\hbar\omega_0)$. After Pockrand and Swalen (1978).

in the SPP dispersion curve, which is caused by an absorptive surface layer, is a pronounced structure in the energy region of absorption. Depending on the width of the isotropic absorption band, these authors obtain a more or less pronounced double back bending in the dispersion curve, the low energy back bending designating the energetic position of the absorption maximum. As the authors further note, double back bending is only expected for films with isotropic dielectric constants. Anisotropy may lead to a single back bending in such a way that for absorption parallel to the surface a single structure at low energy is observed, while for a transition moment perpendicular to the surface a single back bending at high energy is observed. In both cases the corresponding second structure, which is clearly seen in the isotropic case only, is damped away.

We will demonstrate the usefullness of SPP excitation in characterizing optical properties of surface layers for the system pyridine–silver. Recently this system has received much attention in Raman spectroscopy, since the pyridine molecule of the Ag electrode surface can exhibit an enhancement in Raman intensity of about 10^6 compared to the Raman intensity of the pyridine molecule in solution (see e.g. Jeanmaire and VanDuyne 1977). This strong enhancement was usually found only when the Ag electrode was subjected to a so-called electrochemical activation cycle in the presence of pyridine and in a chloride ion containing electrolyte. This cycle consists of a potential scan to anodic values, where Ag is oxidized and precipitated as AgCl and a subsequent reduction to metallic Ag on the cathodic scan. The total amount of Ag transferred back and forth for observing the Raman enhancement typically ranged in the equivalent from submonolayers to several tens of monolayers (Pettinger et al. 1978). The SPP excitation experiment can reveal what happens to the electrode surface during such a treatment. In fig. 13 we see the dispersion curves for Ag in 0.1 M NaCl + 10^{-2} M pyridine before and after the activation cycle (Tadjeddine and Kolb 1980b). It becomes evident that such an electrochemical treatment causes the formation of a pyridine–Ag surface complex, which has absorption bands in the near IR. The existence of such bands around 800 to 900 nm has also been shown by external reflectance measurements (Pettinger et al. 1978). The result demonstrates that the properties of an adsorbate can differ substantially from those of the same entity in solution since pyridine is nonabsorbing throughout the whole visible wavelength region. The strong interaction with Ag, which is introduced by the dissolution and redeposition of Ag (perhaps by forming Ag adatoms) causes new absorption bands to appear at much lower energies, close to those wavelengths where the Raman experiments are done. This could suggest that the existence of resonance Raman effects has to be considered for the interpretation of the enhancement.

8. *The Use of Field Enhancements in SPP Excitation*

Classical field strength calculations using Maxwell's equations (Hansen 1968) reveal that an enormous field enhancement occurs at the metal–ambient interface in the case of SPP excitation under ATR conditions. A typical intensity ($\sim E^2$) distribution for a 500 Å thick Ag film on glass (Kretschmann configuration) is reproduced in fig. 14 (Pettinger et al. 1979). Here, a total enhancement of about 150 with respect to the intensity of the incoming electromagnetic wave is obtained, mostly in the direction normal to the surface (z-direction).

This enhancement has been used for detection of overlayers and a more accurate determination of their properties. In fig. 15 the result of a study on the electrochemical oxide formation on gold by a combined ellipsometric – SPP/ATR experiment is shown (Chao et al. 1977a, b). While for the nonresonant case the formation and reduction of the surface oxide on gold is barely seen, this process can be followed in detail when SPP excitation is achieved. This is simply done by changing slightly the angle of incidence until resonance occurs. This is an interesting phenomenon, since by a very small angle variation (into and out of the resonance) the field distribution of the probing electromagnetic wave is changed considerably and hence information from different regions of the system under study may be sampled.

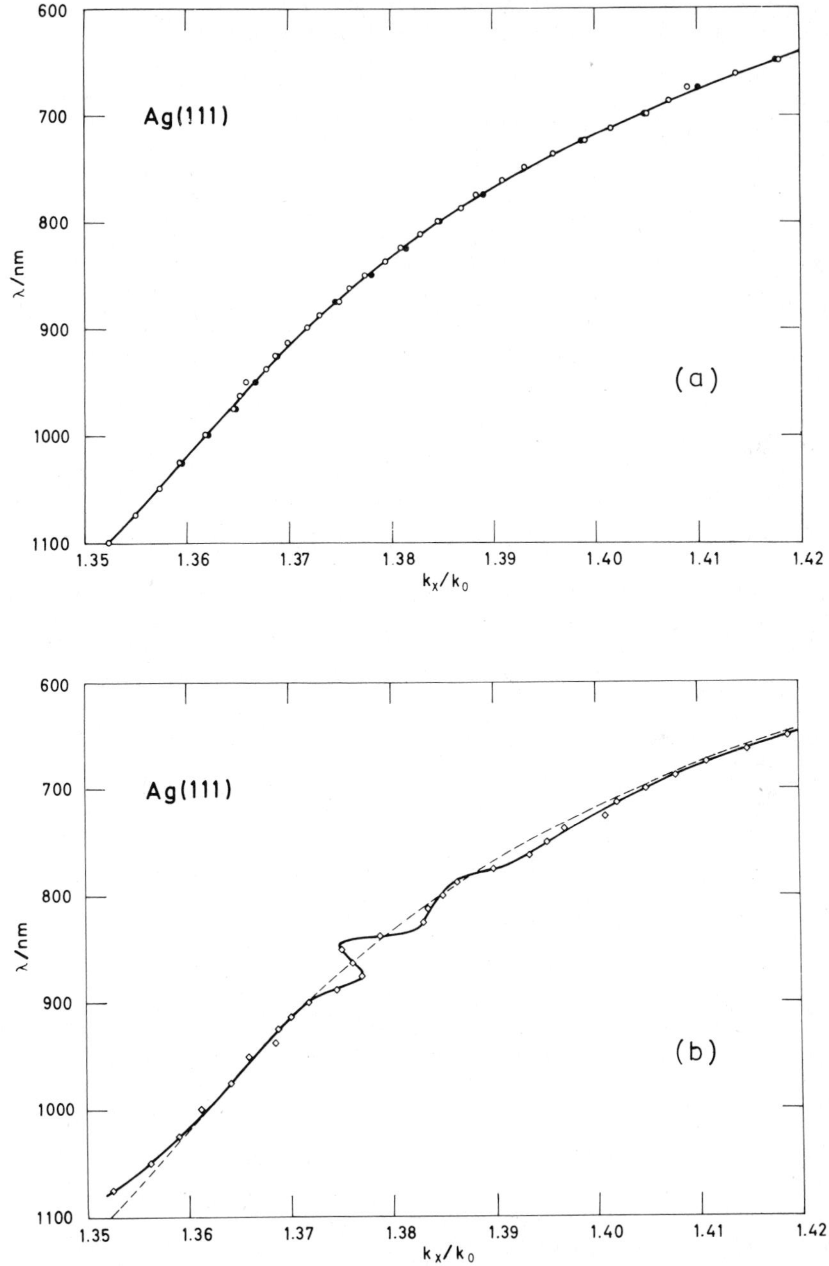

Fig. 13: Dispersion curves for a Ag(111) electrode in 0.1 M NaCl: (a) (–○–○–○–) before activation; (–●–●–●–) after one activation cycle, no pyridine present; (b) after 6 activation cycles in the presence of pyridine. After Tadjeddine and Kolb (1980b).

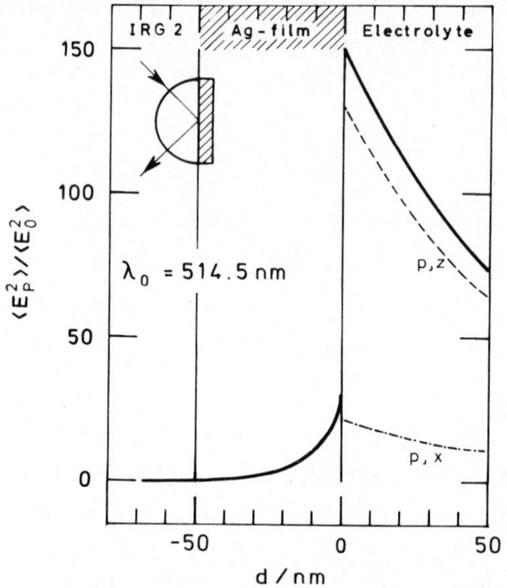

Fig. 14. Time-averaged and normalized square of the electric field strength for p-polarized light as a function of distance d from the electrode–electrolyte interface under SPP resonance conditions: $\varphi_1 = 50°$, $n_{ATR} = 1.906$, $n_{H_2O} = 1.336$, $d_{Ag} = 500$ Å. The components parallel (x) and perpendicular (z) to the surface are also shown.

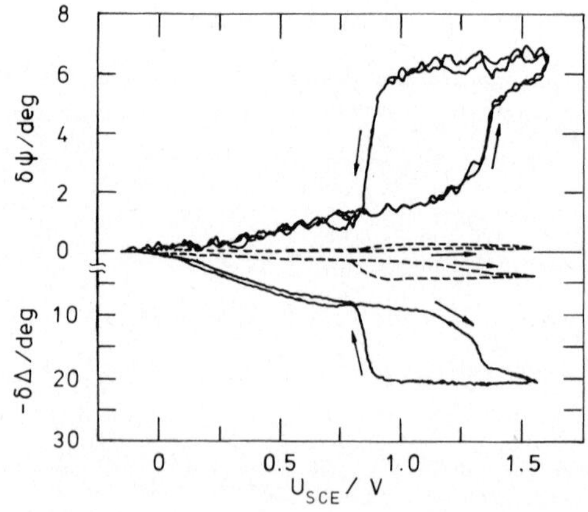

Fig. 15. Variation of phase Δ and azimuth ψ for the gold–electrolyte interface with electrode potential at fixed angles of incidence: (——) with and (---) without SPP excitation; 0.5 M H_2SO_4; $\lambda = 609.3$ nm. After Chao et. al. (1977b).

Another field in spectroscopy, where intensity problems are often encountered, is Raman spectroscopy. Because of the extremely small intensity of normal Raman scattering, signals from smooth surfaces are often difficult to obtain in reasonably good quality. Chen et al. (1976) first suggested the use of field enhancement in SPP excitation for Raman studies. To my knowledge, so far no one has obtained Raman spectra by means of SPP enhanced fields, which would otherwise have been very difficult or impossible to obtain. To demonstrate the signal enhancement, when SPPs are excited, measurements were carried out with the pyridine–Ag system, where the field enhancement is superimposed onto the commonly observed and often discussed system-specific enhancement. In fig. 16 the Raman signal at $1010\,cm^{-1}$ for the ring-breathing vibration of pyridine on Ag is shown as a function of the angle of incidence (Pettinger et al. 1979). The occurrence of the SPP excitation within a small angular range around 50° is clearly seen by an increase in Raman intensity. The disappointingly low enhancement factor, which misses the theoretically predicted one by an order of magnitude, may be caused by several effects, one of them certainly being the strong damping of the SPPs (as indicated by the relatively broad half width of the resonance). The reason for this may be found in either a non-optimum Ag film thickness or the absorptive properties of the pyridine–Ag surface layer (see previous section) or in both.

Recently, Girlando, Philpott et al. (1980) also demonstrated the use of SPP enhanced fields in Raman spectroscopy by studying thin organic polymer films on a silver grating, where SPP excitation occurs via coupling

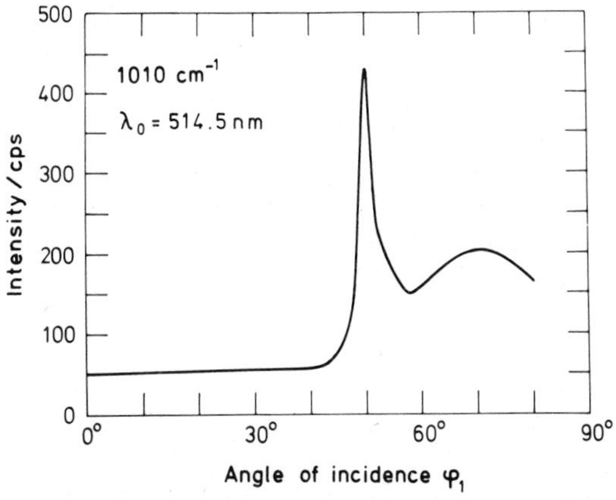

Fig. 16. Angle dependence of the Raman intensity at $1010\,cm^{-1}$ from pyridine on Ag after electrochemical activation (ATR configuration); SPP excitation occurs at $\varphi_1 = 50°$. After Pettinger et al. (1979).

through the grating vector. An enhancement factor of 30 with respect to the off-resonance case was reported.

9. SPP Excitation by External Reflection

In an external reflectance experiment, light line and SPP dispersion curve do not intersect and hence direct SPP excitation by light is not possible because of lack of momentum conservation. When the surface or interface, however, supplies a momentum source SPP excitation becomes possible. In return, it is hoped that from the study of SPP excitation under these circumstances information is gained on the structure and properties of such a surface. In the following we will discuss three possible momentum sources on surfaces: random roughness, regular steps and adsorbates. Since the influence of SPP excitation on the reflectivity under external reflection conditions is often very small, modulation spectroscopy is usually employed for detection of minute changes in the reflected power.

9.1 SPP Excitation on Rough Surfaces

SPP excitation by surface roughness is conveniently studied to a high degree of accuracy by electroreflectance at the metal–electrolyte interface. Electroreflectance (ER), that is the change in reflectance upon variation of the electrode potential, is a very surface-sensitive tool, as the static electric field which is modulated with the electrode potential is screened within the very first layer of the metal (Thomas–Fermi screening length). Care has to be taken that the potential is modulated only within the so-called double layer charging region, where no electrochemical reactions take place and the double layer behaves like a condensor, which is charged and discharged. The ER of metals has been described extensively elsewhere (McIntyre 1973, Kolb 1977) and will not be discussed in this context.

The ER spectra for a perfectly flat Ag(111) surface exhibit a smooth background around 3.5 eV, where SPP excitation is expected to occur since random roughness will mostly favor excitation at large k-values (Kolb and Kötz 1977b). This smooth background is shown as dotted line in the spectra in fig. 17 recorded for s- and p-polarized light at $\varphi_1 = 45°$. Then the Ag(111) surface is roughened by simply dissolving and redepositing some Ag during an anodic potential cycle. The occurrence of SPP excitation suddenly appears as a dip in the static reflectance spectrum (see insert in fig. 17a) and more clearly by a derivative-like structure in the ER spectra around 3.5 eV (see arrow). The derivative-like structure in the ER indicates already that the shallow reflectance minimum due to SPP excitation is shifted in energy by the potential modulation, similar to the case of ATR

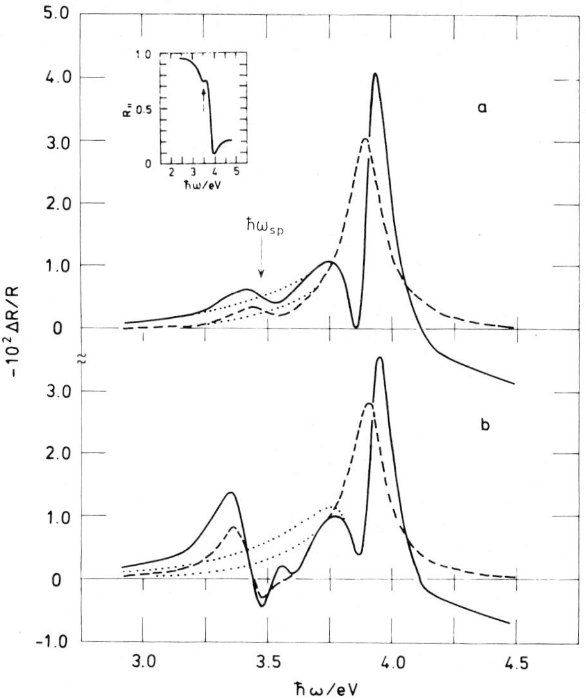

Fig. 17. Electroreflectance spectra of a slightly rough (a) and a more strongly roughened (b) Ag(111) surface for p (——) and s (---) polarization; 0.5 M $NaClO_4$ (pH 2); −0.5 to 0.0 V; $\varphi_1 = 45°$. The dotted lines represent the curves for the initially flat surface. Insert: reflectance spectrum of a rough Ag electrode in 0.5 M $NaClO_4$: $\varphi_1 = 45°$; p-polarization. After Kolb and Kötz (1977b).

conditions as discussed in sect. 6. We indeed find again a linear shift of $\hbar\omega_{sp}$ with electrode potential for charging the Ag electrode positively, while at potentials negative of the pzc this shift ceases. The shift $\partial\hbar\omega_{sp}/\partial U$ at positive potentials (with respect to the pzc) is about −0.2 eV/V (Kolb and Kötz 1977b) and hence roughly one order of magnitude larger than that found at small k values by ATR measurements. The larger shift in $\hbar\omega_{sp}$ with potential at large k values can be understood by the smaller penetration depth of these SPPs, which therefore are more sensitive to surface changes.

When the roughening is increased, the structure due to SPP excitation around 3.5 eV becomes more complicated and will usually exhibit a splitting of the derivative-like structure (fig. 17b) (Kolb and Kötz 1977b). This splitting has been tentatively explained by Kretschmann et al. (1979) by a splitting in the SPP dispersion relation for a statistically rough surface.

9.2 SPP Excitation on Stepped Surfaces

The natural roughness of a polycrystalline Ag surface supplies a large spectrum of k values for momentum conservation during SPP excitation (see e.g., Raether, ch. 9). Hence, excitation occurs over an extended k and energy range in the respective dispersion curve.

In the search for a more well defined momentum source we prepared a stepped single crystal surface, assuming that the terrace width supplies the desired k in one direction only. We used an Ag(110) surface, which was tilted by 3° against the (110) plane, with the steps along the $[00\bar{1}]$ direction (Kötz and Kolb 1978). The terrace width should be about 60 Å, assuming steps of atomic height. This would correspond to $k = 0.1\ \text{Å}^{-1}$. The normal-incidence ER spectra of this surface are shown in fig. 18; they were recorded for $e \perp [1\bar{1}0]$ and $e \| [1\bar{1}0]$, which coincides with e being parallel and perpendicular to the steps (just for clarity: moving perpendicular to the steps would lead up and down the stairs). The difference in the two spectra in the SPP excitation region around 3.5 eV (350 nm) is striking. For $e \| [1\bar{1}0]$ the structure due to SPPs is extremely sharp and pronounced, while for the other direction SPP excitation is hardly seen at all in $\Delta R/R$. In the first case, the excitation direction is such that the steps can supply a well-defined k value, while in the second case a more or less perfectly flat

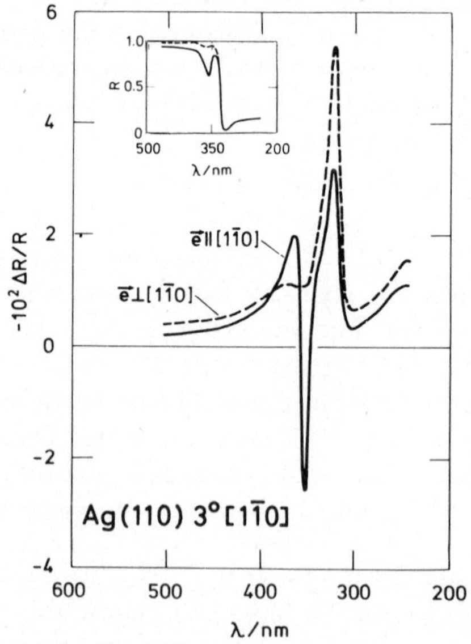

Fig. 18. Normal-incidence electroreflectance spectra and static reflectance spectra (insert) of a stepped Ag(110) surface: 0.5 M $NaClO_4$; −0.5 to 0.0 V. The steps lead in $[1\bar{1}0]$ direction. After Kötz and Kolb (1978).

surface is seen as one looks down the rails. The same behavior is also seen in the static reflectance spectra for the two polarization directions at normal incidence (see insert in fig. 18). In the first case a sharp dip is observed, indicative of strong SPP excitation, which is largely absent for the other direction. The ER spectra outside the energy range for SPP excitation are nearly identical to those for a perfect Ag(110) surface and exhibit the characteristic anisotropy in $\Delta R/R$ for the two main crystallographic directions (Furtak and Lynch 1975, Tadjeddine et al. 1980). It may not be surprising that the atomic rows in the Ag(110) surface along $[1\bar{1}0]$ are not found to act as a grid of atomic dimensions for SPP excitation.

9.3 Adsorbate Induced SPP Excitation

It is well known from LEED experiments in the gas phase that many adsorbates in the submonolayer range form superstructures on the surface, that is, they adsorb at regular lattice sites. The same is believed to be true for some adsorbates on electrode surfaces. E.g. extensive studies gave evidence that metal adsorbates in submonolayer amount on foreign metal surfaces form regular structures (Bewick and Thomas 1975, Schultze and Dickertmann 1976, Kolb et al. 1979). Such a system is Pb on Ag(111), where comparative studies for deposition in UHV and electrodeposition in an electrochemical cell suggested that in both cases the same growth behavior occurs (Takayanagi et al. 1980). At low coverages ($0 < \theta < 0.2$) random adsorption of Pb is found, followed by the formation of a regular superstructure of the type $(\sqrt{3} \times \sqrt{3})R30°$ which gradually disappears when the monolayer is completed as a hexagonal close packed layer. The superstructure has its maximum coverage at $\theta = \frac{1}{2}$.

It seems an appealing idea to use the regular structure of adlayers as a source for well-defined momentum in enabling SPP excitation. The facilitation of SPP excitation by the adlayer could give us structural clues on the adlayer itself. Figure 19 shows spectra of normalized reflectance differences, $\Delta R/R$, for Ag(111) electrodes with various coverages of Pb. The normalized reflectance difference is defined as (McIntyre and Aspnes 1971):

$$\frac{\Delta R}{R} = \frac{R(\theta) - R(\theta = 0)}{R(\theta = 0)},$$

and describes the change in reflectance of the bare substrate due to adsorption or film formation. For very low Pb coverages and for the completed monolayer we do not find any structure in $\Delta R/R$ in the energy range of interest (i.e. around 350 nm), $\Delta R/R$ reflecting changes due to the different optical properties of the overlayer only. However, at medium coverages, where the superstructure is formed, we observe a pronounced

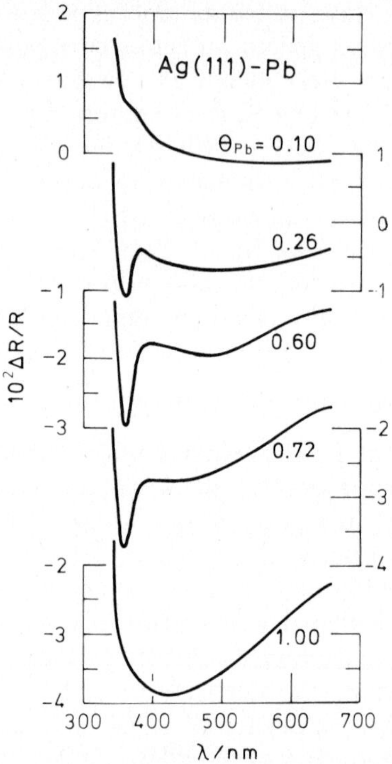

Fig. 19. Normalized reflectance differences, $\Delta R/R$, for a Ag(111) electrode due to Pb deposition as a function of wavelength; normal incidence. After Kolb and Rath (1980).

dip at 350 nm, obviously due to SPP excitation. Since the intensity of this dip correlates well with the coverage dependence of the $(\sqrt{3} \times \sqrt{3})R30°$ structure, we believe that it is the Pb in this structure which allows SPP excitation. It is not yet clear, which k values are supplied by such an adlayer, nor do we know whether other conditions, such as a certain polarizability of the adatoms, may in addition be a necessary requirement. We have also observed similar effects for Pb on Ag(110), although less pronounced. More experimental data are necessary to safely establish this effect; however, we are confident that this is a promising way for obtaining information on the superstructure formation in adlayers at electrode–electrolyte interfaces.

Finally it should be mentioned that small metal clusters on a perfectly flat foreign metal surface can induce SPP excitation via external reflection in very much the same way as "own" roughness. This has been demonstrated by SPP excitation on a flat Ag(111) electrode surface, which was covered with small 3-dimensional Cu clusters (Kötz and Kolb 1980).

10. Conclusion and Outlook

The use of SPPs in the study of the electrode–electrolyte interface seems indeed a promising way of obtaining new information on bare and adsorbate covered surfaces and on the electrochemical interface in general. The method is applicable in-situ, is non-destructive and yields information which can be specific on an atomic or molecular level and hence is complementary in many respects to classical electrochemical techniques with their ability to yield exact thermodynamic data.

The problems which seem most rewarding to pursue with this method are manyfold. A detailed study of the potential influence on $\hbar\omega_{sp}$ should give answers towards the electronic structure of the double layer and our knowledge there is still rudimentary. This includes also the study of the electronic properties of adsorbates and electrochemically formed surface compounds. Adsorbate induced SPP excitation could be another promising field, if the promise holds that we obtain structural information on adlayers. And finally, the study of well prepared single crystal surfaces by SPPs for evaluation of surface optical properties is still at an initial state and it should eventually yield information about surface properties which are considerably more accurate than that obtained by normal optical techniques.

Acknowledgements:

A great deal of the ATR experiments, done in the author's laboratory, was carried out by Dr. A. Tadjeddine during several visits as a guest at the Fritz-Haber-Institut; his skill and enthusiasm lead to a very pleasant and fruitful cooperation. I am also indebted very much to Professor F. Forstmann for many stimulating discussions.

References

Bewick, A. and B. Thomas, 1975, J. Electroanal. Chem. **65**, 911.
Bockris, J.O'M., M.A. Devanathan and K. Muller, 1963, Proc. R. Soc. **A274**, 55.
Bockris, J.O'M. and A.K.N. Reddy, 1970, Modern Electrochemistry (Plenum Press, New York).
Chao, F., M. Costa and A. Tadjeddine, 1977, J. Physique (Paris) **38**, C5-97.
Chao, F., M. Costa, A. Tadjeddine, F. Abelès, T. Lopez-Rios and M.L. Theye, 1977, J. Electroanal. Chem. **83**, 65.
Chen, Y.J., W.P. Chen and E. Burstein, 1976, Phys. Rev. Lett. **36**, 1207.
Forstmann, F. and H. Stenschke, 1978, Phys. Rev. **B17**, 1489.
Furtak, T.E. and D.W. Lynch, 1975, Phys. Rev. Lett. **35**, 960.
Gerischer, H., D.M. Kolb and J.K. Sass, 1978, Adv. Phys. **27**, 437.

Girlando, A., M.R. Philpott, D. Heitmann, J.D. Swalen and R. Santo, 1980, J. Chem. Phys. **72**, 5187.
Gordon, J.G. and S. Ernst, 1980, Surf. Sci. **101**, 499.
Hansen, W.N., 1968, J. Opt. Soc. Am. **58**, 380.
Harrison, J.A., J.E.B. Randles and D.J. Schiffrin, 1973, J. Electroanal. Chem. **48**, 359.
Jeanmaire, D.L. and R.P. VanDuyne, 1977, J. Electroanal. Chem. **84**, 1.
Kloos, T., 1968, Z. Physik **208**, 77.
Kolb, D.M., 1977, J. Physique (Paris) **38**, C5-167.
Kolb, D.M., 1978, Advances in Electrochemistry and Electrochemical Engineering, Vol. 11, eds., H. Gerischer and C.W. Tobias (Wiley, New York) p. 125.
Kolb, D.M., M. Przasnyski and H. Gerischer, 1974, J. Electroanal. Chem. **54**, 25.
Kolb, D.M. and R. Kötz, 1977a, Surf. Sci. **64**, 698.
Kolb, D.M. and R. Kötz, 1977b, Surf. Sci. **64**, 96.
Kolb, D.M., R. Kötz and K. Yamamoto, 1979, Surf. Sci. **87**, 20.
Kolb, D.M. and D.L. Rath, 1981, in preparation.
Kötz, R., 1978, Thesis, Technical University Berlin.
Kötz, R., D.M. Kolb and J.K. Sass, 1977, Surf. Sci. **69**, 359.
Kötz, R. and D.M. Kolb, 1978, Z. Physik. Chem. N.F. **112**, 69.
Kötz, R. and D.M. Kolb, 1980, Surf. Sci. **97**, 575.
Krane, K.J. and H. Raether, 1976, Phys. Rev. Lett. **37**, 1355.
Kretschmann, E., 1971, Z. Physik **241**, 313.
Kretschmann, E., 1978, Opt. Commun. **26**, 41.
Kretschmann, E., T.L. Ferrell and J.C. Ashley, 1979, Phys. Rev. Lett. **42**, 1312.
Lang, N.D., 1971, Phys. Rev. **B4**, 4234.
Lopez-Rios, T., F. Abelès and G. Vuye, 1978, J. Physique (Paris) **39**, 645.
McIntyre, J.D.E., 1973, Surf. Sci. **37**, 658.
McIntyre, J.D.E. and D.E. Aspnes, 1971, Surf. Sci. **24**, 417.
Orlowski, R. and H. Raether, 1976, Surf. Sci. **54**, 303.
Orlowski, R., P. Urner and D.L. Hornauer, 1979, Surf. Sci. **82**, 69.
Otto, A., 1968, Z. Physik **216**, 398.
Otto, A., 1976, Optical Properties of Solids – New Developments, ed., B.O. Seraphin (North-Holland, Amsterdam) p. 677.
Parsons, R., 1961, Advances in Electrochemistry and Electrochemical Engineering, Vol. 1, ed., P. Delahay and C.W. Tobias (Interscience, New York) p.1.
Pettinger, B., U. Wenning and D.M. Kolb, 1978, Ber. Bunsenges. Phys. Chem. **82**, 1326.
Pettinger, B., A. Tadjeddine and D.M. Kolb, 1979, Chem. Phys. Lett. **66**, 544.
Pockrand, I. and J.D. Swalen, 1978, J. Opt. Soc. Am. **68**, 1147.
Pockrand, I., J.D. Swalen, R. Santo, A. Brillante and M.R. Philpott, 1978, J. Chem. Phys. **69**, 4001.
Raether, H., 1977, Physics of Thin Films, Vol. 9, ed., G. Hass (Academic Press, New York) p. 145.
Raether, H., 1980, Springer Tracts in Modern Physics, Vol. 88 (Springer, Berlin) chap. 10.
Ritchie, R.H., 1973, Surf. Sci. **34**, 1.
Schultze, J.W. and D. Dickertmann, 1976, Surf. Sci. **54**, 489.
Sparnaay, M.J., 1972, The Electrical Double Layer (Pergamon Press, Oxford).
Tadjeddine, A. and D.M. Kolb, 1980a, Proc. 4th Int. Conf. on Solid Surfaces (Cannes) Vol. I, p. 615.
Tadjeddine, A. and D.M. Kolb, 1980b, J. Electroanal. Chem. **111**, 119.
Tadjeddine, A. and D.M. Kolb, 1981, to be published.
Tadjeddine, A., D.M. Kolb and R. Kötz, 1980, Surf. Sci. **101**, 277.
Takayanagi, K., D.M. Kolb, K. Kambe and G. Lehmpfuhl, 1980, Surf. Sci. **100**, 407.

Trasatti, S., 1977, Advances in Electrochemistry and Electrochemical Engineering, Vol. 10, ed., H. Gerischer and C.W. Tobias (Wiley, New York) p. 213.
Valette, G. and A. Hamelin, 1973, J. Electroanal. Chem. **45**, 301.
Vetter, K.J., 1967, Electrochemical Kinetics (Academic Press, New York).
Yeager, E., 1977, J. Physique (Paris) **38**, C5 - 1.

Chapter 9

Surface Plasmons and Roughness*

H. RAETHER

Institut für Angewandte Physik
Universität Hamburg
2 Hamburg 36,
F.R. Germany

*We are here using the customary description of a class of surface excitations as surface plasmons, independent whether retardation plays a role or not. An alternative description as polaritons or surface polaritons used frequently in this volume is, of course, valid.

Surface Polaritons
Edited by
V.M. Agranovich and D.L. Mills

Contents

1. Introductory remarks on roughness . . . 333
2. Influence of roughness on the dispersion relation of surface plasmons . . . 335
    2.1. Experimental results . . . 336
    2.2. Theoretical remarks . . . 341
3. Coupling of photons with surface plasmons via roughness . . . 343
    3.1. Theoretical results on light emission . . . 344
    3.2. Experimental results on light emission – determination of roughness . . . 349
    3.3. Experiments with enhanced roughness . . . 354
    3.4. Exterior and interior roughness . . . 358
    3.5. Deficit of the reflected intensity due to surface plasmons . . . 362
    3.6. Measurement of roughness with the help of the electron microscope . . . 368
4. Scattering of light without excitation of surface plasmons . . . 369
    4.1. General remarks . . . 369
    4.2. Results in the region of visible light – determination of roughness . . . 370
    4.3. Scattering experiments with X-rays . . . 377
    4.4. Discussion of the scattering equations . . . 378
5. Rather rough surfaces; limits of present theory . . . 381
    5.1. Depolarization of the emitted radiation . . . 381
    5.2. Wavelength dependence of $|s(\Delta k)|^2$ . . . 384
6. Plasmons on surfaces of periodic profile . . . 384
    6.1. Dispersion relation of surface plasmons on sinusoidally perturbed surfaces . . . 385
    6.2. Influence of higher harmonics . . . 391
    6.3. Coupling of surface plasmons and of guided light modes on sinusoidal surfaces . . . 393
7. Conclusion . . . 399

References . . . 400

1. Introductory Remarks on Roughness

Considerations in physics assume in general a plane boundary of a solid. There are however a number of problems in which such an idealization is not allowed: the mobility of electrons in a very thin metal film or in a Si MOS device is strongly influenced by the scattering of electrons at the interior rough boundary; reflection of electromagnetic waves at rough surfaces produces diffusely scattered radiation in addition to the specularly reflected beam; this refers to radio waves scattered at the ocean surface, to light reflected at a metal surface, to X-rays reflected at the mirrors of an X-ray telescope or to electrons and atoms reflected at a crystal surface.

In recent years the role of roughness in the coupling of photons with surface plasmons – nonradiative and radiative – has attracted much interest. A number of experiments and theoretical investigations has been reported, e.g., on the emission of light by plasmons propagating along a rough surface, light absorption at rough surfaces or at gratings by the excitation of plasmons, a change of the dispersion relation of surface plasmons (SP) on rough surfaces etc. It is the aim of this article to review these phenomena and to show the common principles.

In the following we concentrate on the small roughness case: the roughness profile shall be described by the profile function $z = S(x, y)$ (see fig. 1), where x, y are the coordinates of a point in the surface and z is measured from $z = 0$, given by a fictitious plane defined as

$$\bar{S} = \bar{z} = \frac{1}{F}\int\int_F S(x, y)\, dx\, dy = 0. \qquad (1)$$

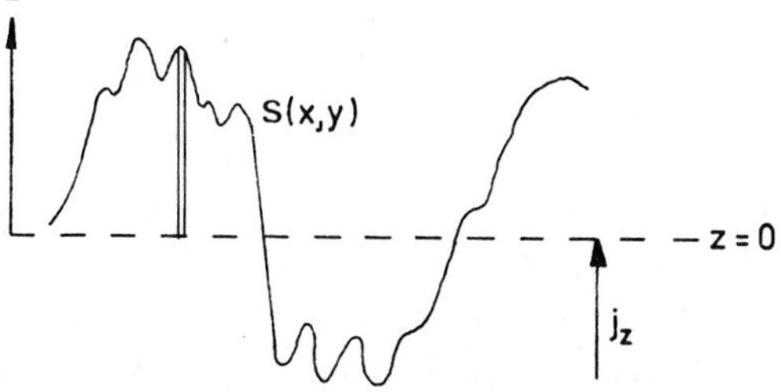

Fig. 1. Schematic representation of the roughness profile $z = S(x, y)$.

The assumption of small roughness means that the mean square root

$$(\overline{S^2})^{1/2} = \delta \tag{2}$$

shall be small compared to the wavelength λ of the incident light:

$$\delta \cdot 2 \cos \theta \ll \lambda \quad \text{or} \quad \delta \ll \lambda, \tag{3}$$

since in general $\cos \theta \sim 1$. If λ is small as in the case of X-rays (sect. 4.3) this condition is fulfilled by using grazing incidence ($\theta \sim 89°$, $\cos \theta \sim 10^{-2}$).

In a first approximation we describe the scattering of light as follows: The polarization currents induced by the incoming field, feed the peaks and valleys, so that they act as Hertzian dipoles radiating light with their typical angular distribution. The contribution of the valleys is taken as negative ($z < 0$). Phase differences along S are neglected. The intensity of the scattered light is therefore determined by $\overline{S^2}$, whereas the angular dependence is given by the lateral distribution of the scattering dipoles over the surface. If these are uniformly spread over the surface, as on a smooth surface, no diffuse scattering is observed.

If we write this distribution function as a Fourier integral each component acts as a sinusoidal grating with the diffraction condition

$$\Lambda(\cos \theta_0 - \cos \theta) = \nu\lambda, \tag{4}$$

Λ is the wavelength of the Fourier spatial component, θ_0, θ the angle of incidence and that of observation, measured against the normal, and ν is a whole number. The roughness has in general a statistical character, so that it has to be described by a stochastic function, the correlation function $G(x, y)$, see eq. (20). The mean distance over which $S(x, y)$ and $S(x - x', y - y')$ are correlated is called the correlation length σ. No correlation means that no intensity is scattered out of the specular reflected beam ($\sigma \to 0$).

In general $G(x, y)$ is an unknown function and has to be derived from experiments. An important exception is, e.g., the case of a sinusoidal grating; here $S(x)$ is a simple function so that comparison of calculated and observed values is possible, see sect. 6.

In the following, results are summarized and compared. In sect. 2 the influence of roughness on the dispersion relation (change of eigenfrequency and damping) is described. This is a rather strong and rather well-understood effect. The important phenomenon is the damping produced by the emission of light due to the decay of the nonradiative SPs into photons via roughness. This is described in sect. 3. This light and its angular distribution contain information of the roughness of the surface and allows one to derive δ and σ. In sect. 4 light scattering is reviewed, which works with frequencies far from the plasmon resonances. This procedure is very interesting for roughness determination. The light scattering experiments allow one to demonstrate the limits of the theoretical conceptions based on single scattering of the plasmons, and show that multiple scattering becomes important with increasing roughness rather early (sect. 5). In the last section (sect. 6) experimental results on sinusoidal gratings are discussed with emphasis on the SP aspects of the subject.

2. *Influence of Roughness on the Dispersion Relation of Surface Plasmons*

The electromagnetic field of surface plasmons (SP) guided at the metal surface has its maximum in the boundary between the metal and the air. It is thus sensitive to changes of the dielectric neighborhood as well as to the geometry of the boundary. The latter means that roughness shall have an influence on the properties of the SP, especially on the dispersion relation (valid for smooth surfaces):

$$k = \frac{\omega}{c}\left(\frac{\epsilon}{\epsilon+1}\right)^{1/2}, \tag{5}$$

with ω the frequency of the SPs and k their wavevector (parallel to the surface). The dielectric function $\epsilon(\omega)$ is complex $(\epsilon' + i\epsilon'')$ and gives, for $\epsilon'' < |\epsilon'|$ (Raether 1977, 1980, Bruns and Raether 1970),

$$k' = \frac{\omega}{c}\left(\frac{\epsilon'}{\epsilon'+1}\right)^{1/2}, \tag{6a}$$

$$k'' = k'\frac{\epsilon''}{2\epsilon'^2}\frac{\epsilon'}{\epsilon'+1}. \tag{6b}$$

The interior damping of the amplitude of the SP is given by

$$\exp(-k''x), \tag{7}$$

with x the direction of propagation*. Both equations are in good agreement with the observed position of the reflection minimum (θ_0) and its shape determined by its angular width $\theta_{1/2}$ (see Kretschmann 1971, Schröder 1981). This interior damping has also been measured by the attenuation length of SPs propagating along a copper surface using two prisms: one to transform light into SPs on the surface, the other to extract light from the SPs having run a certain distance x on the surface, so that eq. (7) can be

*If we introduce into eq. (5) the dielectric function $\epsilon(\omega)$ of the nearly free electron gas, the properties of the nonradiative surface plasmons are described; if we use $\epsilon(\omega)$ of the phonons, we obtain the relations for surface phonons or surface polaritons. The following equations are thus valid for both types of excitations. The experiments described in the following sections are made with surface plasmons, except those mentioned in sect. 3.5.

applied. The agreement of observed and measured values was not satisfying (Schoenwald et al. 1973).

The experiments showed that the behaviour of the SP is different on smooth and rough surfaces: the phase velocity ω/k is reduced with increasing roughness characterized by the root mean square height $(\overline{S^2})^{1/2}$, whereas the damping of the waves increases by virtue of the interior damping due to $\mathrm{Im}(\epsilon)$ (see eq. (6b)). The damping due to roughness is observed by the emission of light, easily visible if the plasmon frequencies lie in a suitable frequency range.

2.1. Experimental Results

The experiments to study these properties were performed with the prism (ATR) method, see fig. 2: Light of intensity I_0 excites plasmons on the air/metal boundary at a special angle θ_0 given by the dispersion relation. The resonance condition is indicated either by a dip in the reflected intensity (I_r) or by a peak of the light emitted by the coupling of SPs with surface roughness (I_e). A great number of experiments has been made with silver films of about 500 Å thickness evaporated on fire-polished quartz plates. The (natural) roughness of these films can be increased by covering

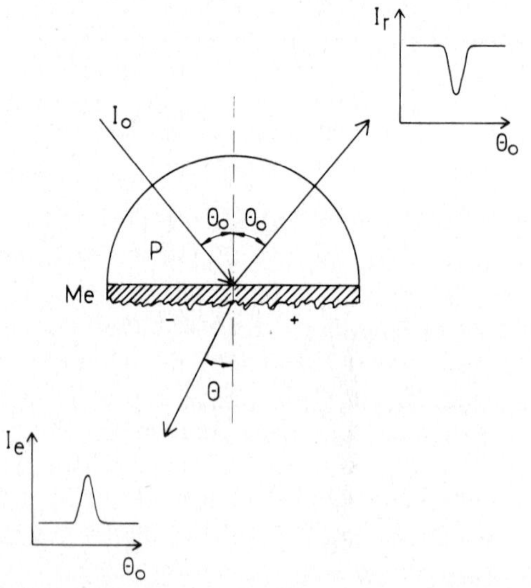

Fig. 2. Using a half cylinder (P) of, e.g., SiO_2 (refractive index $n = 1,5$) plasmons are produced on the surface of the rough metal film (Me). They can be observed by the minimum in the reflected light (I_r) (upper right) or by the maximum of the emitted light (I_e) at a fixed θ (lower left); the intensity is measured with a photomultiplier. Here the plane of observation is parallel to the plane of incidence $\phi = 0°$.

the quartz plates with layers of crystalline material such as CaF_2 or silver of different thickness before the silver is deposited. The roughness obtained is calibrated as will be explained later. The first measurements were done to study the behaviour of the reflection minimum as a function on the roughness of silver and gold surfaces (Hornauer et al. 1974, Braundmeier et al. 1974, Kapitza 1976). The results are the following (fig. 3): the resonance is displaced to higher θ_0 values, or to larger k values, at fixed wavelength (λ) of the incoming light. The wavevector k is related to θ_0 by $k = n(\omega/c) \sin \theta_0$. This means that the phase velocity of the surface plasmons (ω/k) decreases. Further the width of the reflection minimum ($\theta_{1/2}$) increases with the roughness of the silver surface, which had been defined in these first experiments qualitatively by the thickness of a crystalline layer of CaF_2 vaporized on the quartz substrate before the silver film of 500 Å has been deposited. More detailed experiments on rough gold films were published as figs. 4 and 5 display. Figure 4 demonstrates the dispersion relation of SPs for films of different roughness; instead of an $\omega(k)$ plot, the dependence $\lambda(\theta_0)$ at which the minimum of the reflected intensity lies has been chosen, since this is more favourable for demon-

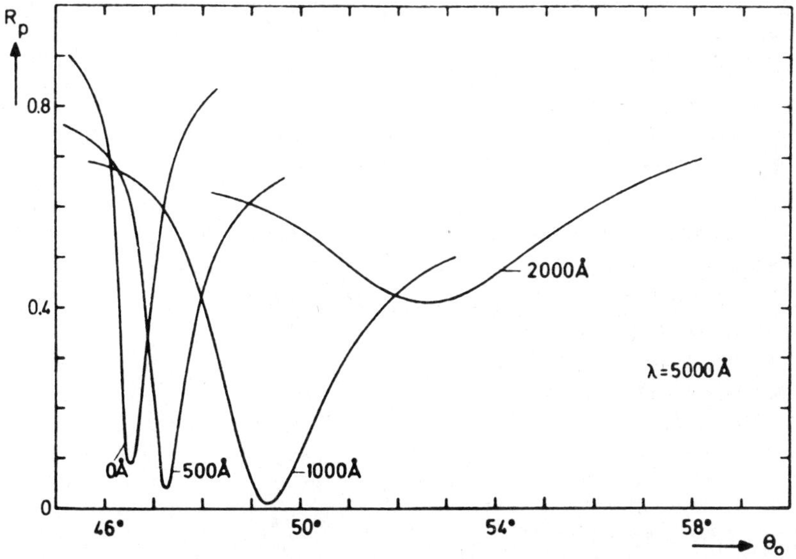

Fig. 3. Reflected (relative) light intensity (R_p) dependence on the angle of incidence θ_0 (λ = 5000 Å). The curve (0 Å) is obtained at a "smooth" silver surface of natural roughness, the other curves at silver films roughened by underlayers of CaF_2 of 500, 1000 and 2000 Å thickness (Hornauer et al. 1974). The arrangement is that of fig 2, upper right. The reflection minimum reaches its deepest value at 1000 Å CaF_2 since the matching condition, internal damping equal to radiation damping, is satisfied (see also fig. 39 and sect. 6.1).

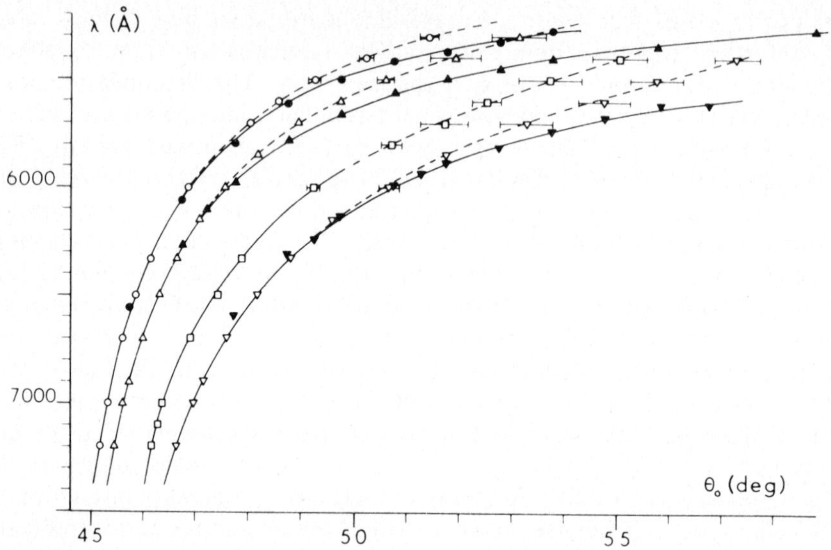

Fig. 4. Dispersion relation of surface plasmons on rough gold surfaces Kapitza (1976): The position of the resonance minimum, given by the ordinate λ on "smooth" ($d(CaF_2) = 0$ Å) and on rougher gold surfaces ($d(CaF_2) > 0$ Å) is plotted as a function of the resonance angle θ_0 (see fig. 2). The dispersion relation is measured in two ways: The full line demonstrates the minimum values obtained at constant θ_0 with varying wavelengths of the light whereas the dashed line represents the data measured at a given light wavelength and varying angles of incidence, see also fig. 17. For the explanation of this difference see sect. 3.3. The result is described as a decrease of the eigenfrequency at θ_0 = const or as an increase of θ_0 at λ = const with increasing roughness. The thickness of the gold film: 700 Å. The thickness $d(CaF_2)$ from left to right: 0 Å, 500 Å, 1000 Å, 2000 Å; similar in fig. 5.

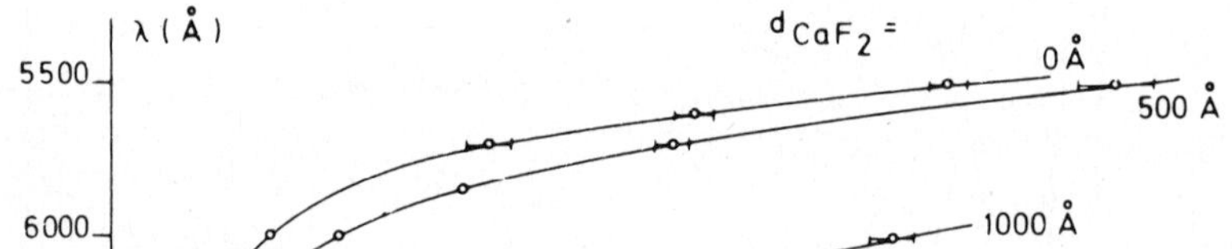

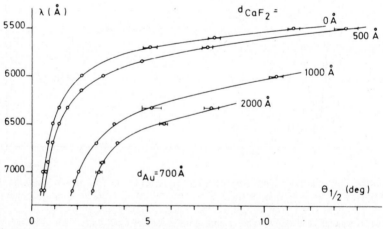

Fig. 5. The increase of the half width ($\theta_{1/2}$) of the resonance minimum on wavelength (λ) for gold films of different roughness Kapitza (1976).

strating the roughness effect. One recognizes the decrease of the phase velocity with larger roughness. Besides this displacement, the half width of the reflection minimum increases, as demonstrated by fig. 5. The curve $d_{CaF_2} = 0$ Å is calculated with the dispersion relation of a smooth silver surface using the dielectric function $\epsilon(\omega)$ of silver from the literature, e.g., by Schröder (1981). It is in agreement with the observed curve, indicating that its roughness can be neglected within the accuracy of these experiments. This film is called a "smooth" one in the following. Table 1 shows the quantitative influence of roughness.

The damping is due to the emission of light, which is not allowed from a smooth surface. This radiation has been found in the first experiments with the prism method in which the silver film has been vaporized directly on the base of the prism, see fig. 6 (Kretschmann and Raether 1968). The strong peak (logarithmic scale!) is displaced with exciting wavelength according to the dispersion relation (eq. 5): the larger the k or θ_0 value, the higher the energy $\hbar\omega$ of the plasmon. Its angular intensity distribution has been measured with a multiplier as a function of the angle θ from the foil normal (see fig. 7) in the arrangement of the lower part of fig. 2. This dependence allows one to obtain information on the roughness of the metal surface.

If the peak of the scattered intensity in fig. 6 is measured, either by varying θ_0 at fixed λ or by varying λ at fixed θ_0 (see fig. 2, lower left) on

Table 1
Values of the observed displacement ($\Delta\theta_0$) of the ATR minimum (θ_0) to larger values and the increase of its half width ($\theta_{1/2}$) due to the roughness of the silver film (500 Å thick) at different SP wavelengths and different roughness.

δ(Å) $d(CaF_2)$(Å)		8–10 500	15–17 1000	~20 1500
	(degree) $\Delta\theta_0$	0,4	1,7	2,5
$\lambda = 5600$ Å	$\theta_{1/2}$	0,4	1	1,8
$\lambda = 5000$ Å	$\Delta\theta_0$	0,7	2,8	4,1
	$\theta_{1/2}$	—	2,4	4
$\lambda = 4400$ Å	$\Delta\theta_0$	1,7	6	8,3
	$\theta_{1/2}$	1	8	11

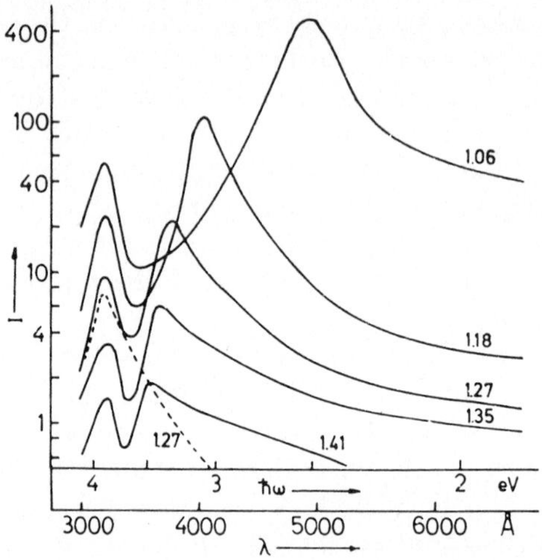

Fig. 6. Relative light intensity emitted from a "smooth" silver surface due to the radiative decay of surface plasmons as function of the wave length. The surface plasmons are excited by incident p-polarized light with the prism method, see fig. 2, lower left. The parameters by the curves are values of $n \sin \theta_0$. The angle θ is kept constant. The dotted line is measured with s-polarized light: its maximum at $\lambda = 3200$ Å is independent of θ_0; it stems from light scattered at the quartz–silver boundary since silver has a maximum of transparency at 3200 Å. The scale is logarithmic (Kretschmann and Raether 1968). The maximum of light intensity is emitted from the silver film via roughness if the dispersion relation is fulfilled. With larger $\sin \theta_0$ or larger k values the frequency at which the maxima are observed approaches the value of $\hbar\omega_s(k \to \infty)$, reflecting thus the dispersion curve.

surfaces of different roughness, (see Raether 1977, fig. 66) the same dependence is measured as observed in figs. 4 and 5, where the reflection minima are recorded; the method of reflection and that of scattering are equivalent sources of information of roughness.

How does one understand these results? The increase of damping with larger roughness (or with δ) can be understood easily if we use the picture of polarization currents excited by the incident electromagnetic field at the surface which radiate as Hertzian dipoles, emitting an intensity proportional to S^2.

The displacement of k to larger values at a given ω (or the increase of ω at a fixed k) is less simple to describe; it has its origin in a double (or higher-order) scattering process: The SPs propagating along a given direction are scattered out of this direction and backscattered again into the original direction. The resulting total plasmon field then has a lower phase velocity. Similarly a photon field in a dielectric of refractive index n propagates with reduced phase velocity c/n outside the eigenfrequencies of the dielectric. Whereas the damping process is produced by a single

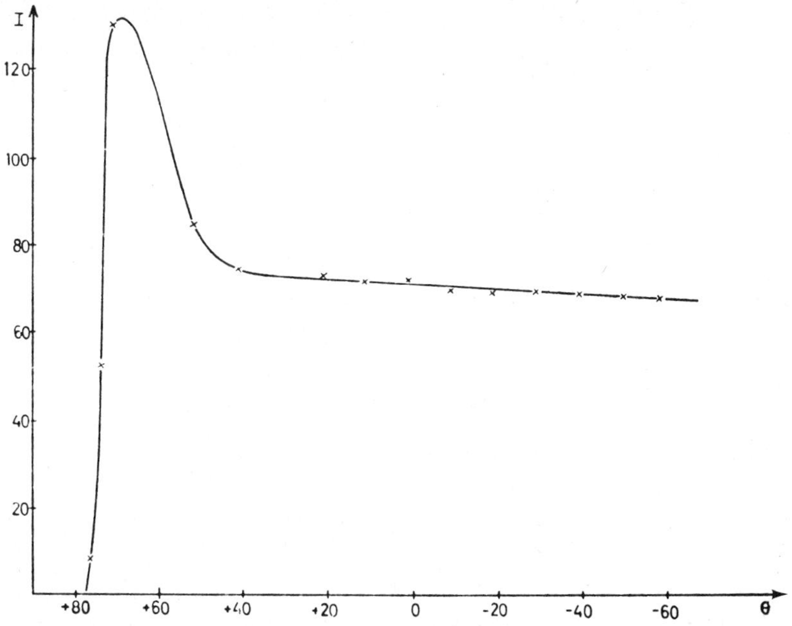

Fig. 7. Angular dependence of light emitted from the surface of a silver film (800 Å thick) if surface plasmons are excited. $\theta = 0$ is the direction of the foil normal. A strong maximum of the light intensity is observed in the forward direction, $\theta > 0$. The angle of incidence is fixed at the resonance angle θ_0 for maximum excitation of plasmons (Bruns and Raether 1968).

scattering process, the change of the dispersion relation is correlated with multiple processes, at least double processes.

In a special case the phase velocity of the SPs can increase. This happens for ω larger than an eigenfrequency ω_0 of the system. This is realized in a metal surface covered with a (monomolecular) dye containing an eigenfrequency ω_0 (Waehling et al. 1978).

2.2. Theoretical Remarks

A theoretical treatment of the phenomena described in sect. 2.1 has been given by Kröger and Kretschmann (1976). These authors, starting from integral equations for the field of the eigenmodes, derive a system of linear functions which is reduced to a set of linear equations, a formalism developed by these authors for the theory of reflectivity at rough surfaces (Kröger and Kretschmann 1970). The determinant of the linear equations has to be zero; regarding terms up to the square of $S(x, y)$, a dispersion relation $\omega(k)$ is obtained. This is different from that of the smooth surface, which is displayed in eqs. (5) or (13).

In the dispersion relation, k has to be replaced by $(k+\overline{\Delta k})$ where $\overline{\Delta k}$ is a complex quantity: its real part describes the displacement, whereas the imaginary part describes the damping of the plasmon mode on the rough surface, or

$$\overline{\Delta k} = \overline{\Delta k'} + \mathrm{i}\overline{\Delta k''}, \tag{8}$$

$$\overline{\Delta k'} = (\omega/c)^3 \int \mathrm{d}^2k\,|s(\Delta k)|^2 A(k, k_0), \tag{9a}$$

$$\overline{\Delta k''} = (\omega/c)^3 \int_0^{2\pi} \mathrm{d}\phi\,|s(\Delta k)|^2 B(k_0, \phi). \tag{9b}$$

For Δk, see eq. (24). For the terms A and B see Kröger and Kretschmann (1976). This result shows that the dispersion relation is changed in such a way that for a given ω the position k of the resonance is displaced to $k+\overline{\Delta k'}$ and its damping is increased by $\overline{\Delta k''}$; in this approximation both are proportional to $\overline{S^2}$.

Maradudin and Zierau (1976) approached the same question with a classical Green's function technique and obtained equivalent results. Another mathematical approach in one dimension is reported by Toigo et al. (1977) to calculate the dispersion relation with the help of "reduced Rayleigh equations". In the limit of small roughness, it can be written as (Kretschmann 1979):

$$D_r(k) = D_s(k) - (\omega/c)^2(\epsilon-1)^2 I(k), \tag{10}$$

$$I(k) = (\omega/c)^2 \int \mathrm{d}k'|s(k-k')|^2 C(k, k'), \tag{11}$$

for

$C(k, k')$ see Kretschmann et al. (1979), with (12)

$$D_s(k) = \epsilon_0 k_{z1} + \epsilon_1 k_{z0}, \tag{13}$$

the dispersion relation of SP on a smooth surface with ϵ_0 the dielectric function of air or vacuum, and ϵ_1 that of the metal; $k_{zi} = (\epsilon_i(\omega/c)^2 - k^2)^{1/2}$ the z component of the wavevector perpendicular to the surface. δ^2 is included in $|s(\Delta k)|^2$.

These equations show that the dispersion relation deviates from that of the smooth surface proportional to δ^2. The results are equivalent to those of Kröger and Kretschmann (1976). A quantum mechanical approach is presented by Hall and Braundmeier Jr. (1978). As just mentioned the imaginary part of Δk represents damping by light emission and by scattering as explained in fig. 34. In different regions of the dispersion relation both contributions are different: for $\omega \to \omega_s$ scattering; for $\omega \ll \omega_s$, radiation damping induced by roughness is more important (Mills 1975) [$\omega_s = \omega(k\to\infty)$, eq. 5].

Looking at table 1 we see that a dependence with nearly δ^2 is observed in the experiments. The quantitative evaluation however, requires a knowledge of $|s\Delta k|^2$, the Fourier transform of the correlation function $G(x, y)$ over the whole Δk region, which is however not available. If one introduces a Gaussian function for $|s(\Delta k)|^2$ (see eq. (28) below), with the parameters δ and σ taken from the visible region, into eqs. (9a and b) one obtains after integration values of the coefficient describing the δ^2 dependencies which are much too small to fit the observations. This demonstrates the following estimate (Raether 1977a): similarly as internal damping is described by k'', (see eq. (6b)) the radiative damping k_{rad} due to roughness can be calculated using eq. (10). Assuming a Gaussian correlation function one obtains for k_{rad}/k'' with $\sigma = 1000$ Å, $\lambda = 4000$ Å, $\epsilon = -4 + \text{i}\ 0,4$ (silver) and $\delta = 10$ Å, a value of $k_{\text{rad}}/k'' = 0.01$. This is a negligible increase of the width $\theta_{1/2}$ of the reflection minimum which disagrees with the observation (see e.g. fig. 5). We have to conclude that the extrapolation of the validity of the Gaussian correlation function is not allowed to large Δk values. It is very probable that at larger Δk values (Λ about 100 Å and less, for Λ see eq. (24)), the crystalline structure of the surface contributes more to the roughness spectrum than the Gaussian function. It should be possible by evaluating the structure of a silver surface photographed with an electron microscope to obtain the value of the correlation function at these large Δk values. With the help of eqs. (9a, b) the comparison with observed data is then possible.

A more favourable situation for evaluating eqs. 9a and 9b is given in the case of sinusoidal gratings. If the behaviour of SPs on sinusoidal gratings is observed, similar changes of the dispersion relation as a function of the amplitude h of the sinusoidal grating is found ($\overline{\Delta k}'$ and $\overline{\Delta k}'' \sim h^2$); since the correlation function in the case of a sinusoidal grating is rather simple to calculate from eqs. 9(a) and (b) a more detailed comparison of observed and calculated values is possible. The good agreement demonstrates that the theory is correct, as we shall see in sect. 6.

3. *Coupling of Photons with Surface Plasmons via Roughness*

The phenomenon of interaction of photons with polaritons (here non-radiative surface plasmons) is of interest not only intrinsically, but also because of the possibility of obtaining information on the roughness structure of the surface. The intensity of the radiation, its spectral and angular distribution is shown in figs. 6 and 7. An approach for calculating this light emission has been reported by E.A. Stern using the concept of polarization currents (Stern 1967). A more general theory has been published by Kröger and Kretschmann (1970) who describe the radiation phenomena in a linear approximation of the Maxwell field (single scattering) as a function of $z = S(x, y)$ so that the intensity of the emitted light becomes a function of $\overline{S^2}$. Kretschmann has derived a number of valuable relations for the experimental verification of the theory (Kretschmann 1972a, 1972b, 1974) and has used them for the first quantitative determination of the roughness parameters δ, σ and $|s(\Delta k)|^2$, see eqs. (27) and (21).

A different detailed approach is given by Maradudin and Mills (1975) and Mills (1975, 1976) who obtain the same final results, apart from a minor correction (a factor of π).

In the following the theoretical background is explained to understand the observed data obtained with SPs. A detailed presentation of the theoretical part will be found in ch. 10 by A.A. Maradudin.

3.1. Theoretical Results on Light Emission

In the experimental arrangement of fig. 2 light excites surface plasmons on the rough air/metal boundary which decay into photons (see also sect. 2.1. The intensity emitted into the solid angle $d\Omega$, ($dI/I_0\, d\Omega$, if I_0 is the incoming p-polarized intensity) is given by Kretschmann (1972a) as

$$\frac{dI}{I_0\, d\Omega} = \frac{1}{4}\left(\frac{\omega}{c}\right)^4 \frac{\sqrt{\epsilon_0}}{\cos\theta_0} |t^p(\theta_0)|^2 |W|^2 |s(\Delta k)|^2. \tag{14}$$

Here θ_0 is the angle of incidence and t^p the Fresnel transmission coefficient t^p_{012} of the layer system: half cylinder P (e.g., quartz) (ϵ_0)/metal (Me) (ϵ_1)/air (ϵ_2). t^p which measures the amplitude of the electric field at the metal–air (or vacuum) interface is given by:

$$t^p_{012} = \frac{t_{21} t_{10} \exp(ik_{1z} d_1)}{1 + r_{21} r_{10} \exp(ik_{1z} d_1)}, \tag{14a}$$

with d_1 the thickness of the metal film. It passes a maximum at $k = k_0$, see fig. 16 (for more detail see Raether (1977)). Its square $|t^p|^2$ measures the enhancement of the field intensity on the surface of metal, which can amount to ~80 in the case of smooth silver for $\lambda = 5461$ Å ($\epsilon_1'' = 0, 45$), see also fig. 16. It increases $\sim\lambda$ to longer wavelengths.

The dipole radiation function $W(\theta)$ determines the angular light intensity distribution produced by the three components of the polarization currents which are induced by the incoming light field. The relation for W, if we observe in the plane of incidence ($\phi = 0°$) and if surface roughness is assumed (see sect. 3.4.), is given by

$$|W|^2 = A(\theta, \epsilon_1) \sin^2\psi \left[\left(\frac{1 + \sin^2\theta}{|\epsilon_1|}\right)^{1/2} - \sin\theta\right]^2, \tag{15}$$

with

$$A(\theta, \epsilon_1) = \frac{|\epsilon_1| + 1}{|\epsilon_1| - 1} \frac{4}{1 + (1/|\epsilon_1|) \operatorname{tg}^2 \theta}. \tag{16}$$

This is an approximation valid for $\epsilon_1'' \ll |\epsilon_1'| (\epsilon_1' < 0)$ and $2\pi d \sqrt{|\epsilon_1|} > \lambda$, conditions which are fulfilled for silver films of $d \geqslant 500$ Å and $\lambda \geqslant 4000$ Å. The exact expressions are found in Kretschmann (1972a). The differences between the exact and the approximate relations are indicated in fig. 10. Here ψ relates the plane of polarization to the plane of observation: for $\psi = 90°$ both planes are parallel (p-polarized), for $\psi = 0°$ both planes are perpendicular (s-polarized). The angle ϕ measures the angles between the plane of incidence, plane 1 in fig. 8, and the plane of observation, plane 2. In fig. 8 we have $\phi = 90°$.

As we see, s-polarized light ($\psi = 0°$) does not exist in this first order approximation, whereas $\psi = 90°$ (E vibrates in the plane of observation)

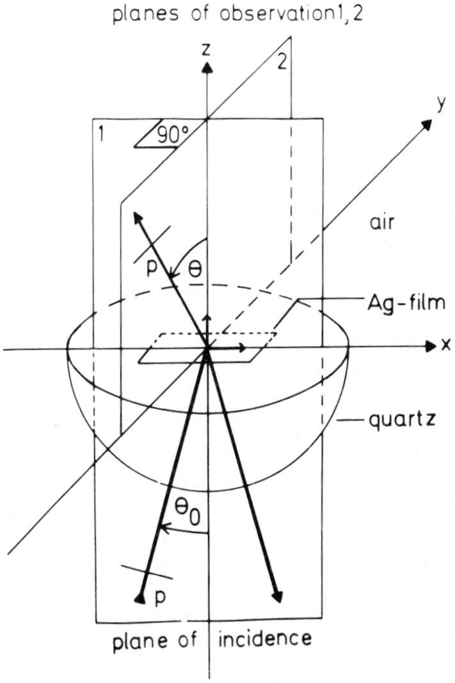

Fig. 8. Experimental arrangement for the observation in the plane (2) perpendicular to the plane of incidence (plane 1); (θ) is the angle of observation and p indicates the direction of polarization. If the polarization (p) lies perpendicular (parallel) to the plane of observation, it is called s-polarization (p-polarization). The two short arrows at the origin denote the x- and z-component of the radiating dipole or of the function $W(\theta)$ (Horstmann 1977).

gives intensity to which both components of the dipole, normal and tangential contribute. Both components lie on the surface (in air) since surface roughness is assumed; for interior roughness, see sect. 3.1. Its radiation pattern, which is reproduced in fig. 9, has a dominant maximum of radiation intensity in the backward direction as the result of interference of the normal and the tangential dipole components. Figure 9 is calculated for $\lambda = 5500$ Å and has its peak at $-55°$; this value as well as the peak height do not change very much with λ.

The model presented in sect. 3 allows one to derive further experiments supporting this model without knowing the roughness function $|s(\Delta k)|^2$ (see eq. 21). Instead of turning the light detector (multiplier or diode) around the light-emitting surface in the plane of incidence (rotation axis = y-axis), the detector can be turned in a plane perpendicular to it around the x-axis ($\phi = 90°$) (see fig. 8 plane 2), p indicates the direction of polarization. Then the dipole function becomes

$$|W|^2 = A(\theta, \epsilon_1)\left|\sin\theta \sin\psi + \cos\theta\cos\psi\left(1 + \frac{\sin^2\theta}{|\epsilon_1|}\right)^{1/2} - \mathrm{i}\sin\theta\frac{\mathrm{tg}\,\theta}{|\epsilon_1|}\right|^2. \tag{17}$$

For $\psi = 0$ (polarization perpendicular to the plane of observation) the detector records the tangential component of the dipole (W_x), see fig. 8. $|W|^2$ can be approximated ($|\epsilon_1| \gg 1$) by

$$|W|^2 \cong \frac{\cos^2\theta}{1 + (\mathrm{tg}^2\,\theta)/|\epsilon_1|}. \tag{18}$$

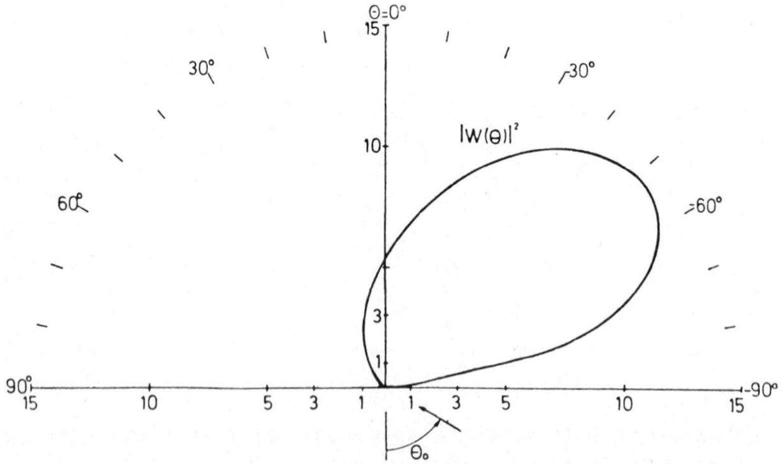

Fig. 9. Calculated polar diagram of the radiation pattern $|W(\theta)|^2$ produced by surface plasmons propagating from right to left; it is the result of the interference of the radiation of a dipole normal (W_z) and a dipole tangential (W_x) to the surface, see eq. 15.

It passes a maximum at $\theta = 0$ and drops to zero at $\theta = 90°$, since a dipole placed on a metallic surface emits no intensity into $\theta = 90°$ direction, the direct and the reflected beam cancel each other.

For $\psi = 90°$ the detector sees the normal component of the dipole (W_z). Its radiation function $|W|^2$ can be approximated by

$$|W|^2 \cong \frac{\sin^2 \theta}{1 + (\mathrm{tg}^2\, \theta)/|\epsilon_1|}. \tag{19}$$

These calculated radiation patterns, see fig. 10 are in good agreement with the observed intensity distribution (Kretschmann 1972a). The agreement confirms the concept of the theory (for deviations see sects. 3.4 and 5).

The last term in eq. (14) describes the roughness spectrum. If we assume a correlation between the quantities $S(x', y')$ at a point (x', y') and $S(x - x', y - y')$ at $(x - x', y - y')$ we introduce a correlation function $G(x, y)$ as follows:

$$G(x, y) = \frac{1}{F} \int_F \mathrm{d}x'\, \mathrm{d}y' S(x', y') S(x' - x, y' - y). \tag{20}$$

Then $|s(\Delta k)|^2$ becomes the Fourier transform of $G(x, y)$, or,

$$|s(\Delta k)|^2 = \frac{1}{(2\pi)^2} \int_F \mathrm{d}x\, \mathrm{d}y\, G(x, y) \exp[-\mathrm{i}(\Delta k_x x + \Delta k_y y)]. \tag{21}$$

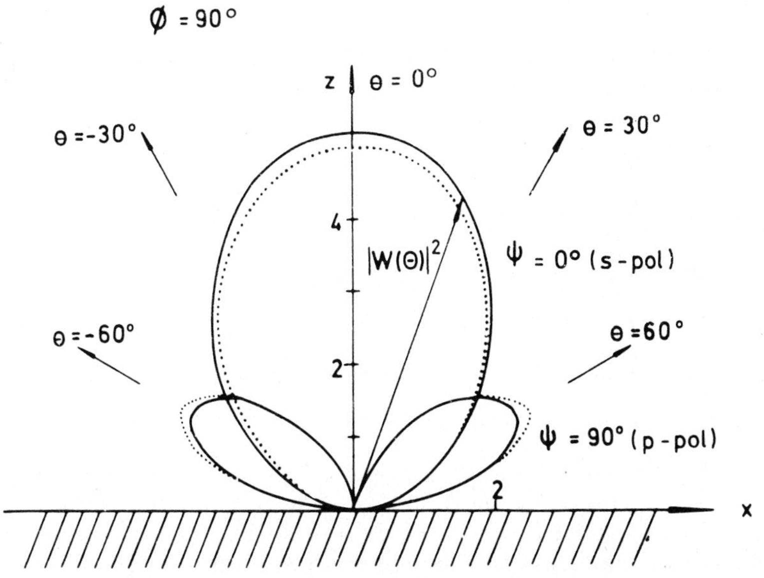

Fig. 10. Polar diagram of the dipole function $|W(\theta)|^2$. Silver foil, 700 Å thick, $\lambda = 5000$ Å. θ is the angle of observation in the plane (2) of fig. 8. The dotted curve represents the dipole function in the approximation mentioned in the text (Kretschmann 1972).

If no correlation exists, i.e., a complete random distribution, we obtain from eq. (20):

$$G(x, y) = 0 \quad \text{for} \quad x \neq 0, y \neq 0. \tag{22}$$

From eq. 20 follows the rms height

$$G(0, 0) = \overline{S^2} = \delta^2, \tag{23}$$

$$|\Delta k| = |k - k_0| = |k_r| = 2\pi/\Lambda, \tag{24}$$

is the roughness vector disposible in the roughness spectrum of the surface and transferred to the tangential component of the incoming photon.

In the case of the prism method with the wavevector of the incident light, $k_0 = \sqrt{\epsilon_0}(\omega/c) \sin \theta_0$ ($\sqrt{\epsilon_0}$ the refractive index of the half cylinder P in fig. 2), the wavevector of the scattered light, $k = (\omega/c) \sin \theta$, and ϕ the angle between the plane of incidence and that of observation, we obtain for Δk

$$\Delta k = (\omega/c)((\sqrt{\epsilon_0} \sin \theta_0)^2 - 2\sqrt{\epsilon_0} \sin \theta_0 \sin \theta \cos \phi + \sin^2 \theta)^{1/2} \tag{24a}$$

It has its largest value at $\phi = 0$ (see fig. 8) and $\theta < 0$:

$$\Delta k = (\omega/c)(\sqrt{\epsilon_0} \sin \theta_0 + \sin \theta). \tag{25}$$

For $\theta = 0$ we have $\Delta k = \sqrt{\epsilon_0}(\omega/c) \sin \theta_0$ as indicated in fig. 14. The maximum of Δk is given by $\Delta k_{\max} \sim 2(2\pi/\lambda)$.

If the experiments yield $|s(\Delta k)|^2$ as function of Δk in a certain range of Δk values it is possible to calculate the rms height

$$\overline{S^2} = 2\pi \int_{\Delta k_1}^{\Delta k_2} \mathrm{d}\,\Delta k \cdot \Delta k |s(\Delta k)|^2. \tag{26}$$

This value of δ can be obtained only in a limited Δk region given by θ_1, θ_2 and λ, see eq. (25).

Very often one assumes a Gaussian distribution

$$G(x, y) = \delta^2 \exp[-(x^2 + y^2)/\sigma^2], \tag{27}$$

with σ the correlation length; its Fourier transform is given by

$$|s(\Delta k)|^2 = (1/4\pi)\sigma^2\delta^2 \exp[-\sigma^2\Delta k^2/4], \tag{28}$$

and allows the calculation of δ and σ from the slope of the $\ln|s(\Delta k)|^2$ versus $(\Delta k)^2$ plot and the value $\ln|s(0)|^2$ which one gets by extrapolation to $(\Delta k) \to 0$. The correlated $\bar{\Lambda}$ is given by $\bar{\Lambda} = \pi\sigma$ derived from $\frac{1}{4}\sigma^2\bar{\Delta}k^2 = 1$.

The use of the Gaussian function is given more by the wish to obtain analytical expressions by performing certain integrals. The experiments can prove the validity of this function only in a limited region of Δk given by the wavelength of light used in the experiment. Equation 25 shows that

$\Delta k_{max} \sim 2(2\pi/\lambda)$. The extrapolation to very large Δk values with the values of δ and σ obtained in the visible region is therefore doubtful as we have seen in sect. 2.

The use of X-rays should allow a comparison of the function $|s(\Delta k)|^2$ and thus of the δ values observed in the lower Δk region with that in the region of large Δk values, see sect. 4.

3.2. *Experimental Results on Light Emission – Determination of Roughness*

Figure 11 shows schematically the experimental arrangement (see, below). The measurements were performed in air in general. Experi-

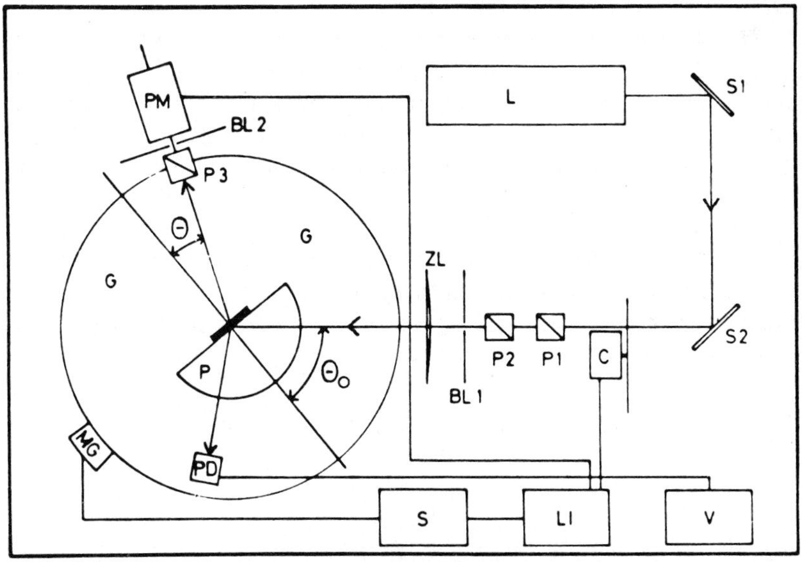

Fig. 11. Diagram of the prism method for measuring the light emission at an angle θ as well as the reflected light. For many purposes a convenient light source (L) is a laser of some m W*. The light beam passes the polarizer ($P_1 + P_2$) and the diaphragm (BL1) and enters the quartz half cylinder P. The metal film deposited on a thin quartz plate is fixed at its base. The SPs are excited at the metal/air boundary. The cylinder lens (ZL) compensates the refraction at the curved surface of the half cylinder. The light emitted into the angle θ and the reflected light (θ_0) are recorded with a photomultiplier or a photodiode. The currents are amplified ((V) amplifier, LI Lock-in) and plotted in S, e.g., as fig. 26 shows. The half cylinder (P) rotates with the goniometer (G); the calibration, see fig. 26, is made by marks in (MG) (Hillebrecht 1979).

*Experiments with a pulsed dye laser ($t_{puls} = 1{,}5\ \mu s$, max power ~ 20 kW, $\lambda = 6000$ Å) on silver and gold films (~ 500 Å thick) had shown that a variation of the density of the primary intensity by about a factor of 10^8 did not influence the position and the shape of the reflection minimum of SP excitation. At ~ 2 kW/cm^2 the films are destroyed (Horstmann 1976).

ments on silver films in a vacuum of some 10^{-7}–10^{-6} Torr (preparation and measurements without exposition of the sample to air) had shown that there is no difference of the data to those obtained in air, except that the position of the reflection minimum displaces slightly just after the condensation of the metal film, apparently due to some recristallization process (Twietmeier 1975). The observed data are evaluated using first-order theory. The intensity is recorded as a function of the scattering angle θ at an angle of incidence θ_0 which fulfills the resonance condition. The relative intensity (dotted line) scattered at the surface of a silver film (thickness ~500 Å) evaporated on a fire-polished (deposition rate of about 10 Å/s) and carefully cleaned suprasil plate is displayed in fig. 12. Comparing it with fig. 9 one notices a strong change of the distribution and the peak direction; it is attributed to the roughness function $|s(\Delta k)|^2$ of this surface. The same is seen comparing fig. 13 (left curve (1)] with fig. 9. This gives the possibility of deriving the roughness function $|s(\Delta k)|^2$ in the region Δk given by the light wavelength and the angles of observation. After it has been demonstrated that the properties of the dipole function $|W|^2$ are in agreement with experimental results, see fig. 10, the roughness function can be deduced with the help of eq. (14). It may be emphasized that the dielectric function $\epsilon(\omega)$ of the metal is separately

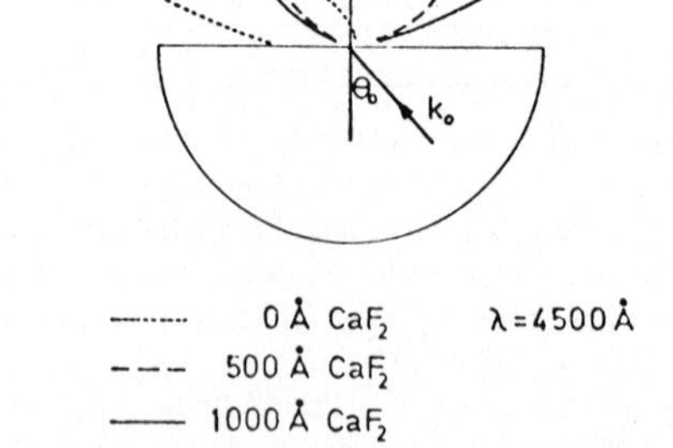

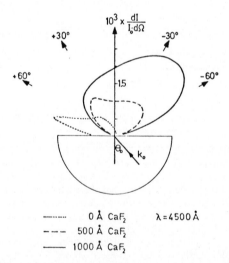

Fig. 12. Observed polar diagram of the normalized p-polarized intensity per solid angle element $d\Omega$: $dI/I_0\,d\Omega$ versus angle of observation θ for $\lambda = 4500$ Å and different thickness of the CaF_2 underlayer; dotted line: 0 Å CaF_2; dashed line: 500 Å CaF_2; full line: 1000 Å CaF_s. A CaF_2 film of 2000 Å thickness gives a diagram of nearly the same shape as that of 1000 Å CaF_2 but about 4 times more intense. θ_0 angle of maximum plasmon excitation, k_0 wavevector of the incoming light (Hornauer 1976).

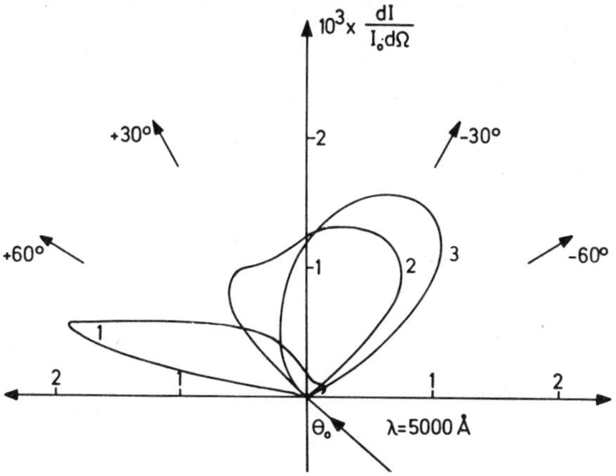

Fig. 13. Observed polar diagram of the p-polarized radiation of a "smooth" silver film, 550 Å thick; $\lambda = 5000$ Å. The intensity pattern of the film (1) (its intensity has been reduced by $\frac{1}{2}$) shows a forward maximum characteristic for the fire-polished suprasil substrates, whereas the substrates of the films (2) and (3) are "supersmooth", apparently of slightly different quality. Roughness parameter: Curve (1) $\delta = 4$–5 Å, $\sigma \simeq 1100$ Å, curve (2) $\delta = 4$–6 Å, $\sigma \simeq 1100$ Å, curve (3) $\delta = 8$ Å, $\sigma \simeq 800$ Å (Hornauer 1977).

determined at the same surface with the prism method, so that $|W|^2$ and $|t_p|^2$ are well known.

Figure 14 demonstrates (on a logarithmic scale) $|s(\Delta k)|^2$ against $(\Delta k)^2$. The curves are practically independent of the wavelength (4–6000 Å). They are interpreted as being composed of two parts (Kretschmann 1974, Hornauer 1976, Orlowsky et al. 1979): a linear part in the region of $1\text{–}3 \times 10^{-3}$ Å^{-1} or a roughness period Λ from ~6000 Å to ~2000 Å and a strongly increasing $|s|^2$ region, sometimes with a maximum, at low Δk-values. The linear part is characteristic of the roughness of the silver surface, whereas the shape at low Δk is due to the wavy surface of the quartz substrate. This conclusion has been verified by several experiments:

(a) Silver films of the same thickness have been vaporized under conditions as equal as possible on quartz (suprasil) substrates of different smoothness, which increases from curve (1) to curve (3), see fig. 13. The peak in the forward direction due to the long undulations of the substrate is no longer dominant and the polar diagram approaches the shape of that in fig. 9. In table 2 values of $|s(\Delta k)|^2$ at small Δk are compared, if the silver film, 500 Å thick, is vaporized on quartz substrates of different smoothness.

(b) Before the silver film is deposited on the fire-polished quartz plate the substrate is covered with a crystalline dielectric layer, e.g., of MgF_2 or CaF_2 of different thickness (Hornauer 1976). The result is seen in fig. 12 which demonstrates that with increasing thickness (and thus roughness) of the CaF_2 underlayer the diagram approaches qualitatively that of fig. 9. The

Table 2
Comparison of values of the roughness spectrum $|s(\Delta k)|^2$ for silver films ($d = 500$ Å) evaporated on "smooth" and "supersmooth" quartz substrates for $(\Delta k)^2 < 0.4 \times 10^{-6}$ Å^{-2} (Orlowsky et al. 1979).

$\Lambda = 2\pi/\Delta k$	$\lvert s(\Delta k)\rvert^2$	
	'supersmooth'	'smooth'
10 000 Å	4×10^6 Å^4	12×10^6 Å^4
15 000 Å	7×10^6 Å^4	35×10^6 Å^4
20 000 Å	1×10^6 Å^4	73×10^6 Å^4

characteristic surface geometry of the substrate is nearly covered by the roughness of the underlayer.

These results support the interpretation of the increase of $|s(\Delta k)|^2$ as an effect of the geometry of the substrate. Since this increase of $|s|^2$ takes place at $\Delta k < 1 \times 10^{-3}$ Å^{-1}, a wavy structure of wavelength of some 10 000 Å and more has been derived. This structure stems from the mechanical process of polishing, (Kretschmann 1972, Orlowsky et al. 1979).

Angular distributions of the same type (forward and backward lobes) on "smooth" silver films have been measured in a similar experimental arrangement by Braundmeier and Hall (1975), Hall and Braundmeier (1978). A disagreement between these results and those of Kretschmann (1972), found by these authors in explaining the backward lobe is due to an error in the computer program by Hall and Braundmeier (1978). Its correction restores agreement. The directions of the lobes are determined by the radiation pattern (backward lobe in fig. 9) and by the roughness function $|s|^2$. The roughness of the silver surface and the surface structure of the substrate determine together the finite pattern as is shown in the experiments described in the following (see sect. 3.3).

The linear part of the $|s|^2$ curve, see fig. 14, is attributed to the roughness of the silver film; it has been evaluated using eq. (14), $\mathrm{d}I/I_0 \cdot \mathrm{d}\Omega$ is given by the observed data whereas the other terms as $|t^{\mathrm{p}}|^2$ and $|W|^2$ have been calculated with the dielectric function $\epsilon(\omega)$ measured at the same specimen using the prism method; these values of $\epsilon(\omega)$ are in good agreement with those obtained with the ellipsometric method, see e.g., the appendix of Schröder (1981). With eq. (26) the value of δ on silver films of about 500 Å thickness vaporized on fire-polished quartz (suprasil) plates has been determined as within 8 Å; it is valid in the region of Δk of $1\text{–}3 \times 10^{-3}$ Å^{-1}.

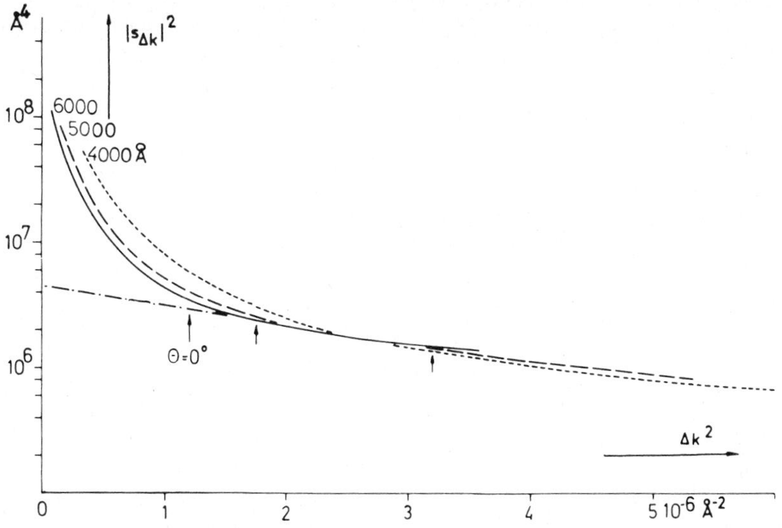

Fig. 14. Experimental roughness spectrum $|s(\Delta k)|^2$ as a function of $(\Delta k)^2$ for a "smooth" silver film of thickness 550 Å (Kretschmann 1974, 1972). The arrows indicate $\theta = 0°$, the observation normal to the surface, at three different wave lengths.

If the observed data are evaluated with the Gaussian correlation function a linear plot results: $\ln|s(\Delta k)|^2$ versus $(\Delta k)^2$. The value of σ is derived from the slope whereas δ comes from $\ln|s(0)|^2$ extrapolated back to $\Delta k = 0$. The figures obtained are $\delta = 7$ Å, in agreement with the above value, and $\sigma = 1180$ Å. The extrapolation to $\Delta k \to 0$ can be justified by the agreement of the values obtained with both procedures. From $\sigma = 1180$ Å a value of $\bar{\Lambda} \sim 3500$ Å follows, see eq. 28.

Values obtained by different authors with the same experimental device, see below, confirmed this value in the Δk range of $1\text{–}3 \times 10^{-3}$ Å^{-1}. If a series of samples are prepared under conditions as similar as possible, the data scatter by about 30% (Orlowsky et al. 1979).

With a different geometry of the prism method a similar figure, $\delta = 5{,}5$ Å, has been obtained on 1000 Å thick silver films (Bodesheim and Otto 1974). Measurements of the radiation emitted into the upper half space of the prism, see fig. 2, see sect. 5.1, allowed to derive the roughness parameters: $\delta = 6$ Å, $\sigma = 1200$ Å on silver films under similar conditions in good agreement with the above data (Simon and Guha 1976).

Measurements of the roughness without SP excitation give values of δ which agree with those obtained with SPs within fluctuations produced by the individual roughness of the different samples, see sect. 4.

The intensity of light scattered from a silver film is reduced by $\approx \frac{1}{3}$ if the quartz substrate is replaced by a cleavage surface of mica. The rms values have, on the average, a tendency to smaller values than those on a quartz substrate (Pschalek 1980).

As mentioned above, light hitting a rough surface is transformed into SPs which are partially absorbed in the metal. A rough surface will thus show additional absorption near 3400 Å in Ag or a peak in Im (ϵ) (Jasperson and Schnatterly 1969). This has been measured as an increase of the temperature of the sample (Kaspar and Kreibig 1977). Reduction of the roughness by annealing diminishes the height of this roughness absorption.

If electrons are used to excite SPs on silver targets, light is emitted by the interaction of nonradiative SPs with the roughness of the surface. Its observed angular distribution (Sauerbrey et al. 1973a) is in agreement with the theoretical concept of surface roughness interaction of Kröger and Kretschmann (1970) and negligible volume roughness as has been shown by Sauerbrey et al. (1973b). The roughness function $|s(\Delta k)|^2$ obtained at Ag films of thickness 10 000 Å with electron excitation (Dobberstein 1970) is different from fig. 14: it increases up to $\Delta k \sim 7 \times 10^{-3}$ Å^{-1}, whereas it decreases monotonically in fig. 14 up to $\Delta k \sim 7 \times 10^{-3}$ Å^{-1} (film thickness ~500 Å, light excitation of SPs). The difference of both results has not been cleared up. There may be a real difference in the surface roughness structure of the thick Ag films in the excitation of the SPs by electrons (broad K-spectrum) or in the different method of the determination of $|s|^2$.

The light scattering method just described which gives detailed information on roughness is very sensitive and should be applicable to surfaces with δ values lower than those presented here presumably by a factor of 5; the limit is given by the smoothness of the samples which can be prepared.

This photon–plasmon coupling has also been studied on silver and gold surfaces immersed in an electrolyte; for details see ch. 8 of D.M. Kolb.

3.3. *Experiments with Enhanced Roughness*

To test this method further the inherent roughness of silver and gold films has been enhanced by depositing a crystalline film of CaF_2, MgF_2, LiF or silver before vaporizing the gold or silver film on the quartz substrate. This technique allows one to increase the roughness by increasing the thickness of this underlayer. In the case of silver underlayers the first thin film (thickness d_1, e.g., 50 Å) has been vaporized on quartz substrates of different temperature (T), e.g., 150°C, to produce more or less strong coalescence. After cooling back to room temperature further evaporation increased the thickness of the film by d_2. d_1 and d_2 are varied with $d_1 + d_2 = \text{const}$, e.g., 350 Å. These conditions are written in the following as 50 Å/350 Å, 150°C. The evaporation rate was 8–10 Å/s. It is the meaning of this experiment that the film of silver (d_2) follows more or less the rough

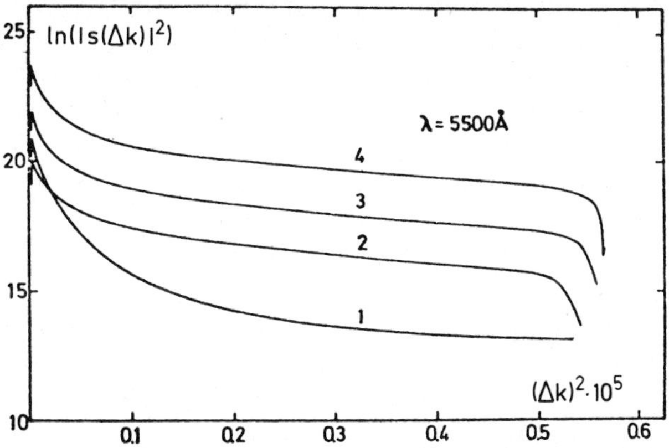

Fig. 15. Roughness function, $\ln|s(\Delta k)|^2$ as a function of Δk^2 of a silver film, 350 Å thick, with different thickness of underlayers of silver: (1) 0 Å/350 Å; $\delta = 7 \pm 2$ Å: (2) 50 Å/350 Å; 200°C; $\delta = 12 \pm 2$ Å: (3) 100 Å/350 Å; 200°C; $\delta = 34 \pm 6$ Å. Evaporation rate of the silver 8–10 Å/s. The correlation length σ is about 1100 Å (Urner 1976).

profile of the first very thin film (d_1) (Orlowsky et al. 1979, Schröder 1969a).

The striking effect of underlayers of different thickness on the angular intensity distribution observed by Hornauer (1976) is seen in fig. 12: the peak turns from the forward direction more and more into the backward direction and the intensity grows in general. The roughness spectrum for $\lambda = 5500$ Å is reproduced in fig. 15 (Urner 1976) which displays (a) the increase of $|s(\Delta k)|^2$ or of the emitted intensity and that of δ^2 with the thickness of the underlayer (b) the reduction of the waviness of the substrate (logarithmic scale!) at small $(\Delta k)^2$ values. The decrease at $(\Delta k)^2 > 5 \times 10^{-6}$ Å^{-2} is supposed to be multiple scattering (Hornauer 1976). The values of δ^2 for different experimental conditions are found in table 3. The reproducibility of values measured on different samples prepared under as similar conditions as possible is rather good and reaches ~30%.

The correlation lengths remain roughly constant, 800 ± 150 Å, and show no trend with the thickness of the underlayer (Hornauer 1976).

The dielectric function ϵ (and thickness d) necessary for the calculation of $|t_{012}^p|^2$ is derived from reflection measurements at the same specimen as used for the roughness measurements by fitting the observed minimum to the theoretical reflection curve. The ϵ values of the "smooth" silver films are identical with those in the literature, see Schröder (1981). Since the reflection minima displace and become flatter with increasing roughness, the calculated $\epsilon(\omega)$ changes (Hornauer 1976), and thus $|t_{012}^p|^2$ changes, as fig. 16 indicates for a silver film of 650 Å thickness (Urner 1976). This procedure takes into account to a certain extent the change of the dispersion

Table 3
Roughness δ of silver films with different underlayers (Orlowsky et al. 1979)

CaF_2 d(Å)	δ(Å)	LiF d(Å)	δ(Å)
0	5	0	5
500	10	580	5
1000	15	2000	5
2000	(28)	d(Ag) = 350 Å	
d(Ag) = 350 Å			
T = 22°C		T = 22°C	

Ag d_1(Å)	δ(Å)	Ag T(°C)	δ(Å)
0	7	22	6
10	12	100	8
100	(34)	150	10
		200	12
$d_1 + d_2 = 350$ Å		$d_1 = 50$ Å	
T = 200°C		$d_1 + d_2 = 350$ Å	

relation D_s, see eq. 13, with increasing roughness.

This effect has the consequence that the field enhancement on rough surfaces is lower than that on a smooth surface. This is of interest for the study of the giant Raman effect of adsorbates on rough metal surfaces.

Very rough surfaces. If the above mentioned experiments with silver underlayers are extended to thicker films of the first silver layer: 200 Å/350 Å, 200°C and 350 Å/350 Å, 200°C the width of the reflection minima increases so much that it is not possible to evaluate it and to derive a value of δ.

If the experiments are performed with thin silver films of 50 Å and 150 Å, condensed on quartz substrates at 200°C, two reflection minima are observed with p-polarized light, at $\lambda = 3600$ Å and 4300 Å; the measurements were done with θ_0 = const and λ variable, see sect. 3.3, Urner (1976). Very probably the surface structure resembles that of a surface covered with half spheres which is better described with a model proposed by Berreman (1967).

Further features of the dispersion relation at rough surfaces. If the dependence of λ_{min}, the wavelength of the reflection minimum on the angle θ_0, is measured, different values are obtained depending on whether the observations are made at fixed λ and variable θ_0 or vice versa (see fig. 17).

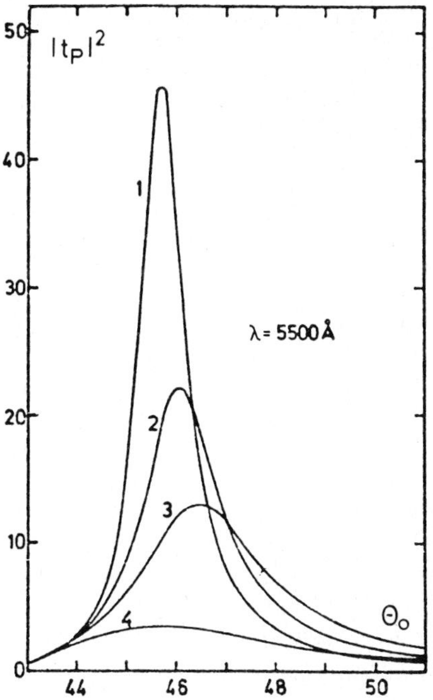

Fig. 16. Transmission factor $|t_{012}^{P}|^2$ as function of the angle of incidence θ_0 for silver films of different roughness. $|t^{P}|^2$ measures the enhancement of the field intensity at the surface. The roughness of the surfaces 1, 2, 3 is given in fig. 15. $\lambda = 5500$ Å (Urner 1976).

Increasing θ_0 further the curve measured at fixed λ bends back to the light line in contrast to that at fixed θ_0. This is a geometrical effect and depends on how one crosses the "dispersion relation valley" which becomes rather broad at shorter wavelengths (Arakawa et al. 1973, Kovener et al. 1976, Orlowsky and Raether 1976). The angle θ_0 at which the curve bends back displaces to lower values of θ_0 with stronger roughness, see fig. 17 (Orlowsky et al. 1979). More detailed observations to find a reproducible relation between rms height δ and this θ_0 for back-bending were not successful. The reason is probably that in the region of the bending back the reflection minima are rather shallow so that the minimum position is not well defined (Waehling 1979).

The branch of the curve of the reflection minima turns back to the light line (L) and returns again. The interband transition of silver at ~3250 Å produces a rather sharp minimum (see inset in fig. 17). Beyond $\lambda = \lambda_s =$ 3400 Å the excitation is a single electron excitation, whereas below λ_s we have the collective excitation of SPs; this is proved by the normal behaviour of the energy density at the boundary air/silver in contrast to the resonant enhancement for $\lambda < \lambda_s$ (see inset of fig. 2 in Orlowsky and Raether 1976). The structure of the reflection curve for $\lambda < \lambda_p = 3280$ Å

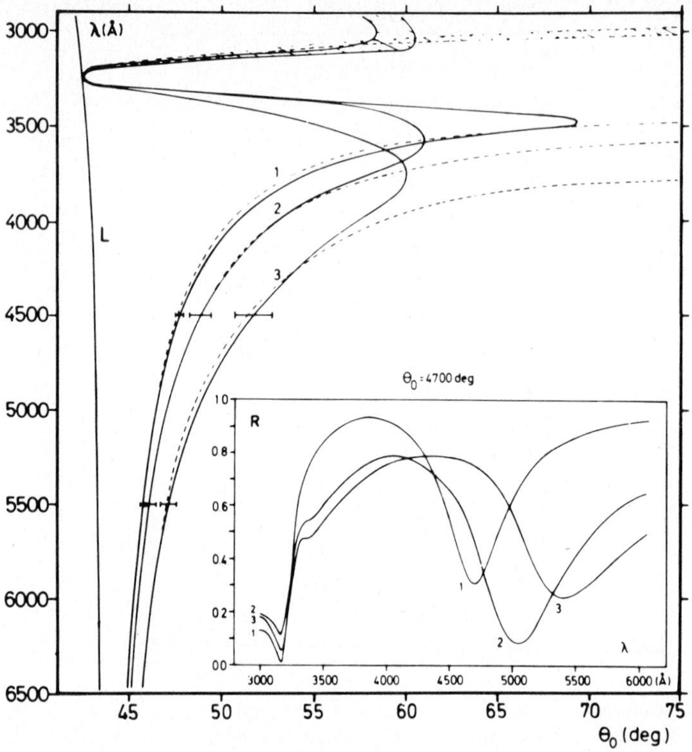

Fig. 17. Dispersion relation of surface plasmons on silver films of different roughness. λ is the wave length at which the minimum of reflection is observed and θ_0 the correlated angle. Dashed lines: derived from $R(\lambda)$ at constant θ_0; full lines: derived from $R(\theta_0)$ at constant λ, see sect. 3.3. Curve (1) "smooth" silver surface (350 Å silver on quartz substrate, see text). Curve (2) and (3) are obtained with underlayers of silver to increase the roughness of the surfaces, the conditions are the same as in fig. 15. L Light line. The displacement of the bending back angle to smaller angles θ_0 with increasing roughness which is due to the increasing damping is remarkable. The inset reproduces the reflection curve $R(\lambda)$ at different roughness.

($\omega_p = 3{,}78$ eV), here is $\epsilon > 0$, can be described by the "dispersion relation" eq. (5) which, for $\epsilon > 0$, is identical with tg $\theta = \sqrt{\epsilon}$, the Brewster condition.

3.4. Exterior and Interior Roughness

The completion of the model by introducing in addition to the dipole component j_z the components parallel to the rough surface allowed one to prove that the experiments with the nonradiative surface plasmons measure the geometric profile of the surface (exterior roughness) and not, or to a negligible extent, interior inhomogenieties of the plasma film. Thus earlier uncertainty in evaluation has been overcome (Stern 1967).

Further, the observed angular distribution of the light intensity could be explained.

The point is that the radiating dipole component j_z has to be placed in the interior of a thin vacuum slit *in* the metal if fluctuations of the dielectric function in the metal occur (and a smooth surface is assumed) whereas this j_z dipole has to be located in a very thin vacuum slit *on* the metal surface if surface roughness is regarded. If the dielectric function outside the metal is equal to 1 (air or vacuum), the dipole is located *on* the surface and one needs no slit. The slit has to be very thin compared to λ in order to be able to neglect multiple reflections (Kröger and Kretschmann 1970). An important contribution to clear this question has been published by Juranek (1970), see also Bedeaux and Vlieger (1973, 1976).

These two positions of the radiating dipole lead to different results, since the normal component of the electric field is discontinuous ($E_z^{\text{ext}}\epsilon^{\text{ext}} = E_z^{\text{int}}\epsilon^{\text{int}}$), whereas the tangential components are continuous. As a consequence the z-component of the dipole W_z has to be replaced by $W_z/|\epsilon|$ if volume scattering is taken into account. This difference in W_z allows one to perform experiments to decide which type of roughness is responsible for the light emission: with wavelengths $\lambda > \lambda_s$, λ_s the wavelength of the SP for $k \to \infty$ is 3400 Å. For Ag the results are:

(1) If one observes, at a fixed angle θ, the light intensity as a function of the angle ψ between the plane of polarization and the plane of observation this dependence should go as (at $\phi = 90°$, see fig. 8):

$$\begin{aligned} &\sim \cos^2(\psi - \theta) \qquad \text{for surface roughness,} \\ &\sim \cos^2 \psi \qquad \text{for volume roughness,} \end{aligned} \tag{29}$$

see Kretschmann (1972). Figure 18 displays the observed values (full curves). The dashed dotted curves are the calculated data (silver film, 650 Å thick, $\lambda = 5000$ Å, $\epsilon^1(\text{Ag}) \sim -10$). In the case of volume roughness the dashed line is expected; the tangential dipole component is much larger compared to the normal component, which is strongly reduced by the factor $1/|\epsilon|$. The rotation of the maximum of the polar diagram into the 45° direction if $\theta = 45°$ indicates that the normal and tangential components are of comparable value, see fig. 18, or that surface scattering dominates.

A disagreement appears insofar as the observed intensity at $\psi - \phi = 90°$ and $\psi = 90°$ does not go to zero; a residual intensity remains. This deviation obviously indicates a breakdown of the first approximation (see sect. 5).

(2) The quotient of the light intensity $I_p(\psi = 90°)/I_s(\psi = 0°)$ (polarization plane parallel ($\psi = 90°$) and perpendicular ($\psi = 0°$) to the plane of observation) as a function of θ are rather different for the two types of roughness for $\theta > 0$, since in case of volume scattering W_z has to be replaced by $W_z/|\epsilon|$ and $|\epsilon|$ is rather large in the λ region of 4–8000 Å in

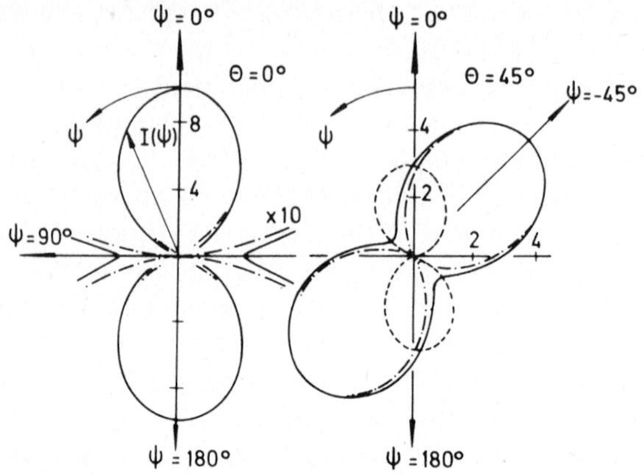

Fig. 18. Polar diagram of the emitted intensity at $\theta = 0°$ and $\theta = 45°$ as a function of the polarization angle ψ (angle between the polarization plane and the plane of observation). Silver film thickness 650 Å, $\lambda = 5000$ Å, $\phi = 90°$. Full line: observed values, normalized to the intensity I ($\psi = 0°$, $\theta = 0°$) equal to 10. Dashed–dotted: theoretical surface emission. Dashed: theoretical volume emission. At $\theta = 0°$ (left) surface and volume intensity distribution are equal (one "sees" only the tangential component of the dipole in fig. 8). At $\theta = 45°$ the normal dipole component is strongly reduced by $1/|\epsilon|$, so that both distributions are different. $|\epsilon|$ is the dielectric function of the bulk (Kretschmann 1972).

silver. In fig. 19 this quotient is shown as a function of θ, demonstrating that surface roughness predominates if light wavelengths of 4–8000 Å are used for the investigation of silver (Kretschmann 1972b). The residual intensity at $\theta = 0°$ in fig. 19(a) is discussed in sect. 5.

Interior roughness. An interesting point is the question under which conditions does volume roughness play the dominant role. This occurs if $\epsilon(\omega)$ has small values, so that the interior field E_z^{int} becomes large. Small ϵ values mean that *radiative* SPs have to be excited. For silver, one needs wavelengths of about 3280 Å ($\hbar\omega_p = 3{,}78$ eV) to excite these plasmons, as experiments have shown (Brambring and Raether 1965, Raether 1977). In this case $|\epsilon|$ becomes small so that the W_z component dominates and W_x and W_y can be neglected. In this simplified picture the light comes from a dipole perpendicular to the metal surface which emits an intensity given by

$$\frac{dI}{I_0\, d\Omega} = \frac{1}{4}\left(\frac{\omega}{c}\right)^2 \left(\frac{1}{\cos\theta_0}\right) |t_p|^2 |W_z|^2 |s(\Delta k)|^2_{\text{Vol}}, \tag{30}$$

with

$$|t_p|^2 |W_z|^2 = \sin^2\theta \sin^2\theta_0 |1 + r_p(\theta)|^2 |\epsilon - 1|^2/\epsilon^4, \tag{31}$$

and $|s(\Delta k)|_{\text{vol}}$ the roughness function of the volume.

Three features of this volume scattering are remarkable:

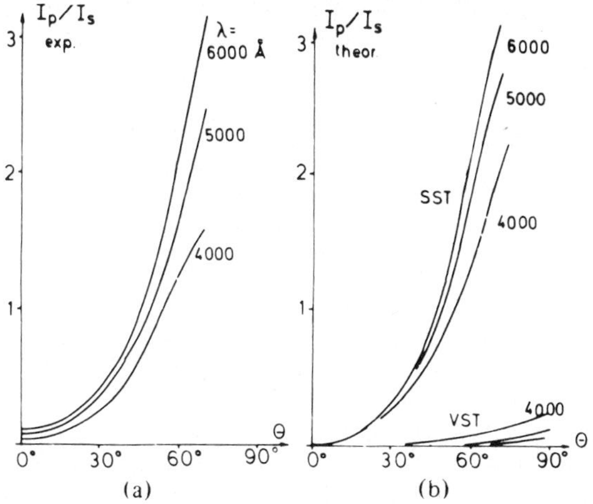

Fig. 19. Experimental (a) and theoretical ratio $I_p/I_s = I(\psi = 90°)/I(\psi = 0°)$, (b) as a function of θ for surface (SST) and volume (VST) emission at different wavelengths (Kretschmann 1972).

(1) At small ϵ or $\omega \sim \omega_p$ the relative intensity passes a maximum. This spectral dependence is shown in fig. 20 (Pokrowsky and Raether 1979), with k left of the light line, $k < \omega/c$ (for more details see Raether (1980)).
(2) The angular distribution of this radiation (plasma resonance radiation) is proportional to $(\sin\theta \sin\theta_0)^2$ and symmetric in θ and θ_0 as has been shown in Schreiber (1968).
(3) The dependence of $I_p(\psi = 90°)/I_s(\psi = 0°)$ on θ has higher values than for surface scattering, since in this case the dipole component $W_z/|\epsilon|$ dominates at the plasma frequency (W_x is small and constant). This is the reverse to fig. 19.

Formally the interior roughness can be understood as caused by a fluctuation of the dielectric function $\epsilon(\omega)$ (Wilems and Ritchie 1965, Kröger and Kretschmann 1970):

$$\epsilon(x, y, z) = \epsilon + \Delta\epsilon(x, y, z). \tag{32}$$

Under certain assumptions a value of $\Delta\epsilon = 0.1$ in the interior of a silver film is derived which reduces to 0,04 after annealing the silver film. For more details see Pokrowsky and Raether (1979).

Experiments on the plasma resonance emission have been made on silver samples treated in different ways to examine the behaviour of the resonance peak: the resonance maximum at $\lambda_p = 3280$ Å is hardly observed (1) on an epitaxially grown silver film (produced by vaporization on a mica substrate at 280°C) (2) on a solidified melt (3) on a silver film produced by vaporization on a substrate of −190°C (Schröder 1969b). In cases (1) and (2) the silver has been annealed, so that the inner roughness is strongly

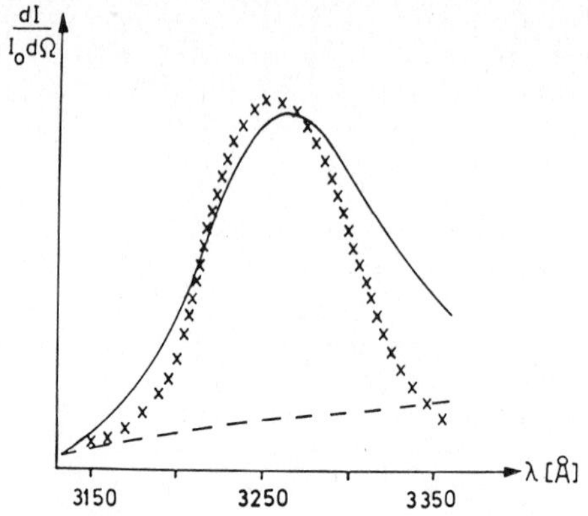

Fig. 20. Comparison of the theoretical intensity of surface emission (dashed line) and of volume emission (full line) with the observed relative intensity (crosses) at a silver film, 500 Å thick, at $\theta_0 = \theta = 30°$. The dependency shows a maximum at the plasma frequency ω_p of silver, $\hbar\omega_p = 3.78$ eV or $\lambda_p = 3250$ Å (Pokrowsky and Raether 1979).

reduced, see also the remark above. In case (3) the crystallites are so small that Im (ϵ) is larger than its normal value near the plasma frequency and thus the factor $1/|\epsilon|^4$, see eq. 31, is reduced (Schröder 1969b).

If volume *and* surface scattering exists, one can in a first approximation assume that both intensities superpose in case that phase relation do not exist.

With these theoretical considerations on interior and exterior roughness together with the experiments the conditions have been precised necessary to evaluate the observed light scattering data.

The calculations assuming normal incidence (or $E_z = 0$) can ignore this problem.

3.5. Deficit of the Reflected Intensity due to Surface Plasmons

In the foregoing section the light emitted from SPs due to roughness into directions different from the specularly reflected beam has been reported. In the following it is shown that the light emitted from SPs into all directions reduces the intensity of the beam specularly reflected at the non-smooth surface, so that an intensity deficit is produced.

If light hits a metal surface such as Al, Ag, Mg etc., in the experiments mostly made at nearly normal incidence, the specularly reflected intensity can show a deficit in the energy region of the nonradiative SPs (Hunter 1964, Williams et al. 1967, Dobberstein et al. 1968) see fig. 21 (Gesell et al. 1973).

The interpretation as a roughness effect is given by two experiments (Jasperson and Schnatterly 1969, Feuerbacher and Steinmann 1969, Stanford and Bennet 1969):

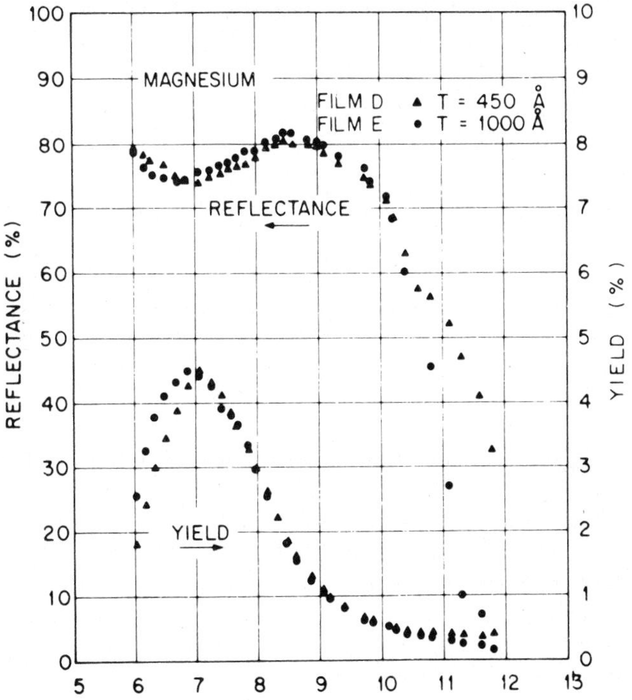

Fig. 21. Reflectance data for magnesium. The minimum at ~7 eV is produced by the excitation of surface plasmons via roughness. The lower curve shows the simultaneously measured yield of photoelectrons (Gesell et al. 1972). $\hbar\omega_s(\mathrm{Mg}) = 7{,}1$ eV; half width 1,4–2 eV.

(a) If the natural roughness of the metal film is increased by depositing underlayers on the substrate before the metal film is evaporated on it (see sect. 3.3) the deficit becomes stronger. This proves that the light energy is dissipated by excitation of SPs via roughness. This excitation appears as a peak in the absorption or as a peak in the imaginary part of the dielectric function (Jasperson and Schnatterly 1969).

(b) If a dielectric layer is deposited *on* the metal surface the position of the dip of the reflected intensity is displaced to lower energies by the depolarizing effect of the layer.

The explanation of the observation of the deficit of the intensity is shown in fig. 22: If light strikes a surface at an angle θ_0, the k vector along the surface is $k = \omega/c \sin\theta_0$. By coupling with roughness k can be changed into $k + k_r$ where $k_r = 2\pi/\Lambda$ (see eq. 24) is the roughness vector. For $k + k_r > \omega/c$ a SP can be created which produces a deficit in the reflected light. The larger the roughness the higher is the deficit. If $k + k_r$ does not

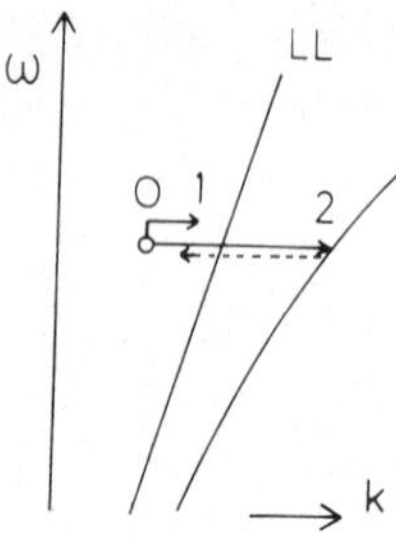

Fig. 22. Light of frequency ω and of wavevector $k = \omega/c \sin\theta_0$ (point (0) left of the light line (LL)) can excite surface plasmons by the roughness vector k_r (arrow) at point (2) of the dispersion curve. The process (01), left of (LL), represents diffusely scattered light. Only single scattering processes are considered, e.g., backscattering from point (2) into the region $k < \omega/c$ (dotted line) is neglected.

cross the light line, the incoming photons are scattered as photons into other directions and absorbed.

This deficit has been calculated for the case of normal incidence by Crowell and Ritchie (1970), Elson and Ritchie (1971, 1974). The value of the calculated reflectance comes out as $R(\omega) = R_0(\omega) + \Delta R_s(\omega)$ with $R_0(\omega)$ the reflectance of an ideal smooth surface and $\Delta R_s = \Delta R' + \Delta R$, (<0). Here $\Delta R'$ describes the deficit by light scattering process 01 in fig. 22, whereas ΔR, the deficit by transformation into SPs. $\Delta R'$ can be obtained by integrating eq. (36) over $d\Omega$ or $d\mathbf{k}$. ΔR for an undamped free electron gas and with the assumption of a Gaussian correlation function has a structure as (Crowell and Ritchie, 1970):

$$-\Delta R(\omega) = \delta^2\sigma^2\left(\frac{\omega}{\sqrt{\omega_s^2 - \omega^2}}\right)^5 A(\omega, \omega_p) \exp\left\{-\frac{\omega^2}{\omega_s^2 - \omega^2}\frac{\sigma^2}{4} B(\omega, \omega_p)\right\} \quad (32a)$$

with

$\omega_s = \omega_p/\sqrt{2}$ and A, B finite functions for $0 \leq \omega < \omega_s$.

The main features are: ΔR disappears with $\omega \to 0$ and with $\omega \to \omega_s$; it increases with δ^2, in agreement with observation.

Further calculations have considered the damping in the metal ($\mathrm{Im}(\epsilon) > 0$) (Kretschmann and Kröger 1975): in fig. 23 the calculated dependence of $\Delta R/R_0$ on $\hbar\omega$ with the damping coefficient $\gamma = 1/\omega_p\tau$ as parameter is shown at a fixed value of σ. The dashed line demonstrates the case without SP excitation in the limit $\sigma \gg \lambda$ or $\lambda/\sigma \to 0$ (scalar case). For $\gamma = 0$ both calculations give the same result. As fig. 23 shows, the influence of γ which leads to rounding the edge at ω_s, see fig. 5 in Gesell et al. (1973), cannot be neglected.

Comparison of the calculated deficit with data measured on different metals: Al (Endriz and Spicer 1971, Daude et al. 1972), Mg (Gesell et al.

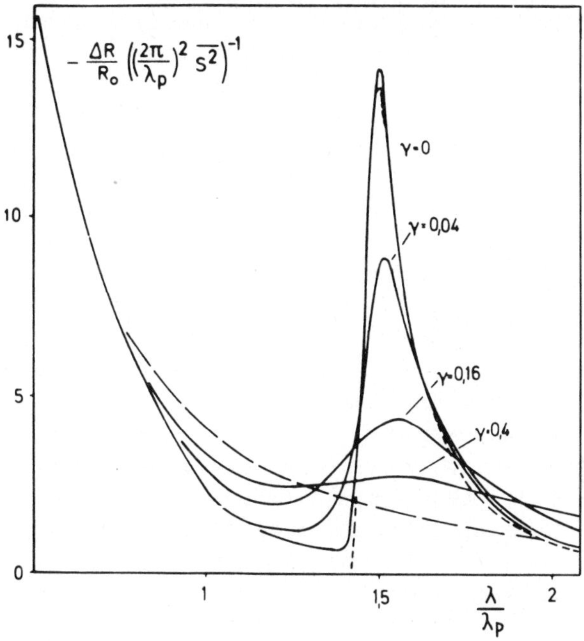

Fig. 23. Calculated relative minimum of reflectance due to the excitation of surface plasmons with γ as parameter and normalized to δ^2 at a free electron gas with damping $\gamma = 1/\omega_p\tau$. A Gaussian correlation function is assumed with a correlation length σ so that $2\pi\sigma/\lambda_p = 2$. The dashed curve: scalar limit $\sigma \to \infty$. With increasing damping the curves approach the curve ($\sigma \to \infty$) (Kretschmann and Kröger 1975).

1973) demonstrate general agreement in shape and in its change with increasing roughness δ; the roughness parameters δ and σ are fitted to the theoretical curves calculated with the assumption of a Gaussian correlation function. Damping has been neglected. The character of these results is thus a more qualitative one; in an exact calculation the change of the dispersion relation with roughness should be regarded too. The result of the observations can be summarized as follows: On clean Al films of 800–1000 Å thickness deposited on float-glass substrates rms heights of 12; 14,7 and 37,7 Å have been obtained using the Gaussian model (the corresponding σ values are 918, 744 and 378 Å resp.) (Endriz and Spicer 1971). These values were obtained on glass substrates overcoated with CaF_2, except the film with $\delta = 12$ Å; the technique of measurement was not sensitive enough to determine values of δ below this figure. Similar measurements on Al films performed at the same time (Daudé et al. 1972) have yielded smaller values. A rms height δ of 8 Å and $\sigma = 320$ Å has been measured at a film 900 Å thick, whereas on rougher films produced by overcoating the substrate with CaF_2, LiF or MgF_2 values of δ and σ are obtained, comparable with the above data. The results of measurements on Mg films of 450–1000 Å thickness on (uncoated) glass substrates using Gaussian model are, e.g., δ: 29 Å ($\sigma = 160$ Å) and 27 Å (180 Å) at ~450 thick films and $\delta = 45$ Å ($\sigma = 375$ Å) and 49 Å (475 Å) at a 1000 Å thick film (Gesell et al. 1973) (see also fig. 21). The roughness of the Mg films seems to be stronger than that of Al and Ag.

Altogether one can say comparing these results with those of sect. 3.3 that a rough agreement exists. "However the actual physical significance of the derived values remains unanswered until the theory is tested by direct measurements of δ and σ" (Gesell et al. 1973). This limitation is due to the integration over $d\Omega$ or $d\boldsymbol{k}$ loosing thus the information about $|s(\Delta k)|^2$. This is not the case if one observes the differential intensity, see e.g. eq. (14).

A similar situation arises if one introduces the mean free path length e.g., of scattering or of radiation damping. In this case one needs the $|s(\Delta k)|^2$ function for the integration over Δk which is not well known.

Under the same assumptions the influence of a dielectric coating of the rough metal surface has been calculated (Mills and Maradudin 1975, Elson 1976). A determination of the thickness of the coating is possible, although it is not very accurate. In this case one has to do with the same relations as in electron energy loss spectroscopy, where the energy position of the SPs is studied with the varying thickness of a coating of the surface (Raether 1977, 1980).

Very rough surfaces. Transmission experiments at 550 Å silver films with p-polarized light ($\theta_0 = 45°$) have shown that the intensity of light of $\lambda > \lambda_s = 3400$ Å is strongly reduced at rough surfaces compared to smooth surfaces. This missing light is partially absorbed, partially scattered into other directions. This has been demonstrated by light scattering experiments at thicker silver films ($\theta_0 = \theta = 30°$, $\phi = 90°$) and p-polarized light; here the intensity at wavelengths $\lambda > \lambda_s = 3400$ Å is much higher at rough surfaces than at smooth surfaces (Schröder 1969c). For the explication see fig. 22.

Surface polaritons on rough surfaces of semiconductors by Raman scattering. The surface polaritons or surface phonons on GaP, GaAs etc. are surface waves in the frequency gap between ω_{LO} and ω_T ($\epsilon < -1$) which lie in the infrared between 20 and 40 meV. These surface waves can be excited by electrons or by light with the prism method, see e.g., Raether (1977, 1980). These SPs can also be observed in the Raman signal as Evans et al. (1973) have reported. A diagram of the experiment is shown in fig. 24 (Ushioda 1981). The incoming light ($\lambda = 5145$ Å) suffers an inelastic scattering process (via nonlinear interaction) by which the photon loses the energy $\hbar\omega_{sp}$ and is in general deflected by the angle θ, creating a SP of energy $\hbar\omega_{sp}$ and of momentum $\hbar k_{sp}$.

In the experiments described above, in which the SP, the polarization

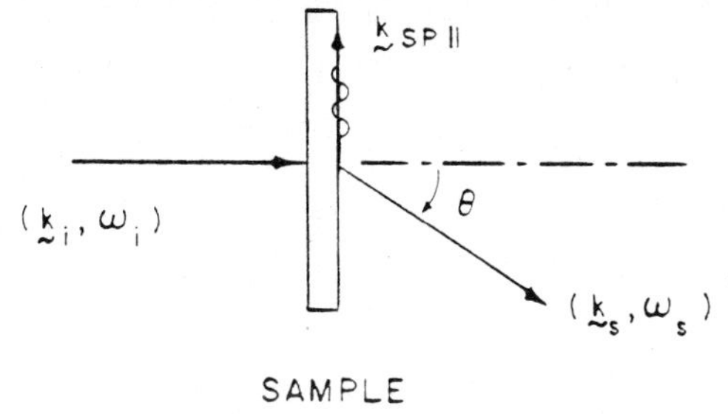

Fig. 24. Schematic representation of Raman scattering to produce surface polaritons (Ushioda 1981).

waves of the nearly free electron gas, are excited by the prism method or grating coupler, the energy of the photon is equal to that of the plasmon; whereas here the polariton energy $\hbar\omega_{sp}$ is small compared with the photon energy $\hbar\omega$ of several eV.

Conservation of energy and momentum are fulfilled if

$$\hbar\boldsymbol{\omega}/c = \hbar\boldsymbol{\omega}_s/c + \hbar\boldsymbol{k}_{sp}, \tag{33}$$

$$\hbar\omega = \hbar\omega_s + \hbar\omega_{sp}, \tag{33a}$$

with $\hbar\omega_s$ the energy of the scattered photon.

Conservation of energy and momentum means in case of plasmons, excited:

(1) by grating coupling (g reciprocal wavevector),

$$\hbar\omega = \hbar\omega_{sp}, \qquad \hbar(\boldsymbol{\omega}/c)\sin\theta_0 + \hbar\boldsymbol{g} = \hbar\boldsymbol{k}_{sp}; \tag{34}$$

(2) by the prism method,

$$\hbar\omega = \hbar\omega_{sp}, \qquad n\hbar(\boldsymbol{\omega}/c)\sin\theta_0 = \hbar\boldsymbol{k}_{sp}, \tag{34a}$$

(n refractive index).

Since the energy loss $\hbar\omega_{sp}$ is small compared to the energy of the incoming photon ($\hbar\omega$), momentum conservation can be written as

$$\hbar(\omega/c)\sin\theta = \hbar k_{sp}. \tag{35}$$

By changing θ and measuring the peak position of the Raman signal the dispersion relation of the SP can be obtained, here θ has been varied between 0° and 4° or k_{sp} between 0 and 10^{-4} Å^{-1}. The situation is the same as with electron energy loss spectroscopy. Figure 25 shows the Raman signal of the SP as well as the LO bulk phonon of GaP. Comparison of the spectrum on smooth (curve C) and on rough surfaces (A) shows that on a rough surface the signal is more intense and its width is on the average larger than on a smooth surface. This is in agreement with the results

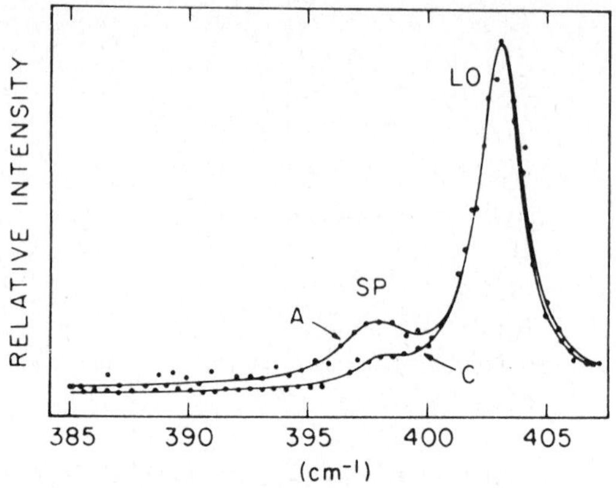

Fig. 25. Raman spectra of surface polaritons (SP) for a rough (A) and a smooth (C) surface of GaP. LO: longitudinal bulk phonon signal (Ushioda et al. 1979).

reported above which have shown that light emission increases with roughness so that the damping of the SP becomes larger.

These results are however rather qualitative: the roughness parameter (δ) has been estimated from the size of the powder used for the final polishing: it is assumed that 0,3 μm powder produces rough surfaces with $\delta = 3000$ Å and 0,05 μm powder produces smoother surfaces with $\delta = 500$ Å. Before drawing any conclusions from such figures which are certainly too large, the sensitivity of the method has to be improved to derive more reliable data of the roughness.

Light emission from junctions. An interesting application of the plasmon–photon coupling via surface discontinuities has been found in tunnel junctions, e.g., Al-Al_2O_3-Indium oxide (doped) (which is transparent in the visible light region) or a thin metal film of Ag or Au. Tunneling electrons driven by a bias voltage excite SPs which couple via electrode roughness (or small particles) with light, so that the whole junction emits light (McCarthy and Lambe 1978, Adams et al. 1979). For theoretical considerations see Rendel et al. (1978), Laks and Mills (1979).

3.6. *Measurement of Roughness with the Electron Microscope*

The light scattering experiments allow to determine the small scale roughness using the first order scattering theory. The experimental limit of the first order approximation using SPs lies at 20–30 Å. Theoretical difficulties have not yet been overcome to calculate second order scattering intensities and thus to evaluate light scattering experiments on rougher surfaces, see sect. 5. Which methods or roughness determination can be applied for rougher surfaces?

One of the powerful techniques to get information about the fine structure of a surface is given by the shadowing method. This method has been applied a long time ago to have an insight into the submicroscopic structure of boundaries of metals and insulators. As one of many examples one can study the correlation of roughness and plasma radiation (Schreiber 1968, Schröder 1969c). Modern instrumentation allows to resolve lateral extension of about 50 Å on a replica of a surface. Shadowing with an angle of about 75° gives a minimum height of 10–20 Å to be detected, so that surface irregularities of this height should be measurable and $\delta = (\overline{S^2})^{1/2}$ of some 10 Å.

The electronmicroscopic method can thus replace the first order light scattering method for roughness heights larger than 20–30 Å.

Measuring point by point the transparency of a photo of a replica with a microdensiometer one can derive, using certain assumptions, the correlation function $G(x)$, as has been realized at Ag (Dobberstein 1970) and Mg (Rasigni et al. 1977). This procedure has been improved so that the function $G(x)$ of silver films (thickness 900 Å) could be determined for $\Delta k < 5 \times 10^{-3}$ Å^{-1}; it has a Gaussian form and gives values of $\delta = 18$ Å and $\sigma = 210$ Å (Rasigni 1981). Light scattering experiments (Heitmann and Permien 1977) yielded $\delta = 8$–10 Å and $\sigma = 2000$ Å at 1000 Å thick films. This difference in the roughness structure is perhaps due to different preparation procedures, e.g., different vaporization rates. Comparison of roughness measurements at the same sample with different methods (light scattering and replica procedure) is necessary to clear this question and to examine the reliability of the methods.

4. *Scattering of Light without Excitation of Surface Plasmons*

4.1. *General Remarks*

If light hits the surface directly without using a prism scattered photons are observed. Here no SP interfere if the surfaces are not too rough. The intensity scattered at silver surfaces is about 10^2–10^3 times lower. If the background is sufficiently reduced, the scattered light can be detected under these conditions, too. It is interesting not only as a proof of the theoretical concept, but also for examining the practicability of this method to measure roughness parameters independent of the condition of SP excitation. The special interest is the possibility of comparing the results of both methods.

Experiments with nonresonant emission have been reported, together with calculations of the emitted intensity by Beaglehole and Hunderi (1970), Hunderi and Beaglehole (1969, 1970), using the concept of field-induced polarization currents (Stern 1967). These authors simplified the problem insofar as they treated the case of normal incidence of the light beam ($\theta_0 = 0°$). The plane of polarization lies either parallel (p-polarized) or perpendicular (s-polarized) to the plane of observation. Similar relations for $\theta_0 = 0$ have been obtained with other methods, leading to equal results, (Maradudin and Mills 1975, Marvin et al. 1975, Elson and Ritchie 1974).

These equations have already been derived earlier for radar measurements in a more general form: finite ϕ, the angle between the plane of incidence and the plane of observation and $\mu \neq 1$, μ the magnetic permeability (Barrick 1970). In all these cases surface roughness has been assumed.

The observation of Beaglehole and Hunderi of the angular distribution of the scattered intensity is in agreement with the calculated dependency, a Gaussian correlation function assumed. Values of δ are not given.

A quantitative evaluation of such scattering experiments in order to obtain the roughness parameters has first been made by Heitmann and Permien (1977), see also Hillebrecht (1980).

4.2. Results in the Region of Visible Light – Determination of Roughness

The experimental arrangement of the nonresonant excitation of light is shown in fig. 27 inset, see also fig. 11. The light beam coming from the air side at an angle of incidence θ_0 (here $\theta_0 = 0$) excites in the rough surface polarization currents which emit light at an angle θ detected by a multiplier. The film, which is deposited on the half cylinder P in fig. 11, could be studied with the prism method, too, by turning the half cylinder by 180° (see fig. 11), so that results in and out of resonance can be compared at the same surface. The relations valid for this arrangement have been given for $\lambda \gg \delta$ and $\Lambda \gg \delta$ as

$$\frac{\mathrm{d}I}{I_0\,\mathrm{d}\Omega} = \frac{1}{4}\left(\frac{\omega}{c}\right)^4 \frac{1}{\cos\theta_0} |W_{ii}|^2 |s(\Delta k)|^2, \tag{36}$$

for s- (W_{ss}) and p- (W_{pp}) polarized light as derived e.g., from Kröger and Kretschmann (1970). W_{ps} and W_{sp} are zero in the plane of incidence ($\phi = 0$) in the single scattering approximation. If ϕ is finite and $\theta_0 \neq 0$, (sp) and (ps) light is expected; corresponding formulas are found in Barrick (1970).

The relations W_{ii} of eq. 36 are

$$W_{pp} = \frac{4(\epsilon - 1)\cos\theta_0 \cos\theta}{(\epsilon\cos\theta_0 + \sqrt{\epsilon - \sin^2\theta_0})(\epsilon\cos\theta + \sqrt{\epsilon - \sin^2\theta})} \times (\epsilon \sin\theta_0 \sin\theta - \sqrt{\epsilon - \sin^2\theta_0}\sqrt{\epsilon - \sin^2\theta}), \tag{37}$$

$$W_{ss} = \frac{4(\epsilon - 1)\cos\theta_0 \cos\theta}{(\cos\theta_0 + \sqrt{\epsilon - \sin^2\theta_0})(\cos\theta + \sqrt{\epsilon - \sin^2\theta})}. \tag{38}$$

It may be added that the denominator of W_{pp} for θ_0 resp. θ in eq. (37) are identical with D_s of eq. (13). Its absolute value put to zero yields eq. (5). In the configuration of fig. 27 inset in which light propagating in a medium with $\epsilon_0 = 1$ hits the metal surface, the wavevector is limited to $k \leq \omega/c$, so that no resonance takes place in contrast to the case $\epsilon_0 > 1$, e.g., as in glass, where eq. (5) can be fulfilled, see configuration in fig. 2.

The W_{pp} term has an interference factor composed of the normal and tangential component. The $W_{ii}(\theta)$ functions are plotted in fig. 27 for $\theta_0 = 0°$; W_{pp} is somewhat dependent on λ: at $\theta = 60°$, $|W_{pp}|^2$ varies by about 20% (between 6500 Å and 4500 Å) $|W_{ss}|^2$ is independent of λ since $|W_{ss}|^2$ can be written, for real ϵ and $\epsilon' < -1$,

$$|\hat{W}_{ss}|^2 = 4\cos\theta\cos\theta_0, \tag{39}$$

dotted line in fig. 27.

Similarly to the case of SP excitation (sect. 3.4 and fig. 19) the quotient

$$dI_{pp}/dI_{ss} = |W_{pp}|^2/|W_{ss}|^2 \tag{40}$$

is independent of the roughness function.

The experimental results are displayed in figs. 26 and 27. Figure 26 reproduces the angular dependence of the pp and the ss intensity for $\theta_0 = 0$ scattered at a silver film of $\delta = 6$–8 Å and 4000 Å thickness. In addition the depolarized relative intensities are shown for comparison with pp and ss intensities. In fig. 27 the good agreement of dI_{pp}/dI_{ss} for "smooth" silver films with the calculated values at $\theta_0 = 0°$ (also at $\theta_0 = 60°$) can be seen. This supports the assumption of the calculation and confirms further that *surface* roughness produces the light emission (see sect. 3.4). With increasing θ, the intensity decreases, so that the error bars of the experimental dI_{pp}/dI_{ss} curve become larger. At rougher surfaces the experimental data of the quotient dI_{pp}/dI_{ss} lie below the theoretical ones, especially at larger θ and for $\delta > 15$ Å (Hillebrecht 1979, Heitman and Permien 1977). Presumably the effect of the double scattering reduces the p intensity in favour of the s intensity.

If the surface has the appropriate roughness, SPs can be created by the incoming light, see sect. 3.5, which in a second scattering process are scattered back into the light circle and increase thus the emitted light. This is assumed to explain the higher observed p intensity compared to the calculated one (see fig. 2, Hunderi and Beaglehole 1970). It shall be pointed out for a discussion of the roughness influence that not the value of δ alone is important but also the character of $|s(\Delta k)|^2$, i.e., in case of a Gauss function, the value of σ; it determines the extension of the $|s(\Delta k)|^2$ in k

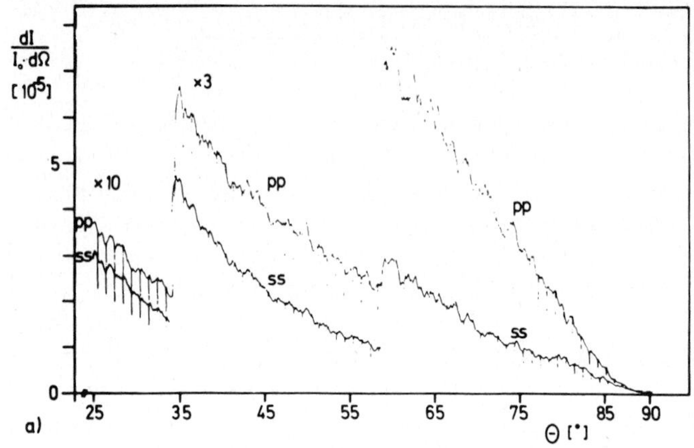

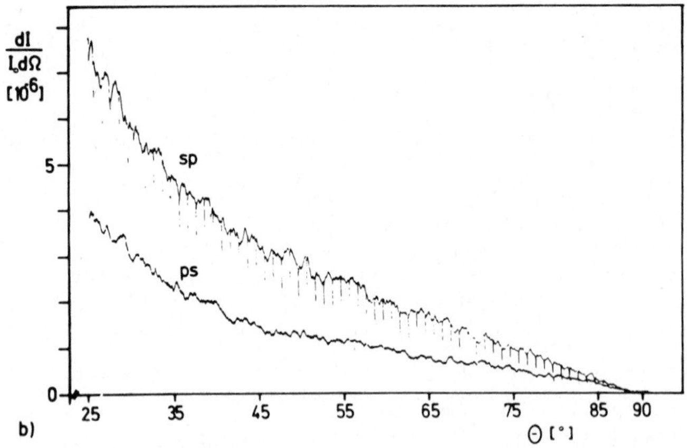

Fig. 26. Observed angular intensity distribution of pp, ss, sp and ps light ($\lambda = 5145$ Å) scattered on a silver film, 4000 Å thick. Normal incidence. The value of dΩ amounts to 8×10^{-5} sr (Hillebrecht 1980).

space and thus the strength of the plasmon–photon coupling. It can thus happen that $\mathrm{d}I_{pp}/\mathrm{d}I_{ss}$ exceeds the calculated (dashed) curve in fig. 27.

A curve similar to fig. 27 (dashed line) has been measured by Mishra and Bray (1977) for $\mathrm{d}I_{pp}/\mathrm{d}I_{ss}$, eq. (40), with light reflected at the ripples on the surface of a solid, e.g., GaAs and CdS produced by bulk-phonon-induced mechanical perturbations. The amplitude δ of the ripples is about one Å and its wavelength $\Lambda \sim 10^4$ Å, so that $\delta/\Lambda \sim 10^{-4}$, which justifies the application of the linear theory. In the above case of a silver surface of "natural roughness" δ/Λ is about a few 10^{-3}; several 10^{-2} seem to be near the limit of the linear theory, see sect. 5.

The observed intensity curves $(\mathrm{d}I/I_0\,\mathrm{d}\Omega)_{ii}$ decrease monotonically with θ: the pp intensities are essentially larger than the ss values, especially at

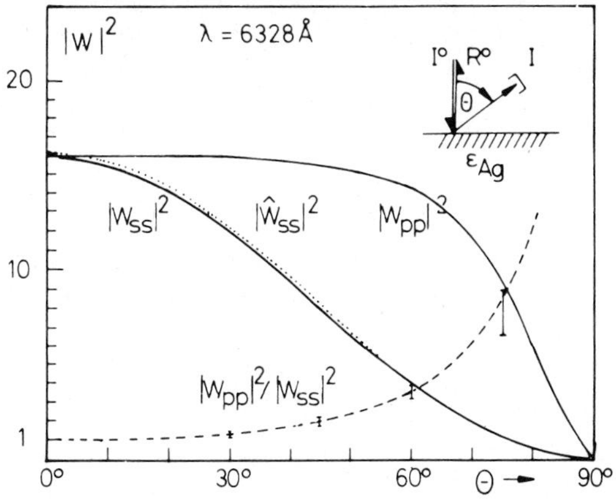

Fig. 27. Calculated dipole functions $|W_{pp}|^2$ and $|W_{ss}|^2$ for normal incidence ($\theta_0 = 0°$) and $\lambda = 6328$ Å (He–Ne laser) for silver ($\epsilon = -17.6 + i0.67$). The dashed curve displays the calculated ratio $|W_{pp}|^2/|W_{ss}|^2$. The error bars indicate experimental values for films of different thickness (500–4000 Å) and different roughness (Heitmann and Permien 1977).

higher θ (see fig. 26). These curves differ from the $W_{ii}(\theta)$ since they are multiplied by the roughness factor $|s|^2$. The function $|s(\Delta k)|^2$ can be determined directly by the same procedure as in sect. 3; fig. 28 shows the $|s(\Delta k)|^2$ curves: $\ln|s|^2$ as a function of Δk^2, at different θ_0 and λ for silver films 500 Å thick. Most of them are derived from the pp curves, their intensity being stronger and thus the error bars of the values obtained are smaller. The curves have the same trend as those deduced from the SP

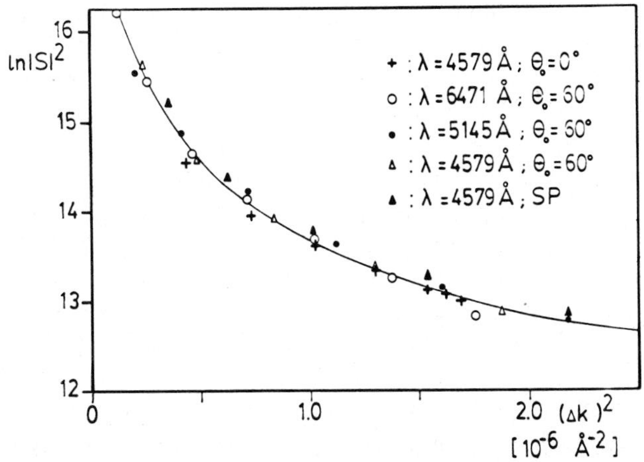

Fig. 28. Roughness functions of a silver film, 500 Å thick, obtained at different angles of incidence θ_0 and wavelengths up to $\Delta k = 1.5 \times 10^{-3}$ Å^{-1}. For comparison data obtained on the same sample using enhancement by SP resonance (Hillebrecht 1980) are shown.

resonance curves: at large Δk a nearly linear region, and at low Δk an increase of $|s(\Delta k)|^2$ by at least a factor of 10. The observations were extended to $\theta_0 = -60°$ to get larger Δk values, the Δk region is given by eq. 25.

The evaluation of the roughness taken from these measurements without exciting SPs (fig. 28) yield δ values of about 3–4 Å (Heitmann and Permien 1977, Hillebrecht 1980). This is in agreement with the figures in sect. 3. Similarly as in sect. 3 the silver films were roughened by underlayers which produce an increase of the roughness value δ, see fig. 35.

Roughness on thicker silver films has been measured: values of δ: 6–8 Å and $\sigma = 2450$ Å were reported on 4000 Å thick films.

As mentioned above the measurements can be repeated at the same surface with the prism method which works with an emitted intensity more than 10^2 times stronger. The results obtained agree with those of the nonresonant method (see fig. 28).

Light scattering experiments as those described above with silver deposited on polished quartz substrates have been published (Elson and Bennett 1979b) but the authors did not give roughness values.

The same experimental and theoretical procedure as just described has been applied to study the roughness structure of evaporated Cu films. However the authors come to the conclusion that the roughness amplitude could not be reliably derived from the data observed at their films. They can however calculate a lower limit of δ for Cu films of 5000 Å thickness on quartz: $\delta > 4{,}4$ Å and Cu on mica: $\delta > 5{,}8$ Å (Jansen and Hoffman 1979).

A more detailed study of the angular distribution of the light (pp, ss, ps and sp) which is scattered from thick silver films (~5000 Å) with CaF_2 underlayers of 500–3000 Å thickness, using normal incidence, has been reported by Sari, Cohen and Scherkoske (Sari et al. 1980). Instead of separating the waviness of the substrate which is concentrated at small Δk values from the roughness of the silver film, see e.g., fig. 14, the $|s(\Delta k)|^2$ function is approximated by two Gaussian correlation functions and fitted to the observed curve with these four parameters. The (δ, σ) values are, e.g., at $\lambda = 5145$ Å for the rough silver films: $\delta = 39$ Å, $\sigma = 790$ Å, in general $\delta = 40$–50 Å, $\sigma = 800$–1200 Å and for the waviness: $\delta = 6$ Å, $\sigma = 7500$ Å. These values are about 2–3 times larger than those obtained on films of similar thickness which have been described above. Perhaps different preparations make these differences. This discrepancy should be examined further.

Parallel to these measurements of the differential scattered intensity, the deficit of the reflectivity ΔR integrated over θ, see sect. 3.5, at different wavelengths has been measured and compared by Bush et al. (1980). The evaluation has been made with the complex ϵ of silver, so that damping is not neglected; rather good agreement is achieved with the roughness data

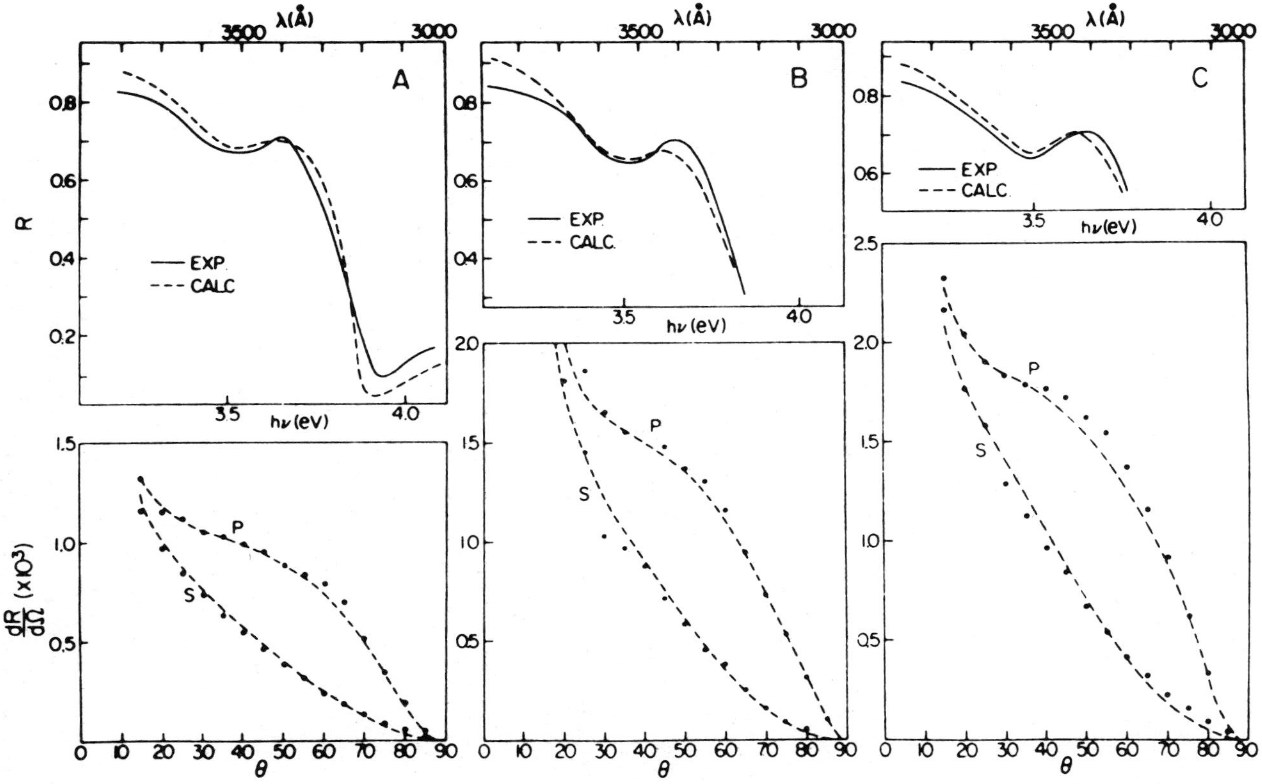

Fig. 29. Scattered light distribution at $\lambda = 4579$ Å (below) and reflectance near the plasma edge of silver on silver foils of increasing roughness; A: $\delta = 41$ Å, $\sigma = 790$ Å. B: $\delta = 50$ Å, $\sigma = 800$ Å. C: $\delta = 52$ Å, $\sigma = 820$ Å. These roughness parameters (δ, σ) give the best fits (dashed lines) to the angular scattering (dots: observed values). To improve agreement at low θ a wavy surface component has been introduced (A: $\delta' = 4$ Å, $\sigma' = 7000$ Å. B: $\delta' = 8$ Å, $\sigma' = 7000$ Å. C: $\delta' = 5$ Å, $\sigma' = 8000$ Å). The upper reflectance calculations use these parameters (Bush et al. 1980). The reflectance curves are similar to fig. 21.

from scattering experiments of the authors, see fig. 29. In these calculations the change of the dispersion relation by roughness, see sect. 2, has been taken into account; this means that the solution k for a smooth surface, eq. (13), has to be replaced by $k+\overline{\Delta k}$ with a real (displacement) and an imaginary (increase of damping) part, see eq. (8). In Sari et al. (1980) the correction of k should be written $k = k_0(1+\Sigma)^{1/2}$; here we are using $k_0+\overline{\Delta k}$, so that $\overline{\Delta k} = \frac{1}{2}k_0\Sigma$. The relations for Σ are evaluated with a Gaussian cor-

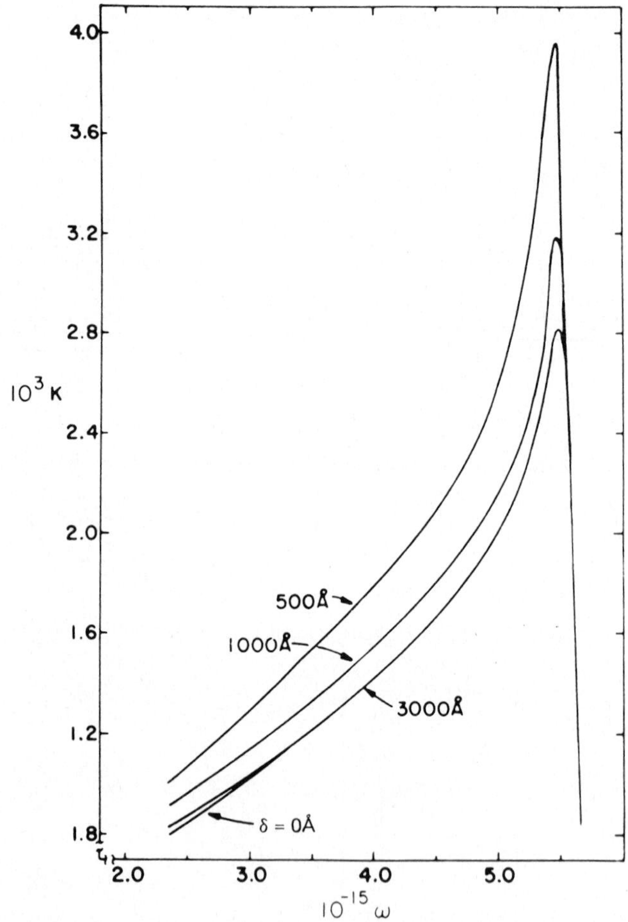

Fig. 30. Comparison of calculated dispersion relations for SPs on silver films of roughness $\delta = 0$ and $\delta = 100$ Å with different correlation lengths σ. The qualitative shape is in good agreement with the observation, see fig. 17 (turned two times). For the quantitative agreement, see text. The curve $\delta = 0$ Å coincides nearly with that for $\delta = 100$ Å and for $\sigma = 3000$ Å since $\delta/\sigma = 1/30$ is sufficiently small (Sari et al. 1980).

relation function. Figure 30 displays the calculated dependency $\omega(k)$ for a silver film of $\delta = 100$ Å and different σ values. The qualitative agreement with the observed data, fig. 17, is evident. The values of $\overline{\Delta k}$ which show displacements into the right sense to larger k values, at $\omega = \text{const.}$, however, are more than 10 times smaller than the observed $\overline{\Delta k}$ values for $\delta = 10$ Å. The comparison is made for $\delta = 10$ Å, since for this roughness the linear approximation is valid; the theoretical results of Sari et al., obtained for $\delta = 100$ Å, are calculated for $\delta = 10$ Å using the relation $\overline{\Delta k} \sim \delta^2$. The same discrepancy is obtained for the imaginary part of Σ. (For small δ/σ ($\sigma = 3000$ Å) the calculated curves agree with the observed data.) The disagreement for the rough surfaces is believed to have its origin in the questionable extrapolation of $|s(\Delta k)|^2$ to large Δk values as already discussed, see sect. 2.2.

At δ of 15–20 Å discrepancies become apparent which very probably are connected with the non-validity of the linear approximation or the single-scattering assumption. These discrepancies are the λ dependence of δ and the observation of s-polarized light with p-polarized incident light and vice versa. These effects are dealt with in sect. 5.

4.3. *Scattering Experiments with X-Rays*

The reflection of X-rays has developed new interest in connection with the production of mirrors of high quality for X-ray telescopes, Wolter type, in X-ray astronomy. Here X-rays are reflected at grazing incidence below the angle of total reflection. Since the surface is not smooth, diffuse scattering takes place which reduces the intensity of the specular beam. This loss of light shall be as small as possible, a maximum of the rms height of $\delta \cong 10$ Å is supportable (Korte and Laine 1979).

Experiments to determine the diffuse scattering have been performed (Korte and Laine 1979, Lenzen 1978, Trumper et al. 1979): A very narrow beam of X-rays (Cu L: 13, 3 Å; Al K α: 8,3 Å) hits the surface at an angle of $\sim 1°$. The intensity of the primary, as well as of the reflected beam is scanned with a very small slit; fig. 31 demonstrates the broadening of the X-ray beam by diffuse scattering.

The evaluation has been done with eqs. (48) and (49):

$$\frac{dI}{I_R \, d\Omega} = 4\left(\frac{\omega}{c}\right)^4 \varphi_0 \varphi^2 |s(\Delta k)|^2. \tag{41}$$

Here I_0 is replaced by I_R, since total reflection takes place. As fig. 31 shows, the scattering angles $\varphi = 90° - \theta$ and $\varphi_0 = 90° - \theta_0$ are of the order of 10 arc min or 10^{-3} rad so that the conditions eq. 50 are fulfilled.

X-ray measurements (Trümper et al. 1979) on the same sample with different wavelengths and different angles φ_0 yield δ values which agree

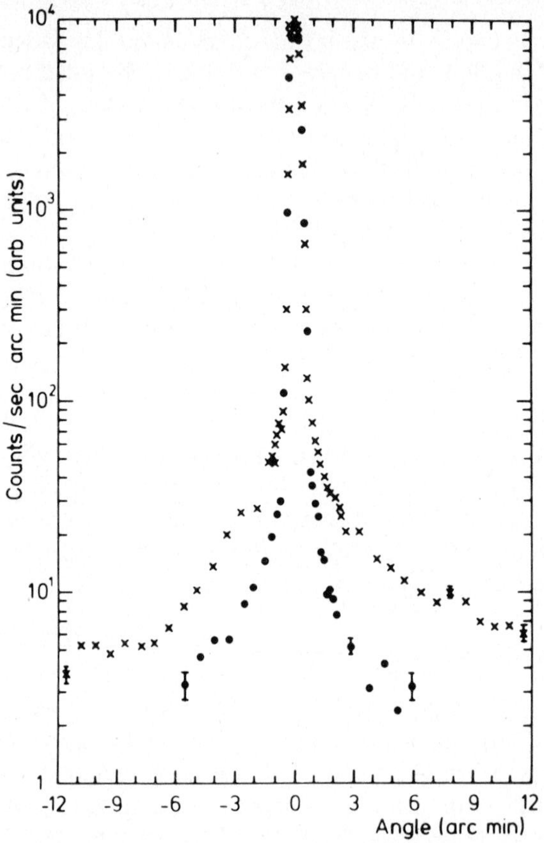

Fig. 31. Profile of an X-ray beam reflected at grazing incidence (1.5°) from a surface of gold (800 Å thick) with a roughness of $\delta = 5.7$ Å (crosses). The dots indicate the primary beam (AlKα: $\lambda = 8.3$ Å) (deKorte and Lainé 1979).

rather well within the error limits. E.g., the δ values measured for a gold-coated quartz surface with $\lambda = 8.3$ Å vary between 3.2 and 2.4 Å at φ_0 between 1.0° and 3.0°. An independent control of this result has been possible by investigating the sample with an electron interference microscope (Lichte 1977). With this instrument a mean roughness height δ of $3.2 \pm 1{,}2$ Å has been measured in good agreement with the above value of $\delta = 2{,}5 \pm 0{,}5$ Å obtained by scattering experiments at the same sample. This consistency provides strong support to the validity of the scattering equations.

4.4. *Discussion of the Scattering Equations*

The scattering eqs. (37, 38) are applied to many fields. Some approximations of interest in this article shall be discussed briefly.

1. (a) surface of a metal: $\epsilon' < 0$ and $|\epsilon'| > 1$, $\epsilon' \gg \epsilon''$, we obtain

$$|W_{ss}|^2 = (4 \cos \theta_0 \cos \theta)^2, \tag{42a}$$

$$|W_{pp}|^2 = [4(1 - \sin \theta_0 \sin \theta)]^2. \tag{42b}$$

In order to obtain eq. 42b $|\epsilon'|$ should be rather large:

$$\sqrt{|\epsilon'|} \cos \theta > 1, \quad \sqrt{|\epsilon'|} \cos \theta_0 > 1. \tag{43}$$

(b) $\theta_0 = 0$ (normal incidence):

$$|W_{ss}|^2 = (4 \cos^2 \theta)^2, \quad |W_{pp}|^2 = 4^2. \tag{44}$$

p- and s-polarized light have different W functions and thus different intensities of reflection.

(c) Limiting case: $\theta \cong \theta_0$,

$$|W_{ss}|^2 = (4 \cos^2 \theta)^2, \quad |W_{pp}|^2 = (4 \cos^2 \theta)^2. \tag{45}$$

The terms for p- and s-polarized light become equal. $\theta \cong \theta_0$ means that all diffraction orders are concentrated near the specular beam; this condition is identical with $\lambda \ll \Lambda$ or with $|\epsilon'| \to \infty$ or with $\lambda \to 0$.

2. Nonmetallic surface: $\epsilon \geqslant 1$,

$$|W_{ss}|^2 = (\epsilon - 1)^2, \quad |W_{pp}|^2 = (\epsilon - 1)^2 \cos^2(\theta + \theta_0). \tag{46}$$

Both terms are different for $\theta_0 = 0$ as well as for $\theta = \theta_0$ (here $|\epsilon'|$ will not reach large values).

3. Grazing incidence, important for X-rays,

$$\theta = 90° - \varphi, \quad \theta_0 = 90° - \varphi_0, \qquad \varphi, \varphi_0 \text{ small}. \tag{47}$$

Equation 38 gives:

$$\text{in case of insulators } (\epsilon \sim 1) \quad W_{ss} = W_{pp} = 4\varphi\varphi_0; \tag{48}$$

$$\text{in case of metals } (|\epsilon| \geqslant 1) \quad W_{ss} = W_{pp} = 4\varphi\varphi_0, \tag{49}$$

$$\text{if} \quad \varphi < \sqrt{|\epsilon - 1|} \quad \text{and} \quad \varphi_0 < \sqrt{|\epsilon - 1|}, \tag{50}$$

e.g., with $|\epsilon - 1| = 10^{-3}$, φ should be smaller than 0,03 rad (~2°).

The calculation of eq. 48, 49 goes as follows:

$$W_{pp} = \frac{4(\epsilon - 1) \sin \varphi_0 \sin \varphi(\epsilon \cos \varphi \cos \varphi_0 - \sqrt{\epsilon - \cos^2 \varphi_0}\sqrt{\epsilon - \cos^2 \varphi})}{(\epsilon \sin \varphi_0 + \sqrt{\epsilon - \cos^2 \varphi_0})(\epsilon \sin \varphi + \sqrt{\epsilon - \cos^2 \varphi})}$$

$$= \frac{4(\epsilon - 1)\varphi_0\varphi(\epsilon - (\epsilon - 1))}{(\epsilon\varphi_0 + \sqrt{\epsilon - 1})(\epsilon\varphi + \sqrt{\epsilon - 1})}. \tag{51}$$

The above conditions eq. (50) yield the result eqs. 48, 49, and similarly for W_{ss}. This shows that the vectorial character of light can be neglected at grazing incidence on metals and nonmetals in the X-ray region (scalar treatment).

In the literature the scalar case (1c) has often been discussed (Beckmann and Spizzichino 1963, Welford 1977). It leads to a simple relation between δ and the integrated intensity if eq. (45) is integrated over $d\Omega$. For incident unpolarized light

$$\left(\frac{dI}{I_0\, d\Omega}\right)_{\text{unpol}} = \frac{1}{2}\left(\frac{dI}{I_0\, d\Omega}\right)_{p} + \frac{1}{2}\left(\frac{dI}{I_0\, d\Omega}\right)_{s}, \tag{52}$$

and assuming a Gaussian roughness function, one obtains (Davies 1954, Bennett and Porteus 1961, Kretschmann and Kröger, 1975):

$$\frac{R_0 - R_s}{R_0} \cong (4\pi\delta/\lambda)^2, \tag{53}$$

($\lambda \gg \delta$) with R_0 the reflected intensity at the smooth surface, R_s the specular reflected intensity at the rough surface. It is only valid in case of $\epsilon \to \infty$, i.e., metallic substances as we have seen.

This simple relation has been used to measure the integrated scattered intensity and to deduce a value of δ. A review of this method and its description has been given elsewhere (Elson et al. 1979a). This method has a qualitative character: As is shown in sect. 3 the light is spread over the half space, so that $\theta \sim \theta_0$ is too rough an approximation, except in case of X-rays (see sect. 4.3). The integration of the scattered intensity brings an important loss of information which one should avoid. For example, as we have seen, the roughness function $\ln |s|^2$ as function of $(\Delta k)^2$ is composed of two parts: the linear part characteristic of the silver and an increasing part at low $(\Delta k)^2$ values originating from the wavy substrate. Integration over $d\Omega$ makes such details of the roughness structure disappear and one obtains in this case too high an average value of δ. This is demonstrated by a comparison of δ values obtained with both methods (Hornauer 1977): The values obtained with the integrated method are about two times higher than those of the differential method; e.g., measurements of the roughness of silver films as those mentioned in sect. 3 yield values of $\delta = 5$–8 Å and $\sigma = 800$–1000 Å whereas $\delta = 14$–16 Å is obtained at these samples with the integrating method. A further disadvantage is that one does not see the limits of the approximations used for the evaluation similar to those mentioned in sect. 5.

A procedure of remarkable capability for the determination of roughness is the interferometric method (Tolanski method, FECO technique). It can determine the rms height δ of a surface with a resolution of nearly 5–10 Å's. However since the lateral resolution is $\gtrsim 10\,000$ Å, this value is averaged over $>1\ \mu$, so that detailed information on the submicroscopic roughness spectrum ($|s(\Delta k)|^2$) is not available.

There exist observations to measure the statistics of surface roughness with the method just mentioned (Elson et al. 1979a). There may be a relation between the roughness data obtained with light scattering and those obtained with FECO technique, but such a relation must not necessarily exist.

5. *Rather Rough Surfaces; Limits of Present Theory.*

In comparing experimental with calculated values discrepancies appeared which are negligible at silver surfaces of natural roughness of $\delta = 3$–6 Å which, however, become stronger with increasing roughness. Very probably they indicate the limits of the single-scattering theory.

5.1. Depolarization of the Emitted Radiation

In experiments at normal incidence ($\theta_0 = 0°$) a small amount, some %, of depolarized light, dI_{sp} and dI_{ps}, has been observed (Beaglehole and Hunderi 1970, Kretschmann 1972b, Hall and Braundmeier 1973, Heitmann and Permien 1977, Hillebrecht 1980). Figure 32 displays results of the ratio dI_{sp}/dI_{pp} (solid lines) and dI_{ps}/dI_{ss} (dashed lines) for $\lambda = 5145$ Å and different roughness from experiments without SP excitation. Absolute values are found in fig. 26. The "smooth" silver film shows a small depolarization, equal for s and p light similar to the 50 Å curve in fig. 32, which increases differently with roughness, see fig. 32. Similar depolarization effects have been observed in experiments with SPs, see figs. 18 and 19 with $\phi = 90°$. In fig. 19 the ratio $dI\ (\psi = 90°)/dI(\psi = 0°)$ should go to zero with $\theta \to 0$, fig. 19b, however there remains a certain residual intensity $I_{pp}/I_{ss}(\theta = 0)$ see fig. 19a. It increases with roughness δ as fig. 33 shows (Horstmann 1977): up to about $dI_p/dI_s = 0{,}3$ to 0,4 the dependence is linear in δ^2; for higher δ values (15–20 Å) the evaluation is not very accurate. Therefore the values of $\theta_{1/2}$ (half width of the reflection minimum) have been taken as abscissa which represent a good quantity to characterize the roughness. The ratio $dI_p/dI_s(\theta = 0°)$ now becomes a linear function of $\theta_{1/2}$ (for $\lambda = 5309$ Å). It is interesting that the underlayer material has an influence on this dependence. Whereas CaF_2 and Ag have a strong roughening effect, LiF as underlayer produces practically no change in the parameters as table 4 shows (Horstmann 1976).

Measurements of the intensity of the s-component ($\psi = 0°$), incident light p-polarized, as a function of θ in the arrangement of fig. 2 for $\lambda = 5500$ Å between $\theta = \pm 70°$, showed that it is symmetric around $\theta = 0°$ with a maximum at $\theta = 0°$, similar to a $\cos^2$ distribution (Urner 1976).

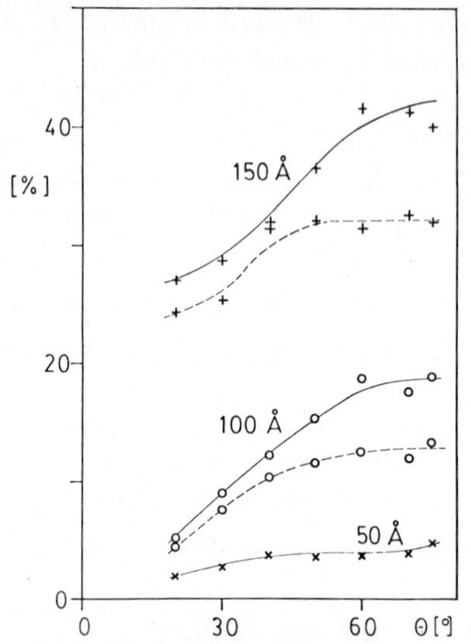

Fig. 32. Ratio of the intensities indicating the depolarization of the scattered light dI_{ps}/dI_{ss} (dashed lines) and dI_{sp}/dI_{pp} (solid lines) at different roughness of the surface of a silver film 500 Å thick. The roughness has been enhanced by underlayers of silver with the following values: Lower curve: 50 Å/500 Å; 150°C: $\delta = 7$ Å. Middle curve: 100 Å/500 Å; 150°C: $\delta = 9$ Å. Upper curve: 150 Å/500 Å; 150°C: $\delta = 22$ Å. For an explanation see sect. 5.1 (Hillebrecht 1980).

Table 4

Roughness data for silver films (350 Å thick) with underlayers of LiF of different thickness (d). θ_0: position, $\theta_{1/2}$: width of the reflection minimum, δ: rms height and $I_p/I_s(\theta = 0°)$: residual light intensity (Horstmann 1977)

d_{LiF}(Å)	θ_0(5500 Å)	$\theta_{1/2}$(5500 Å)	δ(Å)	$I_p/I_s(\theta = 0°)$
0	45.7°	1.39°	4	<0.08
500	45.8°	1.25°	4	<0.08
1000	45.9°	1.33°	4	<0.08
2000	46.0°	1.25°	4	<0.08

The forbidden polarization component can be understood by a double scattering process as is indicated in fig. 34 (right-hand part). By this process a component of the radiating dipole perpendicular to the original polarization plane is produced.

An argument which supports this explanation is as follows (Horstmann 1977): Assuming second-order scattering the (forbidden) $dI_p(0°)$ intensity will be proportional to δ^4, whereas $dI_s(0°)$ is $\sim\delta^2$. It has been shown that $\theta_{1/2} \sim \delta^2$ (see sect. 2), so that it follows that dI_p/dI_s (at 0°) $\propto \theta_{1/2}$ or δ^2. This rough

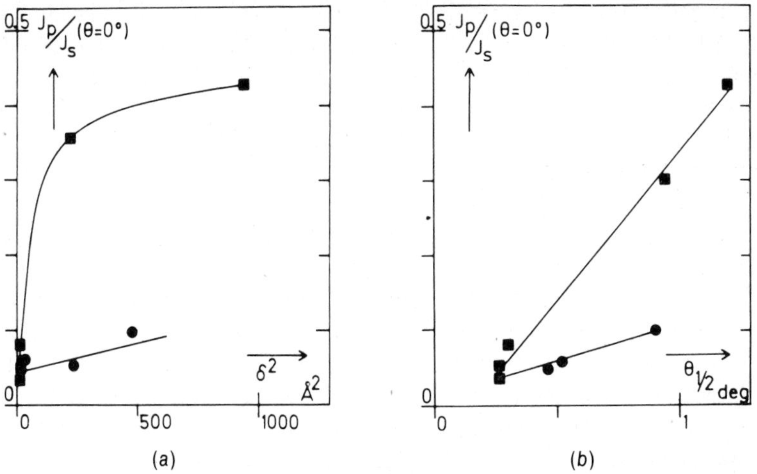

Fig. 33. Rest light intensity ratio dI_p/dI_s ($\theta = 0°$) (a) versus δ^2 and (b) versus the half width of the SP minimum $\theta_{1/2}$ on a silver film, thickness 550 Å, at $\lambda = 5309$ Å with Ag (■) and CaF_2 (●) underlayers of different thickness. The silver underlayers were 50 Å/550 Å, 200°C, $\delta = 10$ Å; 100 Å/550 Å, 200°C, $\delta = 15$ Å and 150 Å/550 Å, 200°C, $\delta = 25$ Å. The CaF_2 underlayers had thicknesses of 500, 1000 and 2000 Å which produce values of $\delta = 7$, 15 and 20 Å (Horstmann 1977). Read in the figures dI_p/dI_s instead of I_p/I_s, similar in fig. 19.

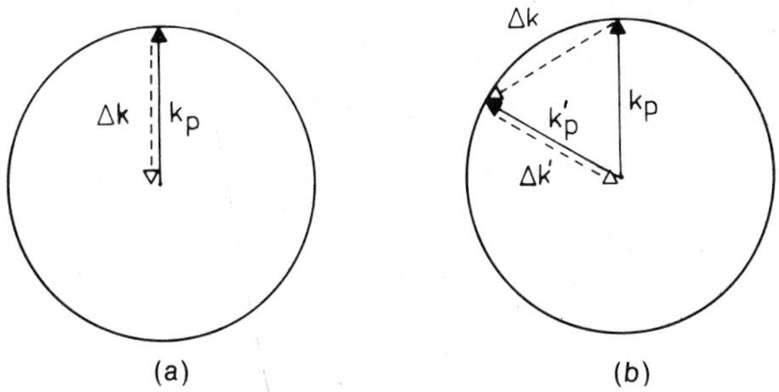

Fig. 34. Scattering process with light emission in z-direction. The circle shows the dispersion relation in the plane (k_x, k_y). Its diameter depends on the frequency of the SP. The correlated light circle always lies in the interior of these circles. At left: single process: the plasmon is cancelled by a Δk transfer and light is emitted perpendicular to the surface (z-direction). At right: scattering of the plasmon by Δk before light is emitted (second order process). This diagram shows that SPs propagating on a rough surface disappear by radiation and scattering processes described by $\Delta k''$ of eq. 9b in addition to the interior damping due to excitation of interband transitions in general.

explanation is confirmed by the observations (see fig. 31a) for $dI_p/dI_s < 0.3$. Detailed calculations have not yet been published.

The scattering process of fig. 34(b) which changes the direction of the wavevector $\boldsymbol{k}_p$ into $\boldsymbol{k}_p' = \boldsymbol{k}_p + \Delta\boldsymbol{k}$ with $|\boldsymbol{k}_p| = |\boldsymbol{k}_p'|$ can be demonstrated in a simple experiment (Braundmeier and Tomaschke 1975, Simon and Guha 1976): A light detector, placed in fig. 8 below the half sphere under the angle θ_0 against the normal of the silver surface and turned around it, records light intensity. It comes from scattered SPs with k_p', lying on the dispersion circle, see fig. 32(b). These SPs with K_p' are transformed into photons by the half sphere itself, see Kröger and Raether (1971). This light cone can be photographed as a light circle. Its intensity radiated into this light cone is of the order of 0,1% of the incoming light, so that I_0 in eq. (14) is not influenced by this effect.

Since the excitation of the SP with K_p' is determined by the density of Δk values available from the roughness spectrum, the light intensity (isotropic roughness assumed) on the circle allows to determine $|s(\Delta k)|^2$ using eq. (14). The roughness parameters measured on a silver film (500 Å thick) were: $\delta = 6$ Å and $\sigma = 1400$ Å in good agreement with (Kretschmann 1972), see also sect. 3.2.

5.2. *Wavelength dependence of* $|s(\Delta k)|^2$

Another result not yet explained by theory has been found in the λ dependence of the roughness function as fig. 35 shows. At low roughness the $|s|^2$ function is independent of the wavelength λ; with increasing roughness, however, $|s(\Delta k)|^2$ shows a splitting at different λ so that at larger δ the determination of δ becomes questionable. It seems that up to $\delta \sim 20$–30 Å the methods described are applicable (Orlowsky et al. 1979). Similar results have been obtained with silver films roughened with silver underlayers and measured with and without SPs; both methods lead to the same results (Hillebrecht 1980). The fact that the deviation is stronger for shorter wavelengths can be due to the stronger and thus multiple scattering of these SPs.

These experiments, especially the measurements of the angular distribution and of the wavelength dependence of the $|s(\Delta k)|^2$ function, demonstrate that this method allows one to get information on the limits of the linear approximation.

6. *Plasmons on Surfaces of Periodic Profile*

If light is diffracted from a reflection grating, rapid variations of its intensity are observed in a rather narrow region of wavelength at constant angle of incidence or vice versa. Two phenomena are responsible for these variations:

(a) a discontinuous change of intensity takes place if the angle θ in the grating eq. (4) (valid for s- and p-polarized light) reaches 90°. The emer-

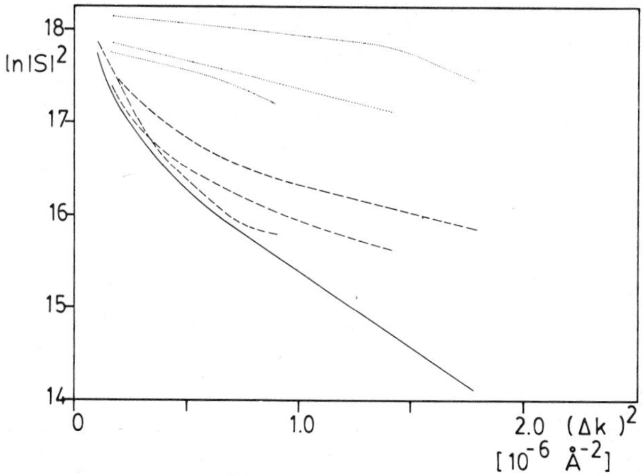

Fig. 35. Roughness functions $|s(\Delta k)|^2$ for silver films (500 Å thick) with different roughness: the lowest curve (full line), $\delta \simeq 7$ Å, shows no dependence on wavelength λ. The dashed curve $\delta \simeq 10$ Å, splits for different λ: 6471 Å (lowest curve) 5145 Å and 4579 Å (highest curve). The same is found for these λ values with $\delta \simeq 22$ Å (dotted curve). It should be noticed that here the wavy structure of the substrate is nearly covered, also see sect. 3 (Hillebrecht 1980).

gence or disappearance of a spectral order at grazing angle θ produces a change of intensity. This effect is more pronounced in a grating which produces only a few diffraction beams of lower order as in a sinusoidal grating.

(b) Using p-polarized light the excitation of SPs can occur as discussed in fig. 22. In the context of this article the second phenomenon is of special interest.

6.1. *Dispersion Relation of Surface Plasmons on Sinusoidally Perturbed Surfaces*

A sinusoidal grating, grating constant a, produces, in addition to the reflected light ($\nu = 0$), diffracted beams of $\nu = \pm 1$. We assume that the grating has shallow grooves so that the higher orders can be neglected. It is now possible that the tangential component of the wavevector of the incoming light, $(\omega/c) \sin \theta_0$, which is polarized perpendicular to the grooves can be increased by the roughness vector $g = 2\pi/a$, so that the dispersion relation of SP

$$k = (\omega/c)\sqrt{\epsilon/(\epsilon+1)} = (\omega/c) \sin \theta_0 + 2\pi/a \tag{54}$$

is fulfilled.

This is demonstrated by fig. 36. Varying θ_0 so that k passes the light line $k = \omega/c$ (LL) a deep intensity minimum is observed due to the excitation of

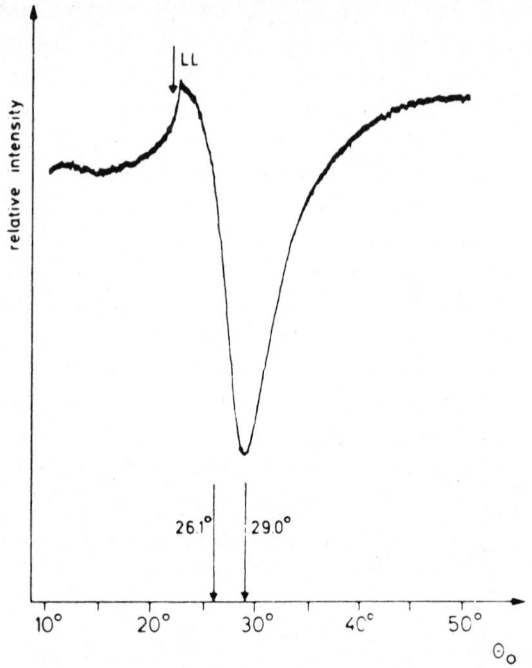

Fig. 36. Light intensity reflected from a sinusoidal grating in zero order as a function of the angle of incidence; p-polarized light. LL light line. The minimum indicates the excitation of surface plasmons. The angle 26,1° calculated for a plane surface is smaller than the observed angle on a sinusoidal grating (Raether 1977, 1980).

SP. This strong change of the reflected intensity is not observed if s-polarized (electric vector perpendicular to the plane of incidence) is used. It shall be noticed that the minimum does appear at a position θ_0 somewhat larger than calculated for a smooth surface. This displacement is due to the geometrical deviation of the surface profile from a plane, as detailed experiments on interference gratings of varying amplitudes have demonstrated. Systematic investigation of the dependence $\omega(k)$ by Pockrand (1974) had a result very similar to that shown on rough surfaces in fig. 4: the position of the resonance minimum (see fig. 36) is displaced to large θ_0 (or k values) and the width increases with larger amplitudes (h) of the interference grating; h varied from 0 to 850 Å, see fig. 37b, c.

The interference gratings were produced by illuminating a thin photoresist film with the interference pattern of a laser beam ($\lambda = 4$–6000 Å). After developing the photoresist with a suitable solvent a grating with a sinusoidal groove profile was obtained which was coated by evaporation with a metal film (Ag, Au, Al, etc.) of ~500 Å thickness. This grating was used as a reflection grating, see fig. 37a.

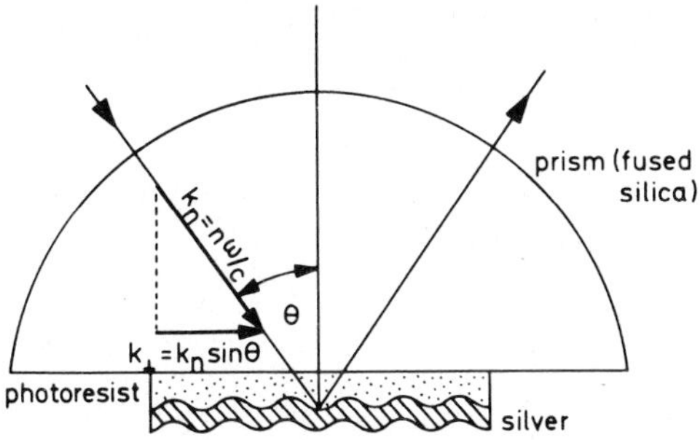

Fig. 37a. Display of the experimental arrangement which differs from the usual ATR set-up in so far as the metal surface has the profile of a sinusoidal grating.

These observations are of interest since they can be interpreted with the theory of Kröger and Kretschmann (1976). Here the Fourier transform of the grating, $S(x) = h \sin 2\pi x/a$, is simple; it consists of two delta functions

$$|s(\Delta k)|^2 = \tfrac{1}{2}\overline{S^2}|\delta(\Delta k - 2\pi/a) + \delta(\Delta k + 2\pi/a)] \tag{55}$$

with $\overline{S^2} = \frac{1}{2}h^2$. This allows one to calculate the displacement of the position and the increase of the half width of the reflection minimum and to compare them with values observed on silver gratings (Pockrand and Raether 1976, 1977). The calculated values θ_0 and $\theta_{1/2}$ dependent on h are in good agreement with the observed data. These experiments have been improved (Rosengart 1978): the gratings were controlled for being sinusoidal, so that the contribution of higher harmonics to the grating profile is reduced (less than 5%, see below) and further the calculations of the position (θ_0) and the width ($\theta_{1/2}$) as function of h were performed with the Rayleigh procedure (the electromagnetic field is represented by an expansion in diffracted orders $S(x) = \nu\Sigma\; S(\nu) \exp i(2\pi/a)\nu x$. It is assumed that this expansion can be continued through the selvedge region $|z| < h$ back to the surface of the grating. Thus the coefficients of the expansion can be determined by the boundary conditions). The agreement is as good as fig. 38 demonstrates for Au and Ag gratings. This shows that the Rayleigh method is a better approximation than the first order approximation.

The just mentioned expansion converges, if $h/a < 0{,}072$, and a one dimensional sinusoidal boundary assumed (Millar 1969, Petit 1966, Hill and Celli 1978). This convergence condition is fulfilled in the experiments with $a = 14\,000$ Å up to $h = 500$ Å ($h/a = 0{,}03$) and nearly with $a = 4417$ Å ($h/a = 0{,}11$). Here the question arises, up to which value of h the first order approximation is valid. The experiments on sinusoidal gratings have shown that the changes of the dispersion relation become measurable with a sufficient accuracy for $h \geqslant 80$ Å, see fig. 37b; thus no statement concerning the validity of the approximation below 80 Å is possible. However such a

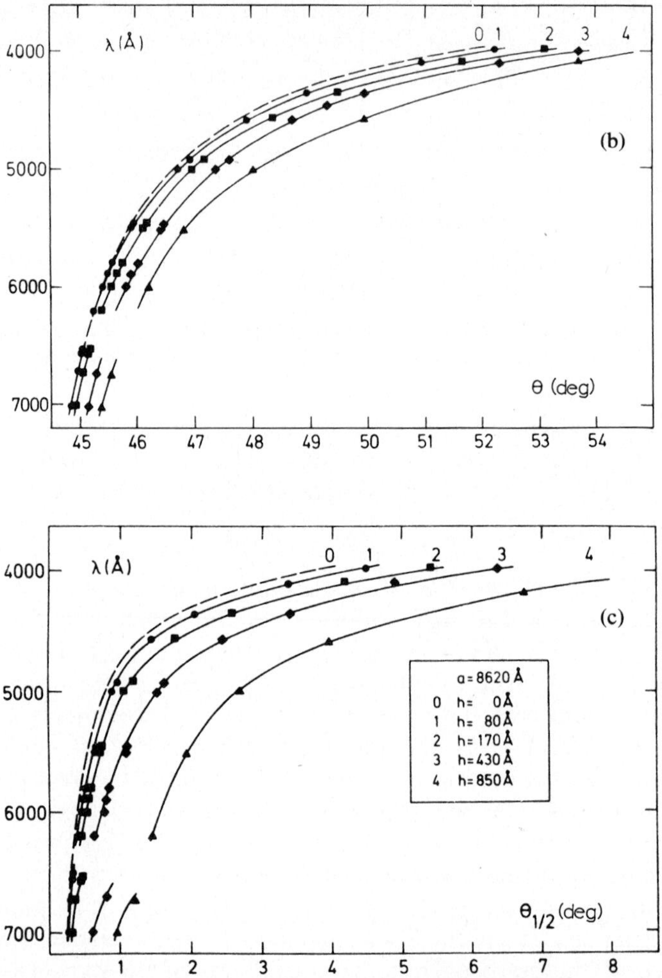

Fig. 37b, c. Demonstration of the dispersion relation measured with the set-up of fig. 37a at gratings of different amplitudes h as parameter, see inset of fig. 37c; (b) shows the displacement of the reflection minimum, (c) its width as function of λ. The grating constant: $a = 8620$ Å, the light wavelength: 6000 Å (Pockrand 1974).

limit can be derived from experiments similar to those displayed in fig. 2, in which the rough surface is replaced by a sinusoidal grating, see fig. 37a. Here the roughness function is well known. If SPs are excited, maximum light is emitted in certain directions: $n(\omega/c)\sin\theta_0 - \nu g = (\omega/c)\sin\theta_\nu$ see eq. (54) if θ_0 fulfills the resonance condition. This intensity which can now be calculated with eq. (55) is compared to the observed one as function of h. The results on a silver grating ($a = 5930$ Å) with different wavelengths

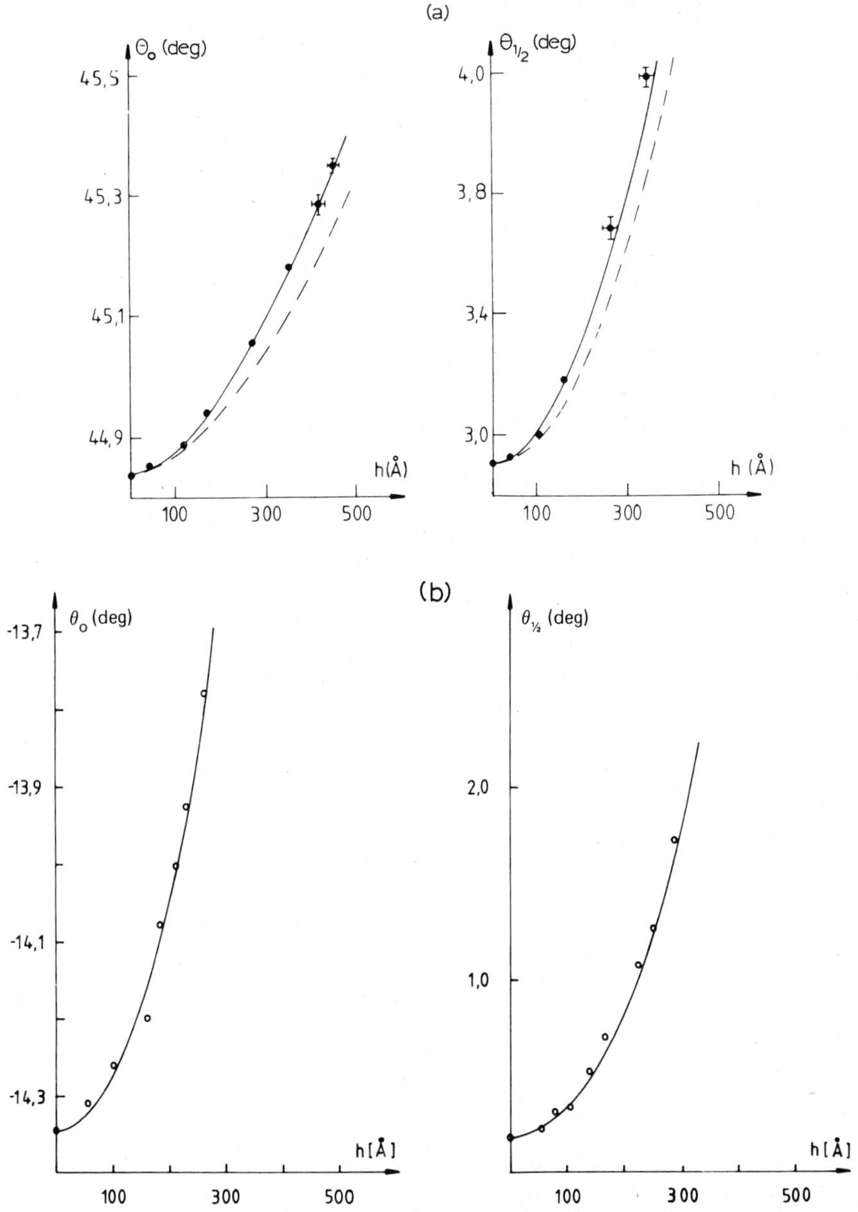

Fig. 38. Displacement of the reflection minimum (θ_0) and its width ($\theta_{1/2}$) as a function of the amplitude h of the grating measured with $\lambda = 5682$ Å. (a): gold, $a = 14\,983$ Å. Dots with error bars: observed values; full line: calculated with a Rayleigh procedure; dashed line: calculated with the approximation eq. (9). (b) silver, $a = 4400$ Å. Dots observed values; full line: calculated with a Rayleigh procedure (Rosengart 1978).

showed that the first order theory holds up to about $h \sim 70$ Å ($\sim 10\%$ deviation) (Heitmann and Raether 1976, Heitmann 1977).

If we compare this limit with the 20–30 Å on rough silver surfaces a difference is apparent. The reason can be that in a sinusoidal profile the values of $S(x)$ never exceed h, whereas a rough surface with a Gaussian roughness function contains naturally values of $S(x)$ larger than $(\overline{\delta^2})^{1/2}$. It is thus understandable that the limit of the above approximation lies higher at a sinusoidal profile than at a rough profile.

Besides the position of the reflection minimum and its width, the height of the reflection minimum is a function of h too; it has a minimum with varying h. The quotient of the intensity in zero order to that of the incoming light: I^0/I_0 can reach zero (see Pockrand 1976, Maystre and Petit 1976, Hutley and Maystre 1976). In fig. 39 the relative intensity $(I^0 + I^{-1})/I_0$ is plotted; it demonstrates how the reflection minimum changes with h similarly to the results on surfaces of different roughness (see fig. 3). Further, fig. 39 shows that the grating with $h \sim 160$ Å absorbs the maximum energy; one can say that the system with $h = 160$ Å fulfills a matching condition as it exists in the prism method for the thickness d of the metal film, see fig. 4 in Kretschmann (1971).

The increase of the width of the reflection minimum approximately proportional to h^2 is correlated with the fact that the intensity, radiated into a diffraction order, increases with h^2. The displacement can be understood with the same arguments as in sect. 2.

Intensity variations in the diffracted beams of a grating were observed a long time ago (Wood 1902). Explanations of these "anomalies" were

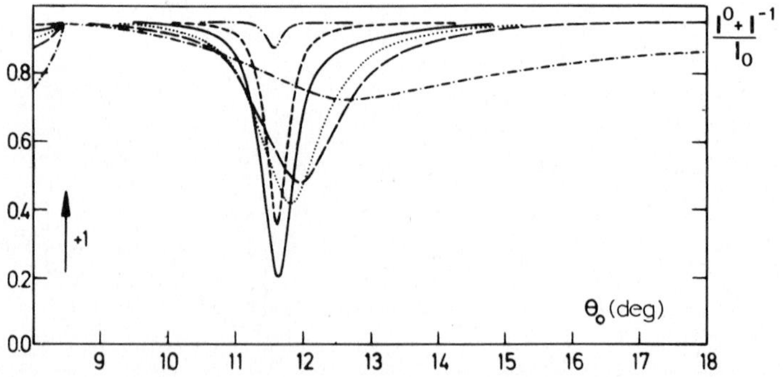

Fig. 39. Reflection of p-polarized light at a thick silver film (~2000 Å) with sinusoidal surface profile as a function of the angle of incidence θ_0 with the amplitude h as parameter. The intensities of the zero and first order are added and normalized to the incoming intensity. The grating constant is 6015 Å and $\lambda = 5145$ Å. h varies between 25 Å and 560 Å: $h = 25$ Å (–···–), $h = 75$ Å (–––), $h = 135$ Å (—), $h = 215$ Å (·····), $h = 315$ Å (––), $h = 550$ Å (–·–·) (Pockrand 1976).

proposed by different authors (Rayleigh 1907, Fano 1941, Twersky 1952, 1962, Hessel and Oliner 1965, Hägglund and Sellberg 1966). A special type of such variations, namely dips in the reflected intensity if p-polarized light is incident perpendicular to the rulings of a commercial grating (Al), has been described by Teng and Stern 1967 and explained as SP effect. Ritchie et al. (1968) observed strong maxima in the reflected light of Al and Au gratings under similar conditions and correlated them with SP excitation too. The maxima do not appear in s-polarized light. Experiments on gratings covered with thin dielectric layers supported this explanation (Cowan and Arakawa, Stanford and Bennett 1969). For further work explaining these anomalies in the context of SPs see Hutley and Bird (1973), McPhedran and Maystre (1974). The reflexion pattern from gratings on semiconductor surfaces (doped In Sb) in the infrared ($\epsilon < -1$) have been investigated theoretically in more detail (Van den Berg and Borburgh 1974).

An explanation using the concept of SPs is identical with solutions of Maxwell's equations. The theoretical work which starts with the Hamiltonian leads to the same results as those obtained with Maxwell's equations, at least in the approximation used to date (Ritchie et al. 1968).

6.2. *Influence of Higher Harmonics*

Earlier measurements on ruled gratings of Al and Au (Ritchie et al. 1968) with a sawtooth-like groove profile showed intensity discontinuities but, as just mentioned, they consist of peaks, and no minima were observed; the interpretation has been left open. Detailed studies of Rothballer (1977) on sinusoidal interference gratings with varying modulation height h had the following result: the zero order reflection ($\nu = 0$) has an intensity minimum if the dispersion relation for SP is fulfilled as described in fig. 34; this has been found to be independent of the amplitude of the grating (shape and depth of the minimum change with h). The higher orders $\nu \geq 1$ behave rather differently: they show a maximum, sometimes a minimum, or they have an oscillator-like shape if the angle θ_0 is scanned. An understanding of these results came from more detailed experiments which studied the shape of the first-order diffraction intensity of a series of gratings which were prepared with constant first harmonic profiles but which contained increasing amplitude of second harmonic components. That is the profile $S(x) = h^{(1)} \cos(2\pi x/a) + h^{(2)} \cos(4\pi x/a)$ is changed by increasing $h^{(2)}$ at constant $h^{(1)}$ (Rosengart and Pockrand 1977).

Experimentally, this varying second-order contribution is obtained by pre-exposing the photoresist so that various curvatures of the characteristic can be reached and the ratio $h^{(2)}/h^{(1)}$ is changed (Pockrand 1974).

The values of $h^{(1)}$ and $h^{(2)}$ were calculated from the intensities of diffracted s-polarized light in the first and second order (I^1/I_0 and I^2/I_0). The

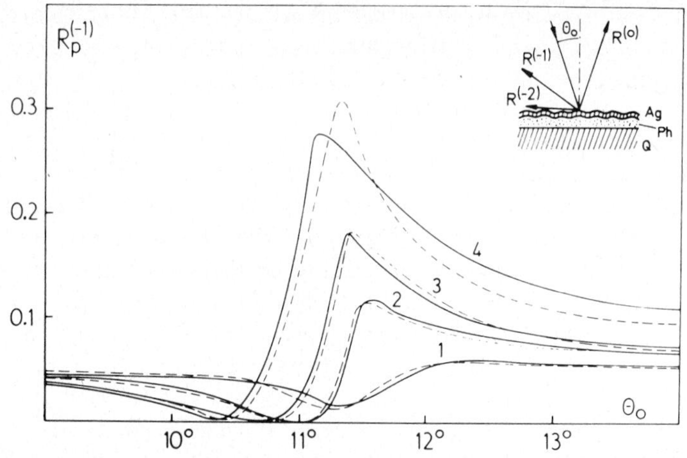

Fig. 40. Relative intensity R_p^{-1} of the diffracted beam ($h = -1$) as a function of θ_0 for a grating $s(x) = h^{(1)}\cos(2\pi x/a) + h^{(2)}\cos(4\pi x/a)$ with a varying $h^{(2)}/h^{(1)}$ and $h^{(1)} = 150$ Å; thickness of the silver film 2000 Å; $\lambda = 5145$ Å; $a = 6000$ Å. Observed values: full lines. Calculated values: dashed lines (1): $h^{(2)}/h^{(1)} = -0.05$; (2) $= -0.11$; (3) $= -0.16$; (4) $= -0.26$. The dielectric function $\epsilon(\omega)$ is taken from silver data for a smooth film (Pockrand and Rosengart 1977).

influence of the increasing value of $h^{(2)}/h^{(1)}$ on the shape of the (-1)st order, $R_p^{(-1)}(\theta_0)$, is shown in fig. 40: at $h^{(2)}/h^{(1)} = 0.05$ the shape of $R_p^{(-1)}(\theta)$ is still that of a well developed minimum which is progressively deformed with increasing $h^{(2)}$, similar to the result mentioned above (Rothballer 1977). The agreement with calculated values using the measured heights $h^{(2)}$ is satisfying. Thus we can say that the minima which are observed in the first order of nearly pure ($h^{(2)}/h^{(1)} < 0.05$) sinusoidal gratings are strongly deformed by the contributions of higher orders, so that maxima are produced. Higher orders are present in gratings of sawtooth-like profile, so that the shape of the diffracted intensity in Ritchie et al. (1968) differs strongly from a minimum as in fig. 36. A calculation of their intensity profile requires knowledge of the amplitudes and phases of the higher orders and a lot of computer time.

If these gratings are covered with thin dielectric films, the dispersion relation is depressed due to depolarization effects (see Pockrand 1976, Raether 1977, 1980 for further literature).

Machined optical surfaces which are periodically corrugated can be controlled by light scattering. They act as a diffraction grating which is superimposed on the natural roughness of the surface. If the periods (Λ) are not too long, diffraction peaks appear on each side of the specular beam; their angular intensity distribution contains information on the details of the corrugation (Church and Zavada 1975).

6.3. *Coupling of Surface Plasmons and of Guided Light Modes on Sinusoidal Surfaces*

In the case of SPs propagating on a periodically perturbed surface their dispersion relation is written:

$$K = (\omega/c)\sqrt{\epsilon/(\epsilon+1)} + \nu g, \tag{56}$$

ν whole numbers. The k-space is thus covered with ν dispersion curves of distance g which cross each other. Figure 41 demonstrates how the interaction of two plasmons k and $k-g$ through scattering at the amplitude of the grating leads to a splitting of the dispersion curves δk at the points of intersection. The width of the gap (δk) is proportional to the Fourier component h_g of the profile. This splitting has been observed on a commercial ruled grating and interpreted as just mentioned (see Ritchie et al. 1968, Wheeler et al. 1976). Detailed measurements on sinusoidal interference gratings (silver, $h = 185$ Å, $a = 6080$ Å, $\lambda = 6328$ Å) of the position of k, δk and of the value of the half width ($k_{1/2}$) are displayed in fig. 42 (compare also with fig. 41). The increase of the half width ($k_{1/2}$) and the change of the k-gap with stronger coupling (see also table 5) are demonstrated. The development of the intensity of the plasmons is seen in fig. 43 and displays how the intensity of the strong branch (originally excited via the prism) decreases, yielding energy into the other branch, and comes back to its original intensity after decoupling.

These data are not obtained in the usual reflection arrangement, but in

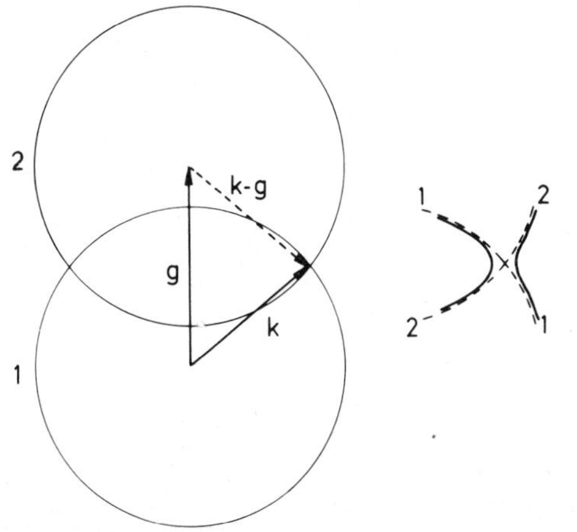

Fig. 41. The two circles represent the dispersion relations of plasmons propagating on a sinusoidal grating ($g = 2\pi/a$ the reciprocal wavevector). The coupling of the plasmons k and $k-g$ increases with the amplitude h of the grating; as a consequence an increasing gap is produced.

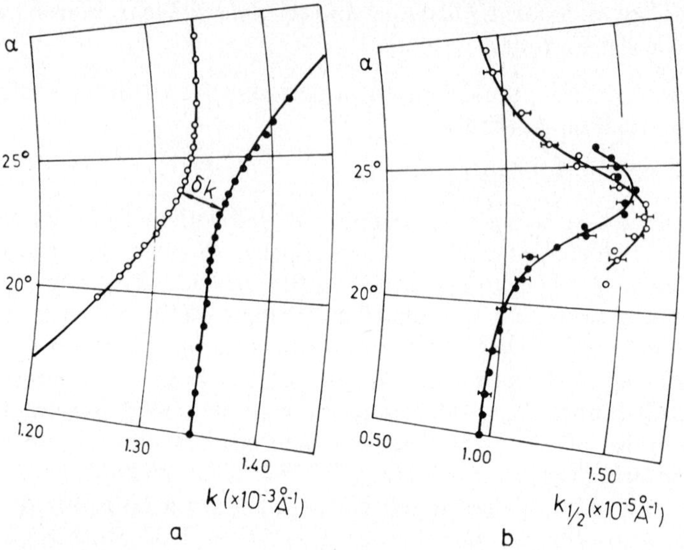

Fig. 42. Observed coupling of two plasmons (1) and (2) via the reciprocal lattice vector g, see fig. 41, characterized by the splitting δk and the half width $k_{1/2}$ as a function of the angle α. $\lambda = 6328$ Å, $a = 6080$ Å and $h = 185$ Å (Pockrand 1978).

Table 5
Coupling strength measured by δk as function of the amplitude h (Pockrand 1979).

h(Å)	$\delta k(\times 10^{-4}$ Å$^{-1})$
200	0.4
400	0.8
600	1.1
800	1.4

the scattering mode of the prism method seen schematically in fig. 2 and fig. 11, so that the SP resonance is measured by the intensity peak of the emitted light instead of a reflection minimum. Furthermore the continuous change of the direction of the g vector has been realized by turning the direction of the rulings around the normal of the grating by an angle α (at $\alpha = 90°$ the plane of incidence is perpendicular to the grooves).

The interaction of SPs with periodic structures as gratings has been calculated by Mills (1977). The gap in the dispersion relation is given by

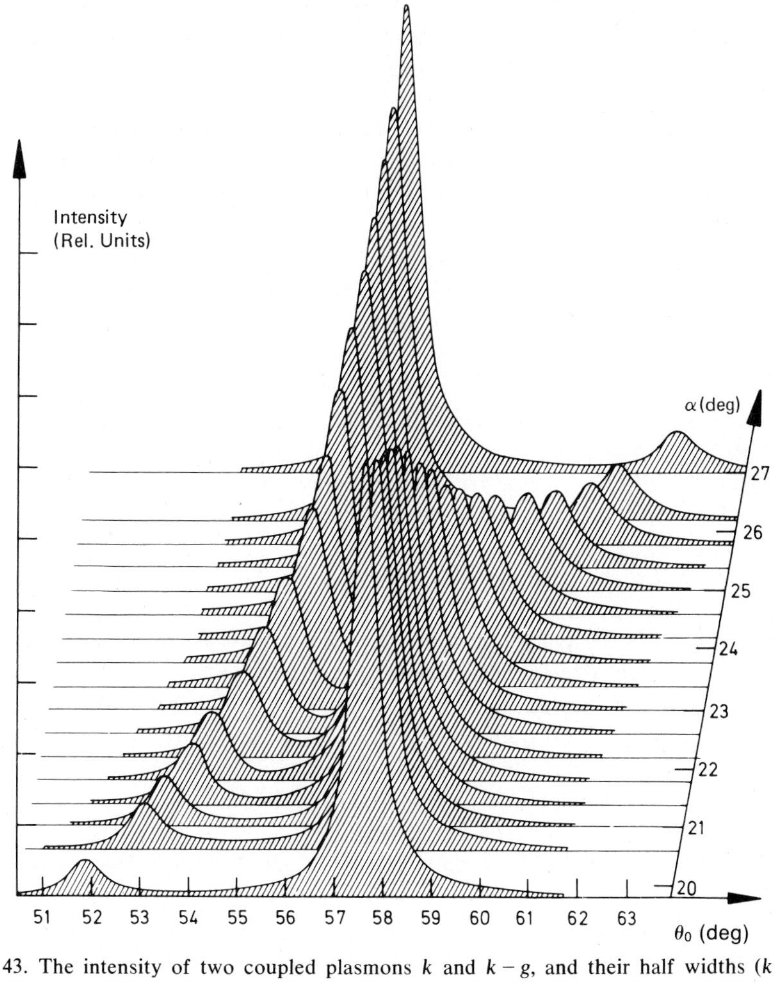

Fig. 43. The intensity of two coupled plasmons k and $k-g$, and their half widths ($k_{1/2}$) as function of the angle α near the gap. θ_0 is the angle of incidence (Pockrand 1978).

$$\delta\omega/\omega \sim \delta k/k = (4k/\sqrt{\epsilon})h \sin^2 \tfrac{1}{2}\phi \tag{57}$$

with $\phi/2 = 90° - \alpha$, $(90 - \alpha)$ the angle between $\boldsymbol{k}$ and $\boldsymbol{g}$. The observed value of $\delta k/k$ taken from fig. 42 is about 0.03, whereas eq. (57) gives with $\lambda = 6300$ Å ($\epsilon' = -18$) and $\alpha = 25°$, a value of 1.8×10^{-1}, not in good agreement.

Coupling of SPs on the wavy metal/air boundary with those on the (plane) metal/photoresist boundary through the silver film has been observed too, if the metal film is not too thick (Pockrand 1975).

The appearance of frequency splitting as observed at gratings has been reported on statistically rough surfaces too: If the intensity of light reflected at a rough surface is measured, two minima, separated by about 0.1 eV, are observed on silver surfaces near the SP frequency ω_s for $k \to \infty$. This has been found in electroreflection experiments performed without a prism coupler (Kötz et al. 1979) and in experiments on radiation emission excited by low energy electrons (Kretschmann et al. 1980, Chung et al. 1980). This splitting ($\sim$0.2 eV) has been observed in normal experiments on Na and K too (Palmer and Schnatterly 1972).

For comparison with theory it has been assumed (Kretschmann et al. 1979, 1980) that the roughness function is peaked around a value Δk_R at large Δk values as has been discussed in sect. 2.2. The $|s(\Delta k)|^2$ function is then approximated by a delta function $\delta(\Delta k - \Delta k_R)$, or several such functions with neighbouring values of Δk_R, so that the integral eq. (11) can be calculated. A splitting of the frequency of about 0.2 eV with values of $(\overline{S^2})^{1/2}/\Lambda_R \sim 1/20(\Delta k_R = 2\pi/\Lambda_R)$ results which gives with $\Lambda_R = 400$ Å a value of $(\overline{S^2})^{1/2} = 20$ Å which is a reasonable number.

Such a peaking of the $|s|^2$ function, see also the remark in Dobberstein (1970), or preference of distances Λ_R and heights $S(x, y)$ may be caused by a preferred average crystal size in the surface.

This frequency splitting produces two branches of the dispersion relation, branches which are nearly horizontal at these large Δk values; their length depends on the number of Δk_R values involved. Thus the frequency gap remains rather well defined and can be separated in the experiments.

The splitting and the shape of the dispersion relation over a larger k range can be made visible at once by using instead of a nearly parallel primary light beam a focussed beam with the focus on the metal film. Since the directions which fulfill the dispersion relation are directions of reduced reflected intensity –it is zero in the optimum case– the light cone is crossed by black lines, demonstrating the "band gaps" (Kretschmann 1978).

If a metal surface is covered with a dielectric film, e.g., LiF, sufficiently thick, so that standing waves in the direction of the film normal are possible, guided light modes can propagate in the dielectric film. These inhomogeneous waves, the SP or TM_0 mode as well as the other TM and TE modes are excited by the prism method (Hornauer and Raether 1973). If the metal/air boundary is a wavy surface, coupling of these modes with each other through the vector $\mathbf{g}$ or multiples of it is observed, e.g., SPs with different TM and TE modes and between each other (inversion of modes) (Pockrand and Raether (1976)). The influence of roughness on light modes is smaller than on SPs ($m_p = 0$) so that the linear theory holds up to values of $\delta \sim 40$ Å. (Hornauer and Raether in press). This is due to the different field distributions: the guided modes have their maximum field in the interior of the film whereas the maximum of SP is at the boundary metal/air. For further references see Galantowicz (1974), Kaminow et al. (1974).

Another type of experiment is performed with sinusoidal dielectric surfaces as displayed schematically by fig. 44. Here the SPs are excited by the prism method on the (smooth) surface of a silver film (500 Å thick) which is covered with a resist film of 1200 Å thickness. The resist film can be modulated up to a depth of 575 Å with the grating constant $a = 6017$ Å. The SP resonance for silver and no resist film lies at $\theta_0 = 34°$ (see fig. 45),

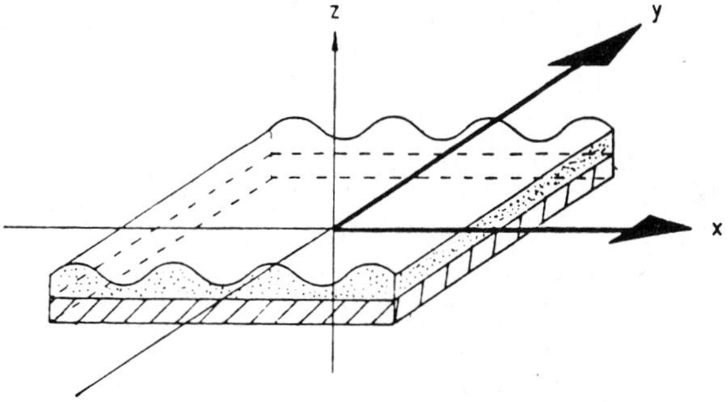

Fig. 44. Scheme of the arrangement of the dielectric grating (resist) on a plane metallic film (silver) (Rosengart 1978).

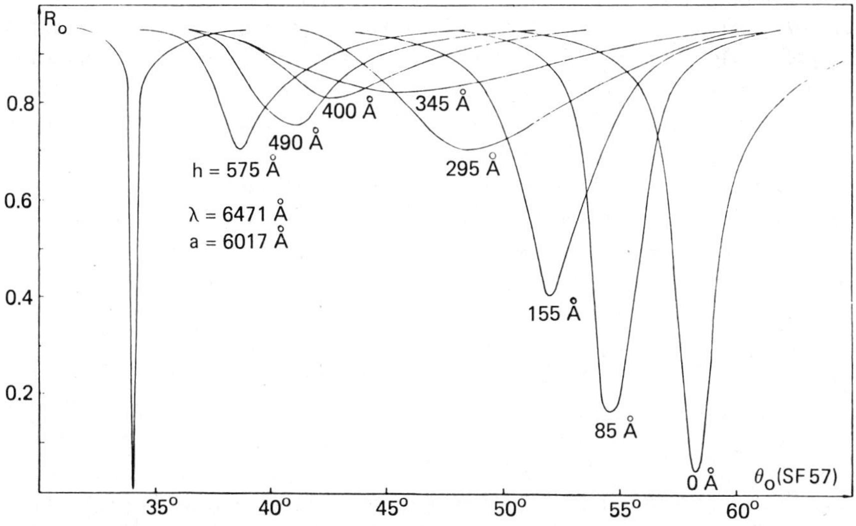

Fig. 45. Reflection intensity observed in the experimental arrangement of fig. 44. The reflection minima displace with increasing modulation depth h. $\lambda = 6471$ Å, $a = 6017$ Å (Rosengart 1978).

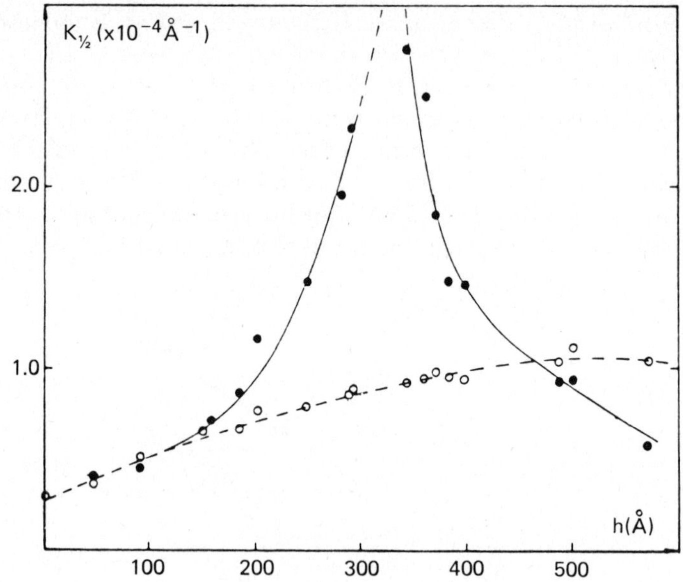

Fig. 46. The half width ($k_{1/2}$) of the reflection minima of fig. 45 as function of h in the k_x and k_y direction (Rosengart 1978).

whereas a smooth resist film of 1200 Å thickness displaces the minimum to 58,1° and increases damping. In case that the SPs propagate parallel to k_x, perpendicular to the grating, fig. 45 is obtained. With increasing modulation the resonance minimum is continuously displaced to lower θ_0 values and approaches θ_0 for silver; parallel to this change the half width goes through a maximum as is indicated in fig. 46. The height of the reflection minimum R increases with modulation indicating mismatching and passes a maximum at $2h \sim 390$ Å; in k_y direction $R_{\min}$ increases continuously. Both curves separate each other at $2h \sim 200$ Å, the same value at which different behaviour of the dependencies in k_x and k_y direction in fig. 46 become visible.

If the grating is turned around the normal of the grating, the position of the minima lie on the dispersion circle (see fig. 34). Due to the spatial periodicity the same situation as in fig. 41 can arise described by the condition $k \cos \alpha = 2\pi/a$; the coupling via the grating ($g = 2\pi/a$) leads to a splitting of the minima; the strength of coupling is given by eq. (57). Similar to the case of a metal a linear increase of the splitting δk with $2h$ has been found up to $2h \sim 400$ Å. Its absolute value is about a factor 8 smaller than that at a metal grating of the same modulation. This is apparently due to the different values of ϵ (Ag) (= 18) and ϵ (resist) (2,2). A theoretical treatment has not been published.

7. *Conclusion*

Roughness is in general an effect regarded as a small correction. In the case of SPs however this is different since the electromagnetic field is strongly concentrated at the boundary. The result is that the dispersion relation $\omega(k)$ is changed already at a relatively small value of the rms roughness height δ: the eigenfrequency is changed, and the damping is increased. The latter is due to light emission produced by coupling of SPs with light via roughness.

To calculate this change of the dispersion relation the knowledge of the roughness function $|s(\Delta k)|^2$ at large Δk values is needed, but not known; this explains the discrepancies between observed and calculated data, obtained by extrapolation of the Gaussian correlation function to large Δk.

Similar change of the dispersion relation of SPs on sinusoidal gratings with its simple $|s(\Delta k)|^2$ function has been observed and explained using the same theoretical concept. The quantitative agreement of calculated and observed values shows that the basic theory is right. The clearer situation in the case of a sinusoidal surface profile allows to observe in more detail the coupling of the SP with the wavevector k and the scattered SP: $k+g$ via the grating and the splitting due to this interaction. Similar splitting has been observed on rough surfaces at wavelengths λ_s near the SP with $k \to \infty$.

The light emitted from SPs on rough surfaces together with its angular distribution using the theoretical results well established in the linear approximation allows to determine the $|s(\Delta k)|^2$ function and shows that the Gauss function is a good approximation in the region of visible light up to Δk values of $\sim 2.5 \times 10^{-3}$ Å^{-1}. Values of ~ 5 Å are measured on evaporated silver films of "natural" smoothness; these measurements were extended to surfaces roughened by underlying material as CaF_2, MgF_2, Ag etc. Increasing values of δ with increasing thickness of the underlying material are measured.

These experiments with enhanced roughness indicate that beyond δ: 20–30 Å the validity of the linear approximation ceases. For lower values of $\delta \sim 5$ Å the light scattering method is also sensitive enough.

Light scattering by roughness without excitation of SPs although less intense can be used to determine the roughness parameters at the same sample used for the measurements with SPs. The results of both experiments agree also the limits of the linear approximation.

It is promising that this light scattering method can be extended to the scattering of X-rays which yields interesting results.

A further step to get an insight into light scattering by roughness has been done by investigating the effect of exterior (surface) roughness and interior roughness as due, e.g., to inhomogenieties of the dielectric constant of the volume of the solid and its different behaviour. It could be verified that the scattering of light takes place at the surface if "nonradiative" SPs are used.

As long as light scattering data cannot be evaluated for $\delta \gtrsim 30$ Å, the electron microscope with its high spatial resolution can be applied using shadowed replicas to determine the roughness spectrum $|s(\Delta k)|^2$ for $\delta \gtrsim 50$ Å.

Instead of exciting SPs by the prism (or ATR) method, light can produce SPs via surface roughness. This is observed, e.g., as a deficit in the specularly reflected light. The roughness parameters are derived from the shape of the deficit measured as function of the wavelength of light assuming a Gaussian function for $|s(\Delta k)|^2$. The sensitivity is in general less than that of the other light scattering methods.

Measurements of roughness have also been performed by integrating the light scattered from the rough surface. The validity of the "scalar theory" (equal behaviour of p- and s-polarised light) is assumed –a rather rough assumption– which simplifies the calculation drastically. The roughness values have thus a more qualitative character.

This review of the experimental and theoretical situation of the roughness problem demonstrates the state of the art. It shows that reliable light scattering methods are available for small roughness. The SPs are important for enhancing the emitted light intensity, so that rather accurate measurements can be made. However it is highly desirable to collect more experimental material to clear the open questions.

Independent on this application the interaction of SPs and photons via roughness which creates radiative damping of the SP, represents an interesting problem in itself. Especially the behaviour of SPs on sinusoidal gratings is an instructive example of the physics of electromagnetic waves –here plasmons– on periodically perturbed surfaces.

I wish to thank for more detailed information, Professors H. Bräuninger (Garching), R. Bray (Purdue), D.G. Hall (Rochester), E. Kretschmann (Hannover), D.L. Mills (Irvine), R. Loudon (Essex), M. and G. Rasigni (Marseille) S.O. Sari (Oregon), S. Ushioda (Irvine). Further I would like to thank Professor Davis Lynch for his reading of the manuscript.

References

Adams, A., J.C. Wyss and P.K. Hansma, 1979, Phys. Rev. Lett., **42**, 912 (there earlier literature).

Arakawa, E.T., M.W. Williams, R.N. Hamm and R.H. Ritchie, 1973, Phys. Rev. Lett. **31**, 1127.

Barrick, D.E., 1970, in: Radar Cross Section Handbook, ed. G. Ruck, Vol. 2, (Plenum Press) p. 706.

Beaglehole, D. and O. Hunderi, 1970, Phys. Rev. **B2**, 309.

Beckmann, P. and A. Spizzichino, 1963, The Scattering of Electromagnetic Waves from Rough Surfaces (Pergamon Press).

Bedeaux, D. and J. Vlieger, 1973, Physica **67**, 55.
Bedeaux, D. and J. Vlieger, 1976, Physica **85A**, 389.
Bennett, H.E. and J.O. Porteus, 1961, JOSA **51**, 123.
Berreman, D.W., 1967, Phys. Rev. **163**, 855.
Bodesheim, J. and A. Otto, 1974, Surf. Sci. **45**, 441.
Brambring, I. and H. Raether, 1965, Phys. Rev. Lett. **15**, 882.
Braundmeier, Jr. A.J. and H.E. Tomaschke, 1975, Opt. Commun. **14**, 99.
Braundmeier, Jr. A.J. and E.T. Arakawa, 1974, J. Phys. Chem. Solid **35**, 517.
Braundmeier, Jr. A.J. and D.G. Hall, 1975, Surf. Sci. **49**, 376.
Bruns, R. and H. Raether, 1970, Z. Phys. **237**, 98.
Bush, I.A., D.K. Cohen, K.D. Scherkoske and S.O. Sari, 1980, JOSA **70**, 1020.
Chung, M.S., T.A. Callcott, E. Kretschmann and E.T. Arakawa, 1980, Surf. Sci. **91**, 245.
Church, E.L. and I.M. Zavada, 1975, Appl. Opt. **14**, 1788.
Cowan, J.J. and E.T. Arakawa, 1970, Z. Phys. **235**, 97.
Crowell, J. and R.H. Ritchie, 1970, JOSA **60**, 794.
Daudé, A., A. Savary and S. Robin, 1972, JOSA **62**, 1.
Davies, H., 1954, Proc. Inst. Ed. Engrs. (London) **101**, 209.
de Korte, P.A.I. and R. Lainé, 1979, Appl. Opt. **18**, 236.
Dobberstein, P., 1970, Phys. Lett. **31A**, 307.
Dobberstein, A., A. Hampe and G. Sauerbrey, 1968, Phys. Lett. **27A**, 256.
Elson, I.M., 1975, Phys. Rev. **B12**, 2541.
Elson, I.M., 1976, JOSA **66**, 682.
Elson, I.M. and R.H. Ritchie, 1971, Phys. Rev. **B4**, 4129.
Elson, I.M. and R.H. Ritchie, 1974, Phys. Status Solidi (b)**62**, 461.
Elson, J.M., H.E. Bennett and J.M. Bennett, 1979a, Appl. Optics and Optical Engineering, Vol. 7, 191 (Academic Press, 1979). In this article those light scattering experiments are not considered which regard the vectorial character of light. The results have thus an approximate character.
Elson, J.M. and J.M. Bennett, 1979b, Opt. Eng. **18**, 116.
Endriz, I.G. and W.G. Spicer, 1971, Phys. Rev. **4**, 4144, 4159.
Evans, D.I., S. Ushioda and I.D. McMullen, 1973, Phys. Rev. Lett. **31**, 369.
Fano, U., 1941, JOSA **31**, 213.
Feuerbacher, B.P. and W. Steinmann, 1969, Opt. Comm. **1**, 81.
Galantowicz, 1974, Appl. Opt. **13**, 2525.
Gesell, T.F., E.T. Arakawa, M.W. Williams and R.N. Hamm, 1973, Phys. Rev. **B7**, 5141.
Hägglund, I. and F. Sellborg, 1966, JOSA **56**, 1031.
Hall, D.G. and A.J. Braundmeier, Jr., 1973, Opt. Comm. **7**, 343.
Hall, D.G. and A.I. Braundmeier, Jr., 1978, Phys. Rev. **B17**, 3808.
Hall, D.G. and A.T. Braundmeier, 1978, Phys. Rev. **B17**, 1557.
Heitmann, D., 1977, Opt. Comm. **20**, 292.
Heitmann, D. and V. Permien, 1977, Opt. Comm. **23**, 131.
Heitmann, D. and H. Raether, 1976, Surf. Sci. **59**, 1.
Hessel, A. and A.A. Oliner, 1965, Appl. Opt. **4**, 1275.
Hill, N.R. and V. Celli, 1978, Phys. Rev. **B17**, 2478.
Hillebrecht, U., 1979, unpublished (Hamburg).
Hillebrecht, U., 1980, J. Phys., D., Appl. Phys. **13**, 1625.
Hornauer, D.L., 1976, Opt. Comm. **16**, 76.
Hornauer, D., 1977, unpublished (Hamburg).
Hornauer, D., H. Kapitza and H. Raether, 1974, J. Phys. **D1**, **L**, 100.
Hornauer, D. and H. Raether, 1973, Opt. Comm. **7**, 297.
Horstmann, C., 1977, Opt. Comm. **21**, 176.

Horstmann, C., 1976, unpublished (Hamburg).
Hunderi, O. and D. Beaglehole, 1969, Opt. Comm. **1**, 101.
Hunderi, O. and D. Beaglehole, 1970, Phys. Rev. **B2**, 321 (theory).
Hunter, W.R., 1964, JOSA **54**, 208.
Hutley, M.C. and V.M. Bird, 1973, Opt. Acta **20**, 771.
Hutley, M.C. and D. Maystre, 1976, Opt. Comm. **19**, 431.
Jansen, F. and R.W. Hoffman, 1979, Surf. Sci. **83**, 313; 1980, J. Vac. Sci. Technol. **17**, (4), 842.
Jasperson, S.N., and S.E. Schnatterly, 1969, Phys. Rev. **188**, 759.
Juranek, H.J., 1970, Zs. Phys. **233**, 324.
Kaminow, I.P., W.L. Mammel and H.P. Weber, 1974, Appl. Opt. **13**, 396.
Kapitza, H., 1976, Opt. Comm. **16**, 73.
Kaspar, W. and U. Kreibig, 1977, Surf. Sci. **69**, 619.
Kötz, R., H.I. Lewerenz and E. Kretschmann, 1979, Phys. Lett. **70A**, 452.
Kovener, G.S., R.W. Alexander Jr. and R.I. Bell, 1976, Phys. Rev. **B14**, 1458.
Kretschmann, E., 1971, Zs. Phys. **241**, 313.
Kretschmann, E., 1972a, Opt. Comm. **5**, 331.
Kretschmann, E., 1972b, Opt. Comm. **6**, 185.
Kretschmann, E., 1974, Opt. Comm. **10**, 356. Proc. Int. Conf. on "Thin Films" Venice (1972).
Kretschmann, E., 1978, Opt. Comm. **26**, 41.
Kretschmann, E., T.A. Callcott and E.T. Arakawa, 1980, Surf. Sci. **91**, 237.
Kretschmann, E., T.L. Ferrell and I.C. Ashley, 1979, Phys. Rev. Lett. **42**, 1312.
Kretschmann, E. and E. Kröger, 1975, JOSA 65, **150**.
Kretschmann, E. and H. Raether, 1967, Zs. Naturf. **22a**, 1623.
Kretschmann, E. and H. Raether, 1968, Zs. Naturf. **A23**, 615.
Kröger, E. and H. Raether, 1971, Z. Phys. **224**, 1.
Kröger, E. and E. Kretschmann, 1970, Zs. Phys. **237**, 1.
Kröger, E. and E. Kretschmann, 1976, phys. stat. sol. (b) **76**, 515.
Laks, B. and D.L. Mills, 1979, Phys. Rev. **B20**, 4962.
Lenzen, R., 1978, Thesis, Tübingen.
Lichte, H., 1977, Thesis, Tübingen.
Maradudin, A.A. and D.L. Mills, 1975, Phys. Rev. **B11**, 1392.
Maradudin, A.A. and W. Zierau, 1976, Phys. Rev. **B14**, 484.
Marvin, A., F. Toigo and V. Celli, 1975, Phys. Rev. 11, 2777. Here a real ϵ is assumed insofar the relations are not as general as eqs. 38 and 37.
Maystre, D. and R. Petit, 1976, Opt. Comm. **17**, 196.
McCarthy, S.L. and J. Lambe, 1978, Appl. Phys. Lett. **33**, 858. (there earlier literature).
McPhedran, R.C. and D. Maystre, 1974, Opt. Acta. **21**, 413.
Millar, R.F., 1969, Proc. Camb. Philos. Sci. **65**, 773.
Mills, D.L., 1975, Phys. Rev. **B12**, 4036.
Mills, D.L., 1976, Phys. Rev. **B14**, 5539.
Mills, D.L., 1977, Phys. Rev. **B15**, 3097.
Mills, D.L. and A.A. Maradudin, 1975, Phys. Rev. **B12**, 2943.
Mishra, S. and R. Bray, 1977, Phys. Rev. Lett. **39**, 222; 1979, Solid State Commun. **32**, 621.
Orlowsky, R. and H. Raether, 1976, Surf. Sci. **54**, 303.
Orlowsky, R., P. Urner and D. Hornauer, 1979, Surf. Sci. **82**, 69.
Palmer, R.E. and S.E. Schnatterly, 1972, Phys. Rev. **B4**, 2329.
Petit, R. and M. Cadilhac, 1966, C.R. Acad. Sci. **B262**, 468.
Pockrand, I., 1974, Phys. Lett. **49A**, 259.
Pockrand, I., 1975, Opt. Comm. **13**, 311.
Pockrand, I., 1976, J. Phys. D, **9**, 2423.
Pockrand, I., 1978, Thesis, Hamburg.

Pockrand, I. and H. Raether, 1976a, Opt. Comm. **17**, 353.
Pockrand, I. and H. Raether, 1976b, Opt. Comm. **18**, 395.
Pockrand, I. and H. Raether, 1977, Appl. Opt. **16**, 1784; 2803.
Pokrowsky, P. and H. Raether, 1979, Surf. Sci. **83**, 423.
Pschalek, N., 1980, unpublished (Hamburg).
Raether, H., 1977a, Nuovo Cim. 39 B.N. **2**, 817.
Raether, H., 1977, Surface Plasma Oscillations and Their Applications, Physics of Thin Films **9**, 145 (Academic Press). Here the properties of SP are briefly described and in Chap. 10 of the book of the author: Excitation of Plasmons and Interband Transitions by Electrons, Springer Tracts, Vol. 88 (1980).
Rasigni, M., G. Rasigni, Ch. T. Hua and J. Pons, 1977, Opt. Lett. **1**, 126.
Rasigni, M., G. Rasigni, J.P. Palmari and A. Llebaria, Phys. Rev. (in press). I have to thank Dr. Rasigni for the data before publication.
Rayleigh, Lord, 1907, Phil. Mag. **4**, 60; 1907, Proc. Roy. Soc. **A79**, 399.
Rendell, R.W., D.J. Scalpino and B. Mühlschlegel, 1978, Phys. Rev. Lett. **41**, 1746.
Ritchie, R.H., E.T. Arakawa, J.J. Cowan and R.N. Hamm, 1968, Phys. Rev. Lett. **21**, 1530.
Rosengart, E.H., 1978, unpublished (Hamburg.).
Rosengart, E.H. and I. Pockrand, 1977, Opt. Lett. **1**, 149.
Rothballer, W., 1977, Opt. Comm. **29**, 429.
Sari, S.O., D.K. Cohen and K.D. Scherkoske, 1980, Phys. Rev. **B21**, 2162.
Sauerbrey, G., E. Wöckel and P. Dobberstein, 1973a, Phys. Status Solidi (b) **60**, 665.
Sauerbrey, G., E. Wöckel and P. Dobberstein, 1973b, Phys. Status Solidi (b) **60**, 845.
Schoenwald, J., E. Burstein and J. Elson, 1973, Solid State Commun. **12**, 185.
Schreiber, P., 1968, Z. Phys. **211**, 257.
Schröder, E., 1969a, Opt. Comm. **1**, 13.
Schröder, E., 1969b, Z. Phys. **222**, 33.
Schröder, E., 1969c, Z. Phys. **225**, 26.
Schröder, U., 1981, Surf. Sci. **102**, 118.
Simon, H.J. and J.K. Guha, 1976, Opt. Comm. **18**, 391.
Stanford, I.L. and H.E. Bennett, 1969, Appl. Opt. **8**, 2556.
Stern, E.A., 1967, Phys. Rev. Lett. **19**, 1321.
Teng, Y.Y. and E.A. Stern, 1967, Phys. Rev. Lett. **19**, 511.
Toigo, F.D., A. Marvin, V. Celli and N.R. Hill, 1977, Phys. Rev. **B15**, 5618.
Trümper, J., B. Aschenbach and H. Bräuninger, 1979, SPIE 184, Space Optics, 12.
Twersky, V., 1952, J. Appl. Phys. **23**, 1099; 1962, JOSA **52**, 145.
Twietmeier, H., 1975, unpublished (Hamburg).
Urner, P., 1976, unpublished (Hamburg).
Ushioda, S., 1981, Progress in Optics, Vol. 19 (North-Holland, Amsterdam).
Ushioda, S., A. Aziza, I.B. Valdez and G. Mattei, 1979, Phys. Rev. **B19**, 4012.
Van den Berg, P.M. and J.C.M. Borburgh, 1974, Appl. Phys. **3**, 55.
Waehling, G., D. Möbius and H. Raether, 1978, Z. Naturf. **A33**, 907.
Waehling, R., 1979, unpublished (Hamburg).
Welford, W.T., 1977, Optic. and Quant. Electronics **9**, 269.
Wheeler, C.E., E.T. Arakawa and R.H. Ritchie, 1976, Phys. Rev. **B13**, 2372.
Wilems, R.E. and R.H. Ritchie, 1965, Phys. Rev. Lett. **15**, 882.
Williams, M.W., E.T. Arakawa and L.C. Emerson, 1967, Surf. Sci. **6**, 127.
Wood, R.W., 1902, Phil. Mag. **4**, 396; 1945, Phys. Rev. **48**, 928.

CHAPTER 10

Interaction of Surface Polaritons and Plasmons with Surface Roughness*

ALEXEI A. MARADUDIN

Department of Physics
University of California
Irvine, CA 92717
U.S.A.

*This research was supported in part by U.S. Army Research Grant #DAAG-29-78-G0108.

Surface Polaritons
Edited by
V.M. Agranovich and D.L. Mills

Contents

1. Introduction 407
2. Underlying concepts and results 410
   2.1. Surface polaritons on a flat surface 411
   2.2. Characterization of a rough surface 417
      2.2.1. A deterministic, periodic surface profile 418
      2.2.2. A statistically rough surface profile 419
3. Interaction of surface polaritons and plasmons with a grating 423
   3.1. Surface polaritons on a grating 424
      3.1.1. Determination of the dispersion relation for surface polaritons on a grating on the basis of Rayleigh's hypothesis 425
      3.1.2. Derivation of exact dispersion relations for surface polaritons on a grating 435
   3.2. Propagation of a surface plasmon over a grating 448
      3.2.1. The Rayleigh hypothesis 448
      3.2.2. Green's theorem 454
   3.3. Interaction of a surface polariton with a Rayleigh wave 459
4. Propagation of a surface polariton over a randomly rough surface 470
   4.1. Dispersion relation for a surface polariton on a randomly rough surface . . . 472
   4.2. Attenuation of surface polaritons by surface roughness 486
   4.3. Interaction of surface plasmons with a rough surface 495
5. Directions for future work 503
References 508

1. Introduction

In recent years there has been considerable interest on the part of theorists and experimentalists alike in the study of surface polaritons (see for example, the reviews by Mills and Burstein 1974; Burstein et al. 1974; Kliewer and Fuchs 1974; Bryksin et al. 1974; Otto 1974, 1976; Bell et al. 1975; Nkoma et al. 1976; Maradudin et al. 1980; and Ushioda 1980). These are electromagnetic waves that propagate along the surface of a dielectric medium and whose amplitudes decay exponentially with increasing distance from the surface into the dielectric and into the vacuum in contact with it. Many of these studies have focused primarily on surface polaritons with frequencies in the infrared, where a variety of resonances in the dielectric, or magnetic, response of the substrate can lead to the satisfaction of the conditions required for surface polariton propagation.

In particular, attention has been given to methods of generating surface polaritons, and to determining their dispersion relation. Their attenuation has been studied experimentally and theoretically, and recently nonlinear interactions of surface polaritons have been studied theoretically and experimentally.

In the great majority of the theoretical studies of these properties of surface polaritons it has been assumed that the surface is perfectly flat, even on an atomic scale. However, even the most carefully prepared surfaces possess some measure of roughness. The only exceptions to this general statement may be those surfaces that are prepared by cleaving a crystal along a favorable plane. It thus becomes a matter of some interest to determine the degree to which surface roughness affects the propagation characteristics of surface polaritons. It will turn out that the interaction of surface polaritons with surface roughness can have consequences of several kinds. The dispersion relation can be altered in significant ways: the surface polariton can be attenuated as it propagates along a rough surface; its frequency can be shifted, and new branches can appear in the dispersion curve that are absent for propagation on a flat surface. The electromagnetic field of a surface polariton, while remaining largely localized to the vicinity of the surface, can acquire a radiative component in the presence of surface roughness.

From a comparison of theoretical and experimental results for the propagation characteristics of a surface polariton on a rough surface it should be possible to obtain useful information characterizing the nature of the surface, its ideality, or its departures therefrom.

It is of interest to point out that after the pioneering work of Cohn (1900), Uller (1903), Weyl (1919), and Zenneck (1909), who discovered surface polaritons theoretically, a good deal of attention was paid them in theoretical studies by Sommerfeld (1909, 1911, 1926), Weyl (1919), and Zenneck (1909), of the propagation of radio waves around the surface of the earth, a rough surface if ever there was one, although in their work the effects of surface roughness were ignored. In contrast with the macroscopic scale that characterizes the roughness of the earth's surface, which can be measured in units of meters to kilometers, it is surface roughness on a microscopic scale, where lengths are measured in tens to hundreds of angstroms, that is proving to be of interest to physicists.

The surface roughness with which surface polaritons interact is generally a consequence of the method by which the surface is prepared, e.g. by mechanical polishing in which grits are used, or by electro polishing, or it is a consequence of the method by which the solid is formed, e.g., by deposition from a vapor onto a substrate. For this type of rough surface the surface profile is not known in any exact sense. We are therefore forced to describe such surfaces by certain statistical properties that in turn depend on certain physical quantities that play the role of the parameters of the theory. The values of these parameters generally have to be obtained from measurements (Dobberstein 1970, Bodesheim and Otto 1974, Kaspar and Kreibig 1977, Jansen and Hoffman 1979), or from comparisons between theoretical predictions and experimental results (Kretschmann 1972a, b, c, Mills and Maradudin 1975). A good deal of this chapter is devoted to the interaction of surface polaritons with such statistically rough surfaces.

However, not all rough surfaces of physical interest are statistically rough. An important category of rough surfaces is that in which the surface profile is described by a deterministic function of the coordinates in the plane of the surface that is also a periodic function of those coordinates. An example of such rough surfaces is provided by diffraction gratings. Such surfaces are of interest for several reasons. They are rough surfaces whose profiles are known. They therefore provide an excellent testing ground for the predictions of theories of rough surface phenomena. This use for gratings has been exploited by several groups, most notably by Raether and his co-workers (Pockrand 1975, Pockrand and Raether 1976 a, b) in studies of the propagation of surface polaritons along such periodically corrugated surfaces.

As was indicated in the opening paragraph of this introduction, surface polaritons are electromagnetic surface waves, i.e., they are solutions of Maxwell's equations in which the effects of retardation – the finiteness of the speed of light – are included. An important subclass of surface polaritons are surface plasmons (Ritchie 1957). These can be viewed as the limiting case of surface polaritons when the speed of light is allowed to become infinitely large. Alternatively, and equivalently, they are solutions of Laplace's equation that propagate along the surface of a dielectric medium, and whose amplitudes decay exponentially with increasing distance from the surface into the dielectric and into the vacuum adjoining it. They are therefore electrostatic surface waves.

Despite the fact that surface plasmons are a limiting case of surface polaritons, in an operational sense they can be said to have an existence of their own, independently of the existence of surface polaritons, since there are experiments that measure them directly. These involve, for example, external probes that have a longitudinal electric field associated with them, such as a beam of electrons, whose wavelength is shorter than the vacuum wavelength of the dipole active excitation in the dielectric medium with which the surface plasmon is associated.

Surface roughness can affect surface plasmons just as it affects surface polaritons. It is often easier, however, to treat the effects of surface roughness on surface plasmons by starting directly from Laplace's equation, rather than by passing to the limit of an infinite speed of light in the corresponding result for a surface polariton. It is the former approach to the interaction of surface plasmons that will be taken in this chapter.

Our emphasis in the discussion of the propagation of surface polaritons and surface plasmons over a rough surface will be on presenting results that are explicit – in analytic form wherever possible. This restricts our discussion of this subject for the most part to slightly rough surfaces, when statistically rough surfaces are being considered, for which perturbation-theoretic methods, and variants thereof, can be employed. In particular, we will make extensive use of the Rayleigh hypothesis (Rayleigh 1907, 1945). This is the assumption that the expressions for the electromagnetic, or electrostatic, fields that are valid in the regions of space above the maximum vertical excursion of the surface profile from flatness, and below the minimum excursion, can be continued into the surface itself. It is known (Petit and Cadilhac 1966, Millar 1971 a, b, 1973) that this hypothesis has a limited applicability to periodically corrugated surfaces: crudely put, the ratio of the maximum amplitude of such a profile to its period cannot be too large for it to be valid, and a definite prescription exists for determining what "too large" is for a given profile (Hill and Celli 1978). No such estimates exist for a statistically rough surface, but it is reasonable to assume that they are comparable to those for periodically corrugated surfaces. In our work we will assume that the use of the Rayleigh hypothesis is valid for the statistically rough surfaces we consider. In contrast, in the case of deterministic, periodic rough surfaces the combination of the known form of the surface profile function and its periodicity enables powerful theoretical-*cum*-computational methods to be used in studying the propagation of surface polaritons across a grating that yield essentially exact results. These methods hold out the promise of being useful for the study of surface polariton propagation along statistically rough surfaces as well, and will be discussed in both contexts in this chapter.

It is our purpose in this chapter to survey existing theoretical work on the interaction of surface polaritons and plasmons with surface roughness, and to present some new results. It has not been our intention, however, merely to present a compilation of results and references to the literature. This article has been written with the deliberate aim of describing the theoretical techniques that have been developed for dealing with such problems in sufficient detail that it can serve as a self-contained guide for anyone interested in carrying out such calculations for himself. The experimental aspects of this subject have been treated in ch. 9.

For an earlier review that treats both the theoretical and experimental aspects of the interaction of surface polaritons and plasmons with surface roughness the reader is referred to the article by Raether (1977).

After a presentation of some useful results concerning surface polaritons on a flat surface, and a discussion of how both kinds of rough surfaces, statistical and deterministic/periodic, will be characterized in this chapter, we turn in sect. 3 to an analysis of the propagation of surface polaritons over a grating. The problem will be discussed first on the basis of the Rayleigh hypothesis, and explicit results valid in the weak roughness limit will be obtained. This will be followed by the presentation of a formally exact formulation of the problem, to illustrate the manner in which an exact solution to the problem can be obtained. The propagation of a surface plasmon over a periodically corrugated surface will then be treated, both on the basis of the Rayleigh hypothesis and by an exact formulation of the problem. Not all gratings are stationary, however, and we conclude section 3 with a discussion of the interaction of a surface polariton with the moving grating created by the passage of a Rayleigh acoustic surface wave along the dielectric medium. Section 4 is devoted to the interaction of surface polaritons and plasmons with a statistically rough surface, with particular emphasis on the attenuation of surface polaritons by surface roughness. We conclude this chapter with a brief discussion of directions the theoretical study of the interaction of surface polaritons and plasmons can take in the future.

2. *Underlying Concepts and Results*

In this chapter we will be concerned exclusively with the effects on surface polaritons and plasmons of roughness on a surface that in the absence of that roughness would be planar.

Consequently, so that the results obtained in this chapter can be contrasted with the better known results for a flat surface (Mills and Burstein 1974, Burstein et al. 1974), and because in the small roughness limit the roughness will be treated as a perturbation on a flat surface, it is useful to present a brief summary of those properties of surface polaritons and plasmons on a flat surface that will be needed in what follows. However, before we turn to a discussion of how the interaction of surface polaritons and plasmons with surface roughness alters the properties they possess on a flat surface, we have to indicate how rough surfaces will be characterized in this chapter.

We take up these two subjects in the present section.

2.1. Surface Polaritons on a Flat Surface

We consider the following physical system. The region $x_3<0$ is filled by a dielectric medium characterized by an isotropic frequency-dependent dielectric constant $\epsilon_a(\omega)$, that we assume is real. The region $x_3>0$ is filled by a second dielectric medium characterized by a real dielectric constant $\epsilon_b(\omega)$. The solutions of Maxwell's equations in this system that are wave-like in directions parallel to the plane $x_3=0$ but whose amplitudes decay exponentially with increasing distance into each medium from the interface are given by

$$\boldsymbol{E}(x,t)=\begin{pmatrix}\hat{k}_1A-\hat{k}_2B\\ \hat{k}_2A+\hat{k}_1B\\ \mathrm{i}k_\parallel A/\alpha_b\end{pmatrix}\exp(\mathrm{i}\,\boldsymbol{k}_\parallel\cdot\boldsymbol{x}_\parallel-\alpha_b x_3-\mathrm{i}\omega t)\qquad x_3\geq 0 \tag{2.1a}$$

$$=\begin{pmatrix}\hat{k}_1C-\hat{k}_2D\\ \hat{k}_2C+\hat{k}_1D\\ -\mathrm{i}k_\parallel C/\alpha_a\end{pmatrix}\exp(\mathrm{i}\boldsymbol{k}_\parallel\cdot\boldsymbol{x}_\parallel+\alpha_a x_3-\mathrm{i}\omega t)\qquad x_3\leq 0 \tag{2.1b}$$

and

$$\boldsymbol{H}(\boldsymbol{x},t)=\begin{pmatrix}\mathrm{i}\dfrac{\omega\epsilon_b(\omega)}{c\alpha_b}\hat{k}_2A-\mathrm{i}\dfrac{c\alpha_b}{\omega}\hat{k}_1B\\ -\mathrm{i}\dfrac{\omega\epsilon_b(\omega)}{c\alpha_b}\hat{k}_1A-\mathrm{i}\dfrac{c\alpha_b}{\omega}\hat{k}_2B\\ \dfrac{ck_\parallel}{\omega}B\end{pmatrix}\exp(\mathrm{i}\boldsymbol{k}_\parallel\cdot\boldsymbol{x}_\parallel-\alpha_b x_3-\mathrm{i}\omega t)\qquad x_3\geq 0 \tag{2.2a}$$

$$= \begin{pmatrix} -\mathrm{i}\dfrac{\omega\epsilon_{\mathrm{a}}(\omega)}{c\alpha_{\mathrm{a}}}\hat{k}_2 C + \mathrm{i}\dfrac{c\alpha_{\mathrm{a}}}{\omega}\hat{k}_1 D \\ \mathrm{i}\dfrac{\omega\epsilon_{\mathrm{a}}(\omega)}{c\alpha_{\mathrm{a}}}\hat{k}_1 C + \mathrm{i}\dfrac{c\alpha_{\mathrm{a}}}{\omega}\hat{k}_2 D \\ \dfrac{ck_\parallel}{\omega} D \end{pmatrix} \exp(\mathrm{i}\boldsymbol{k}_\parallel \cdot \boldsymbol{x}_\parallel + \alpha_{\mathrm{a}} x_3 - \mathrm{i}\omega t) \qquad x_3 \le 0. \tag{2.2b}$$

In these expressions $\boldsymbol{x}_\parallel$ and $\boldsymbol{k}_\parallel$ are two-dimensional position and wave vectors given by

$$\boldsymbol{x}_\parallel = \hat{x}_1 x_1 + \hat{x}_2 x_2, \tag{2.3a}$$

$$\boldsymbol{k}_\parallel = \hat{x}_1 k_1 + \hat{x}_2 k_2, \tag{2.3b}$$

respectively, where $\hat{x}_1$ and $\hat{x}_2$ are two mutually perpendicular unit vectors in the plane $x_3 = 0$. In addition, we have $\hat{k}_\alpha = k_\alpha / k_\parallel$ ($\alpha = 1, 2$). The decay constants $\alpha_{\mathrm{a}}(k_\parallel \omega)$ and $\alpha_{\mathrm{b}}(k_\parallel \omega)$ appearing in eqs. (2.1)–(2.2) are given by

$$\alpha_{\mathrm{a}}(k_\parallel \omega) = \left(k_\parallel^2 - \epsilon_{\mathrm{a}}(\omega)\frac{\omega^2}{c^2}\right)^{1/2} \tag{2.4a}$$

$$\alpha_{\mathrm{b}}(k_\parallel \omega) = \left(k_\parallel^2 - \epsilon_{\mathrm{b}}(\omega)\frac{\omega^2}{c^2}\right)^{1/2}. \tag{2.4b}$$

These constants must be real and positive in order that eqs. (2.1) and (2.2) describe electromagnetic waves localized to the interface between the two dielectric media at the plane $x_3 = 0$. (If we had allowed $\epsilon_{\mathrm{a}}(\omega)$ and $\epsilon_{\mathrm{b}}(\omega)$ to be complex, where quite generally Im $\epsilon_{\mathrm{a,b}}(\omega) > 0$, we would have to require that Re $\alpha_{\mathrm{a,b}} > 0$.)

If we set $B = D = 0$, eqs. (2.1)–(2.2) describe a p-polarized wave, i.e. one whose electric vector is in the sagittal plane – the plane defined by the wave vector and the normal to the surface. If we set $A = C = 0$, eqs. (2.1)–(2.2) describe an s-polarized wave, i.e., one whose electric vector is perpendicular to the sagittal plane.

If we now require the continuity of the tangential components of $\boldsymbol{E}(\boldsymbol{x}, t)$ across the plane $x_3 = 0$, and use the linear independence of $\hat{k}_1$ and $\hat{k}_2$, we find that $A = C$ and $B = D$. If we then use these results together with the requirement that the tangential components of $\boldsymbol{H}(\boldsymbol{x}, t)$ be continuous across the surface $x_3 = 0$, we obtain as the condition that a p-polarized wave exist

$$\left(\frac{\epsilon_{\mathrm{a}}(\omega)}{\alpha_{\mathrm{a}}(k_\parallel \omega)} + \frac{\epsilon_{\mathrm{b}}(\omega)}{\alpha_{\mathrm{b}}(k_\parallel \omega)}\right) A = 0, \tag{2.5}$$

i.e., a p-polarized surface wave can exist only if its frequency and wave vector are related in the manner indicated in this equation. The condition that an s-polarized wave exist is obtained in the same way and is given by

$$(\alpha_a(k_\parallel\omega) + \alpha_b(k_\parallel\omega))\, B = 0. \tag{2.6}$$

We see immediately from eq. (2.6) that because α_a and α_b are both required to be positive in order that we have an electromagnetic wave localized at the plane $x_3 = 0$, the only solution of this equation is $B = 0$. Thus an s-polarized surface wave cannot exist in the structure we are considering here.

The presence of the dielectric constants (which can be negative in certain ranges of ω) in eq. (2.5) has the consequence that a nontrivial ($A \neq 0$) solution of this equation can exist. This requires that

$$\epsilon_a(\omega)/\epsilon_b(\omega) = -\alpha_a(k_\parallel\omega)/\alpha_b(k_\parallel\omega). \tag{2.7}$$

Since α_a and α_b are both required to be positive, we see from this equation that a necessary condition that it has a solution is that either $\epsilon_a(\omega)$ or $\epsilon_b(\omega)$ (but not both) must be negative at the frequency ω of the surface wave. Of the two dielectric media in contact along the interface $x_3 = 0$ in the structure under consideration here, the one whose dielectric constant is negative at the surface wave frequency is termed the *surface active medium.*

The electromagnetic surface wave whose field vectors are given by eqs. (2.1) and (2.2) (with $B = D = 0$ and $A = C$), and whose frequency is obtained from the dispersion relation (2.7), is called a *surface polariton.*

We can solve eq. (2.7) by squaring both sides, using eqs. (2.4), and rearranging the terms. The result is

$$k_\parallel^2 = \frac{\omega^2}{c^2}\frac{\epsilon_a(\omega)\epsilon_b(\omega)}{\epsilon_a(\omega) + \epsilon_b(\omega)}. \tag{2.8}$$

This gives an explicit expression for the magnitude of the wave vector of the surface polariton as a function of its frequency, a relation that can be inverted trivially to yield ω as a function of $k_\parallel$. However, because the method used to obtain eq. (2.8) can introduce spurious solutions, in order that any solution of eq. (2.8) correspond to a surface polariton it must satisfy the conditions that $\alpha_a(k_\parallel\omega)$ and $\alpha_b(k_\parallel\omega)$ both be positive, and that either $\epsilon_\alpha(\omega)$ or $\epsilon_b(\omega)$ be negative at that frequency.

If, instead of working in the regime where the effects of retardation are important, we turn to the case where they are unimportant, i.e. to the electrostatic limit, it is Laplace's equation for the scalar potential that we must solve. The solution of this equation that is wavelike in directions in the plane $x_3 = 0$, but decays exponentially as $|x_3| \to \infty$, is

$$\varphi(\boldsymbol{x}, t) = A \exp(i\boldsymbol{k}_\parallel \cdot \boldsymbol{x}_\parallel - k_\parallel x_3 - i\omega t) \qquad x_3 > 0 \tag{2.9a}$$

$$= B \exp(i\boldsymbol{k}_\parallel \cdot \boldsymbol{x}_\parallel + k_\parallel x_3 - i\omega t, \qquad x_3 < 0. \tag{2.9b}$$

We must require that $\varphi(\boldsymbol{x}, t)$ be continuous across the plane $x_3 = 0$: this ensures the continuity of the tangential components of the corresponding macroscopic electric field $\boldsymbol{E}(\boldsymbol{x}, t) = -\nabla\varphi(\boldsymbol{x}, t)$. This condition yields the result that $A = B$. We next require that $\epsilon\partial\varphi/\partial x_3$ be continuous across the plane $x_3 = 0$: this condition is equivalent to the condition that the normal component of the electric displacement $\boldsymbol{D}(\boldsymbol{x}, t)$ be continuous across this interface. The condition whose satisfaction permits this to be done is

$$\epsilon_a(\omega) + \epsilon_b(\omega) = 0. \tag{2.10}$$

The surface excitation whose potential is given by eq. (2.9) (with $A = B$), and whose frequency is obtained from eq. (2.10) is called a *surface plasmon.*

We again see, from eq. (2.10), that in order that a surface plasmon exist one of the dielectric constants has to be negative.

We also note that the frequency that is the solution of eq. (2.10) is independent of the wave vector $\boldsymbol{k}_\parallel$, unlike the solution of eq. (2.7) or of eq. (2.8). The reason for this ultimately is that in the electrostatic limit there is no characteristic length in the problem. In the electromagnetic case there is one, connected with the finite speed of light. It is the vacuum wavelength of the infrared active elementary excitation in the dielectric medium that contributes a pole at a frequency Ω_0 to its dielectric constant. This is defined by $\lambda_0 = 2\pi/k_0$ where $k_0 = \Omega_0/c$. For values of $k_\parallel$ larger than k_0 the effects of retardation are unimportant: light propagates over the short distances that correspond to such values of $k_\parallel$ essentially instantaneously. For values of $k_\parallel$ smaller than k_0 the finite speed of light has to be taken into account in considering its propagation over such larger distances. The surface polariton dispersion relation has a qualitatively different form in each of these limits, as can be inferred from a comparison of eqs. (2.7) and (2.10) and as will be seen explicitly below.

If we compare eqs. (2.8) and (2.10) we see that the frequency of the surface plasmon is also the limiting frequency of the surface polariton as $k_\parallel \to \infty$.

In what follows we will be concerned only with the case in which madium (a) ($x_3 < 0$) is the dielectric medium, whose dielectric constant is $\epsilon_a(\omega) \equiv \epsilon(\omega)$. Medium (b) ($x_3 > 0$) will be assumed to be vacuum, so that $\epsilon_b(\omega) \equiv 1$. The surface active medium in this case is clearly the dielectric medium.

For illustrative purposes two forms for the dielectric constant $\epsilon(\omega)$ will be used. The first of these,

$$\epsilon(\omega) = \epsilon_\infty + (\epsilon_0 - \epsilon_\infty)\omega_T^2/(\omega_T^2 - \omega^2), \tag{2.11}$$

corresponds to the case of a diatomic, cubic polar crystal, with two ions per primitive unit cell. In eq. (2.11) ϵ_0 and ϵ_∞ are the static and optical frequency dielectric constants, respectively, and ω_T is the frequency of the infinite wavelength transverse optical vibration modes. The second dielectric constant we will use is given by

$$\epsilon(\omega) = \epsilon_\infty(1 - \omega_p^2/\omega^2), \tag{2.12}$$

and represents the contribution to the dielectric constant of a nearly free electron metal or n-type semiconductor from intraband transitions. In eq. (2.12) ϵ_∞ is the background dielectric constant of the material, and ω_p is the electronic plasma frequency, given by $\omega_p^2 = 4\pi ne^2/m^*\epsilon_\infty$, where n is the electron number density, e is the magnitude of the electronic charge, and m^* is the effective mass of the charge carriers.

The frequency $\omega_0(k_\parallel)$ of the surface polariton at a plane dielectric–vacuum interface is given by

$$\omega_0^2(k_\parallel) = \frac{1}{2}\left(\frac{c^2k_\parallel^2}{\epsilon_\infty}(1+\epsilon_\infty) + \omega_L^2\right) - \frac{1}{2}\left[\left(\frac{c^2k_\parallel^2}{\epsilon_\infty}(1+\epsilon_\infty) + \omega_L^2\right)^2 - 4\frac{c^2k_\parallel^2}{\epsilon_\infty}(\omega_T^2 + \epsilon_\infty\omega_L^2)\right]^{1/2} \tag{2.13a}$$

$$\xrightarrow[k_\parallel \to \omega_T/c]{} \omega_T^2 \tag{2.13b}$$

$$\xrightarrow[k_\parallel \to \infty]{} \left(\frac{\epsilon_0 + 1}{\epsilon_\infty + 1}\right)\omega_T^2, \tag{2.13c}$$

for the dielectric constant given by eq. (2.11). Here $\omega_L = (\epsilon_0/\epsilon_\infty)^{1/2}\omega_T$ is the frequency of the infinite wavelength longitudinal optical vibration modes. This dispersion curve is plotted in fig. 1. For the dielectric constant given

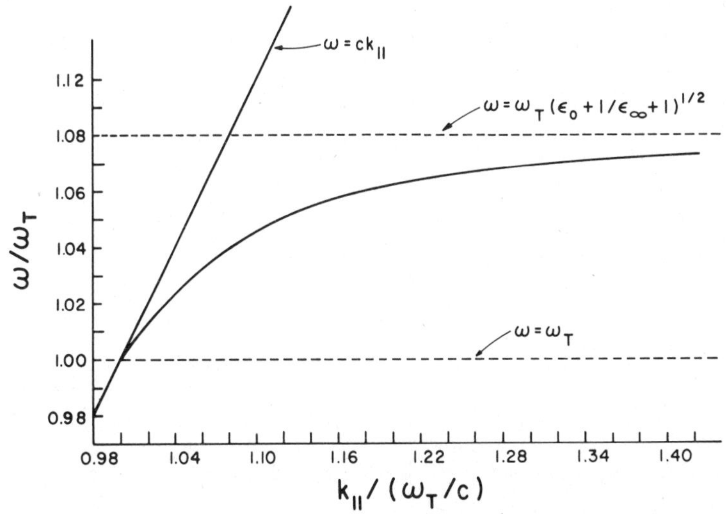

Fig. 1. Dispersion curve for a surface polariton at a flat GaAs–vacuum interface. $\epsilon_0 = 12.9$, $\epsilon_\infty = 10.9$, $\omega_T = 273\ \text{cm}^{-1}$, $\omega_L = 297\ \text{cm}^{-1}$.

by eq. (2.12) the surface polariton frequency takes the form

$$\omega_0^2(k_\parallel) = \frac{1}{2}\left(\frac{c^2k_\parallel^2}{\epsilon_\infty}(1+\epsilon_\infty)+\omega_p^2\right) - \frac{1}{2}\left[\left(\frac{c^2k^2}{\epsilon_\infty}(1+\epsilon_\infty)+\omega_p^2\right)^2 - 4c^2k_\parallel^2\omega_p^2\right]^{1/2} \tag{2.14a}$$

$$\xrightarrow[k_\parallel\to 0]{} \quad c^2k_\parallel^2 \tag{2.14b}$$

$$\xrightarrow[k_\parallel\to\infty]{} \quad \frac{\epsilon_\infty}{\epsilon_\infty+1}\,\omega_p^2\,. \tag{2.14c}$$

This dispersion curve is plotted in fig. 2 for the case that $\epsilon_\infty = 1$ (a simple, free electron metal).

From the plots in these two figures we see that the wave vectors $k_\parallel \cong \omega_T/c$ and $k_\parallel \cong \omega_p/c$, respectively, essentially define the boundary between the regime where the surface polariton dispersion curve is dispersive and the regime where it is not (i.e. the surface plasmon regime), as noted on general grounds above.

In the remainder of this chapter we examine how the results of this subsection are altered when the dielectric–vacuum interface is rough. We begin by describing how rough planar surfaces will be described in this chapter.

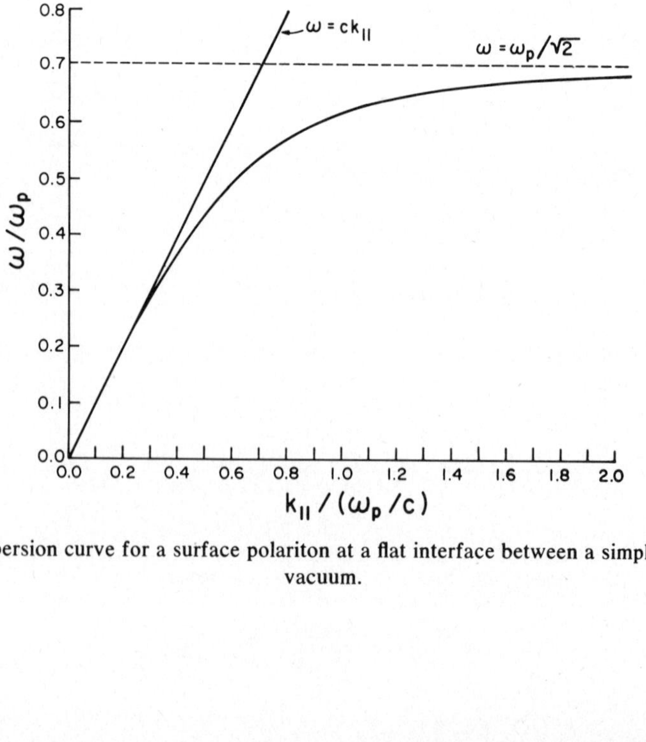

Fig. 2. Dispersion curve for a surface polariton at a flat interface between a simple metal and vacuum.

2.2. Characterization of a Rough Surface

As the planar surface perturbed by the roughness we take the plane $x_3 = 0$. In the presence of roughness the equation defining the surface is

$$x_3 = \zeta(\boldsymbol{x}_\parallel), \tag{2.15}$$

where the real function $\zeta(\boldsymbol{x}_\parallel)$ is called the *surface roughness profile function*. It is a function of the two-dimensional position vector $\boldsymbol{x}_\parallel$, eq. (2.3a).

In what follows we aseume that the region $x_3 > \zeta(\boldsymbol{x}_\parallel)$ is occupied by vacuum, while the region $x_3 < \zeta(\boldsymbol{x}_\parallel)$ is occupied by a dielectric medium, characterized by an isotropic, frequency-dependent dielectric tensor

$$\epsilon_{\alpha\beta}(\omega) = \delta_{\alpha\beta}\epsilon(\omega). \tag{2.16}$$

The region $\zeta_{\min} < x_3 < \zeta_{\max}$ is called the *selvedge region*.

In many problems the unit vector normal to the surface at each point is required. If we further specify this vector to be directed from the dielectric medium into the vacuum, it is given by the normalized gradient of the function $f(\boldsymbol{x}) = x_3 - \zeta(\boldsymbol{x}_\parallel)$, and has the explicit form

$$\hat{n} = \left(-\frac{\partial\zeta(\boldsymbol{x}_\parallel)}{\partial x_1}, -\frac{\partial\zeta(\boldsymbol{x}_\parallel)}{\partial x_2}, 1\right)\left[1 + \left(\frac{\partial\zeta(\boldsymbol{x}_\parallel)}{\partial x_1}\right)^2 + \left(\frac{\partial\zeta(\boldsymbol{x}_\parallel)}{\partial x_2}\right)^2\right]^{-1/2}. \tag{2.17}$$

To avoid ambiguities that might otherwise arise, we will sometimes denote the vector $\hat{n}$ defined by eq. (2.17) by $\hat{n}_+$, and will define the unit vector $\hat{n}_-$ as the negative of $\hat{n}_+$, so that

$$\hat{n}_+ = \hat{n} = -\hat{n}_-. \tag{2.18}$$

It is also often necessary to define two mutually perpendicular unit vectors that are tangent to the surface at each point. There is a great deal of arbitrariness in defining such vectors, since any two mutually perpendicular unit vectors in the plane perpendicular to $\hat{n}$ can be chosen for this purpose. If we choose one of these vectors, $\hat{t}_1$, to be nearly in the x_1-direction, while the second, $\hat{t}_2$, is nearly in the x_2-direction, we obtain the results that

$$\hat{t}_1 = N_1\left(1, 0, \frac{\partial\zeta(\boldsymbol{x}_\parallel)}{\partial x_1}\right) \tag{2.19a}$$

$$\hat{t}_2 = N_2\left(-\frac{\dfrac{\partial\zeta(\boldsymbol{x}_\parallel)}{\partial x_1}\dfrac{\partial\zeta(\boldsymbol{x}_\parallel)}{\partial x_2}}{1 + \left(\dfrac{\partial\zeta(\boldsymbol{x}_\parallel)}{\partial x_1}\right)^2}, 1, \frac{\dfrac{\partial\zeta(\boldsymbol{x}_\parallel)}{\partial x_2}}{1 + \left(\dfrac{\partial\zeta(\boldsymbol{x}_\parallel)}{\partial x_1}\right)^2}\right) \tag{2.19b}$$

$$\cong N_2\left(-\frac{\partial\zeta(\boldsymbol{x}_\parallel)}{\partial x_1}\frac{\partial\zeta(\boldsymbol{x}_\parallel)}{\partial x_2}, 1, \frac{\partial\zeta(\boldsymbol{x}_\parallel)}{\partial x_2}\right), \tag{2.19c}$$

to second order in $\zeta(\boldsymbol{x}_\parallel)$. The normalization factors are

$$N_1 = \left[1 + \left(\frac{\partial \zeta(\boldsymbol{x}_\parallel)}{\partial x_1}\right)^2\right]^{-1/2} \tag{2.20a}$$

$$N_2 = \left[1 + \left(\frac{\partial \zeta(\boldsymbol{x}_\parallel)}{\partial x_1}\right)^2\right]\left[1 + 2\left(\frac{\partial \zeta(\boldsymbol{x}_\parallel)}{\partial x_1}\right)^2 + \left(\frac{\partial \zeta(\boldsymbol{x}_\parallel)}{\partial x_2}\right)^2 + \left(\frac{\partial \zeta(\boldsymbol{x}_\parallel)}{\partial x_1}\right)^4 + \left(\frac{\partial \zeta(\boldsymbol{x}_\parallel)}{\partial x_1}\frac{\partial \zeta(\boldsymbol{x}_\parallel)}{\partial x_2}\right)^2\right]^{-1/2}. \tag{2.20b}$$

Beyond this point we have to differentiate between the case that $\zeta(\boldsymbol{x}_\parallel)$ describes a surface profile that is a deterministic, periodic function of $\boldsymbol{x}_\parallel$, and the case that $\zeta(\boldsymbol{x}_\parallel)$ describes a statistically rough surface profile. We take up each of these cases in turn.

2.2.1. *A Deterministic, Periodic Surface Profile*

The simplest example of a deterministic, periodic surface profile, and the only one to be considered in this chapter, is a planar surface on which a diffraction grating has been ruled (fig. 3). In this case the surface profile function is a function of one coordinate only, that we choose to be x_1. It has the periodicity property

$$\zeta(x_1 + a) = \zeta(x_1), \tag{2.21}$$

where a is the period of the grating. The function $\zeta(x_1)$ can therefore be expanded in a Fourier series,

$$\zeta(x_1) = \sum_{n=-\infty}^{\infty} \hat{\zeta}(n) \exp\left(i\frac{2\pi n}{a}x_1\right), \tag{2.22}$$

where the Fourier coefficients $\{\hat{\zeta}(n)\}$ are given by

$$\hat{\zeta}(n) = \frac{1}{a}\int_0^a dx_1\, \zeta(x_1) \exp\left(-i\frac{2\pi n}{a}x_1\right). \tag{2.23}$$

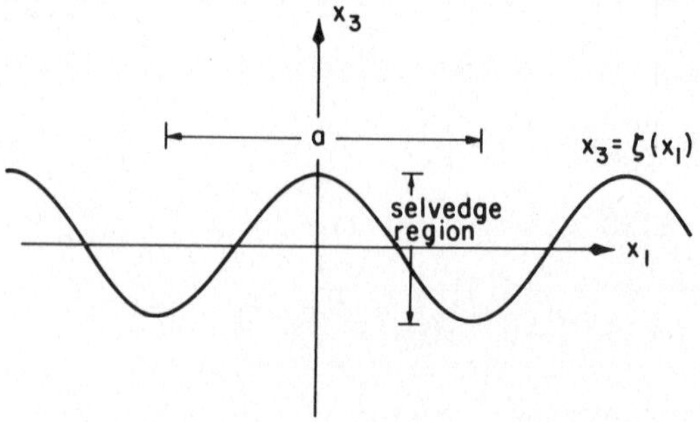

Fig. 3. A typical grating profile.

It is convenient to assume that $\zeta(x_1)$ has been defined in such a way that

$$\hat{\zeta}(0) = \frac{1}{a}\int_0^a \mathrm{d}x_1\,\zeta(x_1) = 0. \tag{2.24}$$

2.2.2. *A Statistically Rough Surface Profile*

In the case of a statistically rough surface the function $\zeta(\boldsymbol{x}_\parallel)$ is generally unknown. Because of this we characterize it by certain statistical properties. Underlying this characterization is the assumption that there is not a single function $\zeta(\boldsymbol{x}_\parallel)$. There is instead an ensemble of realizations of this function. Physical properties associated with a statistically rough surface are to be averaged over this ensemble, and it is assumed that this ensemble average does not differ significantly from the spatial average over a single rough surface. The probability that $\zeta(\boldsymbol{x}_\parallel)$ has a particular value at the point $\boldsymbol{x}_\parallel$ is given by a probability distribution function. An explicit form for this distribution function is usually not required. What are specified instead are the first few of its moments. In common with most theoretical treatments of surface roughness the one presented in this chapter is based on the assumption that $\zeta(\boldsymbol{x}_\parallel)$ is a stationary stochastic process, characterized by the two properties that

$$\langle \zeta(\boldsymbol{x}_\parallel)\rangle = 0 \tag{2.25}$$

$$\langle \zeta(x_\parallel)\zeta(x'_\parallel)\rangle = \delta^2 W(|\boldsymbol{x}_\parallel - \boldsymbol{x}'_\parallel|). \tag{2.26}$$

In eqs. (2.25)–(2.26) the angular brackets $\langle\cdots\rangle$ denote an average over the ensemble of realizations of the function $\zeta(\boldsymbol{x}_\parallel)$. The quantity δ^2 appearing in eq. (2.26) is the mean square departure of the surface from flatness,

$$\langle \zeta^2(\boldsymbol{x}_\parallel)\rangle = \delta^2. \tag{2.27}$$

Equation (2.25) expresses the assumption that the nominal surface of the dielectric medium is the plane $x_3 = 0$. This assumption involves no loss of generality. The fact that the correlation function $W(|\boldsymbol{x}_\parallel - \boldsymbol{x}'_\parallel|)$ depends on $\boldsymbol{x}_\parallel$ and $\boldsymbol{x}'_\parallel$ only through their difference (the stationarity assumption) has the consequence that infinitesimal translational invariance is restored to our rough surface system when physical quantities are averaged over the ensemble of realizations of the function $\zeta(\boldsymbol{x}_\parallel)$. The fact that this correlation function depends on the magnitude of the vector $\boldsymbol{x}_\parallel - \boldsymbol{x}'_\parallel$ is an expression of the assumption that all directions are equivalent on the rough surface. If the medium bounded by the rough surface is isotropic in its properties in the plane $x_3 = 0$, a consequence of this assumption of the equivalence of all directions on the rough surface is that isotropy in the plane $x_3 = 0$ is restored to physical quantities by the averaging process.

Although for many applications it suffices to have knowledge of only the first two moments of the probability distribution function of $\zeta(\boldsymbol{x}_\parallel)$, it is sometimes necessary to specify higher order moments as well. A commonly made assumption that allows this to be done is that the $\{\zeta(\boldsymbol{x}_\parallel)\}$ for different values of $\boldsymbol{x}_\parallel$ are Gaussianly distributed random variables. The consequences of this assumption are that (*i*) the ensemble average of a product of an odd number of $\zeta(\boldsymbol{x}_\parallel)$'s vanishes, e.g.,

$$\langle \zeta(\boldsymbol{x}_\parallel)\zeta)\boldsymbol{x}'_\parallel)\zeta(\boldsymbol{x}''_\parallel)\rangle = 0, \tag{2.28}$$

where $\boldsymbol{x}_\parallel$, $\boldsymbol{x}'_\parallel$, $\boldsymbol{x}''_\parallel$ do not need to be different points; and (*ii*) the average of the product of an even number of $\zeta(\boldsymbol{x}_\parallel)$'s is given by the sum of the products of the averages of the $\zeta(\boldsymbol{x}_\parallel)$'s taken two-by-two different in all possible ways, e.g.,

$$\begin{aligned}\langle \zeta(\boldsymbol{x}_\parallel)\zeta(\boldsymbol{x}'_\parallel)\zeta(\boldsymbol{x}''_\parallel)\zeta(\boldsymbol{x}'''_\parallel)\rangle &= \langle \zeta(\boldsymbol{x}_\parallel)\zeta(\boldsymbol{x}'_\parallel)\rangle\langle \zeta(\boldsymbol{x}''_\parallel)\zeta(\boldsymbol{x}'''_\parallel)\rangle \\ &+ \langle \zeta(\boldsymbol{x}_\parallel)\zeta(\boldsymbol{x}''_\parallel)\rangle\langle \zeta(\boldsymbol{x}'_\parallel)\zeta(\boldsymbol{x}'''_\parallel)\rangle + \langle \zeta(\boldsymbol{x}_\parallel)\zeta(\boldsymbol{x}'''_\parallel)\rangle\langle \zeta(\boldsymbol{x}'_\parallel)\zeta(\boldsymbol{x}''_\parallel)\rangle.\end{aligned} \tag{2.29}$$

Each average of a pair of $\zeta(\boldsymbol{x}_\parallel)$'s on the right hand side of this equation, called a *contraction*, is given by eq. (2.26).

We turn now to an examination of the correlation function $W(|\boldsymbol{x}_\parallel|)$. From eqs. (2.26) and (2.27) it follows that

$$W(0) = 1. \tag{2.30}$$

It is clearly an even function of $\boldsymbol{x}_\parallel$. In addition, $W(|\boldsymbol{x}_\parallel|)$ tends to zero as $|\boldsymbol{x}_\parallel| \to \infty$, because on a statistically rough surface the heights of the roughness at two widely separated points are uncorrelated. We will define the *transverse correlation length* σ as the distance over which $W(|\boldsymbol{x}_\parallel|)$ decreases appreciably.

Thus, the model of statistical surface roughness we are going to use in this chapter depends on two parameters: the root-mean-square departure of the surface from flatness, δ, and the transverse correlation length σ (fig. 4).

For calculational purposes it is convenient, even necessary, to introduce the Fourier transform of the surface roughness profile function:

$$\zeta(\boldsymbol{x}_\parallel) = \int \frac{d^2k_\parallel}{(2\pi)^2} \exp(i\boldsymbol{k}_\parallel \cdot \boldsymbol{x}_\parallel)\, \hat{\zeta}(\boldsymbol{k}_\parallel). \tag{2.31}$$

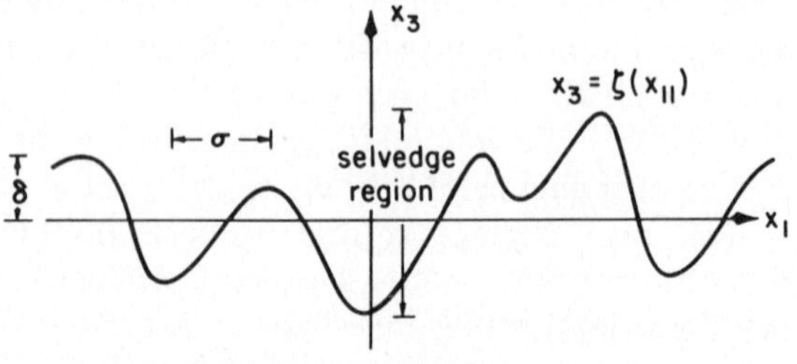

Fig. 4. A randomly rough surface profile.

Here $k_\parallel$ is the two-dimensional wave vector defined by eq. (2.3b). The Fourier coefficients $\{\hat{\zeta}(\boldsymbol{k}_\parallel)\}$ are now random quantities. From the Fourier inversion theorem and the averages given by eqs. (2.25) and (2.26) we find that the statistical properties of these coefficients are characterized by

$$\langle\hat{\zeta}(\boldsymbol{k}_\parallel)\rangle = 0 \tag{2.32}$$

$$\langle\hat{\zeta}(\boldsymbol{k}_\parallel)\hat{\zeta}(\boldsymbol{k}'_\parallel)\rangle = \delta^2 g(k_\parallel)(2\pi)^2\delta(\boldsymbol{k}_\parallel + \boldsymbol{k}'_\parallel). \tag{2.33}$$

For certain applications we require the average $\langle|\hat{\zeta}(\boldsymbol{k}_\parallel)|^2\rangle$. This is found from eq. (2.33) to be

$$\langle|\hat{\zeta}(\boldsymbol{k}_\parallel)|^2\rangle = L_1 L_2 \delta^2 g(k_\parallel), \tag{2.34}$$

where L_1 and L_2 are the dimensions in the x_1- and x_2-directions of the rough portion of the surface $x_3 = 0$. To obtain this result we have used the fact that in two-dimensions $\delta(\boldsymbol{k}_\parallel = 0) = L_1 L_2/(2\pi)^2$. The function $g(k_\parallel)$ appearing in eq. (2.33) is called the *surface structure factor*, and is defined by

$$g(k_\parallel) = \int \mathrm{d}^2x_\parallel \exp(-\mathrm{i}\boldsymbol{k}_\parallel \cdot \boldsymbol{x}_\parallel)\, W(|\boldsymbol{x}_\parallel|) \tag{2.35a}$$

$$= 2\pi \int_0^\infty \mathrm{d}x_\parallel\, x_\parallel J_0(k_\parallel x_\parallel) W(x_\parallel), \tag{2.35b}$$

where $J_0(k_\parallel x_\parallel)$ is a Bessel function. From the inversion formula

$$W(|\boldsymbol{x}_\parallel|) = \int \frac{\mathrm{d}^2k_\parallel}{(2\pi)^2} \exp(\mathrm{i}\boldsymbol{k}_\parallel \cdot \boldsymbol{x}_\parallel)\, g(k_\parallel) \tag{2.36}$$

and eq. (2.30) we see that $g(k_\parallel)$ is normalized according to

$$\int \frac{\mathrm{d}^2k_\parallel}{(2\pi)^2} g(k_\parallel) = 1. \tag{2.37}$$

In the case that the $\{\zeta(\boldsymbol{x}_\parallel)\}$ are regarded as Gaussianly distributed random variables, the Fourier coefficients $\{\hat{\zeta}(\boldsymbol{k}_\parallel)\}$ possess the following properties: (i) the average of the product of an odd number of $\hat{\zeta}(\boldsymbol{k}_\parallel)$'s vanishes, e.g.,

$$\langle\hat{\zeta}(\boldsymbol{k}_\parallel)\hat{\zeta}(\boldsymbol{k}'_\parallel)\hat{\zeta}(\boldsymbol{k}''_\parallel)\rangle = 0, \tag{2.38}$$

and (ii) the average of the product of an even number of $\hat{\zeta}(\boldsymbol{k}_\parallel)$'s is given by the sum of the products of the averages of the $\hat{\zeta}(\boldsymbol{k}_\parallel)$'s taken two-by-two different in all possible ways, e.g.,

$$\begin{aligned}\langle\hat{\zeta}(\boldsymbol{k}_\parallel)\hat{\zeta}(\boldsymbol{k}'_\parallel)\hat{\zeta}(\boldsymbol{k}''_\parallel)\hat{\zeta}(\boldsymbol{k}'''_\parallel)\rangle &= \langle\hat{\zeta}(\boldsymbol{k}_\parallel)\hat{\zeta}(\boldsymbol{k}'_\parallel)\rangle\langle\hat{\zeta}(\boldsymbol{k}''_\parallel)\hat{\zeta}(\boldsymbol{k}'''_\parallel)\rangle \\ &+ \langle\hat{\zeta}(\boldsymbol{k}_\parallel)\hat{\zeta}(\boldsymbol{k}''_\parallel)\rangle\langle\hat{\zeta}(\boldsymbol{k}'_\parallel)\hat{\zeta}(\boldsymbol{k}'''_\parallel)\rangle + \langle\hat{\zeta}(\boldsymbol{k}_\parallel)\hat{\zeta}(\boldsymbol{k}'''_\parallel)\rangle\langle\hat{\zeta}(\boldsymbol{k}'_\parallel)\hat{\zeta}(\boldsymbol{k}''_\parallel)\rangle.\end{aligned} \tag{2.39}$$

Each average of a pair of $\hat{\zeta}(\boldsymbol{k}_\parallel)$'s on the right hand side of this equation is given by eq. (2.33).

In order to obtain explicit, quantitative results for quantities associated with a rough surface we require an explicit, analytic form for the correlation function $W(|\boldsymbol{x}_\parallel|)$. In this chapter we will generally adopt a Gaussian form for $W(|\boldsymbol{x}_\parallel|)$ (see, for example, Elson and Ritchie (1971)),

$$W(|\boldsymbol{x}_\parallel|) = \exp - (x_\parallel^2/\sigma^2), \tag{2.40}$$

where σ is the transverse correlation length. The surface structure factor $g(k_\parallel)$ obtained by combining eqs. (2.35) and (2.40) is

$$g(k_\parallel) = \pi\sigma^2 \exp(-\tfrac{1}{4}\sigma^2 k_\parallel^2). \tag{2.41}$$

Although it possesses the general properties required of $W(|\boldsymbol{x}_\parallel|)$ discussed above, it should be emphazied that there is nothing unique about our choice of the Gaussian form. It has been used to fit experimental data quite successfully, in view of its simplicity (Mills and Maradudin 1975). Moreover, there are experiments (Hillebrecht 1980) that show that $g(k_\parallel)$ is well represented by a Gaussian function for "smooth" silver films (i.e. films for which $\delta < 15$ Å). However, there are also experimental data (Dobberstein 1970, Bennett 1976, Rasigni et al. 1977, Jansen and Hoffman 1979) that suggest that at least under some methods of surface preparation the surface structure factor $g(k_\parallel)$, and therefore also $W(|\boldsymbol{x}_\parallel|)$ is not Gaussian. There is also some evidence that in some cases $W(|\boldsymbol{x}_\parallel|)$ is not a monotonically decreasing function of $|\boldsymbol{x}_\parallel|$ (Dobberstein 1970, Kaspar and Kreibig 1977). Finally, there are experiments that show that in some cases more than one transverse length scale is required (Kaspar and Kreibig 1977).

It is the combination of physical reasonableness and computational ease present in the Gaussian form for $W(|\boldsymbol{x}_\parallel|)$ that prompts its use here. It allows many of the integrals in which $W(|\boldsymbol{x}_\parallel|)$, or equivalently $g(k_\parallel)$, appears to be carried out analytically rather than numerically. This makes it possible to obtain more explicit results than would be the case if some other form for $W(|\boldsymbol{x}_\parallel|)$ were assumed.

However, to avoid tying our results completely to a particular form for $g(k_\parallel)$, we will present them in terms of an arbitrary $g(k_\parallel)$ before proceeding to evaluate them for the particular choice given by eq. (2.41).

It is natural to inquire at this point as to the magnitudes of the parameters δ and σ. Fits of theoretical expressions to experimental results for the reflectivity of super smooth aluminum surfaces yielded values of $\delta =$ 12 Å and $\sigma = 200$ Å (Mills and Maradudin 1975). The values of δ and σ for aluminum films grown on CaF_2 substrates were determined by a similar fitting procedure to be approximately 37 Å and 440 Å, respectively (Daudé et al. 1972). For evaporated copper films σ was found by optical methods to be in the range 200–250 Å, while $\delta > 4.4$ Å for the smoothest sample (Jansen and Hoffman 1979). Horstmann (1977) has obtained values of δ ranging from ~7 to ~20 Å for silver films of 550 Å thickness deposited on CaF_2

films of 500–2000 Å thickness predeposited on fused silica substrates. Silver films obtained by predepositing silver films of 50–150 Å on a heated (200°C) fused silica substrate, and then building them up to a thickness of 550 Å at room temperature, were found to have values of δ ranging from ~10 to ~25 Å. A value of δ of 67 Å was obtained for thick silver films (3 μm thickness) by Kaspar and Kreibig (1977).

The significance of the values of δ and σ just quoted lies in the fact that they are typically many times the value of a typical lattice parameter for a crystal. This means that in studying the effects of surface roughness on physical properties of a solid it should be a good approximation to regard the solid as a continuous medium, ignoring the discreteness of its structure, since the distances over which the surface roughness profile changes significantly are large compared with the scale of that discreteness. This approximation is made in all that follows.

3. *Interaction of Surface Polaritons and Plasmons with a Grating*

It seems appropriate to begin a review of the interaction of surface polaritons and plasmons with surface roughness by considering the particular case of a rough surface that a one-dimensional grating represents. The fact that the surface profile function $\zeta(x_\parallel)$ is a known function of x_1, and is periodic, simplifies the derivation of the surface polariton dispersion relation even if the Rayleigh hypothesis is adopted and, what is more important, permits an exact derivation of the surface polariton dispersion relation to be carried out. The latter does not appear to be possible at the present time for a randomly rough surface. The example of a grating also provides an introduction to the techniques that can be used for the study of surface polaritons on randomly rough surfaces, and the results obtained for a grating help to interpret those obtained for a randomly rough surface.

We begin with a derivation of the dispersion relation for a surface polariton on a grating in sect. 3.1. The derivation is first carried out on the basis of the Rayleigh hypothesis, and the dispersion relation is obtained in a quite explicit form, and its solution analyzed, in the small roughness limit. This is followed by a formally exact formulation of the problem of obtaining the dispersion relation for a surface polariton on a grating that is valid, in principle, without restrictions on the surface profile. In the discussion in sect. 3.1, we follow largely the treatment of Toigo et al. (1977), with the difference that we consider only the propagation of surface polaritons over a grating, while these authors were primarily concerned with the refraction of an incident electromagnetic wave from a grating.

In sect. 3.2 we repeat to a large extent the analysis of sect. 3.1 for the propagation of surface plasmons over a grating. The dispersion relation is obtained on the basis of the Rayleigh hypothesis, and is solved in the small roughness limit. We follow this discussion by presenting an exact formulation of this problem that is free from the limitations present in the use of the Rayleigh hypothesis.

The gratings studied in sect. 3.1 and 3.2 are stationary structures. In sect. 3.3 we outline a treatment, due to Mills (1977), of the interaction of a surface polariton with a moving grating. The latter is produced by the time-dependent corrugation of a solid surface as a Rayleigh elastic surface wave (Rayleigh 1887, Landau and Lifshitz 1970) propagates along the surface of the solid. This case is of interest not only because of the effects that become possible due to the fact that the surface profile varies periodically with time, but also because the surface profile due to a Rayleigh wave is known with greater accuracy than it is for a grating ruled on a surface. This affords, in principle, a more accurate comparison between theory and experiment than is possible with a static grating. The analysis in this section is based on the Rayleigh hypothesis, which should be quite accurate in this context.

3.1. Surface Polaritons on a Grating

There appears to be two general methods by which the dispersion relation for a surface polariton propagating over a grating can be obtained. The first is to study the scattering of a plane electromagnetic wave from such a surface and to determine the frequency at which the amplitude of the scattered (or transmitted) wave has a pole. This approach has been followed by Toigo et al. (1977). [The scattering of plane electromagnetic waves from gratings ruled on dielectric surfaces had earlier been studied theoretically on the basis of Green's theorem by Maystre and his colleagues (Maystre 1972, 1973, 1974, Maystre and Vincent 1972). However, these authors were not specifically concerned with the propagation of surface polaritons across such gratings.] The second is to seek directly the eigenmodes of the electromagnetic field in the vicinity of a grating whose amplitudes decay with increasing distance normal to the plane of the grating. It is this approach we will take in the present chapter.

The system we consider is the one that was described in sect. 2.2: vacuum fills the region $x_3 > \zeta(x_1)$, while the region $x_3 < \zeta(x_1)$ is filled with a dielectric medium characterized by an isotropic, frequency-dependent dielectric constant $\epsilon(\omega)$.

We assume a p-polarized surface polariton propagating in the x_1-direction. The only nonzero components of the associated electromagnetic field are $H_2(x_1, x_3 | \omega)$ and

$$E_1(x_1, x_3 \mid \omega) = -\frac{ic}{\epsilon\omega}\frac{\partial}{\partial x_3} H_2(x_1, x_3 \mid \omega) \tag{3.1}$$

$$E_3(x_1, x_3 \mid \omega) = \frac{ic}{\epsilon\omega}\frac{\partial}{\partial x_1} H_2(x_1, x_3 \mid \omega), \tag{3.2}$$

where ϵ denotes the dielectric constant of the medium in which the electric field is being calculated, and a time dependence of the fields given by $\exp(-i\omega t)$ has been assumed.

The magnetic field component $H_2(x_1, x_3 \mid \omega)$ satisfies the wave equations

$$\left(\frac{\partial^2}{\partial x_1^2} + \frac{\partial^2}{\partial x_3^2} + \frac{\omega^2}{c^2}\right) H_2(x_1, x_3 \mid \omega) = 0 \qquad x_3 > \zeta(x_1), \tag{3.3}$$

$$\left(\frac{\partial^2}{\partial x_1^2} + \frac{\partial^2}{\partial x_3^2} + \epsilon(\omega)\frac{\omega^2}{c^2}\right) H_2(x_1, x_3 \mid \omega) = 0 \qquad x_3 < \zeta(x_1), \tag{3.4}$$

in each of the regions of its definition. The boundary conditions satisfied by $H_2(x_1, x_3 \mid \omega)$ at the corrugated surface $x_3 = \zeta(x_1)$ follow from the continuity of the tangential components of $\boldsymbol{H}(\boldsymbol{x} \mid \omega)$ and $\boldsymbol{E}(\boldsymbol{x} \mid \omega)$ across this surface. They are given by

$$H_2(x_1, x_3 \mid \omega)\Big|_{x_3=\zeta(x_1)_-} = H_2(x_1, x_3 \mid \omega)\Big|_{x_3=\zeta(x_1)_+} \tag{3.5a}$$

$$\frac{1}{\epsilon(\omega)}\frac{\partial}{\partial n} H_2(x_1, x_3 \mid \omega)\Big|_{x_3=\zeta(x_1)_-} = \frac{\partial}{\partial n} H_2(x_1, x_3 \mid \omega)\Big|_{x_3=\zeta(x_1)_+} \tag{3.5b}$$

where $\partial/\partial n$ is the normal derivative at the surface $x_3 = \zeta(x_1)$:

$$\frac{\partial}{\partial n} = \left[1 + \left(\frac{\partial \zeta}{\partial x_1}\right)^2\right]^{-1/2} \left[\frac{\partial}{\partial x_3} - \frac{\partial \zeta}{\partial x_1}\frac{\partial}{\partial x_3}\right]. \tag{3.6}$$

We begin by showing how eqs. (3.3)–(3.5) can be solved on the basis of the Rayleigh hypothesis. Within this approximation it is possible to obtain some explicit results analytically, without recourse to numerical computation. Following the derivation of the surface polariton dispersion relation by Rayleigh's method we outline determinations of this dispersion relation by methods based on Green's theorem, that are formally exact, and hence are free from the limitations present in the use of Rayleigh's hypothesis.

3.1.1. *Determination of the Dispersion Relation for Surface Polaritons on a Grating on the Basis of Rayleigh's Hypothesis*

The solutions of eqs. (3.3)–(3.4) outside the selvedge region that vanish as $|x_3| \to \infty$ can be written in the forms

$$H_2(x_1, x_3 \mid \omega) = \sum_{p=-\infty}^{\infty} A_p(k\omega) \exp\left[\mathrm{i}\left(k + \frac{2\pi p}{a}\right) x_1\right] \exp[-\alpha_p(k\omega)x_3] \qquad x_3 > \zeta_{\max} \tag{3.7a}$$

$$= \sum_{p=-\infty}^{\infty} B_p(k\omega) \exp\left[\mathrm{i}\left(k + \frac{2\pi p}{a}\right) x_1\right] \exp[\beta_p(k\omega)x_3] \qquad x_3 < \zeta_{\min}, \tag{3.7b}$$

where k is the wave vector of the surface polariton, and

$$\alpha_p(k\omega) = \left[\left(k + \frac{2\pi p}{a}\right)^2 - \frac{\omega^2}{c^2}\right]^{1/2}; \qquad \left(k + \frac{2\pi p}{a}\right)^2 > \frac{\omega^2}{c^2} \tag{3.8a}$$

$$= -\mathrm{i}\left[\frac{\omega^2}{c^2} - \left(k + \frac{2\pi p}{a}\right)^2\right]^{1/2}; \qquad \left(k + \frac{2\pi p}{a}\right)^2 < \frac{\omega^2}{c^2}, \tag{3.8b}$$

while

$$\beta_p(k\omega) = \left[\left(k + \frac{2\pi p}{a}\right)^2 - \epsilon(\omega)\frac{\omega^2}{c^2}\right]^{1/2} \qquad \mathrm{Re}\,\beta_p(k\omega) > 0 \quad \mathrm{Im}\,\beta_p(k\omega) < 0. \tag{3.9}$$

The solutions (3.7) possess the Bloch property,

$$H_2(x_1 + a, x_3 \mid \omega) = \exp(\mathrm{i}ka) H_2(x_1, x_3 \mid \omega), \tag{3.10}$$

required by the periodicity of our system in the x_1-direction.

We now invoke the Rayleigh hypothesis for the determination of the coefficients $\{A_p(k\omega)\}$ and $\{B_p(k\omega)\}$. In the present context this hypothesis consists of the assumption that the solutions (3.7) which are exact outside the selvedge region, can be continued in to the surface itself, and substituted into the boundary conditions (3.5). When this is done we obtain the pair of equations

$$\sum_{p=-\infty}^{\infty} \{-\exp[-\alpha_p(k\omega)\zeta(x_1)] \exp[\mathrm{i}k_p x_1] A_p(k\omega) + \exp[\beta_p(k\omega)\zeta(x_1)] \exp[\mathrm{i}k_p x_1]\, B_p(k\omega)\} = 0 \tag{3.11a}$$

$$\sum_{p=-\infty}^{\infty} \left\{\left[\alpha_p(k\omega) + \mathrm{i}k_p \frac{\partial \zeta}{\partial x_1}\right] \exp[-\alpha_p(k\omega)\zeta(x_1)] \exp[\mathrm{i}k_p x_1]\, A_p(k\omega) + \frac{1}{\epsilon(\omega)}\left[\beta_p(k\omega) - \mathrm{i}k_p \frac{\partial \zeta}{\partial x_1}\right] \exp[\beta_p(k\omega)\zeta(x_1)] \exp[\mathrm{i}k_p x_1]\, B_p(k\omega)\right\} = 0, \tag{3.11b}$$

where we have simplified the notation by defining

$$k_p \equiv k + 2\pi p/a. \tag{3.12}$$

To proceed farther it is convenient to introduce the expansions

$$\exp[-\alpha_p(k\omega)\zeta(x_1)] = \sum_{n=-\infty}^{\infty} I_n^{(p)}(k\omega)\exp[\mathrm{i}(2\pi n/a)x_1] \tag{3.13a}$$

$$\exp[\beta_p(k\omega)\zeta(x_1)] = \sum_{n=-\infty}^{\infty} J_n^{(p)}(k\omega)\exp[\mathrm{i}(2\pi n/a)x_1], \tag{3.13b}$$

where the Fourier coefficients are formally defined by

$$I_n^{(p)}(k\omega) = (1/a)\int_0^a \exp[-\alpha_p(k\omega)\zeta(x_1)]\exp[-\mathrm{i}(2\pi n/a)x_1]\,\mathrm{d}x_1 \tag{3.14a}$$

$$J_n^{(p)}(k\omega) = (1/a)\int_0^a \exp[\beta_p(k\omega)\zeta(x_1)]\exp[-\mathrm{i}(2\pi n/a)x_1]\,\mathrm{d}x_1. \tag{3.14b}$$

It follows from these results that

$$\frac{\partial\zeta}{\partial x_1}\exp[-\alpha_p(k\omega)\zeta(x_1)] = -\frac{1}{\alpha_p(k\omega)}\sum_{n=-\infty}^{\infty}\left(\frac{2\pi \mathrm{i}n}{a}\right) I_n^{(p)}(k\omega)\exp\left(\mathrm{i}\frac{2\pi n}{a}x_1\right) \tag{3.15a}$$

$$\frac{\partial\zeta}{\partial x_1}\exp[\beta_p(k\omega)\zeta(x_1)] = \frac{1}{\beta_p(k\omega)}\sum_{n=-\infty}^{\infty}\left(\frac{2\pi \mathrm{i}n}{a}\right) J_n^{(p)}(k\omega)\exp\left(\mathrm{i}\frac{2\pi n}{a}x_1\right). \tag{3.15b}$$

When the expansions (3.13) and (3.15) are employed in eqs. (3.11), and the Fourier coefficient in each of the resulting equations is equated to zero, the following pair of coupled homogeneous equations for the coefficients $\{A_n(k\omega)\}$ and $\{B_n(k\omega)\}$ is obtained:

$$\sum_{p=-\infty}^{\infty}\{I_{r-p}^{(p)}(k\omega)A_p(k\omega) - J_{r-p}^{(p)}(k\omega)B_p(k\omega)\} = 0 \tag{3.16a}$$

$$\sum_{p=-\infty}^{\infty}\left\{\frac{\omega^2/c^2 - k_r k_p}{\alpha_p(k\omega)} I_{r-p}^{(p)}(k\omega)A_p(k\omega) + \frac{\epsilon(\omega)(\omega^2/c^2) - k_r k_p}{\epsilon(\omega)\beta_p(k\omega)} J_{r-p}^{(p)}(k\omega)B_p(k\omega)\right\} = 0. \tag{3.16b}$$

The dispersion relation for surface polaritons on a grating is obtained by equating to zero the determinant of the coefficients in these equations.

It is difficult, however, to deal with a pair of matrix equations, each of infinite order, even by purely numerical methods, when truncation of each set of equations would be employed, and convergence would be tested by increasing the number of terms in the expansions (3.7) in some systematic way. In the present case it is possible to obtain a single matrix equation for the coefficients $\{A_p(k\omega)\}$; alternatively, a single matrix equation can be obtained for the coefficients $\{B_p(k\omega)\}$. The dispersion relation for surface polaritons can then be obtained from either of these single sets of equations. We now show how this can be done.

For this purpose we return to the boundary conditions in the forms given by eqs. (3.11). We multiply eq. (3.11a) by $[\beta_r(k\omega)+ ik_r(\partial\zeta/\partial x_1)]\exp[\beta_r(k\omega)\zeta(x_1)-ik_rx_1]$ and integrate the resulting equation with respect to x_1 over the interval $(0, a)$. We next multiply eq. (3.11b) by $-\epsilon(\omega)\exp[\beta_r(k\omega)\zeta(x_1)-ik_rx_1]$ and integrate the resulting equation with respect to x_1 over the interval $(0, a)$. We add the two equations obtained in this way, with the result that

$$\begin{aligned}\sum_{p=-\infty}^{\infty}(1/a)\int_0^a dx_1\,\{&[-(\beta_r(k\omega)+\epsilon(\omega)\alpha_p(k\omega))-i(k_r+\epsilon(\omega)k_p)(\partial\zeta/\partial x_1)]\\ &\times\exp[(\beta_r(k\omega)-\alpha_p(k\omega))\zeta(x_1)]\exp[-i(k_r-k_p)x_1]\,A_p(k\omega)\\ &+[(\beta_r(k\omega)-\beta_p(k\omega))+i(k_r+k_p)(\partial\zeta/\partial x_1)]\\ &\times\exp[(\beta_r(k\omega)+\beta_p(k\omega))\zeta(x_1)]\exp[-i(k_r-k_p)x_1]B_p(k\omega)\}=0. \quad (3.17)\end{aligned}$$

The second integral can be shown to vanish with the aid of an integration by parts of the term containing $(\partial\zeta/\partial x_1)$ and the definition (3.9). The first integral can be simplified somewhat by integrating by parts the term containing $(\partial\zeta/\partial x_1)$. The result is the following equation for the coefficients $\{A_p(k\omega)\}$ alone:

$$\begin{aligned}\sum_{p=-\infty}^{\infty}(1-\epsilon(\omega))&\frac{\beta_r(k\omega)\alpha_p(k\omega)-k_rk_p}{\beta_r(k\omega)-\alpha_p(k\omega)}\\ &\times\frac{1}{a}\int_0^a dx_1\exp[(\beta_r(k\omega)-\alpha_p(k\omega))\zeta(x_1)]\\ &\times\exp[-i(k_r-k_p)x_1]\,A_p(k\omega)=0. \quad (3.18)\end{aligned}$$

In a similar fashion an equation for the $\{B_p(k\omega)\}$ alone can also be obtained. To achieve this we multiply eq. (3.11a) by $[\alpha_r(k\omega)-ik_r(\partial\zeta/\partial x_1)]\exp[-\alpha_r(k\omega)\zeta(x_1)-ik_rx_1]$, and eq. (3.11b) by $\exp[-\alpha_r(k\omega)\zeta(x_1)-ik_rx_1]$; we integrate each of the resulting equations with respect to x_1 over the interval $(0, a)$; and finally we add the two equations so obtained. The result is

$$\begin{aligned}\sum_{p=-\infty}^{\infty}\frac{1}{a}\int_0^a dx_1\,\Big\{&\Big[-(\alpha_r(k\omega)-\alpha_p(k\omega))+i(k_r+k_p)\frac{\partial\zeta}{\partial x_1}\Big]\\ &\times\exp[-(\alpha_r(k\omega)+\alpha_p(k\omega))\zeta(x_1)]\exp[-i(k_r-k_p)x_1]\,A_p(k\omega)\\ &+\Big[\Big(\alpha_r(k\omega)+\frac{\beta_p(k\omega)}{\epsilon(\omega)}\Big)-i\Big(k_r+\frac{k_p}{\epsilon(\omega)}\Big)\frac{\partial\zeta}{\partial x_1}\Big]\\ &\times\exp[-(\alpha_r(k\omega)-\beta_p(k\omega))\zeta(x_1)]\exp[-i(k_r-k_p)x_1]\,B_p(k\omega)\Big\}=0. \quad (3.19)\end{aligned}$$

In this case it is the first integral that can be shown to vanish if the term containing $(\partial\zeta/\partial x_1)$ is integrated by parts and the definition (3.8) is used. The second integral can be simplified through an integration by parts, and the resulting equation for the $\{B_p(k\omega)\}$ takes the form:

$$\sum_{p=-\infty}^{\infty}\left(1-\frac{1}{\epsilon(\omega)}\right)\frac{k_r k_p-\alpha_r(k\omega)\beta_p(k\omega)}{\alpha_r(k\omega)-\beta_p(k\omega)}$$
$$\times\frac{1}{a}\int_0^a dx_1\exp[-(\alpha_r(k\omega)-\beta_p(k\omega))\zeta(x_1)]$$
$$\times\exp[-i(k_r-k_p)x_1]\,B_p(k\omega)=0. \tag{3.20}$$

The dispersion relation for surface polaritons on a grating can therefore be obtained from the requirement that the determinant of the coefficients in either the set of equations (3.18) or (3.20) vanishes.

A calculation of the dispersion relation in this fashion would require the determination of the $r-p$ Fourier coefficient of the function $\exp[-(\alpha_r(k\omega)-\beta_p(k\omega))\zeta(x_1)]$ for the particular function $\zeta(x_1)$ chosen. This determination can always be carried out numerically. If the function $\zeta(x_1)$ is a simple sinusoid, e.g., if

$$\zeta(x_1)=\zeta_0\cos(2\pi x_1/a), \tag{3.21}$$

then

$$(\exp-(\alpha_r(k\omega)-\beta_p(k\omega))\zeta(x_1))_{r-p}=(-1)^{r-p}I_{r-p}((\alpha_r-\beta_p)\zeta_0), \tag{3.22}$$

where $I_n(z)$ is a modified Bessel function. If $\zeta(x_1)$ is represented by a sequence of straight line segments, the required Fourier coefficient can also be obtained analytically. Finding the zeros of the determinantal equation that is the dispersion relation for surface polaritons on a grating has to be done with the aid of a computer.

However, in the limit of small roughness the dispersion relation can be obtained in an explicit form, and an approximate analytic solution of it obtained. For this purpose we use eq. (3.18) for the $\{A_p(k\omega)\}$. The small roughness limit consists of making the approximation that

$$(1/a)\int_0^a dx_1\exp[(\beta_r(k\omega)-\alpha_p(k\omega))\zeta(x_1)]\exp[-i(k_r-k_p)x_1]$$
$$\approx(1/a)\int_0^a dx_1\{1+(\beta_r(k\omega)-\alpha_p(k\omega))\zeta(x_1)\}\exp[-i(2\pi/a)(r-p)x_1]$$
$$=\delta_{rp}+(\beta_r(k\omega)-\alpha_p(k\omega))\hat{\zeta}(r-p). \tag{3.23}$$

When this result is substituted into eq. (3.18), the latter takes the form

$$A_r(k\omega)=-\frac{\beta_r(k\omega)-\alpha_r(k\omega)}{k_r^2-\beta_r(k\omega)\alpha_r(k\omega)}\sum_{p=-\infty}^{\infty}\hat{\zeta}(r-p)(k_r k_p-\beta_r(k\omega)\alpha_p(k\omega))A_p(k\omega). \tag{3.24}$$

The coefficient $A_0(k\omega)$ describes a surface polariton propagating along a flat surface. We assume it is of O(1). Then, for $r \neq 0$, we obtain from eq. (3.24) the result that

$$A_r(k\omega) = -\frac{\beta_r(k\omega) - \alpha_r(k\omega)}{k_r^2 - \beta_r(k\omega)\alpha_r(k\omega)} \hat{\zeta}(r)\,[k_r k - \beta_r(k\omega)\alpha_0(k\omega)]A_0(k\omega) + O(\zeta^2). \tag{3.25}$$

The equation for $A_0(k\omega)$ is (recalling that $\hat{\zeta}(0) = 0$)

$$\frac{k^2 - \beta_0(k\omega)\alpha_0(k\omega)}{\beta_0(k\omega) - \alpha_0(k\omega)} A_0(k\omega) + \sum_{p\neq 0} \hat{\zeta}(-p) \times [kk_p - \beta_0(k\omega)\alpha_p(k\omega)]A_p(k\omega) + O(\zeta^2) = 0. \tag{3.26}$$

Substitution of eq. (3.25) into eq. (3.26) yields the dispersion relation for surface polaritons in the small roughness limit in the form

$$\frac{k^2 - \beta_0(k\omega)\alpha_0(k\omega)}{\beta_0(k\omega) - \alpha_0(k\omega)} = \sum_{p(\neq 0)} |\hat{\zeta}(p)|^2 \frac{\beta_p(k\omega) - \alpha_p(k\omega)}{k_p^2 - \beta_p(k\omega)\alpha_p(k\omega)} \times (kk_p - \beta_0(k\omega)\alpha_p(k\omega))(k_p k - \beta_p(k\omega)\alpha_0(k\omega)). \tag{3.27}$$

This result can be put into a more transparent form with the use of the identities

$$\beta_p^2(k\omega) - \alpha_p^2(k\omega) = (1 - \epsilon(\omega))(\omega^2/c^2) \tag{3.28a}$$

$$[k_p^2 - \beta_p(k\omega)\alpha_p(k\omega)][\beta_p(k\omega) + \alpha_p(k\omega)] = (\epsilon(\omega)\alpha_p(k\omega) + \beta_p(k\omega))(\omega^2/c^2). \tag{3.28b}$$

With the use of these results we obtain finally the dispersion relation in the form

$$\epsilon(\omega)\alpha_0(k\omega) + \beta_0(k\omega) = (1 - \epsilon(\omega))^2 \sum_{p(\neq 0)} |\hat{\zeta}(p)|^2 \times \frac{(kk_p - \beta_0(k\omega)\alpha_p(k\omega))(k_p k - \beta_p(k\omega)\alpha_0(k\omega))}{\epsilon(\omega)\alpha_p(k\omega) + \beta_p(k\omega)}. \tag{3.29}$$

The solution of eq. (3.29) gives the dispersion curve, $\omega = \omega(k)$, for a surface polariton on a grating.

We can obtain an approximate solution of this equation in the following way. We note that the dispersion relation in the absence of surface corrugations, i.e. for a flat surface, is

$$\epsilon(\omega)\alpha_0(k\omega) + \beta_0(k\omega) = 0. \tag{3.30}$$

We denote the solution of this equation by $\omega = \omega_0(k)$. We next note that if (k, ω) is a point on the flat surface dispersion curve (k_p, ω) will also be, provided that

$$k^2 = k_p^2 = (k + 2\pi p/a)^2,$$

i.e. that,

$$k = -\pi p/a, \qquad p = \pm 1, \pm 2, \pm 3, \dots \tag{3.31}$$

The values of k given by eq. (3.31) coincide with the boundaries of the Brillouin zones for our one-dimensional system. When k equals the value given by eq. (3.31) the pth term in the sum on the right hand side of eq. (3.29) diverges as ω approaches $\omega_0(k)$. Thus for $k = -(\pi_p/a)$ we keep only the pth term in the sum and obtain for the dispersion relation

$$\epsilon(\omega)\alpha_0(k\omega) + \beta_0(k\omega) \cong (1-\epsilon(\omega))^2|\hat{\zeta}(p)|^2 \times \frac{(k^2 + \alpha_0(k\omega)\beta_0(k\omega))^2}{\epsilon(\omega)\alpha_0(k\omega) + \beta_0(k\omega)}, \qquad k = -\frac{\pi p}{a}, \tag{3.32}$$

where we have used the fact that for $k = -(\pi p/a)$, $\alpha_p(k\omega) = \alpha_0(k\omega)$, and $\beta_p(k\omega) = \beta_0(k\omega)$. Equation (3.32) can be rewritten in the form

$$\epsilon(\omega)\alpha_0(k\omega) + \beta_0(k\omega) = \pm|1-\epsilon(\omega)|\,|\hat{\zeta}(p)| \times (k^2 + \alpha_0(k\omega)\beta_0(k\omega)), \qquad k = -\pi p/a. \tag{3.33}$$

The content of eq. (3.33) is that a gap opens up in the surface polariton dispersion curve at $k = -(\pi p/a)$, and the values of $\omega(k)$ on the two sides of the gap are given by the solutions of eq. (3.33). If we denote the solution of eq. (3.33) by $\omega(k) = \omega_0(k) \pm \Delta\omega(k)$, we find that to lowest order in $|\hat{\zeta}(k)|$:

$$\omega(k) = \omega_0(\pi p/a) \pm \tfrac{1}{2}\omega_G(p), \tag{3.34a}$$

where

$$\omega_G(p) = 2|\hat{\zeta}(p)|(|\epsilon(\omega)|-1)\left.\frac{k^2 + \alpha_0(k\omega)\beta_0(k\omega)}{\left|\frac{\partial}{\partial\omega}(\epsilon(\omega)\alpha_0(k\omega) + \beta_0(k\omega))\right|}\right|_{\substack{k=\pi p/a \\ \omega=\omega_0(\pi p/a)}}$$

$$= 4|\hat{\zeta}(p)|\frac{\omega^2}{c}\frac{|\epsilon(\omega)|}{(|\epsilon(\omega)|-1)^{3/2}}\left[1 + \frac{\omega\epsilon'(\omega)}{2|\epsilon(\omega)|(|\epsilon(\omega)|-1)}\right]^{-1}\Bigg|_{\omega=\omega_0(\pi p/a)} \tag{3.34b}$$

In writing eqs. (3.34) we have used the fact that the frequency of the surface polariton for $k = -(\pi p/a)$ is the same as for $k = +(\pi p/a)$.

One can go further and obtain the form of the dispersion curve in the vicinity of the gap at $k = -(\pi p/a)$. If we write $k = \kappa_p + \Delta k$, where $\kappa_p = -(\pi p/a)$, and $\omega(k) = \omega_p(\kappa_p) + \Delta\omega(k)$, we find that as $\Delta k \to 0$

$$\omega(k) = \omega_0(\pi p/a) \pm [A^2(p)(c\Delta k)^2 + \tfrac{1}{4}\omega_G^2(p)]^{1/2}, \tag{3.35}$$

where the upper sign corresponds to $\Delta k > 0$, the lower to $\Delta k < 0$, and

$$A(p) = \frac{(|\epsilon(\omega)| - 1)^{1/2}}{|\epsilon(\omega)|}\left[1 + \frac{\omega\epsilon'(\omega)}{2|\epsilon(\omega)|(|\epsilon(\omega)| - 1)}\right]^{-1}\Bigg|_{\omega=\omega_0(\pi p/a)}. \tag{3.36}$$

Thus, this derivation of the dispersion curve for a surface polariton on a grating by the Rayleigh method in the small roughness limit shows that a gap is opened up on the flat surface dispersion curve at values of $k = -\pi p/a$ for each value of p for which $\hat{\zeta}(p)$ is nonzero. This is depicted qualitatively in fig. 5.

It might be thought that because the right hand side of eq. (3.29) is of second order in $\hat{\zeta}(p)$ consistency would have required that we keep terms of second order in $\hat{\zeta}(p)$ in the expansion on the right hand side of eq. (3.23). This, however, is not the case. For, if the terms of second order in $\hat{\zeta}(p)$ are kept on the right hand side of eq. (3.23), the dispersion relation that replaces eq. (3.29) is, to second order in $\hat{\zeta}(p)$,

$$\begin{aligned}\epsilon(\omega)\alpha_0(k\omega) + \beta_0(k\omega) = (1-\epsilon(\omega))^2 \sum_{p(\neq 0)} |\hat{\zeta}(p)|^2 \\ \times \frac{(kk_p - \beta_0(k\omega)\alpha_p(k\omega))(k_pk - \beta_p(k\omega)\alpha_0(k\omega))}{\epsilon(\omega)\alpha_p(k\omega) + \beta_p(k\omega)} \\ -\tfrac{1}{2}(\epsilon(\omega)\alpha_0(k\omega) + \beta_0(k\omega)) \\ \times (\beta_0(k\omega) - \alpha_0(k\omega))^2 \sum_{p(\pm 0)} |\hat{\zeta}(p)|^2.\end{aligned} \tag{3.37}$$

It is readily seen that because the second term on the right hand side of eq. (3.37) is proportional to the term on the left hand side of this equation, it cannot affect the dispersion relation to $O(|\hat{\zeta}(p)|^2)$: it first contributes to the dispersion relation to $O(|\hat{\zeta}(p)|^4)$.

Up to this point the discussion in this subsection has dealt with the

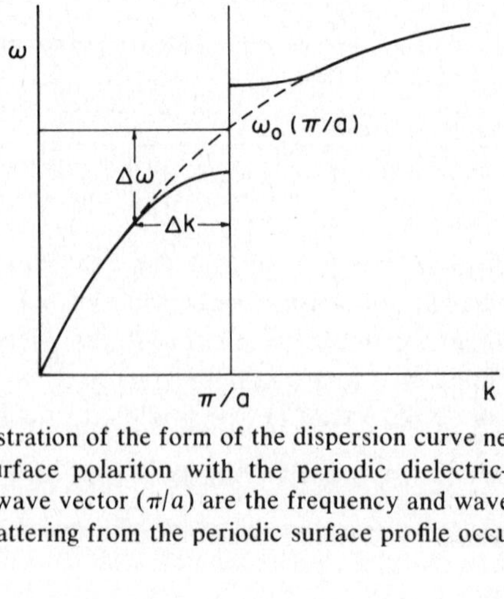

Fig. 5. A schematic illustration of the form of the dispersion curve near the gap opened up by the interaction of a surface polariton with the periodic dielectric–vacuum interface. The frequency $\omega_0(\pi/a)$ and wave vector (π/a) are the frequency and wave vector for which Bragg scattering from the periodic surface profile occurs.

determination of the dispersion relation for surface polaritons on a grating. However, the results obtained here can be used for the discussion of a complementary problem, viz., the diffraction of a surface polariton as it propagates over a grating. We conclude this subsection with a brief consideration of this topic.

We have already pointed out above that the terms with $m = 0$ in eqs. (3.7) describe a p-polarized surface polariton propagating in the x_1-direction along a *flat* dielectric–vacuum interface. According to the discussion in sect. 2.1 this requires that

$$A_0(k\omega) = B_0(k\omega)$$

in eqs. (3.7), and that the frequency ω and wave vector k of the surface polariton be related by the dispersion relation

$$k = \frac{\omega}{c}\left(\frac{\epsilon(\omega)}{\epsilon(\omega)+1}\right)^{1/2} \tag{3.38}$$

in a frequency range where $\epsilon(\omega) < 0$, provided that $k > \omega/c$.

In the small roughness limit eqs. (3.18) and (3.20) can be used to solve for the coefficient $\{A_p(k\omega)\}$ and $\{B_p(k\omega)\}$ for $p \neq 0$ in terms of $A_0(k\omega)$. To first order in $\hat{\zeta}(p)$ we have that

$$A_p(k\omega) = -\frac{(1-\epsilon(\omega))}{\epsilon(\omega)\alpha_p + \beta_p}\,\hat{\zeta}(p)(k_pk_0 - \beta_p\alpha_0)\,A_0(k\omega), \tag{3.39a}$$

$$B_p(k\omega) = -\frac{(1-\epsilon(\omega))}{\epsilon(\omega)\alpha_p + \beta_p}\,\hat{\zeta}(p)(k_pk_0 - \alpha_p\beta_o)\,A_0(k\omega). \tag{3.39b}$$

The first order diffracted field is therefore obtained from eqs. (3.7) as

$$H_2^{(1)}(x_1, x_3 \,|\, \omega) = -(1-\epsilon(\omega))A_0(k\omega)\sum_{p=-\infty}^{\infty}\hat{\zeta}(p) \times \frac{k_pk_0 - \beta_p\alpha_0}{\epsilon(\omega)\alpha_p + \beta_p}\exp(ik_px_1 - \alpha_px_3) \qquad x_3 > \zeta_{\max} \tag{3.40a}$$

$$= -(1-\epsilon(\omega))\,A_0(k\omega)\sum_{p=-\infty}^{\infty}\hat{\zeta}(p)\frac{k_pk_0 - \alpha_p\beta_0}{\epsilon(\omega)\alpha_p + \beta_p} \times \exp(ik_px_1 + \beta_px_3) \qquad x_3 < \zeta_{\min}. \tag{3.40b}$$

We now note from the definition, eq. (3.9), that since $\epsilon(\omega) < 0$ for a point on the flat surface dispersion curve given by eq. (3.38), $\beta_p(k\omega)$ must always be real and positive, for all p, for ω and k satisfying the flat surface dispersion relation. Therefore, in the approximation given by eq. (3.40b), and indeed in all higher approximations, the diffracted waves in the substrate cannot be radiative: they are always localized to the interface.

The situation is different for the diffracted waves in the vacuum given, in first approximation, by eq. (3.40a). In this case it is quite possible for ω and

k to be related by eq. (3.38), when $k > \omega/c$, and still have

$$(k + 2\pi p/a)^2 < \omega^2/c^2, \tag{3.41}$$

for some value(s) of p. We note, however, that if these conditions are satisfied by some p, they cannot be for $-p$. Consequently, at most only one of the pair of first order diffracted fields, defined by p and $-p$, can be radiative.

It is straightforward to determine the portions of the flat surface dispersion curve that give rise to diffracted fields that are radiative in nature. Equation (3.41) can be rewritten as

$$\omega/c > |k + 2\pi p/a|. \tag{3.42}$$

Consequently, if one plots the (infinite) set of straight lines $\omega/c = |k + 2\pi p/a|$ on a flat surface dispersion curve, the portions of that curve lying to the right of the light line $\omega = ck$ (for $k > 0$) that also lie above the curves $\omega = |k + 2\pi p/a|$ will correspond to incident surface polaritons that will radiate into the vacuum above the grating due to their interaction with the periodic surface profile. This is depicted in figs. 6a and 6b for two different grating periods a. The same flat surface dispersion curve has been used in plotting these figures as is shown in fig. 2. The radiative portions of the curve are denoted by the heavy lines in both curves.

Incident surface polaritons belonging to these portions of the dispersion curve will be damped by radiating their energy into the vacuum. The energy conservation method used in sect. 4.2 to obtain the damping of a surface polariton by roughness on a randomly rough surface can be used here to obtain the mean free path of a surface polariton on a grating. We do not present such a calculation here, however, and content ourselves with merely pointing out its possibility.

All of the results presented up to this point have been obtained on the basis of the Rayleigh hypothesis. However, it is known now that the Rayleigh hypothesis has a limited range of applicability, although results obtained by its use may have an asymptotic nature even when they formally diverge (Goodman 1977, Hill and Celli 1978). It is known not to be valid if the surface profile function $\zeta(x_1)$ is not an analytic function of x_1 (Millar 1969, Hill and Celli 1978). Thus it is invalid for profiles that possess sharp curves, for example. Even when $\zeta(x_1)$ is an analytic function of x_1, the Rayleigh hypothesis is valid only when the surface profile is sufficiently smooth, i.e. when, crudely spaking, the slope of the profile at each point is sufficiently small. Methods are now available (Hill and Celli 1978) for determining the limits of validity of Rayleigh hypothesis that place the preceding qualitative remarks on a quantitative footing. The conclusion one is led to is that the Rayleigh hypothesis breaks down for surfaces that are strongly corrugated.

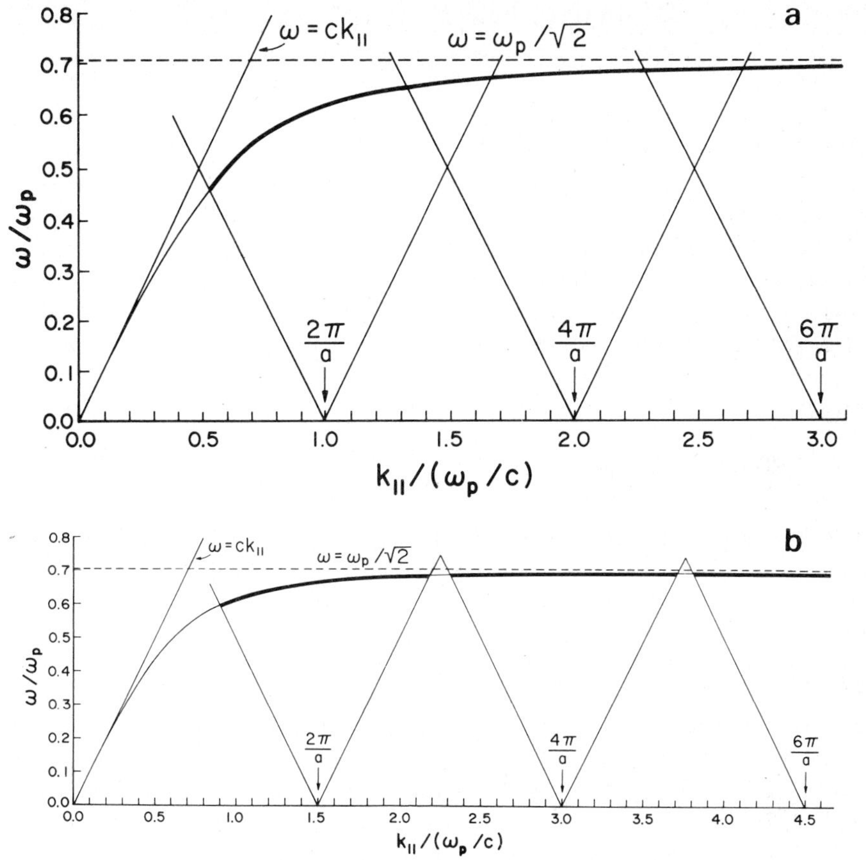

Fig. 6. A depiction of the regions of the dispersion curve for a surface polariton at a flat dielectric–vacuum interface in which it will radiate energy into the vacuum due to its interaction with a periodic surface profile, for two different periods of the profile.

Since not all the periodically corrugated surfaces one is likely to encounter in practice are sufficiently smooth that the Rayleigh hypothesis is valid for them, it is desirable to have alternative ways of studying the propagation of surface polaritons over a grating that are not subject to the limitations of methods based on the Rayleigh hypothesis. Such methods have been developed in recent years, and we turn to a discussion of them in the following subsection.

3.1.2. Derivation of Exact Dispersion Relations for Surface Polaritons on a Grating

The starting point for the derivation of an exact dispersion relation for surface polaritons on a grating is Green's theorem (see, for example,

Jackson 1962). If $u(x)$ and $\upsilon(x)$ are arbitrary scalar fields defined in the volume V bounded by the closed surface Σ, this theorem states that

$$\int_V (u\nabla^2\upsilon - \upsilon\nabla^2 u)\, \mathrm{d}^3x = \int_\Sigma \left(u\frac{\partial\upsilon}{\partial n} - \upsilon\frac{\partial u}{\partial n}\right) \mathrm{d}S, \tag{3.43}$$

where $\partial/\partial n$ is the normal derivative at the surface Σ, directed outward from inside the volume V.

We consider first the case that the volume V is the vacuum above the dielectric, $x_3 > \zeta(x_1)$. In this case the surface Σ consists of two parts: the surface S of the dielectric defined by the equation $x_3 = \zeta(x_\parallel)$, and a hemisphere in the upper half space, $S^{(+\infty)}$, whose radius is allowed to become infinite. We seek the function $H_2^>(x_1x_3\,|\,\omega)$ that satisfies the equation

$$\left(\frac{\partial^2}{\partial x_1^2} + \frac{\partial^2}{\partial x_3^2} + \frac{\omega^2}{c^2}\right) H_2^>(x_1x_3\,|\,\omega) = 0 \qquad x_3 > \zeta(x_1) \tag{3.44}$$

and vanishes exponentially rapidly as $x_3 \to +\infty$. To this end we introduce the Green's function $G_0(x_1x_3\,|\,x_1'x_3')$ that satisfies the equation

$$\left(\frac{\partial^2}{\partial x_1^2} + \frac{\partial^2}{\partial x_3^2} + \frac{\omega^2}{c^2}\right) G_0(x_1x_3\,|\,x_1'x_3') = -4\pi\delta(x_1 - x_1')\delta(x_3 - x_3') \tag{3.45}$$

and outgoing wave or decaying exponential boundary conditions at infinity. We multiply eq. (3.45) from the left by $H_2^>(x_1x_3\,|\,\omega)$, and subtract from the resulting equation the equation that is obtained by multiplying eq. (3.44) from the left by $G_0(x_1x_3\,|\,x_1'x_3')$. When the result is integrated over the volume V and eq. (3.43) is used, we obtain the pair of integral equations

$$\left.\begin{matrix} x_3 > \zeta(x_1) & H_2^>(x_1x_3\,|\,\omega) \\ x_3 < \zeta(x_1) & 0 \end{matrix}\right\} = \frac{1}{4\pi}\int_S \Big(G_0(x_1x_3\,|\,x_1'x_3') \tag{3.46a}$$

$$\times \frac{\partial}{\partial n_-'} H_2^>(x_1'x_3'\,|\,\omega) - \frac{\partial}{\partial n_-'} G_0(x_1x_3\,|\,x_1'x_3')\, H_2^>(x_1'x_3'\,|\,\omega)\Big)\, \mathrm{d}s_1'. \tag{3.46b}$$

Equation (3.46b) expresses the extinction theorem (Wolf 1973): the field and its normal derivative on the surface act as sources that "extinguish" the field in the dielectric medium. Since we seek a function $H_2^>(x_1x_3\,|\,\omega)$ that vanishes as $x_3 \to +\infty$ there is no contribution to the integral over the surface Σ from the integration over the surface of the hemisphere of infinite radius, $S^{(+\infty)}$. In writing eq. (3.46) we have utilized the symmetry of $G_0(x_1x_3\,|\,x_1'x_3')$ expressed by

$$G_0(x_1x_3\,|\,x_1'x_3') = G_0(x_1'x_3'\,|\,x_1x_3). \tag{3.47}$$

That $G_0(x_1x_3\,|\,x_1'x_3')$ possesses this symmetry property follows from the explicit expressions for this function given by

$$G_0(x_1x_3 \mid x_1'x_3') = \mathrm{i}\pi H_0^{(1)}((\omega/c)[(x_1-x_1')^2+(x_3-x_3')^2]^{1/2}) \tag{3.48a}$$

$$= \int_{-\infty}^{\infty} \mathrm{d}q \, \frac{1}{\alpha(q\omega)} \exp[\mathrm{i}q(x_1-x_1') - \alpha(q\omega)\,|x_3-x_3'|], \tag{3.48b}$$

where $H_0^{(1)}(z)$ is a Hankel function (Abramowitz and Stegun 1968), and

$$\alpha(q\omega) = (q^2-\omega^2/c^2)^{1/2} \qquad q^2 > \omega^2/c^2 \tag{3.49a}$$

$$= -\mathrm{i}(\omega^2/c^2-q^2)^{1/2} \qquad q^2 < \omega^2/c^2. \tag{3.49b}$$

We now turn to the case that the volume V is the dielectric medium, $x_3 < \zeta(x_1)$. In this case the surface Σ consists of the surface S of the dielectric medium and a hemisphere of infinite radius in the lower half space, $S^{(-\infty)}$. In this case we seek the function $H_2^<(x_1x_3 \mid \omega)$ that satisfies the equation

$$\left(\frac{\partial^2}{\partial x_1^2} + \frac{\partial^2}{\partial x_3^2} + \epsilon(\omega)\frac{\omega^2}{c^2}\right) H_2^<(x_1x_3 \mid \omega) = 0 \qquad x_3 < \zeta(x_1), \tag{3.50}$$

and vanishes exponentially rapidly as $x_3 \to -\infty$. We introduce the Green's function $G_\epsilon(x_1x_3 \mid x_1'x_3')$ that satisfies the equation

$$\left(\frac{\partial^2}{\partial x_1^2} + \frac{\partial^2}{\partial x_3^2} + \epsilon(\omega)\frac{\omega^2}{c^2}\right) G_\epsilon(x_1x_3 \mid x_1'x_3') = -4\pi\delta(x_1-x_1')\delta(x_3-x_3') \tag{3.51}$$

and outgoing wave or decaying exponential boundary conditions at infinity. This Green's function has the representations

$$G_\epsilon(x_1x_3 \mid x_1'x_3') = \mathrm{i}\pi H_0^{(1)}(\epsilon(\omega)^{1/2}(\omega/c)[(x_1-x_1')^2+(x_3-x_3')^2]^{1/2}) \qquad \epsilon(\omega) > 0 \tag{3.52a}$$

$$= 2K_0(|\epsilon(\omega)|^{1/2}(\omega/c)[(x_1-x_1')^2+(x_3-x_3')^2]^{1/2}) \qquad \epsilon(\omega) < 0, \tag{3.52b}$$

where $K_0(z)$ is a modified Bessel function. A Fourier integral representation of this Green's function is

$$G_\epsilon(x_1x_3 \mid x_1'x_3') = \int_{-\infty}^{\infty} \mathrm{d}q \, \frac{1}{\beta(q\omega)} \exp[\mathrm{i}q(x_1-x_1') - \beta(q\omega)\,|x_3-x_3'|] \tag{3.53}$$

where

$$\beta(q\omega) = (q^2-\epsilon(\omega)(\omega^2/c^2))^{1/2} \qquad \mathrm{Re}\,\beta(q\omega) > 0, \quad \mathrm{Im}\,\beta(q\omega) < 0. \tag{3.54}$$

When we apply Green's theorem to the space occupied by the dielectric medium, we obtain the pair of integral equations

$$\left.\begin{matrix} x_3 > \zeta(x_1) & 0 \\ x_3 < \zeta(x_1) & H_2^<(x_1x_3 \mid \omega) \end{matrix}\right\} = \frac{1}{4\pi}\int_S \bigg(G_\epsilon(x_1x_3 \mid x_1'x_3') \tag{3.55a}$$

$$\times \frac{\partial}{\partial n_+'} H_2^<(x_1'x_3' \mid \omega) - \frac{\partial}{\partial n_+'} G_\epsilon(x_1x_3 \mid x_1'x_3') H_2^<(x_1'x_3' \mid \omega)\bigg)\,\mathrm{d}s_1'. \tag{3.55b}$$

Equation (3.55a) expresses the extinction theorem in this case. If we take into account the continuity conditions (3.5) we can rewrite this pair of equations in the form

$$\left.\begin{matrix} x_3 > \zeta(x_1) & 0 \\ x_3 < \zeta(x_1) & H_2^{<}(x_1x_3 \mid \omega) \end{matrix}\right\} = -\frac{1}{4\pi}\int_S \Big(\epsilon(\omega) G_\epsilon(x_1x_3 \mid x_1'x_3') \tag{3.56a}$$

$$\times \frac{\partial}{\partial n'} H_2^{>}(x_1'x_3' \mid \omega) - \frac{\partial}{\partial n_-'} G_\epsilon(x_1x_3 \mid x_1'x_3') H_2^{>}(x_1'x_3' \mid \omega)\Big)\, ds_1'. \tag{3.56b}$$

Equations (3.46) and (3.56) are not all independent, however. A complete set of equations for obtaining the values of $H_2^{>}(x_1x_3 \mid \omega)$ and $\partial H_2^{>}(x_1x_3 \mid \omega)/\partial n_-$ on the surface S can be obtained by letting $x_3 \to \zeta(x_1)+$ in eq. (3.46a) and letting $x_3 \to \zeta(x_1)-$ in eq. (3.56b). The left hand side becomes $H_2^{>}(x_1, \zeta(x_1))$ in each case, in view of the continuity condition (3.5a), and we obtain the pair of equations

$$H_2^{>}(x_1\zeta(x_1) \mid \omega) = \frac{1}{4\pi} \lim_{x_3 \to \zeta(x_1)+} \int_S \Big(G_0(x_1x_3 \mid x_1'x_3') \frac{\partial}{\partial n_-'} H_2^{>}(x_1'x_3' \mid \omega)$$

$$- \frac{\partial}{\partial n_-'} G_0(x_1x_3 \mid x_1'x_3')\, H_2^{>}(x_1'x_3' \mid \omega)\Big)\, ds_1' \tag{3.57a}$$

$$H_2^{>}(x_1\zeta(x_1) \mid \omega) = -\frac{1}{4\pi} \lim_{x_3 \to \zeta(x_1)-} \int_S \Big(\epsilon(\omega) G_\epsilon(x_1x_3 \mid x_1'x_3')$$

$$\times \frac{\partial}{\partial n_-'} H_2^{>}(x_1'x_3' \mid \omega) - \frac{\partial}{\partial n_-'} G_\epsilon(x_1x_3 \mid x_1'x_3')\, H_2^{>}(x_1'x_3' \mid \omega)\Big)\, ds_1'. \tag{3.57b}$$

Alternatively, we can obtain a second pair of equations for determining the values of $H_2^{>}(x_1x_3 \mid \omega)$ and $\partial H_2^{>}(x_1x_3 \mid \omega)/\partial n_-$ on the surface S by letting $x_3 \to \zeta(x_1)-$ in eq. (3.46b) and letting $x_3 \to \zeta(x_1)+$ in eq. (3.56a). In this way we obtain

$$0 = \frac{1}{4\pi} \lim_{x_3 \to \zeta(x_1)-} \int_S \Big(G_0(x_1x_3 \mid x_1'x_3') \frac{\partial}{\partial n_-'} H_2^{>}(x_1'x_3' \mid \omega)$$

$$- \frac{\partial}{\partial n_-'} G_0(x_1x_3 \mid x_1'x_3')\, H_2^{>}(x_1'x_3' \mid \omega)\Big)\, ds_1' \tag{3.58a}$$

$$0 = \frac{-1}{4\pi} \lim_{x_3 \to \zeta(x_1)+} \int_S \Big(\epsilon(\omega) G_\epsilon(x_1x_3 \mid x_1'x_3') \frac{\partial}{\partial n_-'} H_2^{>}(x_1'x_3' \mid \omega)$$

$$- \frac{\partial}{\partial n_-'} G_\epsilon(x_1x_3 \mid x_1'x_3')\, H_2^{>}(x_1'x_3' \mid \omega)\Big)\, ds_1'. \tag{3.58b}$$

The two sets of equations (3.57) and (3.58) are equivalent, however. To see this, and at the same time obtain a pair of equations that is better suited for numerical calculations than either eqs. (3.57) or eqs. (3.58), we evaluate the limits in these equations explicitly. This has to be done carefully, because of the singularity possessed by the Green's functions $G_0(x_1x_3 | x_1'x_3')$ and $G_\epsilon(x_1x_3 | x_1'x_3')$ when $x_1 = x_1'$ and $x_3 = x_3'$.

In the succeeding discussion we follow the analysis of Meecham (1956). The singularity in each Green's function is given by

$$G(x_1x_3 | x_1'x_3') = -2 \ln[(x_1 - x_1')^2 + (x_3 - x_3')^2]^{1/2}$$
$$+ \text{constant} + \text{terms that vanish as } x_{1,3} \to x_{1,3}', \tag{3.59}$$

where G on the left hand side stands for either G_0 or G_ϵ.

We first consider the integral

$$I_1^\pm = \lim_{x_3 \to \zeta(x_1)\pm} \int_S G(x_1x_3 | x_1'x_3')\, \psi(x_1'x_3')\, ds_1' \tag{3.60}$$

where $\psi(x_1x_3)$ is an arbitrary function. We break up this integral into two contributions

$$I_1^\pm = \lim_{x_3 \to \zeta(x_1)\pm} \int_{S-S_0} G(x_1x_3 | x_1'x_3')\, \psi(x_1'x_3')\, ds_1'$$
$$+ \lim_{x_3 \to \zeta(x_1)\pm} \int_{S_0} G(x_1x_3 | x_1'x_3')\, \psi(x_1'x_3')\, ds_1' = I_{1a}^\pm + I_{1b}^\pm. \tag{3.61}$$

Here S_0 is a portion of the surface that is disposed symmetrically about the point $(x_1, \zeta(x_1))$, and whose length will ultimately be allowed to go to zero. We assume that the length of this portion, 2ϵ, is sufficiently small that the surface can be regarded as a straight line segment within it that, in general, is inclined with respect to the x_1-axis. In the second of these integrals we replace G by its singular part, given by eq. (3.59), and transform to coordinates ξ_1 and ξ_3 that are tangent to and normal to the surface S at the point $(x_1, \zeta(x_1))$ (see fig. 7). We assume that $\psi(x_1x_3)$ is sufficiently smoothly varying along the surface in the vicinity of the point $(x_1, \zeta(x_1))$ that we can evaluate it at this point and take it outside the integral. In this way we find that

$$I_{1b}^\pm \simeq 2\psi(x_1, \zeta(x_1)) \lim_{\substack{\xi_3 \to 0\pm \\ \epsilon \to 0}} \int_{-\epsilon}^{\epsilon} \ln[\xi_1^2 + \xi_3^2]^{1/2}\, d\xi_1$$

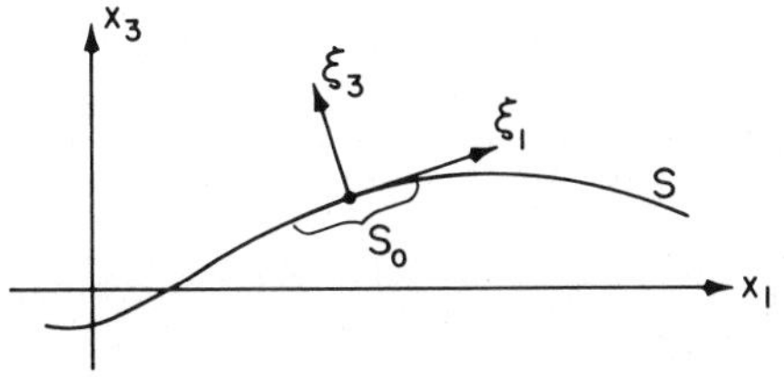

Fig. 7. A portion of a corrugated surface in the vicinity of the point where the normal derivative of the Green's function has a singularity.

$$= -4\psi(x_1, \zeta(x_1)) \lim_{\epsilon\to 0} \int_0^{\epsilon} \ln \xi_1 \, d\xi_1 = 0. \tag{3.62}$$

It follows that we can write the integral $I_1^{\pm}$ as

$$I_1^{\pm} = \int_S G(x_1\, \zeta(x_1) \,|\, x_1'x_3')\psi(x_1'x_3')\, ds_1'. \tag{3.63}$$

The situation is different when we consider the integral

$$I_2^{\pm} = \lim_{x_3\to\zeta(x_1)\pm} \int_S \frac{\partial}{\partial n_-'} G(x_1x_3 \,|\, x_1'x_3')\, \psi(x_1'x_3')\, ds_1' \tag{3.64}$$

$$\begin{aligned} &= \lim_{x_3\to\zeta(x_1)\pm} \int_{S-S_0} \frac{\partial}{\partial n_-'} G(x_1x_3 \,|\, x_1'x_3')\, \psi(x_1'x_3')\, ds_1' \\ &\quad + \lim_{x_3\to\zeta(x_1)\pm} \int_{S_0} \frac{\partial}{\partial n_-'} G(x_1x_3 \,|\, x_1'x_3')\, \psi(x_1'x_3')\, ds_1' \\ &= I_{2a}^{\pm} + I_{2b}^{\pm}. \end{aligned} \tag{3.65}$$

Proceeding as in the evaluation of $I_{1b}^{\pm}$, we obtain for $I_{2b}^{\pm}$ the result that

$$\begin{aligned} I_{2b}^{\pm} &\sim -2\psi(x_1, \zeta(x_1)) \lim_{\substack{\xi_3\to 0\pm \\ \epsilon\to 0}} \int_{-\epsilon}^{\epsilon} \frac{\xi_3 d\xi_1}{\xi_1^2 + \xi_3^2} \\ &= -4\psi(x_1, \zeta(x_1)) \lim_{\substack{\xi_3\to 0\pm \\ \epsilon\to 0}} \operatorname{sgn} \xi_3 \int_0^{\epsilon/|\xi_3|} \frac{du}{1+u^2} \\ &= \mp 2\pi\psi(x_1, \zeta(x_1)). \end{aligned} \tag{3.66}$$

The integral $I_2^{\pm}$ can therefore be written as

$$I_2^{\pm} = P \int \frac{\partial}{\partial n_-'} G(x_1\, \zeta(x_1) \,|\, x_1'x_3')\, \psi(x_1'x_3')\, ds_1' \mp 2\pi\psi(x_1, \zeta(x_1)), \tag{3.67}$$

where P indicates that the Cauchy principal value of the integral is to be evaluated.

When we use the results given by eqs. (3.63) and (3.67) in the pair of equations (3.57) the latter take the form

$$\begin{aligned} H_2^{>}(x_1, \zeta(x_1) \,|\, \omega) &= \frac{1}{2\pi} \int_S G_0(x_1\, \zeta(x_1) \,|\, x_1'x_3') \frac{\partial}{\partial n_-'} H_2^{>}(x_1'x_3' \,|\, \omega)\, ds_1' \\ &\quad - \frac{1}{2\pi} P \int_S \frac{\partial}{\partial n_-'} G_0(x_1\, \zeta(x_1) \,|\, x_1'x_3')\, H_2^{>}(x_1'x_3' \,|\, \omega)\, ds_1' \end{aligned} \tag{3.68a}$$

$$\begin{aligned} H_2^{<}(x_1\zeta(x_1) \,|\, \omega) &= -\frac{1}{2\pi} \int_S \epsilon(\omega) G_\epsilon(x_1\, \zeta(x_1) \,|\, x_1'x_3') \frac{\partial}{\partial n_-'} H_2^{<}(x_1'x_3' \,|\, \omega)\, ds_1' \\ &\quad + \frac{1}{2\pi} P \int_S \frac{\partial}{\partial n_-'} G_\epsilon(x_1\zeta(x_1) \,|\, x_1'x_3')\, H_2^{<}(x_1'x_3' \,|\, \omega)\, ds_1'. \end{aligned} \tag{3.68b}$$

If, instead, we use the results of eqs. (3.63) and (3.67) in the pair of equations (3.58), we obtain exactly the same pair of equations as is given by eqs. (3.68).

We can convert the integrals on the right hand side of eqs. (3.68) into integrals over the surface $x_3' = 0$ by noting that

$$\mathrm{d}s_1' = \left[1 + \left(\frac{\mathrm{d}\zeta(x_1')}{\mathrm{d}x_1'}\right)^2\right]^{1/2} \mathrm{d}x_1', \tag{3.69}$$

and setting $x_3' = \zeta(x_1')$ in the integrands. We also have that

$$\frac{\partial}{\partial n_-'} = \left[1 + \left(\frac{\mathrm{d}\zeta(x_1')}{\mathrm{d}x_1'}\right)^2\right]^{-1/2} \left(\frac{\mathrm{d}\zeta(x_1')}{\mathrm{d}x_1'}\frac{\partial}{\partial x_1'} - \frac{\partial}{\partial x_3'}\right). \tag{3.70}$$

Taking these results into account we rewrite eqs. (3.68) finally as

$$\begin{aligned} H(x_1 | \omega) = &\frac{1}{2\pi}\int G_0(x_1\, \zeta(x_1) | x_1' \zeta(x_1')) L(x_1' | \omega)\, \mathrm{d}x_1' \\ &- \frac{1}{2\pi} P \int \left[\frac{\mathrm{d}\zeta(x_1')}{\mathrm{d}x_1'} \frac{\partial G_0(x_1 \zeta(x_1) | x_1' x_3')}{\partial x_1'} \right. \\ &\left. - \frac{\partial G_0(x_1 \zeta(x_1) | x_1' x_3')}{\partial x_3'}\right]_{x_3' = \zeta(x_1')} H(x_1' | \omega)\, \mathrm{d}x_1', \end{aligned} \tag{3.71a}$$

$$\begin{aligned} H(x_1 | \omega) = &-\frac{1}{2\pi}\int \epsilon(\omega) G_\epsilon(x_1 \zeta(x_1) | x_1' \zeta(x_1')) L(x_1' | \omega)\, \mathrm{d}x_1' \\ &+ \frac{1}{2\pi} P \int \left[\frac{\mathrm{d}\zeta(x_1')}{\mathrm{d}x_1'} \frac{\partial G_\epsilon(x_1 \zeta(x_1) | x_1' x_3')}{\partial x_1'} \right. \\ &\left. - \frac{\partial G_\epsilon(x_1 \zeta(x_1) | x_1' x_3')}{\partial x_3'}\right]_{x_3' = \zeta(x_1')} H(x_1' | \omega)\, \mathrm{d}x_1', \end{aligned} \tag{3.71b}$$

where

$$H(x_1 | \omega) = H_2^>(x_1 \zeta(x_1) | \omega), \tag{3.72a}$$

$$L(x_1 | \omega) = \left[1 + \left(\frac{\mathrm{d}\zeta(x_1)}{\mathrm{d}x_1}\right)^2\right]^{1/2} \frac{\partial}{\partial n_-} H_2^>(x_1 x_3 | \omega)\bigg|_{x_3 = \zeta(x_1)}. \tag{3.72b}$$

Equations (3.71) constitute a pair of coupled, linear integral equations for the functions $H(x_1 | \omega)$ and $L(x_1 | \omega)$. The condition that this pair of equations have nontrivial solutions yields the dispersion relation for surface polaritons on a grating.

The analysis that leads from eqs. (3.57) or (3.58) to eqs. (3.71) seems necessary. In a purely numerical solution of eqs. (3.57) or (3.58), which appears to be the only way in which these equations can be solved, it might be overlooked that the vicinity of the point $(x_1, 0, \zeta(x_1))$ on the corrugated surface contributes as much to the integrals containing the normal derivative of the Green's function as does the entire rest of the surface.

We now express $H(x_1 | \omega)$ and $L(x_1 | \omega)$ in forms that display the Bloch property these functions possess, viz.

$$H(x_1 \mid \omega) = e^{ikx_1} u(k\omega \mid x_1), \qquad L(x_1 \mid \omega) = e^{ikx_1} v(k\omega \mid x_1), \tag{3.73a}$$

where k is the wave vector of the surface polariton, and $u(k\omega \mid x_1)$ and $v(k\omega \mid x_1)$ are periodic functions of x_1,

$$u(k\omega \mid x_1 + a) = u(k\omega \mid x_1), \qquad v(k\omega \mid x_1 + a) = v(k\omega \mid x_1). \tag{3.73b}$$

When these expressions are substituted into eqs. (3.71), together with the Fourier integral representations of $G_0(x_1x_3 \mid x_1'x_3')$ and $G_\epsilon(x_1x_3 \mid x_1'x_3')$, it is found that $u(k\omega \mid x_1)$ and $v(k\omega \mid x_1)$ are the solutions of the following pair of equations:

$$u(k\omega \mid x_1) = \frac{1}{a} P \int_{-a/2}^{a/2} dy_1 M_{11}(k\omega \mid x_1y_1) u(k\omega \mid y_1) + \frac{1}{a} \int_{-a/2}^{a/2} dy_1 M_{12}(k\omega \mid x_1y_1)\, v(k\omega \mid y_1), \tag{3.74a}$$

$$u(k\omega \mid x_1) = \frac{1}{a} P \int_{-a/2}^{a/2} dy_1 M_{21}(k\omega \mid x_1y_1) u(k\omega \mid y_1) + \frac{1}{a} \int_{-a/2}^{a/2} dy_1 M_{22}(k\omega \mid x_1y_1) v(k\omega \mid y_1), \tag{3.74b}$$

where

$$M_{11}(k\omega \mid x_1y_1) = \sum_{m=-\infty}^{\infty} \exp[i \frac{2\pi m}{a}(x_1 - y_1)] \frac{\exp[-\alpha_m(k\omega)\,|\zeta(x_1) - \zeta(y_1)|]}{\alpha_m(k\omega)} \times \left\{ ik_m \frac{d\zeta(y_1)}{dy_1} + \alpha_m(k\omega)\,\mathrm{sgn}(\zeta(x_1) - \zeta(y_1)) \right\}, \tag{3.75a}$$

$$M_{12}(k\omega \mid x_1y_1) = \sum_{m=-\infty}^{\infty} \exp[i \frac{2\pi m}{a}(x_1 - y_1)] \frac{\exp[-\alpha_m(k\omega)\,|\zeta(x_1) - \zeta(y_1)|]}{\alpha_m(k\omega)}. \tag{3.75b}$$

$$M_{21}(k\omega \mid x_1y_1) = -\sum_{m=-\infty}^{\infty} \exp[i \frac{2\pi m}{a}(x_1 - y_1)] \frac{\exp[-\beta_m(k\omega)\,|\zeta(x_1) - \zeta(y_1)|]}{\beta_m(k\omega)} \times \left\{ ik_m \frac{d\zeta(y_1)}{dy_1} + \beta_m(k\omega)\,\mathrm{sgn}(\zeta(x_1) - \zeta(y_1)) \right\}, \tag{3.75c}$$

$$M_{22}(k\omega \mid x_1y_1) = -\sum_{m=-\infty}^{\infty} \exp[i \frac{2\pi m}{a}(x_1 - y_1)] \frac{\exp[-\beta_m(k\omega)\,|\zeta(x_1) - \zeta(y_1)|]}{\beta_m(k\omega)}. \tag{3.75d}$$

The sums over m in the expressions for these kernels converge exponentially rapidly for $\zeta(x_1) \neq \zeta(y_1)$ because of the presence of the absolute values $|\zeta(x_1) - \zeta(y_1)|$ in the summands.

A way of solving eqs. (3.74) appears to be the RR' method of García and Cabrera (1978), devised originally for the solution of an integral equation in the theory of the scattering of waves from a periodic hard surface. In this method the integrals over y_1 are replaced by discrete summations over a dense set of equally spaced points:

$$\frac{1}{a}\int_{-a/2}^{a/2} f(y_1)\,\mathrm{d}y_1 \to (1/2N) \sum_{n=-N+1}^{N} f(na/2N). \tag{3.76}$$

The variable x_1 assumes the same values, i.e. $x_1 \to ma/2N$, $m = -N+1, \ldots, N$. The kernels $M_{ij}(k\omega \mid x_1 y_1)$ go over into "discretized" kernels $M_{ij}(k\omega \mid mn)$. The diagonal elements $M_{ij}(k\omega \mid nn)$ require special treatment because the series (3.75) for these elements diverge as $\Sigma_m(1/|m|)$. García and Cabrera (1978) show how they are to be evaluated. In the present case additional care has to be taken because of the presence of the Cauchy principal value integrals in eqs. (3.74).

In this way the pair of coupled integral equations, eqs. (3.47), become a pair of $2N \times 2N$ homogeneous matrix equations. The solvability condition for this pair of matrix equations yields the dispersion relation for surface polaritons on a grating. Convergence of the solution is tested by replacing N by $N+1$, say, in the matrix forms of eqs. (3.74) and seeing if the solutions change only within some preassigned limit. This process is continued until the solution does not change by more than this limit. A determination of this dispersion relation by the method outlined here has yet to be carried out.

More manageable equations from which the dispersion relation for surface polaritons on a grating can be obtained result if the observation points are taken on planes $x_3 = $ const. that lie outside the selvedge region, i.e. for $x_3 > \max \zeta(x_1)$ or $x_3 < \min \zeta(x_1)$. This is due to the fact that $x_3 - x_3'$ in eqs. (3.48b) and (3.53) does not change sign in either of these cases so that it is unnecessary to take the absolute value. We begin by rewriting eqs. (3.46) and (3.56) in terms of $H(x_1 \mid \omega)$ and $L(x_1 \mid \omega)$:

$$\left.\begin{matrix} x_3 > \zeta(x_1) & H_2^>(x_1, x_3 \mid \omega) \\ x_3 < \zeta(x_1) & 0 \end{matrix}\right\} = \frac{1}{4\pi}\int \mathrm{d}x_1' \Big\{ G_0(x_1x_3 \mid x_1'x_3')\Big|_{x_3'=\zeta(x_1')} \tag{3.77a}$$

$$\times L(x_1' \mid \omega) - \left(\frac{\mathrm{d}\zeta}{\mathrm{d}x_1'}\frac{\partial}{\partial x_1'} - \frac{\partial}{\partial x_3'}\right) G_0(x_1x_3 \mid x_1'x_3')\Big|_{x_3'=\zeta(x_1')} H(x_1' \mid \omega)\Big\}, \tag{3.77b}$$

$$\left.\begin{matrix} x_3 > \zeta(x_1) & 0 \\ x_3 < \zeta(x_1) & H_2^<(x_1, x_3 \mid \omega) \end{matrix}\right\} = -\frac{1}{4\pi}\int \mathrm{d}x_1' \Big\{ \epsilon(\omega) G_\epsilon(x_1x_3 \mid x_1'x_3')\Big|_{x_3'=\zeta(x_1')} \tag{3.78a}$$

$$\times L(x_1' \mid \omega) - \left(\frac{\mathrm{d}\zeta}{\mathrm{d}x_1'}\frac{\partial}{\partial x_1'} - \frac{\partial}{\partial x_3'}\right) G_\epsilon(x_1x_3 \mid x_1'x_3')\Big|_{x_3'=\zeta(x_1')} H(x_1' \mid \omega)\Big\}. \tag{3.78b}$$

Then, if we take eq. (3.77b) for $x_3 < \min \zeta(x_1)$ and eq. (3.78a) for $x_3 > \max \zeta(x_1)$, and use the Fourier representations (3.48b) and (3.53) for the Green's functions we obtain convenient equations for $H(x_1 \mid \omega)$ and $L(x_1 \mid \omega)$:

$$\frac{1}{a}\int_{-a/2}^{a/2} dy_1 \exp[-ik_m y_1 - \alpha_m(k\omega)\zeta(y_1)] \times \{L(y_1 \mid \omega) - (\alpha_m(k\omega) - ik_m\zeta'(y_1))H(y_1 \mid \omega)\} = 0 \text{ for each } m \quad (3.79a)$$

$$\frac{1}{a}\int_{-a/2}^{a/2} dy_1 \exp[-ik_m y_1 + \beta_m(k\omega)\zeta(y_1)] \times \{\epsilon(\omega)L(y_1 \mid \omega) + (\beta_m(k\omega) + ik_m\zeta'(y_1))H(y_1 \mid \omega)\} = 0 \text{ for each } m. \quad (3.79b)$$

It is clear by construction that the correct boundary values of the field $H_2(x_1, x_3 \mid \omega)$ and its normal derivative, $H(x_1 \mid \omega)$ and $L(x_1 \mid \omega)$, satisfy the extinction theorem equations (3.79). By an argument due to Toigo et al. (1977) it can be shown that these equations define $H(x_1 \mid \omega)$ and $L(x_1 \mid \omega)$ uniquely in the sense that eqs. (3.79) have only the trivial solution $H(x_1 \mid \omega) = L(x_1 \mid \omega) \equiv 0$, except for the particular frequencies where undamped surface polaritons occur. The argument runs as follows. The right hand side of eqs. (3.77) and (3.78), for any square-integrable $H(x_1' \mid \omega)$ and $L(x_1' \mid \omega)$, represent functions of x_1 and x_3 that are analytic in each of these variables and are solutions of the Helmholtz equations (3.3) and (3.4), except at $x_3 = \zeta(x_1)$. Every solution $H(x_1 \mid \omega)$, $L(x_1 \mid \omega)$ of eqs. (3.79) satisfies the extinction theorem given by eq. (3.77b) for all $x_3 < \zeta(x_1)$ and the extinction theorem given by eq. (3.78a) for all $x_3 > \zeta(x_1)$, by the principle of analytic continuation, even though eqs. (3.79) were established by using the extinction theorem outside the selvedge region. Then eqs. (3.77a) and (3.78b) give the field component $H_2(x_1, x_3 \mid \omega)$ everywhere, with the correct matching conditions, and represent the unique solution of the Helmholtz equations (3.3) and (3.4), in view of the uniqueness theorem for the Helmholtz equation (Messiah 1965). Thus $H(x_1 \mid \omega)$ and $L(x_1 \mid \omega)$ must be uniquely determined by eqs. (3.79), in the sense given above, to within functions of measure zero that do not change the right hand sides of eqs. (3.77), and (3.78).

A possible way of solving the pair of equations (3.79) is by the RG method of García and Cabrera (1978). In this method integration over y_1 is again replaced by summation according to eq. (3.76). The wave vector k_m is given $2N$ values symmetrically disposed about the origin (k itself can be restricted to the interval $-(\pi/a) < k \le (\pi/a)$), viz. $m = -N+1, \ldots, N$. This procedure results in a pair of $2N \times 2N$ homogeneous matrix equations, the solvability condition for these equations yields the dispersion relation for surface polaritons on a grating. Convergence of the results obtained is again tested by increasing N until the solutions do not change by more than some preassigned amount.

The RG method so far has not been applied to the solution of eqs. (3.79). However, when it has been applied to the solution of a similar integral equation in the theory of the scattering of waves from a corrugated hard surface (García and Cabrera 1978) it seems to give convergent solutions only for rather small corrugation strengths (e.g. the ratio ζ_0/a for the sinusoidal profile $x_3 = \zeta_0 \cos(2\pi x_1/a)$), that are nevertheless larger than the corrugation strength beyond which the Rayleigh hypothesis is no longer valid. The results obtained also depend critically on the number of points $2N$ used.

An alternative way of solving eqs. (3.79) is provided by expanding $H(y_1 | \omega)$ and $L(y_1 | \omega)$ in Fourier series according to

$$H(y_1 | \omega) = \sum_{n=-\infty}^{\infty} e^{ik_n y_1} \hat{H}_n(k\omega) \tag{3.80a}$$

$$L(y_1 | \omega) = \sum_{n=-\infty}^{\infty} e^{ik_n y_1} \hat{L}_n(k\omega). \tag{3.80b}$$

When these expansions are substituted into eqs. (3.79) and integrations by parts are used to simplify the terms containing $\zeta'(y_1)$, the Fourier coefficients $\hat{H}_n(k\omega)$ and $\hat{L}_n(k\omega)$ are found to obey the simple equations

$$\sum_{n=-\infty}^{\infty} I_{m-n}^{(m)}(k\omega)\left\{\frac{\omega^2/c^2 - k_m k_n}{\alpha_m(k\omega)} \hat{H}_n(k\omega) + \hat{L}_n(k\omega)\right\} = 0 \tag{3.81a}$$

$$\sum_{n=-\infty}^{\infty} J_{m-n}^{(m)}(k\omega)\left\{\frac{\epsilon(\omega)(\omega^2/c^2) - k_m k_n}{\epsilon(\omega)\beta_m(k\omega)} \hat{H}_n(k\omega) - \hat{L}_n(k\omega)\right\} = 0. \tag{3.81b}$$

To solve eqs. (3.81) one can keep only $2N$ equations for N values of k_m and solve for the first N coefficients $\hat{H}_n(k\omega)$ and $\hat{L}_n(k\omega)$. Such a calculation has been carried out very recently by Laks et al. (1980) for two different surface profiles: the sinusoidal profile given by eq. (3.21) and a symmetric sawtooth profile given by

$$\zeta(x_1) = h + (4h/a)x_1; \qquad -a/2 \le x_1 \le 0, \tag{3.82a}$$

$$= h - (4h/a)x_1; \qquad 0 \le x_1 \le a/2. \tag{3.82b}$$

A free electron dielectric constant, $\epsilon(\omega) = 1 - (\omega_p^2/\omega^2)$ was used in these calculations. The convergence of the calculations was quite good. For the sinusoidal profile convergence for the surface polariton frequencies was achieved for values of ζ_0/a up to 0.6, the largest value of this ratio considered. For the largest values of ζ_0/a the zeros of determinants of up to 52×52 matrices were being sought. Convergence of the calculations for the sawtooth profile was somewhat slower, and calculations were carried out for values of h/a up to only 0.2.

A typical result of their calculations is shown in fig. 8 for the sinusoidal profile with $\zeta_0/a = 0.2$, for $0 \le k \le \pi/a$. Several branches of the dispersion curve were obtained for this profile. These can be regarded, in the reduced zone scheme, as the result of the folding back into the first Brillouin zone the pieces of the dispersion curve from the second, third, . . . , Brillouin zones, obtained in the extended zone scheme. The number of branches and

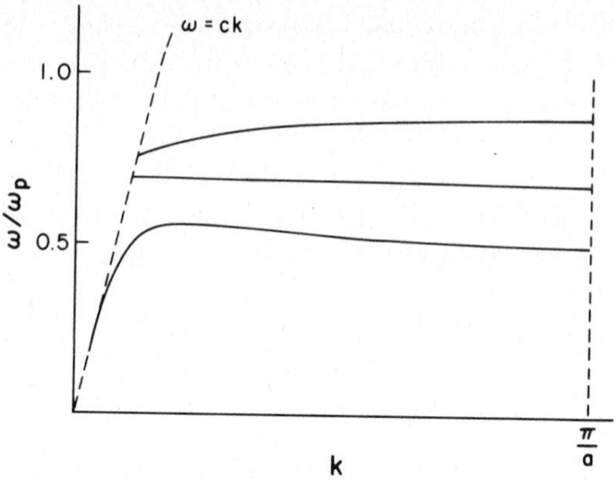

Fig. 8. The dispersion curve for surface polaritons on a grating characterized by the sinusoidal profile $\zeta(x_1) = \zeta_0 \cos(2\pi x_1/a)$ and a free electron form for the dielectric constant, $\epsilon(\omega) = 1 - (\omega_p^2/\omega^2)$. In the present case $\zeta_0/a = 0.2$ (Laks et al. 1980).

their range in k are limited by the fact that certain portions of the surface polariton dispersion curve for a flat surface correspond to modes that are radiative in the presence of a grating on that surface, as has been discussed in sect. 3.1.1, and illustrated in figs. 5.

The results of Laks et al. (1980) are encouraging in that they demonstrate the feasibility of carrying out calculations of dispersion curves for surface polaritons on rather strongly corrugated surfaces. However, their significance goes beyond this, as the following argument shows.

Equations (3.81) should be compared with eqs. (3.16), which were obtained on the basis of the Rayleigh hypothesis. We can write eqs. (3.16) in the form:

$$\sum_{n=-\infty}^{\infty} \begin{pmatrix} I^{(n)}_{m-n} & -J^{(n)}_{m-n} \\ \dfrac{\omega^2/c^2 - k_m k_n}{\alpha_n} I^{(n)}_{m-n} & \dfrac{\epsilon(\omega)(\omega^2/c^2) - k_m k_n}{\epsilon(\omega)\beta_n} I^{(n)}_{m-n} \end{pmatrix} \begin{pmatrix} A_n \\ B_n \end{pmatrix} = 0, \tag{3.16}$$

$$\sum_{n=-\infty}^{\infty} \begin{pmatrix} I^{(m)}_{m-n} & \dfrac{\omega^2/c^2 - k_m k_n}{\alpha_m} I^{(m)}_{m-n} \\ -J^{(m)}_{m-n} & \dfrac{\epsilon(\omega)(\omega^2/c^2) - k_m k_n}{\epsilon(\omega)\beta_m} J^{(m)}_{m-n} \end{pmatrix} \begin{pmatrix} \hat{L}_n \\ \hat{H}_n \end{pmatrix} = 0. \tag{3.83}$$

We see that the matrix kernels in these equations are the transposes of each other, provided that $I^{(m)}_{m-n} = I^{(m)}_{n-m}$ and $J^{(m)}_{m-n} = J^{(m)}_{n-m}$. If we write the limits of integration in eqs. (3.13) as $(-a/2, a/2)$ rather than as $(0, a)$, which is permitted by the periodicity of the integrand, we see that these conditions are satisfied if $\zeta(x_1)$ is an even function of x_1, i.e., if $\zeta(-x_1) = \zeta(x_1)$.

The uniqueness of the solutions of eq. (3.79) have important consequences for the Rayleigh method, as exemplified by eqs. (3.11). Indeed, Toigo et al. (1977) conclude that eqs. (3.11) are generally valid, but solutions by simple Fourier expansions such as those that lead to eqs. (3.16) may diverge, at least for some range of strengths of the surface corrugation. The same can be said of eqs. (3.79) and the system of eqs. (3.81), or equivalently (3.83), obtained from them by Fourier expansion. We can however, make stronger statements on the basis of the numerical results of Laks et al. (1980), at least for surface profile functions $\zeta(x_1)$ that are even functions of x_1. According to the result obtained in the preceding paragraph, the solutions of the equations of the Rayleigh method, eqs. (3.11), by Fourier expansions converge for the same range of strengths of the surface corrugation as do the solutions of the equations obtained by the extinction theorem form of Green's theorem, eqs. (3.79), by Fourier expansions, and this range of corrugation strengths exceeds that for which the Rayleigh hypothesis has been found to be valid. In particular, Fourier series solutions of either eqs. (3.11) or (3.79) converge even for a nonanalytic surface profile, viz., the symmetric sawtooth profile. This would not be expected on the basis of the usual arguments concerning the validity of the Rayleigh hypothesis (Millar 1969, Hill and Celli 1978). The range of values of ζ_0/a and h/a for which the Fourier series solutions converge is not known at the present time, and may be limited. Indeed, there is a suggestion that such solutions converge only for a finite range of values of h/a in the case of the symmetric sawtooth profile. This follows from the results of analogous calculations of the dispersion curve for surface plasmons on a grating that will be discussed at the end of sect. 3.2.2. On the basis of the latter results one can also suspect that Fourier series solutions of the reduced equations of the Rayleigh method given by eqs. (3.18) and (3.20) may converge for smaller corrugation strengths than Fourier series solutions of the full set of equations (3.11) obtained by the Rayleigh method, perhaps only for corrugation strengths for which the Rayleigh hypothesis is valid.

One is therefore led to conclude that at least for gratings defined by a symmetric profile function, the solution of the full set of equations (3.11) obtained on the basis of the Rayleigh hypothesis by Fourier expansions yields convergent results for corrugation strengths for which the Rayleigh hypothesis is no longer valid, and for profiles for which it also is invalid.

It is possible that solving eqs. (3.79) by methods other than the RG method or expansions in Fourier series will yield results valid for an arbitrary surface profile function. Given the simplicity of eqs. (3.79) in comparison with eqs. (3.75)–(3.76), the search for such methods of solution should be worthwhile.

3.2. Propagation of a Surface Plasmon Over a Grating

A surface plasmon propagating over a grating has its dispersion relation modified from what it is for a flat surface, just as a surface polariton does. However, as we will see, there are qualitative differences between the two cases. In this section we study the propagation of surface plasmons over a grating, initially by the Rayleigh method, and then by the use of Green's theorem.

The results of the present section can be obtained as the limits of the results of the preceding section when the speed of light is allowed to become infinite. However, because of its comparative simplicity, the analysis presented here has an interest of its own, as well as in juxtaposition with the discussion presented in sect. 4.3 concerning the propagation of a surface plasmon over a statistically rough surface.

For simplicity we consider a surface plasmon propagating perpendicularly to the grooves of a grating, whose profile is given by the equation $x_3 = \zeta(x_1)$. The region defined by $x_3 > \zeta(x_1)$ is vacuum, while the region $x_3 < \zeta(x_1)$ is filled by a dielectric medium characterized by an isotropic, frequency-dependent dielectric tensor $\epsilon_{\mu\nu}(\omega) = \delta_{\mu\nu}\epsilon(\omega)$.

In many respects the derivation of the dispersion relation for surface plasmons on a grating mimics that presented for surface polaritons in the preceding section. The chief difference is that in the case of surface plasmons we seek the solution of Laplace's equation

$$\left(\frac{\partial^2}{\partial x_1^2} + \frac{\partial^2}{\partial x_3^2}\right) \varphi(x_1 x_3 \mid \omega) = 0, \tag{3.84}$$

for a p-polarized surface plasmon propagating in the x_1-direction, that vanishes as $|x_3| \to \infty$, and satisfies the boundary conditions

$$\varphi(x_1 x_3 \mid \omega)\Big|_{x_3=\zeta(x_1)-} = \varphi(x_1 x_3 \mid \omega)\Big|_{x_3=\zeta(x_1)+} \tag{3.85a}$$

$$\epsilon(\omega)\frac{\partial}{\partial n}\varphi(x_1 x_3 \mid \omega)\Big|_{x_3=\zeta(x_1)-} = \frac{\partial}{\partial n}\varphi(x_1 x_3 \mid \omega)\Big|_{x_3=\zeta(x_1)+}. \tag{3.85b}$$

3.2.1. The Rayleigh Hypothesis

We begin by solving the problem on the basis of the Rayleigh hypothesis. The solution of eq. (3.84) that vanishes as $x_3 \to +\infty$ can be written for $x_3 > \zeta_{\max}$ in the form

$$\varphi^{>}(x_1 x_3 \mid \omega) = \sum_{p=-\infty}^{\infty} A_p \exp[ik_p x_1 - \alpha_p(k) x_3]; \qquad x_3 > \zeta_{\max}, \tag{3.86}$$

where

$$k_p \equiv k + 2\pi p/a, \qquad \alpha_p(k) = |k + 2\pi p/a|. \tag{3.87}$$

Similarly, the solution of eq. (3.84) that vanishes as $x_3 \to -\infty$ can be written for $x_3 < \zeta_{\min}$ in the form

$$\varphi^{<}(x_1x_3 \mid \omega) = \sum_{p=-\infty}^{\infty} B_p \exp[ik_px_1 + \alpha_p(k)x_3]; \qquad x_3 < \zeta_{\min}. \tag{3.88}$$

Both solutions possess the Bloch property

$$\varphi(x_1 + a, x_3 \mid \omega) = e^{ika}\ \varphi(x_1x_3 \mid \omega), \tag{3.89}$$

as they must.

We now assume that the solutions (3.86) and (3.88) can be continued in to the surface itself, and apply the boundary conditions (3.85). The expansion coefficients $\{A_p\}$ and $\{B_p\}$ are thereby found to satisfy the pair of coupled homogeneous equations

$$\sum_{p=-\infty}^{\infty} \{-e^{-\alpha_p(k)\zeta(x_1)}\, e^{ik_px_1}\, A_p + e^{\alpha_p(k)\zeta(x_1)}\, e^{ik_px_1}\, B_p\} = 0, \tag{3.90a}$$

$$\sum_{p=-\infty}^{\infty} \left\{\left[\alpha_p(k) + ik_p \frac{d\zeta(x_1)}{dx_1}\right] e^{-\alpha_p(k)\zeta(x_1)}\, e^{ik_px_1}\, A_p \right.$$
$$\left. + \epsilon(\omega)\left[\alpha_p(k) - ik_p \frac{d\zeta(x_1)}{dx_1}\right] e^{\alpha_p(k)\zeta(x_1)}\, e^{ik_px_1}\, B_p\right\} = 0. \tag{3.90b}$$

With the aid of the expansions:

$$\exp[-\alpha_p(k)\zeta(x_1)] = \sum_{n=-\infty}^{\infty} I_n^{(p)}(k) \exp(i\frac{2\pi n}{a} x_1) \tag{3.91a}$$

$$\exp[\alpha_p(k)\zeta(x_1)] = \sum_{n=-\infty}^{\infty} J_n^{(p)}(k) \exp(i\frac{2\pi n}{a} x_1), \tag{3.91b}$$

eqs. (3.90) can be transformed into the pair:

$$\sum_{p=-\infty}^{\infty} \{-I_{r-p}^{(p)}(k)\, A_p + J_{r-p}^{(p)}(k)\, B_p\} = 0, \tag{3.92a}$$

$$\sum_{p=-\infty}^{\infty} \left\{\frac{k_rk_p}{\alpha_p(k)} I_{r-p}^{(p)}(k)\, A_p + \epsilon(\omega)\frac{k_rk_p}{\alpha_p(k)} J_{r-p}^{(p)}(k)\, B_p\right\} = 0. \tag{3.92b}$$

The dispersion relation for surface plasmons on a grating is obtained by equating to zero the determinant of the coefficients in these equations.

However, it is possible to obtain equations for the coefficients $\{A_p\}$ and $\{B_p\}$ separately, and these lead to simpler dispersion relations. If we multiply eq. (390a) by $[\alpha_r(k) + ik_r\, d\zeta(x_1)/dx_1] \exp[\alpha_r(k)\zeta(x_1) - ik_rx_1]$, multiply eq. (3.90b) by $-\epsilon^{-1}(\omega) \exp[\alpha_r(k)\zeta(x_1) - ik_rx_1]$, add the resulting equations, and integrate their sum on x_1 over the interval $(0, a)$, we obtain an equation for the $\{A_p\}$ alone:

$$\sum_{p=-\infty}^{\infty} \frac{1}{a}\int_0^a \mathrm{d}x_1 \{[\epsilon(\omega)\alpha_r(k)+\alpha_p(k)]+\mathrm{i}[\epsilon(\omega)k_r+k_p]\frac{\mathrm{d}\zeta(x_1)}{\mathrm{d}x_1} \times \exp[(\alpha_r(k)-\alpha_p(k))\zeta(x_1)]\exp[-\mathrm{i}(k_r-k_p)x_1]\}A_p=0. \tag{3.93}$$

In a similar fashion we can obtain an equation for the $\{B_p\}$ alone. We multiply eq. (3.90a) by $[\alpha_r(k)-\mathrm{i}k_r\,\mathrm{d}\zeta(x_1)/\mathrm{d}x_1]\exp[-\alpha_r\zeta(x_1)-\mathrm{i}k_rx_1]$, multiply eq. (3.90b) by $\exp[-\alpha_r(k)\zeta(x_1)-\mathrm{i}k_rx_1]$, add the resulting equations, and integrate the sum on x_1 over the interval $(0, a)$. In this way we obtain the equation

$$\sum_{p=-\infty}^{\infty} \frac{1}{a}\int_0^a \mathrm{d}x_1 \{[\alpha_r(k)+\epsilon(\omega)\alpha_p(k)]-\mathrm{i}[k_r+\epsilon(\omega)k_p]\frac{\mathrm{d}\zeta(x_1)}{\mathrm{d}x_1} \times \exp[-(\alpha_r(k)-\alpha_p(k))\zeta(x_1)]\exp[-\mathrm{i}(k_r-k_p)x_1]\}B_p=0. \tag{3.94}$$

We can now use eq. (3.93) to obtain the dispersion relation for surface plasmons on a grating in the small roughness limit. For this purpose it is convenient to rewrite it in the form

$$\sum_{p=-\infty}^{\infty} \left\{\delta_{rp}(\epsilon(\omega)+1)\,\alpha_r(k)+(1-\delta_{rp})(1-\epsilon(\omega)) \times \frac{\alpha_r(k)\alpha_p(k)-k_rk_p}{\alpha_r(k)-\alpha_p(k)}\frac{1}{a}\int_0^a \mathrm{d}x_1 \exp[(\alpha_r(k)-\alpha_p(k))\zeta(x_1)] \times \exp[-\mathrm{i}(k_r-k_p)x_1]\right\}A_p=0. \tag{3.95}$$

To first order in $\zeta(x_1)$ this equation takes the form

$$(\epsilon(\omega)+1)\alpha_r(k)\,A_r=-(1-\epsilon(\omega))\sum_{p(\neq r)}\hat{\zeta}(r-p)[\alpha_r(k)\alpha_p(k)-k_rk_p]\,A_p. \tag{3.96}$$

We again use the fact that A_0 describes the propagation of a surface plasmon over a flat surface, and hence take it to be of O(1), while A_r for $r\neq 0$ is at least of $\mathrm{O}(\hat{\zeta})$. In this way we obtain for the dispersion relation to $\mathrm{O}(|\hat{\zeta}|^2)$ the result that

$$(\epsilon(\omega)+1)^2=2(1-\epsilon(\omega))^2\sum_{p(\neq 0)}|\hat{\zeta}(p)|^2[\alpha_0(k)\alpha_p(k)-kk_p]. \tag{3.97}$$

The dispersion relation for surface plasmons on a flat surface according to eq. (2.10) is

$$\epsilon(\omega)+1=0. \tag{3.98}$$

Consequently, if we denote the solution of eq. (3.98) by ω_0, independent of k, the solution of eq. (3.97) can be written in the form

$$\omega(k)=\omega_0+\Delta\omega(k), \tag{3.99a}$$

where

$$\Delta\omega(k) = \pm\frac{2\sqrt{2}}{\epsilon'(\omega_0)} f(k), \tag{3.99b}$$

with

$$f(k) = \left\{\sum_{p(\neq 0)} |\hat{\zeta}(p)|^2[\alpha_0(k)\alpha_p(k) - kk_p]\right\}^{1/2}$$
$$= f(-k). \tag{3.99c}$$

Thus, we see from eqs. (3.99) that the dispersion curve for a surface plasmon on a grating consists of two branches for each branch possessed by the dispersion curve for a flat surface. This is in contrast with the result obtained for surface polaritons in the preceding section, where it was found that a similar splitting of the dispersion curve occurred only at special values of k, equal to $\pm\pi p/a$, that correspond to the boundaries of the one-dimensional Brillouin zones associated with the one-dimensional periodicity of the grating.

The difference between these two cases can be understood on the basis of the following argument. The splitting of the dispersion curve associated with a periodic structure occurs at those values of k for which a point on the flat surface dispersion curve is degenerate with another point on the dispersion curve separated from it by a translation vector of the lattice reciprocal to the one defining the periodicity of the given structure. Since the dispersion curve for a surface polariton is dispersive and even in k, $\omega(-k) = \omega(k)$, this degeneracy can occur only at isolated values of k for which $-k$ equals $k + 2\pi p/a$ for some p, and this condition defines the values of k that correspond to the Brillouin zone boundaries. In the case of a surface plasmon, however, while the flat surface dispersion curve is even in k it is also dispersionless, i.e., it is flat. This means that every point on it is degenerate with every other point on it. That is, the point on it corresponding to an *arbitrary* value of k is degenerate with points on it corresponding to values of k given by $k + 2\pi p/a$ for *all* p. Thus a gap opens up at every point of the flat surface dispersion curve, and it is thereby split into two branches in the small roughness limit. The splitting of the dispersion curve for a surface plasmon on a grating into two branches is therefore ultimately connected to the infinite degeneracy of the corresponding dispersion curve for a flat surface.

As an example, we consider the case of the profile given by

$$\zeta(x_1) = \zeta_0 \cos(2\pi x_1/a). \tag{3.100}$$

The only nonvanishing Fourier components of this function are

$$\hat{\zeta}(1) = \hat{\zeta}(-1) = \tfrac{1}{2}\zeta_0. \tag{3.101}$$

The function $f(k)$ defined by eq. (3.99c) in this case becomes

$$f(k) = \zeta_0 (k/2)^{1/2}(2\pi/a - k)^{1/2}; \qquad 0 \le k \le 2\pi/a, \tag{3.102a}$$

$$= 0; \qquad 2\pi/a \le k. \tag{3.102b}$$

The problem of obtaining the dispersion relation of a surface plasmon on a grating can be formulated as an eigenvalue problem that offers the possibility of going beyond the small roughness limit in a simple fashion. Returning to eq. (3.95), we rewrite it in the form

$$\sum_{p=-\infty}^{\infty} M_{rp}(k)A_p = \frac{\epsilon(\omega)+1}{\epsilon(\omega)-1} A_r, \tag{3.103}$$

where the matrix $M(k)$ has the elements

$$M_{rp}(k) = \frac{\alpha_r(k)\alpha_p(k) - k_r k_p}{\alpha_r(k)[\alpha_r(k) - \alpha_p(k)]} \frac{1}{a}\int_0^a \mathrm{d}x_1 \exp[(\alpha_r(k) - \alpha_p(k))\zeta(x_1)]$$
$$\times \exp[-\mathrm{i}\frac{2\pi}{a}(r-p)x_1]; \qquad r \ne p, \tag{3.104a}$$

$$= 0; \qquad r = p. \tag{3.104b}$$

Thus, if we denote the eigenvalues of the matrix $M(k)$ by $\{\lambda_s(k)\}$, where s merely labels the distinct eigenvalues, the dispersion relation for a surface polariton on a grating can be written as

$$\frac{\epsilon(\omega)+1}{\epsilon(\omega)-1} = \lambda_s(k), \tag{3.105}$$

and leads to a separate branch for each distinct eigenvalue.

Solutions of the eigenvalue problem posed by eqs. (3.104)–(3.105) have been carried out recently by Glass and Maradudin (1980) for the sinusoidal profile, eq. (3.21), and the symmetric sawtooth profile, eq. (3.82). A free electron dielectric constant, $\epsilon(\omega) = 1 - (\omega_p^2/\omega^2)$, was used in these calculations. In the case of the sawtooth profile, it proved to be impossible to obtain convergent results for the eigenvalues $\{\lambda_s(k)\}$. This is not altogether surprising. It is known (Millar 1969) that the Rayleigh hypothesis is invalid for a surface profile function $\zeta(x_1)$ that is not an analytic function of x_1.

For the sinusoidal profile, on the other hand, good convergence for the eigenvalues $\{\lambda_s(k)\}$ could be obtained for values of ζ_0/a up to about 0.1, with matrices $M(k)$ of up to 38×38 in size. The larger the value of ζ_0/a the larger the matrix M that had to be diagonalized. If a $2N \times 2N$ matrix is used to approximate $M(k)$, then $2N$ eigenvalues $\{\lambda_s(k)\}$ are obtained. These eigenvalues occur in pairs, $\pm\lambda_s(k)$, and are periodic in k with period $2\pi/a$. In the small roughness limit, the largest pair of eigenvalues is proportional to ζ_0/a; the next largest pair is proportional to $(\zeta_0/a)^3$; the next largest to $(\zeta_0/a)^5$; and so on. It turns out to be difficult to calculate the smaller eigenvalues with high accuracy without going to large matrices: the rate of convergence of the eigenvalues decreases as their magnitudes decrease.

In fig. 9 is plotted a typical result for $\Delta\omega(k) = \omega(k) - \omega_0$, where $\omega(k)$ is the frequency of a surface plasmon on a grating and $\omega_0 = 2^{-1/2}\omega_p$ is the frequency of a surface plasmon on a flat surface. In these calculations a value of $\zeta_0/a = 0.07$ was assumed. The curves shown were obtained from the diagonalization of a 38×38 matrix approximation to $M(k)$. Only the branches corresponding to the four largest eigenvalues $\{\lambda_s(k)\}$ are plotted in this figure. The remaining branches are too close to the line $\Delta\omega(k) = 0$ to be depicted graphically.

We now turn to methods for obtaining the dispersion relation for surface plasmons on a grating that are not subject to the limitations of the Rayleigh hypothesis. Some additional comments concerning the numerical results of Glass and Maradudin (1980) just described will be made following the presentation of these methods.

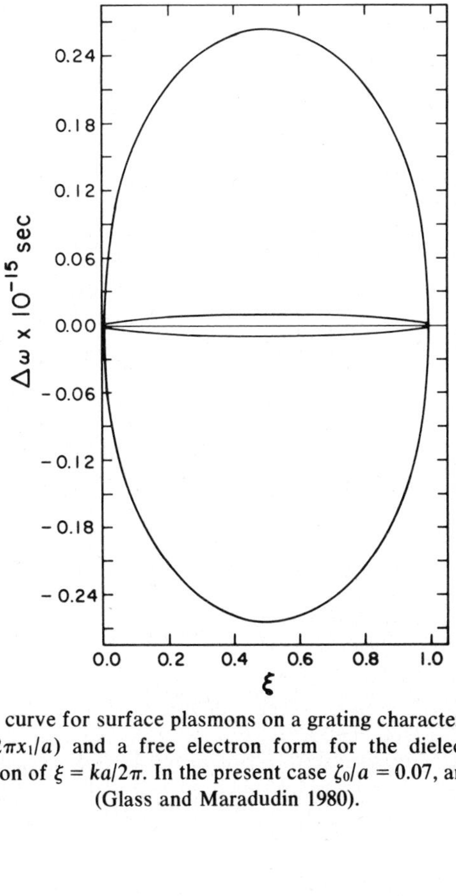

Fig. 9. The dispersion curve for surface plasmons on a grating characterized by the sinusoidal profile $\zeta(x_1) = \zeta_0 \cos(2\pi x_1/a)$ and a free electron form for the dielectric constant, $\epsilon(\omega) = 1 - (\omega_p^2/\omega^2)$, as a function of $\xi = ka/2\pi$. In the present case $\zeta_0/a = 0.07$, and $\omega_p = 3.699 \times 10^{15}\ \mathrm{s}^{-1}$ (Glass and Maradudin 1980).

3.2.2. Green's Theorem

There may arise situations in which it is desired to know the dispersion relation for surface plasmons on a grating that is rougher than those for which the Rayleigh hypothesis is valid. In this subsection we sketch out how the dispersion relation can be obtained in such cases.

Our starting point is Green's theorem as given by eq. (3.43). If we apply it to the vacuum above the dielectric medium, we obtain the pair of equations:

$$\begin{matrix} x_3 > \zeta(x_1) & \varphi^{>}(x_1, x_3 \mid \omega) \\ x_3 < \zeta(x_1) & 0 \end{matrix} \Bigg\} = -\frac{1}{4\pi}\int_S \Bigg\{\frac{\partial}{\partial n'_-} G(x_1x_3 \mid x'_1x'_3)$$

$$\times \varphi^{>}(x'_1, x'_3 \mid \omega) - G(x_1x_3 \mid x'_1x'_3) \quad (3.106a)$$

$$\times \frac{\partial}{\partial n'_-} \varphi^{>}(x'_1, x'_3 \mid \omega)\Bigg\} ds'_1, \quad (3.106b)$$

where the Green's function $G(x_1x_3 \mid x'_1x'_3)$ is the solution of

$$\nabla^2 G(x_1x_3 \mid x'_1x'_3) = -4\pi\delta(x_1 - x'_1)\delta(x_3 - x'_3), \quad (3.107)$$

subject to exponentially decaying or outgoing wave conditions at infinity, in its Fourier representation. An explicit expression for $G(x_1x_3 \mid x'_1x'_3)$ is provided by

$$\begin{aligned} G(x_1x_3 \mid x'_1x'_3) &= -2\ln[(x_1 - x'_1)^2 + (x_3 - x'_3)^2]^{1/2} \\ &= G(x'_1x'_3 \mid x_1x_3). \end{aligned} \quad (3.108)$$

The Fourier integral representation of this function is

$$G(x_1x_3 \mid x'_1x'_3) = \int_{-\infty}^{\infty} dq \, \frac{\exp[iq(x_1 - x'_1) - |q|\,|x_3 - x'_3|]}{|q|}. \quad (3.109)$$

If we now apply Green's theorem to the region occupied by the dielectric medium, we obtain the following pair of equations:

$$\begin{matrix} x_3 > \zeta(x_1) & 0 \\ x_3 < \zeta(x_1) & \varphi^{<}(x_1, x_3 \mid \omega) \end{matrix} \Bigg\} = -\frac{1}{4\pi}\int_S \Bigg\{\frac{\partial}{\partial n'_+} G(x_1x_3 \mid x'_1x'_3)$$

$$\times \varphi^{<}(x'_1, x'_3 \mid \omega) - G(x_1x_3 \mid x'_1x'_3) \quad (3.110a)$$

$$\times \frac{\partial}{\partial n'_+} \varphi^{<}(x'_1, x'_3 \mid \omega)\Bigg\} ds'_1. \quad (3.110b)$$

If we make use of the boundary conditions (3.85) as well as the fact that $\partial/\partial n_+ = -\partial/\partial n_-$, we can transform eqs. (3.110) into the form:

$$\left.\begin{array}{ll} x_3>\zeta(x_1) & 0 \\ x_3<\zeta(x_1) & \varphi^{<}(x_1, x_3\,|\,\omega) \end{array}\right\} = \frac{1}{4\pi}\int_S \left\{\frac{\partial}{\partial n'_-} G(x_1x_3\,|\,x'_1x'_3)\right. \tag{3.111a}$$

$$\times\, \varphi^{>}(x'_1, x'_3\,|\,\omega) - \frac{1}{\epsilon(\omega)} G(x_1x_3\,|\,x'_1x'_3)$$

$$\left.\times \frac{\partial}{\partial n'_-}\varphi^{>}(x'_1, x'_3\,|\,\omega)\right\} \mathrm{d}s'_1. \tag{3.111b}$$

In just the same way as this was done in sect. 3.1.2 in going from eqs. (3.46) and (3.56) to eqs. (3.68) we can obtain the following pair of homogeneous equations for the values of $\varphi^{>}(x_1, x_3\,|\,\omega)$ and $(\partial/\partial n_-)\,\varphi^{>}(x_1, x_3\,|\,\omega)$ on the surface S:

$$\varphi^{>}(x_1, \zeta(x_1)\,|\,\omega) = -\frac{1}{2\pi} P\int_S \frac{\partial}{\partial n'_-} G(x_1\zeta(x_1)\,|\,x'_1x'_3)\,\varphi^{>}(x'_1, x'_3\,|\,\omega)\,\mathrm{d}s'_1$$

$$+\frac{1}{2\pi}\int G(x_1\zeta(x_1)\,|\,x'_1x'_3)\frac{\partial}{\partial n'_-}\varphi^{>}(x'_1x'_3\,|\,\omega)\,\mathrm{d}s'_1, \tag{3.112a}$$

$$\varphi^{>}(x_1, \zeta(x_1)\,|\,\omega) = \frac{1}{2\pi} P\int_S \frac{\partial}{\partial n'_-} G(x_1\zeta(x_1)\,|\,x'_1x'_3)\,\varphi^{>}(x'_1, x'_3)\,\mathrm{d}s'_1$$

$$-\frac{1}{\epsilon(\omega)}\frac{1}{2\pi}\int_S G(x_1\zeta(x_1)\,|\,x'_1x'_3)\frac{\partial}{\partial n'_-}\varphi^{>}(x'_1x'_3\,|\,\omega)\,\mathrm{d}s'_1. \tag{3.112b}$$

When we convert the integrals on the right hand sides of these equations into integrals over the surface $x'_3 = 0$, they take the forms:

$$H(x_1\,|\,\omega) = -\frac{1}{2\pi} P\int\left[\frac{\mathrm{d}\zeta(x'_1)}{\mathrm{d}x'_1}\frac{\partial G(x_1\zeta(x_1)\,|\,x'_1x'_3)}{\partial x'_1}\right.$$

$$\left.-\frac{\partial G(x_1\zeta(x_1)\,|\,x'_1x'_3)}{\partial x'_3}\right]_{x'_3=\zeta(x'_1)} H(x'_1\,|\,\omega)\,\mathrm{d}x'_1$$

$$+\frac{1}{2\pi}\int G(x_1\zeta(x_1)\,|\,x'_1\zeta(x'_1))L(x'_1\,|\,\omega)\,\mathrm{d}x'_1, \tag{3.113a}$$

$$H(x_1\,|\,\omega) = \frac{1}{2\pi} P\int\left[\frac{\mathrm{d}\zeta(x'_1)}{\mathrm{d}x'_1}\frac{\partial G(x_1\zeta(x_1)\,|\,x'_1x'_3)}{\partial x'_1}\right.$$

$$\left.-\frac{\partial G(x_1\zeta(x_1)\,|\,x'_1x'_3)}{\partial x'_3}\right]_{x'_3=\zeta(x'_1)} H(x'_1\,|\,\omega)\,\mathrm{d}x'_1$$

$$-\frac{1}{\epsilon(\omega)}\frac{1}{2\pi}\int G(x_1\zeta(x_1)\,|\,x'_1\zeta(x'_1))L(x'_1\,|\,\omega)\,\mathrm{d}x'_1, \tag{3.113b}$$

where here and in the remainder of this section,

$$H(x_1\,|\,\omega) = \varphi^{>}(x_1, \zeta(x_1)\,|\,\omega), \tag{3.114a}$$

$$L(x_1\,|\,\omega) = \left[1+\left(\frac{\mathrm{d}\zeta(x_1)}{\mathrm{d}x_1}\right)^2\right]^{1/2}\frac{\partial}{\partial n_-}\varphi^{>}(x_1x_3\,|\,\omega)\bigg|_{x_3=\zeta(x_1)}. \tag{3.114b}$$

Equations (3.113) are a pair of coupled, linear, homogeneous integral equations for $H(x_1 \mid \omega)$ and $L(x_1 \mid \omega)$. The condition that this pair of equations have nontrivial solutions yields the dispersion relation for surface plasmons on a grating.

We next write $H(x_1 \mid \omega)$ and $L(x_1 \mid \omega)$ in Bloch form:

$$H(x_1 \mid \omega) = e^{ikx_1} u(k\omega \mid x_1), \qquad L(x_1 \mid \omega) = e^{ikx_1} \upsilon(k\omega \mid x_1), \tag{3.115}$$

where u and υ are periodic in x_1 with period a. When eqs. (3.115) are used in eqs. (3.113), together with eq. (3.114), it is found that $u(k\omega \mid x_1)$ and $\upsilon(k\omega \mid x_1)$ are the solutions of the following pair of equations:

$$u(k\omega \mid x_1) = \frac{1}{a} P \int_{-a/2}^{a/2} dy_1 M_{11}(k \mid x_1 y_1) u(k\omega \mid y_1) + \frac{1}{a} \int_{-a/2}^{a/2} dy_1 M_{12}(k \mid x_1 y_1) \upsilon(k\omega \mid y_1), \tag{3.116a}$$

$$u(k\omega) \mid x_1) = -\frac{1}{a} P \int_{-a/2}^{a/2} dy_1 M_{11}(k \mid x_1 y_1) u(k\omega \mid y_1) - \frac{1}{\epsilon(\omega)} \frac{1}{a} \int_{-a/2}^{a/2} dy_1 M_{12}(k \mid x_1 y_1) \upsilon(k\omega \mid y_1), \tag{3.116b}$$

where

$$M_{11}(k \mid x_1 y_1) = \sum_{m=-\infty}^{\infty} \exp\left[i\frac{2\pi m}{a}(x_1 - y_1)\right] \frac{\exp[-\alpha_m(k)|\zeta(x_1) - \zeta(y_1)|]}{\alpha_m(k)} \times \left[ik_m \frac{d\zeta(y_1)}{dy_1} + \alpha_m(k) \operatorname{sgn}(\zeta(x_1) - \zeta(y_1))\right], \tag{3.117a}$$

$$M_{12}(k \mid x_1 y_1) = \sum_{m=-\infty}^{\infty} \exp\left[i\frac{2\pi m}{a}(x_1 - y_1)\right] \frac{\exp[-\alpha_m(k)|\zeta(x_1) - \zeta(y_1)|]}{\alpha_m(k)}. \tag{3.117b}$$

From the results given by eqs. (3.116) we find that the pair of equations for $u(k\omega \mid x_1)$ and $\upsilon(k\omega \mid x_1)$ can be reduced to a single homogeneous integral equation for $u(k\omega \mid x_1)$:

$$\frac{1+\epsilon(\omega)}{1-\epsilon(\omega)} u(k\omega \mid x_1) = \frac{1}{a} P \int_{-a/2}^{a/2} dy_1 M_{11}(k \mid x_1 y_1) u(k\omega \mid y_1). \tag{3.118}$$

Thus, if we denote the eigenvalues of the integral equation

$$\lambda f(x_1) = \frac{1}{a} P \int_{-a/2}^{a/2} dy_1 M_{11}(k \mid x_1 y_1) f(y_1) \tag{3.119}$$

by $\{\lambda_s(k)\}$, where $s = 1, 2 \ldots$ labels the distinct eigenvalues, the dispersion curve for surface plasmon consists of several branches $\omega = \omega_s(k)$, where $\omega_s(k)$ is the solution of

$$\frac{1+\epsilon(\omega)}{1-\epsilon(\omega)} = \lambda_s(k), \qquad s = 1, 2, \ldots \tag{3.120}$$

The solution of eq. (3.119) can be carried out by any of the methods discussed in connection with the solution of eqs. (3.75)–(3.76). This has not been done as yet.

Just as in the case of the surface polariton on a grating, more manageable equations from which the dispersion relation for surface plasmons on a grating can be obtained follow from taking the observation points on planes $x_3 =$ constant that lie outside the selvedge region. Thus, we start with eqs. (3.106) and (3.111), written in terms of $H(x_1|\omega)$ and $L(x_1|\omega)$,

$$\begin{array}{ll} x_3 > \zeta(x_1) & \varphi^>(x_1, x_3|\omega) \\ x_3 < \zeta(x_1) & 0 \end{array} \Big\} = -\frac{1}{4\pi}\int dx_1' \left\{\left(\frac{d\zeta}{dx_1'}\frac{\partial}{\partial x_1'} - \frac{\partial}{dx_3'}\right)\right. \quad \begin{array}{r}(3.121a)\\(3.121b)\end{array}$$

$$\left.\times G(x_1x_3|x_1'x_3')\Big|_{x_3'=\zeta(x_1')} H(x_1'|\omega) - G(x_1x_3|x_1'x_3')\Big|_{x_3'=\zeta(x_1')} L(x_1'|\omega)\right\},$$

$$\begin{array}{ll} x_3 > \zeta(x_1) & 0 \\ x_3 < \zeta(x_1) & \varphi^<(x_1, x_3|\omega) \end{array} \Big\} = \frac{1}{4\pi}\int dx_1' \left\{\left(\frac{d\zeta}{dx_1'}\frac{\partial}{\partial x_1'} - \frac{\partial}{\partial x_3'}\right)\right. \quad \begin{array}{r}(3.122a)\\(3.122b)\end{array}$$

$$\times G(x_1x_3|x_1'x_3')\Big|_{x_3'=\zeta(x_1')} H(x_1'|\omega) - \frac{1}{\epsilon(\omega)}$$

$$\left.\times G(x_1x_3|x_1'x_3')\Big|_{x_3'=\zeta(x_1')} L(x_1'|\omega)\right\}.$$

Then, if we use eq. (3.121b) for $x_3 < \min \zeta(x_1)$ and eq. (3.122a) for $x_3 > \max \zeta(x_1)$, together with the Fourier representation for the Green's function, eq. (3.109), we obtain convenient equations for $H(x_1|\omega)$ and $L(x_1|\omega)$:

$$\frac{1}{a}\int_{-a/2}^{a/2} dy_1\, e^{-ik_m y_1}\, e^{-\alpha_m(k)\zeta(y_1)} \{[ik_m\zeta'(y_1) - \alpha_m(k)]H(y_1|\omega) + L(y_1|\omega)\} = 0$$

$$\text{for each } m \qquad (3.123a)$$

$$\frac{1}{a}\int_{-a/2}^{a/2} dy_1\, e^{-ik_m y_1}\, e^{\alpha_m(k)\zeta(y_1)} \Big\{[ik_m\zeta'(y_1) + \alpha_m(k)]H(y_1|\omega)$$

$$+\frac{1}{\epsilon(\omega)} L(y_1|\omega)\Big\} = 0 \qquad \text{for each } m. \qquad (3.123b)$$

The discussion following eqs. (3.79) applies to these equations and to the methods for their solution as well. If we expand $H(y_1|\omega)$ and $L(y_1|\omega)$ in Fourier series, as in eqs. (3.80), the equations for the Fourier coefficients $\{\hat{H}_n(k\omega)\}$ and $\{\hat{L}_n(k\omega)\}$ take the simple forms

$$\sum_{n=-\infty}^{\infty} I_{m-n}^{(m)}(k) \left\{\frac{k_m k_n}{\alpha_m(k)} \hat{H}_n(k\omega) - \hat{L}_n(k\omega)\right\} = 0 \qquad (3.124a)$$

$$\sum_{n=-\infty}^{\infty} J_{m-n}^{(m)}(k)\left\{\epsilon(\omega)\frac{k_m k_n}{\alpha_m(k)}\hat{H}_n(k\omega)+\hat{L}_n(k\omega)\right\}=0, \tag{3.124b}$$

where $I_n^{(m)}(k)$ and $J_n^{(m)}(k)$ have been defined in eqs. (3.91).

We can again write the equations obtained by the Rayleigh method, eqs. (3.92), and the equations (3.124) obtained from the extinction equations in forms where comparisons can be made:

$$\sum_n \begin{pmatrix} -I_{m-n}^{(n)}(k) & J_{m-n}^{(n)}(k) \\ \dfrac{k_m k_n}{\alpha_n(k)} I_{m-n}^{(n)}(k) & \epsilon(\omega)\dfrac{k_m k_n}{\alpha_n(k)} J_{m-n}^{(n)}(k) \end{pmatrix} \begin{pmatrix} A_n(k\omega) \\ B_n(k\omega) \end{pmatrix} = 0 \tag{3.125a}$$

$$\sum_n \begin{pmatrix} -I_{m-n}^{(m)}(k) & \dfrac{k_m k_n}{\alpha_m(k)} I_{m-n}^{(m)}(k) \\ J_{m-n}^{(m)}(k) & \epsilon(\omega)\dfrac{k_m k_n}{\alpha_m(k)} J_{m-n}^{(m)}(k) \end{pmatrix} \begin{pmatrix} \hat{L}_n(k\omega) \\ \hat{H}_n(k\omega) \end{pmatrix} = 0. \tag{3.125b}$$

We see that the coefficient matrices are transposes of each other, provided that

$$I_{n-m}^{(n)}(k) = I_{m-n}^{(n)}(k), \quad J_{n-m}^{(n)}(k) = J_{m-n}^{(n)}(k), \tag{3.126}$$

i.e., provided $\zeta(-x_1)=\zeta(x_1)$. The consequences of this fact, together with the uniqueness of the solutions to the extinction equations (3.120), for the solutions of the equations of the Rayleigh method, eqs. (3.92), parallel those for the corresponding equations of the preceding subsection for surface polaritons.

Equations (3.124) have been solved numerically recently by Glass and Maradudin (1980), for the sinusoidal surface profile function, eq. (3.21), and for the symmetric sawtooth profile function, eq. (3.82). A free electron dielectric constant, $\epsilon(\omega)=1-(\omega_p^2/\omega^2)$, was used in these calculations. For the former profile, calculations were carried out for values of ζ_0/a up to 0.4. It is not known yet whether this represents a limit on the computational method. For values of ζ_0/a for which the calculations based on the Rayleigh hypothesis, as exemplified by eqs. (3.104)–(3.105), converge, the results obtained by both approaches are identical. The numerical calculations based on eqs. (3.104)–(3.105), however, are much easier to carry out than are those based on eqs. (3.124).

Convergent results for the frequencies of surface plasmons on the sawtooth grating were obtained from eqs. (3.124) for values of h/a up to about 0.04.

For both profiles increasing the ratio ζ_0/a or h/a required increasing the size of the matrix of the coefficients on the left hand side of eqs. (3.124) to obtain accurate values of the surface plasmon frequencies. This was also the case if it was desired to obtain the frequencies of the branches of the plasmon dispersion curve close to the value $\omega = 2^{-1/2}\,\omega_p$ corresponding to a flat surface. Matrices as large as 32×32 were used in these calculations.

The results of Glass and Maradudin (1980) are of interest because they show that the solution of the reduced equation (3.93), obtained from the full set of equations of the Rayleigh method (3.90), by an expansion in Fourier series is convergent, for analytic (sinusoidal) surface profiles, for values of the corrugation strengths that extend up to, and even slightly exceed, the value for which the Rayleigh hypothesis is known to become invalid. It is not convergent, except perhaps in an asymptotic sense, for a nonanalytic (symmetric sawtooth) profile. On the other hand, when the full set of equations of the Rayleigh method (3.90) is solved by expansions in Fourier series the resulting dispersion relation is the same as that obtained by solving the extinction theorem equations (3.124) also by Fourier series expansions, for even profile functions. Convergent results are then obtained for corrugation strengths much larger than that for which the Rayleigh hypothesis is valid, for the sinusoidal profile, and convergent results are also obtained for moderately large amplitude sawtooth profiles. The conclusion would seem to be that the use of the equations of Rayleigh's method in their unreduced form, even when they are solved by expansions in Fourier series, yields convergent results for the frequencies of surface plasmons on gratings whose amplitudes are larger than those for which the Rayleigh hypothesis is valid, but the use of the reduced form of these equations with Fourier series expansions yields convergent results only in those cases where the Rayleigh hypothesis is valid.

Thus, in this and the preceding subsection we have shown how the dispersion relation for surface polaritons and plasmons on a grating can be obtained. We have presented explicit results for these dispersion relations in the small roughness limit on the basis of the Rayleigh hypothesis, and have shown the qualitative nature of their solutions. Exact dispersion relations, free from the limitations present in the use of the Rayleigh method, have been obtained by the use of Green's theorem. Results obtained by numerical solution of several of the equations derived here have been reported. More, however, remains to be done before the most efficient methods for obtaining dispersion curves are known. This represents one of the areas for future work on this problem.

3.3. *Interaction of a Surface Polariton With a Rayleigh Wave*

Not all gratings with which a surface polariton can interact are stationary. A particularly interesting example of the interaction of a surface polariton with a moving grating is provided by the interaction of a surface polariton with a Rayleigh surface (acoustic) wave. Through this interaction a surface polariton can be Bragg-scattered, or a portion of its energy can be caused to be scattered into the vacuum above the crystal, depending on the kinematics of the interaction process. Thus, the use of Rayleigh waves may prove to be an important means for detecting or manipulating surface polaritons. In addition, a Rayleigh wave produces a (moving) grating whose profile is essentially exactly known. In what follows we present an analysis of the interaction of surface polaritons with Rayleigh waves due to Mills (1977).

Surface polaritons may couple to a Rayleigh wave by two different mechanisms. The Rayleigh wave creates a moving sinusoidal diffraction grating on the surface that deflects the fields set up by the surface polariton. In addition, through the electro-optic effect, it modulates the dielectric constant of the substrate in a periodic fashion. We consider both of these mechanisms here.

In contrast with the discussions in the precedng two subsections, in the present treatment we assume that the crystal lies in the upper half-space, $x_3 > 0$, and the vacuum fills the lower half-space, $x_3 < 0$.

In the dielectric medium the dielectric tensor modulated by the passage of the Rayleigh wave over it takes the form

$$\epsilon_{\alpha\beta}(x, t) = \delta_{\alpha\beta}\epsilon(\omega) + \delta\epsilon_{\alpha\beta}(x, t) \tag{3.127a}$$

$$= \delta_{\alpha\beta}\epsilon(\omega) + \sum_{\mu\nu} k_{\alpha\beta\mu\nu}\, e_{\mu\nu}(x, t), \tag{3.127b}$$

to lowest order in the elastic strains $\{e_{\mu\nu}(x, t)\}$ created by the Rayleigh wave, where $k_{\alpha\beta\mu\nu}$ is the photoelastic tensor of the medium. The strain field set up by a Rayleigh wave of a wave vector $\mathbf{Q}_\parallel$ and frequency Ω can be written quite generally as

$$e_{\mu\nu}(x, t) = \sum_j e_{\mu\nu}^{(j)} \exp[i(\mathbf{Q}_\parallel \cdot \mathbf{x}_\parallel - \Omega t)] \exp(-\alpha_j x_3) + \text{c.c.} \tag{3.128}$$

In general the strain field is a linear superposition of as many as three exponentially decaying disturbances. This is indicated by the sum on j in eq. (3.128). The decay constants $\{\alpha_j\}$ and the amplitudes $\{e_{\mu\nu}^{(j)}\}$ can be obtained from the theory of elasticity. If any of the $\{\alpha_j\}$ is complex, it is chosen with $\text{Re}(\alpha_j) > 0$. Combining eqs. (3.127) and (3.128) we can write the modulation of the dielectric tensor as

$$\delta\epsilon_{\alpha\beta}(x, t) = \sum_j \Delta_{\alpha\beta}^{(j)} \exp[i(\mathbf{Q}_\parallel \cdot \mathbf{x}_\parallel - \Omega t)] \exp(-\alpha_j x_3) + \text{c.c.} \tag{3.129a}$$

where

$$\Delta_{\alpha\beta}^{(j)} = \sum_{\mu\nu} k_{\alpha\beta\mu\nu}\, e_{\mu\nu}^{(j)}. \tag{3.129b}$$

It is now necessary to obtain the solutions of the Maxwell equations for the electromagnetic fields in the medium, where the dielectric tensor varies in space and time in the manner just indicated. At the same time the fields in the vacuum must be obtained, and finally the fields in these two regions of space must be matched across the undulating surface caused by the presence of the Rayleigh wave.

The macroscopic electric field in the vacuum and in the medium can be written in Bloch form as

$$\boldsymbol{E}^{<}(\boldsymbol{x}, t) = \sum_{m=-\infty}^{\infty} \boldsymbol{\mathscr{E}}^{(m)}(x_3) \exp[\mathrm{i}\boldsymbol{k}_{\parallel}(m) \cdot \boldsymbol{x}_{\parallel} - \mathrm{i}\omega_m t], \tag{3.130a}$$

$$\boldsymbol{E}^{>}(\boldsymbol{x}, t) = \sum_{m=-\infty}^{\infty} \boldsymbol{E}^{(m)}(x_3) \exp[\mathrm{i}\boldsymbol{k}_{\parallel}(m) \cdot \boldsymbol{x}_{\parallel} - \mathrm{i}\omega_m t], \tag{3.130b}$$

respectively, where

$$\boldsymbol{k}_{\parallel}(m) = \boldsymbol{k}_{\parallel} + m\boldsymbol{Q}_{\parallel}, \tag{3.131a}$$

$$\omega_m = \omega + m\Omega. \tag{3.131b}$$

Here, and in what follows we use script letters to denote components of the electromagnetic field in the vacuum, and block letters to denote the components of the field in the medium.

The corresponding magnetic fields are given by

$$\boldsymbol{H}^{<}(\boldsymbol{x}, t) = \sum_{m=-\infty}^{\infty} \boldsymbol{\mathscr{H}}^{(m)}(x_3) \exp[\mathrm{i}\boldsymbol{k}_{\parallel}(m) \cdot \boldsymbol{x}_{\parallel} - \mathrm{i}\omega_m t] \tag{3.132a}$$

$$\boldsymbol{H}^{>}(\boldsymbol{x}, t) = \sum_{m=-\infty}^{\infty} \boldsymbol{H}^{(m)}(x_3) \exp[\mathrm{i}\boldsymbol{k}_{\parallel}(m) \cdot \boldsymbol{x}_{\parallel} - \mathrm{i}\omega_m t] \tag{3.132b}$$

where

$$\begin{aligned} \boldsymbol{\mathscr{H}}^{(m)}(x_3) = \frac{c}{\mathrm{i}\omega_m} \Big\{ \hat{x}_i \Big[\mathrm{i}k_2(m)\mathscr{E}_3^{(m)}(x_3) - \frac{\mathrm{d}}{\mathrm{d}x_3} \mathscr{E}_2^{(m)}(x_3) \Big] \\ + \hat{x}_2 \Big[\frac{\mathrm{d}}{\mathrm{d}x_3} \mathscr{E}_1^{(m)}(x_3) - \mathrm{i}k_1(m)\, \mathscr{E}_3^{(m)}(x_3) \Big] \\ + \hat{x}_3 [\mathrm{i}k_1(m)\, \mathscr{E}_2^{(m)}(x_3) - \mathrm{i}k_2(m)\, \mathscr{E}_1^{(m)}(x_3)] \Big\} \end{aligned} \tag{3.133a}$$

$$\begin{aligned} \boldsymbol{H}^{(m)}(x_3) = \frac{c}{\mathrm{i}\omega_m} \Big\{ \hat{x}_1 \Big[\mathrm{i}k_2(m)E_3^{(m)}(x_3) - \frac{\mathrm{d}}{\mathrm{d}x_3} E_2^{(m)}(x_3) \Big] \\ + \hat{x}_2 \Big[\frac{\mathrm{d}}{\mathrm{d}x_3} E_1^{(m)}(x_3) - \mathrm{i}k_1(m)\, E_3^{(m)}(x_3) \Big] \\ + \hat{x}_3 [\mathrm{i}k_1(m)\, E_2^{(m)}(x_3) - \mathrm{i}k_2(m)\, E_1^{(m)}(x_3)] \Big\}. \end{aligned} \tag{3.133b}$$

Explicit equations for the electric field amplitude vectors $\boldsymbol{\mathscr{E}}^{(m)}(x_3)$ and $\boldsymbol{E}^{(m)}(x_3)$ are obtained by substituting the expansions (3.130a) and (3.130b) into Maxwell's equations in the vacuum and dielectric medium, respectively. In the former case the components of $\mathscr{E}^{(m)}(x_3)$ satisfy the equations

$$\sum_{\beta} L^{(0)}_{\alpha\beta}(\mathbf{k}_{\|}(m), \omega_m \mid x_3)\, \mathscr{E}^{(m)}_{\beta}(x_3) = 0, \tag{3.134}$$

where $L^{(0)}_{\alpha\beta}(\mathbf{k}_{\|}, \omega \mid x_3)$ is the matrix differential operator

$$L^{(0)}_{\alpha\beta}(k_{\|}, \omega \mid x_3) = \delta_{\alpha\beta}\left[\frac{\mathrm{d}^2}{\mathrm{d}x_3^2} - k_{\|}^2 + \frac{\omega^2}{c^2}\right] - \left[(1-\delta_{\alpha 3})\mathrm{i}k_{\alpha} + \delta_{\alpha 3}\frac{\mathrm{d}}{\mathrm{d}x_3}\right]\left[(1-\delta_{\beta 3})\mathrm{i}k_{\beta} + \delta_{\beta 3}\frac{\mathrm{d}}{\mathrm{d}x_3}\right]. \tag{3.135}$$

The solutions of eqs. (3.134)–(3.135) can be written as

$$\mathscr{E}^{(m)}(x_3) = \mathscr{E}^{(m)}_p \mathscr{E}^{(m)}_p(x_3) + \mathscr{E}^{(m)}_s \mathscr{E}^{(m)}_s(x_3), \tag{3.136a}$$

where

$$\mathscr{E}^{(m)}_p(x_3) = \left[\hat{x}_1 \cos\varphi_m + \hat{x}_2 \sin\varphi_m - \mathrm{i}\hat{x}_3 \frac{k_{\|}(m)}{\alpha_0(m)}\right] \mathrm{e}^{\alpha_0(m)x_3} \tag{3.136b}$$

$$\mathscr{E}^{(m)}_s(x_3) = [-\hat{x}_1 \sin\varphi_m + \hat{x}_2 \cos\varphi_m]\, \mathrm{e}^{\alpha_0(m)x_3}. \tag{3.136c}$$

In eqs. (3.136) φ_m is the angle between the vector $\mathbf{k}_{\|}(m)$ and the x_1-axis, and

$$\alpha_0(m) = (k_{\|}^2(m) - \omega_m^2/c^2)^{1/2} \qquad k_{\|}(m) > \omega/c. \tag{3.137a}$$

$$= -\mathrm{i}(\omega_m^2/c^2 - k_{\|}^2(m))^{1/2} \qquad k_{\|}(m) < \omega/c. \tag{3.137b}$$

The content of eq. (3.137b) is that the interaction of a surface polariton with a Rayleigh wave can give rise to diffracted waves that can radiate energy into the vacuum away from the dielectric–vacuum interface.

In the expansion of the fields given by eqs. (3.130) and (3.132) the terms with $m = 0$ describe a surface polariton propagating along the flat interface between vacuum and a dielectric medium whose dielectric tensor is not being modulated by the Rayleigh wave. If we assume that this incident wave propagates in the $+x_1$-direction, this means that

$$\mathbf{k}_{\|}(0) \equiv \mathbf{k}_{\|} = \hat{x}_1 k_{\|}, \tag{3.138}$$

and that the amplitude $\mathscr{E}^{(0)}(x_3)$ is

$$\mathscr{E}^{(0)}(x_3) = \mathscr{E}^{(0)}_p \left[\hat{x}_1 - \mathrm{i}\hat{x}_3 \frac{k_{\|}(0)}{\alpha_0(0)}\right] \mathrm{e}^{\alpha_0(0)x_3}. \tag{3.139}$$

The heart of the present calculation is the determination of the vectors $\{\boldsymbol{E}^{(m)}(x_3)\}$ that describe the electric field in the dielectric medium over which the Rayleigh wave is propagating. When the expression given by eq. (3.130b) is substituted into Maxwell's equation for a medium whose dielectric tensor is given by eqs. (3.127) and (3.129), it is found that the amplitudes $\{\boldsymbol{E}^{(m)}(x_3)\}$ obey the system of coupled equations

$$\sum_\beta L_{\alpha\beta}(\boldsymbol{k}_\|(m), \omega_m \mid x_3)\, E_\beta^{(m)}(x_3) + \frac{\omega_m^2}{c^2} \sum_\beta \sum_j \Delta_{\alpha\beta}^{(j)}\, \mathrm{e}^{-\alpha_j x_3}\, E_\beta^{(m-1)}(x_3)$$

$$+ \frac{\omega_m^2}{c^2} \sum_\beta \sum_j \Delta_{\alpha\beta}^{(j)*}\, \mathrm{e}^{-\alpha_j^* x_3}\, E_\beta^{(m+1)}(x_3) = 0. \tag{3.140}$$

The matrix differential operator $L_{\alpha\beta}(\boldsymbol{k}_\|, \omega \mid x_3)$ is defined by

$$L_{\alpha\beta}(\boldsymbol{k}_\|, \omega \mid x_3) = \delta_{\alpha\beta}\left[\frac{\mathrm{d}^2}{\mathrm{d}x_3^2} - k_\|^2 + \epsilon(\omega)\frac{\omega^2}{c^2}\right]$$

$$-\left[(1-\delta_{\alpha 3})\, ik_\alpha + \delta_{\alpha 3}\frac{\mathrm{d}}{\mathrm{d}x_3}\right]\left[(1-\delta_{\beta 3})\, ik_\beta + \delta_{\beta 3}\frac{\mathrm{d}}{\mathrm{d}x_3}\right], \tag{3.141}$$

and the last two terms on the left hand side of eq. (3.140) account for the modulation of the dielectric constant of the medium by the Rayleigh wave.

Because the terms with $m = 0$ in eqs. (3.130) and (3.132) describe a surface polariton propagating along a flat dielectric–vacuum interface, we take the field amplitude $\boldsymbol{E}^{(0)}(x_3)$ to be of O(1) in the sense that although it also contains contributions of $O(\Delta^2)$, $O(\Delta^4)$, etc., it is nonvanishing as $\Delta_{\alpha\beta}^{(j)} \to 0$. The field amplitudes $\boldsymbol{E}^{(\pm 1)}(x_3)$ are therefore of $O(\Delta)$, and contain higher order contributions of $O(\Delta^3)$, $O(\Delta^5)$, etc. Similarly, $\boldsymbol{E}^{(\pm 2)}(x_3)$ are of $O(\Delta^2)$, and so on. In what follows we will need only $\boldsymbol{E}^{(0)}(x_3)$ and $\boldsymbol{E}^{(\pm 1)}(x_3)$, the latter to no higher than first order in $\Delta_{\alpha\beta}^{(j)}$, the former to second order.

Because of their coupling to the amplitudes $\{\boldsymbol{E}^{(m)}(x_3)\}$ through the boundary conditions, we can make similar estimates of the coefficients $\{\mathscr{E}_p^{(m)}\}$ and $\{\mathscr{E}_s^{(m)}\}$ in eq. (3.136a). In particular, $\mathscr{E}_p^{(0)}$ is of O(1), while $\mathscr{E}_p^{(\pm 1)}$ and $\mathscr{E}_p^{(\pm 1)}$ are of $O(\Delta)$.

To lowest order in $\Delta_{\alpha\beta}^{(j)}$ the amplitude $\boldsymbol{E}^{(0)}(x_3)$ satisfies the equation

$$\sum_\beta L_{\alpha\beta}(\boldsymbol{k}_\|(0), \omega_0 \mid x_3)\, E_\beta^{(0)}(x_3) = 0. \tag{3.142}$$

We assume that $\boldsymbol{E}^{(0)}(x_3)$ describes a (p-polarized) surface polariton that propagates in the $\hat{x}_1$-direction. The solution of eq. (3.142) in this case is

$$\boldsymbol{E}^{(0)}(x_3) = E_\beta^{(0)}\left[\hat{x}_1 + i\hat{x}_3\frac{k_\|(0)}{\alpha(0)}\right]\mathrm{e}^{-\alpha(0)x_3}, \tag{3.143}$$

where, in general,

$$\alpha(m) = [k_\|^2(m) - \epsilon(\omega_m)(\omega_m^2/c^2)]^{1/2}, \qquad \mathrm{Re}\,\alpha(m) > 0. \tag{3.144}$$

In particular, the decay constant $\alpha(0)$ is required to be real and positive, on the assumption that $\epsilon(\omega)$ is real.

From eq. (3.140) we see that the amplitude $\boldsymbol{E}^{(+1)}(x_3)$ satisfies the inhomogeneous equation

$$\sum_{\beta} L_{\alpha\beta}(\boldsymbol{k}_{\|}(+1), \omega_{+1} \,|\, x_3)\, E_{\beta}^{(+1)}(x_3) = -\frac{\omega_{+1}^2}{c^2} E_p^{(0)} \sum_j \lambda_{(p)\alpha}^{(0)j}\, \mathrm{e}^{-[\alpha_j+\alpha(0)]x_3} \tag{3.145}$$

where, in general,

$$\lambda_{(\sigma)\alpha}^{(m)j} = \sum_{\beta} \Delta_{\alpha\beta}^{(j)}\, n_{(\sigma)\beta}^{(m)}. \tag{3.146}$$

In eq. (3.146) σ is a polarization index that assumes the values p and s, and we have defined the vectors $\boldsymbol{n}_{(p)}^{(m)}$ and $\boldsymbol{n}_{(s)}^{(m)}$ as

$$\boldsymbol{n}_{(p)}^{(m)} = \hat{x}_1 \cos\varphi_m + \hat{x}_2 \sin\varphi_m + \mathrm{i}\hat{x}_3\, k_{\|}(m)/\alpha(m) \tag{3.147a}$$

$$\boldsymbol{n}_{(s)}^{(m)} = -\hat{x}_1 \sin\varphi_m + \hat{x}_2 \cos\varphi_m. \tag{3.147b}$$

Similarly, the amplitude $\boldsymbol{E}_{(x_3)}^{(-1)}$ satisfies the equation

$$\sum_{\beta} L_{\alpha\beta}(\boldsymbol{k}_{\|}(-1), \omega_{-1} \,|\, x_3)\, E_{\beta}^{(-1)}(x_3) = -\frac{\omega_{-1}^2}{c^2} E_p^{(0)} \sum_j \bar{\lambda}_{(p)\alpha}^{(0)j}\, \mathrm{e}^{-[\alpha_j^*+\alpha(0)]x_3}, \tag{3.148a}$$

where

$$\bar{\lambda}_{(\sigma)\alpha}^{(m)j} = \sum_{\beta} \Delta_{\alpha\beta}^{(j)*}\, n_{(\sigma)\beta}^{(m)}. \tag{3.148b}$$

Note that $\bar{\lambda}_{(\sigma)\alpha}^{(m)j}$ is not the complex conjugate of $\lambda_{(\sigma)\alpha}^{(m)j}$.

The general solution of eq. (3.145) is the sum of a particular integral and a complementary function, and is given by

$$\boldsymbol{E}^{(+1)}(x_3) = [E_p^{(+1)}\, \boldsymbol{n}_{(p)}^{(+1)} + E_s^{(+1)}\, \boldsymbol{n}_{(s)}^{(+1)}]\, \mathrm{e}^{-\alpha(+1)x_3} + E_p^{(0)} \sum_j \sum_{\beta} \mathscr{L}_{\alpha\beta}^{(j)-1}(\boldsymbol{k}_{\|}(+1), \omega_{+1})\lambda_{(p)\beta}^{(0)j}\, \mathrm{e}^{-[\alpha_j+\alpha(0)]x_3} \tag{3.149}$$

where, in general, the matrix $\mathscr{L}^{(j)-1}(\boldsymbol{k}_{\|}(m), \omega_m)$ is given by

$$\mathscr{L}^{(j)-1}(\boldsymbol{k}_{\|}(m), \omega_m) = \frac{1}{\epsilon(\omega_m)\{\alpha^2(m) - [\alpha_j + \alpha(m-1)]^2\}}$$
$$\times \begin{pmatrix} \epsilon(\omega_m)\dfrac{\omega_m^2}{c^2} - k_1^2(m) & -k_1(m)k_2(m) & -\mathrm{i}k_1(m)[\alpha_j + \alpha(m-1)] \\ -k_1(m)k_2(m) & \epsilon(\omega_m)\dfrac{\omega_m^2}{c^2} - k_2^2(m) & -\mathrm{i}k_2(m)[\alpha_j + \alpha(m-1)] \\ -\mathrm{i}k_1(m)[\alpha_j + \alpha(m-1)] & -\mathrm{i}k_2(m)[\alpha_j + \alpha(m-1)] & \epsilon(\omega_m)\dfrac{\omega_m^2}{c^2} + [\alpha_j + \alpha(m-1)]^2 \end{pmatrix}. \tag{3.150}$$

In accordance with what has been said above, the coefficients $E_p^{(+1)}$ and $E_s^{(+1)}$ must be understood to be of $O(\Delta)$.

The solution of eq. (3.148) can be obtained from the result given by eqs. (3.149)–(3.150) by the replacements $+1 \rightarrow -1$, $\alpha_j \rightarrow \alpha_j^*$, and $\Delta_{\alpha\beta}^{(j)} \rightarrow \Delta_{\alpha\beta}^{(j)*}$.

With the preceding results in hand we can return to eq. (3.140) and obtain the correction to the field amplitude $\boldsymbol{E}^{(0)}(x_3)$ of second order in $\Delta_{\alpha\beta}^{(j)}$. When this is done, the result can be expressed as

$$\begin{aligned}\boldsymbol{E}^{(0)}(x_3) &= E_p^{(0)}\left[\hat{x}_1 + \mathrm{i}\hat{x}_3 \frac{k_\parallel(0)}{\alpha(0)}\right] \mathrm{e}^{-\alpha(0)x_3} \\ &+ \sum_j \sum_\sigma \sum_\beta E_\sigma^{(-1)} \mathscr{L}_{\alpha\beta}^{(j)-1}(\boldsymbol{k}_\parallel(0), \omega_0)\, \lambda_{(\sigma)\beta}^{(-1)j}\, \mathrm{e}^{-[\alpha_j^* + \alpha(-1)]x_3} \\ &+ \sum_j \sum_\sigma \sum_\beta E_\sigma^{(+1)} \bar{\mathscr{L}}_{\alpha\beta}^{(j)-1}(\boldsymbol{k}_\parallel(0), \omega_0) \bar{\lambda}_{(\sigma)\beta}^{(+1)j}\, \mathrm{e}^{-[\alpha_j + \alpha(+1)]x_3},\end{aligned} \qquad (3.151)$$

where $\bar{\mathscr{L}}_{\alpha\beta}^{(j)-1}(\boldsymbol{k}_\parallel(0), \omega_0)$ is obtained from $\mathscr{L}_{\alpha\beta}^{(j)-1}(\boldsymbol{k}_\parallel(0), \omega_0)$ by the replacements $-1 \rightarrow +1$, $\alpha_j \rightarrow \alpha_j^*$.

The second and third terms on the right hand side of eq. (3.151) are both of $O(\Delta^2)$.

It should be pointed out that in substituting into eq. (3.140) the amplitudes $\boldsymbol{E}^{(\pm 1)}(x_3)$ given by eq. (3.149) and the discussion following eq. (3.150) we have used only the complementary function and have omitted the particular integral. This is due to the fact that the latter will not contribute the resonant denominators that give rise to the gaps in the dispersion curve that are of interest to us (Mills 1980).

The expressions for the electric field components (3.130) contain four sets of unknown coefficients, viz. $\{\mathscr{E}_p^{(m)}\}$, $\{\mathscr{E}_s^{(m)}\}$, $\{E_p^{(m)}\}$, and $\{E_s^{(m)}\}$. As these coefficients are obtained from the boundary conditions, it is necessary to use four independent boundary conditions. These are provided by the requirements that the tangential components of the electric field and the tangential components of the magnetic field be continuous across the dielectric–vacuum interface corrugated by the passage of the Rayleigh wave along it.

The instantaneous position of the surface is given to first approximation by

$$x_3 = u_3(\boldsymbol{x}_\parallel 0, t) = u_\perp \exp[\mathrm{i}(\boldsymbol{Q}_\parallel \cdot \boldsymbol{x}_\parallel - \Omega t)] + u_\perp^* \exp[-\mathrm{i}(\boldsymbol{Q}_\parallel \cdot \boldsymbol{x}_\parallel - \Omega t)]. \qquad (3.152)$$

It is this function that in the present context plays the role of the surface profile function $\zeta(\boldsymbol{x}_\parallel)$ introduced in sect. 2.2.

When eqs. (3.130) are substituted into the two equations that express the continuity of the tangential components of the electric field,

$$\hat{t}_{1,2} \cdot \boldsymbol{E}^{<}(\boldsymbol{x}, t)\Big|_{x_3=u_3} = \hat{t}_{1,2} \cdot \boldsymbol{E}^{>}(\boldsymbol{x}, t)\Big|_{x_3=u_3}, \qquad (3.153)$$

where the unit vectors $\hat{t}_1$ and $\hat{t}_2$ are defined by eqs. (2.19)–(2.20), and both sides are expanded to first order in $u_\perp$ and $u_\perp^*$, we obtain the conditions that

$$E_1^{(m)}(0) - u_\perp\left[\left(\frac{\partial E_1^{(m-1)}}{\partial x_3}\right)_0 + iQ_1E_3^{(m-1)}(0)\right] + u_\perp^*\left[\left(\frac{\partial E_1^{(m+1)}}{\partial x_3}\right)_0 - iQ_1E_3^{(m+1)}(0)\right]$$
$$= \mathscr{E}_1^{(m)}(0) + u_\perp\left[\left(\frac{\partial \mathscr{E}_1^{(m-1)}}{\partial x_3}\right)_0 + iQ_1\mathscr{E}_3^{(m-1)}(0)\right]$$
$$+ u_\perp^*\left[\left(\frac{\partial \mathscr{E}_1^{(m+1)}}{\partial x_3}\right)_0 - iQ_1\mathscr{E}_3^{(m+1)}(0)\right], \tag{3.154}$$

$$E_2^{(m)}(0) + u_\perp\left[\left(\frac{\partial E_2^{(m-1)}}{\partial x_3}\right)_0 + iQ_2E_3^{(m-1)}(0)\right] + u_\perp^*\left[\left(\frac{\partial E_2^{(m+1)}}{\partial x_3}\right)_0 - iQ_2E_3^{(m+1)}(0)\right]$$
$$= \mathscr{E}_2^{(m)}(0) + u_\perp\left[\left(\frac{\partial \mathscr{E}_2^{(m-1)}}{\partial x_3}\right)_0 + iQ_2\mathscr{E}_3^{(m-1)}(0)\right]$$
$$+ u_\perp^*\left[\left(\frac{\partial \mathscr{E}_2^{(m+1)}}{\partial x_3}\right)_0 - iQ_2\mathscr{E}_3^{(m+1)}(0)\right]. \tag{3.155}$$

The two equations that express the requirement that tangential components of $\boldsymbol{H}(x, t)$ be continuous across the corrugated surface are obtianed by replacing $E_\alpha^{(m)}$ and $\mathscr{E}_\alpha^{(m)}$ in eqs. (3.154)–(3.155) by $H_\alpha^{(m)}$ and $\mathscr{H}_\alpha^{(m)}$, respectively.

There is a technical point concerning the use of these boundary conditions that needs to be made here. It has been assumed that the surface corrugated by the Rayleigh wave has been frozen at one instant of time, and the standard Maxwell boundary conditions have been applied across the resulting surface. In fact, however, any point on the surface is moving with a velocity whose magnitude is $\sim\Omega u_3$. When the boundary conditions are applied across a moving surface, all the fields should be transformed to the rest frame of the point on the surface where the boundary conditions are to be applied, and then the tangential components of the transformed $\boldsymbol{E}$ and $\boldsymbol{H}$ fields should be matched across the boundary (Landau and Lifshitz 1960). Mills (1977) has estimated that the procedure used here introduces errors of the order of $\Omega/\omega \sim c_R/c_S$, where c_R and c_S are the phase velocities of the Rayleigh wave and surface polariton, respectively, when the wavelengths of the two waves are comparable. The ratio c_R/c_S under typical circumstances is $\sim 10^{-5}$, so that the procedure followed here leads to negligible error.

The use of the preceding boundary conditions, together with the expressions for the electric field vectors $\{\boldsymbol{\mathscr{E}}^{(m)}(x_3)\}$ and $\{\boldsymbol{E}^{(m)}(x_3)\}$ given by eqs. (3.136)–(3.138) and (3.149)–(3.151), makes it possible to obtain expressions for the amplitudes of the first order diffracted fields $\boldsymbol{\mathscr{E}}_{p,s}^{(\pm1)}(x_3)$ and $\boldsymbol{E}_{p,s}^{(\pm1)}(x_3)$ in terms of the amplitudes of the incident wave, $\boldsymbol{\mathscr{E}}_p^{(0)}(x_3)$ and $\boldsymbol{E}_p^{(0)}(x_3)$. The resulting expressions have been given by Mills (1977), who also discusses the conditions under which both fields ($m = \pm 1$) are localized near the medium–vacuum interface, as are those of the incident surface polariton, as well as the conditions under which one of the diffracted beams is localized near the surface while the other describes a radiating wave that carries energy away from the surface. In view of the emphasis in this chapter on the effects of surface roughness on the dispersion curve of the surface polariton, we will not pursue this point further here, and refer the interested reader to the paper by Mills (1977) where this topic is treated in some detail.

We are now in a position to discuss the dispersion relation for a surface polariton when it propagates on a surface with a Rayleigh wave impressed on it. However, rather than obtaining the complete dispersion curve, we will focus on the gaps that are introduced into it near those values of the wave vector $\boldsymbol{k}_{\|}(0)$ where Bragg reflections from the periodic surface structure can occur. As we have seen in the preceding two subsections a perturbation theory linear in the amplitude of the surface corrugation is sufficient to yield the magnitudes of these gaps and the form of the dispersion curve near the gaps to second order in the surface corrugation amplitude. We have also seen that in this approximation there are as many gaps as there are nonzero Fourier coefficients of the surface roughness profile function. In the present case, as we can see from eq. (3.152), we expect two gaps. These occur at values of the wave vector $\boldsymbol{k}_{\|}(0)$ such that $|\boldsymbol{k}_{\|}(0)-\boldsymbol{Q}_{\|}| = k_{\|}(0)$ and $|\boldsymbol{k}_{\|}(0)+\boldsymbol{Q}_{\|}| = k_{\|}(0)$. This is because for wave vectors in the near vicinity of the position of the gap the dominant contribution to the shift in the dispersion curve from its flat surface form comes from a process in which the surface polariton of wave vector $\boldsymbol{k}_{\|}(0)$ is scattered by the surface corrugations into an intermediate state with wave vectors $\boldsymbol{k}_{\|}(\pm 1)$, and is then scattered back into the initial state by a second interaction with the periodic surface structure. This is a second order process that is in fact a sequence of two first order processes, to each of which the theory developed here applies.

To obtain the form of the dispersion curve in the vicinity of the wave vector $\boldsymbol{k}_{\|}(0) = \hat{x}_1 k_{\|}(0)$ that satisfies the condition $(\boldsymbol{k}_{\|}(0)-\boldsymbol{Q}_{\|})^2 = k_{\|}^2(0)$, viz. $k_{\|}(0) = \frac{1}{2}Q_{\|}^2/Q_1 \equiv k_{\|}^{(s)}$, we proceed as follows. The boundary conditions (3.154)–(3.155), and the corresponding conditions for the magnetic field components, are used to obtain a pair of equations that express the first order amplitudes $\mathscr{E}_p^{(-1)}$ and $E_p^{(-1)}$ in terms of the zero order amplitudes $\mathscr{E}_p^{(0)}$ and $E_p^{(0)}$. These same boundary conditions can be used to obtain a pair of equations for $\mathscr{E}_p^{(0)}$ and $E_p^{(0)}$ in terms of $\mathscr{E}_p^{(-1)}$ and $E_p^{(-1)}$. [It is not necessary to include the effects of the first order coefficients $\mathscr{E}_s^{(-1)}$ and $E_s^{(-1)}$ in this calculation, because they lead to terms in the dispersion relation that are not resonant as $k_{\|}(0) \to k_{\|}^{(s)}$]. Elimination of $\mathscr{E}_p^{(-1)}$ and $E_p^{(-1)}$ from these two pairs of equations yields two homogeneous linear equations for $\mathscr{E}_p^{(0)}$ and $E_p^{(0)}$. The solvability conditions for this pair of equations gives the frequency of the surface polariton for a given value of $\boldsymbol{k}_\parallel(0)$:

$$\epsilon(\omega)\frac{\alpha_0(0)}{\alpha(0)}+1=\Delta\left[\epsilon(\omega)\frac{\alpha_0(-1)}{\alpha(-1)}+1\right]^{-1}, \tag{3.156}$$

where Δ is of second order in $u_\perp$ and $\Delta^{(j)}_{\alpha\beta}$, and a complicated function of $\epsilon(\omega)$, $\alpha_0(0)$, $\alpha(0)$, $\alpha_0(-1)$, $\alpha(-1)$, α_j, $\boldsymbol{n}^{(0)}_{(\sigma)}$ and $\boldsymbol{n}^{(-1)}_{(\sigma)}$.

The vanishing of the left hand side of eq. (3.156) is the dispersion relation for surface polaritons on a flat surface.

For values of $k_\parallel(0)$ in the vicinity of the value where $k_\parallel(0)=k_\parallel(-1)$, i.e. for $k_\parallel(0)\cong k_\parallel^{(s)}$, the denominator of the expression on the right hand side of eq. (3.156) is small for values of ω close to those on the flat surface dispersion curve at the same value of the wave vector, as is the term on the left hand side of this equation. We therefore denote the frequency of the surface polariton on a flat surface corresponding to $k_\parallel^{(s)}$ by ω_s, and set

$$\omega=\omega_s+\delta\omega \tag{3.157a}$$

$$k_\parallel(0)=k_\parallel^{(s)}+\delta k_\parallel, \tag{3.157b}$$

so that

$$k_\parallel(-1)=k_\parallel^{(s)}+\cos\varphi^{(s)}_{-1}\,\delta k_\parallel, \tag{3.157c}$$

where $\varphi^{(s)}_{-1}$ is the angle between $\boldsymbol{k}_\parallel(-1)$ and the x_1-axis when $|\boldsymbol{k}_\parallel(-1)|=k_\parallel^{(s)}$. We expand both $\epsilon(\omega)(\alpha_0(0)/\alpha(0))+1$ and $\epsilon(\omega)(\alpha_0(-1)/\alpha(-1))+1$ about the point $(k_\parallel^{(s)},\omega_s)$ to first order in $\delta k_\parallel$ and $\delta\omega$, and evaluate Δ at $(k_\parallel^{(s)},\omega_s)$, because it is already a small quantity. In this way eq. (3.156) is transformed into

$$\left(\delta\omega-\frac{\omega_s}{\lambda_s k_\parallel^{(s)}}\delta k_\parallel\right)\left(\delta\omega-\frac{\omega_s}{\lambda_s k_\parallel^{(s)}}\cos\varphi^{(s)}_{-1}\,\delta k_\parallel\right)=\frac{|\epsilon(\omega_s)|^2\omega_s^2\Delta_s}{(\epsilon^2(\omega_s)-1)^2\lambda_s^2}, \tag{3.158}$$

where

$$\lambda_s=\lambda(\omega_s)=\left\{1+\frac{\omega}{2|\epsilon(\omega)|\,(|\epsilon(\omega)|-1)}\frac{\partial\epsilon(\omega)}{\partial\omega}\right\}_{\omega=\omega_s}, \tag{3.159}$$

and

$$\Delta_s=\frac{k_\parallel^{(s)2}}{|\epsilon(\omega_s)|}\left|2u_\perp(\epsilon(\omega_s)-1)\sin^2\tfrac{1}{2}\varphi^{(s)}_{-1}\right.$$
$$\left.+\sum_j\frac{1}{(2\alpha+\alpha_j)}\sum_{\alpha\beta}n^{(0)*}_{(p)\alpha}\,\Delta^{(j)}_{\alpha\beta}\,n^{(-1)}_{(p)\beta}\right|^2, \tag{3.160}$$

where $\alpha(-1)=\alpha(0)\equiv\alpha$ is evaluated at $k_\parallel^{(s)}$ and ω_s, as is $\alpha_0(-1)=\alpha_0(0)\equiv\alpha_0$, and $\boldsymbol{n}^{(0,-1)}_{(p)}$. It follows immediately from Eq. (3.158) that

$$\delta\omega=\frac{(|\epsilon(\omega_s)|-1)^{1/2}}{\lambda_s|\epsilon(\omega_s)|^{1/2}}\{c\delta k_\parallel\cos^2\tfrac{1}{2}\varphi^{(s)}_{-1}\pm[(c\delta k_\parallel)^2\sin^4\tfrac{1}{2}\varphi^{(s)}_{-1}+g^2]^{1/2}\} \tag{3.161a}$$

where

$$g^2 = \frac{\omega_s^2 |\epsilon(\omega_s)|^3 \Delta_s}{(|\epsilon(\omega_s)| - 1)^3 (|\epsilon(\omega_s)| + 1)^2}, \tag{3.161b}$$

and we have used eq. (2.8). The + sign is to be chosen for $\delta k_\| > 0$ and the minus sign for $\delta k_\| < 0$.

An important feature of the treatment presented here is the inclusion of the effect of the modulation of the dielectric constant of the substrate by the elasto-optic effect, as displayed in eqs. (3.127)–(3.129). Mills (1977) has considered the relative importance of the elasto-optic coupling and the effect of the modulation of the surface profile in producing the gap in the dispersion curve. He concludes that when the substrate is an insulator, both interactions contribute to the gap on roughly equal footing, although in a particular case one or the other of the two effects may dominate.

In the case of the propagation of a surface polariton on a metal substrate at frequencies in the 5–10 μm range, the situation is less clear. At first glance it appears as if the elasto-optic coupling might be rather unimportant under these conditions. This is because $|\epsilon| \gg 1$ in the infrared (see eq. (2.12)) and the term proportional to $(\epsilon - 1)\, u_\perp$ in the expression for Δ_s, eq. (3.160), becomes very large. Also, $\alpha = |\epsilon|^{1/2} k_\| \gg k_\|$ under these conditions, so the prefactor $(2\alpha + \alpha_j)^{-1}$ becomes small compared to $k_\|^{-1}$. Both of these effects act to enhance greatly the importance of the surface corrugation-induced interaction over that provided by the elasto-optic effect. However, just as the dielectric constant is strongly dependent on frequency in the infrared, and increases dramatically in magnitude as ω increases, so do the elements of the photoelastic tensor. Bennett et al. (1971) have calculated the magnitude and frequency dependence of the elements of the photo-elastic tensor for a simple model of aluminum, for frequencies in the visible. They find that $\omega^2 k_{\alpha\beta\mu\nu}(\omega)$ increases dramatically as ω decreases. Consequently, as $\omega \to 0$ it appears as if the divergence in $k_{\alpha\beta\mu\nu}(\omega)$ may be stronger than in the dielectric constant $\epsilon(\omega)$. If this is so, and their results are extrapolated to infrared frequencies, then the elements of the photo-elastic tensor of the metal may become large enough for the second contribution to Δ_s in eq. (3.160) to become comparable to the first.

At the present time it is extremely difficult to obtain information about the magnitude and frequency dependence of the photoelastic tensor of metals in frequency regimes where the skin depth is small and the reflectivity is close to unity. The results presented here, together with the experimental study of the nonlinear interaction of surface polaritons with Rayleigh waves, may provide a unique opportunity to obtain information about these constants. If the gap opened up in the surface polariton dispersion curve could be measured, then comparison of the result with eq. (3.160) and (3.161) would provide information about these constants. There is considerable flexibility in carrying out such measurements, because the angle between the directions of propagation of the surface polariton and the Rayleigh wave can be varied.

In order to carry out such measurements it will be necessary to propagate both high frequency Rayleigh waves and surface polaritons on a metal surface in a frequency regime where Bragg scattering can occur. At the present time all surface polariton propagation experiments in the infrared have employed the 10.6 μm radiation from CO_2 lasers, and current Rayleigh wave technology does not allow propagation of Rayleigh waves with frequencies high enough for Bragg scattering to occur. The use of sources of infrared radiation of longer wavelength may allow the gap in the dispersion curve to be studied.

A different manifestation of the interaction of surface polaritons with Rayleigh waves than that discussed here has been studied by Talaat et al. (1975). These authors used a Kretschmann prism configuration (Kretschmann and Raether 1968, Kretschmann 1972c)(a geometry that differs from the one assumed here), and observed the radiation emitted by a surface polariton that crosses a Rayleigh wave.

One can expect more experimental studies of this interaction in the future.

4. *Propagation of a Surface Polariton over a Randomly Rough Surface*

When a surface polariton propagates over a randomly rough surface, the problem of obtaining its dispersion relation is a great deal more difficult than it is for propagation over a deterministic, periodic surface. The difficulties arise from the necessity of obtaining the electric field in the vacuum–dielectric system sufficiently explicitly as a function of the surface profile function $\zeta(\boldsymbol{x}_{\|})$, which is unknown, that averaging of the result over the ensemble of realizations of the surface roughness can be carried out. In contrast, in the case of a deterministic, periodic surface profile, the values of the field and of its normal derivative on the surface do not need to be obtained explicitly as function of $\zeta(\boldsymbol{x}_{\|})$: the latter enters the kernels of the integral equations that determine the dispersion relation, and it is sufficient that the solutions, that in general have to be obtained numerically, are obtained only implicitly as functions of $\zeta(\boldsymbol{x}_{\|})$, since no averaging has to be carried out in this case. In practice this means that at the present time the dispersion relation for a surface polariton on a randomly rough surface has been obtained only in the small roughness limit by perturbative methods based essentially on the Rayleigh hypothesis.

The first determinations of the dispersion relation for surface polaritons on a randomly rough surface were published in 1976. Kröger and Kretschmann (1976) presented a derivation based on their earlier work (1970) on the reflectivity of randomly rough surfaces. Maradudin and Zierau (1976) obtained the dispersion relation by obtaining the Green's tensor for the matrix differential operator that enters the Maxwell equation for the macroscopic electric field in the system of vacuum and dielectric separated by a randomly rough interface. When the Green's tensor is averaged over the ensemble of realizations of the surface roughness its pole gives the surface polariton dispersion curve. A year later Toigo et al. (1977) presented a derivation of the dispersion relation for a surface polariton on a grating that utilized the Rayleigh hypothesis and the extinction theorem. Their work was the basis for the discussion in sect. 3.1.1, and yielded the result given by eq. (3.29). They then indicated how this result could be used to obtain the corresponding dispersion relation for a surface polariton on a randomly rough surface, and pointed out its equivalence to the result of Maradudin and Zierau. A detailed derivation of the dispersion relation of Toigo et al. (1977) can be found in the unpublished dissertation of Hill (1978). Subsequently Kretschmann et al. (1979) reported they had verified the equivalence of the dispersion relation of Kröger and Kretschmann (1976) to that of Toigo et al. (1977), and Sari et al. (1980) stated that they, too, have verified the equivalence of the dispersion relations of Toigo et al. and of Maradudin and Zierau. The situation at the present time is therefore that despite the differences among the approaches to the calculation, and in the forms of the result, the dispersion relations for a surface polariton on a randomly rough surface obtained by Kröger and Kretschmann, Maradudin and Zierau, and Toigo et al. are all equivalent.

In contrast with the work cited above, which is purely classical in nature, in a recent paper Hall and Braundmeier Jr. (1978) have obtained the surface polariton dispersion relation for a rough surface by a quantum mechanical calculation. A single particle Green's function is defined for a surface polariton propagating on a rough surface and is related to its flat surface counterpart through Dyson's equation. The surface polariton dispersion relation is then obtained from an approximation to the proper self-energy. No comparison of the result obtained with those of the earlier, classical treatments has been made.

In the present section we study the propagation of a surface polariton over a randomly rough surface by a Green's function approach. This method, which is described in sect. 4.1 is related to the one used by Maradudin and Zierau, but is directed specifically to the determination of the mean electric field in our vacuum–dielectric system, rather than to the determination of the Green's tensor for the Maxwell differential operator. It is also computationally simpler than the approach of Maradudin and

Zierau in its use of projection operator techniques for obtaining the equation satisfied by the mean electric field, whereas Maradudin and Zierau used a diagramatic perturbation theory approach in their work. In the present formulation the "mathematical ambiguities" in the work of Maradudin and Zierau noted by Toigo et al. (1977) have been removed, we believe.

One of the advantages of the Green's function approach, in our view, is that it allows a unified treatment of such problems as the scattering and absorption of electromagnetic radiation by a randomly rough surface, and the attenuation of a surface polariton by surface roughness. In sect. 4.2 the latter problem is treated in a simple fashion by this approach, following the work of Mills (1975). Finally, in sect. 4.3 the dispersion relation for surface plasmons on a randomly rough surface is obtained in the small roughness limit. Although this result can be obtained from the result obtained in sect. 4.1 in the limit as the speed of light $c \to \infty$, it is as easy, and of interest in its own right, to obtain the result by a calculation that proceeds directly from Laplace's equation, which is the approach we will take here.

The dispersion relation for surface plasmons on a randomly rough surface was first obtained by the former approach by Kretschmann et al. (1979). Soon after, it was obtained by Rahman and Maradudin (1980b) from the pole of a certain Green's function arising in their earlier work on the electrostatic image potential in the presence of surface roughness (Rahman and Maradudin, 1980a). The derivation presented here differs from each of the earlier treatments, and is much more direct, in our view.

4.1. Dispersion Relation for a Surface Polariton on a Randomly Rough Surface

In this section we derive the dispersion relation for surface polaritons on a randomly rough surface by a Green's function method based on the work of Maradudin and Zierau (1976). We believe this method is algebraically simple, and the integral equation to which it gives rise can be applied to the solution of the problem of the refraction of an electromagnetic wave by such a vacuum–dielectric interface, (Maradudin and Mills, 1975) and to the attenuation of a surface polariton by surface roughness (Mills, 1975). The latter problem will be discussed on the basis of this approach in the following subsection.

In this treatment it is assumed that the region $x_3 > \zeta(\mathbf{x}_\parallel)$ is occupied by vacuum, and the region $x_3 < \zeta(\mathbf{x}_\parallel)$ is occupied by a dielectric medium, characterized by an isotropic, frequency-dependent dielectric constant $\epsilon(\omega)$. The dielectric constant for this system is position-dependent, and has the form

$$\epsilon(\boldsymbol{x}, \omega) = \theta(x_3 - \zeta(\boldsymbol{x}_\|)) + \epsilon(\omega)\theta(\zeta(\boldsymbol{x}_\|) - x_3), \tag{4.1}$$

where $\theta(x_3)$ is Heaviside's unit step function. This expression can be rewritten as the sum of the dielectric constant for a system of vacuum separated from dielectric by a plane interface at the surface $x_3 = 0$, and a correction due to the surface roughness:

$$\epsilon(\boldsymbol{x}; \omega) = \epsilon_0(x_3; \omega) + \Delta\epsilon(\boldsymbol{x}; \omega), \tag{4.2}$$

where

$$\epsilon_0(x_3; \omega) = \theta(x_3) + \epsilon(\omega)\theta(-x_3) \tag{4.3}$$

$$\Delta\epsilon(\boldsymbol{x}; \omega) = (\epsilon(\omega) - 1)\, U(\boldsymbol{x}) \tag{4.4}$$

with

$$U(\boldsymbol{x}) = \theta(\zeta(\boldsymbol{x}_\|) - x_3) - \theta(-x_3). \tag{4.5}$$

A surface, even a rough surface, represents a static scatterer of an electromagnetic wave. Thus, if in Maxwell's equation,

$$\nabla \times \nabla \times \boldsymbol{E} = -\frac{1}{c^2}\frac{\partial^2}{\partial t^2}\boldsymbol{D}, \tag{4.6}$$

we substitute

$$\boldsymbol{E}(\boldsymbol{x}; t) = \boldsymbol{E}(\boldsymbol{x}; \omega)\, \mathrm{e}^{-\mathrm{i}\omega t}, \tag{4.7a}$$

$$\boldsymbol{D}(\boldsymbol{x}; t) = \boldsymbol{D}(\boldsymbol{x}; \omega)\, \mathrm{e}^{-\mathrm{i}\omega t}, \tag{4.7b}$$

and use the relation

$$\boldsymbol{D}(\boldsymbol{x}; \omega) = \epsilon(\boldsymbol{x}; \omega)\boldsymbol{E}(\boldsymbol{x}; \omega), \tag{4.8}$$

the equation for the amplitude $\boldsymbol{E}(\boldsymbol{x}; \omega)$ takes the form

$$\sum_\beta \left\{\epsilon_0(x_3; \omega)\frac{\omega^2}{c^2}\delta_{\alpha\beta} - \frac{\partial^2}{\partial x_\alpha \partial x_\beta} + \delta_{\alpha\beta}\nabla^2\right\} E_\beta(\boldsymbol{x}; \omega)$$
$$= -\frac{\omega^2}{c^2}(\epsilon(\omega) - 1)U(\boldsymbol{x})E_\alpha(\boldsymbol{x}; \omega). \tag{4.9}$$

To solve eq. (4.9) we introduce the Green's function $D_{\alpha\beta}(\boldsymbol{x}, \boldsymbol{x}'; \omega)$ that satisfies the differential equation (Maradudin and Mills, 1975):

$$\sum_\beta \left\{\epsilon_0(x_3; \omega)\frac{\omega^2}{c^2}\delta_{\alpha\beta} - \frac{\partial^2}{\partial x_\alpha \partial x_\beta} + \delta_{\alpha\beta}\nabla^2\right\} D_{\beta\gamma}(\boldsymbol{x}, \boldsymbol{x}'; \omega)$$
$$= 4\pi\delta_{\alpha\gamma}\delta(\boldsymbol{x} - \boldsymbol{x}'), \tag{4.10}$$

as well as the usual Maxwell boundary conditions at the plane $x_3 = 0$, and outgoing wave or exponentially decaying boundary conditions as $|x_3| \to \infty$. With the aid of this Green's function the differential equation (4.9) can be converted into an integral equation,

$$E_\alpha(\boldsymbol{x};\omega) = E_\alpha^{(0)}(\boldsymbol{x};\omega) - \frac{\omega^2}{4\pi c^2}(\epsilon(\omega)-1) \times \sum_\beta \int \mathrm{d}^3x' D_{\alpha\beta}(\boldsymbol{x},\boldsymbol{x}';\omega)U(\boldsymbol{x}')E_\beta(\boldsymbol{x}';\omega). \tag{4.11}$$

In eq. (4.11) $E_\alpha^{(0)}(\boldsymbol{x};\omega)$ is the solution of the corresponding homogeneous differential equation. Depending on the context, this field represents an electromagnetic wave incident on a flat vacuum–dielectric interface from either the vacuum or the dielectric side, or a surface polariton propagating along a flat vacuum–dielectric interface.

In the light of the definition (4.5) of the function $U(\boldsymbol{x})$, the integral over $\boldsymbol{x}'$ in eq. (4.11) can be rewritten in the form

$$\int \mathrm{d}^3x' D_{\alpha\beta}(\boldsymbol{x},\boldsymbol{x}';\omega)U(\boldsymbol{x}')E_\beta(\boldsymbol{x}';\omega) = \int \mathrm{d}^2x'_\parallel \Big\{ \theta(\zeta(\boldsymbol{x}'_\parallel)) \int_0^{\zeta(\boldsymbol{x}'_\parallel)} \mathrm{d}x'_3\, D_{\alpha\beta}(\boldsymbol{x},\boldsymbol{x}';\omega)E_\beta(\boldsymbol{x}';\omega) - \theta(-\zeta(\boldsymbol{x}'_\parallel)) \int_{\zeta(\boldsymbol{x}'_\parallel)}^0 \mathrm{d}x'_3\, D_{\alpha\beta}(\boldsymbol{x},\boldsymbol{x}';\omega)\,E_\beta(\boldsymbol{x}';\omega)\Big\}. \tag{4.12}$$

We recall that $D_{\alpha\beta}(\boldsymbol{x},\boldsymbol{x}';\omega)$ considered as a function of x'_3 has its discontinuities at the surface $x'_3=0$; on the other hand, the electric field $E_\beta(\boldsymbol{x}';\omega)$ considered as a function of x'_3 has its discontinuities at the surface $x'_3=\zeta(\boldsymbol{x}'_\parallel)$. Thus in the first of the integrals over x'_3 in eq. (4.12) the Green's function is that appropriate to the region $x'_3>0$; the electric field is that appropriate to the region $x'_3<\zeta(\boldsymbol{x}'_\parallel)$, i.e., it is the field in the dielectric. In the second integral the Green's function is that appropriate to the region $x'_3<0$; the electric field is that appropriate to the region $x'_3>\zeta(\boldsymbol{x}'_\parallel)$, i.e., it is the field in the vacuum. Thus, in the small roughness limit, in which the right hand side of eq. (4.12) is sought to the first order in the roughness profile function $\zeta(\boldsymbol{x}'_\parallel)$, we have that

$$\int \mathrm{d}^3x' D_{\alpha\beta}(\boldsymbol{x},\boldsymbol{x}';\omega)U(\boldsymbol{x}')E_\beta(\boldsymbol{x}';\omega) = \int \mathrm{d}^2x'_\parallel \zeta(\boldsymbol{x}'_\parallel)\,\{\theta(\zeta(\boldsymbol{x}'_\parallel))\,D_{\alpha\beta}(\boldsymbol{x},\boldsymbol{x}'_\parallel 0+;\omega)E_\beta(\boldsymbol{x}'_\parallel 0-;\omega) + \theta(-\zeta(\boldsymbol{x}'_\parallel))D_{\alpha\beta}(\boldsymbol{x},\boldsymbol{x}'_\parallel 0-;\omega)E_\beta(\boldsymbol{x}'_\parallel 0+;\omega)\} + \mathrm{O}(\zeta^2). \tag{4.13}$$

This method of obtaining eq. (4.13) is due essentially to Agarwal (1976).

When the result given by eq. (4.13) is substituted into eq. (4.11), the integral equation for the electric field amplitude takes the form

$$E_\alpha(\boldsymbol{x};\omega) = E_\alpha^{(0)}(\boldsymbol{x};\omega) - \frac{\omega^2}{4\pi c^2}(\epsilon(\omega)-1)\sum_\beta \int d^2x'_\parallel \zeta(\boldsymbol{x}'_\parallel)$$
$$\times\{\theta(\zeta(\boldsymbol{x}'_\parallel))D_{\alpha\beta}(\boldsymbol{x}, \boldsymbol{x}'_\parallel 0+;\omega)E_\beta(\boldsymbol{x}'_\parallel 0-;\omega)$$
$$+\theta(-\zeta(\boldsymbol{x}'_\parallel))D_{\alpha\beta}(\boldsymbol{x}, \boldsymbol{x}'_\parallel 0-;\omega)E_\beta(\boldsymbol{x}'_\parallel 0+;\omega\}. \tag{4.14}$$

This expression can be simplified in the following way. We first note the saltus conditions satisfied by the elements of the Green's function tensor,

$$D_{\alpha 1}(\boldsymbol{x}, \boldsymbol{x}'_\parallel 0+;\omega) = D_{\alpha 1}(\boldsymbol{x}, \boldsymbol{x}'_\parallel 0-;\omega) \tag{4.15a}$$

$$D_{\alpha 2}(\boldsymbol{x}, \boldsymbol{x}'_\parallel 0+;\omega) = D_{\alpha 2}(\boldsymbol{x}, \boldsymbol{x}'_\parallel 0-;\omega) \tag{4.15b}$$

$$D_{\alpha 3}(\boldsymbol{x}, \boldsymbol{x}'_\parallel 0+;\omega) = \epsilon(\omega)D_{\alpha 3}(\boldsymbol{x}, \boldsymbol{x}'_\parallel 0-;\omega). \tag{4.15c}$$

We next consider the relation between the electric fields $\boldsymbol{E}(\boldsymbol{x}_\parallel 0+;\omega)$ and $\boldsymbol{E}(\boldsymbol{x}_\parallel 0-;\omega)$. We use the fact that the saltus conditions satisfied by the electric field at the surface $x_3 = \zeta(\boldsymbol{x}_\parallel)$ are

$$\hat{t}_1\cdot \boldsymbol{E}^>(\boldsymbol{x}_\parallel, \zeta(\boldsymbol{x}_\parallel);\omega) = \hat{t}_1\cdot \boldsymbol{E}^<(\boldsymbol{x}_\parallel, \zeta(\boldsymbol{x}_\parallel);\omega), \tag{4.16a}$$

$$\hat{t}_2\cdot \boldsymbol{E}^>(\boldsymbol{x}_\parallel, \zeta(\boldsymbol{x}_\parallel);\omega) = \hat{t}_2\cdot \boldsymbol{E}^<(\boldsymbol{x}_\parallel, \zeta(\boldsymbol{x}_\parallel);\omega), \tag{4.16b}$$

$$\hat{n}\cdot \boldsymbol{E}^>(\boldsymbol{x}_\parallel, \zeta(\boldsymbol{x}_\parallel);\omega) = \epsilon(\omega)\hat{n}\cdot \boldsymbol{E}^<(\boldsymbol{x}_\parallel, \zeta(\boldsymbol{x}_\parallel);\omega), \tag{4.16c}$$

where the subscripts $\gtrless$ denote the fields in the regions $x_3 \gtrless \zeta(\boldsymbol{x}_\parallel)$, respectively. The vectors $\hat{t}_1$, $\hat{t}_2$, and $\hat{n}$ are defined in eqs. (2.19a), (2.19b), and (2.17), respectively. When we expand both sides of each of these equations to first order in $\zeta(\boldsymbol{x}_\parallel)$, they become

$$E_1(\boldsymbol{x}_\parallel, 0+;\omega) + \zeta(\boldsymbol{x}_\parallel)\frac{\partial}{\partial x_3}E_1^>(\boldsymbol{x};\omega)\bigg|_{x_3=0+} + \frac{\partial\zeta(\boldsymbol{x}_\parallel)}{\partial x_1}E_3(\boldsymbol{x}_\parallel, 0+;\omega)$$
$$= E_1(\boldsymbol{x}_\parallel, 0-;\omega) + \zeta(\boldsymbol{x}_\parallel)\frac{\partial}{\partial x_3}E_1^<(\boldsymbol{x};\omega)\bigg|_{x_3=0-} + \frac{\partial\zeta(\boldsymbol{x}_\parallel)}{\partial x_1}E_3(\boldsymbol{x}_\parallel, 0-;\omega), \tag{4.17a}$$

$$E_2(\boldsymbol{x}_\parallel, 0+;\omega) + \zeta(\boldsymbol{x}_\parallel)\frac{\partial}{\partial x_3}E_2^>(\boldsymbol{x};\omega)\bigg|_{x_3=0+} + \frac{\partial\zeta(\boldsymbol{x}_\parallel)}{\partial x_2}E_3(\boldsymbol{x}_\parallel, 0+;\omega)$$
$$= E_2(\boldsymbol{x}_\parallel, 0-;\omega) + \zeta(\boldsymbol{x}_\parallel)\frac{\partial}{\partial x_3}E_2^<(\boldsymbol{x};\omega)\bigg|_{x_3=0-} + \frac{\partial\zeta(\boldsymbol{x}_\parallel)}{\partial x_2}E_3(\boldsymbol{x}_\parallel, 0-;\omega), \tag{4.17b}$$

$$E_3(\boldsymbol{x}_\parallel, 0+;\omega) + \zeta(\boldsymbol{x}_\parallel)\frac{\partial}{\partial x_3}E_3^>(\boldsymbol{x};\omega)\bigg|_{x_3=0+} - \frac{\partial\zeta(\boldsymbol{x}_\parallel)}{\partial x_1}E_1(\boldsymbol{x}_\parallel, 0+;\omega)$$
$$-\frac{\partial\zeta(\boldsymbol{x}_\parallel)}{\partial x_2}E_2(\boldsymbol{x}_\parallel, 0+;\omega) = \epsilon(\omega)\bigg\{E_3(\boldsymbol{x}_\parallel, 0-;\omega)$$
$$+\zeta(\boldsymbol{x}_\parallel)\frac{\partial}{\partial x_3}E_3^<(\boldsymbol{x};\omega)\bigg|_{x_3=0-} - \frac{\partial\zeta(\boldsymbol{x}_\parallel)}{\partial x_1}E_1(\boldsymbol{x}_\parallel, 0-;\omega)$$
$$-\frac{\partial\zeta(\boldsymbol{x}_\parallel)}{\partial x_2}E_2(\boldsymbol{x}_\parallel, 0-;\boldsymbol{\omega})\bigg\}. \tag{4.17c}$$

It follows from these results that

$$E_1(\boldsymbol{x}_\parallel, 0+; \omega) = E_1(\boldsymbol{x}_\parallel, 0-; \omega) + O(\zeta), \tag{4.18a}$$

$$E_2(\boldsymbol{x}_\parallel, 0+; \omega) = E_2(\boldsymbol{x}_\parallel, 0-; \omega) + O(\zeta), \tag{4.18b}$$

$$E_3(\boldsymbol{x}_\parallel, 0+; \omega) = \epsilon(\omega) E_3(\boldsymbol{x}_\parallel, 0-; \omega) + O(\zeta). \tag{4.18c}$$

Since the second term on the right hand side of eq. (4.14) is already explicitly of $O(\zeta)$, then to this order the use of eqs. (4.15) and (4.18) allows us to write the integral equation for the electric field amplitudes as

$$E_\alpha(\boldsymbol{x}; \omega) = E_\alpha^{(0)}(\boldsymbol{x}; \omega) - \frac{\omega^2}{4\pi c^2}(\epsilon(\omega) - 1) \sum_\beta \int d^2x'_\parallel \, \zeta(\boldsymbol{x}'_\parallel) \times D_{\alpha\beta}(\boldsymbol{x}, \boldsymbol{x}'_\parallel, 0+; \omega) E_\beta(\boldsymbol{x}'_\parallel, 0-; \omega). \tag{4.19}$$

The inhomogeneous integral equation (4.19) will be used in the next subsection in a determination of the attenuation of surface polaritons by surface roughness. In the present subsection we are interested in the free oscillations of our vacuum–dielectric system that are localized in the vicinity of the interface. Consequently, we drop the inhomogeneous term $E_\alpha^{(0)}(\boldsymbol{x}; \omega)$ from eq. (4.19), and for the rest of this subsection study the homogeneous integral equation

$$E_\alpha(\boldsymbol{x}; \omega) = -\frac{\omega^2}{4\pi c^2}(\epsilon(\omega) - 1) \sum_\beta \int d^2x'_\parallel \, \zeta(\boldsymbol{x}'_\parallel) \times D_{\alpha\beta}(\boldsymbol{x}, \boldsymbol{x}'_\parallel, 0+; \omega) E_\beta(\boldsymbol{x}'_\parallel, 0-; \omega). \tag{4.20}$$

The system of dielectric separated from vacuum by a planar interface possesses infinitesimal translational invariance in directions parallel to the surface. This fact has the consequence that the Green's function $D_{\alpha\beta}(\boldsymbol{x}, \boldsymbol{x}'; \omega)$ can be Fourier analyzed in the coordinates parallel to the plane $x_3 = 0$ according to

$$D_{\alpha\beta}(\boldsymbol{x}, \boldsymbol{x}'; \omega) = \int \frac{d^2k_\parallel}{(2\pi)^2} \exp[i\boldsymbol{k}_\parallel \cdot (\boldsymbol{x}_\parallel - \boldsymbol{x}'_\parallel)] \, d_{\alpha\beta}(\boldsymbol{k}_\parallel \omega \mid x_3 x'_3). \tag{4.21}$$

Because our vacuum–dielectric system also possesses isotropy in the plane $x_3 = 0$, the Fourier coefficients $\{d_{\alpha\beta}(\boldsymbol{k}_\parallel \omega \mid x_3 x'_3)\}$ can be expressed in terms of simpler coefficients $\{g_{\alpha\beta}(k_\parallel \omega \mid x_3 x'_3)\}$ that depend on the two-dimensional wave vector $\boldsymbol{k}_\parallel$ only through its magnitude. The relation between these coefficients is

$$d_{\alpha\beta}(\boldsymbol{k}_\parallel \omega \mid x_3 x'_3) = \sum_{\mu\nu} g_{\mu\nu}(k_\parallel \omega \mid x_3 x'_3) S_{\mu\alpha}(\hat{k}_\parallel) S_{\nu\beta}(\hat{k}_\parallel), \tag{4.22}$$

where $S(\hat{k}_\parallel)$ is the 3×3 real, orthogonal matrix

$$S(\hat{k}_\parallel) = \begin{pmatrix} \hat{k}_1 & \hat{k}_2 & 0 \\ -\hat{k}_2 & \hat{k}_1 & 0 \\ 0 & 0 & 1 \end{pmatrix} \tag{4.23}$$

with $\hat{k}_\alpha = k_\alpha/k_\parallel$ ($\alpha = 1, 2$). The coefficients $\{g_{\alpha\beta}(k_\parallel\omega \mid x_3x_3')\}$ have been tabulated by Maradudin and Mills (1975).

To solve eq. (4.20) we introduce the definitions

$$\int d^2x_\parallel \exp[-i\boldsymbol{k}_\parallel \cdot \boldsymbol{x}_\parallel]\, E_\alpha(\boldsymbol{x};\omega) = E_\alpha(\boldsymbol{k}_\parallel\omega \mid x_3) \tag{4.24a}$$

$$F_\alpha(\boldsymbol{k}_\parallel\omega \mid x_3) = \sum_\beta S_{\alpha\beta}(\hat{\boldsymbol{k}}_\parallel)E_\beta(\boldsymbol{k}_\parallel\omega \mid x_3). \tag{4.24b}$$

When these relations are used in eq. (4.20) the following equation is obtained for the function $F_\alpha(\boldsymbol{k}_\parallel\omega \mid x_3)$:

$$F_\alpha(\boldsymbol{k}_\parallel\omega \mid x_3) = -\frac{\omega^2}{16\pi^3c^2}(\epsilon(\omega)-1)\sum_{\beta\gamma}\int d^2q_\parallel\, g_{\alpha\beta}(k_\parallel\omega \mid x_3\, 0+) \times M_{\beta\gamma}(\boldsymbol{k}_\parallel;\boldsymbol{q}_\parallel)F_\gamma(\boldsymbol{q}_\parallel\omega \mid 0-), \tag{4.25}$$

where

$$M_{\beta\gamma}(\boldsymbol{k}_\parallel;\boldsymbol{q}_\parallel) = \sum_\nu S_{\beta\nu}(\hat{\boldsymbol{k}}_\parallel)\hat{\zeta}(\boldsymbol{k}_\parallel-\boldsymbol{q}_\parallel)S^{-1}_{\nu\gamma}(\hat{\boldsymbol{q}}_\parallel), \tag{4.26}$$

and $\hat{\zeta}(\boldsymbol{Q}_\parallel)$ is the Fourier transform of $\zeta(\boldsymbol{x}_\parallel)$, defined by eq. (2.31).

Equation (4.25) holds for all x_3. It therefore holds in particular for $x_3 = 0-$. In this way we obtain the homogeneous matrix integral equation satisfied by $F_\alpha(\boldsymbol{k}_\parallel\omega \mid 0-)$:

$$F_\alpha(\boldsymbol{k}_\parallel\omega \mid 0-) = -\frac{\omega^2}{16\pi^3c^2}(\epsilon(\omega)-1)\sum_{\beta\gamma} g^{(0)}_{\alpha\beta}(k_\parallel\omega) \times\int d^2q_\parallel\, M_{\beta\gamma}(\boldsymbol{k}_\parallel;\boldsymbol{q}_\parallel)F_\gamma(\boldsymbol{q}_\parallel\omega \mid 0-). \tag{4.27}$$

In writing eq. (4.27) we have introduced the notation

$$g^{(0)}_{\alpha\beta}(k_\parallel\omega) \equiv g_{\alpha\beta}(k_\parallel\omega \mid 0-\, 0+). \tag{4.28}$$

Since we do not know the surface profile function $\zeta(\boldsymbol{x}_\parallel)$, only its statistical properties that have been discussed in sect. 2, rather than solving eq. (4.27) for $F_\alpha(\boldsymbol{k}_\parallel\omega \mid 0-)$ we will solve it for its value averaged over the ensemble of realizations of the surface profile function. This is equivalent to solving for the mean field in our vacuum–dielectric system.

The integral equation satisfied by $\langle F_\alpha(\boldsymbol{k}_\parallel\omega \mid 0-)\rangle$ can be obtained in the following way. We introduce the smoothing operator P that averages everything it acts on over the ensemble of realizations of $\zeta(\boldsymbol{x}_\parallel)$:

$$Pf \equiv \langle f\rangle. \tag{4.29}$$

We also introduce the operator $Q = 1 - P$. Application of P to both sides of eq. (4.27), using the fact that F_γ on the right hand side can be written as

$(P+Q)F_\gamma$, and noting that $PM_{\beta\gamma}(\boldsymbol{k}_\|;\boldsymbol{q}_\|)\equiv 0$ because $M(\boldsymbol{k}_\|;\boldsymbol{q}_\|)$ is linear in $\hat{\zeta}(\boldsymbol{k}_\|-\boldsymbol{q}_\|)$, leads to the result that

$$PF_\alpha(\boldsymbol{k}_\|\omega\,|\,0-) = -\frac{\omega^2}{16\pi^3c^2}(\epsilon(\omega)-1)\sum_{\beta\gamma} g^{(0)}_{\alpha\beta}(k_\|\omega) \times \int \mathrm{d}^2q_\| PM_{\beta\gamma}(\boldsymbol{k}_\|;\boldsymbol{q}_\|)QF_\gamma(\boldsymbol{q}_\|\omega\,|\,0-). \tag{4.30}$$

In the same way we find that $QF_\alpha(\boldsymbol{k}_\|\omega\,|\,0-)$ satisfies the equation

$$QF_\alpha(\boldsymbol{k}_\|\omega\,|\,0-) = -\frac{\omega^2}{16\pi^3c^2}(\epsilon(\omega)-1)\sum_{\beta\gamma} g^{(0)}_{\alpha\beta}(k_\|\omega) \times \int \mathrm{d}^2q_\| QM_{\beta\gamma}(\boldsymbol{k}_\|;\boldsymbol{q}_\|)PF_\gamma(\boldsymbol{q}_\|\omega\,|\,0-), \tag{4.31}$$

where the term $QM_{\beta\gamma}QF_\gamma$ has been omitted from the right hand side of this equation as of second order in $\hat{\zeta}(\boldsymbol{Q}_\|)$. When eq. (4.31) is substituted into eq. (4.30) we obtain the homogeneous, matrix integral equation satisfied by $\langle F_\alpha(\boldsymbol{k}_\|\omega\,|\,0-)\rangle$:

$$\langle F_\alpha(\boldsymbol{k}_\|\omega\,|\,0-)\rangle = \delta^2\frac{\omega^4}{64\pi^4c^4}(\epsilon(\omega)-1)^2\sum_{\beta} g^{(0)}_{\alpha\beta}(k_\|\omega) \times \sum_{\mu\nu\gamma}\int \mathrm{d}^2q_\|\, g(|\boldsymbol{k}_\|-\boldsymbol{q}_\||)\, R_{\beta\mu}(\hat{k}_\|;\hat{q}_\|)\, g^{(0)}_{\mu\nu}(q_\|\omega) \times R^{-1}_{\nu\gamma}(\hat{q}_\|;\hat{k}_\|)\langle F_\gamma(\boldsymbol{k}_\|\omega\,|\,0-)\rangle, \tag{4.32}$$

where we have used eq. (2.21), and

$$R_{\mu\nu}(\hat{k}_\|;\hat{q}_\|) = \sum_\lambda S_{\mu\lambda}(\hat{k}_\|)S^{-1}_{\lambda\nu}(\hat{q}_\|) \tag{4.33a}$$

$$R^{-1}_{\mu\nu}(\hat{q}_\|;\hat{k}_\|) = \sum_\lambda S_{\mu\lambda}(\hat{q}_\|)S^{-1}_{\lambda\nu}(\hat{k}_\|). \tag{4.33b}$$

We see that eq. (4.32) is a matrix equation for the components of the vector $\langle \boldsymbol{F}(\boldsymbol{k}_\|\omega\,|\,0-)\rangle$. This is in contrast with eq. (4.27) for the components of the vector $\boldsymbol{F}(\boldsymbol{k}_\|\omega\,|\,0-)$, which is a matrix *integral* equation. The comparative simplicity of eq. (4.32) is due to the fact that averaging over the ensemble of realizations of the surface roughness has restored infinitesimal translational invariance in the plane $x_3=0$ to our system.

The dispersion relation for surface polaritons on a randomly rough surface is obtained by equating to zero the determinant of the coefficients in eq. (4.32). Before this is done eq. (4.32) can be simplified considerably. From the results of Maradudin and Mills (1975) the functions $\{g^{(0)}_{\alpha\beta}(k_\|\omega)\}$ defined by eq. (4.28) can be written in the form

$$g^{(0)}(k_\parallel\omega)=\frac{4\pi c^2}{\omega^2}\frac{1}{\epsilon(\omega)\alpha_0(k_\parallel\omega)+\alpha(k_\parallel\omega)}$$

$$\times\begin{pmatrix}\alpha_0(k_\parallel\omega)\alpha(k_\parallel\omega) & 0 & -ik_\parallel\alpha(k_\parallel\omega)\\ 0 & \alpha_0(k_\parallel\omega)\alpha(k_\parallel\omega)-k_\parallel^2 & 0\\ -ik_\parallel\alpha_0(k_\parallel\omega) & 0 & -k_\parallel^2\end{pmatrix}$$

$$\equiv\frac{4\pi c^2}{\omega^2}\frac{1}{\epsilon(\omega)\alpha_0(k_\parallel\omega)+\alpha(k_\parallel\omega)}m(k_\parallel\omega), \tag{4.34}$$

where

$$\alpha_0(k_\parallel\omega)=(k_\parallel^2-\omega^2/c^2)^{1/2} \qquad k_\parallel>\omega/c$$
$$=-i(\omega^2/c^2-k_\parallel^2)^{1/2} \qquad k_\parallel<\omega/c \tag{4.35}$$

$$\alpha(k_\parallel\omega)=[k_\parallel^2-\epsilon(\omega)(\omega^2/c^2)]^{1/2} \qquad \mathrm{Re}\,\alpha>0, \mathrm{Im}\,\alpha<0. \tag{4.36}$$

In obtaining eq. (4.34) we have used the useful identity

$$[k_\parallel^2-\alpha_0(k_\parallel\omega)\alpha(k_\parallel\omega)][\alpha_0(k_\parallel\omega)+\alpha(k_\parallel\omega)]=(\omega^2/c^2)[\epsilon(\omega)\alpha_0(k_\parallel\omega)+\alpha(k_\parallel\omega)]. \tag{4.37}$$

When the result given by eq. (4.34) is substituted into eq. (4.32), the sums over β, γ, μ, ν are carried out, and the angular integration is examined carefully, it is found that the components $\langle F_1(\boldsymbol{k}_\parallel\omega\,|\,0-)\rangle$ and $\langle F_3(\boldsymbol{k}_\parallel\omega\,|\,0-)\rangle$ are decoupled from $\langle F_2(\boldsymbol{k}_\parallel\,|\,0-)\rangle$, and we obtain the pair of equations

$$\begin{pmatrix}\langle F_1(\boldsymbol{k}_\parallel\omega\,|\,0-)\rangle\\ \langle F_3(\boldsymbol{k}_\parallel\omega\,|\,0-)\rangle\end{pmatrix}=\delta^2\frac{(\epsilon(\omega)-1)^2}{\epsilon(\omega)\alpha_0+\alpha}\int\frac{d^2q_\parallel}{(2\pi)^2}\frac{g(|\boldsymbol{k}_\parallel-\boldsymbol{q}_\parallel|)}{\epsilon(\omega)\alpha_0'+\alpha'}$$

$$\times\begin{pmatrix}m_{11}m_{22}'+am_{13}m_{31}'+a^2q_\parallel^2m_{11} & m_{13}m_{33}'+am_{11}m_{13}'\\ m_{31}m_{22}'+am_{33}m_{31}'+a^2q_\parallel^2m_{31} & m_{33}m_{33}'+am_{31}m_{13}'\end{pmatrix}\begin{pmatrix}\langle F_1(\boldsymbol{k}_\parallel\omega\,|\,0-)\rangle\\ \langle F_3(\boldsymbol{k}_\parallel\omega\,|\,0-)\rangle\end{pmatrix} \tag{4.38}$$

$$\langle F_2(\boldsymbol{k}_\parallel\omega\,|\,0-)\rangle=\delta^2\frac{(\epsilon(\omega)-1)^2}{\epsilon(\omega)\alpha_0+\alpha}\int\frac{d^2q_\parallel}{(2\pi)^2}\frac{g(|\boldsymbol{k}_\parallel-\boldsymbol{q}_\parallel|)}{\epsilon(\omega)\alpha_0'+\alpha'}$$
$$\times m_{22}(m_{11}'-a^2q_\parallel^2)\langle F_2(\boldsymbol{k}_\parallel\omega\,|\,0-)\rangle. \tag{4.39}$$

The primes in these equations denote that the wave vector argument in the corresponding functions is $q_\parallel$, and

$$a\equiv\hat{k}_\parallel\cdot\hat{q}_\parallel=\cos\theta, \tag{4.40}$$

where θ is the angle between the vectors $\boldsymbol{k}_\parallel$ and $\boldsymbol{q}_\parallel$. This partial decoupling of the equation for the components of $\langle \boldsymbol{F}(\boldsymbol{k}_\parallel\omega\,|\,0-)\rangle$ is due to the fact that the average over the ensemble of realizations of the surface roughness restores isotropy in the plane $x_3=0$ to our system.

The dispersion relations that are obtained from eqs. (4.38) and (4.39) can be written in the following forms:

$$\epsilon(\omega)\alpha_0 + \alpha = \delta^2(1-\epsilon(\omega))^2 \int \frac{d^2q_\parallel}{(2\pi)^2} \frac{g(|\boldsymbol{k}_\parallel - \boldsymbol{q}_\parallel|)}{\epsilon(\omega)\alpha_0' + \alpha'}$$
$$\times [(k_\parallel q_\parallel - \alpha_0\alpha' \cos\theta)(k_\parallel q_\parallel - \alpha\alpha_0' \cos\theta)$$
$$- \alpha_0\alpha(q_\parallel^2 - \alpha_0'\alpha') \sin^2\theta], \qquad (4.41)$$

$$\alpha_0 + \alpha = \delta^2(1-\epsilon(\omega))^2 \frac{\omega^2}{c^2} \int \frac{d^2q_\parallel}{(2\pi)^2} g(|\boldsymbol{k}_\parallel - \boldsymbol{q}_\parallel|) \frac{q_\parallel^2 \cos^2\theta - \alpha_0'\alpha'}{\epsilon(\omega)\alpha_0' + \alpha'}. \qquad (4.42)$$

In the limit of a flat surface ($\delta = 0$) eq. (4.42) is the dispersion relation that is obtained when s-polarized surface polaritons are sought. In that case the equation $\alpha_0 + \alpha = 0$ has no solution since both α_0 and α are required to be real and positive (for a real dielectric constant) in order that the electromagnetic wave be exponentially localized to the vacuum dielectric interface. If the right hand side of eq. (4.42) is positive, the possibility of solutions of this equation exists. These solutions would be the dispersion curves for roughness-induced, s-polarized surface polaritons, that have no counterpart for a flat surface. We will not consider this equation further here.

Equation (4.41) is the dispersion relation for a p-polarized surface polariton propagating over a randomly rough surface. When the Gaussian assumption (2.41) for $g(k_\parallel)$ is made and the integration over the angle θ is carried out, one obtains the dispersion relation given by eq. (3.23) of the paper by Maradudin and Zierau (1976).

We can rewrite eq. (4.41) in a slightly more compact and symmetric form:

$$\epsilon(\omega)\alpha_0 + \alpha = \delta^2(1-\epsilon(\omega))^2 \int \frac{d^2q_\parallel}{(2\pi)^2} \frac{g(|\boldsymbol{k}_\parallel - \boldsymbol{q}_\parallel|)}{\epsilon(\omega)\alpha_0' + \alpha'}$$
$$\times [(k_\parallel q_\parallel \cos\theta - \alpha_0\alpha')(k_\parallel q_\parallel \cos\theta - \alpha\alpha_0')]. \qquad (4.43)$$

To show the equivalence of the results given by eqs. (4.41) and (4.43) we evaluate the difference between the terms in brackets in the two expressions:

$$[(k_\parallel q_\parallel - \alpha_0\alpha' \cos\theta)(k_\parallel q_\parallel - \alpha\alpha_0' \cos\theta) - \alpha_0\alpha(q_\parallel^2 - \alpha_0'\alpha') \sin^2\theta]$$
$$- [(k_\parallel q_\parallel \cos\theta - \alpha_0\alpha')(k_\parallel q_\parallel \cos\theta - \alpha\alpha_0')]$$
$$= (k_\parallel^2 - \alpha_0\alpha) q_\parallel^2 \sin^2\theta = \frac{\omega^2}{c^2} \frac{\epsilon(\omega)\alpha_0 + \alpha}{\alpha_0 + \alpha} q_\parallel^2 \sin^2\theta, \qquad (4.44)$$

where we have used eq. (4.37). This quantity is of $O(\delta^2)$ in view of eqs. (4.41) or (4.43), and consequently does not contribute to the surface polariton dispersion relation to $O(\delta^2)$.

The dispersion relation (4.43) is the one obtained by Toigo et al. (1977), in the form presented by Kretschmann et al. (1979).

The reason that it suffices to retain only the term linear in $\zeta(\boldsymbol{x}_\parallel)$ in the kernel of the integral equation (4.19) in order to obtain the dispersion relation correct to second order in $\zeta(\boldsymbol{x}_\|)$ is the same here as it was in sects. 3.1 and 3.2. The term of second order in $\zeta(\boldsymbol{x}_\|)$ in the kernel of eq. (4.19) can be shown to yield a contribution to the right hand side of the dispersion relation (4.41) that is proportional to the left hand side of this equation. It therefore affects the dispersion relation first to fourth order in $\zeta(\boldsymbol{x}_\|)$.

If we define

$$F(k_\|\omega) = \epsilon(\omega)\alpha_0 + \alpha \tag{4.45}$$

$$G(k_\|\omega) = \int \frac{d^2q_\|}{(2\pi)^2} \frac{g(|\boldsymbol{k}_\| - \boldsymbol{q}_\||)}{\epsilon(\omega)\alpha_0' + \alpha'} \times [(k_\| q_\| - \alpha_0\alpha' \cos\theta)(k_\| q_\| - \alpha\alpha_0' \cos\theta) - \alpha_0\alpha(q_\|^2 - \alpha_0'\alpha')\sin^2\theta], \tag{4.46}$$

(note that these functions differ from those denoted by the same symbols in the paper of Maradudin and Zierau (1976)), and denote by $\omega_0(k_\|)$ the solution of $F(k_\|\omega) = 0$, the solution of eq. (4.41) correct to $O(\delta^2)$ is given by

$$\omega(k_\|) = \omega_0(k_\|) + \delta^2 \left(\frac{(1-\epsilon(\omega))^2 G(k_\|\omega)}{(d/d\omega)F(k_\|\omega)} \right)_{\omega=\omega_0(k_\|)}. \tag{4.47}$$

A surface polariton propagating along a perfectly smooth surface of a pure crystal is attenuated by processes that can be called intrinsic. By this we mean the dissipative processes present in the bulk of the solid, i.e., the processes that give rise to the imaginary part of the dielectric tensor. In insulators and semiconductors, for frequencies in the infrared, these can be the anharmonic interactions of the normal modes of vibrations; in semiconductors and metals these can be interband electronic transitions. An expression for the attenuation length of the surface polariton in this case can be obtained by inserting the complex dielectric constant of the material into the equation $F(k_\|\omega) = 0$ relating the frequency of the surface polariton to the magnitude of its wave vector, and obtaining from it the imaginary part of the wave vector of the surface polariton $k_\|^{(I)}$. The attenuation length for energy flow is then $(2k_\|^{(I)})^{-1}$.

However, in the presence of surface roughness a surface polariton can be attenuated even if the dielectric constant $\epsilon(\omega)$ is real. We will see this explicitly below. This is the situation we are focusing attention on here. Therefore, to separate the attenuation of surface polaritons that has its origin in the surface roughness from that which has its origin in the dissipative processes present in the bulk of the material, we will assume $\epsilon(\omega)$ to be real everywhere except in the denominator $\epsilon(\omega)\alpha_0' + \alpha'$ in the integrand on the right hand side of eq. (4.41) (or eq. (4.43)). The retention of the small, positive imaginary part of the dielectric constant serves only to define the manner in which the pole in the integrand at the wave vector of the plane interface surface polariton, i.e., the value of $q_\parallel$ for which $\epsilon(\omega)\alpha_0' + \alpha'$ vanishes for $\omega = \omega_0(k_\parallel)$, is to be treated in the evaluation of the integral over $q_\parallel$.

When this is done the frequency $\omega_0(k_\parallel)$ of a surface polariton at a plane vacuum–dielectric interface is purely real. Consequently, we can rewrite eq. (4.47) in the form

$$\omega(k_\parallel) = \omega_0(k_\parallel) + \Delta(k_\parallel) - \mathrm{i}\Gamma(k_\parallel), \tag{4.48}$$

where $\Delta(k_\parallel)$ gives the shift in the frequency of the surface polariton due to surface roughness, while $\Gamma(k_\parallel)$ is related to the inverse lifetime of this mode owing to surface roughness. If we separate $G(k_\parallel\omega)$ into its real and imaginary parts according to

$$G(k_\parallel\omega) = G^{(1)}(k_\parallel\omega) - \mathrm{i}G^{(2)}(k_\parallel\omega), \tag{4.49}$$

we obtain for these quantities

$$\Delta(k_\parallel) = \delta^2 \frac{(1-\epsilon(\omega_0))^2}{[(\mathrm{d}/\mathrm{d}\omega)F(k_\parallel\omega)]_{\omega=\omega_0}} G^{(1)}(k_\parallel\omega_0) \tag{4.50a}$$

$$\Gamma(k_\parallel) = \delta^2 \frac{(1-\epsilon(\omega_0))^2}{[(\mathrm{d}/\mathrm{d}\omega)F(k_\parallel\omega)]_{\omega=\omega_0}} G^{(2)}(k_\parallel\omega_0). \tag{4.50b}$$

The quantity $\Gamma(k_\parallel)$ is the inverse of the lifetime of the amplitude of the electric field of the surface polariton. Since the energy transported by the surface polariton is proportional to the square of this amplitude, the inverse of the lifetime of the surface polariton $\tau(k_\parallel)$ is given by $2\Gamma(k_\parallel)$,

$$1/\tau(k_\parallel) = 2\Gamma(k_\parallel). \tag{4.51}$$

The attenuation length of the surface polariton $l(k_\parallel)$ is the distance over which the energy of the polariton decays to $1/e$ of its initial value. It is obtained by multiplying its lifetime by the energy transport velocity of the surface polariton $V_\mathrm{E}(k_\parallel)$:

$$l(k_\parallel) = V_\mathrm{E}(k_\parallel)\tau(k_\parallel). \tag{4.52}$$

In the absence of damping $V_\mathrm{E}(k_\parallel)$ is equal to the group velocity of the surface polariton. The latter is given by

$$\left(\frac{\partial\omega}{\partial k_\parallel}\right)_{\omega_0} = V_\mathrm{E}(k_\parallel) = c^2 \frac{k_\parallel}{\omega_0} \frac{[\epsilon(\omega_0)+1]^2}{\epsilon(\omega_0)[\epsilon(\omega_0)+1] + \frac{1}{2}\omega_0\epsilon'(\omega_0)}. \tag{4.53}$$

At the same time we have the result that

$$\begin{aligned}\left(\frac{\mathrm{d}}{\mathrm{d}\omega}F(k_\parallel\omega)\right)_{\omega_0} &= \frac{1}{c}\frac{|\epsilon(\omega_0)|+1}{|\epsilon(\omega_0)|\,[|\epsilon(\omega_0)|-1]^{1/2}}(\epsilon(\omega_0)[\epsilon(\omega_0)+1] + \tfrac{1}{2}\omega_0\epsilon'(\omega_0)) \\ &= \frac{ck_\parallel}{\omega_0}\frac{[|\epsilon(\omega_0)|+1][|\epsilon(\omega_0)|-1]^{3/2}}{|\epsilon(\omega_0)|\,V_\mathrm{E}(k_\parallel)},\end{aligned} \tag{4.54}$$

where we have used the fact that $\epsilon(\omega)$ is negative at the frequency of the surface polariton. With the results given by eqs. (4.53)–(4.54) we can rewrite eqs. (4.50) and (4.52) as

$$\Delta(k_\parallel) = \delta^2 \frac{\omega_0}{ck_\parallel} \frac{|\epsilon(\omega_0)|[|\epsilon(\omega_0)|+1]}{[|\epsilon(\omega_0)|-1]^{3/2}} V_{\mathrm{E}}(k_\parallel) G^{(1)}(k_\parallel\omega), \tag{4.55}$$

$$\Gamma(k_\parallel) = \delta^2 \frac{\omega_0}{ck_\parallel} \frac{|\epsilon(\omega_0)|[|\epsilon(\omega_0)|+1]}{[|\epsilon(\omega_0)|-1]^{3/2}} V_{\mathrm{E}}(k_\parallel) G^{(2)}(k_\parallel\omega_0), \tag{4.56}$$

$$\frac{1}{l(k_\parallel)} = \delta^2 \frac{2\omega_0}{ck_\parallel} \frac{|\epsilon(\omega_0)|[|\epsilon(\omega_0)|+1]}{[|\epsilon(\omega_0)|-1]^{3/2}} G^{(2)}(k_\parallel\omega_0). \tag{4.57}$$

In evaluating $G(k_\parallel\omega)$ from eq. (4.46) it is necessary to express the integral over $q_\parallel$ as the sum of an integral $G^<(k_\parallel\omega)$ over the interval $(0, \omega/c)$ and an integral $G^>(k_\parallel\omega)$ over the interval $(\omega/c, \infty)$. This is because in the former interval α_0 and α are real for ω and $k_\parallel$ in the region of $(\omega, k_\parallel)$-space in which surface polaritons exist ($k_\parallel > \omega/c$), α' is real, but α_0' is pure imaginary, $\alpha_0' = -\mathrm{i}\hat{\alpha}_0'$, as can be seen from eq. (4.35). In the latter interval all of the decay constants α_0, α, α_0', α' are real, but the integrand has a pole at the value of $q_\parallel$ for which $\epsilon(\omega)\alpha_0' + \alpha'$ vanishes, i.e., at the wave vector of the surface polariton of frequency ω on a flat surface. Each contribution, $G^<(k_\parallel\omega)$ and $G^>(k_\parallel\omega)$, therefore has to be treated separately.

It is found that $G^<(k_\parallel\omega)$ can be written as

$$G^<(k_\parallel\omega) = G^<_{(1)}(k_\parallel\omega) - \mathrm{i}G^<_{(2)}(k_\parallel\omega), \tag{4.58}$$

where

$$G^<_{(1)}(k_\parallel\omega) = \frac{1}{4\pi^2}\int_0^{\omega/c} \mathrm{d}q_\parallel q_\parallel \int_0^{2\pi} \mathrm{d}\theta \frac{g(|\boldsymbol{k}_\parallel - \boldsymbol{q}_\parallel|)}{\epsilon^2(\omega)\hat{\alpha}_0'^2 + \alpha'^2} \times \{\alpha'[k_\parallel^2 q_\parallel^2 - k_\parallel q_\parallel \alpha_0\alpha' \cos\theta - \alpha_0\alpha q_\parallel^2 \sin^2\theta] - \epsilon(\omega)\alpha\hat{\alpha}_0'^2[k_\parallel q_\parallel \cos\theta - \alpha_0\alpha']\}, \tag{4.59a}$$

$$G^<_{(2)}(k_\parallel\omega) = -\frac{1}{4\pi^2}\int_0^{\omega/c} \mathrm{d}q_\parallel q_\parallel \int_0^{2\pi} \mathrm{d}\theta \frac{g(|\boldsymbol{k}_\parallel - \boldsymbol{q}_\parallel|)}{\epsilon^2(\omega)\hat{\alpha}_0'^2 + \alpha'^2} \times \{\alpha\alpha'\hat{\alpha}_0'[k_\parallel q_\parallel \cos\theta - \alpha_0\alpha'] + \epsilon(\omega)\hat{\alpha}_0'[k_\parallel^2 q_\parallel^2 - k_\parallel q_\parallel \alpha_0\alpha' \cos\theta - \alpha_0\alpha q_\parallel^2 \sin^2\theta]\}. \tag{4.59b}$$

The imaginary part of $G^<(k_\parallel\omega)$ is due to the roughness-induced scattering of a surface polariton into radiative modes in the vacuum.

In evaluating $G^>(k_\parallel\omega)$ we use the identity

$$\frac{1}{\epsilon(\omega)\alpha_0' + \alpha'} = \frac{\epsilon(\omega)\alpha_0' - \alpha'}{\epsilon^2(\omega) - 1} \frac{1}{q_\parallel + k_{\mathrm{sp}}(\omega)} \frac{1}{q_\parallel - k_{\mathrm{sp}}(\omega)}, \tag{4.60}$$

where $k_{sp}(\omega)$ is defined by

$$k_{sp}^2(\omega) = \frac{\omega^2}{c^2}\frac{\epsilon(\omega)}{\epsilon(\omega)+1}. \tag{4.61}$$

Equation (4.60) displays explicitly the pole in the integrand of $G^>(k_\|\omega)$. To define the way in which this pole is circled in carrying out the integration over $q_\|$ we use the fact that $\epsilon(\omega)$ is complex in general, $\epsilon(\omega) = \epsilon_1(\omega) + i\epsilon_2(\omega)$, where $\epsilon_2(\omega) > 0$. Thus we find that

$$k_{sp}(\omega) = k_{sp}^{(1)}(\omega) + ik_{sp}^{(2)}(\omega), \tag{4.62}$$

where

$$k_{sp}^{(1)}(\omega) = \frac{\omega}{c}\frac{[\epsilon_1(\omega)(\epsilon_1(\omega)+1)+\epsilon_2^2(\omega)]^{1/2}}{[(\epsilon_1(\omega)+1)^2+\epsilon_2^2(\omega)]^{1/2}} \tag{4.63a}$$

$$k_{sp}^{(2)}(\omega) = \frac{\omega}{2c}\frac{\epsilon_2(\omega)}{[(\epsilon_1(\omega)+1)^2+\epsilon_2^2(\omega)]^{1/2}[\epsilon_1(\omega)(\epsilon_1(\omega)+1)+\epsilon_2^2(\omega)]^{1/2}}, \tag{4.63b}$$

and we have assumed that $|\epsilon_2(\omega)| \ll 1$. Consequently $k_{sp}^{(2)}(\omega)$ is very small and we make the approximation that

$$\frac{1}{q_\| - k_{sp}(\omega)} \cong \frac{1}{[q_\| - k_{sp}^{(1)}(\omega)]_p} + i\pi\delta(q_\| - k_{sp}^{(1)}), \tag{4.64}$$

where $1/(x)_p$ denotes the principal part of $1/x$. Having served the purpose of defining the manner in which the singularity in the integrand of $G^>(k_\|\omega)$ at $q_\| = k_{sp}(\omega)$ is to be treated, the recognition that $\epsilon(\omega)$ is complex can be forgotten in all that follows. We will treat $\epsilon(\omega)$ as real, and will take $k_{sp}(\omega)$ as given by eq. (4.61) with this real $\epsilon(\omega)$.

It follows then that $G^>(k_\|\omega)$ can be written in the form

$$G^>(k_\|\omega) = G^>_{(1)}(k_\|\omega) - iG^>_{(2)}(k_\|\omega), \tag{4.65}$$

where

$$\begin{aligned} G^>_{(1)}(k_\|\omega) = {} & \frac{1}{4\pi^2}\frac{1}{[\epsilon^2(\omega)-1]}\int_{\omega/c}^{\infty} dq_\| q_\| \int_0^{2\pi} d\theta\, g(|\boldsymbol{k}_\| - \boldsymbol{q}_\||)\frac{\epsilon(\omega)\alpha\alpha_0' - \alpha'}{[q_\|^2 - k_{sp}^2(\omega)]_p} \\ & \times [(k_\| q_\| - \alpha_0\alpha' \cos\theta)(k_\| q_\| - \alpha\alpha_0' \cos\theta) \\ & - \alpha_0\alpha(q_\|^2 - \alpha_0'\alpha')\sin^2\theta], \end{aligned} \tag{4.66a}$$

$$\begin{aligned} G^>_{(2)}(k_\|\omega) = {} & \frac{\omega}{4\pi c}\frac{|\epsilon(\omega)|}{[\epsilon^2(\omega)-1][|\epsilon(\omega)|-1]^{1/2}} \\ & \times \int_{\omega/c}^{\infty} dq_\|\, \delta(q_\| - k_{sp}(\omega)) \int_0^{2\pi} d\theta\, g(|\boldsymbol{k}_\| - \boldsymbol{q}_\||) \\ & \times [(k_\| q_\| - \alpha_0\alpha' \cos\theta)(k_\| q_\| - \alpha\alpha_0' \cos\theta)]. \end{aligned} \tag{4.66b}$$

The imaginary part of $G^>(k_\|\omega)$ arises from the roughness-induced scattering of a surface polariton into other surface polariton modes.

The functions $G^{(1)}(k_\|\omega)$ and $G^{(2)}(k_\|\omega)$ defined by eq. (4.49) are clearly given by

$$G^{(1)}(k_\|\omega) = G^{<}_{(1)}(k_\|\omega) + G^{>}_{(1)}(k_\|\omega), \tag{4.67a}$$

$$G^{(2)}(k_\|\omega) = G^{<}_{(2)}(k_\|\omega) + G^{>}_{(2)}(k_\|\omega). \tag{4.67b}$$

Numerical evaluations of $\Delta(k_\|)$, $\Gamma(k_\|)$, and $l^{-1}(k_\|)$ for NaCl and aluminum have been carried out on the basis of the preceding expressions by Maradudin and Zierau for the Gaussian form of $g(k_\|)$ given by eq. (2.41). One of their results, for aluminum, is shown in fig. 10. The dielectric constant $\epsilon(\omega)$ was chosen to have the form

$$\epsilon(\omega) = 1 - \omega_p^2/\omega^2, \tag{4.68}$$

with $\omega_p = 1.2018 \times 10^5\ \mathrm{cm}^{-1}$. The values of δ and σ used were $\delta = \sigma = 2500$ Å, so that the surface is a rough one. In fact, the value of δ/σ is so large that the small amplitude theory developed in this section is undoubtedly invalid. However, the use of such values for δ and σ makes the effects of surface roughness on the surface polariton dispersion curve much more apparent by exaggerating them than they would be otherwise. Form the figure we see that for a wavelength of 2.1 μm($k_\| = 3 \times 10^{-4}$ Å^{-1})$l^{-1}(k_\|) \cong 4 \times 10^2\ \mathrm{cm}^{-1}$, so that $l(k_\|) \cong 2.5 \times 10^{-3}$ cm.

In all cases studied $\Delta(k_\|)$ was found to be negative, i.e., surface roughness depresses the frequency of a surface polariton below its value in the absence of surface roughness.

The numerical calculations also show that the dominant contributions to $G^{(1)}(k_\|\omega)$ and $G^{(2)}(k_\|\omega)$ come from $G^{>}_{(1)}(k_\|\omega)$ and $G^{>}_{(2)}(k_\|\omega)$, respectively, i.e.,

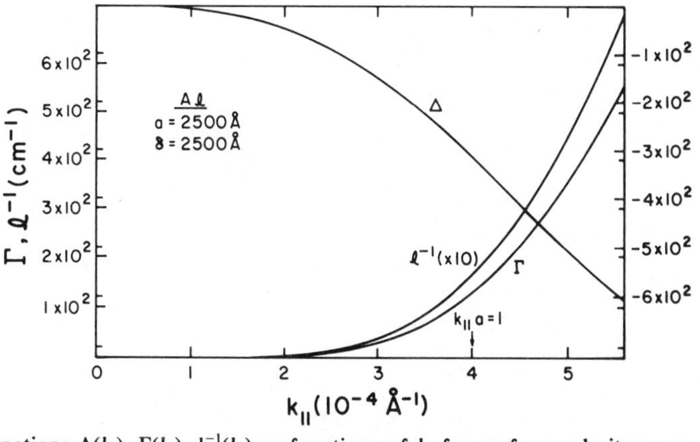

Fig. 10. Functions $\Delta(k_\|)$, $\Gamma(k_\|)$, $l^{-1}(k_\|)$ as functions of $k_\|$ for surface polaritons on a rough Al surface. Note that the graphical value of $l^{-1}(k_\|)$ is to be multiplied by 10 (Maradudin, A. A. and W. Zierau, 1976, Phys. Rev. **B14**, 484).

from the parts of the integrals over $q_\parallel$ that come from the region $q_\parallel > \omega/c$ and correspond to the roughness-induced scattering of a surface polariton into other surface polariton states.

Thus, we have presented a derivation of the dispersion relation for a surface polariton on a randomly rough surface. In the small roughness limit, we have shown its equivalence to the dispersion relation obtained by Toigo et al. (1977), and have discussed its solution and the properties of that solution. The results have been presented for a general form of the surface structure factor $g(k_\parallel)$, so that one can use them together with results for $g(k_\parallel)$ obtained in independent experiments, as well as with analytic forms of the type of eq. (2.28), to calculate the quantities $\Delta(k_\parallel)$, $\Gamma(k_\parallel)$, and $l^{-1}(k_\parallel)$ for a given surface.

We now turn to a determination of $l^{-1}(k_\parallel)$ by an alternative approach that emphasizes its origins in the roughness-induced scattering of a surface polariton as it propagates along a rough surface.

4.2. *Attenuation of Surface Polaritons by Surface Roughness*

The derivation of the surface polariton dispersion curve in the presence of surface roughness presented in the preceding subsection yields the attenuation of the surface polariton as well as the shift in its frequency. If it is only the attenuation of surface polaritons by surface roughness that is of interest, a more direct approach to its determination is provided by a modification of the Green's function approach of the preceding subsection, combined with an argument based on energy conservation (Mills, 1975).

We have seen in the preceding subsection that as a surface polariton propagates along a rough surface it is attenuated by having its energy scattered out of the incident beam through its collisions with the hills and valleys of the surface. The scattering is partly into radiative modes in the vacuum and partly into other surface polariton states. It is natural, therefore, to regard obtaining the roughness-induced attenuation of a surface polariton as a problem in scattering theory, with the roughness playing the role of the scatterer. This is the point of view that will be taken in the present subsection.

The starting point for the present calculation is the integral equation for the electric field in the system of dielectric separated from vacuum by a rough surface derived in the preceding subsection:

$$E_\alpha(\boldsymbol{x};\omega) = E_\alpha^{(0)}(\boldsymbol{x};\omega) - \frac{\omega^2}{4\pi c^2}(\epsilon(\omega)-1)\sum_\beta \int d^2x'_\parallel \zeta(\boldsymbol{x}'_\parallel) \times D_{\alpha\beta}(\boldsymbol{x}, \boldsymbol{x}'_\parallel 0+;\omega) E_\beta(\boldsymbol{x}'_\parallel, 0-;\omega). \tag{4.17}$$

In the present context $E_\alpha^{(0)}(\boldsymbol{x};\omega)$ is the electric field of a surface polariton of frequency ω propagating along a flat vacuum–dielectric interface. We write this field in the form

$$E_\alpha^{(0)}(\boldsymbol{x};\omega) = E_\alpha^{(0)}(\boldsymbol{k}_\|^{(0)}\omega \mid x_3)\exp[\mathrm{i}\boldsymbol{k}_\|^{(0)}\cdot\boldsymbol{x}_\|], \tag{4.69}$$

where $\boldsymbol{k}_\|^{(0)}$ is the wave vector of the incident surface polariton. For a surface polariton propagating in the $+x_1$-direction the associated electromagnetic field amplitudes $\boldsymbol{E}^{(0)}(\boldsymbol{x};\omega)$ and the similarly defined $\boldsymbol{H}^{(0)}(\boldsymbol{x};\omega)$ are

$$\boldsymbol{E}^{(0)}(\boldsymbol{x};\omega) = E^{(0)}\left(1, 0, \frac{\mathrm{i}k^{(0)}}{\beta_0}\right)\exp[\mathrm{i}k^{(0)}x_1 - \beta_0 x_3] \qquad x_3 > 0 \tag{4.70a}$$

$$= E^{(0)}\left(1, 0, -\frac{\mathrm{i}k^{(0)}}{\beta}\right)\exp[\mathrm{i}k^{(0)}x_1 + \beta x_3] \qquad x_3 < 0, \tag{4.70b}$$

$$\boldsymbol{H}^{(0)}(\boldsymbol{x};\omega) = E^{(0)}\left(0, -\frac{\mathrm{i}\omega}{c\beta_0}, 0\right)\exp[\mathrm{i}k^{(0)}x_1 - \beta_0 x_3] \qquad x_3 > 0 \tag{4.71a}$$

$$= E^{(0)}\left(0, -\frac{\mathrm{i}\omega}{c\beta_0}, 0\right)\exp[\mathrm{i}k^{(0)}x_1 + \beta x_3] \qquad x_3 < 0. \tag{4.71b}$$

The decay constants β_0 and β are real and positive and are defined by

$$\beta_0 = \left(k^{(0)2} - \frac{\omega^2}{c^2}\right), \quad \beta = \left(k^{(0)2} - \epsilon(\omega)\frac{\omega^2}{c^2}\right)^{1/2}. \tag{4.72}$$

The wave vector $k^{(0)}$ is related to the frequency ω by the surface polariton dispersion relation

$$\epsilon(\omega)\beta_0 + \beta = 0 \tag{4.73a}$$

or, equivalently,

$$k^{(0)2} = \frac{\omega^2}{c^2}\frac{\epsilon(\omega)}{\epsilon(\omega)+1}. \tag{4.73b}$$

Consequently, from eq. (4.73a) and the requirement that β_0 and β both be positive we see that the incident surface polariton can exist only subject to the conditions that $k_\|^{(0)} > \omega/c$ and $\epsilon(\omega) < 0$.

We first consider the scattered electromagnetic field in the vacuum above the dielectric medium. When we make use of the Fourier representation of $\zeta(\boldsymbol{x}_\|)$ given by eq. (2.31) and the representation of $D_{\alpha\beta}(\boldsymbol{x}, \boldsymbol{x}'; \omega)$ given by eqs. (4.21) and (4.22)–(4.23), together with eq. (4.69), we find that the scattered electric field in first Born approximation can be written in the form

$$E_\alpha^{(s)}(x,\omega) = -\frac{\omega^2}{16\pi^3c^2}(\epsilon(\omega)-1)\int \mathrm{d}^2k_\|\, \mathrm{e}^{\mathrm{i}\boldsymbol{k}\cdot\boldsymbol{x}}\,\hat{\zeta}(\boldsymbol{k}_\| - \boldsymbol{k}_\|^{(0)})\bar{\lambda}_\alpha(\boldsymbol{k}_\|\boldsymbol{k}_\|^{(0)}, \omega), \tag{4.74}$$

where the vector $\boldsymbol{k}$ is given by

$$\boldsymbol{k} = \boldsymbol{k}_{\|} + \hat{x}_3 k_z, \tag{4.75}$$

with

$$k_z = \mathrm{i}\alpha_0(k_{\|}\omega) \tag{4.76}$$

in the notation of the preceding subsection (see eq. (4.35)). The vector $\bar{\lambda}_\alpha(\boldsymbol{k}_{\|}\boldsymbol{k}_{\|}^{(0)}, \omega)$ is defined by

$$\bar{\lambda}_\alpha(\boldsymbol{k}_{\|}\boldsymbol{k}_{\|}^{(0)}, \omega) = \sum_{\alpha'} \lambda_{\alpha'}(\boldsymbol{k}_{\|}\boldsymbol{k}_{\|}^{(0)}, \omega) S_{\alpha'\alpha}(\hat{k}_{\|}), \tag{4.77}$$

where the matrix $S(\hat{k}_{\|})$ has been defined in eq. (4.23), and where

$$\lambda_\alpha(\boldsymbol{k}_{\|}\boldsymbol{k}_{\|}^{(0)}, \omega) = \sum_{\beta} \hat{g}_{\alpha\beta}(k_{\|}\omega \mid +) \mathscr{E}_\beta^{(0)}(\boldsymbol{k}_{\|}\boldsymbol{k}_{\|}^{(0)}\omega \mid -). \tag{4.78}$$

In writing these results we have used the result that for $x_3 > 0$

$$g_{\alpha\beta}(k_{\|}\omega \mid x_3 0+) = \mathrm{e}^{\mathrm{i}k_z x_3} \hat{g}_{\alpha\beta}(k_{\|}\omega \mid +), \tag{4.79}$$

where the tensor $\hat{g}_{\alpha\beta}(k_{\|}\omega \mid +)$ is given by (Maradudin and Mills, 1975)

$$\hat{g}(k_{\|}\omega \mid +) = \frac{4\pi c^2}{\omega^2} \frac{1}{\epsilon(\omega)\alpha_0 + \alpha} \begin{pmatrix} \alpha_0\alpha & 0 & \mathrm{i}\epsilon(\omega)k_{\|}\alpha_0 \\ 0 & \alpha_0\alpha - k_{\|}^2 & 0 \\ \mathrm{i}k_{\|}\alpha & 0 & -\epsilon(\omega)k_{\|}^2 \end{pmatrix}. \tag{4.80}$$

The decay constant $\alpha(k_{\|}\omega)$ has been defined in eq. (4.36). The function $\mathscr{E}_\alpha^{(0)}(k_{\|}k_{\|}^{(0)}\omega \mid -)$ entering eq. (4.78) is

$$\mathscr{E}_\alpha^{(0)}(\boldsymbol{k}_{\|}\boldsymbol{k}_{\|}^{(0)}\omega \mid -) = \sum_{\beta} S_{\alpha\beta}(\hat{k}_{\|}) E_\beta^{(0)}(\boldsymbol{k}_{\|}^{(0)}\omega \mid 0-). \tag{4.81}$$

The scattered field given by eq. (4.74) can now be evaluated in the asymptotic region $|\boldsymbol{x}| \to \infty$ by the method of stationary phase. In carrying through this procedure it must be recognized that two distinct domains of $\boldsymbol{k}_{\|}$ exist. When $k_{\|} < \omega/c$, the quantity k_z is real. This portion of the integral over $\boldsymbol{k}_{\|}$ in eq. (4.74) describes the field radiated away from the surface into the vacuum. When $k_{\|} > \omega/c$, k_z is pure imaginary. This portion of the integral over $\boldsymbol{k}_{\|}$ in eq. (4.74) describes an electromagnetic field localized to the surface, that propagates parallel to it. Roughly speaking, this portion of the integral describes the roughness-induced scattering of the incident surface polariton into other surface polaritons.

If in obtaining the asymptotic form of the contribution to the integral in eq. (4.74) from the range $0 < k_{\|} < \omega/c$ we denote the scalar product $\boldsymbol{k} \cdot \boldsymbol{x}$ by $f(\boldsymbol{k}_{\|}, \boldsymbol{x})$, the stationary points of this function, defined by the conditions $(\partial f/\partial k_\alpha) = 0$ $(\alpha = 1, 2)$ are given by

$$k_1^{(s)} = (\omega/c)(x_1/x), \quad k_2^{(s)} = (\omega/c)(x_2/x). \tag{4.82}$$

Evaluating slowly varying functions of $\boldsymbol{k}_{\|}$ at $\boldsymbol{k}_{\|}^{(s)}$ and removing them outside the integral sign, we find that the asymptotic form of the radiated field in the vacuum is given by

$$E_\alpha^{(s)}(\boldsymbol{x}, \omega) \sim -\frac{\omega^2}{16\pi^3 c^2}(\epsilon(\omega)-1)\hat{\zeta}(\boldsymbol{k}_\|^{(s)} - \boldsymbol{k}_\|^{(0)})$$
$$\times \bar{\lambda}_\alpha(\boldsymbol{k}_\|^{(s)}\boldsymbol{k}_\|^{(0)}, \omega) \exp(\mathrm{i}\omega/cx) \int_{-\infty}^{\infty} \mathrm{d}k_1 \int_{-\infty}^{\infty} \mathrm{d}k_2 \exp(-\mathrm{i}/2\, \boldsymbol{k}_\|^T A \boldsymbol{k}_\|), \qquad (4.83)$$

where the matrix A is

$$A = \frac{cx}{\omega x_3^2}\begin{pmatrix} x_1^2 + x_3^2 & x_1 x_2 \\ x_1 x_2 & x_2^2 + x_3^2 \end{pmatrix}. \qquad (4.84)$$

The remaining integrals in eq. (4.83) are readily evaluated. The radiative part of the scattered electric field obtained in this way is

$$\boldsymbol{E}^{(s)}(\boldsymbol{x}, \omega) = \mathrm{i}\frac{\omega^3(\epsilon(\omega)-1)}{8\pi^2 c^3}\bar{\boldsymbol{\lambda}}(\boldsymbol{k}_\|^{(s)}\boldsymbol{k}_\|^{(0)}, \omega)\zeta(\boldsymbol{k}_\|^{(s)} - \boldsymbol{k}_\|^{(0)}) \cos\theta_s \frac{\mathrm{e}^{\mathrm{i}kx}}{x}, \qquad (4.85)$$

In writing this result we have introduced the magnitude of the wave vector of the scattered radiation, $k = \omega/c$, and have replaced a factor of x_3/x by $\cos\theta_s$, where θ_s is the polar angle of the wave vector $\boldsymbol{k}$ of the scattered wave, i.e., it is the polar angle of the vector $\boldsymbol{x}$.

In obtaining the contribution to the asymptotic form of the scattered field (4.74) from the range $k_\| > \omega/c$ we have to proceed somewhat differently. In this case k_z is pure imaginary and we rewrite eq. (4.74) as

$$E_\alpha^{(s)}(\boldsymbol{x}, \omega) = -\frac{\omega^2}{16\pi^3 c^2}(\epsilon(\omega)-1)\int_{\omega/c}^{\infty} \mathrm{d}k_\| k_\| \int_0^{2\pi} \mathrm{d}\theta'\, \mathrm{e}^{\mathrm{i}\boldsymbol{k}_\|\cdot\boldsymbol{x}_\|}$$
$$\times \mathrm{e}^{-\alpha_0 x_3}\hat{\zeta}(\boldsymbol{k}_\| - \boldsymbol{k}_\|^{(0)})\bar{\lambda}_\alpha(\boldsymbol{k}_\|\boldsymbol{k}_\|^{(0)}, \omega). \qquad (4.86)$$

Because this expression describes a scattered wave in the vacuum that is localized to the surface, we assume x_3 fixed and examine the limit as $|\boldsymbol{x}_\|| \to 0$. If we write $\boldsymbol{x}_\|$ and $\boldsymbol{k}_\|$ in the forms

$$\boldsymbol{x}_\| = x_\|(\cos\theta, \sin\theta), \qquad \boldsymbol{k}_\| = k_\|(\cos\theta', \sin\theta'), \qquad (4.87)$$

we can replace $\boldsymbol{k}_\| \cdot \boldsymbol{x}_\|$ in the integrand of eq. (4.86) by $k_\| x_\| \cos(\theta - \theta') \cong k_\| x_\| = \frac{1}{2}k_\| x_\|(\theta' - \theta)^2$, since for large $x_\|$ the dominant contribution to the integral over θ' comes from values of θ' in the immediate vicinity of the stationary point $\theta' = \theta$. The angular integration can then be carried out with the result that

$$E_\alpha^{(s)}(\boldsymbol{x}, \omega) \sim -\frac{\omega^2}{c^2}\frac{(\epsilon(\omega)-1)}{2(2\pi)^{5/2}}\frac{\mathrm{e}^{-\mathrm{i}\pi/4}}{x_\|^{1/2}}\int_{\omega/c}^{\infty} \mathrm{d}k_\| k_\|^{1/2}\, \mathrm{e}^{\mathrm{i}k_\| x_\|}\, \mathrm{e}^{-\alpha_0 x_3}$$
$$\times \hat{\zeta}(\boldsymbol{k}_\| - \boldsymbol{k}_\|^{(0)})\bar{\lambda}_\alpha(\boldsymbol{k}_\|\boldsymbol{k}_\|^{(0)}, \omega). \qquad (4.88)$$

The vector $\boldsymbol{k}_\|$ in this expression must now be understood to be given by

$$\mathbf{k}_\parallel = k_\parallel(\cos\theta, \sin\theta) = \hat{x}_\parallel k_\parallel, \tag{4.89}$$

i.e. it is directed along the vector $\boldsymbol{x}_\parallel$.

It is now possible to obtain the form of the vector $\bar{\boldsymbol{\lambda}}(\boldsymbol{k}_\parallel \boldsymbol{k}_\parallel^{(0)}, \omega)$. When we use eqs. (4.77)–(4.81), together with eqs. (4.23) and (4.89), we find that

$$\bar{\lambda}_1(\boldsymbol{k}_\parallel \boldsymbol{k}_\parallel^{(0)}, \omega) = \hat{x}_1 \bar{\lambda} \tag{4.90a}$$

$$\bar{\lambda}_2(\boldsymbol{k}_\parallel \boldsymbol{k}_\parallel^{(0)}, \omega) = \hat{x}_2 \bar{\lambda} \tag{4.90b}$$

$$\bar{\lambda}_3(\boldsymbol{k}_\parallel \boldsymbol{k}_\parallel^{(0)}, \omega) = \frac{\mathrm{i}k_\parallel}{\alpha_0} \bar{\lambda} \tag{4.90c}$$

where

$$\bar{\lambda} = \frac{4\pi c^2}{\omega^2} \frac{\alpha_0}{\epsilon(\omega)\alpha_0 + \alpha} [\alpha \hat{x}_1 E_1^{(0)}(\boldsymbol{k}_\parallel^{(0)}\omega \mid 0-) + \mathrm{i}\epsilon(\omega) k_\parallel E_3^{(0)}(k_\parallel^{(0)}\omega \mid 0-)]. \tag{4.91}$$

In writing eqs. (4.90)–(4.91) we have kept only the contributions to $\bar{\lambda}_\alpha(\boldsymbol{k}_\parallel \boldsymbol{k}_\parallel^{(0)}, \omega)$ that have the denominator $\epsilon(\omega)\alpha_0 + \alpha$, i.e. we have neglected the 22 elements of the matrix $\hat{g}_{\alpha\beta}(k_\parallel\omega \mid +)$, eq. (4.80), in view of eq. (4.37). The latter terms have no pole in the complex $k_\parallel$-plane, while, according to eqs. (4.60) and (4.62), the former do.

The latter fact means that we can evaluate the integral over $k_\parallel$ in eq. (4.88) by the use of the contour shown in fig. 11. In the limit as the radius of the circular portion of the contour tends to infinity the integral we seek is given by $2\pi\mathrm{i}$ times the residue at the pole of the integrand plus the contribution from the integral along the vertical segment of the contour from ω/c to $\omega/c + \mathrm{i}\infty$. The latter contribution, however, is found to be proportional to $x_\parallel^{-3}$ for large $x_\parallel$, and thus does not contribute to the radiation field.

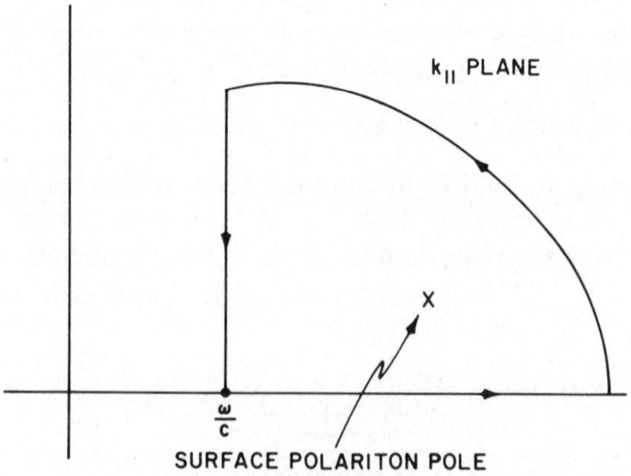

Fig. 11. Contour used to evaluate the contribution to the scattered field from surface polaritons.

The expression for the scattered electric field in this case is therefore found to be

$$E^{(s)}(x, \omega) = [\hat{x}_1 \cos\varphi_s + \hat{x}_2 \sin\varphi_s + \mathrm{i}\hat{x}_3\, k_\|/\alpha_0]\, E^{(s)} \frac{\mathrm{e}^{\mathrm{i}k_\| x_\| - \alpha_0 x_3}}{x_\|^{1/2}}, \tag{4.92}$$

$$E^{(s)} = \frac{(-\mathrm{e}^{-\mathrm{i}\pi/4})}{(2\pi)^{1/2}} \hat{\zeta}(k_\| - k_\|^{(0)}) \frac{\alpha_0 \alpha k_\|^{1/2}}{|\epsilon(\omega)| - 1} \left[\mathrm{i}\frac{\alpha}{k_\|} \cos\varphi_s \times E_1^{(0)}(k_\|^{(0)}\omega \mid 0-) + |\epsilon(\omega)| E_3^{(0)}(k_\|^{(0)}\omega \mid 0-) \right]. \tag{4.93}$$

In writing this expression we have used the notation that

$$\cos\varphi_s = \hat{x}_1 = x_1/x_\|, \qquad \sin\varphi_s = \hat{x}_2 = x_2/x_\|, \tag{4.94}$$

where φ_s is the aximuthal angle of the wave vector $\boldsymbol{k}$ of the scattered wave. The wave vector $\boldsymbol{k}_\|$ in this case is in the direction of observation $\boldsymbol{x}_\|$, and its magnitude, $k_{sp}(\omega)$, is found from eq. (4.61) or equivalently eq. (4.73), i.e., $\boldsymbol{k}_\|$ is the wave vector of the surface polariton created in the scattering process.

The time-averaged Poynting vector associated with each of the electric fields (4.85) and (4.92) is readily evaluated. If $(\mathrm{d}^2E^{(R)}/\mathrm{d}\Omega\mathrm{d}t)\,\mathrm{d}\Omega$ be the energy per unit time radiated into the vacuum, into the solid angle range $\mathrm{d}\Omega$ about the scattering direction, we have that

$$\frac{\mathrm{d}^2E^{(R)}}{\mathrm{d}\Omega\mathrm{d}t} = \frac{\omega^4(1+|\epsilon(\omega)|)}{32\pi^3c^3} \cos^2\theta_s |\hat{\zeta}(k_\| - k_\|^{(0)})|^2 \times \left\{ \frac{|\mathrm{i}(|\epsilon(\omega)| + \sin^2\theta_s)^{1/2} \cos\varphi_s E_1^{(0)}(k_\|^{(0)}\omega \mid -) + |\epsilon(\omega)| \sin\theta_s E_3^{(0)}(k_\|^{(0)}\omega \mid -)|^2}{|\epsilon(\omega)| \cos^2\theta_s + \sin^2\theta_s} + \sin^2\varphi_s |\mathrm{E}_1^{(0)}(\mathrm{k}_\|^{(0)}\omega \mid -)|^2 \right\}. \tag{4.95}$$

In eq. (4.95) the term proportional to $\sin^2\varphi_s$ has its origin in roughness-induced radiation into a final state of s-polarization, and the other term in braces has its origin in radiation into a final state of p-polarization.

Now let the total energy per unit time carried by the surface polariton field in the vacuum above the crystal in the angular range between φ_s and $\varphi_s + \mathrm{d}\varphi_s$ be denoted by $(\mathrm{d}^2E^{(sp>)}/\mathrm{d}\varphi_s\mathrm{d}t)\mathrm{d}\varphi_s$. The time average of the Poynting vector in the vacuum above the crystal is parallel to $\boldsymbol{k}_\|$. If we denote its magnitude by $\mathcal{S}_>$, then

$$\frac{\mathrm{d}^2E^{(sp>)}}{\mathrm{d}\varphi_s\mathrm{d}t} = x_\| \int_0^\infty \mathrm{d}x_3\, \mathcal{S}_> = \frac{\omega}{32\pi^2\alpha_0} \frac{k_\|^2\alpha^2}{(|\epsilon(\omega)| - 1)^2} |\hat{\zeta}(k_\| - k_\|^{(0)})|^2 \times \left| \mathrm{i}\frac{\alpha}{k_\|} \cos\varphi_s E_1^{(0)}(k_\|^{(0)}\omega \mid -) + |\epsilon(\omega)| E_3^{(1)}(k_\|^{(0)}\omega \mid -) \right|^2. \tag{4.96}$$

We have so far considered only the energy flow stored in the electromagnetic field in the vacuum above the crystal. There is also a contribution from the electromagnetic field in the crystal. There is no radiative component to the electromagnetic field in the crystal, because the dielectric constant of the crystal is negative at the frequency of the surface polariton, but there is a surface polariton component. Its contribution to the energy flow stored in the electromagnetic field in the crystal is obtained in the same way as eq. (4.96) is obtained, and is given by

$$\frac{d^2E^{(sp<)}}{d\varphi_s dt} = -\frac{1}{|\epsilon(\omega)|^2}\frac{d^2E^{(sp>)}}{d\varphi_s dt}. \tag{4.97}$$

Thus, the total rate at which energy flows in the surface polariton field is

$$\frac{d^2E^{(sp)}}{d\varphi_s dt} = \left(1-\frac{1}{|\epsilon(\omega)|^2}\right)\frac{d^2E^{(sp>)}}{d\varphi_s dt} = \frac{\omega^4}{32\pi^2c^3}\frac{|\epsilon(\omega)|(|\epsilon(\omega)|+1)}{(|\epsilon(\omega)|-1)^{5/2}}|\hat{\zeta}(\boldsymbol{k}_\parallel - \boldsymbol{k}_\parallel^{(0)})|^2$$
$$\times |\mathrm{i}|\epsilon(\omega)|^{1/2}\cos\varphi_s E_1^{(0)}(\boldsymbol{k}_\parallel^{(0)}\omega\,|\,-) + |\epsilon(\omega)|\,E_3^{(0)}(\boldsymbol{k}_\parallel^{(0)}\omega\,|\,-)|^2. \tag{4.98}$$

The energy stored in the incident wave is

$$\frac{dE^{(0)}}{dt} = L_2\frac{1}{16\pi}\frac{c^2}{\omega}\frac{[|\epsilon(\omega)|-1]^2[|\epsilon(\omega)|+1]}{|\epsilon(\omega)|^{3/2}}|E^{(0)}|^2, \tag{4.99}$$

where L_2 is the length of the rough portion of the surface in the x_2-direction. Equation (4.99) is obtained by multiplying the x_1-component of the time-averaged Poynting vector of the incident surface polariton (the only nonzero component) by $L_2\,dx_3$ and integrating the result over all x_3.

We can now obtain the total energy per unit time $dE^{(T)}/dt$ radiated by the surface polariton as it moves over the rough surface. This is achieved by first averaging eqs. (4.95) and (4.98) over the ensemble of realizations of the surface roughness with the use of eq. (2.33), then integrating the results over the appropriate portion of solid angle (φ_s ranges from $-\pi$ to $+\pi$, while θ_s ranges from 0 to $\pi/2$ in eq. (4.95); φ_s ranges from $-\pi$ to $+\pi$ in eq. (4.98)). Then with the aid of eq. (4.99) one obtains the relation

$$\frac{dE^{(T)}}{dt} = \frac{L_1}{l}\frac{dE^{(0)}}{dt}. \tag{4.100}$$

Here L_1 is the distance traveled by the surface polariton as it passes over the rough region of the surface. The ratio $(dE^{(T)}/dt)/L_1(dE^{(0)}/dt)$ is the energy lost per unit of distance traveled by the surface polariton. This is the inverse of the mean free path l of the wave. We therefore come to the result that

$$1/l = 1/l^{(R)} + 1/l^{(sp)}, \tag{4.101}$$

where

$$\frac{1}{l^{(\mathrm{R})}} = \frac{\delta^2\omega^5}{2\pi^2c^5}\frac{|\epsilon|^{3/2}}{(|\epsilon|-1)^2}\int_{-\pi}^{\pi} \mathrm{d}\varphi_\mathrm{s} \int_0^{\pi/2} \mathrm{d}\theta_\mathrm{s} \cos^2\theta_\mathrm{s} \sin\theta_\mathrm{s}\, g(|\boldsymbol{k}_\parallel - \boldsymbol{k}_\parallel^{(0)}|)$$

$$\times\left(\sin^2\varphi_\mathrm{s} + \frac{|(|\epsilon| + \sin^2\theta_\mathrm{s})^{1/2}\cos\varphi_\mathrm{s} - |\epsilon|^{1/2}\sin\theta_\mathrm{s}|^2}{|\epsilon|\cos^2\theta_\mathrm{s} + \sin^2\theta_\mathrm{s}}\right), \tag{4.102}$$

$$\frac{1}{l^{(\mathrm{sp})}} = \frac{2}{\pi}\frac{\delta^2\omega^5}{c^5}\frac{|\epsilon|^{7/2}}{(|\epsilon|-1)^{9/2}}\int_{-\pi}^{\pi} \mathrm{d}\varphi_\mathrm{s}\, g(|\boldsymbol{k}_\parallel - \boldsymbol{k}_\parallel^{(0)}|)\sin^4\tfrac{1}{2}\varphi_\mathrm{s}. \tag{4.103}$$

The quantity $l^{(\mathrm{R})-1}$ is the contribution to the inverse attenuation length from roughness-induced radiation into the vacuum, while $l^{(\mathrm{sp})-1}$ is that from roughness-induced scattering of the surface polariton into other surface polariton states.

Equation (4.103) can be evaluated in closed form if the Gaussian form for $g(k_\parallel)$, eq. (2.41), is used (Mills 1975), and the integration over φ_s in eq. (4.102) can be carried out analytically, although the integral over θ_s has to be carried out numerically. However, a simple approximation that yields analytic results for both $l^{(\mathrm{R})-1}$ and $l^{(\mathrm{sp})-1}$ is valid in many circumstances in the infrared range of frequencies. Here the wavelength of the surface polariton will often be large compared with the transverse correlation length σ. In this case $\sigma|\boldsymbol{k}_\parallel - \boldsymbol{k}_\parallel^{(0)}| \ll 1$, and $g(|\boldsymbol{k}_\parallel - \boldsymbol{k}_\parallel^{(0)}|)$ can be replaced by $g(0)$. In this limit we obtain the results that (Mills 1975):

$$\frac{1}{l^{(\mathrm{sp})}} = \frac{3}{2}\frac{\delta^2\omega^5}{c^5}\frac{|\epsilon|^{7/2}}{[|\epsilon|-1]^{9/2}}g(0), \tag{4.104}$$

and

$$\frac{1}{l^{(\mathrm{R})}} = \frac{1}{l_\mathrm{s}^{(\mathrm{R})}} + \frac{1}{l_\mathrm{p}^{(\mathrm{R})}}, \tag{4.105}$$

where

$$\frac{1}{l_\mathrm{s}^{(\mathrm{R})}} = \frac{\delta^2\omega^5}{6\pi c^5}\frac{|\epsilon|^{3/2}}{[|\epsilon|-1]^2}g(0) \tag{4.106}$$

$$\frac{1}{l_\mathrm{p}^{(\mathrm{R})}} = \frac{\delta^2\omega^5}{2\pi c^5}\frac{|\epsilon|^{3/2}}{[|\epsilon|-1]^3}\left[-\tfrac{2}{3}|\epsilon| - \tfrac{1}{3} + \frac{3|\epsilon|^2}{|\epsilon|-1}\left(1 - \frac{\tan^{-1}[(|\epsilon|-1)^{1/2}]}{(|\epsilon|-1)^{1/2}}\right)\right]. \tag{4.107}$$

The quantity $1/l_\mathrm{s}^{(\mathrm{R})}$ is the contribution to $1/l^{(\mathrm{R})}$ that comes from radiation into final states of s-polarization, and $1/l_\mathrm{p}^{(\mathrm{R})}$ is the contribution that comes from radiation into final states of p-polarization.

In the limit that $|\epsilon| \gg 1$, the preceding results simplify to

$$1/l^{(\mathrm{sp})} = \tfrac{3}{2}(\delta^2\omega^5/c^5)(g(0)/|\epsilon|) \tag{4.108}$$

$$1/l_\mathrm{s}^{(\mathrm{R})} = (1/6\pi)(\delta^2\omega^5/c^5)(g(0)/|\epsilon|^{1/2}) \tag{4.109}$$

$$1/l_p^{(R)} = (7/6\pi)(\delta^2\omega^5/c^5)(g(0)/|\epsilon|^{1/2}). \tag{4.110}$$

These results apply, e.g., to surface polaritons at a metal–vacuum interface when $\omega \ll \omega_p$, so that $\omega \approx ck_\parallel$ here, and to a semiconductor–vacuum interface when ω is close to ω_T. One sees from these results that in this limit the most effective process for attenuating the surface polariton is the roughness-induced radiation into the vacuum above the substrate. The attenuation length in this case is

$$l = (3\pi/4)(c^5/\delta^2\omega^5)(|\epsilon|^{1/2}/g(0)). \tag{4.111}$$

In the case that $|\epsilon| \approx 1$, i.e., when the frequency of the surface polariton is close to its asymptotic value given by $\epsilon(\omega)+1=0$, we have that

$$1/l^{(sp)} = \tfrac{3}{2}(\delta^2\omega^5/c^5)[g(0)/(|\epsilon|-1)^{9/2}], \tag{4.112}$$

$$1/l_s^{(R)} = (1/6\pi)(\delta^2\omega^5/c^5)[g(0)/(|\epsilon|-1)^2], \tag{4.113}$$

$$1/l_p^{(R)} = (11/30\pi)(\delta^2\omega^5/c^5)[g(0)/(|\epsilon|-1)^2]. \tag{4.114}$$

From these results it is seen that when ϵ is near -1, the surface polariton is damped predominantly by scattering into other surface polariton states. Calculations based on the full expressions (4.102)–(4.103) show that scattering into other surface polariton states dominates radiation into the vacuum for $|\epsilon| \lesssim 7$.

The validity of the results given by eqs. (4.112)–(4.114) requires that $ck_\parallel \gg \omega$ so that $|\epsilon|$ is close to unity, and that $k_\parallel\sigma \ll 1$, i.e., that the wavelength of the surface polariton be long compared to the transverse correlation length.

Thus a scattering theory approach, combined with an energy conservation argument, provides a simple and direct determination of the attenuation of surface polaritons by surface roughness. It appears to lead to simpler expressions to work with than does solving the surface polariton dispersion relation, as was done in the preceding subsection. That the two approaches are equivalent, however, has been remarked on by Kröger and Kretschmann (1976).

The results obtained in this section are based on the assumption of the small roughness approximation, in which only the leading nonzero term in $\zeta(\mathbf{x}_\parallel)$ has been retained in the kernel of the integral equations, eqs. (4.19) and (4.20). In the absence of any criterion for the range of validity of the Rayleigh hypothesis for a randomly rough surface one must look elsewhere for information concerning the range of validity of the theoretical results presented here. In recent measurements of light scattered from rough surfaces of silver films Hillebrecht (1980) has found that first order theory can be applied to surfaces for which $\delta \leq 15$ Å. For roughness characterized by values of $\delta > 15$ Å the Fourier transform of the roughness correlation function $g(k_{\|})$ obtained from his experimental data shows a dependence on the wavelength of the incident light.

The effects of surface roughness on surface polaritons in GaP surfaces have been studied by means of Raman scattering by Ushioda et al. (1979). The surface polariton linewidth was observed to broaden as the surface is roughened, and the dispersion curve is depressed below the theoretical value calculated for a flat surface. Both the observed increase in damping and the frequency shift are well described by the theories of these effects outlined in this section. The values of δ used in fitting their experimental data ranged from 500 Å to 3000 Å, while the values of σ ranged from 1 to 10 μm. These results suggest, in contrast with the result of Hillebrecht (1980), that the limits of validity of the small roughness theory are determined by the interplay between δ and σ and not by the value of δ alone.

4.3. Interaction of Surface Plasmons with a Rough Surface

A surface plasmon can be regarded as the limiting case of a surface polariton when the effects of the retardation of the Coulomb interaction can be neglected. This is the case when the wavelength of the surface polariton parallel to the surface is short compared with the vacuum wavelength of the principal infrared active excitation in the medium over which the surface polariton propagates. This limit can be achieved experimentally, and surface plasmons have been observed and their properties studied as surface excitations in their own right, and not as limiting cases of surface polaritons.

The propagation of a surface plasmon along a randomly rough surface is of considerable interest since its study offers the possibility of providing information about the roughness of the surface. In this section we obtain the dispersion relation for a surface plasmon on a statistically rough surface. It can be obtained as a limiting case of the dispersion curve for a surface polariton on the same kind of surface. Indeed, the dispersion relation for surface plasmons on such a surface was first obtained theoretically by Kretschmann et al. (1979) from the large $k_{\|}$ limit of the dispersion relation for surface polaritons propagating along a statistically slightly rough surface (see sect. 4.1). However, it can be obtained directly and simply from electrostatic, as opposed to electromagnetic, considerations (Rahman and Maradudin 1980b). Such a derivation is of interest in itself, to the extent that surface plasmons are independent surface excitations. Consequently, the present subsection is devoted to the presentation of this derivation.

It follows from the preceding comments that a surface plasmon is an electrostatic surface "wave". In the electrostatic limit the macroscopic electric field is obtainable from a scalar potential that satisfies Laplace's equation. In the present discussion we assume that the region $x_3 < \zeta(\boldsymbol{x}_\parallel)$ is filled with a dielectric characterized by an isotropic, frequency-dependent dielectric tensor $\epsilon_{\mu\nu}(\omega) = \delta_{\mu\nu}\epsilon(\omega)$. The region $x_3 > \zeta(\boldsymbol{x}_\parallel)$ is vacuum. The solution of Laplace's equation that vanishes as $|x_3| \to \infty$ outside the selvedge region can be written in the forms

$$\varphi(\boldsymbol{x}) = \int \frac{\mathrm{d}^2 q_\parallel}{(2\pi)^2} A(\boldsymbol{q}_\parallel) \exp[\mathrm{i}\boldsymbol{q}_\parallel \cdot \boldsymbol{x}_\parallel - q_\parallel x_3] \qquad x_3 > \zeta_{\max} \tag{4.115a}$$

$$= \int \frac{\mathrm{d}^2 q_\parallel}{(2\pi)^2} B(\boldsymbol{q}_\parallel) \exp[\mathrm{i}\boldsymbol{q}_\parallel \cdot \boldsymbol{x}_\parallel + q_\parallel x_3] \qquad x_3 < \zeta_{\max}. \tag{4.115b}$$

On the assumption that these expressions can be continued in to the surface, the unknown coefficients $A(\boldsymbol{q}_\parallel)$ and $B(\boldsymbol{q}_\parallel)$ are to be determined with the aid of the boundary conditions. The latter are

$$\varphi(\boldsymbol{x})\Big|_{x_3=\zeta(\boldsymbol{x}_\parallel)+} = \varphi(\boldsymbol{x})\Big|_{x_3=\zeta(\boldsymbol{x}_\parallel)-}, \tag{4.116a}$$

$$\hat{n} \cdot \nabla\varphi(\boldsymbol{x})\Big|_{x_3=\zeta(\boldsymbol{x}_\parallel)+} = \epsilon(\omega)\hat{n} \cdot \nabla\varphi(\boldsymbol{x})\Big|_{x_3=\zeta(\boldsymbol{x}_\parallel)-}. \tag{4.116b}$$

It can be pointed out here that there is no loss of generality involved in our choice of a vacuum–dielectric interface in the present discussion. In the more general case in which the region $x_3 > \zeta(\boldsymbol{x}_\parallel)$ is filled with a dielectric characterized by a dielectric constant $\epsilon_2(\omega)$, while the region $x_3 < \zeta(\boldsymbol{x}_\parallel)$ is filled with a dielectric characterized by a dielectric constant $\epsilon_1(\omega)$, the dispersion curve for surface plasmons can be obtained from the result obtained here by the replacement of the dielectric constant $\epsilon(\omega)$ by the ratio $\epsilon_1(\omega)/\epsilon_2(\omega)$.

We now substitute eqs. (4.115) into eqs. (4.116) and expand the exponentials $\exp(\pm q_\parallel \zeta(\boldsymbol{x}_\parallel))$ resulting from this step in powers of $q_\parallel \zeta(\boldsymbol{x}_\parallel)$ up to second order terms. When the Fourier decomposition of $\zeta(\boldsymbol{x}_\parallel)$ given by eq. (2.31) is used in the resulting equations, and the Fourier coefficients on both sides of these equations are equated, the following pair of homogeneous equations for $A(\boldsymbol{q}_\parallel)$ and $B(\boldsymbol{q}_\parallel)$ is obtained:

$$\begin{pmatrix} 1 & -1 \\ 1 & \epsilon(\omega) \end{pmatrix} \begin{pmatrix} A(\boldsymbol{q}_\parallel) \\ B(\boldsymbol{q}_\parallel) \end{pmatrix} - \int \frac{\mathrm{d}^2 k_\parallel}{(2\pi)^2} \hat{\zeta}(\boldsymbol{q}_\parallel - \boldsymbol{k}_\parallel) k_\parallel \begin{pmatrix} 1 & 1 \\ \hat{\boldsymbol{k}}_\parallel \cdot \hat{\boldsymbol{q}}_\parallel & -\epsilon(\omega)\hat{\boldsymbol{k}}_\parallel \cdot \hat{\boldsymbol{q}}_\parallel \end{pmatrix} \begin{pmatrix} A(\boldsymbol{k}_\parallel) \\ B(\boldsymbol{k}_\parallel) \end{pmatrix}$$

$$+ \int \frac{\mathrm{d}^2 k_\parallel}{(2\pi)^2} \int \frac{\mathrm{d}^2 Q_\parallel}{(2\pi)^2} \hat{\zeta}(\boldsymbol{q}_\parallel - \boldsymbol{k}_\parallel - \boldsymbol{Q}_\parallel) \zeta(\boldsymbol{Q}_\parallel)$$

$$\times \begin{pmatrix} \tfrac{1}{2}k_\parallel^2 & -\tfrac{1}{2}k_\parallel^2 \\ \dfrac{k_\parallel}{q_\parallel}(\boldsymbol{k}_\parallel \cdot \boldsymbol{q}_\parallel - \tfrac{1}{2}k_\parallel^2 - \boldsymbol{k}_\parallel \cdot \boldsymbol{Q}_\parallel) & \epsilon(\omega)\dfrac{k_\parallel}{q_\parallel}(\boldsymbol{k}_\parallel \cdot \boldsymbol{q}_\parallel - \tfrac{1}{2}k_\parallel^2 - \boldsymbol{k}_\parallel \cdot \boldsymbol{Q}_\parallel) \end{pmatrix} \begin{pmatrix} A(\boldsymbol{k}_\parallel) \\ B(\boldsymbol{k}_\parallel) \end{pmatrix} = 0. \tag{4.117}$$

We have remarked in the introduction that in the case of a statistically rough surface we do not know the surface roughness profile function $\zeta(\mathbf{x}_\parallel)$: we only know certain statistical properties of this function, expressed by eqs. (2.25)–(2.26) or, equivalently by eqs. (2.32)–(2.33). In such a case we seek not the scalar potential $\varphi(\mathbf{x})$ itself, but rather its value averaged over the ensemble of realizations of the surface roughness profile function $\zeta(\mathbf{x}_\parallel)$. In view of eqs. (4.115) this means that we seek not the coefficients $A(\mathbf{q}_\parallel)$ and $B(\mathbf{q}_\parallel)$ themselves, but rather their averaged values $\langle A(\mathbf{q}_\parallel)\rangle$ and $\langle B(\mathbf{q}_\parallel)\rangle$.

It is possible to convert the pair of equations (4.117) for $A(\mathbf{q}_\parallel)$ and $B(\mathbf{q}_\parallel)$ into a pair of equations for $\langle A(\mathbf{q}_\parallel)\rangle$ and $\langle B(\mathbf{q}_\parallel)\rangle$, that can then be solved readily. If we denote the two-component vector $(A(\mathbf{q}_\parallel), B(\mathbf{q}_\parallel))$ by $\mathbf{V}$, the matrix integral equation (4.117) can be rewritten symbolically in the form

$$(L_0 + \delta L)\mathbf{V} = 0 \tag{4.118}$$

where L_0 is a deterministic, 2×2 matrix, while δL is a stochastic, 2×2 matrix integral operator. We next introduce the smoothing operator P that averages everything it acts on over the ensemble of realizations of the surface roughness profile function, e.g.,

$$P\mathbf{A} \equiv \langle \mathbf{A}\rangle. \tag{4.119}$$

We also introduce the operator Q defined by

$$Q \equiv 1 - P. \tag{4.120}$$

We now apply P to eq. (4.118) to obtain

$$(L_0 + P\delta L)P\mathbf{V} + P\delta LQ\mathbf{V} = 0, \tag{4.121}$$

where we have used the identity $V = PV + QV$. We next apply Q to eq. (4.118), and find that

$$(L_0 + Q\delta L)Q\mathbf{V} = -Q\delta LP\mathbf{V}. \tag{4.122}$$

When we solve eq. (4.122) for $Q\mathbf{V}$ and substitute the result into eq. (4.121) we find that the equation satisfied by $P\mathbf{V} \equiv \langle \mathbf{V}\rangle$ is

$$\{L_0 + \langle \delta L(L_0 + Q\delta L)^{-1}\rangle L_0\}\langle \mathbf{V}\rangle = 0. \tag{4.123}$$

In the present case we can write that

$$\delta L = L_1 + L_2, \tag{4.124}$$

where the subscripts denote the order of each term in $\zeta(\mathbf{x}_\parallel)$. To second order in $\zeta(\mathbf{x}_\parallel)$ the equation for $\langle \mathbf{V}\rangle$ becomes

$$\{L_0 + \langle L_2\rangle - \langle L_1 L_0^{-1} L_1\rangle\}\langle \mathbf{V}\rangle = 0. \tag{4.125}$$

When eq. (4.125) is combined with the result given by eq. (2.21), we obtain the equation satisfied by $\langle A(\mathbf{q}_\parallel)\rangle$ and $\langle B(\mathbf{q}_\parallel)\rangle$ in the form

$$\left[(1+\tfrac{1}{2}\delta^2 q_\parallel^2)\begin{pmatrix}1 & -1\\ 1 & \epsilon(\omega)\end{pmatrix} - \frac{\delta^2 q_\parallel}{\epsilon(\omega)+1}\begin{pmatrix} a_{11}(q_\parallel\omega) & a_{12}(q_\parallel\omega)\\ a_{21}(q_\parallel\omega) & a_{22}(q_\parallel\omega)\end{pmatrix}\right]\begin{pmatrix}\langle A(\boldsymbol{q}_\parallel)\rangle\\ \langle B(\boldsymbol{q}_\parallel)\rangle\end{pmatrix} = 0 \tag{4.126}$$

where

$$a_{11}(q_\parallel\omega) = \int \frac{d^2k_\parallel}{(2\pi)^2}\, g(|\boldsymbol{q}_\parallel - \boldsymbol{k}_\parallel|)k_\parallel[\epsilon(\omega) - 1 + 2\hat{q}_\parallel\cdot\hat{k}_\parallel], \tag{4.127a}$$

$$a_{12}(q_\parallel\omega) = \int \frac{d^2k_\parallel}{(2\pi)^2}\, g(|\boldsymbol{q}_\parallel - \boldsymbol{k}_\parallel|)k_\parallel[\epsilon(\omega) - 1 - 2\epsilon(\omega)\hat{q}_\parallel\cdot\hat{k}_\parallel], \tag{4.127b}$$

$$a_{21}(q_\parallel\omega) = \int \frac{d^2k_\parallel}{(2\pi)^2}\, g(|\boldsymbol{q}_\parallel - \boldsymbol{k}_\parallel)k_\parallel[2\epsilon(\omega)\hat{q}_\parallel\cdot\hat{k}_\parallel - (\epsilon(\omega)-1)(\hat{q}_\parallel\cdot\hat{k}_\parallel)^2] \tag{4.127c}$$

$$a_{22}(q_\parallel\omega) = \int \frac{d^2k_\parallel}{(2\pi)^2}\, g(|\boldsymbol{q}_\parallel - \boldsymbol{k}_\parallel|)k_\parallel[2\epsilon(\omega)\hat{q}_\parallel\cdot\hat{k}_\parallel + \epsilon(\omega)(\epsilon(\omega)-1)(\hat{q}_\parallel\cdot\hat{k}_\parallel)^2]. \tag{4.127d}$$

We note that in contrast with the starting equation for $A(\boldsymbol{q}_\parallel)$ and $B(\boldsymbol{q}_\parallel)$, eq. (4.117), which is a matrix integral equation, the equation (4.126) for $\langle A(\boldsymbol{q}_\parallel)\rangle$ and $\langle B(\boldsymbol{q}_\parallel\rangle$ is simply a matrix equation. This is because the average over the ensemble of realizations of the surface roughness restores translational invariance to our vacuum–dielectric system, through the stationarity of the two-point correlation function as expressed by eq. (2.26). That the matrix elements $a_{\mu\nu}(q_\parallel\omega)$ depend on the wave vector $\boldsymbol{q}_\parallel$ only through its magnitude is due to the fact that the averaging process also restores isotropy in the plane $x_3 = 0$ to our system, because the two-point correlation function (2.26) depends on the difference vector $\boldsymbol{x}_\parallel - \boldsymbol{x}'_\parallel$ only through its magnitude.

The dispersion relation for surface plasmons in the presence of surface roughness is obtained from the solvability conditions for the system of equations (4.126). Equating to zero the determinant of the coefficients in this system of equations, with the aid of the result that

$$[A + \delta A| = |A| + |A| \operatorname{Tr} A^{-1}\delta A + O((\delta A)^2), \tag{4.128}$$

we obtain the sought dispersion relation to $O(\delta^2)$ in the form

$$\epsilon(\omega) + 1 - \frac{\delta^2 q_\parallel}{\epsilon(\omega)+1}[\epsilon(\omega)a_{11}(q_\parallel\omega) + a_{21}(q_\parallel\omega) - a_{12}(q_\parallel\omega) + a_{22}(q_\parallel\omega)] = 0. \tag{4.129}$$

Equation (4.129) can be rearranged into the equation

$$\epsilon(\omega) + 1 = \pm\,\delta q_\parallel^{1/2}[\epsilon(\omega)a_{11}(q_\parallel\omega) + a_{21}(q_\parallel\omega) - a_{12}(q_\parallel\omega) + a_{22}(q_\parallel\omega)]^{1/2}. \tag{4.130}$$

In the absence of surface roughness the dispersion relation for surface plasmons at a vacuum–dielectric interface is

$$\epsilon(\omega)+1=0. \tag{4.131}$$

We denote the solution of this equation by $\omega=\omega_0$, where ω_0 is a frequency that is independent of $q_\parallel$. In general, ω_0 is complex because the dielectric constant $\epsilon(\omega)$ is complex in general. If we write the solution of eq. (4.130) as

$$\omega(q_\parallel)=\omega_0+\Delta\omega(q_\parallel), \tag{4.132}$$

then it follows from eq. (4.130) that the surface roughness-induced shift in the frequency of a surface plasmon to lowest order in δ is given by

$$\Delta\omega(q_\parallel)=\pm\frac{\delta q_\parallel^{1/2}}{\epsilon'(\omega_0)}\left[\epsilon(\omega_0)a_{11}(q_\parallel\omega_0)+a_{21}(q_\parallel\omega_0)-a_{12}(q_\parallel\omega_0)+a_{22}(q_\parallel\omega_0)\right]^{1/2}. \tag{4.133}$$

From this result and eq. (4.132) we see that every branch of the surface plasmon dispersion curve for a flat vacuum–dielectric interface is split into two branches by surface roughness.

To obtain a quantitative result for the dependence of $\omega(q_\parallel)$ on $q_\parallel$ we must assume some expression for the surface structure factor $g(k_\parallel)$. We choose the Gaussian form given by eq. (2.41). With this choice the matrix elements $a_{\mu\nu}(q_\parallel\omega)$ defined by eqs. (4.127) can be evaluated in closed form, with the result that

$$a_{11}(q_\parallel\omega)=(4/\sigma\xi^3)[(\epsilon(\omega)-1)\mathcal{I}_0(\xi)+2\mathcal{I}_1(\xi)], \tag{4.134a}$$

$$a_{12}(q_\parallel\omega)=(4/\sigma\xi^3)[(\epsilon(\omega)-1)\mathcal{I}_0(\xi)-2\epsilon(\omega)\mathcal{I}_1(\xi)], \tag{4.134b}$$

$$a_{21}(q_\parallel\omega)=(4/\sigma\xi^3)[-\tfrac{1}{2}(\epsilon(\omega)-1)\mathcal{I}_0(\xi)+2\epsilon(\omega)\mathcal{I}_1(\xi)-\tfrac{1}{2}(\epsilon(\omega)-1)\mathcal{I}_2(\xi)], \tag{4.134c}$$

$$a_{22}(q_\parallel\omega)=(4/\sigma\xi^3)[\tfrac{1}{2}\epsilon(\omega)(\epsilon(\omega)-1)\mathcal{I}_0(\xi)+2\epsilon(\omega)\mathcal{I}_1(\xi)+\tfrac{1}{2}\epsilon(\omega)(\epsilon(\omega)-1)\mathcal{I}_2(\xi)], \tag{4.134d}$$

with

$$\xi=q_\parallel\sigma, \tag{4.135}$$

and

$$\begin{aligned}\mathcal{I}_n(\xi)&=\mathrm{e}^{-\xi^2/4}\int_0^\infty \mathrm{d}u\,u^2\,\mathrm{e}^{-u^2/\xi^2}I_n(u)\\&=\frac{\pi^{1/2}}{16}\mathrm{e}^{-\xi^2/8}\xi^5\left\{I_{n/2-1}\left(\frac{\xi^2}{8}\right)+\left[1+\frac{4}{\xi^2}-\frac{4n}{\xi^2}\right]I_{n/2}\left(\frac{\xi^2}{8}\right)\right\},\end{aligned} \tag{4.136}$$

where $I_\nu(x)$ is a modified Bessel function of the first kind.

When we combine eqs. (4.132), (4.133), and (4.134) we find that in the presence of surface roughness the frequency of the surface plasmon is given by

$$\omega_{\pm}(\xi) = \omega_0 \pm \frac{2\sqrt{2}}{\epsilon'(\omega_0)}\frac{\delta}{\sigma} f(\xi), \tag{4.137}$$

where

$$f(\xi) = \frac{1}{\xi}[3\mathscr{I}_0(\xi) - 4\mathscr{I}_1(\xi) + \mathscr{I}_2(\xi)]^{1/2} \tag{4.138}$$

$$= \left(\frac{3\sqrt{\pi}}{4}\right)^{1/2} \xi^{1/2}\left[1 - \frac{2}{3\sqrt{\pi}}\xi + \cdots\right] \qquad \xi \ll 1 \tag{4.139a}$$

$$= \left(\frac{3}{2}\right)^{1/2}\left[\frac{1}{\xi} + \frac{1}{\xi^3} + \cdots\right] \qquad \xi \gg 1. \tag{4.139b}$$

A plot of the function $f(\xi)$ is given in fig. 12.

The fact that the splitting of the surface plasmon dispersion curve vanishes both as $\xi \to 0$ and $\xi \to \infty$ is not difficult to understand. In the former limit the wavelength of the surface plasmon is much longer than the transverse correlation length, and the surface plasmon does not "see" the roughness over which it propagates. In the latter limit the wavelength of the surface plasmon is much shorter than the transverse correlation length, and the surface plasmon follows the roughness adiabatically. In either case

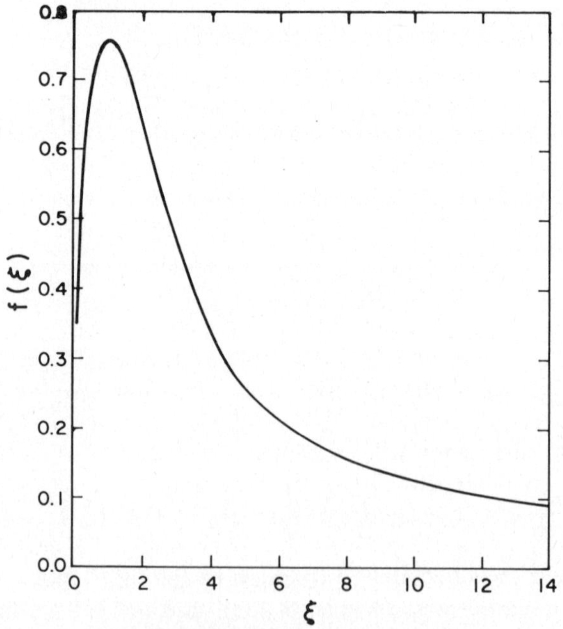

Fig. 12. The function $f(\xi)$ entering the expression for the frequency of a surface plasmon on a randomly rough surface, eq. (4.137).

the surface plasmon sees a flat surface, globally in the former case, and locally in the latter, and its frequency is that for a flat surface. For intermediate values of ξ, e.g. for $\xi \approx 1$ when the wavelength of the surface plasmon is comparable to the transverse correlation length, a kind of resonant interaction of the surface plasmon with the roughness occurs, and the splitting of its dispersion curve goes through a maximum.

We also see that the damping of surface plasmons in the presence of surface roughness is due to the imaginary part of $\epsilon(\omega)$, that in turn contributes an imaginary part to ω_0 and to the roughness-induced frequency shift given by the second term on the right hand side of eq. (4.133). This is in contrast with the situation when the effects of retardation are taken into account. As we have seen in the preceding two subsections, a surface polariton can be damped even in the limit as $\text{Im}\,\epsilon(\omega) \to 0$ by two mechanisms: the surface polariton can radiate energy into the vacuum, and it may be scattered by the surface roughness into other surface polariton states. Such dynamical, or radiative, processes are not possible in the electrostatic limit. The attenuation of the surface plasmon in this limit therefore arises only from the dissipative processes present in the bulk of the dielectric medium that give rise to the imaginary part of the dielectric constant $\epsilon(\omega)$.

The physical reason for the splitting of the surface plasmon dispersion curve has been indicated by Kretschmann et al. (1979). A rough surface can be regarded as a superposition of diffraction gratings, each with its own spacing, amplitude, and orientation in the x_1x_2-plane which vary continuously from one grating to the next. As we have seen in sect. 3.1, each grating can split the surface plasmon dispersion relation. This splitting occurs at a given frequency on the dispersion curve if two degenerate surface plasmons with different wave vectors can couple through the wave vector of the grating. Since the dispersion relation for a surface plasmon given by eq. (4.131) is flat, i.e., it depends neither on the magnitude nor direction of the wave vector $\boldsymbol{k}_\parallel$, all wave vectors entering the Fourier decomposition of $\zeta(\boldsymbol{x}_\parallel)$ couple two degenerate surface plasmons with different wave vectors, and split the dispersion curve thereby.

Although this explanation is correct, we feel that it is useful to present an alternative way of looking at the origin of the splitting that can also be applied to the discussion of other physical phenomena in which surface roughness plays a role (Rahman and Maradudin, 1980a).

It was shown by Rahman and Maradudin (1980a) that the effects of surface roughness on the electrostatic image potential can be reproduced by a simple model in which the surface roughness is replaced by a thin layer of dielectric material straddling the plane $x_3 = 0$, whose dielectric constant $\epsilon_s(\omega)$ is intermediate between that of the vacuum above it and that of the dielectric medium below it. The thickness of the layer is L, and it occupies the region $-\alpha L < x_3 < (1-\alpha)L$ with $0 < \alpha < 1$. The values of the parameters ϵ_s, L, α obtained by Rahman and Maradudin (1980a) are

$$\epsilon_s(\omega) = \tfrac{1}{2}(\epsilon(\omega)+1), \tag{4.140a}$$

$$L = 3\pi^{1/2}\delta(\delta/\sigma), \tag{4.140b}$$

$$\alpha = \tfrac{2}{3}. \tag{4.140c}$$

The dispersion relation for plasmons in this three layer structure is readily found to be

$$(\epsilon(\omega)+\epsilon_s(\omega))(\epsilon_s(\omega)+1)+(\epsilon(\omega)-\epsilon_s(\omega))(\epsilon_s(\omega)-1)\,\mathrm{e}^{-2k_\parallel L} = 0. \tag{4.141}$$

Because the film of dielectric constant $\epsilon_s(\omega)$ is thin, we expand this dispersion relation to first order in L. When the values of the parameters given by eq. (4.140) are substituted into the resulting dispersion relation it takes the form

$$(\epsilon(\omega)+1)^2 - \tfrac{3}{2}\pi^{1/2}(\epsilon(\omega)-1)^2(\delta^2/\sigma^2)\xi = 0, \tag{4.142}$$

the solution of which is

$$\omega(q_\parallel) = \omega_0 \pm \frac{\sqrt{6}\pi^{1/4}}{\epsilon'(\omega_0)}\frac{\delta}{\sigma}\xi^{1/2}. \tag{4.143}$$

This is exactly the result given by eq. (4.137) when the small ξ expression for $f(\xi)$ given by eq. (4.139a) is substituted into it.

The physical reason for the splitting in the present case is the presence of two interfaces in the three-layer system: the interface between vacuum and the thin layer of dielectric constant $\epsilon_s(\omega)$ at $x_3 = \frac{1}{3}L$: and the interface between the thin layer of dielectric constant $\epsilon_s(\omega)$ and the substrate of dielectric constant $\epsilon(\omega)$ at $x = -\frac{2}{3}L$. A surface plasmon can be associated with each interface because at each interface one of the two dielectric media in contact across it is surface active (i.e. has a negative dielectric constant), while the other is surface inactive (i.e. has a positive dielectric constant. This is the condition for the occurrence of a surface plasmon.

Recent experimental work (Palmer and Schnatterly, 1971; Kötz et al., 1979) on the propagation of surface plasmons over a rough planar surface of a metal shows the splitting of the surface plasmon dispersion curve discussed in this subsection. In fig. 13 is shown a plot of essentially the imaginary part of the dielectric constant of sodium measured by Palmer and Schnatterly (1971). The structure near 4.0 eV is due to surface plasmons, and the uppermost curve has the structure of a double peak, the separation of which is 0.2 eV. Similar results were obtained by these authors for potassium. The observed structure has subsequently been attributed to roughness-induced splitting of the surface plasmon dispersion curve (Kretschmann et al. 1979). A separation of the peaks of this mag-

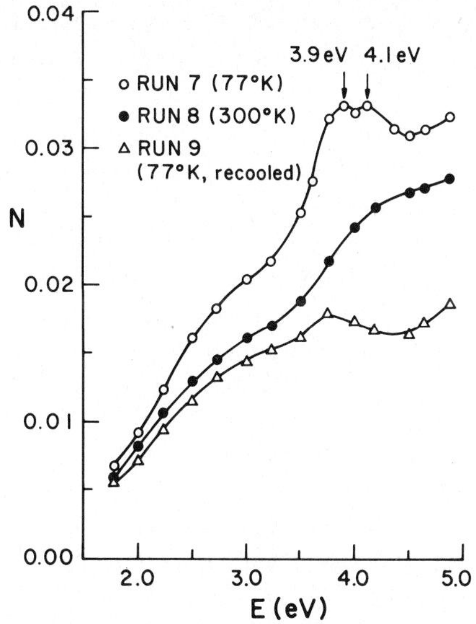

Fig. 13. *N* for sodium. *N* is proportional to the imaginary part of the dielectric constant. The structure near 4.0 eV is due to surface plasmons, (Palmer, R. E. and S. E. Schnatterly, 1971, Phys. Rev. **B4**, 2329).

nitude can be explained by the existing theories (Kretschmann et al. 1979, Rahman and Maradudin 1980b) with a reasonable value of δ/σ, viz. $\delta/\sigma \cong 0.3$.

We have presented here a simple theory of the effects of surface roughness on the dispersion relation for surface plasmons that demonstrates the roughness induced splitting of that dispersion relation observed in recent experiments, and yields quite explicit expressions for the frequencies of each of its two branches. Agreement between theory and experiment can be achieved for reasonable values of the parameters characterizing the surface roughness in a Gaussian model of the roughness. These results suggest that the effects of surface roughness on the surface plasmon dispersion relation can be sufficiently large that they need to be accounted for in precision studies of optical, or more generally electromagnetic, properties of solids in the vicinity of the surface plasmon resonance.

5. Directions for Future Work

In this chapter we have presented a summary of the existing theoretical research into the interaction of surface plasmons and polaritons with surface roughness. It seems appropriate to conclude with a brief discussion of the directions for future work indicated by what has been, or has not been, accomplished to date.

One of the highest priorities for work in this area in the immediate future is the carrying out of additional numerical calculations of the dispersion relation for surface polaritons on a deterministic, periodic surface that go beyond the small roughness limit. We have described in sect. 3 possible ways in which such calculations could be carried out on the basis of the equations obtained through the use of Green's theorem and have presented some results of very recent calculations based on several of the dispersion relations derived there. These methods are essentially those that have been used in the related, but simpler, problem of the scattering of a scalar plane wave from a periodically corrugated hard wall. The results of Laks et al. (1980) and of Glass and Maradudin (1980) are encouraging in showing that the solution of the dispersion relations obtained from the extinction theorem through expansions in Fourier series can yield dispersion curves for surface polaritons and plasmons on rather rough gratings with good accuracy. The limits of applicability of this approach, however, are not known at the present time, particularly for the case of gratings characterized by surface profile functions that are not analytic functions of x_1. These limitations may be fundamental, in the sense that there may be something akin to a finite radius of convergence for the expansion of the fields in Fourier series, they may be due to the capacity of the computers that are employed in the calculations, or both types of limitations may be present. The numerical solution of the dispersion relations presented by eqs. (3.74)–(3.75) and (3.117)–(3.120) should be carried out. The aim of such studies would be to determine the form of the dispersion relation and the method for solving it that yields accurate dispersion curves for the widest range of surface profile functions, and for the largest values of the strength of the corrugation. It may well be the case that one can tailor the approach to the particular problem being studied. We have already seen that eqs. (3.104)–(3.105) are easier to solve than eqs. (3.124) for the sinusoidal profile, eq. (3.21), for the range of values of ζ_0/a where both sets of equations are valid. For the sawtooth profile, eq. (3.82), however, there is really no choice: of the two computational methods just mentioned, only the one based on eqs. (3.124) can be used.

Perhaps new, more powerful methods have to be developed for dealing with the equations that are obtained in seeking the surface polariton dispersion relation. In particular, in view of their comparatively simple form it would seem very worthwhile to have methods for solving the equations resulting from the use of the extinction theorem, eqs. (3.79a) and (3.79b), in addition to the RG method (García and Cabrera 1978), or expansion in Fourier series (Laks et al. 1980) that either converge more rapidly, are suitable for use with strongly corrugated surfaces, or both. For example, in the related problem of scattering of a scalar plane wave from a corrugated hard wall the inhomogeneous integral equation that is obtained through the use of the extinction theorem can be solved very efficiently by an expansion of the unknown function in increasing orders of $\zeta(x_1)$ in which the successive terms are obtained in a recursive fashion (Lopez et al. 1978). This method is attractive in that it does not require the inversion or diagonalization of any matrix. In this way the scattering problem has been solved for surface profiles for which the Rayleigh method breaks down completely. A variant of this method suited to dealing with the pair of coupled homogeneous integral equations (3.79a) and (3.79b) would be welcome indeed.

It would also be useful to relax the assumption that the surface polariton or plasmon is incident normally on the grooves of the grating in obtaining the dispersion relation for these excitations. In addition to the generality, such an extension of the theory would give rise to, it would also be of interest for the following reason. It has been shown recently (Dobrzynski and Maradudin 1972) that surface plasmons can exist in the vicinity of the apex of a dielectric wedge surrounded by vacuum, whose associated electromagnetic fields are wavelike along the apex of the wedge, but decay essentially exponentially with increasing distance from the apex of the wedge, and from its surface, both into the wedge and into the vacuum surrounding it. For an infinitely sharp wedge the frequencies of these electrostatic edge modes form a continuous spectrum. However, it has been shown (Davis 1976, Eguiluz and Maradudin 1976) that rounding off the tip of the wedge makes discrete the spectrum of frequencies of the resulting edge modes. These frequencies fall in the neighborhood of the frequency of the surface plasmon on a flat surface, and it is possible that a surface polariton propagating on a grating with a wave vector $\boldsymbol{k}_{\|}$ that has a component parallel to the grooves can display these edge modes in its frequency spectrum. It has been suggested recently (Burstein and Chen 1981) that the existence of a closely spaced set of edge modes on a grating may play a role in explaining the giant enhancement of the Raman signal from molecules adsorbed on a metal surface (Jeanmaire and Van Duyne 1977).

There is a need to obtain the dispersion relation for surface polaritons on a randomly rough surface to higher than second order in $\zeta(\boldsymbol{x}_{\|})$. A recent experimental study of both the scattering from and reflectance of roughened silver foils (Sari et al. 1980) shows that the position of the surface polariton resonance in the reflectance measurements is given well by a theory that includes the dispersion relation for surface polaritons to second order in $\zeta(\boldsymbol{x}_{\|})$, although some small residual discrepancies remain. The values of the parameters characterizing the surface roughness in this work were $\delta = 40$–50 Å and $\sigma = 800$–1200 Å, so that the use of a dispersion relation appropriate to the small roughness limit would seem to be valid in this case. However, the experimental results of Sari et al. (1980) on much rougher surfaces yielded discrepancies between theory and experiment that these authors take to indicate a limitation in the lowest order scattering theory results, that could imply a corresponding limitation in the lowest order determination of the surface polariton dispersion relation. In such calculations the use of the Green's function approach described in sect. 4.1, extended to provide a kernel of higher than first order in $\zeta(\boldsymbol{x}_{\|})$ in the integral equation (4.20), would eliminate much of the tedious algebra encountered in other approaches to this problem.

All of the discussions of the effects of surface roughness on surface plasmons and polaritons to date have presumed that the dielectric substrate is characterized by an isotropic dielectric tensor. While this assumption applies to many cases of practical interest, e.g. to all crystals of cubic symmetry, there are many crystals of lower symmetry that are of interest and that cannot be described by an isotropic dielectric tensor. In the case of a deterministic, periodic surface profile the Rayleigh method described in sect. 2 can be used to obtain the surface polariton dispersion relation with some increase in algebraic complexity over that required for an optically isotropic medium: two, or three, decay constants for the electromagnetic field in the medium are required, depending on its symmetry. It would be worthwhile to try to obtain such dispersion relations with the use of Green's theorem, but in the absence of such calculations for isotropic media, they seem much more remote for anisotropic media. In the case of randomly rough surfaces the use of the Green's function technique seems to be the simplest for obtaining the surface polariton dispersion relation in the small roughness limit. At the present time the electromagnetic Green's functions required for such calculations have been obtained for a gyrotropic dielectric medium, such as is represented by a III–V compound semiconductor of cubic symmetry, possessing free carriers, in a constant magnetic field applied along the outward normal to a (001) surface (Maradudin and Wallis 1980), in addition to those available for an isotropic medium (Maradudin and Mills 1975).

All of the discussion in this chapter has been based on the assumption of a local relation between the electric displacement vector $\boldsymbol{D}(x, t)$ and the electric field $\boldsymbol{E}(x, t)$. It has been found, however, that the interpretation of certain experimental data on the optical properties of bounded media requires the assumption of a nonlocal, integral relation between $\boldsymbol{D}(\boldsymbol{x}, t)$ and $\boldsymbol{E}(x, t)$,

$$D_\mu(\boldsymbol{x}, t) = \sum_\nu \int \mathrm{d}^3x' \int \mathrm{d}t' \epsilon_{\mu\nu}(\boldsymbol{x}, \boldsymbol{x}'; t - t') E_\nu(\boldsymbol{x}', t'), \tag{5.1}$$

where $\epsilon_{\mu\nu}(\boldsymbol{x}, \boldsymbol{x}'; t - t')$ is the dielectric tensor of the bounded medium (see, for example, Fischer and Lagois, 1979). The existence of a relation between $\boldsymbol{D}$ and $\boldsymbol{E}$ that is nonlocal in the spatial coordinates is usually referred to as a consequence of *spatial dispersion* in the medium. It is to be expected that the surfaces of crystals in which spatial dispersion manifests itself in the optical properties are rough to some extent. An unsolved problem is that of obtaining the dispersion relation for surface polaritons propagating along the surface of a spatially dispersive medium. It is known that the dispersion curve for surface polaritons on the flat surface of a spatially dispersive medium differs from that for a nonspatially dispersive medium (Maradudin and Mills, 1973, Horing, 1977). In particular, surface polaritons are damped in the presence of spatial dispersion, due to Landau damping, when in the absence of spatial dispersion they are not, e.g., in the surface plasmon regime of short wavelengths. The interplay of surface roughness and spatial dispersion can be expected to lead to some interesting results, since each alone does.

Finally, we note that although there have been studies of surface plasmons and polaritons associated with nonplanar surfaces, e.g., with cylindrical or spherical surfaces (Englman and Ruppin 1968a, b; Ruppin and Englman 1968), the surfaces considered have been prefectly smooth. In connection with studies of the optical properties of small metal particles embedded in dielectric matrices (Genzel et al. 1975), studies of the effects of surface roughness on surface plasmons and polaritons in rough spherical particles would be useful. In the only work of this kind of which we are aware, (Lee Jr. et al. 1980) it has been shown that the frequencies of plasmons in a dielectric sphere separated from vacuum by a rough surface are split into two levels each, reminiscent of the roughness-induced splitting of the surface plasmon dispersion curve associated with a flat surface discussed in sect. 3.2 and 4.3. Such an effect might well be observable in the absorption spectra of such systems.

The preceding list of topics that, in our view, deserve study in the future is not intended to be comprehensive. It should serve to indicate, however, that the subject is far from exhausted, and that much good work remains to be done.

Acknowledgements

I am grateful to Professor V. Celli and Dr. A. Marvin for very helpful correspondence concerning the equivalence of the results of Toigo et al. (1977) and of Maradudin and Zierau (1976) for the dispersion relation of surface polaritons on a randomly rough surface. I am also grateful to Professor D. L. Mills for helpful discussions concerning the contents of sect. 3.1.2, 3.3 and 4.2, and to Dr. N. Glass and Dr. B. Laks for discussions about numerical methods for obtaining dispersion curves for surface plasmons and polaritons on a grating. Finally, it is a pleasure to thank Professor H. Raether for discussion and correspondence concerning virtually every topic in this chapter.

References

Abramowitz, M. and I.A. Stegun, 1968, Handbook of Mathematical Functions (Dover, New York) p. 358.

Agarwal, G.S., 1976, Phys. Rev. **B14**, 846.

Bell, R.J., R.W. Alexander, C.A. Ward and I.L. Tyler, 1975, Surf. Sci. **48**, 253.

Bennett, B.I., A.A. Maradudin and L.R. Swanson, 1971, Ann. Phys. (N.Y.) **71**, 357.

Bennett, J.M., 1976, Appl. Opt. **15**, 2705.

Bodesheim, J. and A. Otto, 1974, Surf. Sci. **45**, 441.

Bryksin, V.V., D.N. Mirlin and Yu.A. Firsov, 1974, Usp. Fiz. Nauk. **113**, 29 [English translation: Soviet Physics-Uspekhi **17**, 305 (1974)].

Burstein, E. and C.Y. Chen, 1981, in Proc. VIIth Intern. Conf. on Raman Spectroscopy, ed. W.F. Murphy (North-Holland, New York) p. 346.

Burstein, E., A. Hartstein, J. Schoenwald, A.A. Maradudin, D.L. Mills and R.F. Wallis, 1974, Surface Polaritons – Electromagnetic Waves at Interfaces, in: *Polaritons*, eds. E. Burstein and F. de Martini (Pergamon Press, New York) p. 89.

Cohn, 1900, Das Electromagnetische Feld (Leipzig)-cited in Stratton J.A., 1941, Electromagnetic Theory (McGraw-Hill, New York) p. 584.

Daudé, A., A. Savary and S. Robin, 1972, J. Opt. Soc. Am. **62**, 1.

Davis, L.C., 1976, Phys. Rev. **B14**, 5523.

Dobberstein, P., 1970, Phys. Lett. **31A**, 307.

Dobrzynski, L. and A.A. Maradudin, 1972, Phys. Rev. **B6**, 3810.

Eguiluz, A. and A.A. Maradudin, 1976, Phys. Rev. **B14**, 5526.

Elson, J.M. and R.H. Ritchie, 1971, Phys. Rev. **B4**, 4129.

Englman, R. and R. Ruppin, 1968a, J. Phys. **C1**, 614.

Englman, R. and R. Ruppin, 1968b, J. Phys. **C1**, 1515.

Fischer, B. and J. Lagois, 1979, Surface Exciton Polaritons, in: Topics in Current Physics, vol. **14**: Excitons, ed. K. Cho (Springer-Verlag, Berlin) ch. 4.

García, N. and N. Cabrera, 1978, Phys. Rev. **B18**, 576.

Genzel, L., T.P. Martin and U. Kreibig, 1975, Z. Phys. **B21**, 339.

Glass, N.E. and A.A. Maradudin, 1980 (to be published).

Hall, D.G. and A.J. Braundmeier Jr., 1978, Phys. Rev. **B17**, 3808.

Hill, N.R., 1978, Ph.D. dissertation, University of Virginia (unpublished).

Hill, N.R. and V. Celli, 1978, Phys. Rev. **B17**, 2478.

Hillebrecht, F.U., 1980, J. Phys. **D13**, 1625.

Horing, N.J.M., 1977, in Proc. 7th Intern. Vacuum Congr. and 3rd. Intern. Conf. on Solid Surfaces, eds. R. Dobrozemsky, F. Rüdenauer, F.P. Viebock and A. Breth (Berger, Vienna) p. 419.

Horstmann, C., 1977, Opt. Commun. **21**, 173.

Jackson, J.D., 1962, Classical Electrodynamics (Wiley, New York) pp. 14–15.

Jansen, F. and R.W. Hoffman, 1979, Surf. Sci. **83**, 313.

Jeanmaire, D.L. and R.P. Van Duyne, 1977, J. Electroanal. Chem. **84**, 1.

Kaspar, W. and U. Kreibig, 1977, Surf. Sci. **69**, 619.
Kliewer, K.L. and R. Fuchs, 1974, Advan. Chem. Phys. **27**, 355.
Kötz, R., H.J. Lewerenz and E. Kretschmann, 1979, Phys. Lett. **70A**, 452.
Kretschmann, E., 1972a, Int. Conf. on "Thin Films", Venice; see also, Opt. Commun. **10** (1974) 356.
Kretschmann, E., 1972b, Opt. Commun. **5**, 331.
Kretschmann, E., 1972c, Opt. Commun. **6**, 185.
Kretschmann, E., T.L. Ferrell and J.C. Ashley, 1979, Phys. Rev. Lett. **42**, 1312.
Kretschmann, E. and H. Raether, 1968, Z. Naturforsch. **23a**, 2135.
Kröger, E. and E. Kretschmann, 1970, Z. Phys. **237**, 1.
Kröger, E. and E. Kretschmann, 1976, Phys. Status Solidi **76**, 515.
Landau, L.D. and E.M. Lifshitz, 1960, Electrodynamics of Continuous Media (Pergamon, New York) p. 243.
Landau, L.D. and E.M. Lifshitz, 1970, Theory of Elasticity (Pergamon, New York) p. 109.
Laks, B., D.L. Mills and A.A. Maradudin, 1980 (to be published).
Lee, Jr., R.G., A.A. Maradudin and B. Mühlschlegel, 1980 (to be published).
Lopez, C., F.J. Yndurian and N. Garcia, 1978, Phys. Rev. **B18**, 970.
Maradudin, A.A. and D.L. Mills, 1973, Phys. Rev. **B7**, 2787.
Maradudin, A.A. and D.L. Mills, 1975, Phys. Rev. **B11**, 1392.
Maradudin, A.A. and R.F. Wallis, 1980, to appear in the Int. J. of Raman Spectrosc.
Maradudin, A.A., R.F. Wallis and L. Dobrzynski, 1980, Handbook of Surfaces and Interfaces, ed. L. Dobrzynski (Garland STPM Press, New York), vol. 3, Surface Phonons and Polaritons.
Maradudin, A.A. and W. Zierau, 1976, Phys. Rev. **B14**, 484.
Maystre, D., 1972, Opt. Commun. **6**, 50.
Maystre, D., 1973, Opt. Commun. **8**, 216.
Maystre, D., 1974, Nouv. Rev. Opt. **5**, 65.
Maystre, D. and P. Vincent, 1972, Opt. Commun. **5**, 327.
Meecham, W.C., 1956, J. Ration. Mech. Anal. **5**, 323.
Messiah, A., 1965, Mechanique Quantique (Dunod, Paris) Vol. 1, p. 83.
Millar, R.F., 1969, Proc. Cambridge. Philos. Soc. **65**, 773.
Millar R.F., 1971a, Proc. Cambridge. Philos. Soc. **69**, 175.
Millar, R.F., 1971b, Proc. Cambridge. Philos. Soc. **69**, 217.
Millar, R.F., 1973, Radio Sci. **8**, 785.
Mills, D.L., 1975, Phys. Rev. **B12**, 4036.
Mills, D.L., 1977, Phys. Rev. **B15**, 3097.
Mills, D.L., 1980, private communication.
Mills, D.L. and E. Burstein, 1974, Rep. Prog. Phys. **37**, 817.
Mills, D.L. and A.A. Maradudin, 1975, Phys. Rev. **B12**, 2943.
Nkoma, J., R. Loudon and D.R. Tilley, 1974, J. Phys. **C7**, 3547.
Otto, A., 1974, Adv. in Solid State Phys., Festkorperprobleme XIV, 1.
Otto, A., 1976, Spectroscopy of Surface Polaritons by Attenuated Total Reflection, in: Optical Properties of Solids New Developments, ed. B.O. Seraphin (North-Holland, Amsterdam) p. 677.
Palmer, R.E. and S.E. Schnatterly, 1971, Phys. Rev. **B4**, 2329.
Petit, R. and M. Cadilhac, 1966, C.R. Acad. Sci. **B262**, 468.
Pockrand, I., 1975, Opt. Commun. **13**, 311.
Pockrand, I. and H. Raether, 1976a, Opt. Commun. **17**, 353.
Pockrand, I. and H. Raether, 1976b, Opt. Commun. **18**, 395.
Raether, H., 1977, Surface Plasma Oscillations and Their Applications, in: Physics of Thin Films, vol. 9 (Academic Press, New York) p. 145.
Rahman, T.S. and A.A. Maradudin, 1980a, Phys. Rev. **B21**, 504.

Rahman, T.S. and A.A. Maradudin, 1980b, Phys. Rev. **B21**, 2137.
Rasigni, M., G. Rasigni, C.T. Hua and J. Pous, 1977, Opt. Lett. **1**, 126.
Rayleigh, Lord, 1887, Proc. Lond. Math. Soc. **17**, 4.
Rayleigh, Lord, 1907, Phil. Mag. **14**, 70.
Rayleigh, Lord, 1945, Theory of Sound, 2nd ed. (Dover, New York) vol. II, p. 89.
Ritchie, R.H., 1957, Phys. Rev. **106**, 874.
Ruppin, R. and R. Englman, 1968, J. Phys. **C1**, 630.
Sari, S.O., D.K. Cohen and K.D. Scherkoske, 1980, Phys. Rev. **B21**, 2162.
Sommerfeld, A., 1909, Ann. Physik **28**, 665.
Sommerfeld, A., 1911, Jahrbuch d. drahtl. Telegraphie **4**, 157.
Sommerfeld, A., 1926, Ann. Physik **81**, 1135.
Talaat, H., W.P. Chen, E. Burstein and J. Schoenwald, 1976, IEEE Transactions on Sonics and Ultrasonics **SU-23**, 188.
Toigo, F., A. Marvin, V. Celli and N.R. Hill, 1977, Phys. Rev. **B15**, 5618.
Uller, 1903, Rostock Dissertation–cited in: Stratton, J.A., 1941, Electromagnetic Theory (McGraw-Hill, New York) p. 584.
Ushioda, S., 1980, Light Scattering Spectroscopy of Surface Electro-Magnetic Waves in Solids, in: Progress in Optics, ed. E. Wolf (North-Holland, Amsterdam) (to appear).
Ushioda, S., A. Aziza, J.B. Valdez and G. Mattei, 1979, Phys. Rev. **B19**, 4012.
Ward, C.A., R.J. Bell, R.W. Alexander, G.S. Kovener and I. Tyler, 1974, Appl. Opt. **13**, 2378.
Weyl, H., 1919, Ann. Physik **60**, 481.
Wolf, E., 1973, A Generalized Extinction Theorem and its Role in Scattering Theory, in: Coherence and Quantum Optics, eds. L. Mandel and E. Wolf (Plenum, New York) p. 339.
Zenneck, J., 1909, Ann. Physik, **28**, 665.

CHAPTER 11

Scattering of Surface Polaritons by Order Parameter Fluctuations near Phase Transition Points

V.M. AGRANOVICH, V.E. KRAVTSOV and T.A. LESKOVA

Institute of Spectroscopy
USSR Academy of Sciences
Troitsk, Moscow Oblast 142092
USSR

Surface Polaritons
Edited by
V.M. Agranovich and D.L. Mills

Contents

1. Introduction 513
2. General expression for the intensity of scattered radiation; certain special cases . 514
3. Surface polariton scattering near a superconducting transition point 520
4. Surface polariton scattering near phase transition points in thin films; reasonance . . amplification of scattering 525
5. Conclusion 530
References 531

1. Introduction

At the present time light scattering by order-parameter fluctuations forms the basis for one of the most powerful experimental methods of investigating critical phenomena in condensed media (Bendow et al. 1976, Birman et al. 1979). Such scattering makes it possible to obtain abundant information on the dynamics of crystal lattices in the region of structural phase transitions and, in particular, to determine in many cases the intensity of fluctuations and their dependence on temperature. All the experiments conducted up to the present time, however, dealt with the bulk scattering of radiation. Such a method of investigation has exceptionally limited applicability when we are concerned with phase transitions in nontransparent media or when such transitions occur on the surface. Under such conditions it may prove promising in investigating scattering by the order-parameter fluctuations to use surface waves (surface polaritons, SPs) propagating along the surfaces or interfaces of media (Agranovich 1976, Agranovich and Leskova 1977), rather than the bulk light waves previously employed.

When a surface polariton is scattered, either a surface polariton may be formed in the final state or a bulk polariton. Since bulk radiation is precisely what can be registered experimentally in employing ordinary detectors, the aim of the theory developed in the present chapter consists primarily in determining the intensity of the radiation. The scattering of surface polaritons may be due not only to fluctuations of the order parameter, but to roughness (irregularities) of the surface as well. This circumstance should be kept in mind in arranging experimental investigations of the scattering of surface polaritons by order-parameter fluctuations because it imposes special requirements in preparing the specimens.

As indicated by estimation (see below), the scattering cross section for surface polaritons by order-parameter fluctuations is comparatively small. Hence, to obtain sufficiently intensive scattered radiation fluxes, the fluxes of surface polaritons must be extremely intensive too. Intensive fluxes of surface polaritons can evidently be produced today only by making use of powerful pulsed lasers. In this connection, along with the task of preparing suitable specimens, we are also faced with the problem of the laser strength of the surfaces and of estimating the maximum pumping intensities that do not lead to the destruction of the surfaces or to appreciable changes in their properties.

Surface polaritons of the infrared range, propagating along the surfaces of metals, have the maximum path length (of the order of several centimeters) as has been found from experimental investigations conducted in recent years. Hence metallic surfaces are evidently the most suitable objects to use in investigating the scattering of surface polaritons by order-parameter fluctuations. Under these conditions, the use of surface polaritons makes it possible to investigate, not only phase transitions in the metallic substrate, but also those in sufficiently thin dielectric films deposited on the surface of the metal. Owing to the quasi-two-dimensionality of the system in the latter case, the order-parameter fluctuations may be especially great.

A few words are in order on the arrangement of the material in the present chapter. The next section presents a detailed derivation of the expression for the intensity of bulk radiation produced in scattering surface polaritons by arbitrary fluctuations of the permittivity (dielectric constant) and magnetic permeability of a medium that is not necessarily homogeneous throughout its thickness.

Employing a superconducting transition as an example, sect. 3 investigates temperature dependence of the intensity of surface polariton scattering when the SPs are propagating along the surface of a massive specimen undergoing a phase transition. This section also presents an estimate of the maximum permissible pumping power at which a destruction of the superconducting state does not yet occur.

The general theory developed in sect. 2 is made use of in sect. 4 to determine the intensity of scattered radiation under conditions when the phase transition occurs in a transition layer or in a thin film deposited on a massive substrate. An interesting case, specially discussed in this section, has the frequency of the surface polariton close to that of the optical vibrations in the film. Under such conditions, as shown by Agranovich (1979), resonance amplification of the surface polariton scattering by fluctuations of the order parameter should be observed.

In the Conclusion (sect. 5), the feasibility is considered of experimentally observing the effects discussed in this chapter.

2. *General Expression for the Intensity of Scattered Radiation; Certain Special Cases*

Assume that a medium with the refractive index $n_1^{(0)} = \sqrt{(\mu_1^{(0)}\epsilon_1^{(0)})}$ occupies the half-space $z > 0$, and a medium with the refractive index $n_2^{(0)} = \sqrt{(\mu_2^{(0)}\epsilon_2^{(0)})}$ occupies the half-space $z < 0$. In this case (see ch. 2), the field of the surface polariton propagating along the x-axis is of the form

$$E_x^{(0)} = \mathscr{E}_0\, e^{-\kappa(z)|z|}\, e^{ik_0x} \equiv E_x^{(0)}(z)\, e^{ik_0x},$$

$$E_z^{(0)} = i\, \mathscr{E}_0 \frac{k_0}{\kappa(z)} \operatorname{sign} z\, e^{-\kappa(z)|z|} e^{ik_0x} \equiv E_z^{(0)}(z)\, e^{ik_0x}, \tag{1}$$

where

$$\kappa^2(z) = k_0^2 - \frac{\omega^2}{c^2} n_0^2(z), \quad n_0 = \begin{cases} n_1^{(0)}, & z > 0, \\ n_2^{(0)}, & z < 0. \end{cases}$$

The dispersion law of the surface polariton in this case is found from the expression

$$k_0^2 = \frac{\omega^2}{c^2} \frac{(n_1^{(0)})^2 (n_2^{(0)})^2}{(n_1^{(0)})^2 + (n_2^{(0)})^2}. \tag{2}$$

In particular, if medium 1 is a vacuum and medium 2 is a metal with $\mu = 1$ and $|\epsilon| \gg 1$, then from eq. (2) we readily obtain ($\epsilon = \epsilon' + i\epsilon''$):

$$k_0 \approx \frac{\omega}{c}\left(1 - \frac{\epsilon'}{2|\epsilon|^2} + i\frac{\epsilon''}{2|\epsilon|^2}\right), \tag{3}$$

$$\kappa(z>0) \equiv \kappa \approx \frac{\omega}{c}\frac{1}{|\epsilon|}\left[\left(\frac{|\epsilon| - \epsilon'}{2}\right)^{1/2} + i\left(\frac{|\epsilon| + \epsilon'}{2}\right)^{1/2}\right], \tag{4}$$

$$\kappa(z<0) \equiv \kappa_0 \approx \frac{\omega}{c}\left[\frac{\epsilon''}{|\epsilon|}\left(\frac{|\epsilon| + \epsilon'}{2}\right)^{1/2} - \frac{\epsilon'}{|\epsilon|}\left(\frac{|\epsilon| - \epsilon'}{2}\right)^{1/2}\right]$$
$$- \frac{i\omega}{c}\left[\frac{\epsilon''}{|\epsilon|}\left(\frac{|\epsilon| - \epsilon'}{2}\right)^{1/2} + \frac{\epsilon'}{|\epsilon|}\left(\frac{|\epsilon| + \epsilon'}{2}\right)^{1/2}\right]. \tag{5}$$

Note that the length L_x of propagation of the surface polariton, when Joule heat losses are taken into account, is large (as is evident from eq. (3)) when $|\epsilon| \gg 1$, even in the case when $\epsilon'' \gg |\epsilon'|$. It is determined from the relation

$$L_x = \frac{2c}{\omega}\frac{|\epsilon|^2}{\epsilon''}. \tag{6}$$

For instance, if $\omega/c = 10^3\ \text{cm}^{-1}$ and we have the typical value $|\epsilon| \sim \epsilon'' \approx 10^3$ for most metals in this spectral range, the propagation length $L_x \approx 2$ cm.

Another characteristic feature of surface waves propagating along a metal–vacuum boundary is the large penetration of the field into the vacuum compared to the wavelength:

$$\kappa^{-1} \sim \frac{c}{\omega}\sqrt{|\epsilon|}$$

and the small depth of penetration into the metal:

$$\kappa^{-1} \sim \frac{c}{\omega} \frac{1}{\sqrt{|\epsilon|}}.$$

Consequently, the main part of the energy of the surface wave is concentrated in the vacuum. This, precisely, is what leads to the small difference between dispersion law (3) and $\omega = ck$.

The field determined by eq. (1) is a solution of the Maxwell equations for a perfectly smooth interface between the media, which has a refractive index independent of the coordinates.

If there is roughness on the interface or the refractive index of the media is a function of the coordinates, as is the case when fluctuations are taken into account, the surface polariton is scattered.

Let $\epsilon_{1,2} = \epsilon_{1,2}^{(0)} + \delta\epsilon_{1,2}(r)$ and $\mu_{1,2} = \mu_{1,2}^{(9)} + \delta\mu_{1,2}(r)$. Then from the Maxwell equations we readily obtain

$$\text{curl curl } \boldsymbol{E} - \frac{\omega^2}{c^2} \mu\epsilon \boldsymbol{E} = \frac{\nabla\mu}{\mu} \cdot \text{curl } \boldsymbol{E}, \tag{7}$$

where $\mu\epsilon = n_0^2 + \delta(\mu\epsilon)$.

This equation can be rewritten in integral form, putting $\boldsymbol{E} = \boldsymbol{E}^{(0)} + \boldsymbol{E}^{(s)}$, where $\boldsymbol{E}^{(s)}$ is the field of the scattered wave. Thus

$$E_i(r) = E_i^{(0)}(r) + \int d^3r' D_{ij}(r, r') \left[\frac{\omega^2}{c^2} \delta(\mu\epsilon) \boldsymbol{E} + \frac{\nabla\mu}{\mu} \text{curl } \boldsymbol{E} \right]_j (r'). \tag{8}$$

The Green function $D_{ij}(r, r')$ satisfies the equation

$$\text{curl curl } D - \frac{\omega^2}{c^2} n_0^2 D = \delta(r - r')$$

and the Maxwell boundary conditions at the interface, whereas as $|z| \to \infty$ it contains only outgoing waves: $D(z \to \pm\infty, z') \sim \exp(\pm ikz)$, or decreases exponentially.

In the first order of perturbation theory with respect to $\delta(\mu\epsilon)$, we obtain from eq. (8) the following expression for the field of the scattered wave:

$$E_i^{(s)}(r) = \int d^3r' D_{ij}^{(0)}(r, r') \left[\frac{\omega^2}{c^2} \delta(\mu\epsilon) \boldsymbol{E}^{(0)} + \frac{\nabla\mu}{\mu} \text{curl } \boldsymbol{E}^{(0)} \right]_j (r'). \tag{9}$$

In contrast to $D_{ij}(r, r')$, the Green function $D_{ij}^{(0)}(r, r')$ satisfies the Maxwell boundary conditions at a plane interface $z = 0$.

We note that for the magnetic field of the scattered wave we can write an expression analogous to eq. (9) and obtained from it by the substitutions $E \to H$ and $\epsilon \to \mu$. In a case when $|\nabla\mu/\mu| \ll |\nabla\epsilon/\epsilon|$ it proves convenient to employ expression (9), whereas, in the case of the inverse relationship between these quantities, the expression for the magnetic field $H^{(s)}$ should be resorted to.

Let us now derive the expression for the energy flux of radiation scattered in a vacuum. The time-averaged value of the Poynting vector is of the form:

$$S_k = \frac{c^2}{16\pi\omega}(E_i^{(s)*}\hat{p}_k E_i^{(s)} - E_i^{(s)*}\hat{p}_i E_k^{(s)} + \text{h.c.}), \tag{10}$$

where $\hat{p}_k = -\mathrm{i}\nabla_k$.
Substituting eq. (9) into (10) we obtain after simple but cumbersome transformations,

$$\begin{aligned}\langle S_k\rangle = \frac{c^2}{8\pi\omega}\int\frac{\mathrm{d}^2k_\parallel}{(2\pi)^2}\int \mathrm{d}z'\int \mathrm{d}z'' R_{jj'ss'}(\boldsymbol{k}_\parallel - \boldsymbol{k}_0, z', z'')\\ \times E_s^{(0)}(z')E_{s'}^{(0)*}(z'')k_k d_{lj}(\boldsymbol{k}_\parallel, z, z')d_{lj'}^*(\boldsymbol{k}_\parallel, z, z''),\end{aligned} \tag{11}$$

$$\begin{aligned}R_{jj'ss'} = \left(\frac{\omega}{c}\right)^4 K\delta_{js}\delta_{j's'} + \left(\frac{\omega}{c}\right)^2[L_s\nabla_j\delta_{j's'} + L_{s'}^+\nabla_{j'}\delta_{js} - (L_{i'}^+\nabla_{i'} + L_i\nabla_i)\delta_{js}\delta_{j's'}]\\ + M_{ii'}\nabla_i\nabla_{i'}\delta_{js}\delta_{j's'} + M_{ss'}\nabla_j\nabla_{j'} - M_{is'}\nabla_i\nabla_{j'}\delta_{js} - M_{si'}\nabla_{i'}\nabla_j\delta_{j's'},\end{aligned}$$

where

$$\langle\delta(\mu\epsilon)\delta^*(\mu\epsilon)\rangle = K,$$

$$\left\langle\delta^*(\mu\epsilon)\frac{\nabla_i\mu}{\mu}\right\rangle = L_i,$$

$$\left\langle\delta(\mu\epsilon)\left(\frac{\nabla_i\mu}{\mu}\right)^*\right\rangle = L_i^+,$$

$$\left\langle\left(\frac{\nabla_i\mu}{\mu}\right)\left(\frac{\nabla_k\mu}{\mu}\right)^*\right\rangle = M_{ik}.$$

The angle brackets $\langle\ \ \rangle$ denote averaging over the assembly of values of $\delta\epsilon(r)$ and $\delta\mu(r)$.

Besides, the following notation has been introduced in eq. (11):

$$\nabla_{x'} = \nabla_x = \mathrm{i}k_0, \quad \nabla_{y'} = \nabla_y = 0, \quad \nabla_z = -\kappa(z')\,\mathrm{sign}\, z',$$

$$\nabla_{z'} = -\kappa(z'')\,\mathrm{sign}\, z'',$$

$$D_{ik}^{(0)}(r, r') = \int\frac{\mathrm{d}^2k_\parallel}{(2\pi)^2}d_{ik}(\boldsymbol{k}_\parallel, z, z')\exp[\mathrm{i}\boldsymbol{k}_\parallel(\boldsymbol{\rho} - \boldsymbol{\rho}')].$$

Expression (11) gives the value of the total energy flux of the scattered field. The scattering process with the formation of surface polaritons in the final state makes a contribution to eq. (11), which is related to integration with respect to $|k_\parallel| > \omega/c$. This part of the flux decreases exponentially at great distances from the surface into the vacuum when $z \gg \kappa^{-1}$. The energy flux of the radiation scattered in the vacuum that we are interested in is determined by integrating over small values of $|k_\|| < \omega/c$ and is independent of z at large values of z.

Next we shall consider the problem of surface polariton scattering with the formation of bulk radiation in a vacuum. We shall assume that the medium with a fluctuating dielectric constant (permittivity) and $\mu = 1$ is bounded by the plane $z = 0$. Thus

$$\epsilon(\omega, r) = \begin{cases} 1, & z > 0, \\ \epsilon(\omega) + \delta\epsilon(\omega, r), & z < 0. \end{cases}$$

It follows then from eq. (11) that the energy flux of the radiation scattered in a vacuum is readily found from the relation

$$\langle S_l \rangle = \frac{c}{8\pi}\left(\frac{\omega}{c}\right)^3 \int_{|k_\||<\omega/c} \frac{d^2 k_\|}{(2\pi)^2} k_l \int_{-\infty}^{0} dz' \int_{-\infty}^{0} dz'' K(\boldsymbol{k}_\| - \boldsymbol{k}_0, z', z'') \times d_{ij}(k_\|, z, z') d_{ik}^{*}(k_\|, z, z'') E_j^{(0)}(z') E_k^{(0)*}(z''). \tag{12}$$

The intensity of the radiation scattered in a vacuum along the total length of propagation of a surface polariton is $W = \langle S_z \rangle L_x L_y$, where L_y is the width of the excitation region of the surface wave. The intensity of the initial surface wave at real values of $\epsilon(\omega) < 0$ is determined by the expression

$$W_0 = L_y \frac{c^2}{16\pi\omega} \frac{|\epsilon + 1|^2 (1 - \epsilon)}{|\epsilon|^{3/2}} \mathscr{E}_0^2. \tag{13a}$$

For a complex value of $\epsilon(\omega)$, the general expression for W_0 is quite cumbersome. In the limit, however, $|\epsilon| \gg 1$, the expression is simplified, and is reduced to the following:

$$W_0 = L_y \frac{c^2}{16\pi\omega} \frac{|\epsilon|^2}{(|\epsilon| - \epsilon')^{1/2}} \mathscr{E}_0^2 \sqrt{2}. \tag{13b}$$

It proves convenient in eq. (12) to go over from integration with respect to $\boldsymbol{k}_\|$ to integration with respect to the angles in accordance with the relation: $d^2\boldsymbol{k}_\| = (\omega/c)^2 \sin\theta \cos\theta \, d\theta \, d\varphi$. In this case $k_\| = (\omega/c)\sin\theta$ and $k_z = (\omega/c)\cos\theta$.

Taking into account the above-mentioned changes of variables, we obtain from eqs. (12) and (13) the following expression for the ratio W/W_0 when $|\epsilon| \gg 1$:

$$\frac{W}{W_0} = \frac{\sqrt{2}}{2\pi}\left(\frac{\omega}{c}\right)^5 \frac{(|\epsilon| - \epsilon')^{1/2}}{|\epsilon|^2} L_x \int_{-\pi}^{\pi} \frac{d\varphi}{2\pi} \int_0^{\pi/2} d\theta \sin\theta \cos^2\theta F(\varphi, \theta) \times \int_{-\infty}^{0} dz' \int_{-\infty}^{0} dz'' K(\varphi, \theta, z', z'') \exp[(\kappa(k_\|) + \kappa_0)z'] \exp[(\kappa(k_\|)^* + \kappa_0^*)z'']. \tag{14}$$

The following notation is introduced in eq. (14):

$$\kappa^2(k_\|) = k_\|^2 - \frac{\omega^2}{c^2}\epsilon(\omega),$$

$$F = \mathscr{E}_0^{-2}(\omega/c)^2 \sum_i |\Lambda_i|^2, \tag{15}$$

where, when $z > 0$ and $z' < 0$, $d_{ij}(k_\|, z, z')E_j^{(0)}(z') = \exp(ik_z z) \times \exp[(\kappa(k_\|) + \kappa_0)z']\Lambda_i$.

Making use of the expression for d_{ij} obtained by Maradudin and Mills (1975a), we find for F

$$F = \frac{|\kappa(k_\|)\cos\varphi + (k_0/\kappa_0)k_\||^2}{|\kappa(k_\|) - ik_z\epsilon|^2} + \sin^2\varphi \frac{(\omega/c)^2}{|\kappa(k_\|) - ik_z|^2}, \tag{16}$$

which, in the limit $|\epsilon| \gg 1$, leads to

$$F(\varphi, \theta) = |\epsilon|^{-1}(\sin^2\varphi + \cos^2\varphi/\cos^2\theta). \tag{17}$$

Note that when $|\epsilon| \gg 1$, $|k_\| - k_0| \leq 2\omega/c$. At the same time, $K(\boldsymbol{p})$ strongly depends on $\boldsymbol{p}$ when $|\boldsymbol{p}| \gtrsim \xi^{-1}$, where $\xi(T)$ is the correlation length. In approaching the temperature of the second-order phase transition, the correlation radius $\xi(T)$ increases at the rate $\xi(T) \sim \xi_0|1 - T/T_c|^{-1/2}$. But for all known phase transitions in the infrared region of frequencies $\xi_0 \ll c/\omega$, so that $\xi(T)$ becomes of the order of the wavelength c/ω only in the very narrow region where $|1 - T/T_c| \sim (\xi_0\omega/c)^2$. Specifically, for a phase transition to the superconducting state, characterized by a large value of $\xi_0 \approx 10^{-4}$ cm, $(\xi_0\omega/c)^2 \approx 10^{-2}$ for frequencies $\omega/c \approx 10^3\,\mathrm{cm}^{-1}$. Of interest, therefore, from the physical point of view, is the region $\xi(T) \ll c/\omega$, in which the dependence of K on $\boldsymbol{p}$ can be ignored. Then integration with respect to angles in eq. (14) can be readily carried out and we obtain

$$\frac{W}{W_0} = \frac{\sqrt{2}}{3\pi}\left(\frac{\omega}{c}\right)^5 \frac{(|\epsilon| - \epsilon')^{1/2}}{|\epsilon|^3} L_x \int_{-\infty}^0 dz' \int_{-\infty}^0 dz'' K(z', z'') \times \exp(2\kappa_0 z')\exp(2\kappa_0^* z''). \tag{18}$$

Next we shall find the form of the correlation $K(z', z'')$ for the region of applicability of the Ginzburg–Landau theory for second-order phase transitions (Landau and Lifshits 1976). We shall assume here that $\delta\epsilon(\boldsymbol{r}) = A|\eta|^2$, where η is the order parameter corresponding to the given second-order phase transition. The probability distribution P of the given spatial configuration of order parameter $\eta(\boldsymbol{r})$ is described by the Ginzburg–Landau functional $\Phi\{\eta\}$

$$P \sim \exp\left(-\frac{\Phi\{\eta\}}{T}\right),$$

$$\Phi\{\eta\} = \int d^3r[at|\eta|^2 + g\,|\partial\eta/\partial r|^2 + \tfrac{1}{2}b|\eta|^4], \tag{19}$$

where

$$t = (T - T_c)/T_c.$$

It should be pointed out that the surface contribution to the energy has been omitted in eq. (19). In particular, in the transition of the metal contacting the vacuum to the superconducting state, the correction to the correlator, owing to surface energy, is small when $T_c/\epsilon_F \ll 1$, where ϵ_F is Fermi energy.

For the correlator $K(z', z'')$ above the transition point, where the term $|\eta|^4$ can be neglected, standard calculations based on the function (19) yield the following value:

$$K(z', z'') = |A|^2 \frac{T_c^2}{16g^2} \int \frac{d^2k_\parallel}{(2\pi)^2} \frac{[\exp(-\gamma|z' - z''|) + \exp(-\gamma|z' + z''|)]^2}{\gamma^2}, \tag{20}$$

where $\gamma = \sqrt{(k_\parallel^2 + \xi^{-2}(T))}$ and $\xi^{-2}(T) = at/g$.

Substituting eq. (20) into (18) and integrating as required, we obtain for W/W_0 above the transition point:

$$\frac{W}{W_0} = \frac{\sqrt{2}}{96\pi^2} \frac{(|\epsilon| - \epsilon')^{1/2}}{|\epsilon|^2\epsilon''} \left(\frac{\omega}{c}\right)^2 \frac{|A|^2 T_c^2}{g^2} \Big[\ln|1 + \xi\kappa_0|$$
$$- \ln|1 + 2\xi\kappa_0'| - \frac{\epsilon'}{\epsilon''} \arctan\left(\frac{\xi\kappa_0''}{1 + \xi\kappa_0'}\right)\Big], \tag{21}$$

where $\kappa_0 = \kappa_0' + i\kappa_0''$ is determined according to eq. (5).

Below the transition point, simple calculations yield the following expression for the correlator $K(z', z'')$:

$$K(z', z'') = \frac{\sqrt{2}}{2} |A|^2 \frac{T_c\xi}{g} |\eta_0|^2 [\exp(-\xi^{-1}|z' - z''|\sqrt{2})$$
$$+ \exp(-\xi^{-1}|z' + z''|\sqrt{2})], \tag{22}$$

where $|\eta_0|^2 = at/b$.

In this case

$$\frac{W}{W_0} = \frac{1}{3\pi} \left(\frac{\omega}{c}\right)^2 \frac{(|\epsilon| - \epsilon')^{1/2}}{|\epsilon|^2\epsilon''} |A|^2 |\eta_0|^2 \frac{T_c\xi}{g} \frac{(\xi|\kappa_0|)^2}{|1 + \kappa_0\xi\sqrt{2}|^2} \frac{1 + 2\kappa_0'\xi\sqrt{2}}{\kappa_0'\xi\sqrt{2}}. \tag{23}$$

3. *Surface Polariton Scattering near a Superconducting Transition Point*

Let us now consider as an example the scattering of surface polaritons near a superconducting transition point. Here the order parameter η is a gap Δ in the spectrum of single-electron excitations. Fluctuations of gap Δ contribute, in general, both to the permittivity (dielectric constant) ϵ of the metal and to its permeability. It can be shown that in a pure superconductor the cross sections of scattering by fluctuations of ϵ and μ are of the same order of magnitude. In this case, however, most metals have an anomalous skin effect in the infrared frequency region. For this reason, the discussion given above is inapplicable unless the effects of spatial dispersion are taken into account.

Moreover, as was shown by Gorkov (1959), the coefficient of "rigidity" $g \sim l$ and decreases with the mean free path l of the electrons. This leads to more highly developed fluctuations in a "dirty" superconductor in which $T_c\tau/\hbar \ll 1$, where $\tau = l/v_F$, the time of free path of the electrons.

For this reason we shall limit our discussion to the more important case, practically, of a "dirty" superconductor in the region of the normal skin-effect, in which $l \ll |\kappa_0|^{-1}$. It can be shown that under these conditions the fluctuations of μ make a negligibly small contribution to the scattering cross section, which can thus be determined by eqs. (21) and (23).

To find the dependence of the dielectric constant on the order parameter, use was made of a method developed by Abrikosov et al. (1963) and Abrikosov and Gorkov (1958). As a result, the following expression was obtained for the coefficient $A(\omega)$:

$$|A(\omega)|^2 = 4(\epsilon'')^2 \ln^2\left(\frac{\omega\hbar}{T_c}\right)\frac{1}{(\hbar\omega)^4}. \tag{24}$$

When $\omega\tau \ll 1$, eq. (24) coincides with the results obtained by Abrikosov and Gorkov (1958).

Let us analyse the temperature dependence in various limiting cases of the scattering cross section for a surface polariton when bulk radiation is produced. First we shall discuss the region of small values of $t = |1 - T/T_c|$, when $\xi|\kappa_0| \gg 1$. It follows from eq. (23) that in this region the temperature dependence of W below the transition point is determined by the quantity $\xi|\eta_0|^2 \sim \sqrt{t}$. It is evident that notwithstanding the intensified fluctuations, the scattering cross section decreases due to the decrease of the order parameter.

Above the transition point, eq. (21) shows a logarithmic growth of the cross section as $t \to 0$.

We point out that the theory developed here is restricted to the fluctuation region and is valid only when $1 \gg t \gtrsim (T_c^2 b^2)/(ag^3)$ (see Landau and Lifshits 1976).

On the other hand, it follows from eqs. (21) and (23) that the scattering cross section below the transition point is greater than that above the transition point everywhere outside the fluctuation region, which is extremely narrow ($T_c^2 b^2/ag^3 \approx 10^{-16}$) in a superconductor. Consequently, in the following we shall consider in detail only the temperature region below T_c.

In a "dirty" superconductor the correlation length $\xi = \xi_0/\sqrt{t}$, where $\xi_0 = [(\pi/24)(v_F \hbar l/T_c)]^{1/2}$. But $|\kappa_0| \sim \sqrt{|\epsilon|} \sim \sqrt{l}$. Thus, the quantity $\xi_0|\kappa_0| \sim l$ and it may be sufficiently small with a large impurity concentration. In this case it proves reasonable to consider the limit $\xi|\kappa_0| \ll 1$. In this limit the scattering cross section is found to be proportional to $\xi^2|\eta_0|^2$ and independent of t.

The qualitative form of the temperature dependence of the scattering cross section is shown in fig. 1.

It is an essential fact that with a sufficiently large amount of impurities, when

$$\xi_0|\kappa_0| \lesssim 1 \tag{25a}$$

and

$$\omega\tau \lesssim 1, \tag{25b}$$

The quantity W in the plateau region ($W = W_m$) is independent of the free path length l. Substituting into eq. (23) the values of a, ξ_0 and $|\Delta_0|^2$ found by Gorkov (1959), e.g.

$$a = N(0), \quad |\Delta_0|^2 = \frac{8\pi^2}{7\zeta(3)} T_c^2 t, \quad \xi_0^2 = \frac{\pi}{24}\frac{v_F \hbar l}{T_c},$$

and taking eqs. (25a) and (25b) into account, we readily obtain

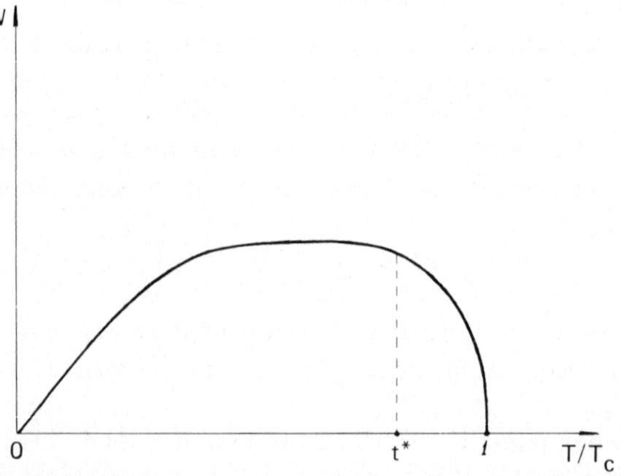

Fig. 1. Temperature dependence of the cross section of surface polariton scattering by the fluctuation of energy gap Δ in a superconductor $(1 - t^* \approx (\xi_0\kappa_0)^2)$.

$$\frac{W_m}{W_0} = \frac{64\pi}{21\zeta(3)} c^{-3}\hbar^{-4} \frac{T_c^3}{\omega N(0)} \ln^2\left(\frac{\omega\hbar}{T_c}\right), \tag{26}$$

where $N(0)$ is the electronic state density at the Fermi level, $\zeta(3)$ is the Riemann zeta function.

Making use of the expression for $\epsilon(\omega)$:

$$\epsilon(\omega) = -\frac{\omega}{\omega + i/\tau} \frac{\omega_p^2}{\omega^2},$$

where ω_p is the frequency of plasma oscillations, it can be shown that if one of the conditions (25a) or (25b) is not satisfied, the maximum value of $W = W_m \sim 1/l$. At given higher values of l, when neither of the conditions (25) is satisfied, $W_m \sim l^{-3/2}$. Hence, eq. (26) yields the maximum attainable scattering cross section for the given substance.

To estimate the quantity W_m/W_0 using eq. (26) we take values typical for niobium alloys: $T_c \approx 10\ \mathrm{K} \approx 1.4 \times 10^{-15}$ erg, $N(0) \approx 0.8 \times 10^{34}\ \mathrm{erg}^{-1}\ \mathrm{cm}^{-3}$. Then, at $\omega = 3 \times 10^{13}\ \mathrm{s}^{-1}$, we obtain $W_m/W_0 \approx 3 \times 10^{-14}$.

Thus, the scattering cross section of surface polaritons by the fluctuations of ϵ near the superconducting transition point is found to be extremely small, and a problem arises of separating it out of all other possible scattering processes. This circumstance imposes strict requirements on the surface quality. Scattering by surface roughness (irregularities) was discussed by Mills (1975). Mills (1975) obtained an expression for the scattering cross section which, when $|\epsilon| \gg 1$, is of the form

$$\frac{W}{W_0} = \frac{4\sqrt{2}}{3\pi} \frac{(|\epsilon| - \epsilon')^{1/2}}{\epsilon''} |\epsilon| \left(\frac{\omega}{c} h\right)^2 \left(\frac{\omega}{c} b\right)^2, \tag{27}$$

where h and b are the average height and width of surface irregularities (roughness), respectively, in a plane so that correlator K in the momentum representation at small momenta is of the form $K = |\epsilon|^2 h^2 b^2$. If for an estimate we take $h = 10^{-7}$ cm, $b = 10^{-6}$ cm, $\omega = 3 \times 10^{13}\ \mathrm{s}^{-1}$ and $|\epsilon| = \epsilon'' = 10^3$, then $W/W_0 = 2 \times 10^{-13}$. Thus, with the specified surface characteristics, which are technologically quite attainable (Endriz and Spicer 1971), the scattering cross sections by fluctuations and by surface irregularities do not differ too greatly. It is to be hoped, therefore, that the characteristic temperature dependence of the cross section in scattering by fluctuations can enable these two contributions to be separated.

Still another difficulty in the experimental observation of the above-mentioned effects arises because of the small value of the scattering cross section for surface polaritons. Since the limit of sensitivity of existing radiation detectors is the value $W \approx 10^{-12}$ W, it is necessary to produce a surface wave flux $W_0 \gtrsim 100$ W. Hence, the necessity arises for estimating the maximum pumping powers at which no appreciable changes yet occur in the properties of the substance being employed.

In this connection, we shall assume, in accordance with Eliashberg (1971), that the electrons acquire energy upon interaction with an electromagnetic field and that the phonons form a thermostat. On this basis, we shall seek the criterion which indicates that the electron energy distribution function is not essentially changed.

The kinetic equation for the electron energy distribution function $n(\epsilon)$ is of the form

$$R[n(\epsilon-\hbar\omega)-2n(\epsilon)+n(\epsilon+\hbar\omega)]=\delta n(\epsilon)/\tau_0, \tag{28}$$

where τ_0 is the energy relaxation time. The left-hand side of eq. (28) expresses the fact that the electrons acquire energy from the field and radiate it in quanta of the amount $\hbar\omega$. To determine the temperature-independent constant R, we shall consider eq. (28) in the limit $\hbar\omega \ll T$. Then

$$\delta n(\epsilon)=\frac{\partial^2 n_0}{\partial\epsilon^2}(\hbar\omega)^2 R\tau_0,$$

where n_0 is the Fermi distribution function. Taking into consideration that

$$\frac{1}{n_0}\frac{\partial^2 n_0}{\partial\epsilon^2}\sim\frac{1}{T^2},$$

we readily find that inequality $\delta n/n_0 \ll 1$ is equivalent to inequality

$$T^2 \gg (\hbar\omega)^2 R\tau_0. \tag{29}$$

On the other hand, a criterion of the type of eq. (29) can be obtained by the following line of reasoning. During the energy relaxation time, the electrons in unit volume acquire energy equal to $E^2\sigma\tau_0$ from the field, where $E^2\sigma$ is the Joule-power losses. Since the electronic heat capacity per unit volume $C \sim N(0)T$, the heating effect is small if

$$T^2 \gg \frac{E^2\sigma}{N(0)}\tau_0. \tag{30}$$

Combining eqs. (29) and (30) we obtain

$$R\sim\frac{E^2\sigma}{N(0)(\hbar\omega)^2}. \tag{31}$$

We are interested, however, in the reverse case when $\hbar\omega \gg T$. Under these conditions, $n(\epsilon-\hbar\omega)\approx 1$, $n(\epsilon+\hbar\omega)\approx 0$ and the required criterion reduces to

$$R\tau_0=\frac{E^2\sigma\tau_0}{N(0)(\hbar\omega)^2}\ll 1. \tag{32}$$

At low temperatures, where the energy relaxation time related to the inelastic collisions of electrons with electrons ($\tau_{el}\sim T^2/\epsilon_F$) and phonons

($\tau_{ph} \sim T^3/\theta_D^2$, where θ_D is the Debye temperature) is very large, we should take as τ_0 the time required for a diffusing electron to emerge from a layer whose thickness is of the order of the depth of penetration of the field $\delta \sim |\kappa_0|^{-1}$. Consequently

$$\tau_0 \sim \delta^2/D,$$

where $D = v_F l/3$ is the diffusion coefficient.

Then from eq. (32) we obtain the following estimate for the maximum field strength E:

$$E_{cr} \sim \hbar\omega/\delta e^*, \tag{33}$$

where e^* is the effective charge of the electron.

The superconducting state can be destroyed, not only as a result of a change in the electron distribution function, but also by the currents produced by the field. The corresponding estimate can be obtained by making use of the investigations by Kulik (1969):

$$E_{cr} \sim \hbar\omega/\xi(T)e^*. \tag{34}$$

Since in the temperature range corresponding to the maximum scattering cross section, $\delta \sim \xi(T)$, estimates (33) and (34) coincide and yield the value $E_{cr} \approx 600$ V/cm at $\delta \approx 3 \times 10^{-5}$ cm. At such amplitudes of the electric field the quantity W_0 determined by eq. (13b), reaches a value of 10^4 W if we assume that the width of the excitation region of the surface wave $L_y \sim 1$ cm. It thus follows from the aforesaid that at large values of $|\epsilon| \sim 10^3$, typical of metals, unfavorable effects appear only at very high pumping powers. Consequently, these effects practically do not restrict the applicability of the theory developed here.

4. *Surface Polariton Scattering near Phase Transition Points in Thin Films; Resonance Amplification of Scattering*

Up to this point we have discussed the scattering of surface polaritons propagating along the surface of a metal that is undergoing a second-order phase transition. The scattering of surface polaritons by the fluctuations of the order parameter can evidently be utilized for investigating critical phenomena in dielectric media as well when these media are applied in the form of a thin film on the surface of a metal.

The scattering cross section for surface polaritons can prove to be considerable, notwithstanding the small thickness of the film, especially in the frequency range close to the frequencies of optical vibrations in a dielectric film. In this case, as will be shown below (see also Agranovich 1979), a resonance amplification of the scattering cross section should occur near the frequencies of both longitudinal and transverse optical vibrations. In the first case, resonance effects result from the growth of the normal component of the electric field in the film, whereas in the second case the resonance multipliers arise from the dependence of the permittivity (dielectric constant) ϵ_1 of the film on the order parameter.

Note that in the following we shall consider dielectric films of the small thickness $d \ll (c/\omega)|\epsilon|^{-1/2}\epsilon_1$, where the propagation length and dispersion of the surface polaritons are not appreciably changed by the presence of the film.

The intensity of scattered radiation in this case, as in the case of a free metallic surface, is proportional to the large propagation length L_x of the surface polariton. Hence the advantage of employing thin films deposited on the surface of a metal. Moreover, the investigation of critical phenomena in thin films and, in particular, the transition from three-dimensional to quasi-two-dimensional behavior of a system is in itself an important fundamental problem.

The flux of radiation scattered in a vacuum is determined by eq. (12). Assuming that the field $E^{(0)}$ of the surface polariton remains unchanged throughout the thickness of the film and making use of the expression for the Green function of a three-layer structure (Maradudin and Mills 1975b) and the expression for the field $E^{(0)}$ in the film,

$$E_x^{(0)} = \mathscr{E}_o, \qquad E_z^{(0)} = \frac{ik_0}{\kappa\epsilon_1}\mathscr{E}_0$$

we obtain, in place of eq. (14):

$$\frac{W}{W_0} = \frac{\sqrt{2}}{2\pi}\left(\frac{\omega}{c}\right)^5 \frac{(|\epsilon| - \epsilon')^{1/2}}{|\epsilon|^2} L_x \int_{-\pi}^{\pi} \frac{d\varphi}{2\pi} \int_0^{\pi/2} d\theta \sin\theta \cos^2\theta$$

$$\times F(\varphi, \theta) \int_0^d dz' \int_0^d dz'' K(\varphi, \theta, z', z''), \tag{35}$$

where

$$F = \frac{|\kappa(k_\parallel)\cos\varphi + (\epsilon/\epsilon_1)^2(k_0/\kappa_0)k_\parallel|^2}{|\kappa(k_\parallel) - ik_z\epsilon|^2} + \frac{(\omega/c)^2}{|\kappa(k_\parallel) - ik_z|^2}\sin^2\varphi. \tag{36}$$

In the region of frequencies close to the frequency of longitudinal optical vibration in the film, when $\epsilon_1(\omega) \to 0$, the expression for F assumes the form

$$F = \frac{|\epsilon|}{|\epsilon_1|^4}\tan^2\theta. \tag{37}$$

Near resonance with the transverse optical vibration in the film, eq. (36)

can be readily transformed in an expression for F that coincides with eq. (17).

To obtain frequency dependence of the scattering cross section in the resonance region in the explicit form, we represent the permittivity (dielectric constant) in the form

$$\epsilon_1 = \epsilon_\infty \prod_i \frac{\omega_{\|i}^2 - \omega^2}{\omega_{\perp i}^2 - \omega^2}. \tag{38}$$

The frequencies $\omega_{\|i}(\eta)$ and $\omega_{\perp i}(\eta)$ are, in general, functions of the order parameter η and can be expanded into series with respect to powers of η:

$$\begin{aligned} \delta\omega_{\|i}^2 &= \alpha_{\|i}\eta + \beta_{\|i}\eta^2 + \cdots, \\ \delta\omega_{\perp i}^2 &= \alpha_{\perp i}\eta + \beta_{\perp i}\eta^2 + \cdots, \end{aligned} \tag{39}$$

where $\delta\omega^2 = \omega^2(\eta) - \omega^2(0)$.

The presence of a linear term in expansion (39) is related to a definite type of symmetry of the high-temperature phase. In particular, in structural phase transitions the values of α_i are nonzero only in crystals that have no inversion center above the transition temperature. An example is the well-known ferroelectric crystal KH_2PO_4 (KDP). In the following we shall discuss specifically the linear relation of $\delta\omega^2$ to the order parameter as in this case the critical features of the scattering cross section are more strongly manifested.

Near the frequency of longitudinal optical vibrations $\omega_{\|i0}$ the variation in the permittivity $\delta\epsilon_1$, due to fluctuations of the order parameter, is of the form

$$\delta\epsilon_1 = -\epsilon_{\infty\|} \frac{\delta\omega_{\|i0}^2}{\omega_{\|i0}^2 - \omega_{\perp i0}^2}, \tag{40}$$

where

$$\epsilon_{\infty\|} = \epsilon_\infty \prod_{i \neq i_0} \frac{\omega_{\|i}^2 - \omega_{\|i0}^2}{\omega_{\perp i}^2 - \omega_{\|i0}^2}.$$

Analogously, near the frequency $\omega_{\perp i0}$ of the transverse optical vibrations in the film we obtain from eq. (38):

$$\delta\epsilon_1 = -\epsilon_{\infty\perp} \frac{\omega_{\|i0}^2 - \omega_{\perp i0}^2}{(\omega^2 - \omega_{\perp i0}^2)^2} \delta\omega_{\perp i0}^2, \tag{41}$$

where

$$\epsilon_{\infty\perp} = \epsilon_\infty \prod_{i \neq i_0} \frac{\omega_{\|i}^2 - \omega_{\perp i0}^2}{\omega_{\perp i}^2 - \omega_{\perp i0}^2}.$$

Substituting the values of F, ϵ_1 and $\delta\epsilon_1$ from eqs. (37) through (41) and (17) into eq. (35), we obtain the following expressions for the scattering cross section of a surface polariton:

$$\frac{W}{W_0}=\frac{\sqrt{2}}{\pi}\frac{(|\epsilon|-\epsilon')^{1/2}}{|\epsilon|\epsilon''}\left(\frac{\omega}{c}\right)^4\frac{|\epsilon|^2}{\epsilon_{\infty\parallel}^2}\frac{(\omega_\perp^2-\omega_\parallel^2)^2\alpha_\parallel^2}{(\omega_\parallel^2-\omega^2)^4}$$

$$\times\int_{-\pi}^{\pi}\frac{d\varphi}{2\pi}\int_0^{\pi/2}d\theta\,\sin^3\theta\int_0^d dz'\int_0^d dz''G(z',z'',\varphi,\theta), \tag{42}$$

$$\frac{W}{W_0}=\frac{2}{\pi}\frac{(|\epsilon|-\epsilon')^{1/2}}{|\epsilon|\epsilon''}\left(\frac{\omega}{c}\right)^4\epsilon_{\infty\perp}^2\frac{(\omega_\perp^2-\omega_\parallel^2)^2\alpha_\perp^2}{(\omega_\perp^2-\omega^2)^4}\int_{-\pi}^{\pi}\frac{d\varphi}{2\pi}$$

$$\times\int_0^{\pi/2}d\theta\,\sin\theta\,(\cos^2\varphi+\sin^2\varphi\cos^2\theta)\int_0^d dz'\int_0^d dz''G(z',z'',\varphi,\theta), \tag{43}$$

where

$$G(z',z'',\varphi,\theta)=\int d^2\rho\langle\eta(0,z')\eta(\rho,z'')\rangle\exp[i(\boldsymbol{k}_\parallel-\boldsymbol{k}_0)\rho].$$

Equation (42) yields the scattering cross section of a surface polariton near the frequency $\omega_\parallel$ of longitudinal optical vibration, and eq. (43), near the frequency $\omega_\perp$ of transverse optical vibration in a dielectric film. It can be seen that when $|\epsilon|\gg\epsilon_{\infty\parallel}\epsilon_{\infty\perp}$, resonance amplification of the scattering cross section near the frequency of longitudinal optical vibration is especially strongly manifested.

Let us assume that as a result of a second-order phase transition no new elements of the continuous symmetry group appear above the transition point, as they do, for instance, in ferroelectric transitions of the order–disorder type. In this case, even purely two-dimensional systems have long-range order below the phase transition point T_c and, outside the fluctuation range of temperatures near T_c, correlator G can be found in the approximation of the Gaussian distribution for the fluctuations η. As an example, we give the form of correlator G for a phase transition that can be described by a real order parameter and the Ginzburg–Landau functional of the form of eq. (19):

$$G(z',z'')=\frac{T_c}{4\gamma g}[\coth(\gamma d)(\cosh\gamma|z'+z''|+\cosh\gamma|z'-z''|)$$
$$-\tanh(\gamma d)(\sinh\gamma|z'+z''|+\sinh\gamma|z'-z''|)], \tag{44}$$

where $\gamma=[(\boldsymbol{k}_\parallel-\boldsymbol{k}_0)^2+\xi^{-2}]^{1/2}$ and d is the thickness of the film. Integrating G with respect to z' and z'', we readily obtain

$$\mathscr{G}=\int_0^d dz'\int_0^d dz''G(z',z'')=\frac{T_c}{4\gamma^3 g}\{\sinh(2\gamma d)[1-\tanh(\gamma d)]$$
$$+(2\gamma d)\tanh(\gamma d)\}. \tag{45}$$

We assume, as previously in sect. 3, that $\xi\omega/c \ll 1$. Then the quantity $\mathscr{G}$ is independent of the angles φ and θ, and is determined in both limiting cases, $d \ll \xi$ and $d \gg \xi$, by the expression

$$\mathscr{G} = T_c\xi^2 d/2g. \tag{46}$$

Substituting eq. (46) into eqs. (42) and (43) we obtain the following expressions for the scattering cross section of surface polaritons:

$$\frac{W}{W_0} = \frac{\sqrt{2}}{3\pi}\frac{(|\epsilon| - \epsilon')^{1/2}}{\epsilon''}\left(\frac{\omega}{c}\right)^4 \frac{|\epsilon|}{\epsilon_{\infty\parallel}^2}\frac{(\omega_\parallel^2 - \omega_\perp^2)^2\alpha_\parallel^2}{(\omega^2 - \omega_\parallel^2)^4}\frac{T_c\xi^2 d}{g}, \tag{47}$$

$$\frac{W}{W_0} = \frac{\sqrt{2}}{3\pi}\frac{(|\epsilon| - \epsilon')^{1/2}}{|\epsilon|\epsilon''}\left(\frac{\omega}{c}\right)^4 \epsilon_{\infty\perp}^2\frac{(\omega_\parallel^2 - \omega_\perp^2)^2\alpha_\perp^2}{(\omega^2 - \omega_\perp^2)^4}\frac{T_c\xi^2 d}{g}. \tag{48}$$

Note that $\xi^2 = g/at$ when $T > T_c$, and that $\xi^2 = g/2at$ when $T < T_c$.

Hence, near the phase transition point the scattering cross section grows in proportion to $|T - T_c|^{-1}$. We should underline again that this law of increase of the cross section is related to the assumption of a Gaussian distribution of the fluctuations. In the fluctuation region, in the immediate vicinity of the transition point, fluctuation interaction plays an essential role. Here, $\xi \sim |t|^{-\nu}$, with ν depending upon the dimensionality of the space and the symmetry of the Ginzburg–Landau functional. With a single-component order parameter the value of $\nu = 0.64$ for a three-dimensional system and $\nu = 1$ for a two-dimensional system (Landau and Lifshits 1976).

As is evident from eqs. (47) and (48), the scattering cross section is proportional to $|t|^{-2\nu}$. Therefore, the measurement of the temperature dependence of the ratio W/W_0 can be used to determine the critical index ν. It would be of interest, of course, to investigate the variation in ν with a reduction in the film thickness d and to follow up the transition from three-dimensional behavior of the system to two-dimensional behavior.

Let us estimate the coefficient α, appearing in eqs. (47) and (48), for a ferroelectric transition. The simplest model Hamiltonian for crystals having no inversion center, which contains a term of cubic anharmonicity, is of the form

$$H = \sum_i \tfrac{1}{2}(M_i\dot{Q}_i^2 + M_i\omega_{\perp i}^2 Q_i^2) + \sum_{i,j,l} b_{ijl}Q_iQ_jQ_l,$$

where Q_i are normal coordinates, M_i is the reduced mass, $\omega_{\perp i}$ is the eigenfrequency of i-mode.

Near the phase transition point the frequency of the soft mode $\omega_{\perp 0} \to 0$. Hence, the corresponding normal coordinate Q_0 is subject to the strongest fluctuations. The frequency shift of the i-mode, related to fluctuations Q_0, is of the form

$$\delta\omega_{\perp i}^2 = \frac{6b_{ii0}}{M_i}Q_0.$$

If we select as the order parameter the polarization vector $\boldsymbol{P}_0 = e\boldsymbol{Q}_0/V_c$, where V_c is the volume of a unit cell and e is the effective charge, then $\alpha_{\perp i}$ is determined by the expression

$$\alpha_{\perp i} = 6b_{ii0}V_c/eM_i. \tag{49}$$

Coefficient $\alpha_{\|i} = \alpha_{\perp i}$ under the condition that the oscillator strength is independent of the order parameter.

We shall assume, as usual, that in its order of magnitude the coefficient of cubic anharmonicity b_{ii0} is equal to

$$b_{ii0} \sim M_i\omega_{\perp i}^2/a, \tag{50}$$

where a is the crystal lattice constant.

Making use of eqs. (49) and (50) we obtain the following estimate for the quantity α_i:

$$f_i \equiv \frac{\alpha_i}{\omega_{\perp i}^2} \sim \frac{a^2}{e} \approx 10^{-6}\ \text{cgse}. \tag{51}$$

To estimate the scattering cross section for surface polaritons by the fluctuations of the order parameter near a ferroelectric transition we use the relation

$$\frac{T_c\xi^2}{g} = \frac{T_c}{at} = \frac{R}{2\pi t}, \tag{52}$$

where the static permittivity $\epsilon_1(0)$ is determined by the Curie law

$$\epsilon_1(0) = \frac{Rk_B^{-1}}{(T - T_c)}.$$

Substituting eqs. (51) and (52) into (47) we obtain a final estimate of the scattering cross section of surface polaritons with the frequency $\omega \approx \omega_\|$. Thus

$$\frac{W}{W_0} \approx \frac{\sqrt{2}}{6\pi^2}(|\epsilon| - \epsilon')^{1/2}\frac{|\epsilon|}{\epsilon''}\frac{1}{\epsilon_{\infty\|}^2}\left(\frac{\omega_\|}{c}\right)^4\left[\frac{\omega_\perp^2}{\omega_\|^2}\left(1 - \frac{\omega_\perp^2}{\omega_\|^2}\right)\right]^2\left(\frac{\omega_\|}{\Gamma}\right)^4\frac{Rd}{t}f^2, \tag{53}$$

where Γ is the attenuation constant of optical vibrations with the frequency $\omega_\|$. Assuming for the purpose of estimation the following values of the parameters in eq. (53): $\epsilon' = -8000$, $\epsilon'' = 4000$, $\epsilon_{\infty\|} = 3$, $\omega_\| = 10^3\ \text{cm}^{-1}$, $(\omega_\perp/\omega_\|)^2 = \frac{1}{2}$, $\Gamma = 10\ \text{cm}^{-1}$, $d = 10^{-6}\ \text{cm}$, $R = 4 \times 10^{-13}\ \text{erg}$, $f = 10^{-6}$ cgse, $|T - T_c| = 1°$ and $T_c = 100$ K, we obtain $W/W_0 = 2 \times 10^{-10}$.

5. Conclusion

The estimate of the scattering cross section of surface polaritons by the fluctuations of the order parameter near the ferroelectric transition in

KDP-type crystals, obtained on the basis of eq. (53), indicates that the effect being discussed could be observed experimentally even at pumping powers $W_0 \approx 1$ W. It should be noted, however, that the above estimate is of a very tentative nature. In particular, the strong dependence of W/W_0 on the frequency $\omega_\parallel$ indicates that the scattering cross section can be substantially increased if the surface polaritons are excited in resonance with the highest high-frequency optical vibrations in the crystal whose frequency may reach several thousand cm^{-1}.

It can be pointed out, in addition, that the intensity of radiation scattered in a vacuum is higher than the limit of the experimentally detectable intensity even at a film thickness d of the order of several Angstroms. This fact enables us to hope that surface polariton scattering may also serve as a method for investigating phase transitions in monatomic films adsorbed on the surface of metals or phase transitions associated with the reconstruction of metallic surfaces.

References

Abrikosov, A.A. and L.P. Gorkov, 1958, Zh. Eksp. Teor. Fiz. **35**, 1558.

Abrikosov, A.A., L.P. Gorkov and I.E. Dzyaloshinskii, 1963, Methods of Quantum Field Theory in Statistical Physics (Prentice-Hall, Englewood Cliffs, New Jersey).

Agranovich, V.M., 1976, JETP Lett. **24**, 588.

Agranovich, V.M., 1979, Light Scattering in Solids, eds. J.L. Birman, H.Z. Cummins and K.K. Rebane (Plenum Publishing Corp., New York) 113.

Agranovich, V.M. and T.A. Leskova, 1977, Solid State Commun. **21**, 1065.

Bendow, B., J.L. Birman and V.M. Agranovich, eds., 1976, Theory of Light Scattering in Condensed Matter (Plenum Press, New York and London).

Birman, J.L., H.Z. Cummins and K.K. Rebane eds., 1979, Light Scattering in Solids (Plenum Publishing Corp., New York).

Eliashberg, G.M., 1971, Zh. Eksp. Teor. Fiz. **61**, 1254.

Endriz, J.G. and W.E. Spicer, 1971, Phys. Rev. **B4**, 4144, 4159.

Gorkov, L.P., 1959, Zh. Eksp. Teor. Fiz. **37**, 1407.

Kulik, I.O., 1969, Zh. Eksp. Teor. Fiz. **57**, 600.

Landau, L.D. and E.M. Lifshits, 1976. Statistical Physics (Nauka Publishers, Moscow) Part I, 3rd edition.

Maradudin, A.A. and D.L. Mills, 1975a, Phys. Rev. **B11**, 1392.

Maradudin, A.A. and D.L. Mills, 1975b, Phys. Rev. **B12**, 2943.

Mills, D.L., 1975, Phys. Rev. **B12**, 4036.

PART III

Nonlinear Interactions and Surface Polaritons

E. BURSTEIN
Y. R. CHEN
F. DEMARTINI
R. LOUDON
Y. R. SHEN
J. E. SIPE
G. I. STEGEMAN
S. USHIODA

CHAPTER 12

Raman Scattering by Surface Polaritons

S. USHIODA and R. LOUDON*

Department of Physics
University of California
Irvine, California 92717
U.S.A.

*Permanent address: Department of Physics, Essex University, Colchester CO4 3SQ, ENGLAND

Surface Polaritons
Edited by
V.M. Agranovich and D.L. Mills

Contents

1. Introduction 537
2. Surface-polariton electric-field fluctuations 538
   2.1 Single-interface surface polaritons 538
   2.2 Double-interface surface polaritons 545
   2.3 Guided-wave polaritons 550
3. Theory of surface-polariton light scattering 553
   3.1. Geometrical considerations 553
   3.2. The surface-polariton cross section 559
4. Experimental method and results 563
   4.1 Single-interface surface polaritons 565
   4.2. Double-interface surface polaritons 568
   4.3. Guided-wave polaritons 574
5. Effects of surface roughness 577
6. Conclusion 584

References 585

1. *Introduction*

Light-scattering spectroscopy has proved to be a fruitful technique for the measurement of dispersion relations and other properties of phonon surface polaritons. We consider only this kind of surface polariton, where the excitation consists of a surface electromagnetic wave coupled to an optic lattice vibration of the active material. Other experimental methods for determining the dispersion relations for the various kinds of surface polariton are covered elsewhere in this volume, including attenuated total reflection (Mirlin, ch. 1), grating couplers (Raether ch. 9), and electron scattering (Ibach and Mills 1982). The present chapter is devoted solely to the experimental and theoretical studies of the Raman scattering properties of surface polaritons.

Almost all the light scattering associated with excitations in solids is produced by the same basic mechanism, in which the monochromatic incident light couples to the dynamic variables of the excitation of interest, generating a frequency-shifted polarization that radiates the scattered light. In thermal equilibrium, the dynamic variables engage in random fluctuations whose statistical properties are determined by the thermal excitation probabilities of the various states of the system. The spectrum of the scattered light is controlled by the frequency spectra, or power spectra, of the relevant random fluctuations. The main aim of experiments is the extraction of information on the excitation power spectra from the distorting effects of the measurement process, and the main aim of theory is the prediction and interpretation of the power spectra.

Surface polaritons are electromagnetic excitations of a sample and their light-scattering properties are mainly controlled by the electric-field fluctuations in the region of the sample surface. The main theoretical tool in calculating the required power spectra is linear response theory. In sect. 2 we consider the linear response theory for the electric-field fluctuations and thereby derive the main properties of the surface polaritons. We treat the simplest case of a single interface in some detail to show the nature of the method; results are also given for the surface modes in double-interface systems and for guided-wave polaritons. In sect. 3 we show how the power spectra are used to obtain expressions for the surface light-scattering cross section and we describe the information that can be obtained by Raman scattering experiments. Since the scattered intensity is extremely weak owing to the small scattering volume involved, various special spectroscopic techniques are needed to observe the Raman spectra. These experimental methods are described in sect. 4, which also reviews the experimental results for single-interface modes, double-interface modes and guided-wave polaritons. In sect. 5 we discuss the effects of surface roughness on the surface-polariton spectra, reviewing theories of the effects and presenting experimental results. Section 6 contains concluding remarks and a comparison of light-scattering spectroscopy with other methods for observing surface polaritons.

2. Surface-Polariton Electric-Field Fluctuations

The purpose of the present section is an account of the way in which linear response theory is used to derive the thermal fluctuations in electric field associated with the surface polaritons. The results provide information on the polarization, frequency dependence, temperature dependence and strength of the fluctuations; they can also be interpreted in terms of the surface-polariton dispersion relation and linewidth. We give rather complete details of the linear response theory in the simplest case of a single interface, but for the more complicated double-interface surface polaritons and the guided-wave polaritons we quote the main results without detailed derivations.

In order to emphasize the physical features of the surface polaritons, we keep the theoretical models as simple as possible. The materials treated are assumed to have isotropic optical and vibrational properties, although in some comparisons with experiment we quote results for anisotropic materials. It is also assumed that the relative permittivities of the materials treated are independent of the electromagnetic wavevector, so that there are no spatial dispersion effects and the usual boundary conditions are sufficient to determine the electromagnetic fields. Spatial dispersion effects on surface polaritons are treated elsewhere in this volume (Agranovich, ch. 5).

The surface-polariton properties derived below are those relevant to the light-scattering experiments. More comprehensive discussions are given by Nkoma et al. (1974) for the single interface, by Mills and Subbaswamy (1980), and elsewhere in this volume (Mirlin, ch. 1).

2.1. Single-Interface Surface Polaritons

The geometrical arrangement assumed in the present subsection is shown in fig. 1. The two isotropic materials have a flat surface of contact taken as the $z = 0$ plane, and different relative permittivities κ_1 and κ_2. The theory

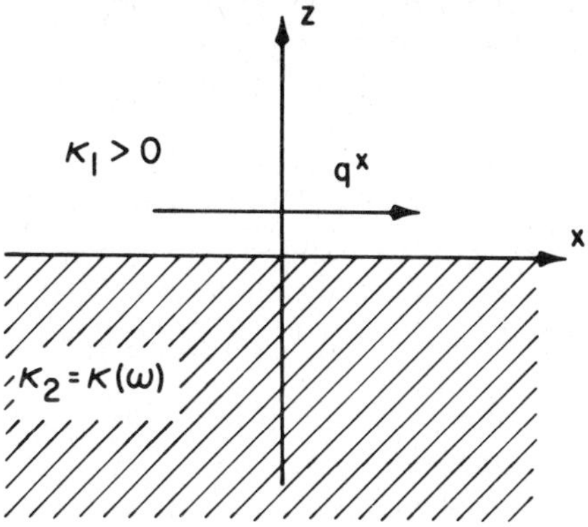

Fig. 1. Single interface geometry.

could be developed in a form that is symmetrical in the properties of the two media. However in the later applications to light-scattering experiments, all of the coupling between electromagnetic field and lattice vibration, and all of the light-scattering interaction takes place in one of the media. We take material 2 to be this active medium, with a relative permittivity that shows a resonance at the far-infrared frequency of a dipole-active optic lattice vibration. Material 1 plays a more passive role, and is often free space in practical measurements; its relative permittivity is generally assumed to be a positive constant. The surface area A and thickness L of medium 2 have dimensions much larger than any characteristic lengths in the material, so that the sample is effectively semi-infinite.

It is convenient to consider fluctuations that have a given wavevector component q^x parallel to the surface; because of the symmetry of the system, the direction of the x-axis in the surface is arbitrary. The electric-field fluctuations with frequency ω have different wavevector z components in the two media, given by

$$q_1^{z2} = \kappa_1(\omega/c)^2 - q^{x2} \quad (1)$$

$$q_2^{z2} = \kappa_2(\omega/c)^2 - q^{x2}. \quad (2)$$

The power spectra of the electric-field fluctuations can be obtained by a rather simple application of linear response theory (Landau and Lifshitz 1969, Barker and Loudon 1972, Forster 1975), whose essential steps we now indicate. Consider the effect of applying a fictitious polarization

$$\boldsymbol{P}\exp(-i\omega t + iq^x x)\delta(z - z')/A \quad (3)$$

on the plane $z = z'$ in material 2 (that is $z' \leq 0$). The energy of interaction between the polarization and the electric field $\boldsymbol{E}(z)\exp(iq^x x)$ in medium 2 is

$$H_{\mathrm{int}} = -\boldsymbol{E}(z') \cdot \boldsymbol{P} \exp(-\mathrm{i}\omega t). \tag{4}$$

It is not difficult with the help of Maxwell's equations and the usual dielectric boundary conditions to calculate the electric and magnetic fields induced in the two media by the three Cartesian components of the polarization eq. (3). The electric-field Green functions, or linear response functions, for medium 2 are then simply obtained from

$$\langle\langle E^i(z); E^j(z')^* \rangle\rangle_{q^x,\omega} = E^i(z)/P^j, \qquad i, j = x, y, z. \tag{5}$$

We consider only the case where the stimulus coordinate z' and the response coordinate z both lie in medium 2, since only these results are needed for the light-scattering calculation. However, the same methods can of course provide the Green functions for cases where z and/or z' lie in medium 1 (the complete results are given by Maradudin and Mills 1975).

The results produced by the calculation outlined above are

$$\begin{aligned} &\langle\langle E^i(z); E^j(z')^* \rangle\rangle_{q^x,\omega} \\ &= \frac{\mathrm{i}/q_2^z}{2A\epsilon_0\kappa_2} \begin{bmatrix} -q_2^{z^2} & q^x q_2^z \\ -q^x q_2^z & q^{x^2} \end{bmatrix} \frac{\kappa_1 q_2^z - \kappa_2 q_1^z}{\kappa_1 q_2^z + \kappa_2 q_1^z} \exp\{-\mathrm{i}q_2^z(z+z')\} \\ &+ \frac{\mathrm{i}/q_2^z}{2A\epsilon_0\kappa_2} \begin{bmatrix} q_2^{z^2} & -q^x q_2^z \operatorname{sgn}(z-z') \\ -q^x q_2^z \operatorname{sgn}(z-z') & q^{x^2} \end{bmatrix} \exp\{\mathrm{i}q_2^z |z-z'|\} \end{aligned} \tag{6}$$

for i, $j = x$, z, where

$$\operatorname{sgn}(z-z') = \begin{cases} 1 & \text{for} \quad z > z' \\ -1 & \text{for} \quad z < z' \end{cases}, \tag{7}$$

and

$$\begin{aligned} &\langle\langle E^y(z); E^y(z')^* \rangle\rangle_{q^x,\omega} \\ &= \frac{\mathrm{i}q_2^2/q_2^z}{2A\epsilon_0\kappa_2} \left\{ \frac{q_2^z - q_1^z}{q_2^z + q_1^z} \exp\{-\mathrm{i}q_2^z(z+z')\} + \exp\{\mathrm{i}q_2^z |z-z'|\} \right\}. \end{aligned} \tag{8}$$

The remaining components of the medium 2 Green function all vanish. Note that the Green functions given above all satisfy the symmetry property (Loudon 1978a):

$$\langle\langle E^i(z); E^j(z')^* \rangle\rangle_{q^x,\omega} = \langle\langle E^j(z'); E^i(z)^* \rangle\rangle_{-q^x,\omega}. \tag{9}$$

It is seen that all the above Green functions have a similar structure, which is similar to that found in other surface problems, for example acoustic fluctuations (Loudon 1978b). The contributions proportional to $\exp\{\mathrm{i}q_2^z|z-z'|\}$ are bulk terms that depend only on the properties of medium 2 and on the direct distance between the stimulus and response coordinates z' and z. The contributions proportional to $\exp\{-\mathrm{i}q_2^z(z+z')\}$ are surface terms that depend on the properties of both media 1 and 2 and on the round-trip distance from z' to z via the surface. Barker (1972) has derived a related but different kind of Green function that describes the response of a dielectric to an imposed surface charge layer.

The electric-field correlation functions or power spectra appropriate to Stokes-component light-scattering measurements are obtained from the Green functions by the fluctuation-dissipation theorem (Loudon 1978a):

$$\langle E^i(z)E^j(z')^*\rangle_{q^x,\omega} = (\mathrm{i}\hbar/2\pi)\{n(\omega)+1\} \times \{\langle\langle E^j(z'); E^i(z)^*\rangle\rangle^*_{q^x,\omega} - \langle\langle E^i(z); E^j(z')^*\rangle\rangle_{q^x,\omega}\}, \quad (10)$$

which reduces in the diagonal case to

$$\langle E^i(z)E^i(z')^*\rangle_{q^x,\omega} = (\hbar/\pi)\{n(\omega)+1\}\,\mathrm{Im}\langle\langle E^i(z); E^i(z')^*\rangle\rangle_{q^x,\omega}. \quad (11)$$

In these equations $n(\omega)$ is the usual Bose–Einstein factor.

Surface polaritons are the electric waves that correspond to the poles in the surface parts of the Green functions, with the conditions

$$\begin{aligned} &\mathrm{Im}\, q_1^z \geq 0 \\ &\mathrm{Im}\, q_2^z \geq 0 \end{aligned} \quad (12)$$

imposed to ensure localization of the excitations close to the $z = 0$ surface (the z dependence of the fields in medium 1 is $\exp(\mathrm{i}q_1^z z)$). Since any imaginary parts in κ_1 and κ_2 must be positive, eqs. (1) and (2) show that eq. (12) implies

$$\begin{aligned} &\mathrm{Re}\, q_1^z \geq 0 \\ &\mathrm{Re}\, q_2^z \geq 0. \end{aligned} \quad (13)$$

The damping of the optical phonons that produce the resonance in the relative permittivity and participate in the surface polaritons is usually small. As a first approximation it is useful to consider the zero-damping limit where κ_1 and κ_2 are real. The wavevector z components given by eqs. (1) and (2) are then either purely real or purely imaginary, q_2^z being real when κ_2 is positive and when

$$q^{x2} < \kappa_2(\omega/c)^2. \quad (14)$$

This is the regime of the bulk polaritons that propagate without attenuation, and eq. (2) is the usual bulk-polariton dispersion relation for medium 2 (Mills and Burstein 1974).

Both wavevector z components are purely imaginary when

$$q^{x2} > \begin{cases} \kappa_1(\omega/c)^2 \\ \kappa_2(\omega/c)^2. \end{cases} \quad (15)$$

It is then clear from (12) that the surface part of the y component Green function (8) shows no pole in the zero-damping limit. However it is seen that the zx-plane Green functions (6) do have surface poles if κ_1 and κ_2

have opposite signs. The surface-polariton electric-field fluctuations are thus confined to the plane determined by the surface component of the fluctuation wavevector and the normal to the interface. The relevant denominator from eq. (6) can be written with the help of eqs. (1) and (2) in the form

$$\frac{1}{\kappa_1 q_2^z + \kappa_2 q_1^z} = \frac{\kappa_1 q_2^z - \kappa_2 q_1^z}{\kappa_1 - \kappa_2} \frac{1}{\kappa_1\kappa_2(\omega/c)^2 - (\kappa_1 + \kappa_2) q^{x2}}, \tag{16}$$

and the condition for a pole is $\omega = \pm\,\omega_{SP}$ where

$$q^{x2} = \frac{\kappa_1\kappa_2}{\kappa_1 + \kappa_2} \frac{\omega_{SP}^2}{c^2}. \tag{17}$$

This is the surface-polariton dispersion relation; the ω_{SP} that satisfies this relation with κ_2 also evaluated at ω_{SP} is the surface polariton frequency for wavevector q^x. The frequency and wavevector can both be real only if

$$\kappa_1\kappa_2 < 0 \quad \text{and} \quad \kappa_1 + \kappa_2 < 0. \tag{18}$$

In order to make further progress it is necessary to make some assumptions about the forms of the relative permittivities. We take κ_1 to be a real positive constant but assume that κ_2 has a transverse resonance at a far-infrared frequency ω_T, with the Lorentz form

$$\kappa_2 \equiv \kappa(\omega) = \kappa_\infty + \frac{(\kappa_0 - \kappa_\infty)\omega_T^2}{\omega_T^2 - \omega^2 - i\omega\Gamma} = \kappa_\infty \frac{\omega_L^2 - \omega^2 - i\omega\Gamma}{\omega_T^2 - \omega^2 - i\omega\Gamma}. \tag{19}$$

Here κ_0 and κ_∞ are the relative permittivities of medium 2 well below and well above the resonance at ω_T, and ω_L is the usual longitudinal frequency:

$$\omega_L = (\kappa_0/\kappa_\infty)^{1/2}\,\omega_T. \tag{20}$$

With the damping parameter Γ ignored for the present, κ_2 is real and the surface-polariton frequencies for which the requirements (18) are satisfied are easily shown to occupy the range given by

$$\omega_T < \omega_{SP} < \omega_M, \tag{21}$$

where the maximum surface-polariton frequency is defined by

$$\omega_M^2 = \frac{\kappa_1 + \kappa_0}{\kappa_1 + \kappa_\infty}\,\omega_T^2. \tag{22}$$

Figure 2 shows the surface polariton dispersion relation; the hatched area is the region occupied by the bulk polaritons in medium 2, where different parts of the region correspond to different q_2^z. There is no overlap between the surface and bulk polariton dispersion relations and there is therefore no possibility of linear coupling between the two kinds of excitation.

The surface polariton poles in the denominator of eq. (16) can be

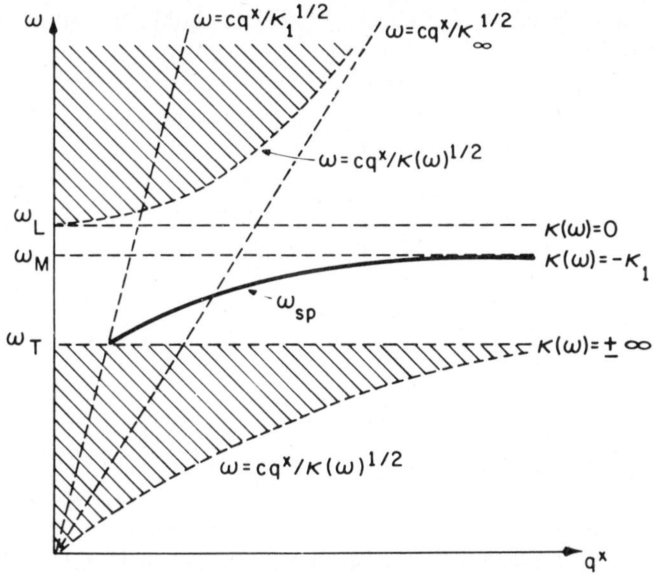

Fig. 2. Dispersion relation for single-interface surface polaritons. The hatched areas show the regions for which medium 2 supports bulk polaritons.

exhibited more explicitly by writing this equation in the form

$$\frac{1}{\kappa_1 q_2^z + \kappa_2 q_1^z} = -\frac{i\pi c}{2\omega} \frac{\kappa_1 q_2^z - \kappa_2 q_1^z}{\kappa_1^2 - \kappa_2^2} \left(\frac{\kappa_1 + \kappa_2}{\kappa_1 \kappa_2}\right)^{1/2} \frac{\partial \omega_{SP}}{\partial q^x} \times \{\delta(\omega - \omega_{SP}) + \delta(\omega + \omega_{SP})\}. \tag{23}$$

It is now a simple matter to evaluate the surface field fluctuations with the help of the fluctuation-dissipation theorems (10) or (11) and the first terms on the right of eq. (6). We give only the results for $z = z'$, where

$$\langle |E^x(z)|^2 \rangle_{q^x,\omega} = \frac{\hbar c}{A\epsilon_0 \omega} \{n(\omega) + 1\} \frac{\kappa_1^2 |q_2^z|^3}{\kappa_2(\kappa_1^2 - \kappa_2^2)} \left(\frac{\kappa_1 + \kappa_2}{\kappa_1 \kappa_2}\right)^{1/2} \frac{\partial \omega_{SP}}{\partial q^x} \times \{\delta(\omega - \omega_{SP}) + \delta(\omega + \omega_{SP})\} \exp(2|q_2^z|z), \tag{24}$$

$$\langle |E^z(z)|^2 \rangle_{q^x,\omega} = (q^x/|q_2^z|)^2 \langle |E^x(z)|^2 \rangle_{q^x,\omega}, \tag{25}$$

$$\langle E^x(z) E^z(z)^* \rangle_{q^x,\omega} = -\langle E^z(z) E^x(z)^* \rangle_{q^x,\omega} = (iq^x/|q_2^z|) \langle |E^x(z)|^2 \rangle_{q^x,\omega}. \tag{26}$$

These expressions give the electric field fluctuation power spectra responsible for the surface-polariton light-scattering cross section. The field penetration depth $1/|q_2^z|$ in medium 2 is typically of the order of a few microns for phonon surface polaritons. The behavior of surface polaritons in the far-infrared frequency region is thus determined entirely by the surface geometry and the relative permittivity within the first few microns of the crystal surface.

The above results are all derived for the limit of zero damping, with the parameter Γ in the relative permittivity eq. (19) set equal to zero. The effect of a small but finite Γ is obtained by developing the denominator in eq. (16) to the first order in the damping (Nkoma and Loudon 1975). The result is the same as eq. (23) except that the delta functions are replaced by Lorentzian lineshapes, for example

$$\delta(\omega - \omega_{SP}) \to \frac{\Gamma(\omega_{SP})/2\pi}{(\omega - \omega_{SP})^2 + [\Gamma(\omega_{SP})/2]^2}, \tag{27}$$

where

$$\Gamma(\omega_{SP}) = \frac{\kappa_1 \omega_{SP}^2 \omega_T^2 (\kappa_0 - \kappa_\infty)\Gamma}{\kappa_\infty(\kappa_1 + \kappa_\infty)(\omega_M^2 - \omega_{SP}^2)^2 + \kappa_1(\kappa_0 - \kappa_\infty)\omega_M^2\omega_T^2}. \tag{28}$$

This is the linewidth appropriate to a scan across the surface polariton dispersion relation at constant wavevector component q^x. The dependence of the linewidth on the surface polariton frequency is illustrated in fig. 3. It is seen that the linewidth has its maximum value at the upper limit of the surface polariton range, where

$$\Gamma(\omega_M) = \Gamma \tag{29}$$

and falls to the value

$$\Gamma(\omega_T) = \kappa_1\Gamma/(\kappa_1 + \kappa_0 - \kappa_\infty) \tag{30}$$

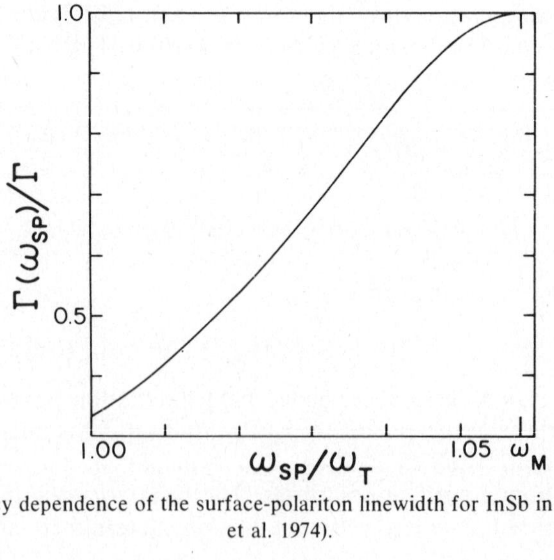

Fig. 3. Frequency dependence of the surface-polariton linewidth for InSb in air (from Nkoma et al. 1974).

at the lower end of the range. It should be stressed that the figure is plotted for a constant damping parameter Γ; in real materials the linewidth may show significant additional variation because of a frequency dependence in the basic damping parameter Γ (Ushioda and McMullen 1972).

There is little experimental information on the frequency linewidths of surface polaritons but their related spatial damping parameters have been measured in propagation experiments (Schoenwald et al. 1973, McMullen 1975). There seems to be reasonable agreement between experiment and theory although there is sometimes additional surface polariton damping when the interface is not perfectly smooth. We discuss the surface-roughness induced damping and frequency shift in sect. 5.

2.2. Double-Interface Surface Polaritons

We now consider the electric-field fluctuations in a double-interface structure of the kind illustrated in fig. 4. The three isotropic materials involved can be all different in general and their relative permittivities are indicated in the figure. Media 1 and 3 are assumed to have real and positive relative permittivities, similar to medium 1 in the single-interface geometry. The surfaces of contact of the materials and the z dimensions of media 1 and 3 are assumed to be of large extent compared to all characteristic lengths associated with the system excitations. However, the thickness d of medium 2 is allowed to take any value; this is the active material for the

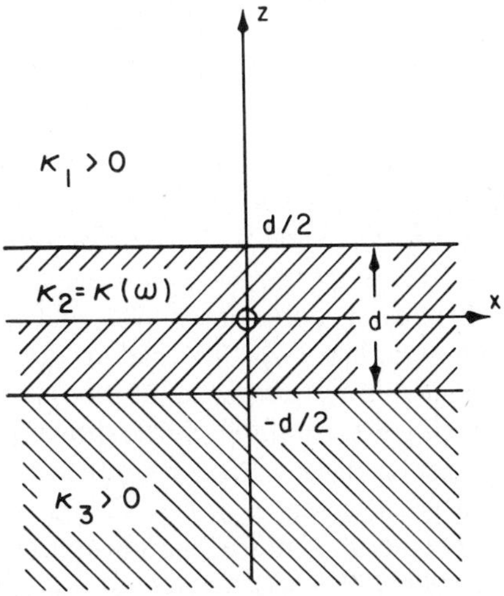

Fig. 4. Double interface geometry.

surface polaritons and its relative permittivity is assumed to have a far-infrared resonance as in eq. (19).

We again consider the electric-field fluctuations that have a given wavevector component q^x parallel to the interfaces; this parallel component of the wavevector is the same in all three media. The wavevector z components are given by equations like (1) and (2), and their real or imaginary characters determine the qualitative natures of the excitations. We define the surface modes to be those whose electric-field amplitudes fall off in media 1 and 3 with increasing distances from the interfaces. The z dependences of the fields thus have the forms

$$\begin{aligned} &\text{medium 1:} \quad \text{field} \sim \exp(\mathrm{i}q_1^z z), \\ &\text{medium 3:} \quad \text{field} \sim \exp(-\mathrm{i}q_3^z z), \end{aligned} \tag{31}$$

where,

$$\begin{aligned} q_1^{z2} &= \kappa_1(\omega/c)^2 - q^{x2}, \\ q_3^{z2} &= \kappa_3(\omega/c)^2 - q^{x2}, \end{aligned} \tag{32}$$

and the surface mode requirements are

$$q^{x2} > \begin{cases} \kappa_1(\omega/c)^2 \\ \kappa_3(\omega/c)^2 \end{cases} \quad \text{and} \quad \begin{cases} \operatorname{Im} q_1^z \geq 0 \\ \operatorname{Im} q_3^z \geq 0 \end{cases}. \tag{33}$$

Now consider the field variations in the slab of material 2. Because of its finite extent in the z direction, both signs of exponent are allowed in the spatial dependence of the field.

$$\text{medium 2:} \quad \text{field} \sim \exp(\mathrm{i}q_2^z z) \quad \text{and} \quad \exp(-\mathrm{i}q_2^z z), \tag{34}$$

where

$$q_2^{z2} = \kappa_2(\omega/c)^2 - q^{x2}. \tag{35}$$

In the limit of zero damping q_2^z can be either purely real or purely imaginary, and both cases fall within the definition of surface modes given in the previous paragraph. The modes that correspond to the field fluctuations with real q_2^z are called *guided-wave polaritons* and their properties are discussed in the following subsection. The modes that correspond to field fluctuations with imaginary q_2^z are called *double-interface surface polaritons* and we consider their properties in the present subsection.

The calculation of the electric-field fluctuations proceeds by the application of linear response theory to the double-interface structure, following the methods outlined in sect. 2.1. The Green functions analogous to eqs. (6) and (8) are considerably more complicated for the three-medium case (Mills and Maradudin 1975) and we do not quote the full results. It is again possible to distinguish bulk and surface contributions to the field fluctuations and our main concern as before is with the dispersion relations and existence conditions for the surface modes. This information is contained in the poles of the surface parts of the Green functions.

We therefore restrict attention to the Green function denominators, analogous to eq. (16) in the single-interface problem. For the electric field components in the zx-plane (TM modes), all the Green functions for pairs of points z and z' that lie in medium 2 have the denominator

$$(\kappa_1 q_2^z + \kappa_2 q_1^z)(\kappa_2 q_3^z + \kappa_3 q_2^z) + \exp(2iq_2^z d)(\kappa_1 q_2^z - \kappa_2 q_1^z)(\kappa_2 q_3^z - \kappa_3 q_2^z), \quad (36)$$

while the analogous Green function for field components parallel to the y-axis (TE modes) has the denominator

$$(q_2^z + q_1^z)(q_3^z + q_2^z) + \exp(2iq_2^z d)(q_2^z - q_1^z)(q_3^z - q_2^z). \quad (37)$$

In these expressions we have adopted the convention

$$\left.\begin{matrix}\mathrm{Re}\, q_2^z \\ \mathrm{Im}\, q_2^z\end{matrix}\right\} \geq 0. \quad (38)$$

The double-interface Green functions reduce to single-interface Green functions in the special cases where medium 2 is made the same as either medium 1 or medium 3. Thus it is seen that the second terms in (36) and (37) vanish in these special cases and the expressions reduce respectively to the denominators that occur in the surface contributions of the Green functions (6) and (8).

The double-interface surface polariton dispersion relation in the zero-damping limit is obtained from the zeros of the Green function denominators for purely imaginary q_2^z. The denominator (37) does not in fact have any zeros in this case, analogous to the lack of poles in the y polarization Green function (8) for the single-interface system. Thus the surface polaritons are TM polarized and their dispersion relation is obtained by setting (36) equal to zero. This result was derived earlier by Mills and Maradudin (1973) (see also Ushioda 1980). The dispersion relation is complicated, and the modes are best understood from plots of their dispersion relations. Figure 5 shows the two branches labelled UM (upper modes) and LM (lower modes) that are obtained when the relative permittivity $\kappa(\omega)$ of the form (19) is again used for κ_2, and κ_3 is assumed larger than κ_1. As before the surface polaritons occur only in the frequency range where κ_2 is negative, and the bulk polaritons of medium 2 occupy the same region as that shown by the shading in fig. 2. The bulk polaritons of media 1 and 3 occupy the regions to the left of the lines $\omega = cq^x/\kappa_1^{1/2}$ and $\omega = cq^x/\kappa_3^{1/2}$ respectively. The LM starts at the intersection of the latter line with the $\omega = \omega_T$ line and asymptotically approaches the frequency determined by $\kappa_2 \equiv \kappa(\omega) = -\kappa_3$. The UM formally starts at the intersection of the lines $\omega = cq^x/\kappa_1^{1/2}$ and $\omega = \omega_T$ and approaches the asymptotic frequency deter-

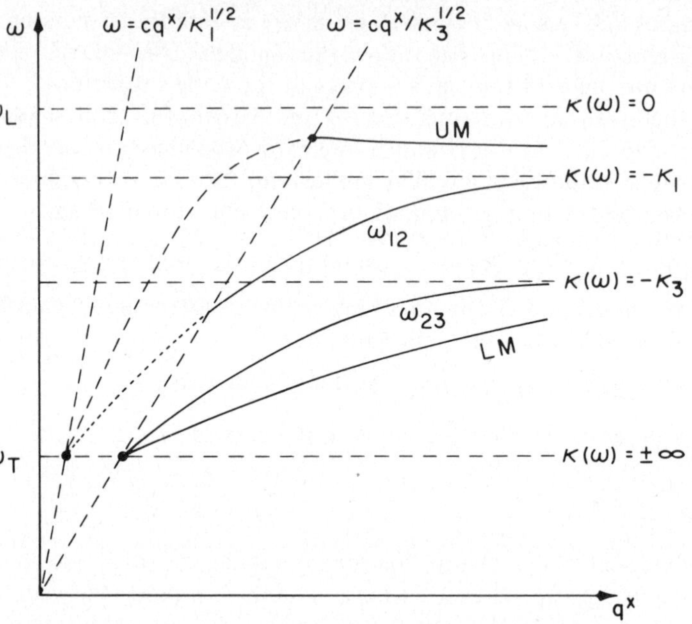

Fig. 5. Dispersion relations for the upper mode (UM) and lower mode (LM) double-interface surface polaritons. The curves labelled ω_{12} and ω_{23} are the thick-slab limiting forms of the UM and LM dispersion relations.

mined by $\kappa_2 \equiv \kappa(\omega) = -\kappa_1$. However the dashed part of the UM line lies in the bulk polariton region for medium 3, and coupling between the modes prevents the formation of true surface polaritons in this region.

The natures of the modes are clarified by considering the thick-slab limit, $|q_2^z|d \to \infty$, where the dispersion relation obtained from (36) simplifies to

$$(\kappa_1 q_2^z + \kappa_2 q_1^z)(\kappa_2 q_3^z + \kappa_3 q_2^z) = 0, \tag{39}$$

equivalent to the pair of relations

$$q^{x2} = \frac{\kappa_1 \kappa_2}{\kappa_1 + \kappa_2} \frac{\omega_{12}^2}{c^2} \tag{40}$$

$$q^{x2} = \frac{\kappa_2 \kappa_3}{\kappa_2 + \kappa_3} \frac{\omega_{23}^2}{c^2}. \tag{41}$$

These are a pair of single-interface surface-polariton dispersion relations similar to eq. (17). The frequencies ω_{12} and ω_{23} that satisfy the pole condition (39), analogous to ω_{SP} in eq. (17), have the wavevector dependence shown by the other pair of curves in fig. 5. The higher frequency mode ω_{12} again has a region of overlap with the bulk polaritons of medium 3, but the effects of coupling between the two kinds of mode tend to zero in the limit of an infinitely thick slab. The excitations on the opposite faces of the slab are independent in this case and the theory of sect. 2.1 applies separately to both. As the slab thickness is now reduced to a value where the second term in (36) is not longer negligible, the two single-interface modes for a given q^x interact, and the mode frequencies ω_{12} and ω_{23} move further apart to the UM and LM curves in fig. 5.

Another special case of interest for some of the experiments discussed later is the symmetrical geometry obtained when medium 3 is the same material as medium 1. The dispersion relation obtained by setting (36) equal to zero factorizes in this case to give separate dispersion relations for the two branches,

$$\kappa_2 q_1^z/\kappa_1 q_2^z = -\tanh(\tfrac{1}{2}|q_2^z|d) \qquad \text{(UM)}, \tag{42}$$

$$\kappa_2 q_1^z/\kappa_1 q_2^z = -\coth(\tfrac{1}{2}|q_2^z|d) \qquad \text{(LM)}. \tag{43}$$

Figure 6 illustrates the dispersion relations for a dielectric slab surrounded by vacuum where $\kappa_1 = \kappa_3 = 1$. Both UM and LM then start at the intersection of the vacuum light line with the line $\omega = \omega_T$, and they have the same asymptotic limit ω_M given by eq. (22). In the limit of a thick slab ($|q_2^z|d \to \infty$), both UM and LM approach the dispersion relation (17) for single-interface surface polaritons, shown by the curve labelled SIM in fig. 6. The symmetric slab modes were discussed by Kliewer and Fuchs (1966).

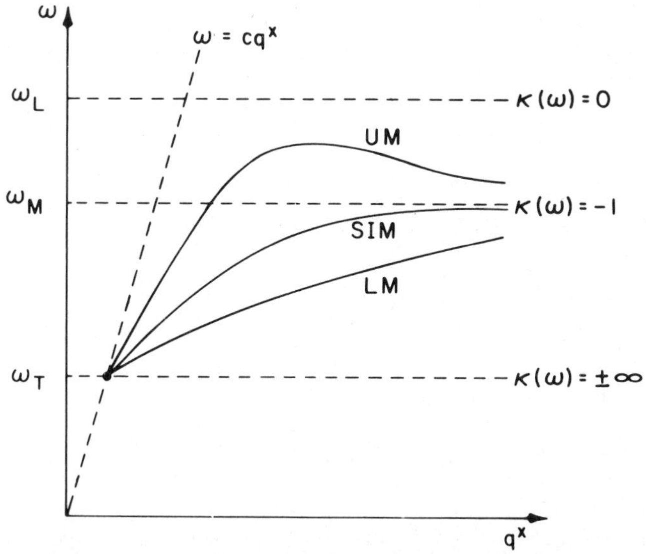

Fig. 6. Double-interface surface-polariton dispersion relations in a symmetric slab geometry. Both curves UM and LM tend to SIM in the thick-slab limit.

The above results are all derived in the limit of a real relative permittivity $\kappa(\omega)$. It is also possible to derive the effect of a nonzero damping Γ in the expression (19) for $\kappa(\omega)$. As in the single-interface geometry, the damping produces a surface polariton linewidth, and the double-interface expression analogous to eq. (28) has been derived and illustrated by Nkoma (1975).

2.3. *Guided-Wave Polaritons*

As defined in the previous subsection, the guided-wave polaritons correspond to the electric-field fluctuations of the double-interface geometry in fig. 4 that decay exponentially away from the interfaces in media 1 and 3 but have an oscillatory behavior with real q_2^z across the slab of material 2. All the electric-field Green functions have poles in the limit of zero damping when q_2^z is assumed real and there exist both TM guided-wave polaritons with dispersion relation obtained by setting (36) equal to zero, and TE guided-wave polaritons with dispersion relation obtained by setting (37) equal to zero.

The guided-wave polariton dispersion relations are quite complicated and we here focus attention on a special case that corresponds to the experimental results discussed later in the review. We suppose that media 1 and 3 are both vacuum so that

$$\kappa_1 = \kappa_3 = 1 \quad \text{and} \quad q_1^{z^2} = q_3^{z^2} = (\omega/c)^2 - q^{x^2}. \tag{44}$$

The TM guided-wave polariton dispersion relation obtained from (36) then factorizes into

$$\kappa_2 = \kappa(\omega) = (q_2^z/|q_1^z|)\tan(\tfrac{1}{2}q_2^z d), \tag{45}$$

and

$$\kappa_2 = \kappa(\omega) = -(q_2^z/|q_1^z|)\cot(\tfrac{1}{2}q_2^z d), \tag{46}$$

and the TE guided-wave polariton dispersion relation obtained from (37) factorizes into

$$1 = (q_2^z/|q_1^z|)\tan(\tfrac{1}{2}q_2^z d) \tag{47}$$

and

$$1 = -(q_2^z/|q_1^z|)\cot(\tfrac{1}{2}q_2^z d). \tag{48}$$

These results were derived by Kliewer and Fuchs (1966).

The dispersion relations (45) to (48) are similar in form to equations that occur in various other calculations, for example the determination of the quantum-mechanical energy levels of a particle in a one-dimensional square-well potential, or the calculation of Love and Sezawa acoustic-wave frequencies in a slab. Physically, the guided-wave polaritons have a standing-wave pattern across the slab as shown in fig. 7. Because the modes produce an electric-field excitation that extends into the vacuum on either side of the slab, their half-wavelength π/q_2^z does not exactly equal an integer divisor of the slab thickness. It can be shown by a graphical solution of the equations that

$$q_2^z = (m + \delta_m)\pi/d, \tag{49}$$

where

$$m = 0, 2, 4, 6, \ldots \quad \text{and} \quad 0 < \delta_m < 1 \tag{50}$$

for the solutions of eqs. (45) and (47), and

$$m = 1, 3, 5, 7, \ldots \quad \text{and} \quad 0 < \delta_m < 1 \tag{51}$$

for the solutions of eqs. (46) and (48). For all the modes,

$$\delta_m \to 0 \quad \text{for} \quad m \to \infty. \tag{52}$$

The dispersion curves for the TM and TE guided-wave polaritons are plotted in fig. 8(a) and fig. 8(b) respectively. They occupy the region to the right of the light line $\omega = cq^x$, the solutions to the left of the line being degenerate with electromagnetic waves in the vacuum outside the slab and therefore not confined to the vicinity of the slab. In the limit of a very thick

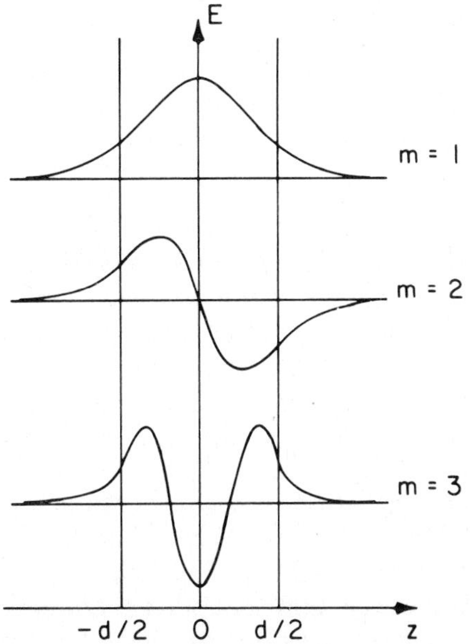

Fig. 7. Electric-field amplitude patterns for the first three guided-wave polaritons.

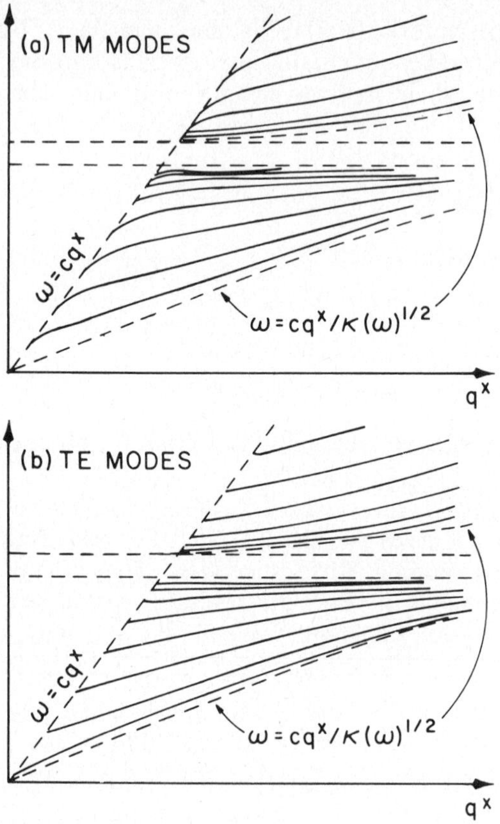

Fig. 8. Dispersion curves of guided wave polaritons in a slab placed in vacuum for (a) transverse magnetic modes and (b) transverse electric modes.

slab, $q_2^z d \to \infty$, the curves in fig. 8 crowd together to form a continuum and the effects of the region outside the slab become unimportant. The polariton curves in this case, including their extension to the left of the light line, just reproduce the shaded regions of fig. 2; the guided-wave polaritons in a thin slab are the remnant of the ordinary propagating bulk polaritons in a thick sample. The guided-wave polaritons are of course of most interest in thin slabs where the mode frequencies for a given q^x are well separated and can be resolved experimentally. We note from fig. 8 that the two kinds of guided-wave polariton dispersion differ in that the TM curves meet the light line tangentially while the TE curves meet the light line at relatively large angles.

This concludes the review of surface polaritons. The topics discussed are those needed to interpret the light-scattering experiments, and many other properties have been omitted, such as coupling of phonons with plasmons, and the effects of magnetic fields on such coupled modes and their associated surface polaritons. Details of these other effects can be found in a review article by Otto (1976) and in the other chapters in the present volume.

3. Theory of Surface-Polariton Light Scattering

The purpose of the present section is an account of the way in which linear response theory provides expressions for the light-scattering cross sections of surface polaritons. The complete derivations are lengthy and the resulting cross sections are complicated. We again give a fuller account of the single-interface case and the main aim is to illustrate the nature of the calculation and present the results that are needed in sect. 4 to interpret experimental spectra for single and double interface structures. More general background information on solid-state light scattering, or Raman scattering, spectroscopy can be found in the books by Cardona (1975) and by Hayes and Loudon (1978).

3.1. Geometrical Considerations

An important feature in the design of a light-scattering experiment is the proper arrangement of the directions of incident and scattered light beams to produce the most favorable coupling to the excitation under investigation. These considerations are particularly important for light scattering by surface excitations.

Suppose that incident light of frequency ω_I and wavevector $\boldsymbol{k}_I$ inside the crystal interacts with an excitation of frequency ω and wavevector $\boldsymbol{q}$ to produce scattered light of frequency ω_S and wavevector $\boldsymbol{k}_S$ inside the crystal. In first-order scattering the process involves the creation or destruction of a single quantum of crystal excitation, corresponding respectively to the Stokes and anti-Stokes components of the scattered light. The probability of the creation process exceeds that of the destruction process in the ratio $\{n(\omega)+1\}/n(\omega)$ and the correspondingly more intense Stokes scattering is that normally observed in the surface-polariton experiments. We consider only this component, and energy and momentum conservation then give

$$\omega_I = \omega_S + \omega, \tag{53}$$

$$\boldsymbol{k}_I = \boldsymbol{k}_S + \boldsymbol{q}. \tag{54}$$

The scattering angle θ is defined as the angle between $\boldsymbol{k}_I$ and $\boldsymbol{k}_S$, as shown in fig. 9.

With the optical frequencies and wavevectors related in the usual way

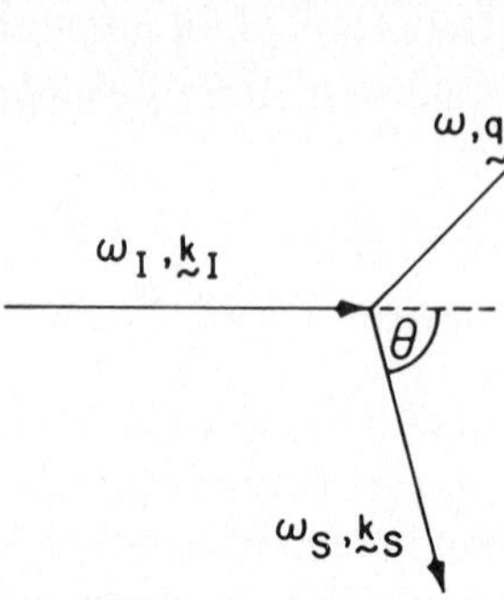

Fig. 9. Stokes light-scattering geometry.

for a material of relative permittivity κ, elimination of the scattered quantities from eqs. (53) and (54) leads to

$$\frac{c^2q^2}{\kappa\omega_I^2} = \frac{\omega^2}{\omega_I^2} + 4\left(1 - \frac{\omega}{\omega_I}\right)\sin^2\frac{\theta}{2}. \tag{55}$$

This expression relates the frequency ω and wavevector magnitude q of those crystal excitations that scatter incident light of frequency ω_I through an angle θ. Figure 10 shows the scans across (ω, q) space made in light-scattering experiments for various values of the scattering angle, where κ is assumed to be independent of the frequency. Since ω is much smaller than the incident frequency ω_I in most experiments, the accessible excitation wavevectors q extend from the light line up to a maximum of order

$$2\omega_I\kappa^{1/2}/c \approx 3\times 10^7\ \mathrm{m}^{-1} \tag{56}$$

for typical values of the parameters.

The scope of light-scattering experiments in the study of surface polaritons is appreciated by considering the superposition of fig. 10 with the appropriate polariton dispersion curves, for example figs. 2, 5, 6 or 8. Although the light lines in these various figures do not necessarily all have the same slopes, it can be seen that most of the surface polariton dispersion curves generally lie within the accessible region for study by light scattering. The more striking variations of surface polariton frequency usually occur at wavevectors of order $10^6\ \mathrm{m}^{-1}$ so that, referring to eq. (56) and fig. 10, small-angle scattering experiments are required to explore the more interesting regions of the polariton dispersion curves.

The above discussion considers a scattering interaction in the interior of the crystal, but there are further features to be taken into account for light scattering at surfaces, where the translational invariance of a crystal of unlimited extent is at least partially removed. In addition, the assumption

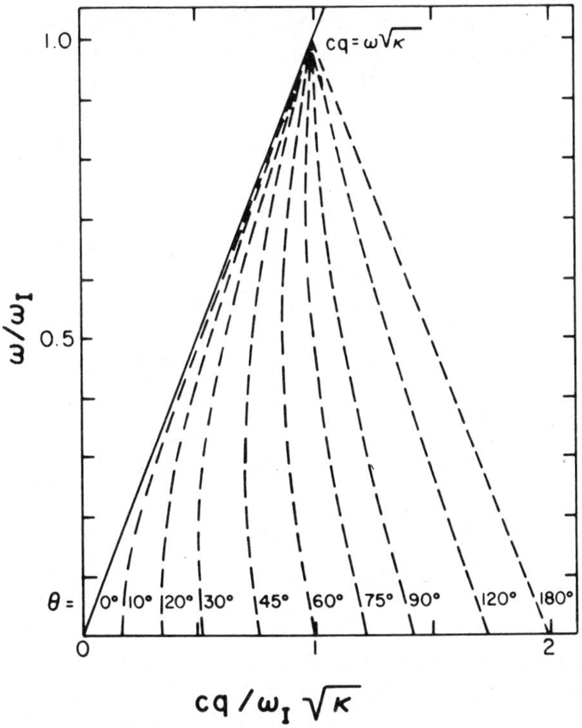

Fig. 10. Accessible region of the frequency ω and wavevector q values of a crystal excitation for study by scattering of incident light of frequency ω_I. The dashed lines show the experimental scans for various scattering angles (from Hayes and Loudon, 1978).

of real optical and excitation wavevectors, underlying the momentum conservation eq. (54), does not hold in most of the experiments on surface polaritons. The occurrence of imaginary wavevector components q^z is discussed in sect. 2; we here briefly discuss the nature of the optical wavevectors.

It is pointed out in sect. 4 that surface-polariton light scattering is weak and is difficult to detect owing to the small scattering angles that must be used. It is therefore essential to take advantage of the enhancement of the cross section that occurs under resonant scattering conditions, sometimes called the resonance Raman effect (see sect. 4.2.3. of Hayes and Loudon (1978) or the more extensive reviews of Martin and Falicov (1975), Richter (1976) and Bendow (1978)). The coupling of incident and scattered light to most crystal excitations occurs via the intermediary of the electrons. The cross sections are particularly large when the incident and/or scattered frequencies are close to electric-dipole transition frequencies of the crystal electronic states, for example interband transitions in semiconductors. These conditions lead of course to an associated absorption of the light, with $\boldsymbol{k}_{\mathrm{I}}$ and $\boldsymbol{k}_{\mathrm{S}}$ now complex wavevectors, that offsets to some extent the resonant effect in the cross section itself. Nevertheless, the net effect in resonant conditions is often a considerable enhancement in the scattering, and any realistic theory must take account of the resulting imaginary parts in the optical wavevectors.

Figure 11 represents the geometrical details of a surface scattering experiment with an incident beam of cross-sectional area a and wavevector $\boldsymbol{k}_{1\mathrm{I}}$ inclined at angle θ_{I} to the interface normal; the area of illuminated surface is $a/\cos\theta_{\mathrm{I}}$. In a single-interface system, the scattered light must be collected from the same surface, and the figure shows its wavevector $\boldsymbol{k}_{1\mathrm{S}}$ inclined at angle θ_{S} to the interface normal. The theory of the surface-scattering cross section has been developed by a number of authors, (Mills et al. 1970, Nkoma and Loudon 1975, Mills et al. 1976, Loudon 1978b) and we here summarize the main results. The differential cross section has the form

$$\frac{\mathrm{d}\sigma}{\mathrm{d}\Omega}=\frac{\kappa_{1}\omega_{\mathrm{I}}\omega_{\mathrm{S}}Aa\cos^{2}\theta_{\mathrm{S}}}{4\pi^{2}c^{2}\cos\theta_{\mathrm{I}}}\left|\frac{\boldsymbol{E}_{1\mathrm{S}}}{\boldsymbol{E}_{1\mathrm{I}}}\right|^{2}, \tag{57}$$

where $\mathrm{d}\Omega$ is the small solid angle of collection of scattered light, $\boldsymbol{E}_{1\mathrm{I}}$ is the electric vector of the incident light, $\boldsymbol{E}_{1\mathrm{S}}$ is the electric vector of the scattered light, and A as before is the total area of the interface. Apart from a geometrical prefactor, which is not of great importance for the present discussion, the cross section is essentially the ratio of the intensities in medium 1 of the scattered and incident light beams. Medium 1 (usually air) has a real relative permittivity, and the optical wavevectors are real.

The frequencies and orientations of the light beams in medium 1 deter-

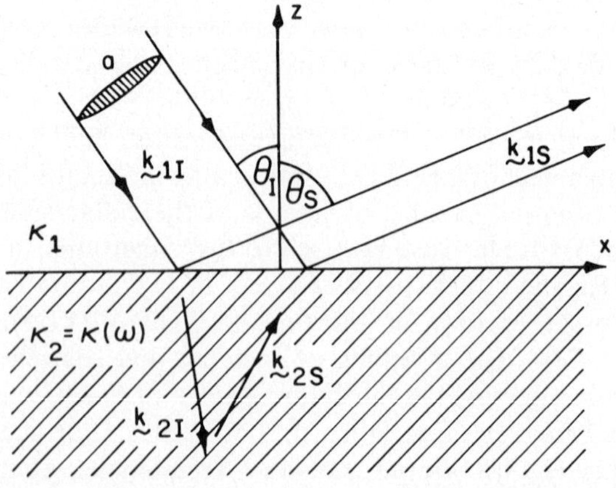

Fig. 11. Arrangement of light beams for backward scattering from a single interface.

mine the properties of the optical wavevectors $\boldsymbol{k}_{2I}$ and $\boldsymbol{k}_{2S}$ in medium 2 where the scattering interactions take place. There is conservation of wavevector parallel to the interface so that

$$k_{1I}^x = k_{2I}^x \quad \text{and} \quad k_{1S}^x = k_{2S}^x; \tag{58}$$

the x components of the wavevectors are therefore real. The complex z components of the wavevectors are determined from

$$k_{2I}^2 = \kappa(\omega_I)\omega_I^2/c^2, \quad \text{and} \quad k_{2S}^2 = \kappa(\omega_S)\omega_S^2/c^2. \tag{59}$$

The effect of resonant scattering conditions in causing absorption of the light beams is represented by the size of the imaginary part of $\kappa(\omega)$. The imaginary part is positive, and for the geometrical arrangement shown in fig. 11 we have

$$\left.\begin{matrix}\operatorname{Re} k_{2I}^z \\ \operatorname{Im} k_{2I}^z\end{matrix}\right\} \le 0, \qquad \left.\begin{matrix}\operatorname{Re} k_{2S}^z \\ \operatorname{Im} k_{2S}^z\end{matrix}\right\} \ge 0. \tag{60}$$

In order to derive an expression for the differential cross section, we need to find a relation between $\boldsymbol{E}_{1S}$ and $\boldsymbol{E}_{1I}$. The scattering process can be considered in three stages. In the first stage, the incident light beam is partially transmitted into medium 2, where its electric vector becomes $\boldsymbol{E}_{2I}$. In the second stage, the incident light couples to a crystal excitation of frequency ω and wavevector $\boldsymbol{Q}$ to form a polarization in medium 2 oscillating at the scattered frequency ω_S given by eq. (53), with the form

$$\boldsymbol{P}_S \exp\{i(\boldsymbol{k}_{2I} - \boldsymbol{Q}) \cdot \boldsymbol{r} - i\omega_S t\}. \tag{61}$$

It should be emphasized that $\boldsymbol{Q}$ is not the same as the wavevector $\boldsymbol{q}$ in eq. (54); although there is momentum conservation parallel to the surface, and we can write

$$Q^x = q^x = k_{2I}^x - k_{2S}^x, \tag{62}$$

there is in general no similar relation perpendicular to the surface where the wavevectors are complex. The third and final stage of the scattering process is the radiation of light by the oscillating polarization $\boldsymbol{P}_S$ in expression (61); the resulting radiated field in medium 1 forms the scattered light beam.

Let us consider for the moment only the third stage. The calculation of the fields generated by the polarization is essentially the same as that employed in the derivation of the electric-field Green functions in sect. 2, except that we now need the field in medium 1 rather than medium 2, and the frequencies now lie in the optical region. The calculation is straightforward and the result is

$$\boldsymbol{E}_{1S} = -\frac{\omega_S \mathbf{g} : \boldsymbol{P}_S \exp(i\boldsymbol{k}_{1S} \cdot \boldsymbol{r})}{\epsilon_0 c(Q^z - k_{2I}^z + k_{2S}^z)}, \tag{63}$$

where g is a 3×3 matrix whose coefficients, given by Mills et al. (1970) and Nkoma and Loudon (1975), express the matching through the interface of the 3 different field-polarization components in media 1 and 2.

It is clear from eq. (63) that the cross section (57) must include a factor

$$\frac{1}{[\mathrm{Re}(Q^z - k_{2\mathrm{I}}^z + k_{2\mathrm{S}}^z)]^2 + [\mathrm{Im}(Q^z - k_{2\mathrm{I}}^z + k_{2\mathrm{S}}^z)]^2}. \tag{64}$$

In the limit where the imaginary parts of the wavevectors all vanish, the factor produces a delta function for the z components, and the combination with eq. (62) then gives the usual wavevector conservation condition similar to eq. (54). This limit is not appropriate to surface-polariton light scattering where the z component of the excitation wavevector may be almost pure imaginary and the z components of the optical wavevectors are complex. Because of the signs of the wavevector components shown in expressions (60), the magnitudes of the optical wavevectors *add* with the same sign in (64) for the backward scattering experiment of fig. 11 where the scattered light emerges from the same surface as that illuminated by the incident light. The real term in the denominator of (64) tends to have a correspondingly large magnitude for the wavevectors employed in surface-polariton light-scattering experiments, leading to a small cross section.

Although the theory outlined above becomes considerably more complicated when applied to the scattering by surface polaritons in a double-interface system, many of the qualitative remarks remain valid, and in particular the cross section still contains the factor (64) (Chen et al. 1975, Nkoma 1975, Mills et al. 1976). The backward-scattering geometry of fig. 11 is still feasible but it is now also possible to have forward scattering in which the scattered light emerges from the opposite surface to that illuminated by the incident light. It is obvious that in forward scattering the signs of the scattered wavevector are opposite to those shown in expressions (60) and the optical wavevector magnitudes *subtract* in the denominator of (64) to give a larger cross section. In actual experimental situations, the forward scattering is 10^2 to 10^3 times more intense than the backward scattering. As will be seen in sect. 4, it has in fact not been possible to observe the backward surface-polariton light scattering from semi-infinite crystals experimentally.

Figure 12 illustrates the geometry used for forward scattering through a film of medium 2 placed in air or vacuum. Since the x components are conserved as in eqs. (58) and (62), the surface wavevector is given by

$$q^x = -k_{\mathrm{S}}^x = (\omega_{\mathrm{S}}/c)\sin\theta \approx (\omega_{\mathrm{I}}/c)\sin\theta \tag{65}$$

for $\omega \ll \omega_{\mathrm{I}}$ and $\boldsymbol{k}_{1\mathrm{I}}$ perpendicular to the film. This relation enables the surface-polariton dispersion curves to be determined from the measured Raman shifts ω at a series of scattering angles θ.

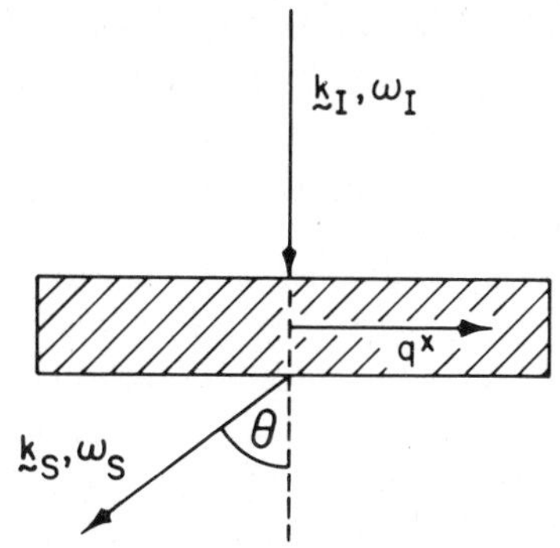

Fig. 12. Experimental geometry for surface-polariton scattering with the incident beam normal to the sample surface.

3.2. The Surface-Polariton Cross Section

In the previous subsection we considered the final stage of the three-stage scattering process and discussed the relative intensities of the forward and backward light scattering by surface polaritons. We now complete the theoretical discussion by including all three stages of the scattering process to obtain an expression for the cross section.

The first experimental results on surface-polariton light scattering were published by Evans et al. (1973). Before this time there had been some work on the theory of the scattering by Ruppin and Englman (1969) and Agranovich and Ginzburg (1972), but the experiments stimulated a large volume of further work. The first detailed explanation of the data was given by Chen et al. (1975); these authors identified the important differences between the forward and backward scattering geometries and showed the forward intensity to be the larger by a factor of 10^2 to 10^3, as discussed in sect. 3.1. It thus became apparent why earlier attempts to observe surface polaritons by backward scattering from opaque crystals had been unsuccessful. The theory also correctly accounted for the observed relative intensities of the bulk and surface polariton light scattering. A more complete version of this theory was given by Mills et al. (1976) (see also Mills and Subbaswamy 1980). Different but equivalent methods have been applied to the surface polariton scattering by Agranovich (1974) and by Agranovich and Leskova (1977). In the outline given below we follow another approach used by Nkoma and Loudon (1975) and Nkoma (1975), but all methods give the same expression for the cross section.

The calculation is essentially a development of the basic cross-section expression (57) into a form that explicitly refers to surface-polariton scattering. We give below some details of the form of the cross section for backward scattering from a semi-infinite medium and indicate the differences that occur in the much more complicated expression for the forward-scattering cross section in the double-interface geometry. The result given in eq. (63) represents the final stage in the scattering process and shows the field radiated by the polarization (61). We now consider the formation of the polarization by coupling of the incident light with the crystal excitation of frequency ω and wavevector $\boldsymbol{Q}$.

The first stage in the scattering process is the transmission of the incident light from medium 1 to medium 2 in the single-interface geometry of fig. 11. The field amplitudes are linearly related,

$$E^j_{2\mathrm{I}} = f^j E^j_{1\mathrm{I}}, \qquad j = x, y, z, \tag{66}$$

where the Fresnel coefficients f^j, listed for example by Nkoma and Loudon (1975), have forms easily obtained from the usual electromagnetic boundary conditions.

The behaviour of the surface polaritons is expressed in terms of the interface electric-field fluctuations treated in sect. 2. However, the polaritons also involve excitation of the far-infrared dipole-active lattice vibration whose transverse frequency ω_T controls the spectral range of the surface polaritons, as in expressions (21) and (22). Let W be the normal coordinate that measures the relative displacement of the positive and negative ions in the optic vibration. Then if Z is the effective charge of the optic mode, the coupling between W and the polariton electric field is governed by the usual Lorentz oscillator equation

$$\ddot{W} + \Gamma\dot{W} + \omega_\mathrm{T}^2 W = Z\boldsymbol{\zeta}\cdot\boldsymbol{E}, \tag{67}$$

where $\boldsymbol{\zeta}$ is the unit vector parallel to the electric-dipole moment of the mode. Thus for fluctuations of frequency ω, the degree of vibrational mode excitation associated with the polariton electric field $\boldsymbol{E}$ can be written

$$W = \beta\boldsymbol{\zeta}\cdot\boldsymbol{E}, \quad \text{where} \quad \beta = Z/(\omega_\mathrm{T}^2 - \omega^2 - \mathrm{i}\omega\Gamma). \tag{68}$$

The surface polariton thus has two coupled dynamical variables W and $\boldsymbol{E}$, and it is important to recognize that both make contributions to the scattering through their interactions with the incident field. The coupling of light to the dynamical variables is caused by their modulation of the optical relative permittivity of medium 2. On a microscopic level, the coupling occurs via virtual excitation of electron–hole intermediate states in a transition amplitude that is of third order in the electron interactions with the photons and optical phonons (Loudon 1963). On a macroscopic or phenomenological level, the coupling between incident light and surface polaritons to form a nonlinear polarization can be written

$$P^i_\mathrm{S} = a^{ij}E^j_{2\mathrm{I}}W^* + b^{ijh}E^j_{2\mathrm{I}}E^{h*}, \tag{69}$$

where we adopt the convention that repeated superscripts are summed over the directions x, y and z. The nature of the coupling coefficients a and b, first introduced by Burstein et al. (1968), and their relation to the microscopic theory of light scattering is fully discussed by Hayes and Loudon (1978); they vary with the incident frequency and contain the resonant effects described above in sect. 3.1.

The scattered field $\boldsymbol{E}_{1S}$ in medium 1 given by eq. (63) is now connected to the incident field $\boldsymbol{E}_{1I}$ in medium 1 and the polariton field $\boldsymbol{E}$ with the help of eqs. (66), (68) and (69). As in previous parts of the calculation, the x components of wavevectors are conserved at the interface but there are in principle contributions to the scattered field from crystal excitations that have any value of the wavevector component Q^z perpendicular to the surface. We assume a polariton electric-field fluctuation $\boldsymbol{E}(Q^z)$ of arbitrary wavevector component and sum over the Q^z to obtain from eq. (63) a total scattered field whose cartesian l component is

$$E^l_{1S} = -\frac{\omega_S E_{1I}\exp(\mathrm{i}\boldsymbol{k}_{1S}\cdot\boldsymbol{r})}{\epsilon_0 c}\sum_{Q^z}\frac{g^{li}f^j\epsilon^j_{1I}(a^{ij}\beta\zeta^h + b^{ijh})E^{h*}(Q^z)}{Q^z - k^z_{2I} + k^z_{2S}}, \tag{70}$$

where $\boldsymbol{\epsilon}_{1I}$ is a unit vector parallel to $\boldsymbol{E}_{1I}$.

The differential cross section is now obtained upon substitution of eq. (70) into eq. (57). The square modulus of the scattered field is clearly proportional to the square modulus of the surface-polariton electric field. The latter is a fluctuating quantity and we indicate its thermal average value by angle brackets as before. The total fluctuation of surface wavevector q^x can be expressed in terms of the power spectrum or frequency spectrum of the fluctuations in the usual way

$$\langle E^h(Q^z)E^{h'}(Q'^z)^*\rangle_{q^x} = \int \mathrm{d}\omega\langle E^h(Q^z)E^{h'}(Q'^z)^*\rangle_{q^x,\omega}. \tag{71}$$

The integral is removed from the cross section by differentiation of both sides of eq. (57) with respect to ω, or equivalently ω_S, to form the spectral differential cross section. The electric field power spectra are obtained from the corresponding Green functions by use of the fluctuation dissipation theorem (10) or (11).

The spectral differential cross section obtained in this way from eq. (57) is

$$\frac{\mathrm{d}^2\sigma}{\mathrm{d}\Omega\mathrm{d}\omega_S} = \frac{\kappa_1\hbar\omega_I\omega_S^3 A a\cos^2\theta_S}{4\pi^3\epsilon_0^2 c^4\cos\theta_I}\{n(\omega)+1\}$$

$$\times\,\mathrm{Im}\sum_{Q^z,Q'^z}\frac{g^{li*}f^{j*}\epsilon^j_{1I}(a^{ij*}\beta\zeta^h + b^{ijh*})\,g^{li'}f^{j'}\epsilon^{j'}_{1I}(a^{i'j'}\beta\zeta^{h'} + b^{i'j'h'})}{(Q^z - k^{z*}_{2I} + k^{z*}_{2S})(Q'^z - k^z_{2I} + k^z_{2S})}$$

$$\times\langle\langle E^h(Q^z); E^{h'}(Q'^z)^*\rangle\rangle_{q^x,\omega}. \tag{72}$$

The remainder of the calculation is straightforward but tedious. The

Green functions in eq. (72) are Fourier transforms of the spatially-dependent electric-field Green functions given in eqs. (6) and (8). With these results substituted, the summations over Q^z and Q'^z in eq. (72) are evaluated by contour integration (Loudon 1978b). The final complete result for the cross section is quite complicated.

A quite simple result is obtained however if attention is restricted to the part of the cross section that is controlled by the surface electric-field fluctuations, that is, the part coming from the first contributions in the Green functions (6) for electric-field components polarized in the xz-plane. Then, in the limit of small damping where q_2^z has only a small real part for ω in the surface-polariton frequency range, the summation in the cross section (72) is approximately equal to

$$-\frac{|\beta a_\mathrm{S}+b_\mathrm{S}|^2}{[\mathrm{Re}(-k_{2\mathrm{I}}^z+k_{2\mathrm{S}}^z)]^2+[\mathrm{Im}(q_2^z-k_{2\mathrm{I}}^z+k_{2\mathrm{S}}^z)]^2}\,\frac{|q_1^z|q_2^2}{|q_2^z|}\,\mathrm{Im}\,\frac{\mathrm{i}}{\kappa_1 q_2^z+\kappa_2 q_1^z}, \tag{73}$$

where

$$|a_\mathrm{S}|^2 = g^{li*}f^{j*}\epsilon_{1\mathrm{I}}^{j}a^{ij*}\zeta^h\epsilon^{h*}g^{li'}f^{j'}\epsilon_{1\mathrm{I}}^{j'}a^{i'j'}\zeta^{h'}\epsilon^{h'} \tag{74}$$

and the vector $\boldsymbol{\epsilon}$ in the zx-plane has components

$$\epsilon^x = -\mathrm{i}q_2^z/q_2, \qquad \epsilon^z = \mathrm{i}q^x/q_2. \tag{75}$$

The definitions of $|b_\mathrm{S}|^2$, $a_\mathrm{S}^*b_\mathrm{S}$ and $a_\mathrm{S}b_\mathrm{S}^*$ are similar to eq. (74).

It should be emphasized that the above discussion is intended to give only an outline of the main steps in the derivations of the single-interface surface-polariton light-scattering cross section; it ignores many of the subtleties covered in the references given at the beginning of the subsection. The main physical predictions of the theory are contained in the factor (73), which is seen to fulfill the properties anticipated in earlier sections. Thus the frequency shift ω and wavevector transfer q^x for maximum light scattering are determined by the same denominator as treated in eqs. (16) and (23); the cross section thus has peaks corresponding to the surface-polariton dispersion relation (17), with widths at constant q^x equal to the surface polariton linewidth (28). The first denominator in (73) is the same as (64), with the imaginary surface-polariton wavevector q_2^z substituted for Q^z. As discussed after (64), the fact that the magnitudes of $k_{2\mathrm{I}}^z$ and $k_{2\mathrm{S}}^z$ are added in the denominator (see (60)) produces a severe reduction in the intensity of backward scattering from a single-interface surface polariton.

The calculation of the corresponding cross section for the double-interface geometry of the slab or film represented in fig. 12 proceeds along similar lines. The added complexity of the optical properties of the double-interface system is accompanied by a substantial lengthening of the expressions that occur in the derivation of the cross section. However, the physical significance of the factors in the cross section remains the same. Thus the final denominator in the single-interface result (73) is replaced by the expressions (36) and (37) whose zeros determine the dispersion relations of the TM modes and the TE modes in the double-interface geometry. The cross section therefore shows peaks when the frequency shift ω and wavevector q^x satisfy the dispersion relations of either the double-interface surface polaritons of sect. 2.2 or the guided-wave polaritons described in sect. 2.3. The latter kind of pole in the cross section of a film was first identified by Subbaswamy and Mills (1978) and the guided-wave polariton dispersion relations were studied experimentally by Valdez et al. (1978). The other important factor in (73), the first denominator, also occurs in the double-interface cross section, and as already discussed in connection with (64), it leads to a considerable enhancement of the cross section for forward scattering over that for backward scattering.

Finally, the strength and symmetry of the scattering for both single and double interface arrangements are controlled by the same coupling coefficients $a^{ij}\zeta^h$ and b^{ijh}. The same coefficients also occur in the cross sections for light scattering by bulk polaritons (Burstein et al. 1968) and for bulk polar lattice vibrations. The symmetries of these coefficients with respect to interchanges of their Cartesian indices i, j and h are tabulated by Hayes and Loudon (1978) for the various crystal structures and vibrational mode characters; this information determines the relation between the intensity of scattering by the same surface polariton for different polarizations of incident and scattered light.

4. Experimental Method and Results

The conceptual scheme for Raman scattering is quite straightforward. One sends a monochromatic beam of light with known frequency ω_I, wavevector $\boldsymbol{k}_I$ and polarization $\boldsymbol{\epsilon}_{1I}$ into a sample and analyzes the corresponding properties ω_S, $\boldsymbol{k}_S$ and $\boldsymbol{\epsilon}_{1S}$ of the scattered light. From momentum and energy conservation (53) and (54) one determines the energy and the momentum of the excitations of the sample that scatter the incident light. The symmetry of the excitation is found from the polarization ($\boldsymbol{\epsilon}_{1I}$, $\boldsymbol{\epsilon}_{1S}$) of the incident and scattered light, and the spectral linewidth Γ of the scattered light measures the lifetime of the scattering excitation. The measured scattered light intensity is determined by the scattering cross section (72) and the efficiency of the spectrometer and detector system.

A schematic diagram of a typical Raman spectroscopy experiment is illustrated in fig. 13. A monochromatic beam of light from a laser is sent through a polarizer and a lens onto the sample. The scattered light emerging at some scattering angle θ is collected by the input optics

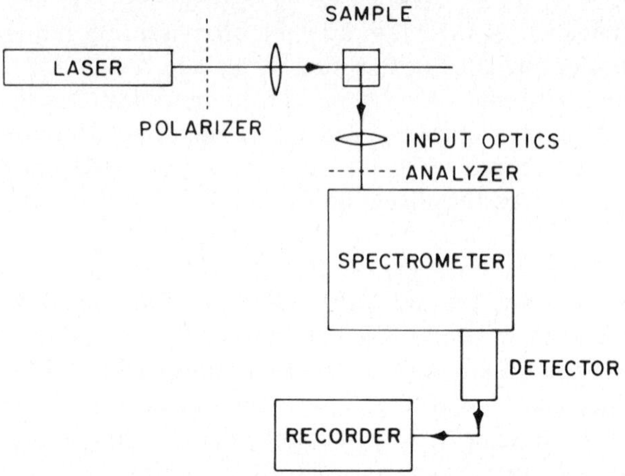

Fig. 13. Schematic diagram of Raman scattering experiment.

consisting of lenses and a polarization analyzer, and energy analyzed by a double-grating spectrometer. The scattered light intensity is measured by a photomultiplier and recorded electronically.

In actual experiments the set-up becomes complex and elaborate, because of the need to detect very weak scattered light in the presence of much stronger elastically scattered light that originates from crystal imperfections. Experimental difficulties involved in Raman scattering experiments can be best appreciated when the following typical situation is considered. A typical Raman signal from surface polaritons is $2 \sim 3$ photon counts/s when the incident laser power is 500 mW CW which corresponds to approximately 10^{18} photons/s in the green. In contrast the intensity of the elastically scattered light corresponds to 10^5–10^6 photons/s only about 100 Å away from the Raman peak. Thus the main effort in Raman scattering experiments is directed toward reduction of the elastically scattered light, minimizing the photomultiplier dark noise and maximizing the Raman signal by collecting as much of the scattered light as possible. For a given photon signal counting rate, the signal-to-noise ratio improves in proportion to the square root of the signal averaging time T. In order to signal-average for a long time, a minicomputer is interfaced to the spectrometer as the experiment controller and data accumulator (Ushioda, Valdez, Ward and Evans 1974). Such a scheme is depicted in fig. 14. In this system the spectrometer is stepped at a fixed interval and the Raman signals are counted for a pre-set period of time. Digital spectrum data are accumulated in the minicomputer memory and later used in digital analysis of the spectrum. A bank of relays under program control can be used to move mirrors, filters and polarizers in and out of the optical path of the experiment.

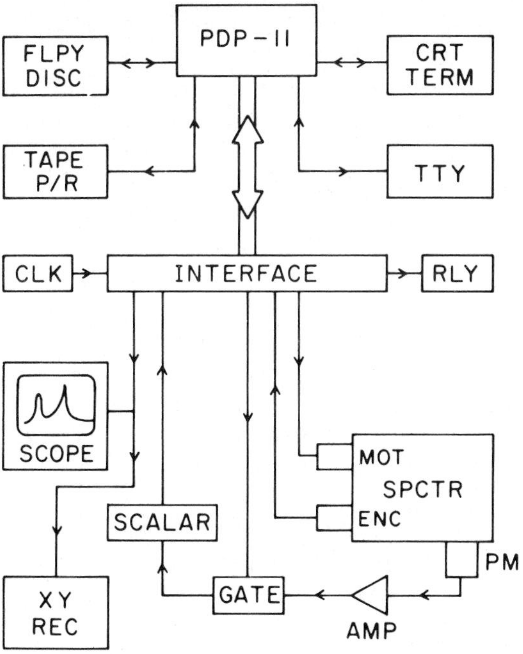

Fig. 14. Computer controlled Raman spectrometer system.

4.1. *Single-Interface Surface Polaritons*

The most straightforward method of measuring the Raman scattering from single-interface surface polaritons would be to measure the backward scattering from the surface of an opaque crystal as illustrated in fig. 11. However, as noted in the preceding section, the Raman scattering intensity in the backward direction is expected to be approximately 10^2 to 10^3 times weaker than the forward scattering intensity. In fact none of the attempts at observing surface polaritons in a backward scattering geometry from a bulk sample have been successful. Single-interface surface polaritons were actually observed in a thin slab of GaP, using a forward scattering geometry depicted in fig. 12 (Valdez and Ushioda 1977). The thickness d of the sample slab was chosen to be large enough so that the thick slab limit ($|q_2^z|d \to \infty$) is applicable. Then the UM and the LM branches, (42) and (43), of the double-interface geometry become degenerate and the dispersion relation is given by eq. (17) appropriate for single-interface surface polaritons. Thin samples are needed, because GaP is strongly absorbing at argon ion laser frequencies (5145 Å and 4880 Å) due to the proximity of the E_0 gap at 2.78 eV (4471 Å) and the indirect gap at 2.25 eV (5524 Å) both at room temperature. By choosing an optimum sample thickness one can take advantage of resonance enhancement of the Raman cross section (Scott et al. 1969, Belle et al. 1973, Weinstein and Cardona 1973) and at the same time satisfy the thick slab limit.

The dielectric properties of GaP are very well characterized near the optical phonon frequency ($\omega_T = 367\ \mathrm{cm}^{-1}$ and $\omega_L = 403\ \mathrm{cm}^{-1}$). In fact GaP is the crystal in which the bulk polaritons were first observed by light scattering (Henry and Hopfield 1965) and in which the surface polariton (SIM) dispersion was measured by the ATR method (Marschall and Fischer 1972).

The sample used by Valdez and Ushioda (1977) was an oriented single crystal of GaP polished down to a thickness of 20 microns. The large parallel faces were the (111) plane and the orientations of $[1\bar{1}0]$ and $[\bar{1}\bar{1}2]$ crystal axes were known in the (111) plane. For shorthand reference purposes we will denote the three crystal axes by $\hat{x}' = [111]$, $\hat{y}' = [1\bar{1}0]$ and $\hat{z}' = [\bar{1}\bar{1}2]$. The bulk Raman tensors for the optical phonon of GaP with respect to these reference axes are given by:

$$R(x') = \frac{R}{\sqrt{3}}\begin{bmatrix} 2 & 0 & 0 \\ 0 & -1 & 0 \\ 0 & 0 & -1 \end{bmatrix}; \qquad R(y') = \frac{R}{\sqrt{3}}\begin{bmatrix} 0 & -1 & 0 \\ -1 & 0 & -\sqrt{2} \\ 0 & -\sqrt{2} & 0 \end{bmatrix};$$

$$R(z') = \frac{R}{\sqrt{3}}\begin{bmatrix} 1 & 0 & -1 \\ 0 & \sqrt{2} & 0 \\ -1 & 0 & -\sqrt{2} \end{bmatrix}, \tag{76}$$

where R is a common factor arising from the scattering matrix elements. These tensors were obtained by rotating the Raman tensor tabulated by Loudon (1964). In terms of a^{ij} and b^{ijh} defined in eq. (69), these tensors can be expressed as

$$R_{ij}(h) = a^{ij}\beta\zeta^h + b^{ijh}; \quad i, j, h = x', y', z'. \tag{77}$$

Figure 15 shows the room temperature Raman spectra of this sample in the frequency region of surface polaritons for several scattering angles. These data were collected using the 5145 Å line of the argon ion laser as the incident light. The incident laser power of approximately 400 mW CW was directed normal to the sample slab in the geometry illustrated in fig. 12. The large peak at $403\ \mathrm{cm}^{-1}$ is the LO phonon peak of bulk GaP and the small peak pointed by an arrow is the surface polariton peak. It is seen that the surface polariton peak shifts to lower frequencies for smaller scattering angles corresponding to smaller values of q^x determined by (65). When the peak position is plotted against the scattering angle as illustrated in fig. 16, we see that the peak frequency follows a smooth dispersion curve. The solid curve in fig. 16 is the theoretical dispersion curve drawn according to eq. (17) for $\kappa_1 = 1$ and κ_2 given by eq. (19) with $\omega_T = 367.3\ \mathrm{cm}^{-1}$, $\omega_L = 403.0\ \mathrm{cm}^{-1}$ and $\kappa_\infty = 9.091$ (Parsons and Coleman 1971). From fig. 16 we see

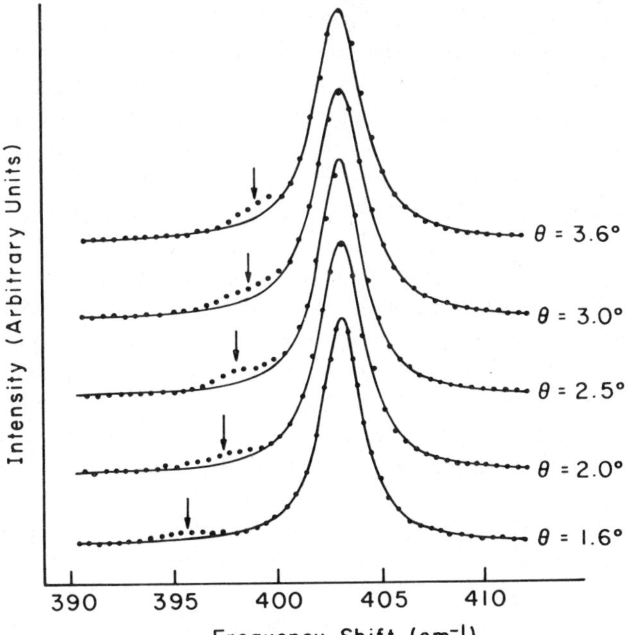

Fig. 15. Raman spectra of surface polaritons in a (111) slab of GaP (20 microns).

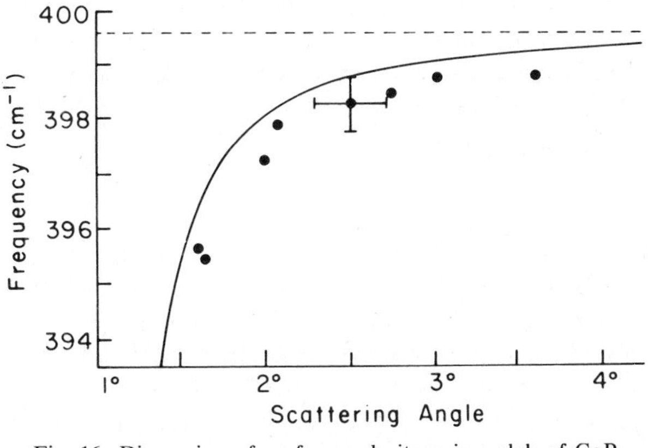

Fig. 16. Dispersion of surface polaritons in a slab of GaP.

that the measured surface polariton dispersion is in good agreement with the theoretical prediction based on the known dielectric parameters of GaP. We note, however, that the data points lie systematically below the theoretical curve. This difference is believed to arise from the presence of roughness of the sample surface, and this point will be discussed further in sect. 5.

In obtaining the data shown in fig. 15 the scattered light with all polarizations was collected. Thus no information on the selection rule was obtained. Moreover, all the light scattered into a cone of the apex angle θ about the direction of the incident beam (normal to the sample surface) was collected. This means that the direction of the surface polariton wavevector q^x in the (111) plane was not specified. In order to determine the Raman scattering selection rule, it is necessary to specify the polarization of the incident and scattered light ($\boldsymbol{\epsilon}_{\mathrm{II}}$, $\boldsymbol{\epsilon}_{\mathrm{IS}}$) as well as the direction of the wavevector q^x with respect to the crystal axes. The measurements to determine the selection rule were made at the scattering angle of $\theta = 2.5°$ (outside crystal) for four configurations of $\boldsymbol{\epsilon}_{\mathrm{II}}$, $\boldsymbol{\epsilon}_{\mathrm{IS}}$ and q^x. The relative intensities measured for the four configurations are tabulated in table 1. The symbols $(y'y')$ and $(y'z')$ signify the polarization directions of the incident and scattered light by ($\boldsymbol{\epsilon}_{\mathrm{II}}$, $\boldsymbol{\epsilon}_{\mathrm{IS}}$), and the orientation of the x', y' and z' axes with respect to the crystallographic axes were defined earlier in this section. The predicted relative intensities based on the bulk Raman tensors (76) and the electric field amplitude for surface polaritons are indicated in parentheses (Ushioda 1980). Since the accuracy of the relative intensity measurements are estimated to be ± 40%, the agreement between the observation and the prediction can be considered satisfactory. Thus we can conclude that the surface polariton scattering selection rule can be found from the bulk Raman tensors. This conclusion is expected on physical grounds, because surface polaritons are macroscopic waves whose wavelength is much greater than the atomic spacing and the penetration depth contains many unit cells of the crystal. So the scattering intensity depends only on the macroscopic properties of the crystal in the same way as bulk excitations.

4.2. *Double-Interface Surface Polaritons*

Raman scattering from double-interface surface polaritons was observed in a thin film of GaAs on a sapphire substrate (Evans et al. 1973). The dispersion branch that was measured in this work was the lower mode

Table 1
Raman scattering selection rule

	Polarization	
q^x direction	(y', y')	(y', z')
$q^x \parallel y'$	1 (1)	2 (2.6)
$q^x \parallel z'$	3 (3.6)	<0.2 (0)

(LM) of fig. 5. The upper mode (UM) lies within the linewidth of the LO phonon of GaAs and it was not possible to resolve the UM surface polariton peak. In a separate and more elaborate experiment Prieur and Ushioda (1975) were able to resolve the peak due to the UM surface polaritons and to obtain the dispersion relation for the same GaAs film on a sapphire substrate.

The sample used for these experiments was grown by Manasevit (Manasevit and Thorsen 1970) by chemical vapor deposition (CVD) of GaAs on the (0001) surface of sapphire. The film was approximately 2500 Å thick and polycrystalline. Thus no data on the Raman scattering selection rule were obtained. The carrier density in the film was sufficiently low so that no effect due to plasmons was observed, and the LO phonon peak appeared at $\omega_L = 292\,\mathrm{cm}^{-1}$ appropriate for the bulk of intrinsic GaAs. The TO phonon frequency of GaAs is $\omega_T = 270\,\mathrm{cm}^{-1}$.

The incident laser used for the measurement of the LM surface polariton scattering was the 4880 Å line of an argon ion laser operating at approximately 400 mW CW. At this wavelength sapphire is completely transparent, but GaAs is quite opaque, because the photon energy (2.77 eV) is far above the energy gap at 1.43 eV. The penetration depth of the 4880 Å light in GaAs is approximately 900 Å calculated from the measured extinction coefficient (Seraphin and Bennett 1967). Thus the film thickness is a little less than three times the penetration depth, so that the scattered light in a near-forward direction can be measured in a transmission scattering geometry. The laser beam was incident on the sample from the sapphire side normal to the surface in the geometry illustrated in fig. 12. The polarization of the incident light was linear and the scattered light polarization was not analyzed.

Figure 17 illustrates the Raman spectra of the GaAs film for different scattering angles θ in the frequency range between the bulk TO and LO phonon peaks at $270\,\mathrm{cm}^{-1}$ and $292\,\mathrm{cm}^{-1}$, respectively. The small peak (pointed by an arrow) between the two bulk phonon peaks was found to be the LM of the double-interface geometry consisting of air, GaAs and sapphire. The scattering angle dependence of the frequency of this peak is plotted in fig. 18. The solid curve is the theoretical dispersion curve for the air/GaAs/sapphire system corresponding to (36). Since sapphire is a uniaxial crystal and the GaAs/sapphire interface is normal to the c-axis, two different relative permittivities $\kappa_{3\parallel}$ (parallel to c-axis) and $\kappa_{3\perp}$ (perpendicular to c-axis) enter the dispersion relation. The relative permittivity for air κ_1 is taken to be unity and the relative permittivity $\kappa(\omega)$ for GaAs is given by eq. (19) with $\kappa_\infty = 11.1$ (Mooradian and McWhorter 1969). The dispersion relation appropriate for the air/GaAs/sapphire system used by Evans et al. (1973) was obtained by Mills and Maradudin (1973), and is given by letting $\kappa_3 = \kappa_{3\perp}$ in (36) and setting the whole expression equal to zero. The

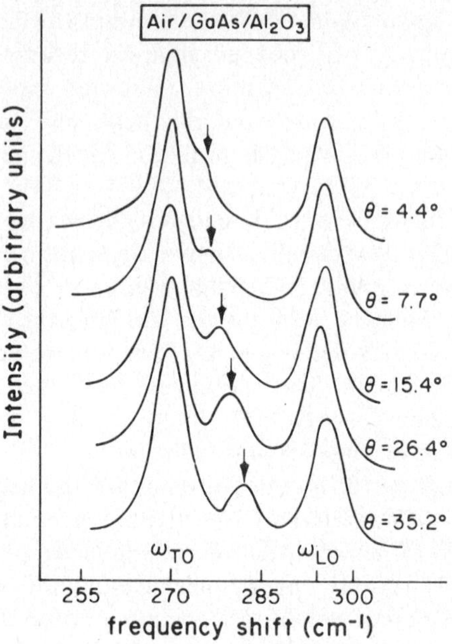

Fig. 17. LM spectra of GaAs film on sapphire.

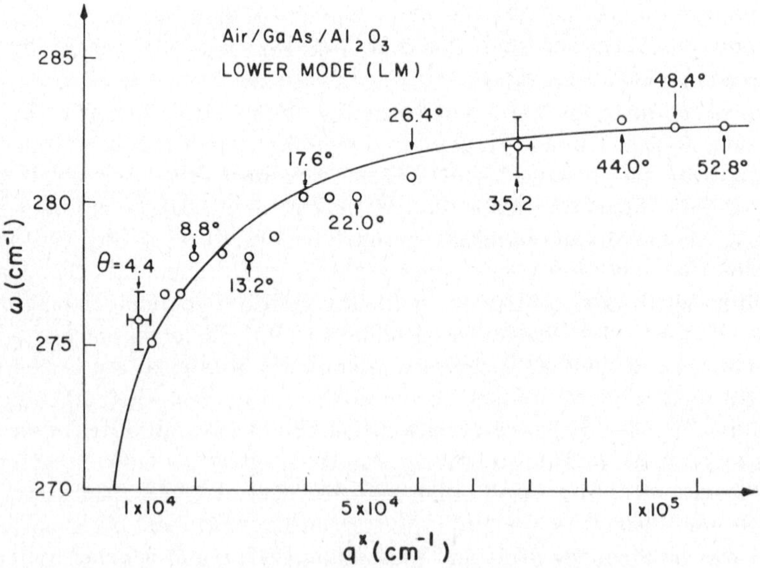

Fig. 18. LM dispersion of GaAs film on sapphire.

expression for q_3^z for this special case becomes:

$$q_3^{z2} = \frac{\kappa_{3\perp}}{\kappa_{3\parallel}}[\kappa_{3\parallel}(\omega/c)^2 - q^{x2}]. \tag{78}$$

The numerical values for the relative permittivities of sapphire used in calculating the dispersion curve of fig. 18 are $\kappa_{3\perp} = 3.2$ and $\kappa_{3\parallel} = 3.1$ (Barker 1963), and the thickness d was set equal to 2500 Å. In fig. 18 we see that the agreement between the measured dispersion and the theoretical prediction based on expression (36) is very good in general. However, at small scattering angles the LM peak starts to merge with the high frequency side of the bulk TO phonon peak, and as a result the frequency determination becomes difficult and inaccurate. The discrepancy between the data and theory seen for the $\theta = 4.4°$ data point results from this kind of difficulty.

In the spectra of fig. 17 only the LM peak can be seen clearly. However, according to theory the UM should be observable between the LM peak and the bulk LO phonon peak as illustrated in fig. 5. If one looks at the spectrum for $\theta = 15.4°$ in fig. 17 for instance, one can recognize an asymmetry in the line shape of the LO phonon peak. It was found that this slight asymmetry results from the presence of the UM scattering on the low frequency side of the LO peak (Prieur and Ushioda 1975). Calculation based on expression (36) shows that the peak position of the UM lies within the linewidth of the LO peak. Thus in order to measure the dispersion of the UM, it was necessary to extract the UM peak position by a detailed analysis of the LO phonon peak. This was done by taking advantage of the large difference in the scattering intensities of surface polaritons for forward and backward directions. While the scattering intensities of the bulk phonons are comparable in both directions, the surface polariton scattering is weaker by a factor of 10^2–10^3 in the backward direction than in the forward direction. Since the scattered photon counting rate in the forward direction is less than 10 cps at the peak, the surface polariton contribution in a backward scattering spectrum is negligible. Thus Prieur and Ushioda (1975) measured the spectra for forward and backward directions and took the difference between the two spectra by using a digital subtraction method. By appropriately adjusting the scale of the forward and backward scattering spectra to null, the difference between them for the bulk phonon peaks, they could find the difference spectra which showed the UM peak clearly. This subtraction procedure is illustrated in fig. 19, which shows the forward and backward scattering spectra as well as the resultant spectrum of the LM and UM after subtraction. By following this data analysis procedure for several scattering angles, Prieur and Ushioda (1975) obtained the dispersion curves for the UM as well as the LM for the 2500 Å film of GaAs on sapphire. The resultant dispersion curves are shown in fig. 20. These measurements were

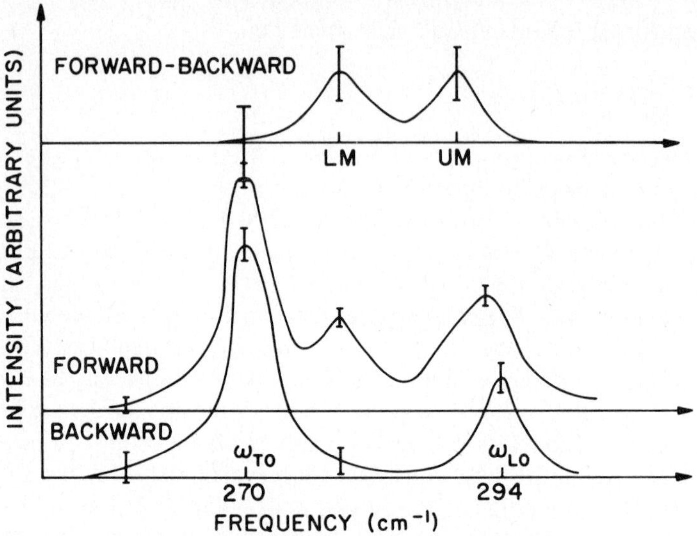

Fig. 19. Extraction of the UM spectrum.

made using the 5145 Å line of the argon ion laser operating at about 400 mW CW. All the data were taken at room temperature, and no polarization analysis was made.

The data points indicated by open circles in fig. 20 were obtained by the method just described. Because of the difficulty with the signal-to-noise problem involved in extracting the difference spectra, they were able to obtain only two data points on the UM dispersion curves as illustrated in fig. 20. However, they could confirm the existence of the UM by another approach. When one examines the behavior of the UM dispersion curve given by setting the denominator (36) equal to zero, one finds that the UM dispersion curve is depressed when the relative permittivity κ_1 is increased by replacing air by an optically more dense material. One also sees that the LM dispersion curve is not affected much by changes in κ_1; this is understandable because the LM originates from the single-interface modes of the GaAs/sapphire interface. The UM on the other hand originates from the single-interface modes of the GaAs/external-medium interface. Thus its dispersion is sensitive to changes in κ_1. Prieur and Ushioda (1975) used this effect in order to shift the UM peak out of the linewidth of the bulk LO peak. This kind of experiment was performed using benzene as medium 1 in contact with the GaAs film, forming a three-layered structure benzene/GaAs/sapphire. Benzene was selected as medium 1 because of its ease of handling and more importantly because its relative permittivity in the far infrared is well characterized (Zelano and King 1970).

The scattering geometry used in these experiments is illustrated in fig. 21

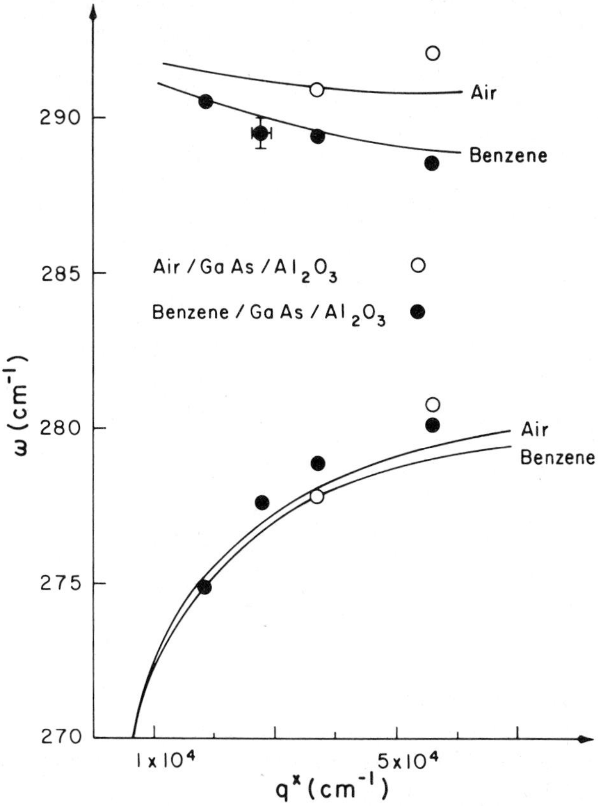

Fig. 20. Dispersion curves for the UM and the LM in air/GaAs/sapphire and benzene/GaAs/sapphire.

where the two optical paths for forward and backward scattering are indicated by the solid line and the dashed line, respectively. The moving mirror that selects the scattering direction is rotated in and out of the laser beam by a solenoid which is controlled by one of the relays interfaced to the minicomputer in fig. 14. Using this scheme the spectra for the two scattering directions can be obtained at once without stepping the grating position of the spectrometer, and thus exact difference spectra can be obtained. The GaAs/sapphire sample was held in an optical cell which can be filled with benzene.

The data points obtained for the benzene/GaAs/sapphire sample are shown in fig. 20 by solid circles. We note that the peak position of the UM is significantly depressed below the UM frequencies for the air/GaAs/sapphire configuration as expected from theory. The LM frequencies are not affected within the accuracy of the measurement. The solid curves for the UM and the LM in fig. 20 are plotted using (36) and appropriate values for

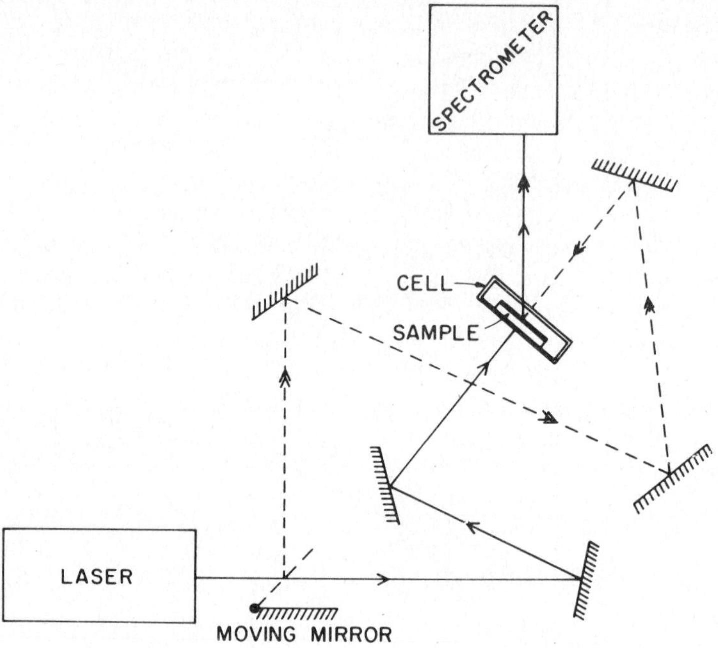

Fig. 21. Scattering geometry for taking the difference spectra between forward and backward scattering.

κ_1, $\kappa_{3\parallel}$, $\kappa_{3\perp}$ and d. Figure 20 shows good agreement between theory and measured dispersion for both air/GaAs/sapphire and benzene/GaAs/sapphire configurations. The sensitivity of the UM dispersion curve to the relative permittivity of the external medium (air or benzene here) is an interesting effect that can be exploited in certain applications.

4.3. *Guided-Wave Polaritons*

The search for Raman scattering from guided-wave polaritons (GWP) was motivated by a theoretical calculation presented by Subbaswamy and Mills (1978). They performed a computer simulation of the Raman spectra of thin films based on the full theoretical expression given by Mills et al. (1976), and found that the peaks due to GWP appear in the spectrum with a comparable strength to that of SIM, LM and UM. As fig. 8 shows the regions of the q_2^x–ω space where GWP are found do not overlap the regions in which other surface modes appear. Thus the GWP were not detected in earlier experiments looking for SIM, LM or UM.

The search for GWP peaks in Raman spectra was performed on thin slabs of GaP varying in thickness between 5 microns to 125 microns. These samples were prepared by polishing a single crystal of GaP and the sample thickness was determined by optical absorption measurements and also by a mechanical gauge. The thickness measured by the two methods agreed with each other within ± 10%. The large sample faces were either (111) or (100).

Typical GWP spectra of a 30 micron thick slab of GaP sample are shown in fig. 22. These spectra were obtained with the incident laser at 5145 Å and at the power level of approximately 100 mW CW. Typical counting rates for GWP peaks were in the range of 0.1 to 1 count/s. In the spectra shown in fig. 22, the bulk mode peaks appear at 367 cm^{-1} and 403 cm^{-1} (TO and LO phonon peaks, respectively), and a series of small peaks are observed below the TO phonon frequency. These peaks were found to be due to GWP. The GWP peaks appear only in the forward scattering geometry shown in fig. 22, and they do not appear at all in the backward scattering

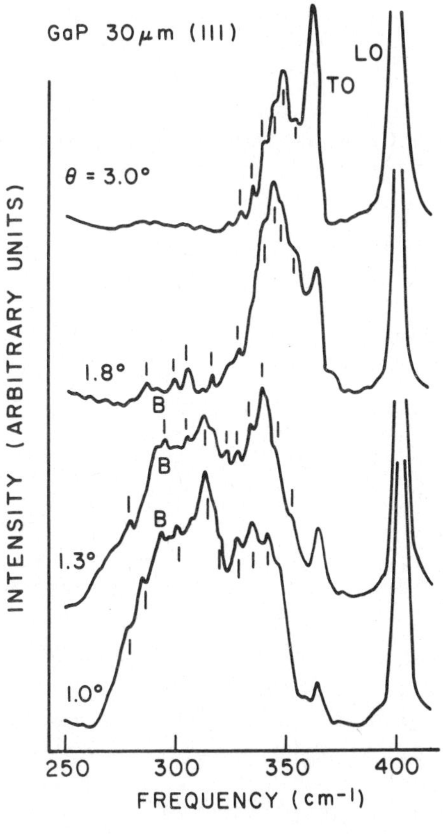

Fig. 22. GWP spectra of a GaP slab.

geometry. The peak labelled by "B" in fig. 22 was seen in the backward scattering geometry also. Thus this peak is not due to GWP. In order to rule out the possibility that some of the peaks seen below the TO phonon frequency are due to second order Raman processes, the sample temperature was lowered to 77 K. The peaks in fig. 22 were observed with similar relative intensities at 77 K, showing that these peaks are not second order Raman features.

Figure 23 illustrates the spectra for different sample thickness and orientations. In a (100) sample GWP are not seen. This result agrees with the prediction given by Subbaswamy and Mills (1978). In a very thick sample (125 microns) all the GWP branches for different m's of (50) and (51) bunch together close to the bulk polariton frequency, and we observe only one broad peak labelled by "π" in fig. 23(b). In the opposite limit of a very thin sample (6 microns), the GWP peaks separate with larger distances between them as seen in fig. 23(c). Also the upper branches of GWP above the LO phonon frequency are seen in fig. 23(c).

The observed frequencies of GWP in a 30 micron (111) slab of GaP are plotted in fig. 24 where the theoretical dispersion curves obtained from eqs. (45) to (48) are also shown. For each angle θ the nearly vertical lines define the trajectory on which eq. (55) is satisfied. Thus the GWP peaks are expected to appear at the crossing points of these curves with the dispersion curves of GWP. We see that the GWP peak positions indicated by circles indeed fall on the crossing points of the two sets of curves. Thus it

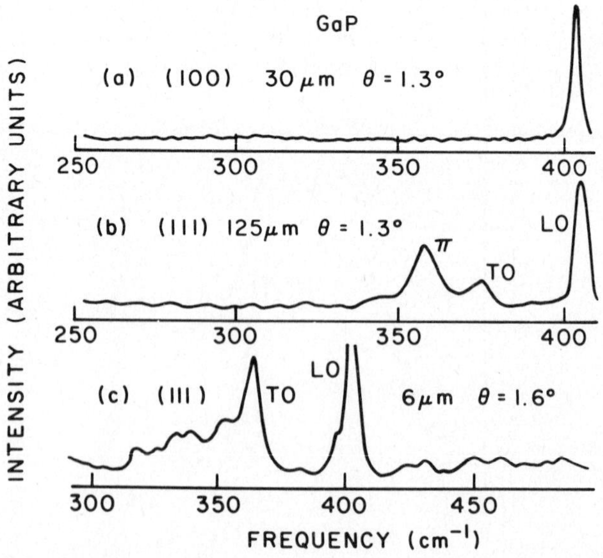

Fig. 23. GWP spectra for samples of different thickness and orientation.

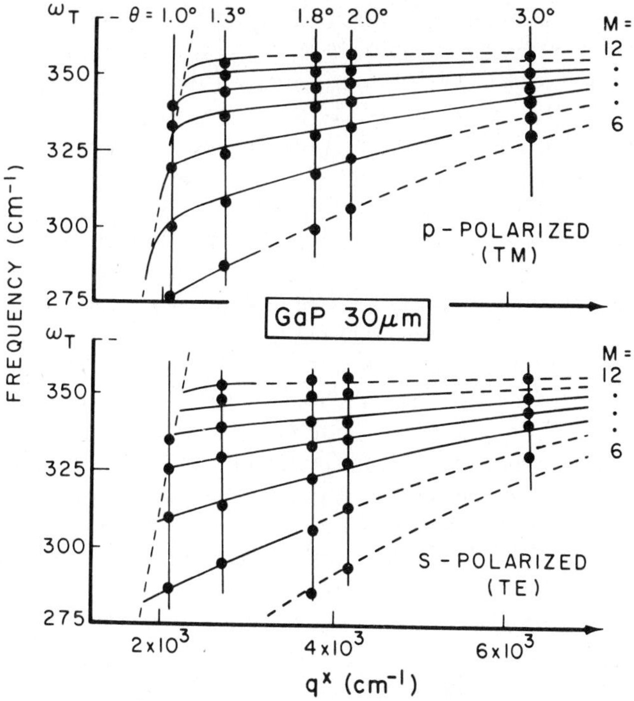

Fig. 24. Dispersion of GWP in a GaP slab.

was found that the peaks seen in fig. 22 are indeed due to GWP and that the dispersion of these peaks agrees well with theoretical predictions.

The portion of the dispersion curve that is shown by dashed curves is the region where the normal component of the wavevector q_2^z cannot be conserved within $\pm\ 2\pi/d$. Thus we expect the scattering intensity to diminish in this portion of the dispersion curve, and indeed only weak scattering is observed in this region. Expression (73) describes the effect of nonconservation of momentum in the direction normal to the surface, and the experimental observation confirms this relationship qualitatively.

This concludes the discussion of experimental work on the various surface polaritons described in sect. 2. All the modes expected to appear in a single-interface geometry and a double-interface geometry have been found by Raman scattering, and their dispersion curves have been found in good agreement with theoretical predictions.

5. *Effects of Surface Roughness*

In the preceding sections we have assumed that the surfaces that support surface polaritons are perfectly smooth and flat. However, the real surfaces that one can produce have varying degrees of roughness associated with the method of preparation and are never perfectly smooth nor flat. The subject of discussion of the present section is the effects that are produced by surface roughness on the behavior of surface polaritons. Although a rigorous theory of the effects of surface roughness becomes quite complex, it is not difficult to anticipate the results on the basis of simple physical considerations. When the surface is not smooth, the surface polaritons that we have discussed are no longer rigorous normal modes of the geometry. Thus we expect that surface polaritons will interact among themselves as well as with the bulk modes of the media on either side of the boundary interface. This effect will allow additional decay and scattering mechanisms for surface polaritons, and increase the linewidth and shift the frequency. The effect of surface roughness on surface polaritons is quite analogous to the effect of anharmonicity on phonons in lattice dynamics. Thus quantitative theories attempt to calculate the contributions of surface roughness to the proper self-energy of surface polaritons. Another problem that needs to be treated is the question of the effect of surface roughness on the light-scattering intensity. So far no theoretical work has been reported on this problem.

Theoretical work dealing with the effects of surface roughness on the dispersion and damping of surface polaritons has been reported by Mills (1975), by Maradudin and Zierau (1976) and by Kröger and Kretschmann (1976). In these studies the effects of rough surfaces are calculated by a perturbation approach keeping only the leading term in the mean square value $\langle S^2 \rangle$ of the surface profile function $z = S(x, y)$. Using the Green functions for Maxwell's equations with a rough boundary, Mills derived the mean free path of surface polaritons in the presence of surface roughness. He found two contributions to the mean free path corresponding to the decay of surface polaritons due to energy loss to other surface polaritons $1/l^{(SP)}$ and to bulk radiation $1/l^{(R)}$. The shift in the frequency of surface polaritons arising from surface roughness was not calculated by Mills. Maradudin and Zierau, starting from the same set of Green functions for rough surfaces, calculated the proper self-energy of the spectral density function for inelastic light scattering by surface polaritons. They obtained the expressions for frequency shift and damping from the real and imaginary parts of the proper self-energy. The approach by Kröger and Kretschmann is quite different from the above two. They derived "transformed boundary conditions" at a rough surface (Kröger and Kretschmann 1970) which incorporate discontinuities in the fields at the boundary arising from the roughness. Then they use a perturbation approach to the dispersion relation of surface polaritons, and obtain a complex shift $\overline{\Delta k}$ in the surface polariton wavevector (q_2^x in the present article). The resulting expression for $\overline{\Delta k}$ is given by:

$$\overline{\Delta k} = \langle S^2\rangle(\omega_{SP}/c)^3\, P\int d^2\boldsymbol{q}\cdot g(\boldsymbol{q}-\boldsymbol{q}_{SP})A(\boldsymbol{q},\boldsymbol{q}_{SP}) + \mathrm{i}\langle S^2\rangle(\omega_{SP}/c)^5\int_0^{2\pi} d\varphi\cdot g(\boldsymbol{q}-\boldsymbol{q}_{SP})B(\boldsymbol{q}_{SP},\varphi) \quad (79)$$

where $A(\boldsymbol{q}, \boldsymbol{q}_{SP})$ and $B(\boldsymbol{q}_{SP}, \varphi)$ are expressions containing the relative permittivity of the active medium $\kappa(\omega)$, the wavevector of the surface polariton $\boldsymbol{q}_{SP}$ and the angle φ between $\boldsymbol{q}$ and $\boldsymbol{q}_{SP}$. $P\int d^2\boldsymbol{q}$ is the Cauchy principal integral over the two dimensional space (q^x, q^y). $g(\boldsymbol{q}-\boldsymbol{q}_{SP})$ is the Fourier transform of the correlation function of the surface profile function $z = S(x, y)$ given by:

$$g(\boldsymbol{q}-\boldsymbol{q}_{SP}) = \frac{1}{\langle S^2\rangle}\int d^2\boldsymbol{r}\, e^{\mathrm{i}(\boldsymbol{q}-\boldsymbol{q}_{SP})\cdot\boldsymbol{r}}\langle S(x, y)S(0, 0)\rangle, \quad (80)$$

where the integration is over the two dimensional space scanned by $\boldsymbol{r} = (x, y)$. In order to obtain numerical values for comparison with experiments, one must assume a suitable form for $g(\boldsymbol{q}-\boldsymbol{q}_{SP})$. A physically reasonable and mathematically manageable form for $g(\boldsymbol{q}-\boldsymbol{q}_{SP})$ is obtained if we assume that the roughness is stochastic with the average height $\langle S\rangle = 0$. Then it is given by:

$$g(\boldsymbol{q}-\boldsymbol{q}_{SP}) = (a/2\pi)^2\exp[-(a/2)^2(\boldsymbol{q}-\boldsymbol{q}_{SP})^2], \quad (81)$$

where a is the transverse correlation distance between peaks and valleys of $z = S(x, y)$. In what follows we will assume the form of $g(\boldsymbol{q}-\boldsymbol{q}_{SP})$ given by eq. (81), but we must remember that this is only a reasonable assumption. On real crystal surfaces $g(\boldsymbol{q}-\boldsymbol{q}_{SP})$ can have many different forms depending on the preparation method and in particular may not be isotropic as it is assumed here. Recent work by Williams and Aspnes (1978) suggests that the form of $g(\boldsymbol{q}-\boldsymbol{q}_{SP})$ is not a simple Gaussian.

If we assume the Gaussian form for $g(\boldsymbol{q}-\boldsymbol{q}_{SP})$, the angular part of the integration in eq. (79) can be performed analytically, and one obtains a result containing modified Bessel functions. The radial part of the integration in the first term of eq. (79) must be performed numerically. The second term in eq. (79) corresponds to the decay of a given surface polariton $(\boldsymbol{q}_{SP}, \omega_{SP})$ into other surface polaritons via interactions mediated by surface roughness, and the result corresponds to $1/l^{(SP)}$ given by Mills. The first term contains both real and imaginary parts and the imaginary part corresponds to the radiative loss of surface polaritons into bulk modes represented by $1/l^{(R)}$ by Mills. The real part of the first term gives a shift in surface polariton frequency. When reasonable values for a corresponding to experimental conditions are assumed, the radiative loss to bulk modes is found to be quite small compared with the loss to other surface polaritons.

The measurements on the effects of surface roughness were made on polished samples of thin GaP slabs, and the surface polariton modes studied were single-interface modes (SIM). Some of the samples were the ones used in the experiments described in sect. 4.1 and the measurement method was identical. The only difference was that the samples were prepared carefully to obtain different surface roughness for each piece. The five samples used in this experiment are listed in table 2 along with the preparation method for each. All samples were polished in stages using finer alumina powder size down to 0.3 micron powder. The final polish for sample A was done using 0.3 micron alumina. Sample B was obtained by annealing sample A for three hours in 10^{-6} Torr vacuum at 500°C. To prepare sample C, sample A was further polished with 0.05 micron alumina powder, and sample D was prepared by annealing sample C under the same condition as for sample B. Sample E was prepared from sample C by etching in a water solution of 0.5 M KCl + 1.0 M $K_3Fe(CN)_6$. Thus samples A and B have a rougher surface than samples C and D. The difference between samples A and C and samples B and D is that residual surface strain is removed in samples B and D by annealing (Evans and Ushioda 1974).

Figure 25 illustrates the difference in the Raman spectra of a "rough" (sample A) and a "smooth" (sample C) surface. In these example spectra it is clear that a "rough" surface produces a higher scattering intensity than a "smooth" surface. However, mere visual inspection of the spectra does not give one a clear idea of the differences in damping and frequency. A numerical data fitting technique was used to deduce ω_{SP} and Γ_{SP} from raw data like the ones shown in fig. 25 (Ushioda et al. 1979). Since absolute calibration of scattering intensities is very difficult, the relative intensity I_{SP}/I_{LO} of the surface polariton intensity to the LO phonon intensity was obtained by the same numerical method. The results for ω_{SP}, Γ_{SP} and I_{SP}/I_{LO} are shown in figs. 26, 27 and 28 for various scattering angles. The estimated uncertainties in these data points are $\omega_{SP} \pm 1\ cm^{-1}$, $\Gamma_{SP} \pm 0.5\ cm^{-1}$ and $I_{SP}/I_{LO} \pm 0.05$.

The dispersion data shown in fig. 26 does not indicate any clear correlation between the frequency ω_{SP} and the surface roughness. However,

Table 2
Samples prepared by different procedures

	Surface preparation	thickness (μm)
Sample A	0.3 μm polish	15
Sample B	0.3 μm polish + annealing	15
Sample C	0.05 μm polish	15
Sample D	0.05 μm polish + annealing	20
Sample E	0.05 μm polish + etching	15

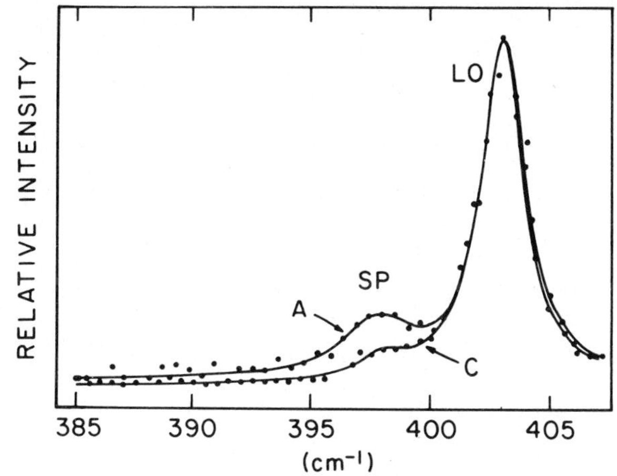

Fig. 25. Surface polaritons at a "rough" and a "smooth" surface.

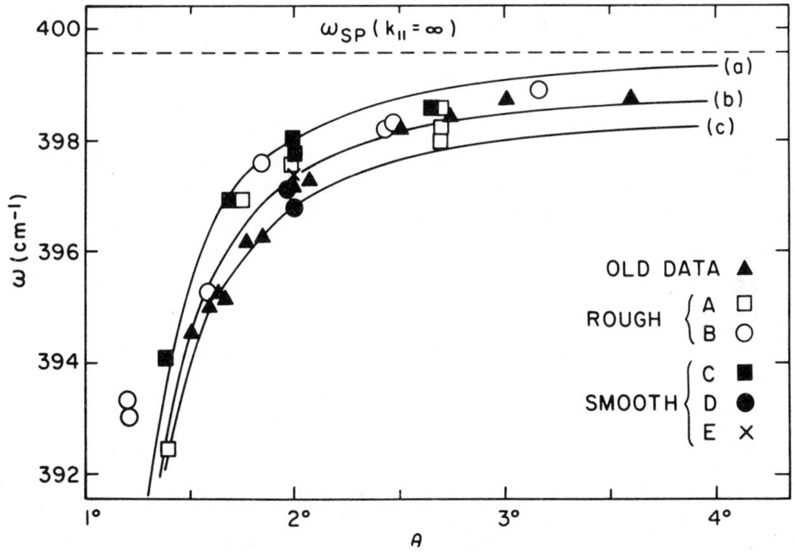

Fig. 26. Dispersion of surface polaritons in different samples; Curve (a) $\kappa_\infty = 9.091$, (b) $\kappa_\infty = 7.8$, (c) $\kappa_\infty = 7.0$.

most of the data points fall 0.5 to 1.0 cm^{-1} below the theoretical dispersion curve (a). The curves (b) and (c) were obtained by varying the optical relative permittivity κ_∞ while keeping ω_L and ω_T fixed at the observed values of 403 cm^{-1} and 367 cm^{-1}, respectively. In fig. 26 it appears that the best fitting is obtained by curve (b) which corresponds to $\kappa_\infty = 7.8$, while curve (a) corresponds to the accepted value of $\kappa_\infty = 9.091$ for the bulk

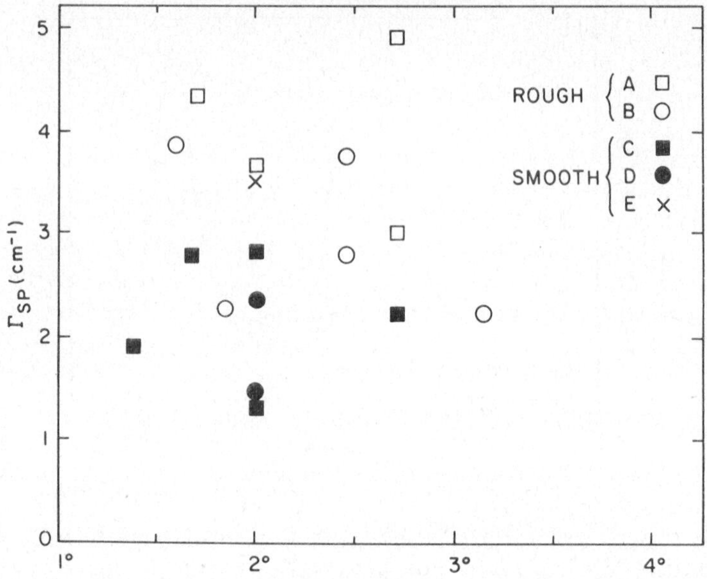

Fig. 27. Damping of surface polaritons in different samples.

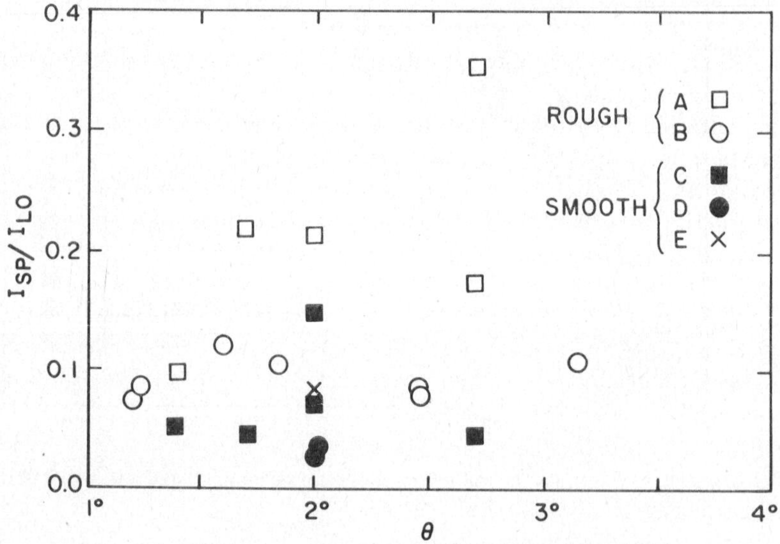

Fig. 28. Raman scattering intensity of surface polaritons in different samples.

crystal of GaP (Parsons and Coleman 1971). Thus the effect of surface roughness can be phenomenologically accounted for by a value of effective surface relative permittivity κ_∞ which is lower than the bulk value. One can understand this effect physically, if one imagines a rough surface region which is only partially filled with a dielectric material because of the presence of ridges and valleys as illustrated in fig. 29. Thus one expects the relative permittivity of the roughness region to be reduced by the filling factor. Evidently the effective relative permittivity of this region is about 15% below the bulk value.

Since the surface roughness is expected to correspond to the size of the final polishing powder, the root-mean-square of the roughness profile $\langle S^2\rangle^{1/2}$ was assumed to be in the range of 500 Å to 3000 Å (0.05 micron to 0.3 micron). Then the frequency shift $\Delta\omega$ due to roughness is estimated to lie in the range of $0.1\ \mathrm{cm}^{-1}$ to $3.6\ \mathrm{cm}^{-1}$ according to (79). This range of $\Delta\omega$ is consistent with the observed deviation of $0.5\ \mathrm{cm}^{-1}$ to $1\ \mathrm{cm}^{-1}$ from the theoretical values of ω_{SP} for a perfectly smooth surface. However, as we can see in fig. 26, no quantitative correlation between $\Delta\omega$ and $\langle S^2\rangle$ was possible because of the lack of quantitative measurement of $\langle S^2\rangle$ and the transverse correlation distance a. The above range of $\Delta\omega$ was obtained for reasonable assumed values of a in the range of 1 to 10 microns.

There is a clear correlation between the surface roughness and the observed surface polariton linewidth Γ_{SP} as seen in fig. 27; a larger linewidth is found for a rougher surface than for a smoother surface. The increase in damping in going from sample A to sample D is of the order of $1 \sim 2\ \mathrm{cm}^{-1}$. This range of shift in Γ_{SP} was found to be consistent with the results found from (79), again assuming $S = 500$ Å to 3000 Å and $a = 1$ micron to 10 microns.

The scattering intensity data in fig. 28 show that rough surfaces produce stronger Raman scattering than smooth surfaces, sometimes by as much as a factor of two. There is no theoretical account of this phenomenon as yet, and no simple physical argument seems to exist on this point. In the case of Raman scattering from adsorbed molecules on metal surfaces, the surface roughness of the metal appears to play a crucial role in enhancing the Raman signal (see, for example, Burstein et al. 1979). The enhancement of I_{SP} observed for roughened GaP surfaces may be related to the giant enhancement observed on metal surfaces, but a systematic theoretical study is needed before the observed effect can be understood.

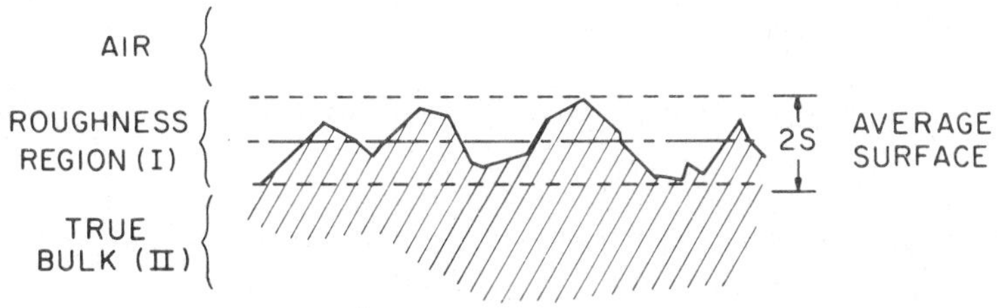

Fig. 29. Schematic profile of a rough surface.

6. Conclusion

In this chapter we have reviewed the essential elements of theory and experiment on the Raman scattering by surface polaritons. All the expected surface polariton dispersion branches in single-interface and double-interface geometries have been found in the Raman spectra of GaAs and GaP samples, and the dispersions of these modes agree with theoretical predictions based on the known values of the relativity premittivity of the sample dielectrics. The experimental results cited here cover all the available reports of Raman scattering by surface polaritons so far, but the review of the theoretical work was restricted to include only the relevant predictions that can be compared with existing experimental results. However, there is more theoretical work dealing with surface polaritons in different kinds of samples and experimental situations (Burstein et. al. 1974). These include papers on surface polaritons in doped semi-conductors with free carriers (for example, Wallis et al. 1974) and surface polaritons associated with current carrying surfaces (Tajima and Ushioda 1978). Also there is a large amount of theoretical and experimental work done on surface plasmons in metals which are observed by the method of attenuated total reflection (ATR) (Otto 1976 and Raether 1977).

As we have seen in this review, the studies of Raman scattering by surface polaritons have been carried out only for a small number of samples and geometries. Most of the past studies of surface polaritons have been done by the ATR method. The light scattering method is difficult because of the low scattering intensity. However, the light scattering probe is more direct than the ATR method in that it measures the surface polariton spectra of an undisturbed surface. In the ATR method a coupler prism must be placed just above the surface and the modes that are actually observed are the normal modes of a double-interface geometry consisting of the coupler prism, the air gap and the sample. Thus the analysis of the resultant absorption spectra is more complex and proper care must be taken to account for the effect of the gap size between the sample and the coupler prism. In contrast the Raman spectra allow direct measurements of undisturbed surface polariton dispersion and damping.

The study of the effects of surface roughness is only at the beginning stage, and so far only qualitative data are available. Experimentally, good methods of quantitative characterization of surface roughness must be developed, and on the theory side different models of surface roughness distribution need to be explored.

Acknowledgements

One of us (S.U.) would like to acknowledge support by a grant from the Air Force Office of Scientific Research; the other of us would like to thank the Department of Physics at Irvine for their hospitality and Professor D. L. Mills for making it possible for us to collaborate in the writing of this article at Irvine.

References

Agranovich, V.M., 1974, Opt. Commun. **11**, 389.
Agranovich, V.M. and V.L. Ginsburg, 1972, Sov. Phys. JETP **34**, 662.
Agranovich, V.M. and T.M. Leskova, 1977, Sov. Phys. Solid State, **19**, 465.
Barker, A.S., 1963, Phys. Rev. **132**, 1474.
Barker, A.S., 1972, Phys. Rev. Lett. **28**, 892.
Barker, A.S. and R. Loudon, 1972, Rev. Mod. Phys. **44**, 18.
Bell, M.I., R.N. Tyte and M. Cardona, 1973, Solid State Commun. **13**, 1833.
Bendow, B., 1978, Springer Tracts in Mod. Phys. **82**, 69.
Burstein, E., C.Y. Chen and S. Lundquist, 1979, Proc. of Second Joint USA–USSR Symposium on Light Scattering in Solids, eds. J. Birman, H. Cummins and K. Rebane (Plenum, New York) p. 479.
Burstein, E., A. Hartstein, J. Schoenwald, A.A. Maradudin, D.L. Mills and R.F. Wallis, 1974, Polaritons, eds. E. Burstein and F. de Martini (Pergamon, New York) p. 89.
Burstein, E., S. Ushioda and A. Pinczuk, 1968, Solid State Commun. **6**, 407.
Cardona, M., ed., 1975, Light Scattering in Solids, Topics in Applied Physics, Vol. **8** (Springer, Berlin).
Chen, Y.J., E. Burstein and D.L. Mills, 1975, Phys. Rev. Lett., **34**, 1516.
Evans, D.J., S. Ushioda and J.D. McMullen, 1973, Phys. Rev. Lett. **31**, 372.
Evans, D.J. and S. Ushioda, 1974, Phys. Rev. **B9**, 1638.
Forster, D., 1975, Hydrodynamic Fluctuations, Broken Symmetry and Correlation Functions (Benjamin, New York).
Hayes, W. and R. Loudon, 1978, Scattering of Light by Crystals (Wiley, New York).
Henry, C.H. and J.J. Hopfield, 1968, Phys. Rev. Lett. **15**, 964.
Ibach, H. and D.L. Mills, 1982, Electron energy loss spectroscopy and surface vibrations (Academic Press, San Francisco) ch. 3.
Kliewer, K.L. and R. Fuchs, 1966, Phys. Rev. **144**, 495.
Kröger, E. and E. Kretschmann, 1970, Z. Phys. **237**, 1.
Kröger, E. and E. Kretschmann, 1976, Phys. Stat. Solidi (b) **76**, 515.
Landau, L.D. and E.M. Lifshitz, 1969, Statistical Physics (Pergamon, Oxford).
Loudon, R., 1963, Proc. Roy. Soc. **A275**, 218.
Loudon, R., 1964, Adv. Phys. **13**, 423.
Loudon, R., 1978a, J. Raman Spect. **7**, 10.
Loudon, R., 1978b, J. Phys. **C11**, 403.
Manasevit, H.M. and A.C. Thorsen, 1970, Met. Trans. **1**, 623.
Maradudin, A.A. and D.L. Mills, 1975, Phys. Rev. **B11**, 1392.
Maradudin, A.A. and W. Zierau, 1976, Phys. Rev. **B14**, 484.
Marschall, N. and B. Fischer, 1972, Phys. Rev. Lett. **28**, 811.

Martin, R. and L.M. Falicov, 1975, Light Scattering Solids, ed., M. Cardona (Springer-Verlag, New York) p. 80.
McMullen, J.D., 1975, Solid State Commun. **17**, 331.
Mills, D.L., 1975, Phys. Rev. **B12**, 4036.
Mills. D.L. and E. Burstein, 1974, Rep. Prog. Phys. **37**, 817.
Mills, D.L. and A.A. Maradudin, 1973, Phys. Rev. Lett. **31**, 372.
Mills, D.L. and A.A. Maradudin, 1975, Phys. Rev. **B12**, 2943.
Mills, D.L. and K.R. Subbaswamy, 1981, Prog. in Opt., Vol. 19, ed., E. Wolf (North-Holland, Amsterdam) p. 45.
Mills, D.L., A.A. Maradudin and E. Burstein, 1970, Ann. Phys. (N.Y.) **56**, 504.
Mills, D.L., Y.J. Chen and E. Burstein, 1976, Phys. Rev. **B13**, 4419.
Mooradian, A. and A.L. McWhorter, 1969, Light Scattering Spectra of Solids, ed., G.B. Wright (Springer, Berlin) p. 297.
Nkoma, J.S. 1975, J. Phys. **C8**, 3919.
Nkoma, J.S. and R. Loudon, 1975, J. Phys. **C8**, 1950.
Nkoma, J.S., R. Loudon and D.R. Tilley, 1974, J. Phys. **C7**, 3547.
Otto, A., 1976, Optical Properties of Solids New Developments, ed. B.O. Seraphin (North-Holland, Amsterdam) p. 678.
Parsons, D.F. and P.D. Coleman, 1971, Appl. Opt. **10**, 1683.
Prieur, J.-Y. and S. Ushioda, 1975, Phys. Rev. Lett. **34**, 1012.
Raether H., 1977, Physics of Thin Films, eds. G. Haas, M.H. Francombe and R.W. Hoffman (Academic Press, New York) Vol. 9, p. 145.
Richter, W., 1976, Springer Tracts in Mod. Phys. **78**, 121.
Ruppin, R. and R. Englman, 1969, Light Scattering Spectra of Solids, ed. G.B. Wright (Springer, Berlin) p. 157.
Schoenwald, J., E. Burstein and J. Elson, 1973, Solid State Commun. **12**, 185.
Scott, J.F., T.C. Damen, R.C.C. Leite and W.T. Silfvast, 1969, Solid State Commun. **7**, 953.
Seraphin, B.O. and H.E. Bennett, 1967, Semiconductors and Semimetals, eds., R.K. Willardson and A.C. Beer (Academic Press, New York) Vol. 3, p. 509.
Subbaswamy, K.R. and D.L. Mills, 1978, Solid State Commun. **27**, 1085.
Tajima, T. and S. Ushioda, 1978, Phys. Rev. **B18**, 1892.
Ushioda, S., 1981, Prog. Opt. Vol. 19, ed., E. Wolf (North-Holland, Amsterdam) p. 139.
Ushioda, S., A. Aziza, J.B. Valdez and G. Mattei, 1979, Phys. Rev. **B19**, 4012.
Ushioda, S. and J.D. McMullen, 1972, Solid State Commun. **11**, 299.
Ushioda, S., J.B. Valdez, W.H. Ward and A.R. Evans, 1974, Rev. Sci. Instrum. **45**, 479.
Valdez, J.B. and S. Ushioda, 1977, Phys. Rev. Lett. **38**, 1098.
Valdez, J.B., G. Mattei and S. Ushioda, 1978, Solid State Commun. **27**, 1089.
Wallis, R.F., J.J. Brion, E. Burstein and H. Hartstein, 1974, Phys. Rev. **B9**, 3424.
Weinstein, B.A. and M. Cardona, 1973, Phys. Rev. **B8**, 2795.
Williams, M.D. and D.E. Aspnes, 1978, Phys. Rev. Lett. **41**, 1667.
Zelano, A.J. and W.T. King, 1970, J. Chem. Phys. **53**, 4444.

CHAPTER 13

Three Wave Nonlinear Interactions Involving Surface Polaritons; Raman Scattering, Light Diffraction and Parametric Mixing

Y. J. CHEN

GTE Laboratories, Inc.
Advanced Technology Laboratory
40 Sylvan Road
Waltham, MA 02254
U.S.A.

E. BURSTEIN

Department of Physics
University of Pennsylvania
Philadelphia, PA 19104
U.S.A.

Surface Polaritons
Edited by
V.M. Agranovich and D.L. Mills

Contents

1. Introduction 589
2. Theory 593
   2.1. General formulation 593
   2.2 Raman scattering 596
   2.3. Nonlinear parametric mixing 598
3. Raman scattering by surface polaritons and parametric generation of surface polaritons 599
   3.1. Raman scattering by surface polaritons 600
   3.2. Parametric generation of surface polaritons 609
4. Enhancement of nonlinear optical interactions using surface polaritons 617
5. Concluding remarks 626
References 627

1. *Introduction*

Surface polaritons are electromagnetic (EM) modes which propagate in a wave-like manner along the interface between two media, but have fields which decrease exponentially in amplitude with distance from the interface. They are interface modes which occur when the dielectric function of one medium, termed the surface active (SA) medium, is negative ($\epsilon_{SA}(\omega) < 0$) and that of the other medium, termed the surface inactive (SI) medium, is positive, and $|\epsilon_{SA}(\omega)| > \epsilon_{SI}(\omega)$. The types of surface polaritons are designated by the nature of the excitation that is involved, i.e., optical phonon surface polariton, exciton surface polariton, etc. (Burstein et al. 1974).

With the advent of the prism coupling methods of Otto (1968) and Kretschmann (1971) which enable one to conveniently couple volume EM waves to surface polaritons, interest in surface excitations and their coupled modes with photons (i.e. surface polaritons) has grown considerably. There is also considerable interest in studying the properties of surface polaritons via nonlinear coupling techniques, and furthermore, in using the special characteristics of surface polaritons to study other surface phenomena (Evans et al. 1973, Simon et al. 1974, 1977, Chen et al. 1979, Gerboshtein et al. 1975, Nkoma 1975, Nkoma and Loudon 1975, Prieur and Ushioda 1975, Chen et al. 1975a, b, 1976a, b, 1977, Chen 1977, Tsang et al. 1979, DeMartini et al. 1976, 1977, Mills et al. 1976, Chen and Burstein 1977, Mills 1977, Bonsall and Maradudin 1978, Fukui et al. 1978, Reinisch et al. 1977. Thusfar the nonlinear optical effects involving surface polaritons, which have been studied experimentally, have involved three wave interactions, such as Raman scattering (Evans et al. 1973, Chen et al. 1975a, Prieur and Ushioda 1975, Mills et al. 1976), light diffraction (Chen et al. 1977), nonlinear parametric mixing (DeMartini et al, 1976, 1977) and second harmonic generation (Simon et al. 1974, 1977). Higher order nonlinear interactions of surface polaritons, specifically four wave nonlinear interactions (e.g., coherent anti-Stokes Raman scattering, etc.) have been proposed (Chen et al. 1976) and were recently observed experimentally (Chen et al. 1979). In this chapter, we focus mainly on the three wave nonlinear interaction processes. Nevertheless, our discussion can be readily generalized to higher order processes.

It should be emphasized that the volume EM waves inside the dielectric media are themselves volume polaritons. In the case of input waves at visible frequencies they involve photon coupled to interband electron–hole (e–h) pair excitations and in special cases coupled with excitons. Both the EM field and the e–h pair content of the polaritons play a role in the nonlinear interaction.

In general, the three wave nonlinear optical interaction can be expressed by a Hamiltonian $\mathcal{H} = -\boldsymbol{P}^{\rm NL}(\omega_3)\cdot\boldsymbol{E}(\omega_3)$, in which $P^{\rm NL}(\omega_3) = \chi^{(2)}(\omega_1, \omega_2)E_1(\omega_1)E_2(\omega_2)$ is the nonlinear polarization generated by two EM waves E_1 and E_2 (Wynne 1974), and $\chi^{(2)}$ is the second order nonlinear susceptibility. The second order nonlinear polarization is given by $P^{\rm NL}(\omega_3) = a(\omega_1, \omega_2)E(\omega_1)u(\omega_2) + b(\omega_1, \omega_2)E(\omega_1)E(\omega_2)$, in which a and b are the lattice deformation and electro-optical nonlinear coefficients due to the lattice displacement $u(\omega_2)$ and the electric field $E(\omega_2)$ of the polariton. Since u and E are related by the equation of motion of the lattice one can, for convenience, express $P^{\rm NL}(\omega_1, \omega_2)$ as $\chi^{(2)}(\omega_1, \omega_2)E(\omega_1)E(\omega_2)$ where

$$\chi^{(2)}(\omega_1, \omega_2) = b(\omega_1, \omega_2) + \gamma(\omega)a(\omega_1, \omega_2) \quad \text{and} \quad \gamma(\omega) = u(\omega_2)/E(\omega_2).$$

In the case of Stokes–Raman scattering by optical phonon surface polaritons, an input EM wave $E_1(\omega_1, \boldsymbol{k}_1)$ (also called the incident EM wave E_i) interacts with an optical phonon surface polariton wave $E_2(\omega_2, \boldsymbol{k}_2)$ to generate $P_3^{\rm NL}$ which radiates an EM wave E_3 (also termed the scattered EM wave E_s). The Stokes–Raman scattering process involves the annihilation of an input polariton E_1, the creation of an output polariton E_3, and the creation of an optical phonon surface polariton E_2, with $\omega_1 = \omega_2 + \omega_3$ and $\boldsymbol{k}_{1\parallel} = \boldsymbol{k}_{2\parallel} + \boldsymbol{k}_{3\parallel}$. Raman scattering by surface polaritons has been reviewed in Ch. 12 by Ushioda and Loudon. In the case of light diffraction by optical phonon surface polaritons, an input EM wave E_1 interacts with an input optical phonon surface polariton wave E_2 (generated for example by prism coupling) to generate an output EM wave E_3. In the light diffraction process an input polariton E_1 is annihilated, the input surface polariton E_2 is either annihilated (+) or created (−) and the output polariton E_3 is created (with $\omega_1 \pm \omega_2 = \omega_3$ and $\boldsymbol{k}_{1\parallel} \pm \boldsymbol{k}_{2\parallel} = \boldsymbol{k}_{3\parallel}$). In the case of parametric difference frequency generation of optical phonon surface polaritons, two input EM waves E_1 and E_2 interact and generate a $P_3^{\rm NL}$ which radiates an optical phonon surface polariton E_3. This process, in which $\omega_1 - \omega_2 = \omega_3$ and $\boldsymbol{k}_{1\parallel} - \boldsymbol{k}_{2\parallel} = \boldsymbol{k}_{3\parallel}$, involves the annihilation of an input polariton E_1, the creation of a polariton E_2 and an optical phonon surface polariton E_3. Finally, in the case of parametric sum frequency generation, two input EM waves, E_1 and E_2, interact to generate an interband e–h pair (or exciton) surface polariton. This process involves the annihilation of the two input polaritons E_1 and E_2 and the creation of a surface polariton E_3 (with $\omega_1 + \omega_2 = \omega_3$ and $\boldsymbol{k}_{1\parallel} + \boldsymbol{k}_{2\parallel} = \boldsymbol{k}_{3\parallel}$).

We note that there can also be processes involving more than one

Table 1

Nonlinear process	Occupation factor	
Raman scattering by surface polaritons	$\bar{n}(\omega_1)(\bar{n}(\omega_2)+1)$	(Stokes process)
	$\bar{n}(\omega_1)\bar{n}(\omega_2)$	(anti-Stokes process)
Light diffraction by surface polaritons	$\bar{n}_1(\omega_1)(\bar{n}(\omega_2)+1)$	(Stokes process)
	$\bar{n}_1(\omega_1)\bar{n}(\omega_2)$	(anti-Stokes process)
Difference-frequency mixing	$\bar{n}_1(\omega_1)(\bar{n}(\omega_2)+1)$	
Sum-frequency mixing	$\bar{n}_1(\omega_1)\bar{n}_2(\omega_2)$	

surface polariton. Thus, one can have two input surface polariton waves interact and generate an output surface polariton wave, etc. In these cases one can make use of prism or grating coupling to detect or generate the surface polariton waves.

The various nonlinear processes differ primarily in two ways: (1) the occupation factor for the input and output waves (see table 1), (2) the nature of the output wave that is to be detected. In the case of Raman scattering by surface polaritons and in the case of light diffraction by surface polaritons, it is an optical frequency polariton that is the output signal, while in the case of parametric difference frequency mixing it is an optical phonon surface polariton that is the output signal.

One can characterize the three wave nonlinear optical interactions involving surface polaritons by dividing them into two groups. In the first group a surface polariton, whose frequency is usually quite different from that of the input waves, is generated by the three wave interaction, as in Raman scattering by surface polaritons and in nonlinear parametric generation of surface polaritons. In the second group one or more surface polaritons are "used" as the input and/or output waves of the nonlinear optical interaction. Examples are the use of e–h pair excitation surface polaritons in nonlinear parametric mixing at a metal-dielectric interface, i.e. in second harmonic generation, and in Raman scattering and/or luminescence by overlayers on metal surfaces, etc.

The purpose of this paper is to present an overview of the three wave nonlinear interaction processes which focusses on the common features and differences among them. In particular, we shall emphasize the special features in the nonlinear processes that result from the evanescent wave nature of the surface polaritons. Thus, as a result of the evanescent character of surface polaritons, the kinematic factors in the matrix elements that control the various nonlinear processes involving them differ from those involving only volume polaritons. Moreover, the resonant generation of the surface polaritons at the surface active interface can lead to a large enhancement of the input and/or output fields at the interface when surface polaritons are used as the input and/or output waves.

In sect. 2, we shall present a general formulation of the nonlinear interactions of EM waves. A Green's function treatment of the nonlinear interactions is employed, since it clearly illustrates the various aspects of the nature of the nonlinear interactions: The Fresnel transfer factors for the input and output waves, the kinematic factor, which is related to the interaction (or coherence) length, and the nonlinear susceptibility. The general formulation treats volume as well as surface polaritons in order to highlight the differences in the kinematic and Fresnel transfer factors.

Section 3 will be devoted to the discussion of Raman scattering by surface polaritons and parametric difference frequency generation of surface polaritons. These two phenomena have conventionally been treated as two different phenomena particularly because in one case the output wave is an optical frequency polariton wave and in the other case, the output wave is an optical phonon surface polariton wave which requires prism (or grating) coupling or nonlinear detection (e.g. Raman scattering). As we shall show both involve the same three wave (two bulk EM waves and a surface polariton wave) interaction matrix elements, and the same kinematic factor. Another common feature of the Stokes–Raman scattering and the parametric difference frequency mixing is that an optical phonon surface polariton is "generated" in both cases. In Stokes–Raman scattering, a quantum of the elementary excitation of the system (an optical phonon surface polariton) and an optical frequency polariton is excited by an input optical frequency polariton. In parametric mixing, the nonlinear medium is "driven" by the two input optical frequency polaritons which determine both the frequency and the wave vector of the $P^{NL}(\omega_3, \boldsymbol{k}_3)$ and the mixing efficiency peaks when both ω_3 and $\boldsymbol{k}_3$ correspond to the frequency and momentum of an optical phonon (surface or volume) polariton. The question of the "free" and "driven" surface polaritons, which arises in a parametric mixing process when frequency and wave vector of the input waves do not match the dispersion relation of the generated elementary excitation, will also be discussed.

In sect. 4 we shall discuss the use of surface polaritons as the input and/or the output waves. The surface polaritons can be coupled into and/or out of the system either nonlinearly (as discussed in sect. 3) or linearly, via grating or prism coupling. Because of the localized nature of the surface polariton the field strengths of both input and/or output surface polaritons at the interface can be greatly enhanced (i.e. the transfer factors of the input and/or output waves are greater than one). The enhancement factor depends on the dielectric functions of both the surface active medium and its adjacent medium. Furthermore since the fields in both the SA medium and the SI medium are enhanced, the elementary excitations (e.g. optical phonons) in either medium can be probed. The examples we will discuss are second harmonic generation at a metal surface, parametric mixing in a nonlinear medium adjacent to a metal surface, and Raman scattering by an overlayer on a metal surface.

2. Theory

2.1. General Formulation

For simplicity, we consider here a system of isotropic media with plane interfaces. The nonlinear interaction, which is characterized by a macroscopic nonlinear polarization $\boldsymbol{P}^{\mathrm{NL}}$, can take place in one or more media. The output EM field $\boldsymbol{E}_0$ generated by the nonlinear interaction of two EM fields $\boldsymbol{E}_1$ and $\boldsymbol{E}_2$ is obtained by solving the Maxwell equation involving P^{NL} in the relevant medium:

$$\left[\nabla\times(\nabla\times)+\epsilon(\omega_0)\frac{\partial^2}{\partial t^2}\right]\boldsymbol{E}_0(\omega_0)=-\frac{4\pi}{c^2}\frac{\partial^2\boldsymbol{P}^{\mathrm{NL}}(\omega_0,\boldsymbol{r})}{\partial t^2}, \tag{1}$$

where $\epsilon(\omega_0)$ is the linear dielectric constant at the output frequency ω_0, and $\omega_0=\omega_1\pm\omega_2$.

The nonlinear wave equation eq. (1) was first treated formally by Bloembergen and Pershan (1962) using a classical EM wave approach which solves the boundary conditions for the EM fields and the Fresnel relations. Later Mills et al. (1970) introduced a Green's function formulism, which emphasized the importance of the transfer factors for EM fields into and out of the nonlinear medium.

The concept of the Green's function method is best illustrated by fig. 1.

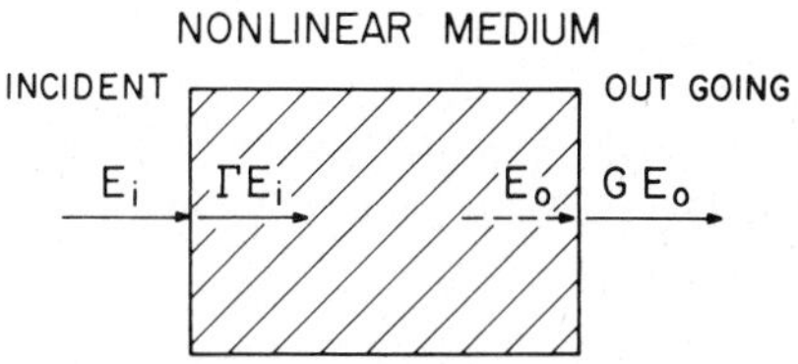

$$E_s(\omega_s)=\left(\frac{\omega_s}{c}\right)^2\int G(z',\omega_s)\,\delta\epsilon^{NL}\prod_i\Gamma(\omega_i\; z')E_i(\omega_i)\,dz'$$

Fig. 1. Schematic diagram for the Green's function treatment of the nonlinear interaction. The Γ transfer functions relate the input EM fields in the nonlinear medium to the incident radiation. The Green's functions G relate the output nonlinear field outside the nonlinear medium to the nonlinear polarization $P^{\mathrm{NL}}=\delta\epsilon^{\mathrm{NL}}\prod_i\Gamma(\omega_i,r')\,E_i(\omega_i)$. The spatiai integration over the entire nonlinear medium yields a kinematic factor which depends on the wave forms of the input and output waves as well as the nature of the nonlinear interaction.

The input EM fields in the nonlinear medium (or media) are related to the incident radiations outside the system by the transfer matrices Γ, which are essentially generalized Fresnel factors. The input waves, $\mathbf{E}_1(\omega_1, \mathbf{r})$ and $\mathbf{E}_2(\omega_2, \mathbf{r})$, interact in the nonlinear medium and excite a nonlinear polarization $\mathbf{P}^{\mathrm{NL}}$. The output wave $\mathbf{E}_0(\omega_0, \mathbf{r})$, which is radiated by $\mathbf{P}^{\mathrm{NL}}(\omega_0, \mathbf{r})$, is governed by the wave equation (1). The wave equation for $\mathbf{E}_0(\omega_0, \mathbf{r})$ can be solved by first obtaining the Green's function $G(\mathbf{r}, \mathbf{r}', t-t')$ of the system:

$$\sum_{\beta}\left[\nabla\times(\nabla\times)+\frac{\epsilon}{c^2}\frac{\partial^2}{\partial t^2}\right]_{\alpha\beta} G_{\beta\xi}(\mathbf{r}, \mathbf{r}', t-t') = 4\pi\delta_{\alpha\xi}\delta(\mathbf{r}-\mathbf{r}')\delta(t-t'), \tag{3}$$

where α, β, ξ represent x, y and z.
The output EM field at $(\mathbf{r}, t)$ can then be written in component form as:

$$E_{0\alpha}(\mathbf{r}, t) = \sum_{\beta}\int\int G_{\alpha\beta}(\mathbf{r}, \mathbf{r}', t-t')\frac{\partial^2 P_{\beta}^{\mathrm{NL}}(\mathbf{r}, t')}{\partial t'^2}\,\mathrm{d}\mathbf{r}'\,\mathrm{d}t'. \tag{4}$$

The Green's function $G(\mathbf{r}, \mathbf{r}', t-t')$ contains a transfer factor similar to the Γ factor which, among other things, relates the output EM field $\mathbf{E}_0(\mathbf{r}, t)$ at $(\mathbf{r}, t)$ which can be outside the nonlinear medium to the nonlinear polarization, $\mathbf{P}^{\mathrm{NL}}(\mathbf{r}', t')$ at $(\mathbf{r}', t')$ inside the nonlinear medium. The spatial integration in eq. (4) over the entire nonlinear medium (or media) yields a kinematic factor which depends on the wave forms of the input and output waves as well as on the nature of the nonlinear interaction.

In this section we shall use the Green's function method to derive the general expressions for the various nonlinear processes. For simplicity we will assume that the input EM waves are plane waves. We can write the Green's function matrix in the following form:

$$G_{\alpha\beta}(\mathbf{r}, \mathbf{r}', t-t') = \int\frac{\mathrm{d}\mathbf{k}_\parallel\,\mathrm{d}\omega}{(2\pi)^3}\exp[\mathrm{i}\mathbf{k}_\parallel(\mathbf{r}-\mathbf{r}')] \times\exp[-\mathrm{i}\omega(t-t')]\mathcal{G}_{\alpha\beta}(\mathbf{k}_\parallel, \omega, z, z'), \tag{5}$$

where $\mathbf{k}_\parallel$ is a two dimensional wave vector which lies in a plane parallel to the surface. The intensity of the output radiation (i.e. the Poynting vector, S) takes the following form:

$$\begin{aligned} S &= \frac{c}{8\pi}\sum_{\alpha}\langle|E_\alpha(\mathbf{r}, t)|^2\rangle \\ &= \frac{c}{8\pi}\sum_{\alpha}\sum_{\beta\beta'}\int\int\mathrm{d}z'\,\mathrm{d}z''\int\mathrm{d}\mathbf{k}_\parallel\,\mathrm{d}\omega\,\mathcal{G}^*_{\alpha\beta'}(\mathbf{k}_\parallel, \omega, z, z'')\mathcal{G}_{\alpha\beta}(\mathbf{k}_\parallel, \omega, z, z') \\ &\quad\times\left(\frac{\omega}{c}\right)^4\int\frac{\mathrm{d}\mathbf{r}_\parallel\,\mathrm{d}t''}{(2\pi)^3}\exp[\mathrm{i}(\mathbf{k}_\parallel\mathbf{r}_\parallel-\omega t'')]\langle P_{\beta'}^{\mathrm{NL}}(\mathbf{r}_\parallel, t'', z'')P_{\beta}^{\mathrm{NL}}(0, 0, z')\rangle_T. \end{aligned} \tag{6}$$

The inversion symmetry in the x–y plane and the time invariance of the nonlinear interaction lead to the conservation of momentum along the x–y plane and the conservation of energy of the nonlinear interaction, i.e., the wave vector component parallel to the plane interfaces $\boldsymbol{k}_{\parallel}$ and the frequency ω are conserved in the nonlinear process. Equation (6) can be rewritten in terms of the momentum along the x–y plane and the frequency of the output radiation, $\boldsymbol{k}_{\parallel}^{\circ}$ and ω_0, as:

$$S = \frac{c}{8\pi}\left(\frac{\omega_0}{c}\right)^4 \sum_{\alpha}\sum_{\beta\beta'} \int\int \mathrm{d}z'\,\mathrm{d}z''\mathcal{G}^{*}_{\alpha\beta'}(\boldsymbol{k}_{\parallel}^{\circ}, \omega_0, z, z'')\mathcal{G}_{\alpha\beta}(\boldsymbol{k}_{\parallel}^{\circ}, \omega_0, z, z')$$
$$\times \langle P^{\mathrm{NL}}_{\beta'}(\boldsymbol{k}_{\parallel}^{\circ}, \omega_0, z'')P^{\mathrm{NL}}_{\beta}(\boldsymbol{k}_{\parallel}^{\circ}, \omega_0, z')\rangle_{\mathrm{T}}. \tag{7}$$

We note that the output intensity is proportional to the correlation function of the nonlinear polarization $\langle P^{\mathrm{NL}}_{\beta'}P^{\mathrm{NL}}_{\beta}\rangle_{\mathrm{T}}$. The integrations over the nonlinear medium lead to a kinematic factor K. As we shall show later, there are, in many cases, drastic differences between the kinematic factor of the nonlinear interactions involving surface polaritons and that involving volume polaritons. Moreover, in cases where more than one interaction geometry is possible (such as in nonlinear parametric mixing where the two input waves can either propagate in the same direction or in opposite directions), each interaction geometry can have a different kinematic factor.

The Green's function G matrix can be readily obtained from eq. (3). For the specific case where $\boldsymbol{k}_{\parallel}$ is along $\hat{\boldsymbol{x}}$, the equations for the components of the G matrix $\mathcal{G}_{\alpha\beta}(\boldsymbol{k}_{\parallel}\hat{\boldsymbol{x}}, \omega, z, z')$ assume the following form:

$$\mathrm{i}k_{\parallel}\frac{\partial}{\partial z}\mathcal{G}_{zx}(z, z') - \left[\frac{\partial^2}{\partial z^2} - \epsilon(z)\frac{\omega^2}{c^2}\right]\mathcal{G}_{xx}(z, z') = 4\pi\delta(z - z'),$$

$$\mathrm{i}k_{\parallel}\frac{\partial}{\partial z}\mathcal{G}_{xx}(z, z') + \left[k_{\parallel}^2 + \epsilon(z)\frac{\omega^2}{c^2}\right]\mathcal{G}_{zx}(z, z') = 0,$$

$$\mathrm{i}k_{\parallel}\frac{\partial}{\partial z}\mathcal{G}_{xz}(z, z') + \left[k_{\parallel}^2 - \epsilon(z)\frac{\omega^2}{c^2}\right]\mathcal{G}_{zz}(z, z') = 4\pi\delta(z - z'),$$

$$\mathrm{i}k_{\parallel}\frac{\partial}{\partial z}\mathcal{G}_{zz}(z, z') - \left[\frac{\partial^2}{\partial z^2} + \epsilon(z)\frac{\omega^2}{c^2}\right]\mathcal{G}_{xz}(z, z') = 0,$$

$$\left[k_{\parallel}^2 - \epsilon(z)\frac{\omega^2}{c^2}\right]\mathcal{G}_{yy}(z, z') - \frac{\partial^2}{\partial z^2}\mathcal{G}_{yy}(z, z') = 4\pi\delta(z, z'). \tag{8}$$

The remaining elements of the G matrix: $\mathcal{G}_{yx}$, $\mathcal{G}_{xy}$, $\mathcal{G}_{yz}$ and $\mathcal{G}_{zy}$ satisfy homogeneous equations and can be taken to be zero.

Since the $\mathcal{G}_{\alpha\beta}$ are the components of the Green's function matrix for the EM field, they satisfy the usual boundary conditions, namely the continuity of the tangential electric and magnetic fields E_{t} and H_{t} (i.e. $\mathcal{G}_{xx}$, $\mathcal{G}_{yy}$, $(\epsilon(z)/k_z^2)\mathcal{G}_{xx}$ and $(\partial/\partial z)\mathcal{G}_{yy}$) and the normal displacement fields D_{n} (i.e. $\epsilon(z)\mathcal{G}_{zx}$ and $\epsilon(z)\mathcal{G}_{zz}$) across the interfaces, where $k_z = (\epsilon(\omega^2/c^2) - k_{\parallel}^2)^{1/2}$ is the normal component of the wave vector. We can rewrite eq. (8) and the corresponding boundary conditions as follows:

$$\text{(i)}\quad \left(\frac{\partial^2}{\partial z^2}+k_z^2\right)\mathcal{G}_{xx}(z,z') = -\frac{4\pi c^2}{\epsilon(z)\omega^2}k_z^2\delta(z-z'),$$

$$\mathcal{G}_{zx} = \frac{\mathrm{i}k_\parallel}{k_z^2}\frac{\partial}{\partial z}\mathcal{G}_{xx}, \tag{9a}$$

with $\mathcal{G}_{xx}$ and $(\epsilon(z)/k_z^2)(\partial\mathcal{G}_{xx}/\partial z)$ continuous across the interfaces.

$$\text{(ii)}\quad \left(\frac{\partial^2}{\partial z^2}+k_z^2\right)\mathcal{G}_{xz} = -\frac{4\pi \mathrm{i}k_\parallel c^2}{\epsilon(z)\omega^2}\frac{\partial}{\partial z}\delta(z-z'),$$

$$\mathcal{G}_{zz} = -\frac{4\pi}{\epsilon(z)}\delta(z-z') + \mathrm{i}\frac{k_\parallel}{k_z^2}\frac{\partial}{\partial z}\mathcal{G}_{xz}(z,z'), \tag{9b}$$

with $\mathcal{G}_{xz}$ and $\epsilon(z)\mathcal{G}_{zz}$ continuous across the interfaces.

$$\text{(iii)}\quad \left(\frac{\partial^2}{\partial z^2}+k_z^2\right)\mathcal{G}_{yy}(z,z') = -4\pi\delta(z-z'), \tag{9c}$$

with $\mathcal{G}_{yy}$ and $(\partial/\partial z)\mathcal{G}_{yy}$ continuous across the interfaces.

Finally, we note that the special case solutions of eq. (9) can be generalized to the case where $\boldsymbol{k}_\parallel = k_x\hat{\boldsymbol{x}} + k_y\hat{\boldsymbol{y}}$ by applying a tensor rotation to $\mathbf{G}(k_\parallel\hat{\boldsymbol{x}},\omega)$:

$$\mathbf{G}(\boldsymbol{k}_\parallel,\omega) = \mathbf{R}\mathbf{G}(k_\parallel\hat{\boldsymbol{x}},\omega)\mathbf{R}^{-1} \tag{10}$$

where

$$\mathbf{R} = \frac{1}{k_\parallel}\begin{pmatrix} k_x & -k_y & 0 \\ k_y & k_x & 0 \\ 0 & 0 & 1 \end{pmatrix}, \qquad k_\parallel = (k_x^2+k_y^2)^{1/2}.$$

2.2. *Raman Scattering*

The nonlinear polarization for a Raman scattering process can be expressed phenomenologically as follows:

$$\boldsymbol{P}_\alpha^{\mathrm{NL}}(\omega_\mathrm{s},\boldsymbol{r}) = \sum_{\beta\xi}[a_{\alpha\beta\xi}u_\xi(\omega_\mathrm{j},\boldsymbol{r}) + b_{\alpha\beta\xi}E_\xi(\omega_\mathrm{j},\boldsymbol{r})]E_\beta(\omega_\mathrm{i},\boldsymbol{r}) \tag{11}$$

where α, β, ξ represent x, y, and z; ω_i, ω_j and $\omega_\mathrm{s} = (\omega_\mathrm{i} \pm \omega_\mathrm{j})$ are frequencies of the incident photon, optical phonon, and scattered photon, respectively;

$$a_{\alpha\beta\xi} = \partial\chi^{(1)}_{\alpha\beta}/\partial u_\xi \quad \text{and} \quad b_{\alpha\beta\xi} \equiv \partial\chi^{(1)}_{\alpha\beta}/\partial E_\xi$$

are the deformation potential and electro-optical Raman tensors due to interaction of the input field $E_\mathrm{i}(\omega_\mathrm{i})$ with the lattice displacement $u(\omega_\mathrm{j})$ and with the macroscopic field $E(\omega_\mathrm{j})$ of the polariton. The higher order Raman tensors such as the wave vector dependent and field induced Raman tensors (Burstein 1972) can also be readily included in eq. (11) when relevant.

The electric field and the lattice displacement of the optical phonon, E and u, are related by the equation of motion of the lattice:

$$M[\omega_T^2 u(\omega) - \omega^2 u(\omega) + \bar{\gamma} u(\omega)] = e^* E(\omega), \tag{12}$$

where M, ω_T, $\bar{\gamma}$ and e^* are the effective mass, the frequency, the damping constant and the effective charge of the optical phonon, respectively. The correlation functions of u and E are related by the fluctuation-dissipation theorem (Abrikosov et al. 1963) as follows:

$$\langle u_\alpha(\omega, z), E_\beta(\omega, z')\rangle_T = \langle E_\beta(\omega, z') u_\alpha(\omega, z)\rangle_T^* = \gamma(\omega)\langle E_\alpha(\omega, z), E_\beta(u, z')\rangle_T$$

and

$$\langle u_\alpha(\omega, z) u_\beta(\omega, z')\rangle_T = \left[\frac{\bar{n}+1}{\bar{N}}\, \delta_{\alpha\beta}\, \mathrm{Im}(\gamma(\omega)) + \gamma^2(\omega)\langle E_\alpha(\omega, z) E_\beta(\omega, z')\rangle_T\right]. \tag{13}$$

where $\bar{n}(\omega)$ is the thermal occupation number of the optical phonon, $\bar{N}$, the number of atom sites in the crystal and $\gamma(\omega) = e^*/[M(\omega_T^2 - \omega^2 - i\bar{\gamma}\omega)]$.

The $\langle E_\alpha E_\beta\rangle_T$ correlation function is related to the Green's function matrix $\mathcal{G}_{\alpha\beta}$ of eq. (9) as follows:

$$\langle E_\alpha(\omega, z) E_\beta(\omega, z')\rangle_T = -(\bar{n}(\omega) + 1) 2\, \mathrm{Im}(\mathcal{G}_{\alpha\beta}(\omega - i\sigma, z, z')), \tag{14}$$

where $\sigma \to 0^+$ is a dummy coefficient.

The input field in the nonlinear medium $E(\omega_i, \boldsymbol{r})$ can be obtained in terms of the incident field E_i by solving the linear wave equation for that system,

$$\sum_\beta [\nabla \times (\nabla \times) - \epsilon(z)(\omega^2/c^2)]_{\alpha\beta} \boldsymbol{E}_\beta(\omega_i) = 0, \tag{15}$$

with boundary conditions that the tangential E components and normal ϵE components are continuous. Thus, E_α can be written as:

$$E_\alpha(\omega_i, \boldsymbol{r}) = \Gamma_{\alpha\beta}(\omega_i, \boldsymbol{r}) E_{i\beta}(\omega_i) \equiv \Gamma_\alpha(\omega_i, \boldsymbol{r}) E_i(\omega_i), \tag{16}$$

where $\Gamma_\alpha(\omega, \boldsymbol{r})$ is the generalized Fresnel matrix for the wave equation (15), defined by

$$\sum_\beta \left[\nabla \times (\nabla \times) + \frac{\epsilon}{c^2}\frac{\partial^2}{\partial t^2}\right]_{\alpha\beta} \Gamma_\beta(\omega, \boldsymbol{r}) = 0. \tag{17}$$

The differential Raman scattering cross section (per unit solid angle and per frequency) can thus be written as

$$\frac{d^2\sigma}{d\omega_0\, d\Omega} = 2(\bar{n}+1)\cos\theta_s\left(\frac{\omega_0}{c}\right)^6 \sum_{\alpha} \sum_{\substack{\beta\xi\eta \\ \beta'\xi'\eta'}} \int\int dz'\, dz''\, \mathcal{G}^*_{\alpha\beta'}(\omega_0, z'')\mathcal{G}_{\alpha\beta}(\omega_0, z')$$

$$\times \Gamma^*_{\xi}(z'')\Gamma_{\xi}(z')\left[\chi^{(2)*}_{\beta'\xi'\eta'}\chi^{(2)}_{\beta\xi\eta}\, \mathrm{Im}(\mathcal{G}_{\eta'\eta}(\omega, \boldsymbol{k}_{\|}, z'', z'))\right.$$

$$\left. + \frac{\delta_{\eta'\eta}}{\bar{n}M}\, a^*_{\beta'\xi'\eta'}a_{\beta\xi\eta}\, \mathrm{Im}\left(\frac{1}{\omega_T^2 - \omega^2 - i\bar{\gamma}\omega}\right)\right], \tag{18}$$

where $\chi^{(2)}_{\beta\xi\eta} = a_{\beta\xi\eta}\gamma(\omega) + b_{\beta\xi\eta}$, and θ_s is the scattering angle.

We note that the second term in eq. (18), which exhibits a "pole" at $\omega = \omega_T$, corresponds to the scattering by a TO phonon. For infrared inactive modes ($e^* = 0$) this is the only existing term. However, since we are interested in the infrared active optical phonons which couple with EM radiation to form the polaritons, the scattering by TO phonon is effectively contained in the first term which has a γ^2 factor (Barker and Loudon 1972). Thus, in the discussion of the next section, we will for convenience omit the second term.

2.3. Nonlinear Parametric Mixing

The nonlinear polarization for a three wave difference-frequency parametric mixing process can be expressed phenomenologically as follows:

$$P^{NL}_{\alpha}(\omega_3, \boldsymbol{r}) = \sum_{\beta\xi} [b_{\alpha\beta\xi} + \gamma(\omega)a_{\alpha\beta\xi}]E_{\beta}(\omega_1)E_{\xi}(\omega_2), \tag{19a}$$

where $\omega_3 = \omega_1 - \omega_2$; **a** and **b** are the lattice deformation potential and the electro-optical nonlinear susceptibilities, respectively; ω_1 and ω_2 are the frequency of the two input waves E_1 and E_2, respectively. In the case of sum-frequency parametric mixing, the output frequency $\omega_3 = \omega_1 + \omega_2$ is far above the lattice oscillation frequency and $\gamma(\omega)$ is small. Thus the nonlinear polarization takes the following form:

$$P^{NL}_{\alpha}(\omega_3, \boldsymbol{r}) = \sum_{\beta\xi} b_{\alpha\beta\xi}E_{\beta}(\omega_1)E_{\xi}(\omega_2). \tag{19b}$$

Again, E_1 and E_2 are related to the incident field strengths E^i_1 and E^i_2 by the relation:

$$E_{j\alpha}(\omega_j, \boldsymbol{r}) = \sum_{\beta} \Gamma_{\alpha\beta}(\omega_j, \boldsymbol{r})E^i_{j\beta}(\omega_i)$$

where $j = 1.2$ for the two input waves; and $\Gamma_{\alpha\beta}(\omega_j, \boldsymbol{r})$ are the generalized Fresnel matrices obtained from the wave equation (17). The output intensity of the three wave parametric mixing takes the following form:

$$S = \bar{N}(\omega_1, \omega_2)\frac{c}{8\pi}\left(\frac{\omega}{c}\right)^4 \sum_{\alpha} \sum_{\substack{\beta\xi\eta \\ \beta'\xi'\eta'}} \iint \mathrm{d}z'\,\mathrm{d}z''\mathcal{G}^*_{\alpha\beta'}(\boldsymbol{k}_\parallel, \omega, z, z'')\mathcal{G}_{\alpha\beta}(\boldsymbol{k}_\parallel, \omega, z, z')$$
$$\times \chi^{(2)}_{\beta'\xi'\eta'}\chi^{(2)}_{\beta\xi\eta}\Gamma^*_{\xi'}(\omega_1, z'')\Gamma^*_{\eta'}(\omega_2, z'')\Gamma_\xi(\omega_1, z')\Gamma_\eta(\omega_2, z''), \tag{20}$$

where $\bar{N}(\omega_1, \omega_2) = \bar{n}(\omega_1)\bar{n}(\omega_2)$, $\omega = \omega_1 + \omega_2$ and $\boldsymbol{k}_\parallel = \boldsymbol{k}_{\parallel 1} + \boldsymbol{k}_{\parallel 2}$ for a sum frequency mixing process, and $\bar{N}(\omega_1, \omega_2) = \bar{n}(\omega_1)(\bar{n}(\omega_2) + 1)$, $\omega = \omega_1 - \omega_2$ and $\boldsymbol{k}_\parallel = \boldsymbol{k}_{\parallel 1} - \boldsymbol{k}_{\parallel 2}$ for a difference-frequency mixing process.

We wish to emphasize that the parametric difference-frequency mixing process, which "generates" a surface polariton, is very similar to the Stokes–Raman scattering process which also generates a surface polariton. The nonlinear susceptibilities of the two processes are the same. The kinematic factors, which correspond to the coupling kinematics of the three waves and are determined by the integration over z' and z'', are basically the same. On the other hand, the two phenomena differ in the nature of the output waves, i.e., the output waves in the Raman scattering are optical frequency polariton waves while the output waves in the difference-frequency parametric generation are optical phonon surface polaritons. The Raman process is a scattering by an elementary excitation, and its scattering cross section is proportional to the imaginary part of the $\langle \boldsymbol{EE}\rangle_{\mathrm{T}}$ correlation function, or the response function of the electric field of the elementary excitation E. The parametric mixing on the other hand, is a process which sets up a driven P^{NL}. The field amplitude of the output wave depends on how well the coupling conditions match the resonance condition of the system. When the coupling conditions, i.e. $(\boldsymbol{k}_\parallel, \omega)$, match the dispersion relation of the surface polariton (or the bulk polariton), the mixing efficiency peaks up, i.e., the transfer functions and/or the kinematic factor are enhanced. Since the input field transfer functions (the Γ matrices) are similar to the output field transfer functions (the $\mathcal{G}_{\alpha\beta}$ matrices), one would expect to get similar field enhancement effects for the Γ matrix when the incident conditions $(\boldsymbol{k}_\parallel, \omega)$ match the dispersion relation of the surface polariton. We shall show in sect. 4, that one can, indeed, take advantage of the enhanced Γ and/or G matrices by coupling to the surface polariton of the system.

3. *Raman Scattering by Surface Polaritons and Parametric Generation of Surface Polaritons*

In this section, we shall discuss the two nonlinear processes, Raman scattering and nonlinear difference frequency mixing in which optical phonon surface polaritons are generated. Our emphasis will be to point out the special features associated with the excitation of surface polaritons by comparing it to the corresponding processes which lead to the excitation of bulk polaritons. Since in both cases the nonlinear interaction takes place in the same nonlinear medium, the input field strengths (e.g., the incident Γ matrices) are the same, and the outgoing transfer factors (e.g., the G matrices) are essentially identical. The major difference between the cross section for the excitation of a surface polariton and that for the excitation of a volume polariton lies in the kinematic factor (Chen et al. 1975a).

3.1. Raman Scattering by Surface Polariton

To simplify the mathematics and to convey more clearly the kinematics of the Raman process involving a surface polariton, we shall first treat the case of a single interface. The discussion will be generalized, without formal detail, to a two-interface (e.g., slab) system in which the middle medium is nonlinear active. Readers who want to see a detailed treatment of the slab case are referred to the papers by Mills et al. (1976) and by Nkoma (1975).

We consider a dielectric-semiconductor (single interface) configuration in which the semiconductor is the only Raman active medium, and either the dielectric or the semiconductor can be the surface active medium. The incident Fresnel transfer functions in the semiconductor ($z'<0$) can be readily obtained as follows:

$$\Gamma_\alpha(\boldsymbol{k}_{\|\mathrm{i}}, \omega_\mathrm{i}, z') = f_\alpha T_{12}(\boldsymbol{k}_{\|\mathrm{i}}, \omega_\mathrm{i}) \exp(-\mathrm{i}k^{\mathrm{i}}_{z2}z'), \tag{21}$$

where $\boldsymbol{k}_{\|\mathrm{i}}$ and ω_i are the wave vector parallel to the interface and the frequency of the incident radiation; $k^{\mathrm{i}}_{zj} \equiv [\epsilon_j(\omega_\mathrm{i})(\omega_\mathrm{i}/c)^2 - k^2_{\|\mathrm{i}}]^{1/2}$ is the normal component of the wave vector in medium j ($j = 1, 2$) and both the real and imaginary part of k^{i}_{zj} are taken to be positive; f_α is the angular coefficient for Γ_α ($\alpha = x$, y and z); and T_{12} is the Fresnel transmission coefficient from medium 1 to medium 2. For transverse magnetic (TM) radiation, $T_{12} = 2k_{z1}(\epsilon_1\epsilon_2)^{1/2}/[\epsilon_1 k_{z2} + \epsilon_2 k_z]$, and for transverse electric (TE) radiation, $T_{12} = 2k_{z1}/[k_{z1} + k_{z2})]$.

There are two possible scattering configurations, as shown in fig. 2. In a backward scattering configuration the output G transfer function is given as follows:

$$\mathcal{G}_{\alpha\beta}(\boldsymbol{k}_{\|\mathrm{s}}, \omega_\mathrm{s}, z, z') = g_{\alpha\beta} T_{21}(\boldsymbol{k}_{\|\mathrm{s}}, \omega_\mathrm{s}) \exp[\mathrm{i}k^{\mathrm{s}}_{z1}z - \mathrm{i}k^{\mathrm{s}}_{z2}z'], \quad \text{for } z>0. \tag{22a}$$

where $g_{\alpha\beta}$ is the angular coefficient for the $\mathcal{G}_{\alpha\beta}$ matrix, s refers to the scattered wave. For a forward scattering configuration the output G transfer function takes the following form:

$$\mathcal{G}_{\alpha\beta}(\boldsymbol{k}_{\|\mathrm{s}}, \omega_\mathrm{s}, z, z') = g_{\alpha\beta} T_{21}(k_{\|\mathrm{s}}, \omega_\mathrm{s}) \exp[\mathrm{i}k^{\mathrm{s}}_{z2}(z'+d) - \mathrm{i}k^{\mathrm{s}}_{z1}(z+d)], \quad \text{for } z<-d, \tag{22b}$$

where d is the thickness of the sample.

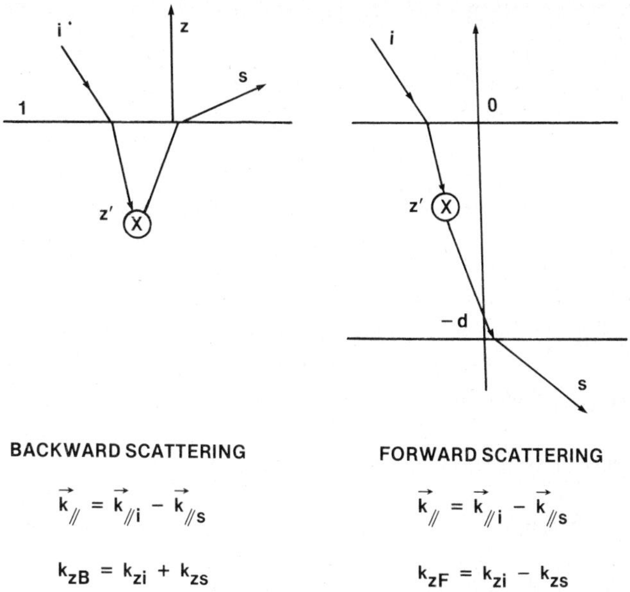

Fig. 2. Diagrams for the forward and backward scattering configuration in a single interface system. Although both scattering configurations have the same amount of wavevector change parallel to the interface, in a forward scattering configuration the change of wavevector normal to the interface is $k_{zF} = k_{zi} - k_{zs}$, while in a backward scattering configuration, the change of wavevector normal to the interface is $k_{zB} = k_{zi} + k_{zs}$.

We should point out that in eqs. (21) and (22) we have neglected the reflected waves at the back interface. As we shall discuss later the presence of the reflected wave makes the scattering kinematics more complex. For now we will assume that the sample is so thick that the second interface can be neglected. Since surface polaritons exist at each surface active interface, there should be an identical surface mode at the $z = -d$ interface. The presence of that surface mode is also neglected. The Raman cross section, given in eq. (18), can be written as

$$\frac{d\sigma}{d\omega_s} \propto \left(\frac{\omega_s}{c}\right)^6 \sum_{\alpha} \sum_{\eta\eta'} [M^*_{\alpha\eta'} M_{\alpha\eta}(I_{\eta'\eta}(\omega - i\sigma) - I^*_{\eta'\eta}(\omega - i\sigma))](\bar{n}(\omega) + 1),$$

$$M_{\alpha\eta} = \sum_{\beta\xi} [\mathcal{G}_{\alpha\beta}(k_{\|s}) f_{\xi}(k_{\|i})(b_{\beta\xi\eta} + \gamma(\omega) a_{\beta\xi\eta}) T_{12}(k_{\|i}) T_{21}(k_{\|s})], \tag{23}$$

where

$$I^{B}_{\eta'\eta}(q_{\|}\omega) = \int \int_{0}^{-d} dz\, dz' \exp(-ik_B z') \exp(ik_B^* z'') \mathcal{G}_{\eta'\eta}(\omega, \boldsymbol{q}_{\|}, z'', z')$$

is the response function of $\langle E_{\eta'} E_\eta \rangle$ for the backscattering configuration and

$$I^{\mathrm{F}}_{\eta'\eta}(q_\| \omega) = \int_0^{-d}\!\!\int \mathrm{d}z''\,\mathrm{d}z' \exp(-\mathrm{i}k_{\mathrm{F}} z') \exp(\mathrm{i}k^*_{\mathrm{F}} z'') \mathcal{G}_{\eta'\eta}(\omega, \boldsymbol{k}_\|, z'', z')$$
$$\times \exp(-2k^{s''}_{z2} d)$$

is the response function of $\langle E_{\eta'} E_\eta \rangle$ for the forward scattering configuration, where $\omega = \omega_\mathrm{i} - \omega_\mathrm{s}$ and $\boldsymbol{q}_\| = \boldsymbol{k}_{\|\mathrm{i}} - \boldsymbol{k}_{\|\mathrm{s}}$ are the frequency and wave vector component parallel to the interface of the optical phonon polaritons; $\sigma \to 0^+$ is a positive infinitesimal which determines the "sign" of $\mathrm{Im}(I_{\eta'\eta}(q_\|, \omega))$; $k_\mathrm{B} \equiv k_{z\mathrm{i}} + k_{z\mathrm{s}}$ and $k_\mathrm{F} = k_{z\mathrm{i}} - k_{z\mathrm{s}}$ are the changes of k_z inside the nonlinear medium in the scattering process for the backward and forward scattering configuration, respectively, and k^* is the complex conjugate of k. We can rearrange eq. (23) and obtain the following simple form for the Raman differential cross section:

$$\frac{\mathrm{d}^2\sigma}{\mathrm{d}\omega_\mathrm{s}\,\mathrm{d}\Omega_\mathrm{s}} \propto -(\bar{n}(\omega) + 1)\left(\frac{\omega_\mathrm{s}}{c}\right)^6 \cos\theta_\mathrm{s} \sum_\alpha \sum_{\eta\eta'} 2\,\mathrm{Im}(M_{\alpha\eta} M^*_{\alpha\eta'} I_{\eta'\eta}(q_\|, \omega - \mathrm{i}\sigma)). \tag{24}$$

Let us now consider a semiconductor with a zincblende crystal structure (e.g., GaAs) with its principal axes along $\hat{x}$, $\hat{y}$ and $\hat{z}$, and with the normal to the interface along $\hat{z}$. Thus, the first order Raman tensors $\mathbf{a}_{\alpha\beta\xi}$ and $\mathbf{b}_{\alpha\beta\xi}$ are nonzero only when α, β, and ξ are mutually orthogonal to each other. We will assume, for convenience, that the incident field is polarized along $\hat{x}$ and the scattered field is polarized along $\hat{y}$ and their corresponding $k_\|$ components are both along $\hat{x}$. For this case the allowed Raman cross section only consists of an I_{zz} term which is associated with the z component electric field of the optical phonon surface (or volume) polaritons.

The Green's function $\mathcal{G}_{zz}(q_\|, \omega, z', z)$ for a single interface configuration takes the form

$$\mathcal{G}_{zz} = \frac{4\pi}{\epsilon_2(\omega/c)^2}\delta(z - z') + \frac{2\pi\mathrm{i}}{q_{z2}} \frac{q_\|^2}{\epsilon^2\omega^2/c^2}$$
$$\times \{[\exp(\mathrm{i}q_{z2}z') + R_{21}\exp(-\mathrm{i}q_{z2}z')]\exp(-\mathrm{i}q_{z2}z)\theta(z' - z)$$
$$+ [\exp(\mathrm{i}q_{z2}z) + R_{21}\exp(-\mathrm{i}q_{z2}z)]\exp(-\mathrm{i}q_{z2}z')\theta(z - z')\},$$
$$\text{for} \quad z < 0 \quad \text{and} \quad z' < 0, \tag{25}$$

where $\theta(z) = 1$ for $z > 0$ and $\theta(z) = 0$ for $z < 0$; and $q_{zj} = (\epsilon_j(\omega)(\omega/c)^2 - q_\|^2)^{1/2}$. The response function I_{zz} can be expressed by the sum of a volume contribution term I^V_{zz} and a surface contribution term I^S_{zz}:

$$I_{zz} = I^\mathrm{V}_{zz} + I^\mathrm{S}_{zz}. \tag{26}$$

We obtain, after some straightforward algebra, the following forms for

I^{V}_{zz} and I^{S}_{zz}, respectively:

(a) *Backward scattering configuration*:

$$I^{V}_{zz} = \frac{4\pi c^2}{i\omega^2\epsilon_2(\omega)} \frac{1}{k_B - k_B^*}\left(1 + \frac{k_\parallel^2}{(q_{z2}+k_B)(q_{z2}-k_B^*)}\right),$$

$$I^{S}_{zz} = \frac{4\pi c^2}{i\omega^2\epsilon_2(\omega)} \frac{\epsilon_1}{\epsilon_1 q_{z2} + \epsilon_2 q_{z1}}\left(\frac{k_\parallel^2}{(q_{z2}+k_B)(q_{z2}-k_B^*)}\right). \tag{27}$$

The I^{V}_{zz} term, for scattering by volume modes, contains a "pole" in the denominator for the excitation of LO phonons $D_L = \epsilon_2(\omega)$, and a "pole" for the excitation of volume polaritons $D_\pi = (q_{z2}+k_B)(q_{z2}-k_B^*)$. I^{S}_{zz} is a "surface correction" term for the scattering response of the volume modes which arises from the fact that the volume modes "reflect" at the interface. When the skin depth of the incident/scattered radiation is not small, I^{S}_{zz} is relatively small compared to that of the I^{V}_{zz} term.

We take special note of the following aspects:

(i) D_π contains a nonresonant factor $(q_{z2}+k_B)$ which, in the case that both the incident/scattered radiations and the phonons are weakly damped, does not appreciably affect the bulk polariton response. However, when the damping of either the incident/scattered radiations or the lattice dielectric functions is large, it can affect the linewidth of the volume polaritons. When the incident/scattered radiations are damped ($k_B = k'_B + ik''_B$) and the lattice dielectric function is undamped (q_z is real), the volume polariton response D_π can be written as $D_\pi = [(q_{z2}^2 - k_B'^2 - k_B''^2) + 2iq_{z2}k''_B]$ and the linewidth of the Raman spectra of the polaritons will be asymmetric (Dresselhaus and Pine 1975).

(ii) The LO phonon response, is determined by $\mathrm{Im}(1/D_L)$, which, in the absence of spatial dispersion, depends only on the damping of the LO phonons and is independent of the damping of the incident/scattered radiations.

(iii) The I^{V}_{zz} term for both LO phonons and volume polaritons has a kinematic factor $K_V = (k_B - k_B^*)^{-1}$, which equals half the skin depth of the incident/scattered radiations or, in the case that the nonlinear medium is transparent (i.e. $k''_B d \ll 1$), to the sample thickness d.

The I^{S}_{zz} term, for scattering by surface polaritons at the 1–2 single interface, contains a factor in the denominator $D_S = \epsilon_2 q_{z1} + \epsilon_1 q_{z2}$. We again take note of the following aspects:

Either the nonlinear semiconductor or the adjacent dielectric can be the surface active medium. When the frequency of the excited surface polariton, ω, is close to the frequency of the TO phonon of the nonlinear medium, $\gamma(\omega)$ is large and as a consequence the lattice displacement contribution to the Raman cross section is large. When ω is far from the TO phonon frequency, $\gamma(\omega)$ is small and the scattering cross section for the surface polaritons is mainly due to the electro-optical term of the nonlinear susceptibility.

When surface polaritons are excited in the absence of polariton damping, $q_{z2} = i\alpha_2$ is purely imaginary. The surface polariton scattering term has a kinematic factor $K_S = \alpha_2^*/|k_B + i\alpha_2|^2$, where α_2^* is a composite attenuation constant of the surface polariton which takes into account the distribution of the energy density of the surface polariton in the adjacent media. For the case of surface polaritons at a semiconductor–air interface, α_2^* is effectively equal to the attenuation constant in the semiconductor α_2.

For the input/output EM waves in the visible, the dielectric constants of a III–V compound semiconductor (e.g., GaAs or GaP) are much larger than unity ($\epsilon(\omega_1)$ and $\epsilon(\omega_2)$ are usually larger than 10) and $k_B' \gg q_\parallel \simeq \alpha_2$ and $k_B' > k_B''$. Thus the presence of k_B' in the denominator of K_S and the fact that it is large in a backward scattering configuration is responsible for the very small scattering cross section of the surface modes relative to that of the volume modes.

We should emphasize that the difference between K_V and K_S is mainly due to the difference in the nature of the volume and surface modes. In the case of Raman scattering by the volume modes the z component of the phonon wave vector, q_{z2}, is predominantly real and can be phase matched in the scattering process by the real part of the z component scattering wave vector k_B. On the other hand, in the case of Raman scattering by the surface polaritons, its z component wave vector, q_{z2}, is predominantly imaginary and therefore cannot be phase matched in the scattering process.

(b) *Forward scattering configuration*:

$$I_{zz}^{V} = \frac{4\pi c^2}{i\omega^2\epsilon_2(\omega)} \frac{1-\exp(-2k_F''d)}{k_F - k_F^*} \left(1 + \frac{q_\parallel^2}{(q_{z2}+k_F)(q_{z2}-k_F)}\right) \exp(-2k_{z2}''d),$$

$$I_{zz}^{s} = \frac{4\pi c^2}{i\omega^2\epsilon_2(\omega)} \frac{\epsilon_1}{\epsilon_1 q_{z2} + \epsilon_2 q_{z1}} \left(\frac{q_\parallel^2}{(q_{z2}+k_F)(q_{z2}-k_F^*)}\right) \exp(-2k_{z2}''d). \tag{28}$$

When the semiconductor is transparent to the incident and the scattered radiations (i.e. when $k_F''d \ll 1$) we can readily write the kinematic factors for the volume and the surface modes as follows:

$$K_V^F \approx d, \qquad K_S^F \approx 1/\alpha_2. \tag{29}$$

Again, the difference in the interaction kinematics of the volume and surface modes shows up in their corresponding kinematic factors. For the volume modes which exist throughout the sample, the kinematic factor,

K_V, equals to the sample thickness. Since the surface polaritons are localized at the interface with an attenuation constant α_2, the corresponding kinematic factor is limited to the penetration depth of the surface polaritons (i.e. to $\delta \sim 1/\alpha_2$).

We can summarize the formulation for the single interface system as follows: For a transparent semiconductor, the kinematic factors of the volume modes in the forward and the backward scattering configurations are nearly the same. However, the kinematic factors of the surface polaritons in the forward and the backward scattering configurations are quite different. The largest kinematical factor for scattering by surface polaritons, $K_S \approx 1/\alpha_2$, can only be obtained in forward scattering configurations. The optimum scattering cross section can be obtained by using a nonlinear medium of thickness $d_0 = 1/\alpha_2$ and incident/scattered frequencies in the resonance absorption region. In particular, when $d = d_0$ and $k''_{zi}d < 1$ $(i = 1, 2)$, the kinematic factor for the Raman scattering by surface polaritons in a forward scattering configuration can be comparable to that for the Raman scattering by volume modes.

We now consider the specific case of a dielectric (3) – semiconductor (2) – air (1) two interface system (as shown in fig. 3) where again, the semiconductor is assumed to be the only Raman active medium. (The formulation for the case in which the dielectric is the Raman active medium is equivalent to that for a single-interface case.) We note that either the semiconductor or the dielectric can be the surface active medium.

When the dielectric is the surface active medium, there is only one surface polariton branch. The dispersion relation of the surface polaritons depends on the dielectric functions of all three media, as well as on the thickness of the semiconductor slab d. When the thickness of the semiconductor is large compared to the penetration depth of the surface polariton ($\alpha_2 d \gg 1$) the surface polariton modes correspond to the 3–2 single interface modes.

When the semiconductor is the surface active medium, there are two surface polariton branches which correspond to the coupled modes of the

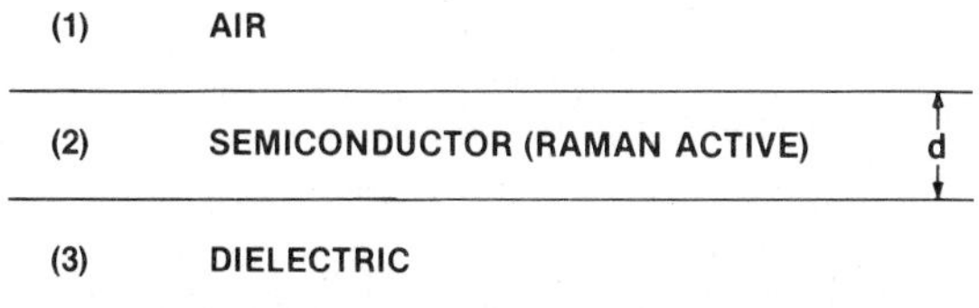

Fig. 3. A dielectric–semiconductor–air two interface system in which the semiconductor "slab" is the Raman active medium.

surface polaritons at the 2–3 and 2–1 single-interface. For a thick slab, i.e. $\alpha_2 d \gg 1$, there is essentially no coupling between the modes at the two surface active interfaces and the two surface polariton branches, Ω_+ and Ω_-, coincide with the uncoupled 2–1 and 2–3 single-interface surface modes. For a thin slab, the coupling between the 2–1 and 2–3 single-interface surface polariton modes pushes the two branches apart. The larger the coupling is (i.e. the smaller $\alpha_2 d$) the larger the shifts in the dispersion are.

The transfer functions for the incident and scattered radiations for the slab system are given as follows:

$$\Gamma_\alpha(k_{\|i}, \omega_i z') = f_\alpha[\Gamma^+(k_{\|i}, \omega_i) \exp(ik^i_{z2}z') + \Gamma^-(k_{\|i}, \omega_i) \exp(-ik^i_{z2}a')],$$

$$\mathcal{G}_{\alpha\beta}(k_{\|s}, \omega_s, z, z') = g_{\alpha\beta}[\mathcal{G}^+(k_{\|s}, \omega_s, z) \exp(+ik^s_{z2}z') + \mathcal{G}^-(k_{\|s}, \omega_s, z) \exp(-ik^s_{z2}z')]. \quad (30)$$

As shown in figs. 4 and 5, there are four scattering processes for either the forward or the backward scattering configurations. In the forward

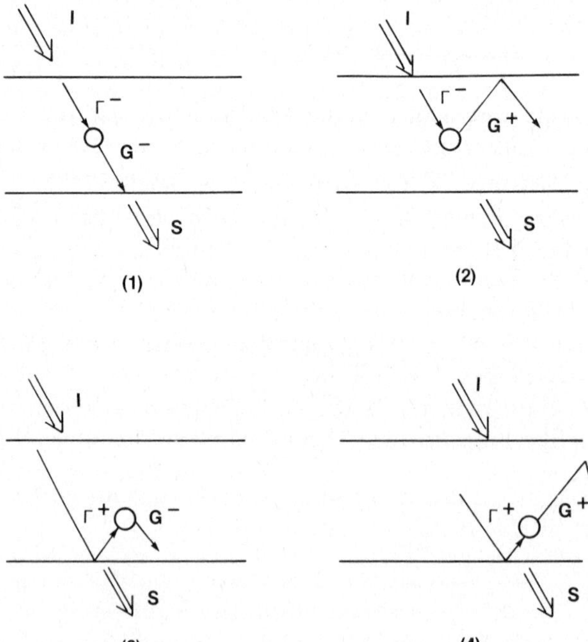

Fig. 4. Scattering processes in the forward scattering configuration for a slab system. Diagrams 1 and 4 are forward scattering processes, in which the change of wavevector normal to the interface is $k_{zF} = k_{zi} - k_{zs}$. Diagrams 2 and 3 are backward scattering processes in which the change of wavevector normal to the interface is $k_{zB} = k_{zi} + k_{zs}$.

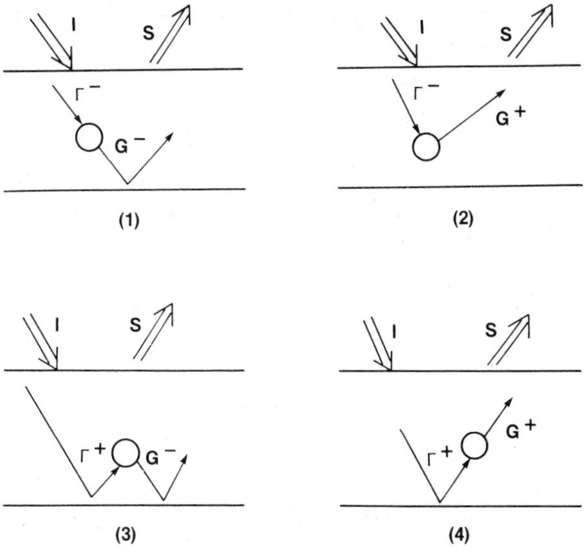

Fig. 5. Scattering processes in the backward scattering configuration for a slab system. Diagrams 1 and 4 are forward scattering processes while diagrams 2 and 3 are backward scattering processes.

scattering configuration (fig. 4) the Raman scattering processes shown in the diagrams 1 and 4 are forward scattering processes, while the Raman scattering processes shown in the diagrams 2 and 3 are backward scattering processes. In the backward scattering configuration (fig. 5) the Raman scattering processes shown in the diagrams 1 and 4 are forward scattering processes while the Raman scattering processes shown in the diagrams 2 and 3 are backward scattering processes. The main difference between a forward and a backward scattering configuration is that the forward scattering process in a forward scattering configuration (diagram 1 in fig. 4) has the shortest path. In all of the other forward scattering processes the path of the incident and/or the scattered radiation are longer by at least a slab thickness d.

The Raman differential cross section for the various scattering processes can be written as follows:

$$\frac{\mathrm{d}^2\sigma}{\mathrm{d}\omega_\mathrm{s}\,\mathrm{d}R_\mathrm{s}}\,\alpha\left(\frac{\omega_\mathrm{s}}{c}\right)^6(\bar{n}(\omega)+1)\cos\theta_\mathrm{s}\sum_{\alpha}\sum_{\substack{\eta\eta'\\ jj'}}\{2\,\mathrm{Im}[M^{(j)}_{\alpha\eta}M^{(j')*}_{\alpha\eta'}I^{(j'j)}_{\eta'\eta}(q_\|,\omega-\mathrm{i}\sigma)]\},$$

$$M^{(j)}_{\alpha\eta}=\sum_{\beta\xi}\mathcal{G}^{(m)}_{\alpha\beta}(k_{\|\mathrm{s}},z)(b_{\beta\xi\eta}+\gamma(\omega)a_{\beta\xi\eta})\Gamma^{(n)}_{\xi}(k_\mathrm{i})],$$

$$I^{(j'j)}_{\eta'\eta}=\int\!\!\int \mathrm{d}z'\,\mathrm{d}z''\exp(-\mathrm{i}k_jz')\exp(\mathrm{i}k^*_{j'}z'')\mathcal{G}_{\eta'\eta}(\omega,q_\|,z'',z'),\qquad(31)$$

where subscripts j and j' designate the 1, 2, 3 and 4 scattering processes; k_j is the change of k_z in the jth scattering process; and the superscripts m and n represent + and − for the forward and backward directions. We note that there are a total of $4 \times 4 = 16$ terms in each $I_{\eta'\eta}$ tensor. Instead of repeating the lengthy algebraic expression given by Mills, et al. (1976), we shall only consider two limiting cases using the coupling kinematics argument discussed above.

(A) "Thick" Slab ($k''_{z2}d > 1$).
As in the single interface configuration, the kinematic factor for scattering by surface polaritons in a forward scattering process is much larger than that in a backward scattering process. Thus, in the forward scattering configuration the Raman cross section of surface polariton is dominated by scattering process 1 shown in fig. 4. Since the latter is also the scattering process with the shortest path it should also be the dominant process for scattering by volume modes.

In the backward scattering configuration the Raman cross section of surface polaritons is dominated by both forward scattering processes 1 and 4 shown in fig. 5. Since these forward scattering processes require a longer propagation path for the incident and/or scattered radiations, they are weaker than that of the forward scattering configuration by a factor of $\exp(-2k''_{z2}d)$. Raman scattering cross section of the volume modes is dominated by the "shortest" scattering process (scattering process 2 shown in fig. 5). Since the kinematic factor for scattering by volume modes in a backward scattering configuration is the same as that in a forward scattering configuration, the Raman cross section of the volume modes, in the backward scattering configuration, is greater than that of the surface polaritons by a factor of $\exp(2k''_{z1}d)$.

We also note the following:

(i) The volume modes in the slab system consist of the LO phonons and volume polaritons of a bulk sample, as well as "slab" (waveguide) modes (Fuchs and Kliewer 1966). The differences in the Raman cross section among them are mainly in the selection rules and in the coupling strength and not in the kinematic factor.

(ii) The interference between the different scattering processes (e.g., between the scattering processes 1 and 2 in a backward scattering configuration) makes it difficult to observe the individual contribution of each scattering process. However as has been shown by Mills et al. (1976), the numerical evaluation of the exact expressions can readily reveal the scattering modes of the system.

(iii) The kinematic factors for the coupled surface polariton modes Ω_+ and Ω_- are roughly the same. However, the Raman intensities of the two surface modes can be somewhat different. This is because of the differences in the magnitude of the lattice displacement contribution which is determined by $\gamma(\omega)$, and because of the differences in the penetration depth. Since Ω_- is somewhat closer to ω_T than Ω_+, $\gamma(\Omega_-)$ is larger than $\gamma(\Omega_+)$. We should also point out that when the dielectric is the surface active medium, the single-interface surface polariton frequency can be quite different from the ω_T in the semiconductor. In that case, the main contribution to the Raman scattering process will be from the electro-optical term.

(B) "Thin" Slab ($k''_{zi}d \ll 1$).
When the slab is very thin one has to take into account, for each scattering configuration, all the possible scattering processes of the volume modes and all the forward scattering processes of the surface modes. When $\alpha_2 d > 1$, the kinematic factor for the forward scattering processes of surface modes is $\sim 1/\alpha_2$. However when $\alpha_2 d \ll 1$ the kinematic factor for the forward scattering processes of surface modes is $\sim d$, the Raman cross sections of the surface modes, for both the forward and the backward scattering configurations, are roughly the same. Furthermore since the kinematic factor for the scattering by volume modes is $\sim d$, there is essentially no difference in the kinematic factors for the volume and the surface modes.

The above discussions of the Raman scattering by surface polaritons can also be applied to that of the light diffraction by optical phonon surface polaritons generated for example by prism coupling (Chen et al. 1977). Light diffraction by an optical phonon surface polariton is similar to parametric mixing of a polariton to an optical phonon surface polariton. The population density of the surface polaritons is increased so that one does not need resonance enhancement to observe the nonlinear interaction. To optimize the effect of light diffraction by surface polaritons in a finite size sample, one again should work with a forward scattering configuration, i.e., the output EM wave is propagating in the same direction as the input EM wave. The frequency shift and the diffracted angle of the output wave yields quantitative information of the dispersion relation of the surface polaritons. Thus, the light diffraction process can be a useful method for studying the properties of surface polaritons.

3.2. *Parametric Generation of Surface Polaritons*

We first consider the difference-frequency mixing of two input EM waves to generate volume and surface optical phonon polaritons in an air (1)–semiconductor (2) single interface system, in which the semiconductor is transparent to both of the input waves and serves as both the nonlinear active as well as the surface active medium. The two input waves can enter into the semiconductor either from the same side, in which the normal

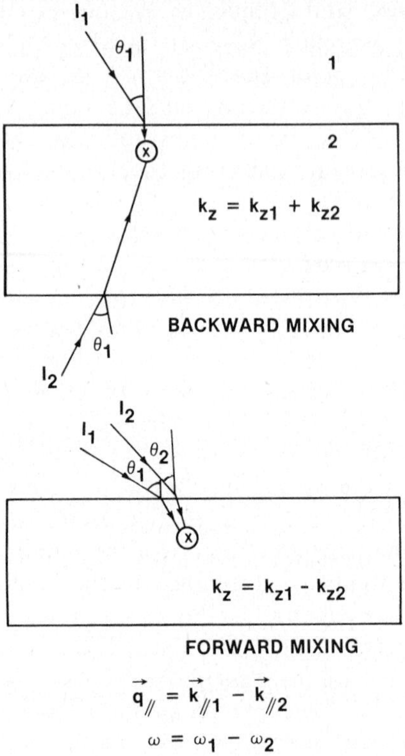

Fig. 6. The forward and backward difference-frequency mixing configurations in a single interface system. In a forward mixing configuration, the change of wavevector normal to the interface is $k_{zF} = k_{z1} - k_{z2}$ while in a backward mixing the configuration the change of wavevector normal to the interface is $k_{zB} = k_{z1} + k_{z2}$.

component mixing wavevector is $k_F = k_{z1} - k_{z2}$, or from the opposite sides, in which the normal component mixing wavevector is $k_B = k_{z1} + k_{z2}$, as shown in fig. 6. The first case will be called a forward mixing configuration and the second case will be called a backward mixing configuration in analogy to the terminologies of our previous discussion of Raman scattering.

The nonlinear polarization of the three wave mixing takes the following form:

$$P_\alpha^{\rm NL}(\boldsymbol{q}_\|, \omega, z') = \sum_{\beta\xi} [(b_{\alpha\beta\xi} + \gamma(\omega)a_{\alpha\beta\xi})f_\beta f_\xi T_{21}(k_{\|1}, \omega_1)T_{21}(k_{\|2}, \omega_2)\exp(\mathrm{i}k_z z')]$$
$$\equiv \boldsymbol{P}^{\rm NL}(\boldsymbol{q}_\|, \omega)\exp(\mathrm{i}k_z z'), \tag{32}$$

where $\boldsymbol{q}_\| = \boldsymbol{k}_{\|1} - \boldsymbol{k}_{\|2}$; $\omega = \omega_1 - \omega_2$; $k_z = k_{z1} + k_{z2}$ for the backward mixing configuration and $k_z = k_{z1} - k_{z2}$ for the forward mixing configuration. For simplicity we assume both $\boldsymbol{k}_{\|1}$ and $\boldsymbol{k}_{\|2}$ are along the $\hat{\boldsymbol{x}}$ axis. The Green's functions inside the semiconductor, which are readily obtained from eq. (3), are given by:

$$\begin{aligned}
\mathcal{G}_{zz} &= \frac{4\pi}{\epsilon_2(\omega^2/c)}\,\delta(z-z') - \frac{2\pi i c^2 q_\|^2}{q_{z2}\omega^2\epsilon_2}\{\exp(-iq_{z2}z')[\exp(iq_{z2}z') \\
&\quad + R_{21}\exp(-iq_{z2}z')]\theta(z'-z) + \exp(-iq_{z2}z') \\
&\quad \times [\exp(iq_{z2}z) + R_{21}\exp(-iq_{z2}z)]\theta(z-z')\}, \\
\mathcal{G}_{xz} &= \frac{2\pi i c^2 q_\|}{\epsilon_2\omega^2}\{[\exp(iq_{z2}z') + R_{21}\exp(iq_{z2}z')]\exp(-iq_{z2}z)\theta(z'-z) \\
&\quad + [-\exp(iq_{z2}z) + R_{21}\exp(-iq_{z2}z)]\exp(-iq_{z2}z')\theta(z-z')\}, \\
\mathcal{G}_{zx} &= \frac{2\pi i c^2 q_\|}{\epsilon_2\omega^2}\{[\exp(iq_{z2}z') - R_{21}\exp(-iq_{z2}z')]\exp(-iq_{z2}z)\theta(z'-z) \\
&\quad + [R_{21}\exp(-iq_{z2}z) - \exp(iq_{z2}z)]\exp(-iq_{z2}z')\theta(z'-z)\}, \\
\mathcal{G}_{xx} &= -\frac{2\pi i c^2 q_{z2}}{\omega^2\epsilon_2}\{[\exp(iq_{z2}z') - R_{21}\exp(-iq_{z2}z')]\exp(-iq_{z2}z)\theta(z'-z) \\
&\quad + [\exp(iq_{z2}z) - R_{21}\exp(-iq_{z2}z)]\exp(-iq_{z2}z')\theta(z-z')\}.
\end{aligned} \tag{33}$$

The x component of the TM mode solution of eq. (4) can be expressed by the sum of a volume contribution term E_x^{V} and a surface contribution term E_x^{S}:

$$\begin{aligned}
E_x(\boldsymbol{q}_\|, \omega, z) &= E_x^{\mathrm{V}}(\boldsymbol{q}_\|, \omega, z) + E_x^{\mathrm{S}}(\boldsymbol{q}_\|, \omega, z), \\
E_x^{\mathrm{V}}(\boldsymbol{q}_\|, \omega, z) &= \frac{1}{k_z + q_{z2}}\frac{4\pi}{\epsilon_2(\omega)}\left\{\frac{1}{k_z - q_{z2}}[\exp(-ik_z z)\right. \\
&\quad \left.\times (P_x^{\mathrm{NL}} q_{z2}^2 + P_z^{\mathrm{NL}} k_z q_\|) - \exp(-iq_{z2}z)q_{z2}(P_x^{\mathrm{NL}} k_z + P_x^{\mathrm{NL}} q_\|)]\right\}, \\
E_x^{\mathrm{S}}(\boldsymbol{q}_\|, \omega, z) &= \frac{1}{k_z + q_{z2}}\frac{4\pi}{\epsilon_2(\omega)}\left\{\frac{\epsilon_1 q_{z2}}{\epsilon_1 q_{z2} + \epsilon_2 q_{z1}}\exp(-iq_{z2}z)(P_x^{\mathrm{NL}} q_{z2} - P_z^{\mathrm{NL}} q_\|)\right\}.
\end{aligned}$$

We note that E_x^{V}, the x field component of the optical phonon polariton generated by the mixing process, peaks when the mixing wave vector k_z matches the polariton wavevector q_{z2}. The first term of E_x^{V}, represents the driven wave which propagates along the driven P^{NL} direction. The second term represents the free wave which propagates as a volume polariton with frequency $\omega = \omega_1 - \omega_2$ and parallel wave vector $\boldsymbol{q}_\| = \boldsymbol{k}_{\|1} - \boldsymbol{k}_{\|2}$. When the phase matching condition is not satisfied in the parametric mixing process (i.e. $k_z \neq q_{z2}$), the free wave will propagate in a different direction from that of the driven wave. When the phase matching condition is satisfied in the parametric mixing process for optical phonon polaritons (i.e. $k_z = q_{z2}$) the free wave and the driven wave coincide and are in phase throughout the sample (i.e. the mixing process is phase matched). When the mixing process is phase matched, the kinematic factor for the mixing to a bulk phonon polariton (which is also termed the coherent length) is equal to the thickness of the sample d.

The expression for E_x^S, which represents the component of the field of the optical phonon surface polariton along x generated by the difference frequency mixing of two input plane waves, involves only a single term. We note moreover that E^S represents a wave which propagates along P^{NL} (i.e. $q_\| = k_{\|1} - k_{\|2}$) as the input waves are plane waves. Thus, the generated surface polariton is a driven wave whose field amplitude peaks as the incident mixing condition $(q_\|, \omega)$ matches the dispersion relation of the surface polariton $\epsilon_1 q_{z2} + \epsilon_2 q_{z1} = 0$. The parametric generation process has a kinematic factor $K_M = 1/(k_z + q_{z2})$ which is identical to that of the Raman scattering process! Thus, the parametric generation efficiency for a forward mixing configuration is much larger than that for a backward mixing configuration.

The above discussion can be readily generalized to a slab configuration as well. When the slab medium is the nonlinear medium, there are again four mixing processes for either a forward mixing configuration or a backward mixing configuration. Of those, two are forward mixing processes and the other two are backward mixing processes. When the slab is transparent to the input EM waves the incident field strengths (e.g. the Γ transfer functions) of the four mixing processes are comparable. Thus the two forward mixing processes which have much larger kinematic factors, dominate. Furthermore, there is very little difference in the generation efficiency for optical phonon surface polaritons between a forward mixing configuration and a backward mixing configuration. In the case that the slab medium is optically absorbing to the input EM waves at the frequencies of the input EM waves, and the slab thickness is comparable to the penetration depth of the input waves, the forward mixing process in a forward mixing configuration, which has the largest input field strengths and the largest kinematic factor, will be the most efficient mixing process. Thus, the parametric generation efficiency for a forward mixing configuration is much larger than that for a backward mixing configuration.

The above discussion is also relevant for the case of nonlinear sum-frequency generation of the interband e–h pair excition (exciton) surface polaritons. However, the kinematic factors for the forward mixing and backward mixing configurations are opposite to that in the nonlinear difference frequency generation (fig. 7). We note that in a nonlinear sum frequency mixing process $\omega = \omega_1 + \omega_2$; $q_\| = k_{\|1} + k_{\|2}$ and $N = \bar{n}_1 \bar{n}_2$. To maximize the kinematic factor, one should use a forward mixing configuration in which the two input waves propagate in opposite directions in order to reduce k_z. Thus, a second harmonic generation process in which only one

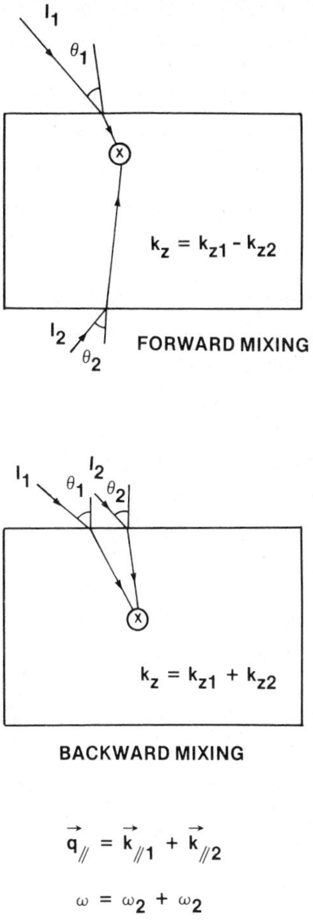

Fig. 7. The forward and backward sum-frequency mixing configurations in a single interface system.

intense beam is used is not an efficient way to generate the sum-frequency surface polaritons.

Since the surface polariton is not a radiative mode, two methods are commonly used to detect the nonlinearly driven surface polariton. In the first method, the surface polariton is *linearly* converted into a volume EM wave via either prism or grating coupling. Both coupling techniques are highly efficient if they are properly used. However, both coupling schemes will disturb the dispersion relation of the surface polariton and should be employed with some caution. In the second approach, the generated surface polariton is probed *nonlinearly*, for example, by Raman scattering. The experimental set-up involves a third "incident" wave. However, one may also encounter a sizeable nonresonant four-wave nonlinear mixing process which does not involve surface polaritons as the intermediate state. The overall nonlinear process, which involves a Raman scattering (light diffraction) by a parametricly generated surface polariton can be expressed as follows:

$$I(\omega_4) \propto n_1(n_2+1)n_3(\omega_4/c)^4|\Gamma_1(\boldsymbol{k}_{\|1}, \omega_1)\Gamma_2(\boldsymbol{k}_{\|2}, \omega_2)\Gamma_3(\boldsymbol{k}_{\|3}, \omega_3) \times G(\boldsymbol{k}_{\|4}, \omega_4)|^2|D(\boldsymbol{q}_{\|}, \omega)|^{-2}K_{\mathrm{M}}(\boldsymbol{q}_{\|}, \omega)K_{\mathrm{R}}(\boldsymbol{k}_4, \omega_4), \quad (35)$$

where $\boldsymbol{k}_{\|4} = \boldsymbol{k}_{\|1} - \boldsymbol{k}_{\|2} + \boldsymbol{k}_{\|3}$; $\boldsymbol{q}_{\|} = \boldsymbol{k}_{\|1} - \boldsymbol{k}_{\|2}$; $D(\boldsymbol{q}_{\|}, \omega) = 0$ is the dispersion relation of the surface polariton of the system; and K_{M} and K_{R} are the kinematic factor of the nonlinear mixing and the Raman scattering process, respectively. Figure 8 shows the optimized geometries for the two, three-wave mixing processes for a single interface system in which both K_{M} and K_{R} are maximized. One should take note of the relationship between the direction of the third wave incident wave and that of the output signal

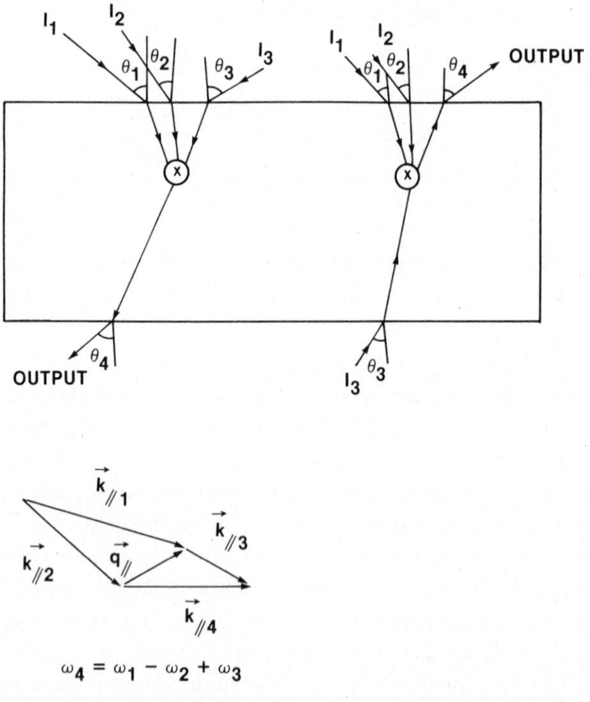

Fig. 8. The form wave mixing configurations in which a surface polariton is excited as the intermediate state. In both mixing configurations the kinematic factors of the two three-wave mixing processes are optimized.

wave. If one of the input wave (e.g. E_1) in the parametric mixing process is also used as the third wave (e.g. the input wave for the Raman scattering process) the output detection of the scattered wave E_4 should be optimally made at the other side of the sample.

As we have noted above when the input waves of the parametric mixing process are infinite plane waves, the generated surface polariton is a driven wave (eq. (34)). However, when the input waves have a finite beam cross section (i.e. when the nonlinear interaction takes place in a finite region) the nonlinear output field of the generated surface polariton has both a "driven" and a "free" term.

We consider a case in which the input EM waves have finite cross section along $\hat{x}$ (the $k_\parallel$ direction). The nonlinear polarization of the difference-frequency mixing process $P^{NL}(x, y)$ thus takes the following form:

$$P^{NL}(x, y) = \phi(x) P^{NL} e^{ik_0 \cdot x},$$

where

$$\phi(x) = 1, \quad 0 \leq x \leq W, \qquad = 0, \quad \text{otherwise}. \tag{36}$$

For convenience, $P^{NL}(x, y)$ can be expressed by the Fourier component $P^{NL}(q)$,

$$P^{NL}(q) \equiv \int P^{NL}(x') e^{-iqx'} dx'. \tag{37}$$

Since for each plane wave component $P^{NL}(q)$, the output plane wave field component of the generated surface polariton is given by eq. (34), i.e.,

$$E(q) \equiv B(q) \frac{P^{NL}(q)}{D(q, \omega)} \equiv P^{NL}(q) \left[\frac{C_1(q)}{q - q_s} + \frac{C_2(q)}{q + q_s} \right], \tag{38}$$

where $D(q_s, \omega) = 0$ is the dispersion relation of the surface polariton for a given ω, and $q_s'' > 0$. The output field amplitude of this generated surface polariton can be written as:

$$E(x) = \int_{-\infty}^{\infty} E(q) e^{iqx} \frac{dq}{2\pi} = \int_{-\infty}^{\infty} P^{NL}(q) \left[\frac{C_1(q)}{q - q_s} + \frac{C_2(q)}{q + q_s} \right] e^{iqx} \frac{dq}{2\pi}. \tag{39}$$

By substituting eq. (37) into eq. (39), we obtain the following expression for the parametricly generated field amplitude:

$$E(x) = P^{\mathrm{NL}}\Big[C_1(q_s)\exp(\mathrm{i}q_s x)\int \phi(x')\exp[\mathrm{i}(k_0 - q_s)x']\,\mathrm{d}x' + C_2(-q_s)\exp(-\mathrm{i}q_s x)\int \phi(x')\exp[\mathrm{i}(k_0 + q_s)x']\,\mathrm{d}x'\Big]. \tag{40}$$

We note that the field amplitude of the generated surface polariton has two terms. The first term represents a surface polariton wave which is driven by the nonlinear polarization $P^{\mathrm{NL}}(q_s)$ and propagates along the driven P^{NL} $(\hat{k}_0)$ direction. The second term represents a surface polariton wave which results from the finite beam size effect, and is propagating opposite to the $\hat{k}_0$ direction. Since $P^{\mathrm{NL}}(q_s\hat{k}_0) \gg P^{\mathrm{NL}}(-q_s\hat{k}_0)$, the second term is small and can be neglected.

By using the form factor $\phi(x')$ of P^{NL} given by eq. (36) and integrating over x' we obtained the following simple result:
for $0 \leqslant x \leqslant W$ (within the driving region)

$$E(x) = P^{\mathrm{NL}}C_1(q_s)\frac{\exp(\mathrm{i}k_0 x) - \exp(\mathrm{i}q_s x)}{\mathrm{i}(k_0 - q_s)}, \tag{41}$$

for $x > W$ (outside the driving region)

$$E(x) = P^{\mathrm{NL}}C_1(q_s)\exp(\mathrm{i}q_s x)\frac{\exp[\mathrm{i}(k_0 - q_s)W] - 1}{\mathrm{i}(k_0 - q_s)}.$$

Within the nonlinear mixing (driving) region ($0 \leqslant x \leqslant W$) the output field of the nonlinear mixing has both a "free propagating" term (of $\exp(\mathrm{i}q_s x)$) and a "driven" term (of $\exp(\mathrm{i}k_0 x)$). Outside the driving region where there is no driven $P^{\mathrm{NL}}(x)$, the output field is entirely that of a "free" wave. We note that eventhough the characteristics (i.e. wave vector and frequency) of the surface polaritons generated outside the driving region is independent of the driving condition, nevertheless the generation efficiency of the nonlinear interaction process depends on the phase matching condition $(k_0 - q_s)$ which peaks when $k_0 - q_s$ is a minimum.

The above discussion can be readily applied to the cases in which the surface polariton is driven linearly by grating or prism coupling using an input beam whose width is small or comparable to the propagation length of the surface polariton. The dispersion relation of the surface polariton in the driving region and that in the free propagating region can be studied via either linear coupling (e.g. prism or grating coupling) or nonlinear coupling (e.g., Raman scattering) techniques. Furthermore, by combining the linear and nonlinear coupling techniques one can study the nonlinear interactions involving surface polaritons in materials with weak nonlinear effects (i.e. at frequencies away from resonance).

4. *Enhancement of Nonlinear Optical Interactions using Surface Polaritons*

As we have shown in the last section the denominators of the input and output transfer functions exhibit poles which correspond to the dispersion relations of surface polaritons at the input and output frequencies, respectively. Thus, the transfer functions can be enhanced by coupling the input and/or output EM waves to surface polaritons. The enhancement of the transfer functions for the input and/or output fields is particularly advantageous in the case of optical phenomena at the surface of metals (or at the surface of any surface active medium where surface polaritons can exist).

A good example is second harmonic generation at a metal (e.g., silver) surface. Since the electromagnetic waves in the metal are evanescent, i.e., they change exponentially with distance from the surface. At frequencies where $|\epsilon'| \gg 1$, the penetration depths of the evanescent waves and that of the surface polaritons are essentially the same. Thus, the kinematic factor for nonlinear interaction in the metal is the same for the evanescent EM waves and the surface polariton waves. However, since surface polaritons are normal modes of the system. The field strengths generated by resonant excitation via prism or grating coupling are considerably larger than that of the evanescent EM waves. It is this aspect which accounts for the enhancement of the Γ transfer functions when surface polaritons are excited by the incident EM waves. Similarly, when the output waves are surface polariton waves, there is a resonant build-up of the field amplitudes. The enhanced fields of the output surface polariton waves can be coupled into volume EM waves either by prism or grating coupling. However, it is the resonant excitation of the surface polaritons by the nonlinear polarization rather than the coupling out of the fields that accounts for the enhancement of the G transfer factors.

In the case of Raman scattering by *very* thin overlayers (e.g., monolayers, etc.) on surface active media, the kinematic factor is proportional to the thickness of the overlayers and is the same for the volume EM and surface polariton incident/scattered waves. Thus, the Raman scattering efficiency can be enhanced by coupling to the surface polariton waves.

There are two prism configurations for linearly coupling of volume and surface EM waves (fig. 9). In the Otto (prism–air gap–metal) configuration, the prism is placed within the surface inactive medium (air). In the Kretschmann (prism–metal film–air) configuration, the prism is placed within the surface active medium (metal). The degree of enhancement of the transfer functions in an Otto configuration depends on the dielectric functions of both the surface active and the surface inactive medium and on the "gap" thickness of the surface inactive medium. In the case of Kretschmann configuration, the degree of enhancement of the transfer functions depends again on the

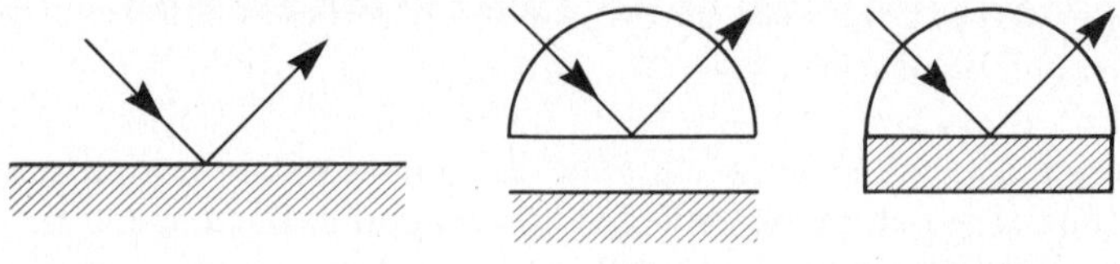

Fig. 9. The external reflection (ER) and the two prism coupling (Otto and Kretschmann) configurations.

dielectric functions of both the SA and SI medium and on the thickness of the surface active medium.

In both prism coupling configurations the presence of the prism introduces a "radiative" damping of the surface polaritons (i.e. the surface polaritons can radiate into the prism). In the Otto configuration, the radiative damping of the volume and surface EM waves is determined by the thickness of the air gap. While in the Kretschmann configuration, the radiative damping is determined by the thickness of the metal film. One obtains an optimum coupling when the radiative damping matches the damping due to the dielectric loss. The Kretschmann configuration is useful in the visible and UV region while the Otto configuration is useful in the IR and far IR region.

Similar considerations play a role in the coupling of volume and surface EM waves by grating. In the grating coupling technique, a grating of wave-vector $\boldsymbol{K}_G$ can diffract (i.e. couple) an incident volume EM wave and convert it into a surface polariton wave via nth order diffraction. The diffraction satisfies the Bragg's condition $\boldsymbol{k}_\parallel^s = \boldsymbol{k}_\parallel^i + n\boldsymbol{k}_G$, where $k_\parallel^i$ and $k_\parallel^s$ are the wavevector parallel to the interface of the incident EM wave and the surface polariton wave respectively; $k_G = 2\pi/a$ is the wavevector of the grating with a periodicity a; and n is an integer corresponding to the diffraction order. The grating coupling technique can also be used to detect surface polaritons by diffracting the surface polaritons into volume EM waves: $\boldsymbol{k}_\parallel^0 = \boldsymbol{k}_\parallel^s - n\boldsymbol{k}_G$. The grating coupling efficiency depends on the amplitude and shape of the grating as well as on the diffraction order. Normally, the lowest order diffraction will yield the highest coupling efficiency. When the amplitude of the grating is large (i.e. greater than 10% of the wavelength) the dispersion relation of the surface polariton can be strongly disturbed (i.e. the dispersion curve exhibits periodic gaps) (Ritchie et al. 1968), and coupling efficiency is reduced.

The enhancement of electric fields that occurs when surface polaritons serve as the input waves in three-wave mixing was first used by Simon and coworkers (1974) to enhance the second harmonic generation at a silver surface in a Kretschmann prism–silver film–air configuration. They also used surface polaritons as the input waves in the second harmonic generation at a silver–quartz crystal interface in a prism–silver film–quartz configuration in which the quartz serves as the primary nonlinear medium. Surface polaritons have also been used to excite the fluorescence of thin overlayers of dye molecules on silver (Gerboshtein et al. 1975, Chen 1977, Chen et al. 1975b). The data obtained by Chen, et al. (1975b) indicated a sizeable enhancement of the fluorescence intensity compared to that observed in a conventional external reflection (ER) configuration. They have also observed a marked enhancement in the Raman scattering by liquid benzene at a silver surface, when using surface polaritons as the excitation radiations in a Kretschmann configuration (Chen et al. 1976b, Chen 1977). Later, Chen et al. (1979) used surface polaritons to enhance the coherent anti-Stokes Raman scattering of liquid benzene at a silver surface.

The surface enhanced Raman scattering by molecules adsorbed at a metal (e.g. silver) surface, (Van Duyne 1979) is another phenomenon where surface polaritons have played a role. The Raman scattering by a monolayer of transparent (e.g., non-resonant) molecules is much too weak to be observed. According to the observation of even the weak signal due to molecules adsorbed at a metal surface is an indication of an appreciable enhancement. In the case of molecules, such as Pyridine, adsorbed on a silver electrode after optimum electrochemical oxidation–reduction cycle, one observes a Raman signal of $\sim 10^5$ cps for 50 mW at 5145 Å which corresponds to an enhancement by a factor of $\sim 10^6$ (Burstein et al. 1981). Although a complete understanding of the surface enhanced Raman scattering by molecules adsorbed on silver has not yet been achieved, there has been considerable experimental and theoretical progress in illucidating the various aspects of the phenomenon. In particular, there is substantial evidence that the short range structures (e.g., bumps, protrusions, etc.) of the submicroscopic surface roughness of the silver electrode exhibit localized plasmon resonance (e.g., localized plasmons). The excitation of the localized plasmons leads to an enhancement of the input and output EM fields (Burstein et al. 1980, Chen and Burstein 1980). The possibility that the giant ($\sim 10^6$) enhancement of the Raman scattering by adsorbed molecules at a silver surface may be due to enhanced EM fields arising from the surface roughness coupling of surface EM waves and volume EM waves has been conjectured but invariably discarded. The Raman scattering by the adsorbed molecules resulting from the use of surface polaritons as the input and output radiations would, even under optimum conditions, only be of the order of 10^3. In the case of a submicroscopically rough silver surface, the surface roughness coupling of volume EM waves and surface EM waves at frequencies in the visible is, in fact, quite weak as indicated by the fact that $(1-R)$ for the silver electrode (where R is the reflectance), which is a measure of the EM field strength, does not differ appreciably from that of a smooth silver surface. On the other hand, the use of gratings to couple the input and output volume EM waves to surface EM waves at the silver–air interface can, as shown by Tsang and coworkers

(1979), lead to a sizeable enhancement ($\sim 10^2$) of the Raman scattering by molecules adsorbed at a silver surface.

Since the field enhancement effect is discussed in a number of chapters in this book, we shall concentrate our discussion mainly on the subject of Raman scattering by overlayers on metals. For a detailed discussion of other nonlinear interactions, such as second harmonic generation, the reader can refer to the related sections in ch. 14 by Shen and deMartini.

For simplicity, we assume that all media are optically isotropic with parallel surfaces. We shall treat in detail only the case of the Kretschmann (prism--metal–dielectric) configuration. The extension to the case of the Otto (prism–dielectric–metal) configuration is straightforward and will only be discussed briefly.

To illustrate the field enhancement effect, we will make a direct comparison of the Raman scattering by an adsorbed layer on a metal surface in the ER and in the Kretschmann scattering configurations. In the ER configuration a TM-polarized radiation is incident at an angle θ_i, with respect to the normal to the metal surface, and the scattered TM-polarized radiation is collected at an angle θ_s (not necessarily collinear with θ_i), both angles being selected to yield a maximum scattering cross section. In the Kretschmann scattering configuration, the "incident" surface polariton is generated by linear coupling with TM-polarized radiation in the prism propagating at an angle $\theta_i = \Theta_{ATR}(\omega_i)$ and the scattered surface polariton (not necessarily collinear with the incident surface polariton) in turn generates a TM-polarized radiation in the prism at an angle $\theta_s = \Theta_{ATR}(\omega_s)$.

The differential Raman cross section of a thin overlayer on a metal surface, which is applicable to both the ER and the Kretschmann scattering configuration, can be readily shown to have the form

$$\frac{d^2I}{d\omega_s\, d\Omega_s} \propto \left(\frac{\omega_s}{c}\right)^4 \sum_a \sum_{\beta\xi\upsilon} |G_{\alpha\beta}(\omega_s, \boldsymbol{k}_{\|s})\delta\epsilon_{\beta\xi}(\omega_j, \boldsymbol{q}_{\|j})\Gamma_{\xi\upsilon}(\omega_i, \boldsymbol{k}_{\|i})\, E_{i\upsilon}|^2 K \cos\theta_s, \tag{42}$$

where α, β, ξ, υ represent x, y, z with the normal of the metal film along z; ω_i, ω_s and $\omega_j = \omega_i - \omega_s$ and $\boldsymbol{k}_{\|i}$, $\boldsymbol{k}_{\|s}$ and $\boldsymbol{q}_{\|j} = \boldsymbol{k}_{\|i} - \boldsymbol{k}_{\|s}$ are the frequencies and the components of the wavevector parallel to the surface of the incident radiation, scattered radiation and vibration modes of the overlayer, respectively; $\delta\epsilon$ is the nonlinear optical tensor of the overlayer; G and Γ are the Fresnel transfer matrix for the scattered and incident radiation, respectively; K is the kinematic factor for the scattering process as defined in sect. 2. In the case of very thin adsorbed layers K reduces simply to Δl, the thickness of the layer.

The components of the Fresnel transfer matrix for the Kretschmann scattering configuration take the following form:

$$\Gamma^{P}_{cfx\upsilon} \approx \Gamma^{0P}_{cbx\upsilon} \qquad \Gamma^{P}_{cfz\upsilon} \approx \Gamma^{0P}_{cbz\upsilon}/\epsilon_f,$$

$$G^{P}_{fc\alpha x} \simeq G^{0P}_{bc\alpha x} \qquad G^{P}_{fc\alpha z} \simeq \epsilon_f G^{0P}_{bc\alpha z},$$

$$\Gamma^{0P}_{cb\xi\upsilon} = f_{\xi\upsilon}(\theta_i) 4k^{i}_{za}k^{i}_{zb}\epsilon_a(\epsilon_b\epsilon_c)^{1/2}\exp(ik^{i}_{za}d)/D,$$

$$G^{0P}_{bc\alpha\beta} = g_{\alpha\beta}(\theta_s) 4k_{za}k_{zb}\epsilon_a(\epsilon_b\epsilon_c)^{1/2}\exp(ik^{s}_{za}d)/D,$$

$$D = (\epsilon_a k_{zc} + \epsilon_c k_{za})(\epsilon_b k_{za} + \epsilon_a k_{za}) - (\epsilon_a k_{zb} - \epsilon_a k_{za})(\epsilon_b k_{za} - \epsilon_a k_{zb})\exp(2ik_{za}z), \tag{43}$$

where the subscripts a, b, c and f represent the metal film, vacuum, prism and overlayer, respectively; ϵ_a is the dielectric constant of the medium a, etc; Γ^0_{ca} and G^0_{bc} are the Fresnel transfer matrices in the absence of the overlayer, which relate the electromagnetic field of the incident and scattered radiation in medium b at the a–b interface to the corresponding field in medium c; $f_{\xi\upsilon}(\theta_i)$ and $g_{\alpha\beta}(\theta_s)$ are simple angular factors; $D = 0$ is the dispersion relation of the surface polariton in the Kretschmann configuration.

For a given ω, the solution of the surface polariton dispersion relation $D(\tilde{g}_s) = 0$ is a complex wavevector $\tilde{g}_s = g'_s + ig''_s$. The condition for optimum coupling angle θ is such that $\epsilon_c^{1/2}(\omega/c)\sin\Theta_{ATR} = g'_s$. The condition for an optimum metal thickness is such that g''_s is minimized.

The components of the Fresnel transfer matrix for the ER scattering configuration take the following form:

$$\Gamma^{E}_{bfx\upsilon} = f_{x\upsilon}(\theta_i)\frac{2\epsilon_b k_{za}}{\epsilon_a k_{zb} + \epsilon_b k_{za}} \qquad \Gamma^{E}_{bfz\upsilon} = \frac{2\epsilon_a k_{bz}}{\epsilon_a k_{zb} + \epsilon_b k_{za}}\frac{f_{z\upsilon}(\theta_i)}{\epsilon_f},$$

$$G^{E}_{fb\alpha x} = g_{\alpha x}(\theta_s)\frac{2\epsilon_b k_{za}}{\epsilon_a k_{zb} + \epsilon_b k_{za}} \qquad G^{E}_{fb\alpha z} \approx g_{\alpha z}(\theta_s)\epsilon_f\frac{2\epsilon_a k_{zb}}{\epsilon_a k_{zb} + \epsilon_b k_{za}}. \tag{44}$$

At the optimum coupling condition Γ^{E}_{zz} and G^{E}_{zz} dominate. Curves giving the dependence of $|\epsilon_f\Gamma_{zz}|^2$ and $|G_{zz}/\epsilon_f|^2\cos\theta_s$ on $k_\parallel$ at $\lambda_i \approx \lambda_s = 6471$ Å for an ER configuration and for a Kretschmann configuration involving a thin overlayer on evaporated silver are shown in fig. 10. We note that in the ER configuration the value of $k_{\parallel i}$ and $k_{\parallel s}$ ranges from zero to $k_0 = \omega/c$, whereas in the Kretschmann configuration the value of $k_\parallel$ extends to $\epsilon_c^{1/2}k_0$.

The curves for the ER configuration are fairly flat. The maximum values of $|\epsilon_f\Gamma^{E}_{bfzz}|^2$ and $|G^{E}_{fbzz}/\epsilon_f|^2\cos\theta_s$, which occur at different values of $k_\parallel$ (corresponding to $\theta^{E}_i \approx 74°$ and $\theta^{E}_s \approx 52°$, respectively), are 2.5 and 1.2, respectively. In the Kretschmann configuration, on the other hand, the curves for $|\epsilon_f\Gamma^{P}_{cfzz}|^2 = |\Gamma^{0P}_{cbzz}|^2$ and $|G^{P}_{fczz}/\epsilon_f|^2\cos\theta_s = |G^{0P}_{bczz}|^2\cos\theta_s$ both exhibit very sharp peaks, at $\theta^{P}_{ATR} \approx 43.6°$, having magnitudes of 250 and 4.0, respectively. The enhancement of the (zz) scattering cross section of the overlayer resulting from the use of surface polaritons is effectively

$$|G^{0P}_{bc}\Gamma^{0P}_{cb}|^2\cos\theta^P/|G^{E}_{bc}\Gamma^{E}_{cb}|^2\cos\theta^E \approx 340.$$

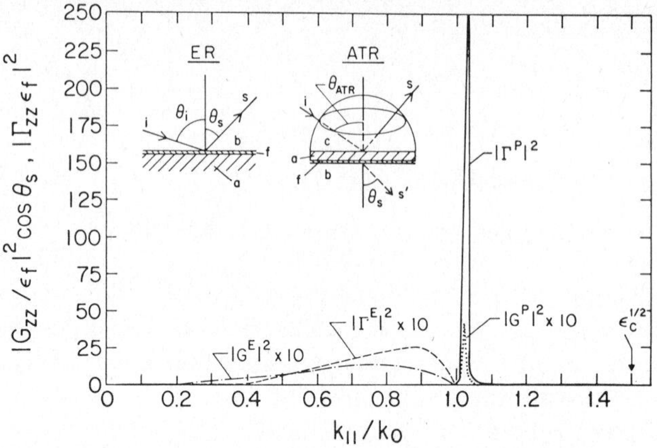

Fig. 10. Curves of $|\Gamma^{E}_{bfzz}\epsilon_f|^2$ and $|G^{E}_{fbzz}/\epsilon_f|^2 \cos\theta_s$ for the ER configuration and of $|\Gamma^{P}_{cfzz}\epsilon_f|^2 = |\Gamma^{0P}_{cbzz}|^2$ and $|G^{P}_{fczz}/\epsilon_f|^2 \cos\theta_s = |G^{0P}_{bczz}|^2 \cos\theta_s$ for the Kretschmann prism configuration versus $k_{\|}/k_0$ for a thin overlayer on silver, based on $\lambda_i \approx \lambda_s = 6471$ Å, $d_0 = 530$ Å, $\tilde{\epsilon}_{Ag} = -19.6 + i0.59$, and $\epsilon_c = 2.25$.

Since the magnitude of $|G^{P}_{fc}/\epsilon_f|^2 \cos\theta^{P}_{ATR}(\approx 4.0)$ is not much greater than $|G^{E}_{fb}/\epsilon_f|^2 \cos\theta^{E}(\approx 1.2)$, one can still achieve an appreciable enhancement in the scattering cross section by using a mixed Kretschmann–ER configuration in which the incident radiation is a surface polariton and the scattered radiation is a volume EM wave in the vacuum. The theoretical enhancement factor resulting from the use of the mixed configuration is approximately 100.

The field enhancement effect has been demonstrated by data on the Raman signals by liquid benzene from a prism–silver film–benzene cell using a Kretschmann prism coupling configuration (Chen et al. 1976a, Chen 1977) as shown in Fig. 11. The incident angle of the incident EM wave is changed by an appropriate rotation of the prism. The reflectivity of the incident EM wave, which has a minimum at the surface plasmon angle, is monitored by a photo cell. The shape and the size of the reflectivity minimum of the incident EM wave serves as an indication of the magnitude of the field enhancement. The Raman signal was also detected in an ER configuration.

The calculated transfer functions for scattering by the vibrational modes of liquid benzene in the Kretschmann configuration are plotted versus incident/output angle in fig. 12. Since the output G transfer factor for an ER scattering configuration is roughly equal to one, the Raman signal I_s is proportional to $|\Gamma|^2 K$. The kinematic factor K (i.e. the interaction length) is equal the $1/2\alpha$, where α is the attenuation constant of the incident surface polariton. Thus, $|\Gamma|^2 \propto I_s/(K|\chi^{(2)}|^2)$. By measuring separately the Raman signal I_{ref} of a known benzene cell of thickness d, under the same scattering condition, we deduce the $|\Gamma|^2$ to be $(I_s/K)(I_{ref}/d) \approx 70$. The result matches the

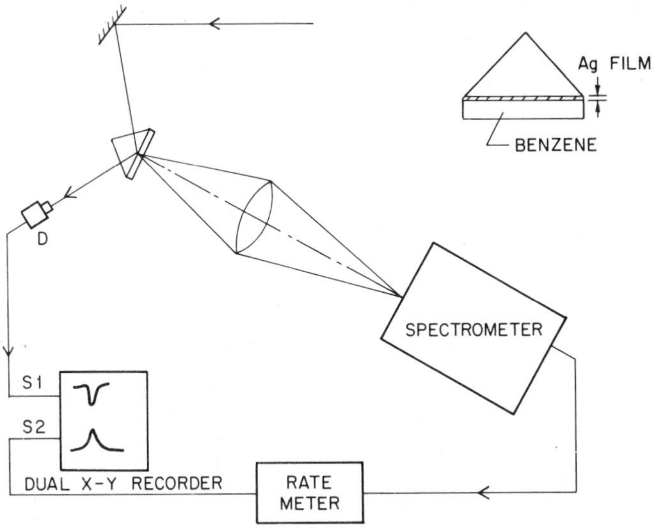

Fig. 11. Experimental set-up for the Raman scattering of liquid benzene in a Kretschmann prism configuration.

theoretical prediction rather well. The experimental set-up shown in fig. 11 can also be used in Coherent Anti-Stokes Raman Scattering (CARS) in liquid benzene (as shown in fig. 13). In this process, two input waves E_1 and E_2 excite the vibrational mode of benzene $\omega = \omega_1 - \omega_2$ and the excited vibrational mode then diffracts the third input wave E_3 into E_4. We note that since only the momentum parallel to the interface is conserved, all three input waves can be coupled into the system as surface polaritons. By properly arranging the angle θ between $k_{\|1}$ and $k_{\|2}$ the excited vibrational mode can also be a surface polariton. Furthermore, by adjusting the incident direction of E_3 (i.e. θ_2), the output CARS signal E_4 can also be a surface polariton. By using the formulation derived in sect. 3 [eq. (35)], the CARS signal can be readily written as follows:

$$I^{\mathrm{CARS}} \propto |\Gamma_1(\boldsymbol{k}_{\|1}, \omega_1)\Gamma_2(\boldsymbol{k}_{\|2}, \omega_2)\Gamma_3(\boldsymbol{k}_{\|3}, \omega_3)G(\boldsymbol{k}_{\|4}, \omega_4)|^2|D(\boldsymbol{q}_{\|}, \omega)|^{-2} \times K_{\mathrm{M}}(\boldsymbol{q}_{\|}, \omega)K_{\mathrm{R}}(\boldsymbol{k}_{\|4}, \omega_4) \approx 5\times 10^6|D(\boldsymbol{q}, \omega)|^{-2}K_{\mathrm{M}}(\boldsymbol{q}_{\|}, \omega)K_{\mathrm{R}}(\boldsymbol{k}_4, \omega_4), \tag{45}$$

where K_{M} and K_{R} are the kinematic factors of the parametric mixing and the Raman scattering process, respectively, and are both proportional to the penetration depth of the input/output surface polaritons.

We should emphasize that the above discussion is based on a nonabsorbing SI medium (i.e. its ϵ'' is very small or zero). If the SI medium is

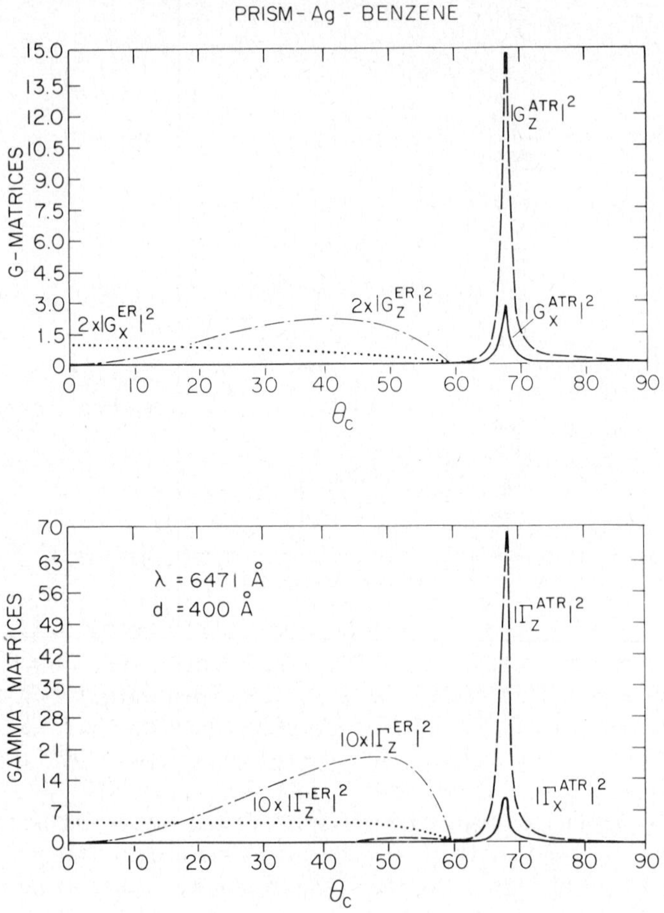

Fig. 12. The Γ and G transfer functions for the Raman scattering of liquid benzene in a benzene–silver ER configuration and in a prism–silver–benzene attenuated total reflection (ATR) configuration.

absorbing, even if it is only a thin layer, the field enhancement effect will be greatly reduced. Figure 14 shows an experimental set-up in which the fluorescent intensities from rhodamine 6G dye coated on the silver film were measured both in a Kretschmann-ER mixed configuration and an ER configuration (Chen et al. 1975b, Chen 1977). Since in both cases the fluorescent intensities were measured in the same way, the field enhancement effect of the $|\Gamma|^2$ factor due to the coupling of the input radiation to a surface polariton can be readily obtained by dividing the fluorescent intensity measured by the Kretschmann–ER mixed configuration with the fluorescent

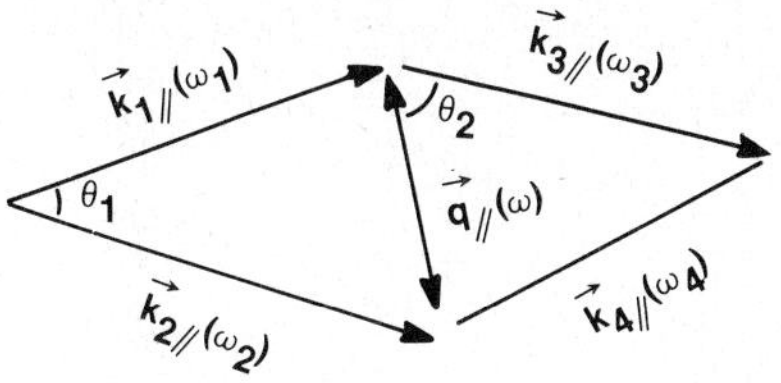

$$\vec{q}_{//} = \vec{k}_{1//} - \vec{k}_{2//} \qquad \vec{k}_{4//} = \vec{k}_{3//} - \vec{q}_{//}$$

$$\omega = \omega_1 - \omega_2 \qquad \omega_4 = \omega_3 + \omega$$

Fig. 13. Diagram for a Coherent Anti-Stokes Raman Scattering (CARS) process.

intensity measured by the ER configuration. The enhancement factors of the fluorescent intensities of rhodamine 6*G* films of various thicknesses are shown in fig. 15 together with corresponding reflection spectrum. Curve 1 is the reflectivity spectrum of a clean silver film and curves 2 to 6 correspond to the reflectivity spectrum of silver film with increasingly thicker coating of

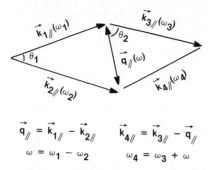

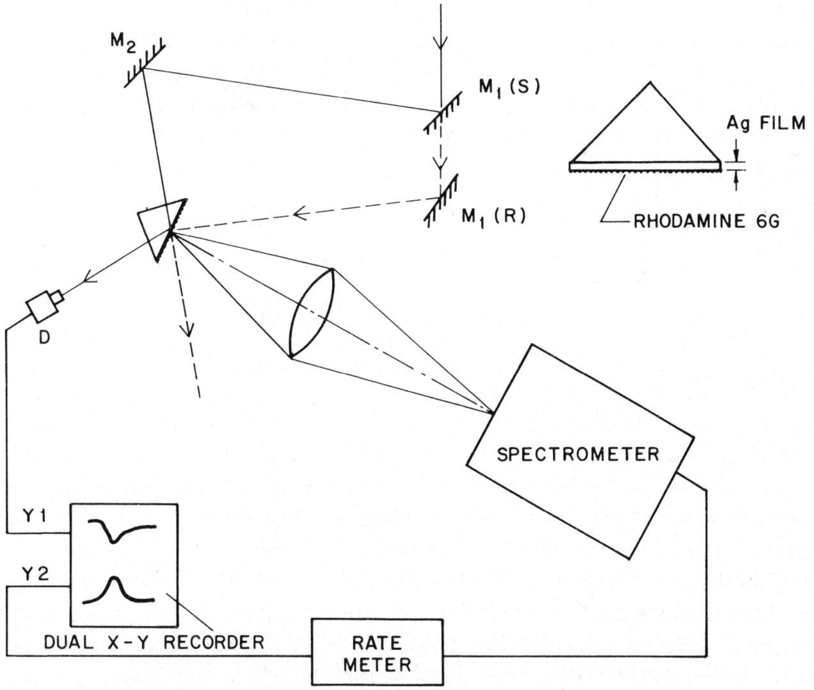

Fig. 14. Experimental set-up for the luminescence measurement of rhodamine 6G dye coated on the silver film in an ER configuration and in a Kretschmann–ER mixed configuration.

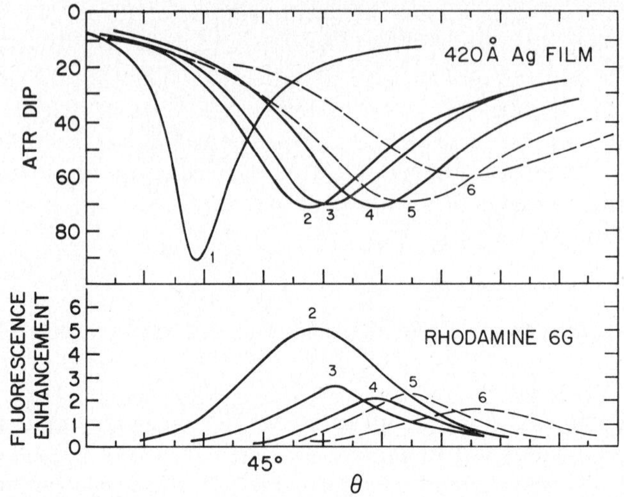

Fig. 15. The enhancement factors of the fluorescent intensities of rhodamine 6G films of various thicknesses by using an input surface polariton together with the corresponding reflectivity spectra.

rhodamine 6*G* film. We note that the largest observed enhancement factor for $|\Gamma|^2$ was only five which is much smaller than the theoretical value of 100 for a nonabsorbing film system. The reduction of the field enhancement factor is due to the large dielectric loss of the rhodamine 6*G* film at the wave length (5145 Å) of the input radiation (i.e. the rhodamine 6*G* is strongly absorbing at 5145 Å).

5. Concluding Remarks

As a result of progress in microstructure fabrication, and particularly in microlithography, it is now possible to fabricate submicron, as well as micron, gratings on a variety of substrates. Such gratings can be used to couple surface and volume polaritons at frequencies in the ultraviolet, as well as in the visible and infrared. More specifically, they can be used as passive surface microstructure amplifiers to enhance the input and output EM fields and a wide variety of nonlinear optical phenomena at the surface of metals, or of any medium which can sustain surface polaritons. Gratings have already been used advantageously in surface enhanced Raman scattering by molecules absorbed on silver. Although the overall grating and prism coupling enhancement factors are comparable in magnitudes, use of grating coupling avoids the spontaneous Raman scattering and luminescence that is invariably present when prism coupling is used. Although the input and output angles in grating couplings are very much more dependent on wavelength than the corresponding angles in prism couplings, this is in fact an advantage since the dependence on wavelength (which corresponds to the Bragg's condition) allows the grating to be used as the wavelength dispersing element. We anticipate that gratings will largely replace prisms as coupling elements for the generation and detection of input and output surface polaritons in nonlinear optical phenomena.

Acknowledgement

We would like to thank D.L. Mills and W.P. Chen for valuable discussions and R. Seymour for his critical reading of this manuscript. This work was supported in part by ONR.

References

Abrikosov, A.A., L.P. Gorkov and I.E. Dzyaloshinskii, 1963, Methods of Quantum Field Theory in Statistical Physics (Prentice-Hall, Englewood Cliffs, NJ).
Barker, A.S. and R. Loudon, 1972, Rev. Mod. Phys. **44**, 18.
Bloembergen, N. and P.S. Pershan, 1962, Phys. Rev. **128**, 606.
Bonsall, L. and A.A. Maradudin, 1978, J. Appl. Phys. **49**, 253.
Burstein, E., 1972, in: Atomic Structure and Properties of Solids (Academic Press, New York) pp. 3–21.
Burstein, E. and C.Y. Chen, 1980, Proc. VIIth Int'l. Conf. on Raman Spectroscopy, Ottawa, 1980, W.F. Murphy, ed. (North-Holland Amsterdam, New York) p. 346.
Burstein, E., W.P. Chen, Y.J. Chen and A. Harstein, 1974, J. Vac. Sci. Technol. **11**, 1004.
Burstein, E., S. Lundquist and D.L. Mills, 1981, The Roles of Roughness, in: Surface Enhanced Raman Scattering, ed., R.K. Chang and T.E. Furtak (Plenum, New York).
Chen, Y.J., 1977, Thesis (Univ. of Pennsylvania, Philadelphia, PA), unpublished.
Chen, Y.J. and E. Burstein, 1977, Nuovo Cimento **39B**, 807.
Chen, C.Y. and E. Burstein, 1980, Phys. Rev. Lett. **45**, 1287.
Chen, Y.J., E. Burstein and D.L. Mills, 1975a, Phys. Rev. Lett. **34**, 1516.
Chen, Y.J., W.P. Chen and E. Burstein, 1975b, Bull. Am. Phys. Soc. **20**, 419.
Chen, Y.J., W.P. Chen and E. Burstein, 1976a, Bull. Am. Phys. Soc. **21**, 338.
Chen, Y.J., W.P. Chen and E. Burstein, 1976b, Phys. Rev. Lett. **36**, 1207.
Chen, Y.J., C.Y. Chen and E. Burnstein, 1977, Bull. Am. Phys. Soc., **22**, 279.
Chen, C.K., A.R.B. deCastro, Y.R. Shen and F. DeMartini, 1979, Phys. Rev. Lett. **43**, 946.
DeMartini, F., G. Giuliani, P. Mataloni, E. Palange and Y.R. Shen, 1976, Phys. Rev. Lett. **37**, 440.
DeMartini, F., M. Colocci, S.E. Kohn and Y.R. Shen, 1977, Phys. Rev. Lett. **38**, 1223.
Dresselhaus, G. and A.S. Pine, 1975, Solid State Commun. **16**, 1001.
Evans, D.J., S. Ushioda and J.D. McMullen, 1973, Phys. Rev. Lett. **31**, 369.
Fuchs, R., and K.L. Kliewer, 1966, Phys. Rev. **150**, 589.
Fukui, M., J.E. Sipe, V.C.Y. So and G.I. Stegeman, 1978, Solid State Commun. **27**, 1265.
Gerboshtein, Y.M., T.A. Merkulov and D.M. Mirlin, 1975, JETP Lett. **22**, 35.
Kretschmann, E., 1971, Z. Physik **241**, 313.

Mills, D.L., 1977, Solid State Commun. **24**, 669.
Mills, D.L., A.A. Maradudin and E. Burstein, 1970, Ann. Phys. (NY) **56**, 504.
Mills, D.L., Y.J. Chen and E. Burstein, 1976, Phys. Rev. **B13**, 4419.
Nkoma, J.S., 1975, J. Phys. **C8**, 3919.
Nkoma, J.S. and R. Loudon, 1975, J. Phys. **C8**, 1950.
Otto, A., 1968, Z. Physik **216**, 398.
Prieur, J.Y. and S. Ushioda, 1975, Phys. Rev. Lett. **34**, 1012.
Reinisch, R., N. Paraire, M. Chapet-Rousseau and S. Laval, 1977, J. Physique 38, 1457.
Ritchie, R.H., E.T. Arakawa, J.J. Cowan and R.N. Hamm, 1968, Phys. Rev. Lett. **21**, 1530.
Simon, H.J., D.E. Mitchell and J.G. Watson, 1974, Phys. Rev. Lett. **33**, 1531.
Simon, H.J., R.E. Benner and J.G. Rako, 1977, Optics Commun. **23**, 245.
Tsang, J.C., J.R. Kirtley and J.A. Bradley, 1979, Phys. Rev. Lett. **43**, 772.
Van Duyne, R. P., 1979, Chemical and Biological Applications of Lasers, Vol. 5, ed., C.B. Moore (Academic Press), Chap. 4.
Wynne, J.J., 1974, Comment Solid State Phys. **6**, 31; 1975, **7**, 7.

CHAPTER 14

Nonlinear Wave Interaction Involving Surface Polaritons

Y. R. SHEN

Department of Physics
University of California
Berkeley, California 94720

Materials and Molecular Research Division
Lawrence Berkeley Laboratory
Berkeley, California 94720
U.S.A.

F. DEMARTINI

Quantum Optics Laboratory
Istituto di Fisica, "G. Marconi"
Università di Roma
00185 Roma
Italy

Surface Polaritons
Edited by
W.M. Agranovich and D.L. Mills

Contents

1. Introduction . . . 631
2. Theory . . . 632
   2.1. Linear excitation of surface polaritons . . . 632
   2.2. Nonlinear excitation of surface polaritons . . . 637
   2.3. Nonlinear interaction of surface polaritons . . . 640
3. Nonlinear excitation of surface polaritons – experimental demonstration . . . 640
4. Second harmonic generation by surface plasmons . . . 647
5. Surface coherent anti-Stokes Raman spectroscopy . . . 653
6. Concluding remarks . . . 658

References . . . 660

1. Introduction

As a subarea of surface physics, surface polaritons have recently attracted much attention (Borstel and Falge 1978). Linear optical properties of surface polaritons in various media have been extensively investigated. Their results and potential applications to surface and material studies are reviewed elsewhere in this book. Nonlinear optical studies involving surface polaritons are, however, very rare. From the physics point of view, the subject is actually a rather interesting one. First, since the surface polaritons are localized to a thin surface layer near an interface, nonlinear optical effects involving surface polaritons have a surface-specific nature. Then, surface nonlinear optics with all interacting waves being surface polaritons is possible, and forms a new branch in the field of nonlinear optics. With tunable laser excitations, surface nonlinear optical spectroscopy can be devised and applications can be envisioned. In this chapter, we shall review work in this area in the past few years. Emphasis will be on the experimental observation in comparison with theoretical prediction.

We shall begin with a theoretical discussion on wave interaction involving surface polaritons in sect. 2. The general theory of linear optical excitation of surface polaritons will first be given. It will then be extended to the case of nonlinear optical excitation with the nonlinear polarization as the driving source (DeMartini and Shen 1976). Actually, generation of surface polaritons by optical mixing of bulk waves can be considered as a special case of nonlinear optical reflection from a surface treated by Bloembergen and Pershan (1962). The general formalism can also be used to describe the generation of bulk and surface waves through mixing of surface polaritons or surface polaritons with bulk waves.

The experimental demonstration of nonlinear excitations of surface polaritons will be reviewed in sect. 3. Two cases will be considered: one with GaP demonstrating nonlinear excitation of surface phonon polaritons by difference-frequency mixing (DeMartini et al. 1976), and the other with ZnO demonstrating second harmonic generation of surface exciton polaritons (DeMartini et al. 1977). It will then be shown that the surface polariton excitation can be probed through either mixing of the surface polaritons with a probe beam or surface roughness scattering. That nonlinear excitation of surface polaritons has some advantages over linear excitation and is useful for the study of surface polaritons will also be discussed.

Second harmonic generation is the simplest nonlinear optical effect, and will be considered in sect. 4 to illustrate the interaction of surface polaritons. Here, surface plasmons will be the subject of discussion since they exist over a wide spectral range covering both the fundamental and the second harmonic. Nonlinearity in the case of a metal–air interface arises from the metal (Simon et al. 1974), while in the case of a metal–dielectric interface may come mainly from the dielectric (Simon et al. 1977). In the former case, a single atomic layer on the surface may be responsible for the observed nonlinearity (Bloembergen et al. 1968). Various aspects of the theory of sect. 2 can be tested out by the experiments of second harmonic generation with surface plasmons.

Coherent anti-Stokes Raman scattering can also be carried out with surface plasmons (Chen et al. 1979). This will be discussed in sect. 5. In general, four-wave mixing of surface plasmons should be observable, and can be used as a spectroscopic technique to probe resonances of a dielectric. The technique has advantages over four-wave mixing spectroscopy in a bulk, and should be most useful for studying molecular overlayers and materials with strong absorption and fluorescence. With picosecond pulse excitation, the signal-to-noise ratio can be enhanced by several orders of magnitude with a resulting sensitivity capable of detecting a submonolayer of adsorbed molecules.

Finally, in sect. 4, we shall speculate on the future progress in the field. Applications to surface physics will be considered in particular. It seems likely that the combined force of surface nonlinear optics and surface physics may open up a new area of exciting interdisciplinary research.

2. *Theory*

2.1. *Linear Excitation of Surface Polaritons*

We consider here a general system of N layers shown in fig. 1. The solution of plane wave propagation in such a medium is governed by the wave equation

$$[\nabla \times (\nabla \times) - (\omega^2/c^2)\epsilon(z)]\boldsymbol{E} = 0 \tag{1}$$

together with the boundary conditions. Assume that each layer is isotropic or cubic, and let the incoming wave be transverse magnetic

$$\boldsymbol{E}_0 = (\hat{x}\mathscr{E}_{0x} + \hat{z}\mathscr{E}_{0z}) \exp \mathrm{i}(k_x x + k_{0z} z - \omega t) \tag{2}$$

with $k_x \mathscr{E}_{0x} = -k_{0z}\mathscr{E}_{0z}$ and $k_0 = (\omega/c)\sqrt{\epsilon_0}$. Then, the reflected wave for $z < z_0$

is

$$\boldsymbol{E}_R = \left(\hat{x} + \hat{z}\frac{k_x}{k_{0z}}\right)\mathscr{E}_{Rx} \exp i(k_x x - k_{0z} z - \omega t). \tag{3}$$

For $z_{j-1} < z < z_j$, the field is the sum of a transmitted wave and a reflected wave, and can be written as

$$\begin{aligned}\boldsymbol{E}_j = \Big\{&\hat{x}[A_j \cos k_{jz}(z - z_{j-1}) + B_j \sin k_{jz}(z - z_{j-1})] \\ &+ \hat{z}\left(\frac{ik_x}{k_{jz}}\right)[-A_j \sin k_{jz}(z - z_{j-1}) + B_j \cos k_{jz}(z - z_{j-1})]\Big\} \exp i(k_x x - \omega t)\end{aligned} \tag{4}$$

with $k_j = (\omega/c)\sqrt{\epsilon_j}$. Finally, for $z > z_N$, there is only an outgoing wave

$$\boldsymbol{E}_T = \left(\hat{x} - \hat{z}\frac{k_x}{k_{Tz}}\right)\mathscr{E}_{Tx} \exp i(k_x x + k_{Tz} z - \omega t). \tag{5}$$

The $2N$ amplitude variables $\mathscr{E}_{Rx}, \ldots, A_j, B_j, \ldots$, and $\mathscr{E}_{Tx}$ are related by the boundary conditions.

$$\begin{gathered}\mathscr{E}_{Rx} - A_1 = -\mathscr{E}_{0x} \\ (\epsilon_0 k_x/k_{0z})\mathscr{E}_{Rx} - (i\epsilon_1 k_x/k_{1z})B_1 = (\epsilon_0 k_x/k_{0z})\mathscr{E}_{0x} \\ \text{- - - - - - - - - - - - - - - - - - -} \\ (\cos k_{jz}d_j)A_j + (\sin k_{jz}d_j)B_j - A_{j+1} = 0 \\ (i\epsilon_j k_z/k_{jz})[-(\sin k_{jz}d_j)A_j + (\cos k_{jz}d_j)B_j] - (i\epsilon_{j+1}k_x/k_{j+1,z})B_{j+1} = 0 \\ \text{- -} \\ (\cos k_{Nz}d_N)A_N + (\sin k_{Nz}d_N)B_N - \mathscr{E}_{Tx} = 0 \\ (i\epsilon_N k_x/k_{Nz})[-(\sin k_{Nz}d_N)A_N + (\cos k_{Nz}d_N)B_N] + (\epsilon_T k_x/k_{Tz})\mathscr{E}_{Tx} = 0\end{gathered} \tag{6}$$

where $d_j = z_j - z_{j-1}$. The above set of equations can be written in the matrix form

$$D\begin{bmatrix}\mathscr{E}_{Rx} \\ \vdots \\ A_j \\ B_j \\ \vdots \\ \mathscr{E}_{Tx}\end{bmatrix} = \begin{bmatrix}-\mathscr{E}_{0x} \\ \dfrac{\epsilon_0 k_x}{k_{0z}}\mathscr{E}_{0x} \\ 0 \\ \vdots \\ 0\end{bmatrix}. \tag{7}$$

Since $k_{jz} = [(\omega/c)^2\epsilon_j - k_x^2]^{1/2}$, so D is a function of k_x. In some cases, one may find that for $k_x = K_x$, the determinant $|D(K_x)|$ vanishes. This actually

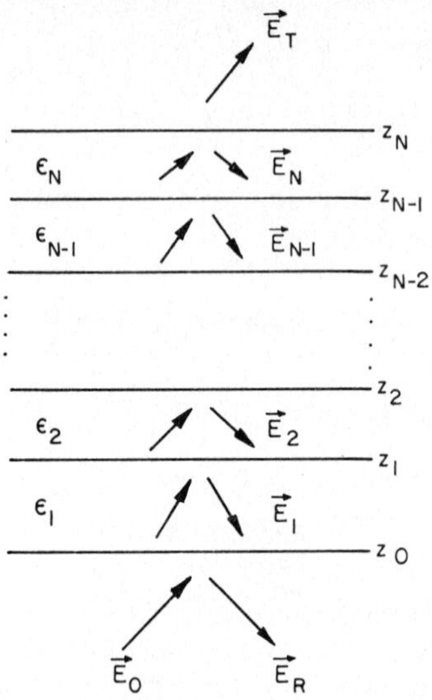

Fig. 1. Electromagnetic wave propagation in a layered medium. Incident wave from $z < z_0$ propagates along $+\hat{z}$.

means that the medium now has an electromagnetic resonance. In other words, in the absence of damping, an electromagnetic field would have generated with a vanishingly small input. If the resonant mode is a guided mode with field strength more or less confined to a particular layer interface, we call it a surface polariton. In general, a layered medium can have several such resonant modes.

We now consider the special case of a single film layer sandwiched between two semi-infinite media.

$$|D| = \begin{vmatrix} -1 & 1 & 0 & 0 \\ 1 & 0 & -iq_0/q_1 & 0 \\ 0 & \cos k_{1z}d_1 & \sin k_{1z}d_1 & -1 \\ 0 & (-i/q_1)\sin k_{1z}d_1 & (i/q_1)\cos k_{1z}d_1 & 1/q_T \end{vmatrix}$$

$$= (\cos k_{1z}d_1/q_1^2 q_T)[i(q_1q_0 + q_1q_T) + (q_1^2 + q_0q_T)\tan k_{1z}d_1], \tag{8}$$

where $q_j = k_{jz}/\epsilon_j = [(\omega/c)^2\epsilon_j - k_x^2]^{1/2}/\epsilon_j$. For real ϵ_j, $|\boldsymbol{D}|$ will vanish only if k_{1z} is imaginary. Let $k_{1z} = -i\beta_1$. We find that for appropriate ϵ_j, we can have $|\boldsymbol{D}| = 0$, which is the dispersion relation of the surface polaritons

$$\tanh \beta_1 d_1 - (q_1q_0 + q_1q_T/q_1^2 + q_0q_T) = 0. \tag{9}$$

If $\beta_1 d_1 \gg 1$, it reduces to

$$(q_1 - q_T)(q_1 - q_0) = 0, \tag{10}$$

or more explicitly in terms of $k_x = K_x$,

$$K_x^2 = \left(\frac{\omega}{c}\right)^2 \frac{\epsilon_1 \epsilon_T}{\epsilon_1 + \epsilon_T}, \tag{11a}$$

$$K_x^2 = \left(\frac{\omega}{c}\right)^2 \frac{\epsilon_1 \epsilon_0}{\epsilon_1 + \epsilon_0}. \tag{11b}$$

These are the familiar dispersion relations for surface polaritons at an interface between two semi-infinite media (Sommerfeld 1909). Physically, $\beta_1 d_1 \gg 1$ corresponds to an optically thick film which effectively decouples the two interfaces so that the two surface polariton modes of eqs. (11a) and (11b) are separately confined to the 0–1 and 1–T interfaces respectively. Note that since k_{1z} is imaginary, eq. (11a) (or (11b)) can be satisfied only if one of the two ϵ's in the equation is negative and the sum of the two ϵ's is also negative. Then, k_{zi} is imaginary and the field drops off exponentially on both sides of the interface. The surface polariton is physically confined to a thin layer of $(k_{z1}^{-1} + k_{zT}^{-1})$ thick at the interface and propagates with a wavevector K_x. In general, $\epsilon = \epsilon' + i\epsilon''$ is complex for a medium, so that $K_x = K_x' + iK_x''$ with K_x' being the wavevector and K_x'' the attenuation constant for the surface polariton. Negative ϵ' arises in media with exciton or phonon reststrahlung bands or in metals below the plasma frequency.

Equation (6) or (7) determines the field amplitudes in the layered medium set up by the incoming wave of eq. (2). When $k_x(\omega) \cong K_x'(\omega)$, the surface polariton is resonantly excited. For the case of a thin film sandwiched between two semi-infinite media, the solution of eq. (6) or (7) is well known (Born and Wolf 1965).

$$\mathscr{E}_{Rx} = \frac{r_{01} + r_{1T}\exp(i2k_{1z}d_1)}{1 + r_{01}r_{1T}\exp(i2k_{1z}d_1)}\mathscr{E}_{0x}$$

$$\mathscr{E}_{Tx} = \frac{t_{01}t_{1T}\exp(ik_{1z}d_1)}{1 + r_{01}r_{1T}\exp(i2k_{1z}d_1)}\left(\frac{k_{Tz}^2\epsilon_0}{k_{0z}^2\epsilon_T}\right)\mathscr{E}_{0x}$$

$$A_1 = \frac{(1 + r_{01})[1 + r_{1T}\exp(i2k_{1z}d_1)]}{1 + r_{01}r_{1T}\exp(i2k_{1z}d_1)}\mathscr{E}_{0x}$$

$$B_1 = \frac{(1 - r_{01})[1 - r_{1T}\exp(i2k_{1z}d_1)]}{1 + r_{01}r_{1T}\exp(i2k_{1z}d_1)}\left(\frac{i\epsilon_0 k_{1z}}{\epsilon_1 k_{0z}}\right)\mathscr{E}_{0x} \tag{12}$$

where r_{01}, r_{1T}, T_{01}, and t_{1T} are the Fresnel coefficients.

$$r_{ij} = (\epsilon_j k_{iz} - \epsilon_i k_{jz})/(\epsilon_j k_{iz} + \epsilon_i k_{jz})$$

$$t_{ij} = 2\sqrt{\epsilon_i \epsilon_j}\, k_i z/(\epsilon_j k_{iz} + \epsilon_i k_{jz}). \tag{13}$$

It is easy to show that $|D| \propto [1 + r_{01}r_{1T}\exp(i2k_{1z}d_1)]$ and hence the relation

$$1 + r_{01}r_{1T}\exp(i2k_{1z}d_1) = 0 \tag{14}$$

is equivalent to eq. (9) representing the dispersion relation of surface polaritons.

In practice, two kinds of geometry are often used to linearly excite surface polaritons in the above sandwich medium. They are shown in fig. 2. For the Otto configuration of fig. 2a (Otto 1968) we have $\epsilon_T' < 0$ and $\epsilon_0' > \epsilon_1' > 0$. The surface polariton that is excited when $k_x(\omega) \cong K_x'(\omega)$ is more or less confined to the 1–T interface. For the Kretschmann configuration of fig. 2b (Kretschmann 1971) we have $\epsilon_1' < 0$ and $\epsilon_0' > \epsilon_T' > 0$. The surface polariton excited is again more or less confined to the 1–T interface. In this latter case, there may also exist a surface polariton mode confined to the 0–1 interface, but since $K_x' > |k_0|$, $|k_T|$, it cannot be linearly excited by the prism coupler in fig. 2. In both cases, when the surface polariton is excited, a reflectivity dip should be observed according to eq. (12). With optimum choice of the film thickness d_1, the reflectivity dip can reach a minimum close to 0. An example is shown in fig. 3.

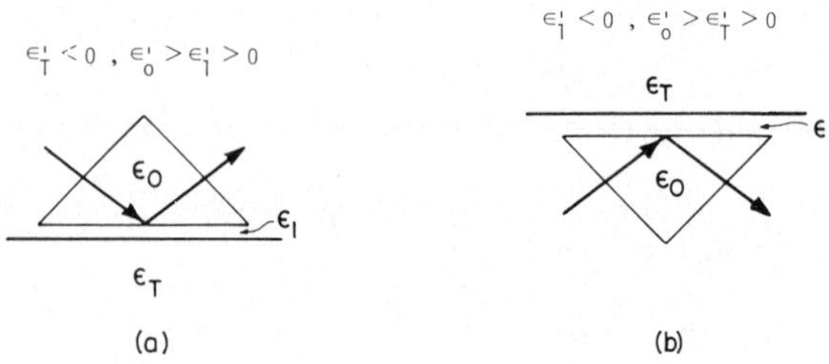

Fig. 2. (a) Otto configuration and (b) Kretschmann configuration for linear excitation of surface polaritons.

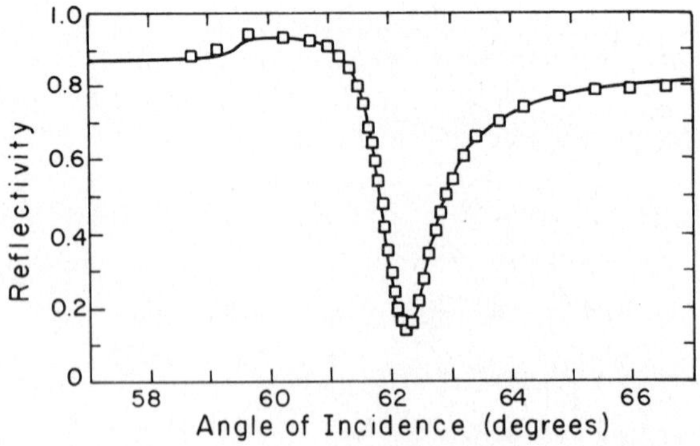

Fig. 3. An example of reflectivity versus incidence angle in the Kretschmann configuration. Here, the sharp dip indicates the resonant excitation of surface plasmons at the silver–liquid crystal interface (after fig. 2(c) of Chu et al. (1980)).

Excitation of surface polaritons is basically the same as excitation of guided optical waves. Thus, instead of prism coupling, a grating on the surface can also be used as the coupler for linear excitation of surface polaritons. Here, the phase mismatch between k_x and K_x' is compensated by the reciprocal lattice vector of the periodic grating in the resonant excitation. Then, surface roughness can also be effective as a coupler for surface polariton excitation since it can be considered as a random grating.

2.2. *Nonlinear Excitation of Surface Polaritons*

Surface polaritons can also be excited through nonlinear optical mixing in the medium (DeMartini 1976). The basic idea is fairly simple. Nonlinear mixing induces a nonlinear polarization $\boldsymbol{P}^{\mathrm{NL}}(\omega, \boldsymbol{k}_s) = \mathcal{P}^{\mathrm{NL}} \exp \mathrm{i}(\boldsymbol{k}_s \cdot \boldsymbol{r} - \omega t)$ in the medium, which, being a collection of oscillating dipoles, acts as a source for generation of the field at ω. If $k_{sx} \cong K_x'(\omega)$, then the surface polariton is resonantly excited. Mathematically, we can treat the problem as an extension of the derivation given in the previous section for linear excitation. First, the wave equation becomes

$$[\nabla \times (\nabla \times) - (\omega^2/c^2)\epsilon(z)]\boldsymbol{E} = (4\pi\omega^2/c^2)\boldsymbol{P}^{\mathrm{NL}}(\omega, \boldsymbol{k}_s). \tag{15}$$

Then, the field is obtained as a sum of homogeneous and particular solutions. Assume $\boldsymbol{E}_0 = 0$. For $z < z_0$

$$\begin{aligned} \boldsymbol{E}_{\mathrm{R}} = {} & (\hat{x} + \hat{z}k_x/k_{0z})\mathscr{E}_{\mathrm{R}x} \exp \mathrm{i}[k_x x - k_{0z}(z - z_0) - \omega t] \\ & + [\hat{x}(\gamma^0_{xx}\mathcal{P}^{\mathrm{NL}}_{0x} + \gamma^0_{xz}\mathcal{P}^{\mathrm{NL}}_{0z}) + \hat{z}(\gamma^0_{zx}\mathcal{P}^{\mathrm{NL}}_{0x} + \gamma^0_{zz}\mathcal{P}^{\mathrm{NL}}_{0z})] \\ & \times \exp \mathrm{i}[k_{s0x}x + k_{s0z}(z - z_0) - \omega t]. \end{aligned} \tag{16a}$$

For $z_{j-1} < z < z_j$.

$$\begin{aligned} \boldsymbol{E}_j = {} & \{\hat{x}[A_j \cos k_{jz}(z - z_{j-1}) + B_j \sin k_{jz}(z - z_{j-1})] \\ & + \hat{z}(\mathrm{i}k_x/k_z)[-A_j \sin k_{jz}(z - z_{j-1}) + B_j \cos k_{jz}(z - z_{j-1})]\} \exp \mathrm{i}(k_x x - \omega t) \\ & + [\hat{x}(\gamma^j_{xx}\mathcal{P}^{\mathrm{NL}}_{jx} + \gamma^j_{xz}\mathcal{P}^{\mathrm{NL}}_{jz}) + \hat{z}(\gamma^j_{zx}\mathcal{P}^{\mathrm{NL}}_{jx} + \gamma^j_{zz}\mathcal{P}^{\mathrm{NL}}_{jz})] \\ & \times \exp \mathrm{i}[k_{sjz}x' + k_{sjz}(z - z_{j-1}) - \omega t]. \end{aligned} \tag{16b}$$

For $z > z_N$

$$\begin{aligned} \boldsymbol{E}_{\mathrm{T}} = {} & (\hat{x} - \hat{z}k_x/k_{\mathrm{T}z})\mathscr{E}_{\mathrm{T}x} \exp \mathrm{i}[k_x x + k_{\mathrm{T}z}(z - z_N) - \omega t] \\ & + [\hat{x}(\gamma^T_{xx}\mathcal{P}^{\mathrm{NL}}_{\mathrm{T}x} + \gamma^T_{xz}\mathcal{P}^{\mathrm{NL}}_{\mathrm{T}z}) + \hat{z}(\gamma^T_{zx}\mathcal{P}^{\mathrm{NL}}_{\mathrm{T}x} + \gamma^T_{zz}\mathcal{P}^{\mathrm{NL}}_{\mathrm{T}z})] \\ & \times \exp \mathrm{i}[k_{s\mathrm{T}x}x + k_{s\mathrm{T}z}(z - z_N) - \omega t]. \end{aligned} \tag{16c}$$

In eq. (16), the $\mathcal{P}^{\mathrm{NL}}$ terms are the particular solutions in various regions obtained from eq. (15). The boundary conditions require that $k_{s\mathrm{R}x} = k_{sjx} = k_{s\mathrm{T}x} = k_x$, and D will be:

$$
D\begin{bmatrix} \mathscr{E}_{\mathrm{R}x} \\ \vdots \\ A_j \\ B_j \\ \vdots \\ \mathscr{E}_{\mathrm{T}x} \end{bmatrix} = \begin{bmatrix} -(\gamma^0_{xx}\mathscr{P}^{\mathrm{NL}}_{0x} + \gamma^0_{xz}\mathscr{P}^{\mathrm{NL}}_{0z}) + (\gamma^1_{zx}\mathscr{P}^{\mathrm{NL}}_{1x} + \gamma^1_{zz}\mathscr{P}^{\mathrm{NL}}_{1z}) \\ \vdots \\ -(\gamma^j_{xx}\mathscr{P}^{\mathrm{NL}}_{jx} + \gamma^j_{xz}\mathscr{P}^{\mathrm{NL}}_{jz})\exp(ik_{sjz}d) + (\gamma^{j+1}_{zx}\mathscr{P}^{\mathrm{NL}}_{j+1,x} + \gamma^{j+1}_{zz}\mathscr{P}^{\mathrm{NL}}_{j+1,z}) \\ -\epsilon_j\left[\gamma^j_{zx}\mathscr{P}^{\mathrm{NL}}_{jx} + \left(\gamma^j_{zz} + \dfrac{4\pi}{\epsilon_j}\right)\mathscr{P}^{\mathrm{NL}}_{jz}\right]\exp(ik_{sjz}d) \\ +\epsilon_{j+1}\left[\gamma^{j+1}_{zx}\mathscr{P}^{\mathrm{NL}}_{j+1,x} + \left(\gamma^{j+1}_{zz} + \dfrac{4\pi}{\epsilon_{j+1}}\right)\mathscr{P}^{\mathrm{NL}}_{j+1,z}\right] \\ \vdots \\ -\epsilon_N\left[\gamma^N_{zx}\mathscr{P}^{\mathrm{NL}}_{Nx} + \left(\gamma^N_{zz} + \dfrac{4\pi}{\epsilon_N}\right)\mathscr{P}^{\mathrm{NL}}_{Nz}\right]\exp(ik_{sNz}d) \\ +\epsilon_{\mathrm{T}}\left[\gamma^{\mathrm{T}}_{zx}\mathscr{P}^{\mathrm{NL}}_{\mathrm{T}x} + \left(\gamma^{\mathrm{T}}_{zz} + \dfrac{4\pi}{\epsilon_{\mathrm{T}}}\mathscr{P}^{\mathrm{NL}}_{\mathrm{T}z}\right)\right] \end{bmatrix} \tag{17}
$$

In comparison with eq. (7) for linear excitation, the nonlinear polarization $\mathscr{P}^{\mathrm{NL}}$ here plays the role of the incoming field $\mathscr{E}_0$ in the linear case. Again, as $k_{sx} \equiv k_x \cong K'_x(\omega)$, the surface polariton is excited.

An important difference between linear and nonlinear excitations should however be noted. In the linear case, it is not possible to excite a surface polariton on a smooth interface between two semi-infinite media because the wavevector $k(\omega)$ of the incoming exciting wave in either medium is always smaller than $K'_x(\omega)$. In the nonlinear case, this becomes possible because optical mixing induces a nonlinear polarization with a wavevector $\boldsymbol{k}_s$ which is the vector sum of the wavevectors of the exciting fields, and in general, one can have $k_{sx} = K'_x$. Here, eq. (17) reduces to

$$
D\begin{bmatrix} \mathscr{E}_{\mathrm{R}x} \\ \mathscr{E}_{\mathrm{T}x} \end{bmatrix} = \begin{bmatrix} -(\gamma^0_{xx}\mathscr{P}^{\mathrm{NL}}_{0x} + \gamma^0_{xz}\mathscr{P}^{\mathrm{NL}}_{0z}) + (\gamma^{\mathrm{T}}_{zx}\mathscr{P}^{\mathrm{NL}}_{\mathrm{T}x} + \gamma^{\mathrm{T}}_{zz}\mathscr{P}^{\mathrm{NL}}_{\mathrm{T}z}) \\ -\epsilon_0\left[\gamma^0_{zx}\mathscr{P}^{\mathrm{NL}}_{0x} + \left(\gamma^0_{zz}\mathscr{P}^{\mathrm{NL}}_{0z} + \dfrac{4\pi}{\epsilon_0}\right)\right] + \epsilon_T\left[\gamma^T_{zx}\mathscr{P}^{\mathrm{NL}}_{\mathrm{T}x} + \left(\gamma^T_{zz}\mathscr{P}^{\mathrm{NL}}_{\mathrm{T}z} + \dfrac{4\pi}{\epsilon_{\mathrm{T}}}\right)\right] \end{bmatrix}. \tag{18}
$$

with

$$
D = \begin{bmatrix} 1 & -1 \\ \dfrac{\epsilon_0 k_x}{k_{0z}} & \dfrac{\epsilon_{\mathrm{T}} k_x}{k_{\mathrm{T}z}} \end{bmatrix}
$$

$$
\begin{aligned} |D| &= (\epsilon_0 k_{\mathrm{T}z} + \epsilon_{\mathrm{T}} k_{0z})/k_{0z}k_{\mathrm{T}z} \\ &= \frac{(\epsilon_{\mathrm{T}} - \epsilon_0)}{k_{0z}k_{\mathrm{T}z}(\epsilon_0 k_{\mathrm{T}z} - \epsilon_{\mathrm{T}} k_{0z})}\left[(\epsilon_{\mathrm{T}} + \epsilon_0)K_x^2 - \frac{\omega^2}{c^2}\epsilon_{\mathrm{T}}\epsilon_0\right]. \end{aligned} \tag{19}
$$

Again, $|D| = 0$ yields the familiar surface polariton dispersion relation

$$K_x^2 = (\omega^2/c^2)\epsilon_T\epsilon_0/(\epsilon_T + \epsilon_0).$$

As an example, we consider the special case where $\boldsymbol{P}_0^{NL} \neq 0$ but $\boldsymbol{P}_T^{NL} = 0$ (DeMartini and Shen 1976). The particular solution of eq. (15) gives

$$\begin{aligned}
\gamma_{xx}^0 &= 4\pi k_{0z}^2/\epsilon_0(k_{0s}^2 - k_0^2),\\
\gamma_{xz}^0 &= -4\pi k_x k_{0sz}/\epsilon_0(k_{0s}^2 - k_0^2),\\
\gamma_{zx}^0 &= -4\pi k_x k_{0sz}/\epsilon_0(k_{0s}^2 - k_0^2),\\
\gamma_{zz}^0 &= -4\pi(k_{0sz}^2 - k_0^2)/\epsilon_0(k_{0s}^2 - k_0^2).
\end{aligned} \tag{20}$$

Then, the solution of eq. (18) yields

$$\begin{aligned}
\mathscr{E}_{Rx} &= \frac{4\pi\epsilon_T k_{0z}}{\epsilon_0(\epsilon_0 k_{Tz} + \epsilon_T k_{0z})(k_{0s}^2 - k_0^2)}\\
&\quad \times \left[\mathscr{P}_{0x}^{NL}(k_{0s}^2 - k_0^2) + \left(\frac{\epsilon_0}{\epsilon_T}k_{Tz} - k_{0sz}\right)(k_{0sz}\mathscr{P}_{0x}^{NL} - k_x\mathscr{P}_{0z}^{NL})\right],\\
\mathscr{E}_{Tx} &= \frac{-4\pi k_{Tz}}{(\epsilon_0 k_{Tz} + \epsilon_T k_{0z})(k_{0s}^2 - k_0^2)}\\
&\quad \times [\mathscr{P}_{0x}^{NL}(k_{0s}^2 - k_0^2) - (k_{0z} + k_{0sz})(k_{0sz}\mathscr{P}_{0x}^{NL} - k_x\mathscr{P}_{0z}^{NL})].
\end{aligned} \tag{21}$$

When $k_x = K_x'$, the real part of $(\epsilon_0 k_{Tz} - \epsilon_T k_{0z})$ vanishes, and hence $\mathscr{E}_R$ and $\mathscr{E}_T$ are resonantly enhanced. The excited surface polariton actually corresponds to only the homogeneous part of eq. (16), i.e.,

$$\begin{aligned}
\boldsymbol{E}_{SP} &= (\hat{x} + \hat{z}k_x/k_{0z})\mathscr{E}_{Rx}\exp \mathrm{i}(k_x x - \omega t) + \beta_0 z \quad \text{for } z < 0\\
&= (\hat{x} + \hat{z}k_x/k_{Tz})\mathscr{E}_{Tx}\exp \mathrm{i}(k_x x - \omega t) - \beta_T z \quad \text{for } z > 0
\end{aligned} \tag{22}$$

where $\beta_0 = -\mathrm{i}k_{0z}$ and $\beta_T = -\mathrm{i}k_{Tz}$.

The surface polariton described here by eq. (22) is a driven wave. In general, there should also exist at the interface free surface polariton waves which are a solution of

$$D\begin{pmatrix}\mathscr{E}_{Rx}\\ \mathscr{E}_{Tx}\end{pmatrix} = 0.$$

As in the bulk case, the amplitude of the free wave is determined by matching the boundary conditions on the interface. For infinite plane wave excitation, the free wave vanishes. For excitation over a finite cross section, on the other hand, the free wave can be important; in particular, it may be the only wave present in regions with no excitation. However, if excitation over a finite cross section can be Fourier decomposed into a set of infinite plane waves, then there is no need to include the free wave in the solution again.

2.3. Nonlinear Interaction of Surface Polaritons

As electromagnetic waves, surface polaritons can also interact nonlinearly. In fact, with linear pulsed laser excitation, the surface polariton intensity being confined to a thin layer can reach unusually high value. The nonlinear polarization induced by mixing of surface polaritons is correspondingly large. Thus, even though $\boldsymbol{P}^{\mathrm{NL}}$ is nonvanishing only in a very thin layer near the interface, the field generated may be readily observable. The fact that $\boldsymbol{P}^{\mathrm{NL}}$ can be greatly enhanced through enhancement of pump field intensities via surface polariton excitations makes the study of nonlinear optical effects on surfaces very appealing.

The output field of surface polariton mixing is again governed by the wave equation in eq. (15). Actually, the general solution of eq. (15) discussed in the previous section is still valid. Consider the case where the surface polaritons can be well approximated as being confined to a single interface at $z = 0$. Then, $\boldsymbol{P}^{\mathrm{NL}}(\omega, \boldsymbol{k}_s)$ can be written as

$$\begin{aligned}\boldsymbol{P}^{\mathrm{NL}}(\omega, \boldsymbol{k}_s) &= \mathcal{P}_0^{\mathrm{NL}} \exp[\mathrm{i}(k_{sx}s - \omega t) + \sigma_0 z] \quad \text{for } z < 0\\ &= \mathcal{P}_{\mathrm{T}}^{\mathrm{NL}} \exp[\mathrm{i}(k_{sx}x - \omega t) - \sigma_{\mathrm{T}} z] \quad \text{for } z > 0.\end{aligned} \tag{23}$$

If $k'_{sx} \cong K'_x(\omega)$, the surface polariton at ω can again be resonantly excited by $\boldsymbol{P}^{\mathrm{NL}}(\omega, \boldsymbol{k}_s)$. This corresponds to a phase-matched generation of surface polariton by mixing of surface polaritons. With all waves involved being surface waves, we have here a true surface nonlinear optical effect. The wavevector component k'_{sx} can also be less than the bulk wavevector k_0 (or/and k_T). In this case, the output is a bulk wave with its direction of propagation defined by the exit angle $\theta = \sin^{-1}(k_{sx}/k_0)$ with respect to the surface normal.

More generally, one can consider the problem of optical mixing of surface polaritons with bulk waves. Since some of the pump fields are surface waves, $\boldsymbol{P}^{\mathrm{NL}}(\omega, \boldsymbol{k}_s)$ induced is again confined to a thin layer near the interface. Equation (15) and its solution discussed in the previous section are clearly also applicable to such a problem. The output can be either surface polariton wave or bulk wave depending on the value of k_{sx}.

3. Nonlinear Excitation of Surface Polaritons – Experimental Demonstration

That surface polaritons can be excited by nonlinear optical mixing has been demonstrated on semiconductor surfaces (DeMartini et al. 1976, 1977). In semiconductors, two types of polariton reststrahlung bands may appear: phonon polariton and exciton polariton. The dielectric constant in the reststrahlung band may become more negative than -1, and hence, according to eq. (11), surface polaritons can exist, at least at the air–semiconductor interface. Many semiconductors possess large second-order optical nonlinearity. It is then possible to induce a second-order nonlinear polarization $\boldsymbol{P}^{(2)}(\omega, \boldsymbol{k}_s)$ in a semiconductor by sum- or difference-frequency mixing. If $k_{sx} \cong K'_x(\omega)$, the surface polariton will be excited.

We consider first the nonlinear excitation of surface phonon polaritons at the air-GaP (110) interface (DeMartini 1976). The infrared dielectric constant of GaP in the frequency region around the $367\,\text{cm}^{-1}$ transverse phonon mode is (Marschall and Fischer 1976):

$$\epsilon(\omega) = \epsilon_\infty + \omega_p^2/(\omega_T^2 - \omega^2 - i\omega_T\Gamma) \tag{24}$$

with $\epsilon_\infty = 9.091$, $\omega_T = 367.3\,\text{cm}^{-1}$, $\omega_p^2 = 1.859\omega_T^2$, and $\Gamma = 1.28\,\text{cm}^{-1}$. In a narrow region with $\omega > \omega_T$, we find $\epsilon(\omega) < -1$. Surface polaritons can then exist with a dispersion relation

$$(K'_x + iK''_x)^2 = (\omega/c)^2\epsilon/(1+\epsilon) \tag{25}$$

plotted in fig. 4. To demonstrate the nonlinear excitation of surface polaritons on GaP, we used the experimental setup shown in fig. 5. A Q-switched ruby laser provided the ω_1 beam at $14403\,\text{cm}^{-1}$. It was then also used to pump two dye laser systems to provide beams at ω_2 and ω_3. In the experiment, ω_3 was fixed at $13333\,\text{cm}^{-1}$ while ω_2 was tuned between 14006 and $14035\,\text{cm}^{-1}$. Then ω_1 and ω_2 beams were used to excite surface polaritons, and the ω_3 beam to probe the excited surface polaritons.

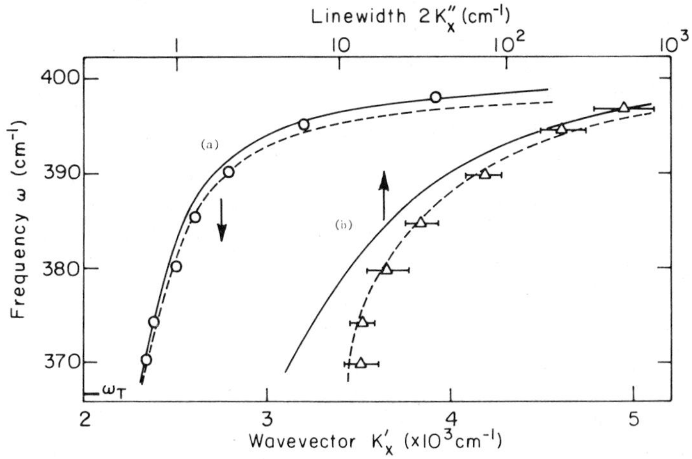

Fig. 4. Dispersion characteristics of surface polaritons at the air–GaP interface. (a) K'_x versus ω, and (b) K''_x versus ω. The solid curves are calculated from the single oscillator model using eqs. (24) and (25). The dashed curves are calculated from the multi-oscillator model of Barker (1968).

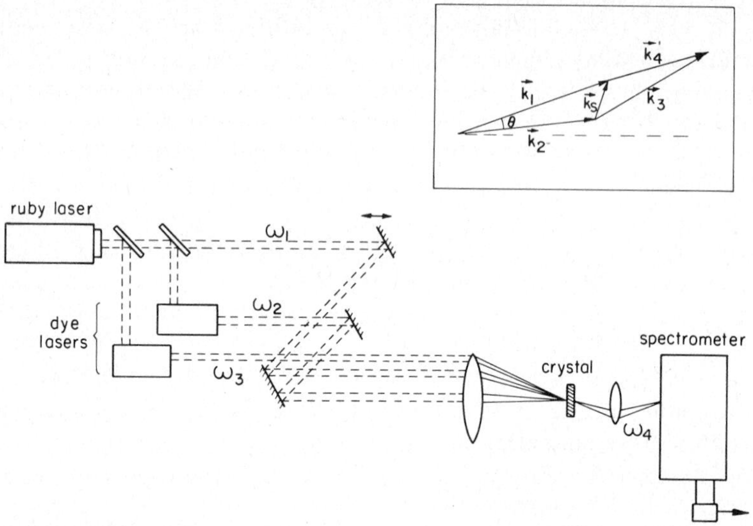

Fig. 5. Experimental setup for nonlinear optical excitation and detection of surface polaritons. The inset shows the wavevector relation for the nonlinear mixing process.

Let $\hat{\hat{x}}$, $\hat{\hat{y}}$, and $\hat{\hat{z}}$ be the crystal axes along [100], with $\hat{\hat{z}} \| \hat{y}$ and $\hat{\hat{x}}$ and $\hat{\hat{y}}$ at 45° with respect to the surface normal $\hat{z}$. We polarized the incoming fields at ω_1 and ω_2 as $\boldsymbol{E} = \hat{\hat{z}} E_1$, and $E_2 \simeq \hat{x} \boldsymbol{E}_2 = (\hat{\hat{x}} + \hat{\hat{y}}) E_2 / \sqrt{2}$. Then, the induced nonlinear polarization in GaP was

$$\boldsymbol{P}_0^{\mathrm{NL}}(\omega, \boldsymbol{k}_s) \cong \chi_{14}^{(2)}(\omega = \omega_1 - \omega_2) \hat{x} E_1 E_2^*, \tag{26}$$

with $\boldsymbol{k}_s = \boldsymbol{k}_1 - \boldsymbol{k}_2$. By adjusting the angle between ω_1 and ω_2 beams, k_{sx} could be varied from 0 to nearly $(k_1 + k_2)$. In our experiment, $\omega = \omega_1 - \omega_2$ was fixed in the range between 368 and 397 cm^{-1} where surface polaritons can exist, and k_{sx} was then varied around $k_{sx} \cong K_x'(\omega)$. It is, of course, also possible to fix k_{sx} and scan $\omega = \omega_1 - \omega_2$ to make $k_{sx} \cong K_x'(\omega)$ for surface polariton excitation. The generated surface polariton field should have the form of E_{SP} described in eqs. (22) and (21) with the subindices 0 and T referring to GaP and air respectively, $\boldsymbol{P}_{0z}^{\mathrm{NL}} \cong 0$, $\epsilon_{\mathrm{T}} = 1$, and $(\epsilon_0 k_{\mathrm{T}z} + k_{0z}) \propto (-k_x + K_x' + \mathrm{i} K_x'')$. We find,

$$\mathscr{E}_{\mathrm{R}x} = \frac{A}{-\Delta k_x + \mathrm{i} K_x''} \mathscr{P}_{0x}^{\mathrm{NL}},$$

$$\mathscr{E}_{\mathrm{T}x} = [k_{\mathrm{T}z}(k_{0sz} k_{0z} + k_{0z}^2)/k_{0z}(k_{0sz} k_{\mathrm{T}z} - k_{0z}^2/\epsilon_0)] \mathscr{E}_{\mathrm{R}x}, \tag{27}$$

where

$$A = -[2\pi(k_{0z} - \epsilon_0 k_{\mathrm{T}z})/K_x' \epsilon_0 (1 - \epsilon_0^2)(k_{0s}^2 - k_0^2)](\epsilon_0 k_{0sz} k_{\mathrm{T}z} - k_{0z}^2),$$

$$\Delta k_x = k_x - K_x'.$$

It is seen that the generated surface polariton intensity ($\propto |E_{SP}|^2$) as a function of the phase mismatch Δk_x is a Lorentzian.

To demonstrate the presence of the generated surface polariton, we used the ω_3 beam as a probe with $\boldsymbol{E}_3$ along $\hat{z}\|\hat{y}$. Optical mixing of the probe beam with the surface polaritons in GaP also induced a second-order nonlinear polarization,

$$\boldsymbol{P}_0^{\mathrm{NL}}(\omega_4) = \chi_{14}^{(2)}(\omega_4 = \omega_3 - \omega)[\hat{x}\mathscr{E}_{\mathrm{R}z}^* + \hat{z}\mathscr{E}_{\mathrm{R}x}^*]\mathscr{E}_3 \times \exp[\mathrm{i}(k_{3x} - k_x)x + (\mathrm{i}k_{3z} - \beta_0)z - \omega_4 t], \quad (28)$$

where $\boldsymbol{k}_3$ is the wavevector of the probe beam in GaP. As expected from eq. (28), the output generated by $\boldsymbol{P}^{\mathrm{NL}}(\omega_4)$ should have TM polarization and should be coherent with its direction fixed by the phase matching relation $k_{1x} - k_{2x} = k_{3x} - k_{4x}$ or equivalently $\boldsymbol{k}_4 = \boldsymbol{k}_3 - (\boldsymbol{k}_1 - \boldsymbol{k}_2)$. This was actually observed in the experiment. The output intensity can be estimated from the general solution for $\boldsymbol{E}(\omega_4)$ in eqs. (16a), (16c), and (21), and should be proportional to $|\boldsymbol{P}_0^{\mathrm{NL}}(\omega_4)|^2$.

$$I(\omega_4, \Delta k_x) \propto |\boldsymbol{P}_0^{\mathrm{NL}}(\omega_4)|^2 = |\overleftrightarrow{\chi}^{(2)}(\omega_4 = \omega_3 - \omega) : \boldsymbol{E}(\omega_3)\boldsymbol{E}_{\mathrm{SP}}(\omega)|^2$$
$$\propto \frac{1}{\Delta k_x^2 + (K_x'')^2}|\chi_{14}^{(2)}(\omega_4 = \omega_3 - \omega)\chi_{14}^{(2)}(\omega = \omega_1 - \omega_2)^* E_1^* E_2 E_3|^2. \quad (29)$$

This explicitly shows that the output signal versus Δk_x should be a Lorentzian with its peak at $\Delta k_x = k_x - K_x'(\omega) = 0$ and its half width equal to K_x''. The experimental results could indeed be fit by eq. (29). Typical examples are given in fig. 6. From the observed half widths of $I(\omega_4, \Delta k_x)$ at various ω, we deduced $K_x'(\omega)$ and $K_x''(\omega)$, as is presented in fig. 4. They are in good agreement with the theoretical curves calculated from the dispersion relation of eq. (25). The attenuation coefficient $K_x''(\omega)$ is more sensitive to the frequency dependence of $\epsilon(\omega)$ than $K_x'(\omega)$. The dashed curves in fig. 4 were calculated with $\epsilon(\omega)$ derived from a multi-oscillator model of Barker (1968) instead of the single-oscillator model used in eq. (24). The experimental results seem to agree better with the multi-oscillator model. Towards small K_x'', the measured half width of $I(\omega_4, \Delta k_x)$ in fig. 4 appears to have a limiting value. This was because when K_x'' was very small, the observed width was dominated by the finite spread of $\boldsymbol{k}$ of the laser beams. With a 50 kW peak power in each of the three laser beams, the peak output was about 0.1 μW at $\omega = 380\ \mathrm{cm}^{-1}$, compared to a theoretical prediction of 0.27 μW.

Nonlinear optical excitation of surface exciton polaritons has also been demonstrated using ZnO (DeMartini et al. 1977). In this case, because excitons only exist at low temperatures, surface exciton polaritons also exist only at low temperatures. In fact, this makes the linear excitation of

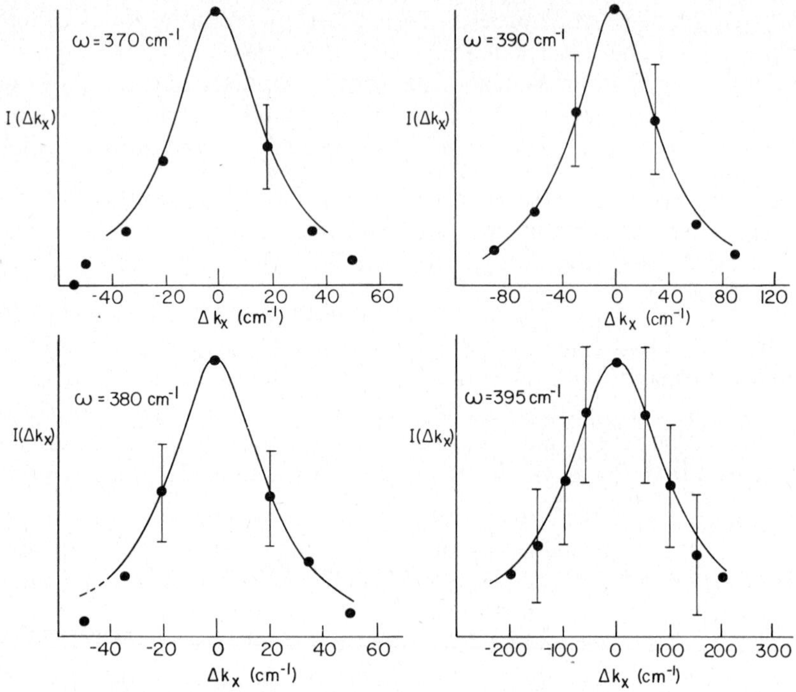

Fig. 6. Examples of experimental data on $I(\omega_4, \Delta k_x)$ versus Δk_x in the GaP case. The solid curves are Lorentzian used to fit the data.

surface exciton polaritons rather difficult (Lagois and Fischer 1976) since the Otto method with critical adjustment of prism spacing from the surface required for optical coupling is not easy to carry out in a helium cryostat. Nonlinear excitation by optical mixing, however, bypasses such difficulty.

ZnO is a uniaxial crystal. It has three prominent excitons, A, B, and C, near the band edge. We consider here only surface polaritons associated with the C exiton. Because of anisotropy of ZnO, the surface polariton dispersion relation for the ZnO–air (or liquid helium) interface in the C exciton reststrahlung band between 3.421 and 3.427 eV is somewhat different from eq. (11) and is given by (Hartstein et al. 1973):

$$K_x^2 = \left(\frac{\omega}{c}\right)^2 \frac{\epsilon_T \epsilon_{0x}(\epsilon_{0x} - \epsilon_T)}{\epsilon_{0x}\epsilon_{0z} - \epsilon_T^2}, \tag{30}$$

with $\epsilon_{bx} < 0$ and $\epsilon_{bz} > \epsilon_T$ or $\epsilon_{bz} < 0$. If the ZnO crystal is oriented with its c-axis in the surface along $\hat{x}$, then since the electric dipole transition for the C exiton is only allowed for polarization parallel to the c-axis, we have, in the reststrahlung band of the exciton,

$$\epsilon_{0x} = \epsilon_\infty - (\epsilon_{00} - \epsilon_\infty)\omega_T^2/[(\omega^2 - \omega_T^2) + i\omega\Gamma],$$

$$\epsilon_{0z} = \epsilon_\infty, \tag{31}$$

following the single-oscillator model and neglecting the anisotropy in ϵ_∞, where ω_T is the transverse exciton frequency, Γ is the damping constant, and $(\epsilon_{00} - \epsilon_\infty)$ is proportional to the exciton oscillator strength. By fitting the experimental dispersion curve with eqs. (30) and (31), the constants ϵ_∞, ϵ_0, ω_T, and Γ can be determined.

The surface exciton polariton here can be excited by sum-frequency mixing, or more simply, by second-harmonic generation (DeMartini et al. 1977). The actual experimental setup used for the observation was shown in fig. 7, and was much simpler than that for the case of GaP. The sample was immersed in superfluid helium. To excite the surface polariton at ω, the dye laser frequency was fixed at $\omega_1 = \frac{1}{2}\omega$, and the beam direction was varied to vary $k_x = 2k_{1x}$ around $K_x'(\omega)$. The excited surface polariton could again be probed through mixing with a probe beam. However, in the present case, since UV photons can be easily detected, the excited surface polaritons could be observed through surface roughness scattering even though the scattering efficiency is usually small for relatively smooth surfaces. Therefore, as shown in fig. 7, the excitation of surface polaritons was monitored simply by a single photomultiplier.

In the experiment, incoming wave of either TE or TM was used. In the former case, the induced second-order nonlinear polarization in ZnO had

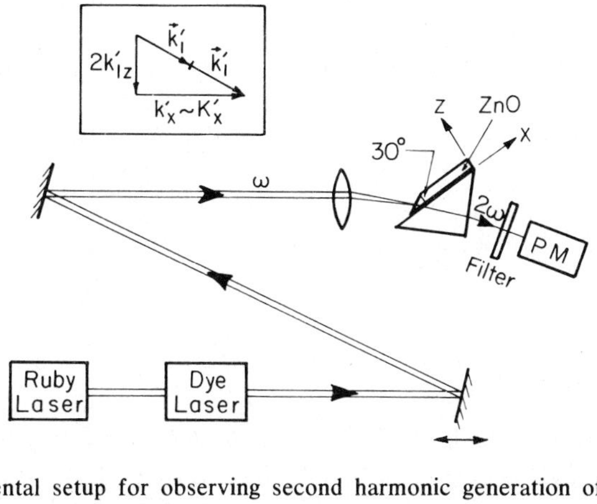

Fig. 7. Experimental setup for observing second harmonic generation of surface exciton–polariton on ZnO. The inset shows the wavevector relation.

only an $\hat{x}$ component

$$P_x^{NL}(\omega) = \chi_{31}^{(2)} E_y^2(\omega_1). \tag{32}$$

In the latter case, the $\hat{x}$-component of $\boldsymbol{P}^{NL}(\omega)$ dominated:

$$P_x^{NL}(\omega) = \chi_{31}^{(2)} E_z^2(\omega_1) + \chi_{33}^{(2)} E_x^2(\omega_1). \tag{33}$$

The surface polariton field generated by $\boldsymbol{P}_x^{NL}(\omega)$ can again be calculated from eqs. (22) and (21). For a 30 ns input pulse with 50 kW peak power focussed to a ~50 μm spot, the surface polariton intensity should correspond to 10^8 photons/pulse. Experimentally, 10^4 photons/pulse was observed, suggesting that the surface roughness coupling efficiency was about 10^{-4}.

As in the case of GaP, the surface polariton intensity $I(\omega)$ versus $\Delta k_x = k_x - K_x'(\omega)$ should be a Lorentzian. This is seen by the fit to the experimental data in fig. 8. The values of $K_x'(\omega)$ and $K_x''(\omega)$ deduced from the fit is plotted in fig. 9. A least square fitting of these data points with eqs. (30) and (31) yielded $\epsilon_\infty = 6.15 \pm 0.01$, $\omega_T = 3.421$ eV, $\epsilon_0 = 6.172 \pm 0.01$, and $\Gamma = 0.25 \pm 0.05$ meV. They can be compared with the following values reported in the literature. $\epsilon_\infty = 6.15$, $\omega_T = 3.4213$ eV, $\epsilon_0 = 6.188$, and $\Gamma =$ 0.5 meV.

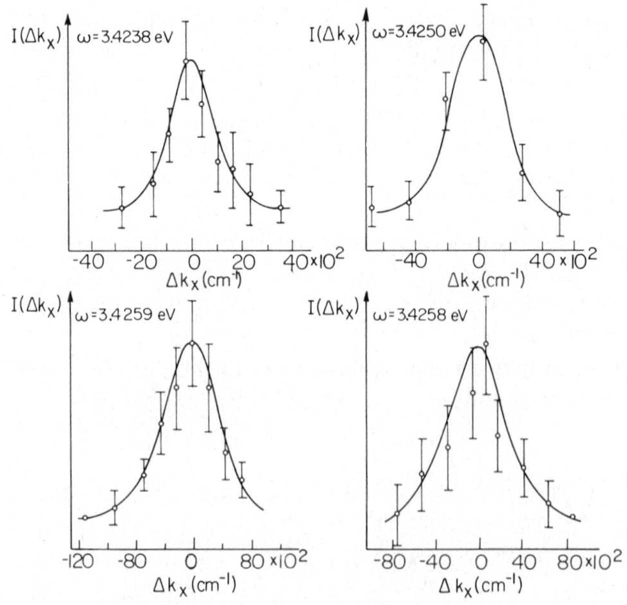

Fig. 8. Examples of experimental results of $I(\omega, \Delta k_x)$ versus Δk_x in the ZnO case. The solid curves are Lorentzian used to fit the data.

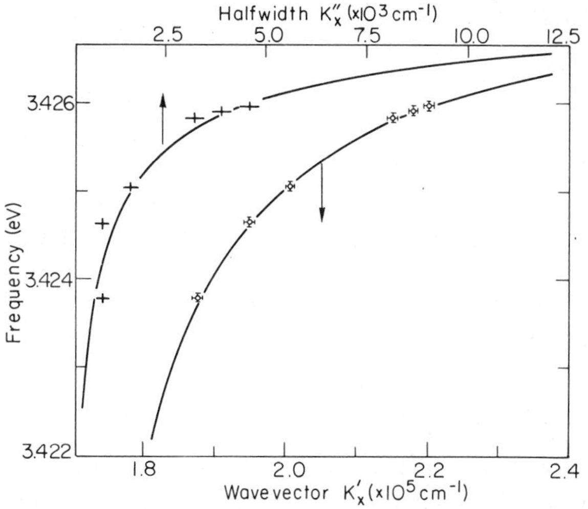

Fig. 9. Dispersion and damping characteristics of surface exciton–polaritons on ZnO. The solid curves are theoretical curves calculated from eqs. (30) and (31).

It is interesting to note that although the data in fig. 8 appear to be quite scattered, the accuracy of determining $K'_x(\omega)$ from the results is excellent. This is because the uncertainty is only $\pm K''_x$, and here $K''_x/K'_x \sim 10^{-2}$. Excitons in the UV are known to be fairly critically dependent of the surface condition. With nonlinear excitation, surface exciton–polaritons can be studied without further surface perturbation aside from original sample preparation. This is a clear advantage over the linear excitation method. Then, the nonlinear excitation method can, in principle, also be used to study surface exciton polaritons under specific surface perturbation.

Unlike linear excitation, the nonlinear excitation method has the advantage that it can be used to excite and study surface polaritons at the interface of two semi-infinite media. Of course, the same method can also be used to study surface polaritons in layered media. In fact, with nonlinear excitation, it may even be possible to excite surface polaritons at an inner interface that cannot be reached by linear excitation.

4. *Second Harmonic Generation by Surface Plasmons*

The dielectric constant of a metal is negative below the plasma frequency. Surface polaritons can therefore exist at a metal–dielectric interface. They are usually called surface plasmons (Borstel and Falge 1978). If the dielectric medium is anisotropic, the surface plasmon dispersion relation is given by

$$K_x^2(\omega) = \left(\frac{\omega}{c}\right)^2 \frac{\epsilon_m \epsilon_\perp (\epsilon_\parallel - \epsilon_m)}{\epsilon_\perp \epsilon_\parallel - \epsilon_m^2} \tag{34}$$

where $\epsilon_m < 0$ is the dielectric constant of the metal with $\epsilon_m^2 > \epsilon_\perp \epsilon_\parallel$, and $\epsilon_\perp > 0$ and $\epsilon_\parallel > 0$ are respectively the dielectric constants of the dielectric medium perpendicular and parallel to the propagation direction of the surface plasmon. If the dielectric medium is isotropic, eq. (34) reduces to

$$K_x^2(\omega) = \left(\frac{\omega}{c}\right)^2 \frac{\epsilon_m \epsilon}{\epsilon_m + \epsilon}. \tag{35}$$

Usually, the Kretschmann method of fig. 2b is used to excite surface plasmons. The metal medium is simply a film of a few hundred Å sandwiched between the prism and the dielectric. The surface plasmon excited is confined more or less to the metal–dielectric interface. Its dispersion is somewhat different from that for two semi-infinite media, and in the case of isotropic dielectric, is given by eq. (9). However, since the metal film can be regarded as optically thick, eq. (34) or (35) is actually a very good approximation of the true dispersion.

With an appropriately chosen metal film thickness in the Kretschmann geometry, the surface plasmon can be optimally excited. A sizeable fraction of the exciting beam power is coupled into the surface plasmon, and hence the surface plasmon intensity is greatly enhanced in comparison with the incoming beam intensity. For example, the enhancement observed in practice can be ~400 for a silver–air interface and ~100 for a silver–dielectric ($\epsilon \sim 2.5$) interface. The strongly enhanced intensity facilitates the study of nonlinear optical processes at the interface.

The simplest nonlinear optical process is the second harmonic generation. It was first demonstrated by Simon and coworkers at the silver–air interface (Simon et al. 1974). The theoretical treatment of the problem follows the general formalism in sect. 2. First, the fundamental surface plasmon field obtained from linear excitation is calculated. Then, the second harmonic output is obtained from the solution of the wave equation discussed in sect. 2 with an induced nonlinear polarization $\boldsymbol{P}^{\mathrm{NL}}(\omega = 2\omega_1)$ in the silver metal as the driving source. Since metal has inversion symmetry, $\boldsymbol{P}^{\mathrm{NL}}(\omega)$ arises only through electric quadrupole and magnetic dipole contribution and is relatively weak (Bloembergen 1968). From symmetry consideration, it can be written as (Bloembergen 1968):

$$\boldsymbol{P}^{\mathrm{NL}}(\omega = 2\omega_1) = \alpha(\boldsymbol{E}_1 \cdot \nabla)\boldsymbol{E}_1 + \beta(\nabla \cdot \boldsymbol{E}_1)\boldsymbol{E}_1 + \gamma \boldsymbol{E}_1 \times \boldsymbol{B}_1, \tag{36}$$

where α, β, and γ are constant coefficients. The first two terms are of electric-quadrupole origin. Since they depend on spatial variation of the fundamental field $\boldsymbol{E}_1$, in uniform metallic media, they are only nonvanishing near the surface within the Thomas–Fermi screening length, i.e., within one or two atomic layers thick. In fact, they can be regarded as arising from electric-dipole nonlinearity of the surface atomic layers that do not have inversion symmetry. The last term of eq. (36) is of magnetic dipole origin, and is nonvanishing throughout the bulk. However, it is usually negligible compared with the electric quadrupole terms. A single layer of atoms without inversion symmetry already has a second-order nonlinearity much larger than that of the magnetic-dipole contribution from a hundred atom layers penetrated by the exciting field, as evidenced by the investigation of Bloembergen et al. (1968).

Even though the nonlinearity is rather weak, second harmonic generation by surface plasmons at the metal–air interface is readily observable with the setup in fig. 10 (Simon et al. 1974). The results generally follow the theoretical predictions of sect. 2. In particular, for co-propagating fundamental surface plasmons, the second harmonic output is a coherent

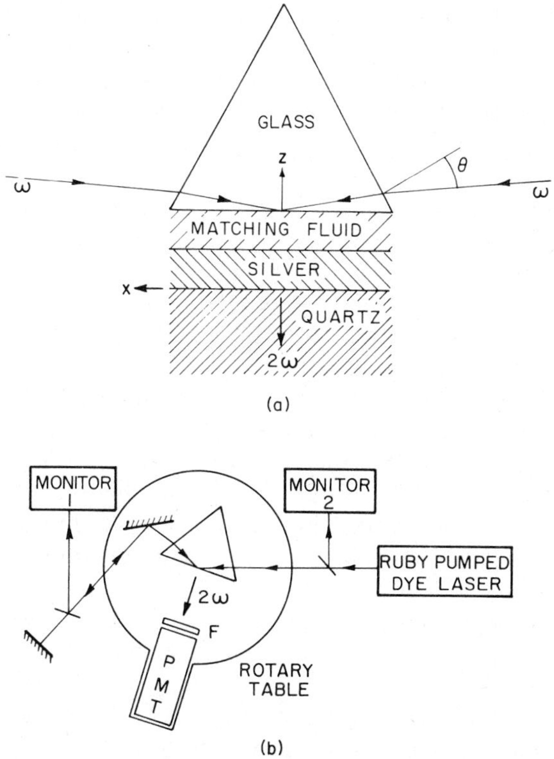

Fig. 10. Experimental setup for observing second harmonic generation by counter-propagating surface plasmons.

beam with its propagation direction specified by $2k_{1x}$ along the surface and its polarization being transverse magnetic. For counter-propagating fundamental surface plasmons, $\boldsymbol{P}^{\mathrm{NL}}(\omega)$ has a zero wavevector component along the surface. The second harmonic output should then be propagating along the normal of the surface (Chen et al. 1979). From symmetry, however, the component of $\boldsymbol{P}^{\mathrm{NL}}(\omega)$ of eq. (36) perpendicular to the surface normal is zero, and hence, no output can be observed (Chen and deCastro). By varying the angle between the two propagating surface plasmons, the coefficients α, β, and γ in eq. (36) can in principle be determined from the observed second harmonic signals.

The second harmonic output can be greatly enhanced if air is replaced by a nonlinear crystal in the above case (Simon et al. 1977). The nonlinearity is now dominated by the crystal instead of the metal. For a quartz–metal interface, for example, the nonlinear susceptibility $\chi^{(2)}$ for quartz is one order of magnitude larger than the effective $\chi^{(2)}$ for metal, so that the output is two orders of magnitude stronger. Then, the symmetry of $\overleftrightarrow{\chi}^{(2)}$ for quartz is also different from that for metal, leading to a different output polarization, even though the output beam direction is still determined by the wavevector component of $\boldsymbol{P}^{\mathrm{NL}}(2\omega)$ along the surface. As a special case, consider the second harmonic generation by counter-propagating surface plasmons at a quartz–silver interface (Chen et al. 1979) with the quartz $\hat{a}$-axis oriented along $\hat{x}$ and the $\hat{b}$-axis along $\hat{z}$. The induced nonlinear polarization in quartz is

$$\boldsymbol{P}^{\mathrm{NL}} = \hat{x}\chi_{11}^{(2)}(\mathscr{E}_{1x}^{+}\mathscr{E}_{1x}^{-} - \mathscr{E}_{1z}^{+}\mathscr{E}_{1z}^{-})\exp(2\beta_1 z - \mathrm{i}2\omega_1 t) \tag{37}$$

where $\boldsymbol{\mathscr{E}}_1^{+}$ and $\boldsymbol{\mathscr{E}}_1^{-}$ are the field amplitudes of the forward and backward fundamental waves respectively. Since $\boldsymbol{P}^{\mathrm{NL}}(\omega)$ along $\hat{x}$ is nonzero, the second harmonic output along the surface normal should be observable. This is indeed the case. Using the experimental setup in fig. 10 with the counter-propagating beam provided by mirror reflection, Chen et al. (1979) observed the second harmonic output along the surface normal. The output beam was highly directional with a beam spread of ~1 mrad. The output intensity as a function of the incidence angle of the exciting laser beam is shown in fig. 11. The observed maximum occurs at the angle where the surface plasmons were optimally excited, and the width of the peak is approximately the width of the surface plasmon resonance observed in the linear reflection curve, similar to the one in fig. 3. The experimental results were in very good agreement with the theoretical calculation following sect. 2, as shown in fig. 11. Even the absolute output power agreed well with the prediction of $\mathbb{P}(\omega) = 5 \times 10^{-25}\mathbb{P}^{+}(\omega_1)\mathbb{P}^{-}(\omega_1)/A$ in cgs units, where A is the beam overlapping area at the interface.

In fig. 12, a typical dispersion curve of surface plasmons at a metal–air or metal–dielectric (transparent) interface is shown. It is seen that $|K_x'(\omega =$

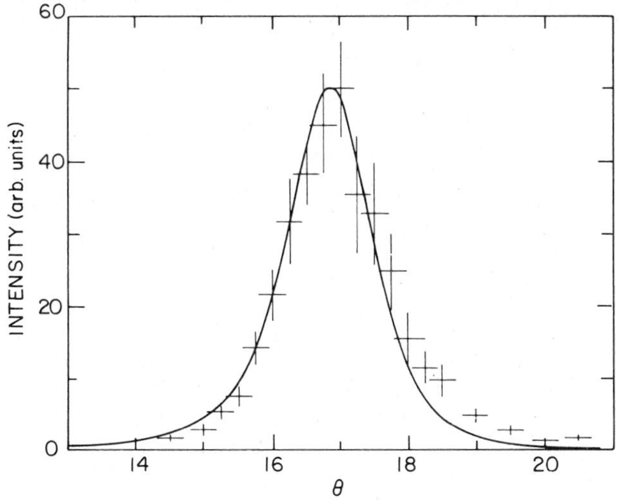

Fig. 11. Second harmonic intensity versus the incidence angle θ (defined in fig. 10).

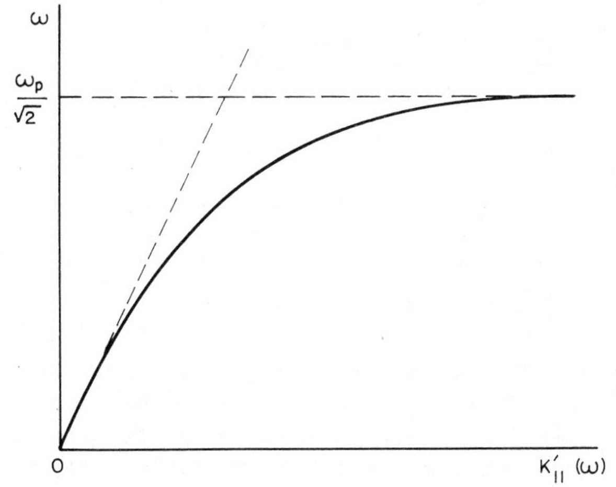

Fig. 12. A typical dispersion curve of surface plasmons.

$2\omega_1)| > |2K_x'(\omega_1)|$, and therefore, phase-matched second harmonic generation of surface plasmons by surface plasmons is clearly impossible. In the case of a single exciting laser beam, if the angle of beam incidence is varied to vary k_{1x}, the second harmonic output should exhibit two separate peaks (Simon et al. 1977). An example is shown in fig. 13 for an aluminum–quartz interface with Nd:YAG laser excitation (DeMartini et al. 1981). The first

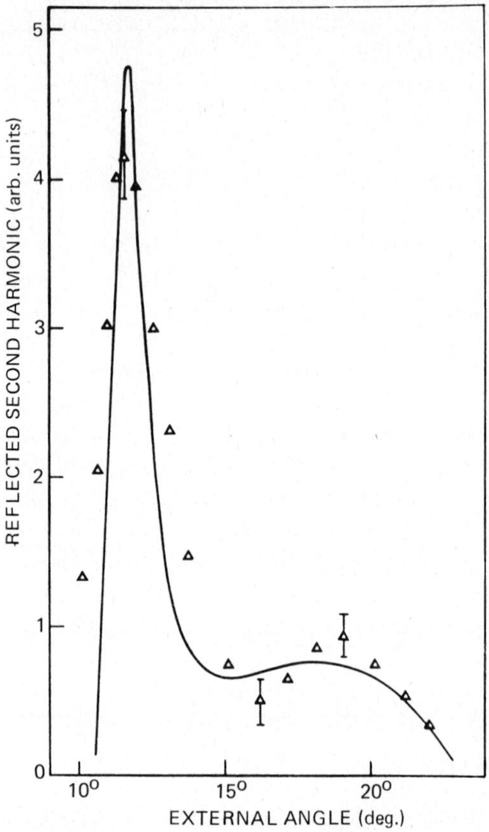

Fig. 13. Second harmonic output at the quartz–aluminum interface as a function of the incidence angle, exhibiting two separate peaks.

peak at smaller angle corresponds to the linear excitation of the fundamental surface plasmons, while the second peak arises because of nonlinear excitation of the second harmonic surface plasmons. Since the second harmonic output is proportional to the square of the resonant excitation profile of the fundamental, but only linearly to that of the second harmonic, the first peak is much sharper and stronger than the second one. The difference becomes larger when $K'_x(\omega) - 2K'_x(\omega_1)$ increases. In fig. 13, the experimental results for the Al–quartz interface are in good accord with the theoretical calculation. Simon et al. (1977) first studied second harmonic generation by surface plasmons at a silver–quartz interface. The surface plasmon dispersion between 1.06 μm and 0.53 μm is appreciably stronger in this case, so that $K'_x(\omega) - 2K'_x(\omega_1)$ is larger. Consequently, the second peak is expected to be 5 orders of magnitude lower than the first peak in a similar plot as fig. 13. Then, because of interference from second harmonic generation in the prism, they were not successful in observing the second peak experimentally. With TE laser excitation, no fundamental surface plasmon can be excited. However, through the nonlinearity of the quartz, the second harmonic surface plasmons can be nonlinearly excited. In this case, only the second peak in fig. 13 should be observed. This has also been demonstrated experimentally (DeMartini et al. 1981).

Third harmonic generation by surface plasmons at a metal–dielectric interface can also be observed (Chen and de Castro). The discussion closely follows that for second harmonic generation. The effect of resonant excitation of fundamental surface plasmons can again be easily seen.

Experiments of harmonic generation by surface plasmons have so far been carried out with visible lasers. On the other hand, the same process in the infrared may be more interesting. First, the attenuation constant K_x'' of the surface plasmons is orders of magnitude less in the infrared. Then, the surface plasmon dispersion is also much less in the infrared so that phase-matching of harmonic generation of surface plasmons by surface plasmons ($K_x'(\omega) = 2K_x'(\omega_1)$) can be approximately satisfied. The harmonic output can then be greatly enhanced, with the possibility of efficient generation of a guided infrared surface wave at a new frequency.

5. Surface Coherent Anti-Stokes Raman Spectroscopy

Third-order nonlinear optical processes with surface plasmons should in general be readily observable, especially if phase-matching can be achieved and the nonlinear susceptibility shows a resonant enhancement. Thus, one expects that coherent anti-Stokes Raman scattering (CARS) at a metal–dielectric interface can be easily observed if both input and output are surface plasmons (Chen et al. 1979). The nonlinearity responsible for the process usually arises from the dielectric medium, and therefore, the resonance enhancement of the nonlinear susceptibility $\overleftrightarrow{\chi}^{(3)}$ is an explicit display of the Raman resonances of the dielectric medium. Physically, surface CARS can also be understood as a two-step process. Two surface plasmon waves at ω_1 and ω_2 with wavevectors $(\boldsymbol{k}_1)_\parallel \simeq \boldsymbol{K}_\parallel'(\omega_1)$ and $(\boldsymbol{k}_2)_\parallel \simeq \mathbf{K}_\parallel'(\omega_2)$ interact nonlinearly at the interface via the nonlinearity of the medium, where $K_\parallel'(\omega)$ describes the surface plasmon dispersion. They first beat with each other and coherently excite a Raman resonance of the medium at $\omega_1 - \omega_2$ with a wavevector $(\boldsymbol{k}_1)_\parallel - (\boldsymbol{k}_2)_\parallel$. Then, the ω_1 wave also acts as a probe beam and mixes with the material excitational wave to generate a nonlinear polarization, and hence a coherent output, at $\omega_a = 2\omega_1 - \omega_2$ with a wavevector component along the interface $(\boldsymbol{k}_a)_\parallel = 2(\boldsymbol{k}_1)_\parallel - (\boldsymbol{k}_2)_\parallel$. Clearly, the output should be a maximum when $(\omega_1 - \omega_2)$ is exactly on resonance $(\omega_1 - \omega_2 = \omega_{ex})$, and when $(k_a)_\parallel = K'_\parallel(\omega_a)$ such that the anti-Stokes surface plasmons at ω_a are resonantly excited (or in other words, generated under the phase-matching condition).

As in the case of bulk CARS (Levenson 1977), if the surface plasmon fields are $\boldsymbol{E}(\omega_1)$ and $\boldsymbol{E}(\omega_2)$, then the nonlinear polarization induced in the dielectric is

$$\boldsymbol{P}^{\mathrm{NL}}(\omega_a) = \overleftrightarrow{\chi}^{(3)}(\omega_a = 2\omega_1 - \omega_2) : \boldsymbol{E}_1(\omega_1)\boldsymbol{E}_1(\omega_1)\boldsymbol{E}_2^*(\omega_2), \tag{38}$$

where the nonlinear susceptibility can be written as the sum of a resonant part and a nonresonant part

$$\overleftrightarrow{\chi}^{(3)}(\omega_a) = \overleftrightarrow{\chi}^{(3)}_{\mathrm{R}} + \overleftrightarrow{\chi}^{(3)}_{\mathrm{NR}}$$

$$\overleftrightarrow{\chi}^{(3)}_{\mathrm{R}} = \frac{\overleftrightarrow{A}}{[(\omega_1 - \omega_2) - \omega_{ex}] + \mathrm{i}\Gamma}. \tag{39}$$

With $\boldsymbol{P}^{\mathrm{NL}}(\omega_a)$ in eq. (38) as the source term for the coherent output at ω_a, the theory formulated in sect. 2 can again be used to describe surface CARS. Since the output is proportional to $|\boldsymbol{P}^{\mathrm{NL}}(\omega_a)|^2$, and hence $|\chi^{(3)}(\omega_a)|^2$, it should clearly exhibit a resonant enhancement when $(\omega_1 - \omega_2)$ approaches ω_{ex}. The output should also be greatly enhanced if first, the input surface plasmon fields $\boldsymbol{E}_1(\omega_1)$ and $\boldsymbol{E}_2(\omega_2)$ are enhanced through resonant excitation (using the Kretschmann geometry), and then the output surface plasmon is also resonantly excited by having $2(\boldsymbol{k}_1)_\parallel - (\boldsymbol{k}_2)_\parallel \cong \boldsymbol{K}'(\omega_a)$. We note that in this case, even though it is a third-order effect, and $\boldsymbol{P}^{\mathrm{NL}}(\omega_a)$ only exists over a very thin layer at the interface, the output may still be easily detectable simply because $|\boldsymbol{P}^{\mathrm{NL}}(\omega_a)|$ can be unusually large through the surface plasmon enhancement of the input fields.

Experimental demonstration of surface CARS has been carried out by Chen et al. (1979). The setup is shown in fig. 14. A Q-switched ruby laser pumping a dye laser system was able to deliver two beams at ω_1 and ω_2, each with 20 mJ/pulse and a linewidth $\leqslant 1\ \mathrm{cm}^{-1}$. The beams excited the surface plasmons through the prism coupler. Their directions could be adjusted to vary $(\boldsymbol{k}_1)_\parallel$ and $(\boldsymbol{k}_2)_\parallel$ and phase-matched generation of anti-Stokes surface plasmons was possible with $2(\boldsymbol{k}_1)_\parallel - (\boldsymbol{k}_2)_\parallel \simeq K'_\parallel(\omega_a)$ as sketched in fig. 14(b). Surface CARS and bulk CARS signals were simultaneously monitored. The latter was used for frequency calibration and normalization.

Figure 15 shows the results of anti-Stokes output versus $(\omega_1 - \omega_2)$ with benzene as the dielectric medium. The resonant peak here corresponds to the $992\ \mathrm{cm}^{-1}$ Raman mode of benzene. The peak is slightly asymmetric, indicating that $\chi^{(3)}_{\mathrm{NR}}$ is nonnegligible. The theoretical curve in fig. 15 was derived using the known values of $\chi^{(3)}_{\mathrm{NR}}$, $\chi^{(3)}_{\mathrm{R}}$, ω_{ex}, and Γ in the literatures. It is seen that the agreement between theory and experiment is excellent.

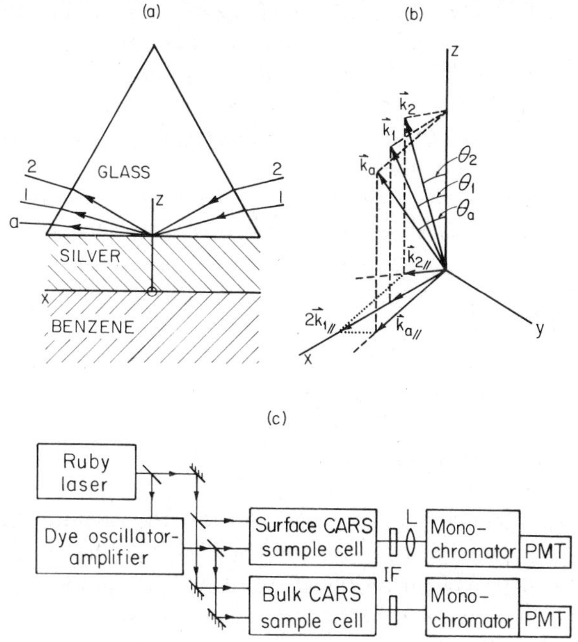

Fig. 14. Experimental setup for surface CARS measurements: (a) the prism–metal–liquid assembly; (b) wavevectors in the glass prism with their components in the xy-plane phase matched; (c) block diagram of the experimental arrangement.

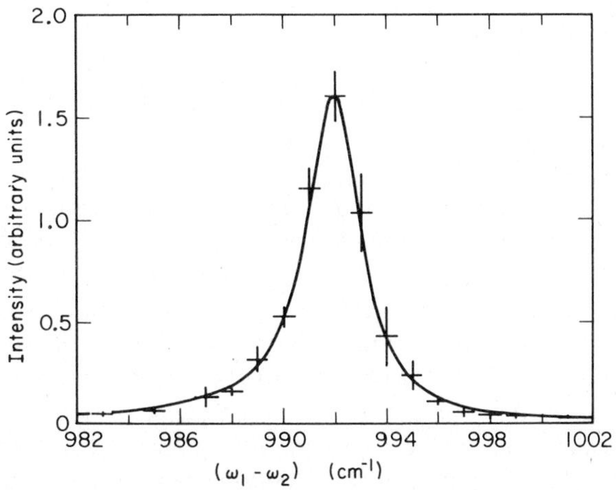

Fig. 15. Anti-Stokes output versus $\omega_1 - \omega_2$ around the 992 cm^{-1} Raman resonance of benzene. The solid curve is a theoretical curve.

That surface CARS can be used as a spectroscopic technique is then obvious. The signal-to-noise ratio detected was very good. At the peak, with 10 mJ/cm^2 pulses of 30 ns pulsewidth, the output was 1.5×10^6 photons/pulse, which was also in good agreement with the theoretical estimate. Since in surface CARS, the output is essentially originated from $\boldsymbol{P}^{\mathrm{NL}}(\omega_a)$ induced by the surface plasmons in a very thin layer (of the order of a reduced wavelength $\lambda/2\pi$) at the interface, the coherent anti-Stokes signal is not expected to decrease appreciably if instead of a semi-infinite dielectric medium, a ~1000 Å dielectric overlayer on the metal surface is used. Thus, surface CARS should be most useful for Raman spectroscopic study of overlayers.

As we mentioned earlier, the output also depends critically on the resonant excitation of surface plasmons. This was demonstrated experimentally by simply rotating the prism assembly about the $\hat{y}$-axis in fig. 14 with respect to the incoming beams. The rotation changed $(\boldsymbol{k}_1)_\parallel$ and $(\boldsymbol{k}_2)_\parallel$, which caused a variation in the resonance conditions of the linear excitation of the surface plasmons at ω_1 and ω_2 and also the nonlinear excitation of the surface plasmons at ω_a. As shown in fig. 16, the output falls rapidly away from the angular position for optimum excitation of all surface plasmons. The experimental data agreed well with the theoretical calculation. In a separate experiment, the surface plasmons at ω_1 and ω_2 were always being optimally excited, but the phase mismatch $\Delta k_\parallel = |2(\boldsymbol{k}_1)_\parallel - (\boldsymbol{k}_2)_\parallel - K'_\parallel(\omega_a)|$ for the resonant excitation of the surface plasmons at ω_a was varied. The output versus $\Delta k_\parallel$ showed a Lorentzian line centered at

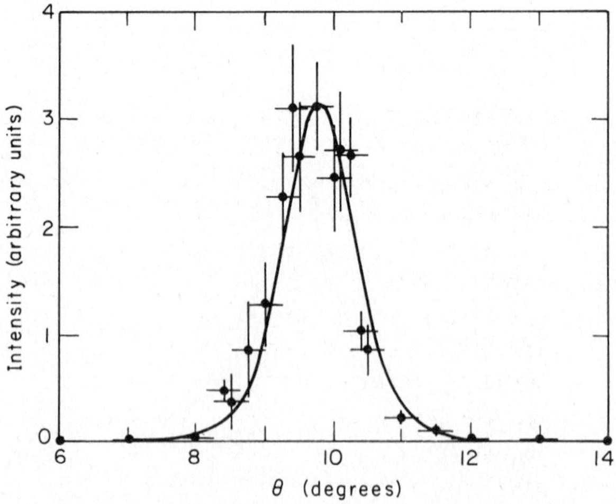

Fig. 16. Anti-Stokes signal versus the angular position of the prism assembly about the $\hat{y}$-axis (fig. 14). θ is the angle between beam 1 and the prism normal.

$\Delta k_{\parallel} = 0$ as expected. Other theoretical predictions for surface CARS were also experimentally verified. For example, the output had a TM polarization characteristic of surface plasmons, and it disappeared with TE excitation. It was also found that the output dependence on the input laser intensities was $I^2(\omega_1)I(\omega_2)$ as predicted.

Aside from the large induced nonlinear polarization and small field penetration depth into the dielectric medium at the interface, surface CARS also has the characteristics of a highly directional output as determined by $(\boldsymbol{k}_a)_{\parallel}$ and a short interaction length. The last aspect results from the high attenuation coefficient $K''_{\parallel}$ of the surface plasmons in the visible. Typically, the attenuation length is $l = 1/K''_{\parallel} \sim 10\,\mu\text{m}$ for surface plasmons at a silver–dielectric interface. The interaction length in surface CARS is therefore only of the order of $10\,\mu\text{m}$. This strong attenuation is the consequence of the large imaginary part in the metal dielectric constant. The absorption of the dielectric medium will not change $K''_{\parallel}$ appreciably as long as its corresponding bulk attenuation length is not much shorter than $10\,\mu\text{m}$. Consequently, the anti-Stokes output will not be seriously affected by the strong absorption of the dielectric medium. This was actually demonstrated experimentally (Chen and de Castro). The short interaction length together with the highly directional anti-Stokes output makes surface CARS particularly suitable for spectroscopic study of materials with strong absorption and luminescence. Since the output comes out from the prism side, the luminescence background is further reduced by the shielding of the metal film separating the prism from the dielectric.

It is interesting to estimate the ultimate sensitivity surface CARS can have for detecting molecular overlayers. As we discussed earlier, the output increases with the input laser intensities as $I^2(\omega_1)I(\omega_2)$. It then seems desirable to focus the exciting beams on the surface as tight as possible even though the signal is also inversely proportional to the excitation area. The maximum laser intensity is, however, limited by the optical damage on the surface. Since the damage usually has a fluence threshold rather than an intensity threshold, the maximum laser intensity can be orders of magnitude higher if picosecond pulses instead of nanosecond pulses are used. Thus, we expect that the ultimate sensitivity of surface CARS can be greatly improved by using picosecond pulse excitation. As an example, we consider surface CARS on a silver–benzene interface. The maximum laser fluence the silver film can withstand is around $25\,\text{mJ/cm}^2$. It is therefore safe to use 10 ps pulses with $10\,\mu\text{J}$/pulse focussed to a spot of $0.15\,\text{mm}^2$. From the experimental results with Q-switched pulses, the output from surface CARS was estimated to be $S \sim 10^{-23}\mathbb{P}^2(\omega_1)\mathbb{P}(\omega_2)T/A^2$ photons/pulse, where the input powers $\mathbb{P}(\omega_1)$ and $\mathbb{P}(\omega_2)$ are in ergs/s, T is the pulsewidth in seconds, and A is the beam overlapping in cm^2. Then, with the 10 ps pulses, the output signal can be

$\sim 10^{11}$ photons/pulse. Reduction of the pulsewidth to 1 ps with the input pulse energy kept unchanged will increase the signal by another two orders of magnitude. We recall that in surface CARS, the signal mainly comes from the nonlinearly induced oscillating dipoles in a surface layer of $\sim \lambda/2\pi$ thick that contains several hundred atomic or molecular layers. If we now have only a single monolayer of molecules on silver, the anti-Stokes output, being coherent, is proportional to the square of the number of molecules radiating, and therefore, should decrease by a factor of 10^4–10^5. Even so, the signal is still as large as 10^6–10^7 photons/pulse and should be easily detected. Thus, we conclude that surface CARS should have an ultimate sensitivity of being able to detect submonolayers of molecules on metals.

6. Concluding Remarks

We have seen in this chapter various different cases of nonlinear wave interaction involving surface polaritons: generation of surface polaritons by mixing of bulk waves and by mixing of bulk and surface waves, generation of bulk waves by mixing of surface polaritons and by mixing of bulk and surface waves, and wave mixing with all input and output waves being surface polaritons. As presented in sect. 2, they can be described by the same general theoretical formalism. Second-order nonlinear optical processes with surface polaritons are strong only if one or both media at the interface lack inversion symmetry. If both media have inversion symmetry, the processes become much weaker. However, they may be more interesting from the surface physics point of view, since in such cases, the nonlinearity may be dominated by the few atomic or molecular layers at the interface. Then, the second-order effects may be used as probes to study the physical properties of these surface layers.

The third-order nonlinear processes involving surface polaritons are generally weaker than the second-order processes. They can, however, become fairly strong when the relevant nonlinear susceptibility undergoes a resonant enhancement. Thus, with picosecond excitation pulses, surface CARS can have a sensitivity of detecting a submonolayer of molecules adsorbed to the surface, and may become a very useful technique for spectroscopic study of adsorbed molecules. The latter is an important problem in surface physics. Electron loss spectroscopy is at present commonly used for measurements of vibrational spectra of adsorbed molecules. It has only a resolution of $\sim 100\,cm^{-1}$, and is not capable of detecting small shifts in the vibrational frequencies due to weak bonding in adsorption. Surface CARS can have a resolution better than $1\,cm^{-1}$, and therefore should yield more valuable information than electron loss spectroscopy. In addition, with picosecond excitation pulses, dynamic behavior or adsorbed molecules can also be studied down to the picosecond time regime. Applications to the detailed studies of adsorption process, molecular reaction, catalytical action, etc. should certainly be considered.

Resonant surface CARS is another interesting area that is yet to be investigated. The additional electronic resonance of the medium seen by either or both input waves should further enhance the output signal, and allow the process to be more selective in specifying the molecules or vibrational modes. This is quite similar to resonant Raman scattering. The resonant surface CARS is particularly attractive in view of the fact that surface CARS is most suited for study of materials with strong absorption and luminescence. CARS is, of course, only a special case of the general four-wave mixing processes. In general, surface four-wave mixing with resonant excitation of materials should also be observable, and can be used for spectroscopic study. For example, two-photon absorption spectra of molecular layers can certainly be obtained by surface four-wave mixing.

Aside from optical mixing, other nonlinear optical effects involving surface polaritons should, in principle, also exist. Stimulated Raman scattering, for example, is one such effect that should be readily observable. In fact, Heritage (1980) has recently reported the observation of stimulated Raman gain from a monolayer of molecules using bulk waves. If surface plasmons are used in the same experiment, the beam intensities at the molecules will be greatly enhanced, and the Raman gain can therefore be orders of magnitude higher. The Raman gain spectroscopy is another potentially useful method for the study of submonolayer of molecules on surfaces.

The Kretschmann geometry for linear excitation of surface plasmons may be most convenient for study of metal–dielectric interfaces, if crystalline metal is not required. For other interfaces, however, the Otto geometry may be more suitable. Technical difficulties still exist in the excitation of surface polaritons, especially when high coupling efficiency is needed, as in the case of a study of nonlinear interaction of surface polaritons. These difficulties can hopefully be overcome if more effort is devoted to the research of surface polaritons.

Acknowledgement

This work was supported by the Division of Materials Sciences, Office of Basic Energy Sciences, U.S. Department of Energy, under contract No. W-7405-ENG-48.

References

Barker, A.S., 1968, Phys. Rev. **165**, 917.
Bloembergen, N., R.K. Chang, S.S. Jha and C.H. Lee, 1968, Phys. Rev. **174**, 813.
Bloembergen, N. and P.S. Pershan, 1962, Phys. Rev. **128**, 606.
Born, M. and E. Wolf, 1965, Principles of Optics, 3rd Ed. (Pergamon Press, New York) p. 61.
Borstel, G. and H.J. Falge, 1978, Appl. Phys. **16**, 211 and references therein.
Chen, C.K. and A.R.B. de Castro, private communications.
Chen, C.K., A.R.B. de Castro and Y.R. Shen, 1979, Opt. Lett. **4**, 393.
Chen, C.K., A.R.B. de Castro and Y.R. Shen, 1979, Phys. Rev. Lett. **43**, 946.
Chen, C.K., A.R.B. de Castro, Y.R. Shen and F. DeMartini, 1979, Phys. Rev. Lett. **43**, 946.
Chu, K.C., C.K. Chen and Y.R. Shen, 1980, Mol. Cryst. Lig. Cryst. **59**, 97.
DeMartini, F., M. Colocci, S.E. Kohn and Y.R. Shen, 1977, Phys. Rev. Lett. **38**, 1223.
DeMartini, F., G. Giuliani, P. Mataloni, E. Palange and Y.R. Shen, 1976, Phys. Rev. Lett. **37**, 440.
DeMartini, F., P. Ristori, E. Santamato and A.C.A. Zammit, 1981, Phys. Rev. **B8**, 3797.
DeMartini, F. and Y.R. Shen, 1976, Phys. Rev. Lett. **36**, 216.
Hartstein, A., E. Burstein, J.J. Brion and R.F. Wallis, 1973, Solid State Commun. **12**, 1083.
Heritage, J.P., 1980, Topical Meeting on Picosecond Phenomena, Cape Cod, Mass., June 18, paper FA 8.
Kretschmann, E., 1971, Z. Phys. **241**, 313.
Lagois, J. and B. Fischer, 1976, Phys. Rev. Lett. **36**, 680.
Levenson, M.D., 1977, Physics Today **30**, 45.
Marschall, N. and B. Fischer, 1972, Phys. Rev. Lett. **28**, 892.
Otto, A., 1968, Z. Phys. **216**, 398.
Simon, H.J., R.E. Benner and J.G. Rako, 1977, Opt. Commun. **23**, 245.
Simon, H.J., D.E. Mitchell and J.G. Watson, 1974, Phys. Rev. Lett. **33**, 1531.
Sommerfeld, A., 1909, Ann. Physik **28**, 665.

CHAPTER 15

Nonlinear Optical Response of Metal Surfaces

J.E. SIPE and G.I. STEGEMAN*

Department of Physics
University of Toronto
Toronto, Ontario M5S 1A7
CANADA

*Present address: Optical Sciences Center, University of Arizona, Tucson, Arizona 85721, U.S.A.

Surface Polaritons
Edited by
V.M. Agranovich and D.L. Mills

Contents

1. Introduction . . . 663
   1.1. Historical review . . . 664
   1.2. The scope of this chapter . . . 666
2. Theory . . . 667
   2.1. Nonlinear response . . . 667
      2.1.1. The hydrodynamic equations . . . 667
      2.1.2. The free electron theory of second harmonic generation . . . 669
      2.1.3. The hydrodynamic theory of second harmonic generation (SHG) . . . 672
      2.1.4. More realistic calculations . . . 675
   2.2. Geometries of interest . . . 676
      2.2.1. Reflection from a free surface . . . 677
      2.2.2. ATR geometries . . . 680
      2.2.3. Surface polariton interactions . . . 684
         2.2.3.1. Contrapropagating incident surface plasmons . . . 686
         2.2.3.2. Second harmonic generation . . . 688
3. Review of experiments . . . 689
   3.1. Reflection experiments . . . 691
   3.2. Attenuated total reflection experiments . . . 696
4. Concluding remarks . . . 698
References . . . 700

1. *Introduction*

The last few years have been witness to a growing interest in nonlinear surface polariton interactions; the mechanisms which lead to these phenomena are rooted in the nonlinear optical response of a metal. Since a reasonable elementary model for a metal is an isotropic free electron gas, which is a highly linear system when excited by electromagnetic waves, the nonlinear mechanisms of interest are amongst the weakest known in nonlinear optics; when compared with the nonlinear mechanisms of a dielectric crystal, they are seen to be of quadrupole and magnetic dipole symmetry. Within the skin depth of the incident electromagnetic field, there are Lorentz forces on free (and to some degree also bound) electrons, as well as interband transitions, both of which lead to nonlinear bulk polarization sources. Further, the breaking of inversion symmetry and the large change in dielectric constant at the metal–vacuum interface introduces strong surface currents within a few Fermi wavelengths of the surface. Because these mechanisms are so very weak, experimental investigations require very high optical power laser sources, and are performed very close to the damage threshold of metal surfaces.

This chapter contains a review of the current state of knowledge concerning the nonlinear optical response of metals. The field has evolved in essentially three stages, the first dealing with an extension of the initial investigations of nonlinear optics to media with inversion symmetry. This period was characterized theoretically by the development of perturbative quantum mechanical calculations which were essentially reduced to an electron gas model for simplicity, and which did not explicitly address the detailed problem of the metal–vacuum interface. The accompanying experiments dealt with the generation of second harmonic radiation obtained on reflection of high power laser radiation from noble metal surfaces. A second stage began with the work of Rudnick and Stern (1971), which started to pose detailed questions about the effect of the metal–vacuum interface on the nonlinear optical response. This dialogue continues as our knowledge of this interface region, and our theoretical attempts to describe it, continue to evolve. The latest rejuvenation of interest was sparked by the application of attenuated total reflection (ATR) techniques to the generation of surface plasmon resonances at a metal–air interface. The resulting second harmonic generation experiments are characterized by coherence length of many wavelengths, and a large reduction in the incident laser power needed for observation.

1.1. Historical Review

The initial interest in the nonlinear optical response of metals was a logical step in the early stages of the evolution of nonlinear optics which followed the invention of the laser. Nonlinear effects were first investigated in media lacking centres of inversion symmetry (quadratic in the fields) and then extended to media with a centre of symmetry (either quadratic or cubic in the fields); a metal is an example of the latter type. In the first two experiments, Brown et al. (1965) and Brown and Parks (1966) reported the generation of second harmonic light on reflection from a silver mirror. The first paper established the existence of the $\boldsymbol{E}(\nabla \cdot \boldsymbol{E})$ nonlinear surface term, which has a large contribution at the boundary due to the discontinuity there of the metal dielectric constant ϵ. With a refinement of experimental technique, the presence of the bulk term $\boldsymbol{E} \times \partial \boldsymbol{H}/\partial t$, which originates in the Lorentz force on the electrons, was confirmed. At the same time Jha (1965, 1966) showed theoretically the importance of the surface by reducing his quantum mechanical formulation of the nonlinear optical response to the limit of a free electron gas. The general laws of nonlinear optical reflection from surfaces of media characterized by arbitrary symmetry were reviewed in a classic book by Bloembergen (1965), and shortly thereafter Bloembergen and Shen (1966a, b) pointed out the importance of the bound electrons (specifically d electrons for silver) in the ion cores near the surface (i.e., in addition to the conduction electrons first considered by Jha (1965)). It was subsequently demonstrated by Jha and Warke (1967) that, except near a resonance associated with interband transitions, the contributions of the bound electrons to the bulk nonlinear terms could be included in terms of the dielectric constants $\epsilon(\omega)$ and $\epsilon(2\omega)$. Intermingled with the ongoing theoretical discussions were reports of further experiments on gold, silver and their alloys by Bloembergen et al. (1966) and Sonnenberg and Heffner (1968). They indeed verified that both the conduction and bound electron effects were important, and obtained improved agreement between experiment and theory. With the exception of a paper by Krivoshchekov and Stroganov (1969) on the variation of the harmonic signal with silver film thickness, this stage of work ended with a classic paper by Bloembergen et al. (1968). Those authors derived fairly general expressions for second harmonic generation in bulk media with inversion symmetry, and applied them with reasonable success to a range of data on metals and semiconductors. At this point, agreement between experiment and theory within a factor of two was being claimed.

A renewal of interest in the nonlinear optical response of a metal was initiated by the work of Rudnick and Stern (1971, 1974). They posed deeper questions about the nature of the surface terms, primarily in the context of a free electron gas. From their analysis they concluded that the surface currents effectively radiate from outside the metal surface and are not shielded by the electron gas with large negative values of ϵ. They also discussed the origin of the surface currents on a more microscopic level than considered previously, and introduced two phenomenological parameters a and b to describe the effects of the details of the surface region on the nonlinear surface currents perpendicular and parallel to the surface respectively. A year before the Rudnick and Stern work, Krivoshchekov and Stroganov (1970) reported measurements of second harmonic generation from sodium and potassium surfaces, and these results provided a stimulus for further theoretical discussions by Wang et al. (1973). They formally included terms in the gradient of the electron density profile in the surface region; in their actual numerical calculations they neglected this term, along with other terms that were divergent or ambiguous, and claimed improved agreement between theory and experiment for the alkali metals. The early quantum mechanical calculations of Jha and coworkers were extended by Bower (1976) to include the effects of interband transitions near their resonance, as well as the effects of surface states. His treatment in its present form is very general and hence very difficult to apply to specific cases. This period of probing more details of the surface region continues today, but more in the context of surface plasmon interactions, to which we now turn.

The present interest in the nonlinear optical response of metals was initiated by the application of total attenuated reflection techniques to the observation of nonlinear surface polariton effects by Simon et al. (1974). When light is incident through a prism onto a prism–metal film–air sample geometry, surface polariton-like resonances are created at the metal–air interface at a specific angle of incidence. The resulting electric fields are enhanced by a factor of 10–100 over the incident fields, and coherence and decay lengths of tens of microns to many centimetres (depending on the frequency) are available for utilization in nonlinear interactions. Using this approach, Simon and coworkers (1974) measured an enhancement of a factor of 30 in the generation of second harmonic radiation on reflection from the prism–silver interface. They subsequently predicted (1975) that large enhancements could also be obtained for alkali metal films and evaluated the film thickness required for optimum results. Chen and Burstein (1977) discussed various experimental geometries for optimizing enhanced phenomena involving surface plasmons. In 1977 Simon et al. reported further experiments on prism–silver film–quartz samples in which nonlinear sources existed also in the dielectric medium quartz. Their results exhibited interference effects between nonlinear sources in the metal and the dielectric, with the quartz's nonlinearities providing a calibration of the metal's nonlinear mechanisms. Sipe et al. (1980a, b) used a hydrodynamic model of an electron gas to analyze these latest experiments with good success; this treatment led to analytical expressions for the Rudnick and Stern (1971) *a* and *b* parameters in terms of an effective plasmon frequency in the surface region. They showed that the Simon et al. (1977) data was insufficient for determining uniquely the value of *a*, but they did suggest extensions to this experiment, as well as a surface reflection experiment which would allow *a* to be measured with some degree of accuracy. In this same time period, the nonlinear interactions between freely propagating surface polariton fields were analyzed by Mills (1977), Fukui and Stegeman (1978a), Fukui et al. (1978b, c, 1979). The Rudnick and Stern nonlinear surface mechanisms were used along with the standard bulk Lorentz force term in the analyses. Mills (1977) considered a case in which long coherence and propagation ($\sim$ cm) lengths were possible but where the frequencies were much less than the plasma frequency, and found that the cross section for second harmonic generation was too small to measure. On the other hand, Fukui and Stegeman (1978a) considered a case in which the frequencies were much closer to the plasma frequency, but where the coherence and attentuation lengths were typically tens of microns: they predicted measurable cross sections. Fukui et al. (1978b) also proposed a guided wave geometry consisting of a thin glass film deposited on a silver surface. They showed that long coherence and decay lengths could be obtained by utilizing various orders of guided modes which resulted in cross sections enhanced by several orders of magnitude. Fukui and coworkers (1978c, d) also analyzed the mixing of oppositely propagating surface plasmons, which leads to the generation of sum frequency fields radiated into the air. Their results indicate that the relative contribution of the various nonlinear mechanisms changes with the relative frequencies of the incident plasmon fields. At present, experimental realization of these proposed geometries is lacking.

1.2. The Scope of this Chapter

Both the theoretical and experimental investigations of the nonlinear optical response of metals will be reviewed in this chapter, with the purpose of setting the stage for future experiments involving nonlinear surface polariton interactions. We shall concentrate solely on those developments which specifically relate to the nonlinearities in the metal. Thus we a priori exclude problems such as surface polariton-enhanced light scattering from monolayers and films (see for example Chen, Chen and Burstein 1976), the nonlinear mixing of surface phonon–plasmon polaritons (Bonsall and Maradudin 1978) and surface plasmon wave phenomena enhanced in the adjacent medium (Chen et al. 1980a, b and DeMartini et al. 1981). A theory for the nonlinear response based on a hydrodynamic model of an electron gas is discussed in sect. 2.1. The nonlinear mechanisms can be studied in various experimental geometries such as reflection from a free surface (sect. 2.2.1), an ATR geometry (sect. 2.2.2) and with freely propagating surface plasmons (sect. 2.2.3). Wherever available, we shall present numerical estimates for the various cross sections involved. In sect. 3, the experiments reported to date will be reviewed with emphasis on how they add to our understanding of nonlinear effects in metals. To date only two experimental geometries have been exploited, namely reflection from free surfaces (sect. 3.1) and ATR methods (sect. 3.2). In sect. 4 we shall summarize the present status of the field and attempt to identify directions which may prove fruitful in the future.

2. Theory

2.1. Nonlinear Response

As mentioned in the introduction, we shall limit ourselves here to a review of treatments of second harmonic generation (SHG) in a bounded electron gas. The simplest microscopic theory which can meaningfully be applied to discuss and calculate such SHG is the so-called "hydrodynamic theory". This theory has recently been used to give a qualitative description of the dynamics of electrons near a surface by Bennett (1970), Rudnick and Stern (1974) and references cited therein, Eguiluz and Quinn (1975, 1976a, b), Sipe (1979) and Sipe et al. (1980a, b); it is described in detail, as applied to linear phenomena, by Eguiluz and Quinn (1976b) and Barton (1979), and its limitations and connections with more accurate microscopic models are discussed by, e.g., Griffin and Kranz (1977) and Eguiluz (1979).

We begin by presenting the basic equations of the hydrodynamic theory and their expansion to second order; this is followed by a discussion of the ambiguities which appear if the simpler, "free electron" theory is used to describe SHG. We then return to the hydrodynamic theory and review its application to SHG, and close with a brief description of more sophisticated calculations.

2.1.1. The Hydrodynamic Equations

In the hydrodynamic theory of an electron gas with a static, positive charge background, the total charge and current densities are:

$$\rho(\boldsymbol{r}, t) = \rho_+(\boldsymbol{r}) + \rho_-(\boldsymbol{r}, t),$$

$$\boldsymbol{j}(\boldsymbol{r}, t) = -en(\boldsymbol{r}, t)\boldsymbol{v}(\boldsymbol{r}, t), \tag{1}$$

where the positive and negative charge densities are given by:

$$\rho_+(\boldsymbol{r}) = en_+(\boldsymbol{r}),$$
$$\rho_-(\boldsymbol{r}, t) = -en(\boldsymbol{r}, t), \tag{2}$$

and $-e$, $n_+(\boldsymbol{r})$ and $n(\boldsymbol{r}, t)$ are respectively the charge on an electron, the positive ion number density, and the electron number density; $\boldsymbol{v}(\boldsymbol{r}, t)$ is the electron velocity field. The densities (1) satisfy the equation of continuity,

$$\nabla \cdot j + \dot{\rho} = 0, \tag{3}$$

and the equation of motion for $\boldsymbol{v}(\boldsymbol{r}, t)$ is Euler's equation for the electron fluid,

$$mn\left[\frac{\partial \boldsymbol{v}}{\partial t} + (\boldsymbol{v} \cdot \nabla)\boldsymbol{v}\right] = -enE - \frac{en}{c}\boldsymbol{v} \times \boldsymbol{B} - \nabla p, \tag{4}$$

where m is the electron mass and $p(\boldsymbol{r}, t)$ is the "quantum pressure". In the Thomas–Fermi theory (cf. e.g. Landau and Lifshitz 1965),

$$p(\boldsymbol{r}, t) = \xi[n(\boldsymbol{r}, t)]^{5/3}, \tag{5}$$

where

$$\xi = (3\pi^2)^{2/3}\hbar^2/5m. \tag{6}$$

A different value results if ξ is set so that the plasmon dispersion relation which follows from the hydrodynamic theory agrees with the long wavelength limit of that dispersion relation as predicted by the random phase approximation (Eguiluz 1979, Barton 1979). The electric and magnetic fields $\boldsymbol{E}$ and $\boldsymbol{B}$ appearing in eq. (4) satisfy the Maxwell equations which, when a polarization potential $\boldsymbol{P}(\boldsymbol{r}, t)$ is introduced by virtue of eq. (3),

$$j = \dot{P}, \qquad \rho = -\nabla \cdot \boldsymbol{P}, \tag{7}$$

takes a form familiar from dielectric theory,

$$\nabla \cdot \boldsymbol{E} = -4\pi\nabla \cdot \boldsymbol{P}, \qquad \nabla \cdot \boldsymbol{B} = 0,$$
$$c\nabla \times \boldsymbol{B} - \dot{\boldsymbol{E}} = 4\pi\dot{\boldsymbol{P}}, \qquad c\nabla \times \boldsymbol{E} + \dot{\boldsymbol{B}} = 0. \tag{8}$$

To calculate the electron response to second order, we expand all fields in the usual way (Bloembergen et al. 1968),

$$n(\boldsymbol{r}, t) = n_0(\boldsymbol{r}) + n_1(\boldsymbol{r}, t) + n_2(\boldsymbol{r}, t) + \cdots,$$
$$\boldsymbol{E}(\boldsymbol{r}, t) = \boldsymbol{E}_0(\boldsymbol{r}) + \boldsymbol{E}_1(\boldsymbol{r}, t) + \boldsymbol{E}_2(\boldsymbol{r}, t) + \cdots, \tag{9}$$
$$\boldsymbol{v}(\boldsymbol{r}, t) = \boldsymbol{v}_1(\boldsymbol{r}, t) + \boldsymbol{v}_2(\boldsymbol{r}, t) + \cdots,$$
$$\boldsymbol{B}(\boldsymbol{r}, t) = \boldsymbol{B}_1(\boldsymbol{r}, t) + \boldsymbol{B}_2(\boldsymbol{r}, t) + \cdots.$$

We find that the zeroth order equation resulting from eqs. (4, 7, 8) gives the usual Thomas–Fermi equations for the equilibrium electron density $n_0(\boldsymbol{r})$, when eqs. (5, 6) are used; the first order equation which results may be written in the form

$$\ddot{\boldsymbol{P}}_1(\boldsymbol{r}, t) - \overleftrightarrow{\boldsymbol{L}}(\boldsymbol{r}) \cdot \boldsymbol{P}_1(\boldsymbol{r}, t) = \frac{e^2 n_0(\boldsymbol{r})}{m} \boldsymbol{E}_1(\boldsymbol{r}, t), \tag{10}$$

(cf. Eguiluz and Quinn 1976b, Sipe 1979, Sipe et al. 1980b), where

$$\overleftrightarrow{\boldsymbol{L}} = -(5/9m)\xi n_0^{-1/3}(\nabla n_0)\nabla + (5/3m)\xi n_0^{2/3}\nabla\nabla, \tag{11}$$

and the second order equation reduces to:

$$\ddot{\boldsymbol{P}}_2(\boldsymbol{r}, t) - \overleftrightarrow{\boldsymbol{L}}(\boldsymbol{r}) \cdot \boldsymbol{P}_2(\boldsymbol{r}, t) = \frac{e^2 n_0(\boldsymbol{r})}{m} \boldsymbol{E}_2(\boldsymbol{r}, t) + \boldsymbol{S}_{\mathrm{f}}(\boldsymbol{r}, t) + \boldsymbol{S}_{\mathrm{p}}(\boldsymbol{r}, t), \tag{12}$$

(Sipe et al. 1980b), where the source terms $\boldsymbol{S}_{\mathrm{f}}(\boldsymbol{r}, t)$ and $\boldsymbol{S}_{\mathrm{p}}(\boldsymbol{r}, t)$ are given by

$$\boldsymbol{S}_{\mathrm{f}} = en_0(\boldsymbol{v}_1 \cdot \nabla)\boldsymbol{v}_1 - e\boldsymbol{v}_1\dot{n}_1 + \frac{e^2}{m} n_1\boldsymbol{E}_1 + \frac{e^2 n_0}{mc} \boldsymbol{v}_1 \times \boldsymbol{B}_1, \tag{13}$$

and

$$\boldsymbol{S}_{\mathrm{p}} = (5e/9m)\xi\nabla(n_1^2 n_0^{-1/3}). \tag{14}$$

The first two terms in eq. (13) are the purely convective sources that always appear in an expansion solution of Euler's equation; the second two terms are cross terms between matter and electromagnetic fields specific to the Lorentz forces appearing on the right hand side of eq. (4); the term (14) is the new second order source due to the presence of the quantum pressure. Of course, the terms appearing in eq. (13) depend implicitly on ξ through the dependence of the linear fields on the value of that parameter.

2.1.2. The Free Electron Theory of Second Harmonic Generation (*SHG*)

We now consider the form these equations and their solutions take in the limit $\xi = 0$, the so-called "free electron" limit in which the quantum pressure is neglected. Writing the first order fields and the second harmonic component of the second order fields as

$$\begin{aligned} f_1(\boldsymbol{r}, t) &= f_1(\boldsymbol{r})\, \mathrm{e}^{-\mathrm{i}\omega t} + \mathrm{cc}, \\ &= 2\mathrm{Re}[f_1(\boldsymbol{r})\, \mathrm{e}^{-\mathrm{i}\omega t}], \end{aligned} \tag{15}$$

and

$$\begin{aligned} f_2(\boldsymbol{r}, t) &= f_2(\boldsymbol{r})\, \mathrm{e}^{-\mathrm{i}\Omega t} + \mathrm{cc}, \\ &= 2\mathrm{Re}[f_2(\boldsymbol{r})\, \mathrm{e}^{-\mathrm{i}\Omega t}], \end{aligned} \tag{16}$$

respectively, where $\Omega = 2\omega$, and dealing first with the first order fields, we find that eq. (10) reduces to

$$\boldsymbol{P}_1(\boldsymbol{r}) = \frac{-e^2 n_0(z)}{m\omega^2}\boldsymbol{E}_1(\boldsymbol{r}). \tag{17}$$

Defining

$$\boldsymbol{D}_1(\boldsymbol{r}) \equiv \boldsymbol{E}_1(\boldsymbol{r}) + 4\pi\boldsymbol{P}_1(\boldsymbol{r}) \equiv \epsilon(\boldsymbol{r};\omega)\boldsymbol{E}_1(\boldsymbol{r}) \tag{18}$$

in the usual way, we find

$$\epsilon(\boldsymbol{r};\omega) = 1 - 4\pi e^2 n_0(z)/m\omega^2, \tag{19}$$

where we have here assumed a jellium model for the positive ions, leading to translational symmetry in the plane of the surface; $\hat{z}$ is set normal to the surface. From eq. (19) we see that, if $n_0(z)$ dropped smoothly from its bulk value to zero, the required continuity of the normal component of $\boldsymbol{D}_1(\boldsymbol{r})$ would lead to a divergence in $\boldsymbol{E}_1(\boldsymbol{r})$ at the plane in which $\epsilon(\boldsymbol{r};\omega)$ vanished. This divergence may be formally removed by including a collision term in Euler's equation (4); if the collisions are characterized by a relaxation time τ, one finds the Drude result: eq. (19) with ω^2 replaced by $\omega(\omega + \mathrm{i}/\tau)$ (cf., e.g., Ashcroft and Mermin 1976). However, the large oscillation in $\boldsymbol{E}_1(\boldsymbol{r})$ at the surface that would then result is completely unphysical, as discussed by Eguiluz and Quinn (1976b). Thus the free electron model is only physically well-defined if the equilibrium density $n_0(z)$ drops from its bulk value to zero as a step function at the surface.

We now turn to eq. (12) (with $\xi = 0$) for the second harmonic polarization potential; using the first order result (17) and

$$\boldsymbol{v}_1 = -(\mathrm{i}e/m\omega)\boldsymbol{E}_1, \tag{20}$$

which follows from eqs. (1, 7, 17), we find

$$\boldsymbol{P}_2(\boldsymbol{r}) = \frac{-e^2 n_0(z)}{m\Omega^2}\boldsymbol{E}_2(\boldsymbol{r}) + \boldsymbol{P}_{\mathrm{NL}}(\boldsymbol{r}), \tag{21}$$

where

$$\boldsymbol{P}_{\mathrm{NL}} = \frac{\mathrm{i}e^3 n_0}{4m^2\omega^3 c}(\boldsymbol{E}_1 \times \boldsymbol{B}_1) + \frac{e^3 n_0}{4m^2\omega^4}(\boldsymbol{E}_1 \cdot \nabla)\boldsymbol{E}_1 - \frac{e^2}{2m\omega^2} n_1 \boldsymbol{E}_1. \tag{22}$$

If for the moment we neglect the $(\boldsymbol{E}_1 \cdot \nabla)\boldsymbol{E}_1$ term in eq. (22), and use the equations

$$n_1 = e^{-1}\nabla \cdot \boldsymbol{P}_1, \tag{23}$$

(cf. eq. 7) and

$$\nabla \cdot \boldsymbol{P}_1 = -(1/4\pi)\nabla \cdot \boldsymbol{E}_1, \tag{24}$$

(cf. eq. 8), we find

$$\boldsymbol{P}_{\mathrm{NL}} = \alpha(\boldsymbol{E}_1 \times \boldsymbol{B}_1) + \beta \boldsymbol{E}_1(\nabla \cdot \boldsymbol{E}_1), \tag{25}$$

where

$$\alpha = \mathrm{i}e^3 n_0/4m^2\omega^3 c,$$
$$\beta = e/8\pi m\omega^2. \tag{26}$$

Equation (25) was derived by Jha (1965). The α and β terms have sometimes been referred to as the "magnetic dipole" and "electric quadrupole" sources respectively, apparently by analogy with corresponding terms in the nonlinear polarization of dielectrics (Adler 1964); of course, there are no bound charges in the model under consideration.

The $(\boldsymbol{E}_1 \cdot \nabla)\boldsymbol{E}_1$ term in eq. (22) strictly vanishes in reflection SHG experiments (see sect. 2.2), but can be non-zero in more complicated geometries; including it, and using eq. (8), we find

$$\boldsymbol{P}_{\mathrm{NL}} = \gamma \nabla(\boldsymbol{E}_1 \cdot \boldsymbol{E}_1) + \beta \boldsymbol{E}_1(\nabla \cdot \boldsymbol{E}_1), \tag{27}$$

instead of eq. (25), where

$$\gamma = e^3 n_0/8m^2\omega^4. \tag{28}$$

We digress here to mention a point that has caused some confusion in the literature (cf. Cheng and Miller 1964, Jha 1965, Bloembergen et al. 1968, Wang et al. 1973). Using eqs. (17, 24), eq. (27) may be written in the form

$$\boldsymbol{P}_{\mathrm{NL}} = \gamma \nabla(\boldsymbol{E}_1 \cdot \boldsymbol{E}_1) + \frac{4\pi e^2 n_0}{m\omega^2} \beta \boldsymbol{E}_1(\nabla \cdot \boldsymbol{E}_1) + \frac{4\pi e^2}{m\omega^2} \beta \boldsymbol{E}_1(\boldsymbol{E}_1 \cdot \nabla n_0), \tag{29}$$

and, if the ∇n_0 term is incorrectly neglected, eq. (29) reduces to the form of eq. (27) but with a much larger term proportional to $\boldsymbol{E}_1(\nabla \cdot \boldsymbol{E})$ at optical frequencies. For a detailed discussion of this matter, we refer to Rudnick and Stern (1971).

Returning now to eqs. (17) and (27), we consider the use of the free electron model to calculate second harmonic generation. In the bulk the electric field is divergenceless, and only the first term in eq. (27) contributes; near the surface, both terms are important. We look first at the component of $\boldsymbol{P}_{\mathrm{NL}}$ parallel to the surface, $\boldsymbol{P}_{\mathrm{NL}}^{\parallel}$; since $\boldsymbol{E}_1^{\parallel}$ is continuous across the surface, using the assumed translational symmetry in the plane of the surface we find

$$\boldsymbol{P}_{\mathrm{NL}}^{\parallel} \simeq \beta \boldsymbol{E}_1^{\parallel}(\partial E_1^z/\partial z), \tag{30}$$

which gives a Dirac delta function at the surface in the limit of a step function equilibrium density profile. Thus there is an effective current sheet radiating at 2ω at the surface, and following back the derivation of eq. (30) it is clear the $\boldsymbol{P}_{\mathrm{NL}}^{\parallel}$ results from the accumulated charge at the surface, signalled by the rapid variation in E_1^z, being driven by the component of the electric field parallel to the surface.

The magnitude of the current sheet due to $\boldsymbol{P}_{\mathrm{NL}}^{\parallel}$ may be calculated in the free electron theory, since eq. (30) may be integrated across the surface. However, for the normal component of $\boldsymbol{P}_{\mathrm{NL}}$ the situation is different. From eq. (27) we find

$$P_{\mathrm{NL}}^{z} = \gamma \frac{\partial}{\partial z}[(E_{1}^{z})^{2}] + \beta E_{1}^{z} \frac{\partial E_{1}^{z}}{\partial z}, \tag{31}$$

which is ambiguous, since both γ and E_{1}^{z} change discontinuously across the surface. Obviously, P_{NL}^{z}, which appears in part because of the rapid variation of the normal field and in part because the response of electrons near the surface need not display inversion symmetry (cf. Rudnick and Stern 1971, Sipe et al. 1980a, b, sect. 2.1.3), is more subtle in nature than $\boldsymbol{P}_{\mathrm{NL}}^{\parallel}$. We note that even if the integral of eq. (31) could be performed, it is not clear whether the current sheet should be placed just inside or just outside the surface. This point has been discussed by Rudnick and Stern (1971, 1974) and Sipe et al. (1980b), and we shall return to it in sect. 2.1.3; it is another ambiguity in the free electron model.

In summary, the free electron theory only leads to physically meaningful results for the first order fields if a step function profile is adopted for the equilibrium electron density; in that limit, however, well-defined predictions for the second harmonic generation cannot be made. Of course, one would certainly expect the free electron theory to be less than accurate in the surface region where the actual fields vary over distances on the order of a few Fermi wavelengths; the point we wish to stress here is that the theory is in a sense somewhat worse: it is inherently ambiguous.

However, for lack of a better theory, many of the researchers cited above have tried to use the free electron theory to describe SHG. This has led to a number of essentially arbitrary "prescriptions" for determining the magnitude and placement of the second harmonic current sheet; in some instances, P_{NL}^{z} has been simply neglected (Wang et al. 1974, Simon et al. 1975). (For a comment on the significance of this, see the discussion following eq. 50). In an important paper, Rudnick and Stern (1971) argued from physical grounds that the second harmonic current sheet should be placed outside the bulk metal, and that its magnitude should be specified in terms of parameters a and b of order unity. These conclusions are borne out to some extent by the application of the hydrodynamic (ξ finite) theory of the electron gas to the calculation of SHG, which we now present following Sipe et al. (1980b).

2.1.3. *The Hydrodynamic Theory of Second Harmonic Generation* (SHG)

Some recent calculations applying the hydrodynamic theory have used the full microscopic equations throughout the medium (Eguiluz and Quinn 1975, 1976a, b). However, as discussed by Sipe (1979), it is permissible to neglect $\overleftrightarrow{L}(\boldsymbol{r})$ in eqs. (10, 12) deep within the metal, since it leads only to corrections of order $(\lambda_F/\lambda)^2 \ll 1$, where λ_F is the Fermi wavelength and λ is the wavelength of light *in vacuo*. Following Sipe et al. (1980b), we place the $z=0$ plane so that in the "bulk" region, $-\infty < z \leq 0$ (the vector $\hat{z}$ is taken to point towards the vacuum), $\overleftrightarrow{L}(\boldsymbol{r})$ may be neglected. The equilibrium electron density then drops to ~ 0 at $z = l$, where $l \sim \lambda_F$; the region $0 < z \leq l$, where $\overleftrightarrow{L}(\boldsymbol{r})$ may not be neglected, is the "selvedge" (Sipe 1979, 1980). Since $l \ll \lambda$, a number of simplifications are possible in the analysis (Sipe et al. 1980b); the approach in fact justifies the concept of a "second harmonic current sheet", and shows how its magnitude should be calculated. We give only the essential results here.

In the bulk the free electron result is found for the second order fields,

$$\boldsymbol{P}_2(\boldsymbol{r}) = \chi(\Omega)\boldsymbol{E}_2(\boldsymbol{r}) + \frac{e^3\bar{n}_0}{8m^2\omega^4}\nabla(\boldsymbol{E}_1(\boldsymbol{r})\cdot\boldsymbol{E}_1(\boldsymbol{r})), \tag{32}$$

(cf. eqs. 21, 27 and 28), where $\Omega = 2\omega$,

$$\begin{aligned}\chi(\omega) &= -e^2\bar{n}_0/m\omega^2,\\ \epsilon(\omega) &\equiv 1 + 4\pi\chi(\omega)\\ &= 1 - 4\pi e^2\bar{n}_0/m\omega^2 \equiv 1 - \omega_p^2/\omega^2,\end{aligned} \tag{33}$$

where $\bar{n}_0$ is the bulk equilibrium electron density and ω_p is the bulk plasma frequency. More complicated equations hold in the selvedge, but the crucial quantity is the integral of the current density over the selvedge,

$$-\mathrm{i}\Omega\boldsymbol{Q}\exp(\mathrm{i}\boldsymbol{K}_\parallel\cdot\boldsymbol{r}) = -\mathrm{i}\Omega\int_0^l \boldsymbol{P}_2(z)\exp(\mathrm{i}\boldsymbol{K}_\parallel\cdot\boldsymbol{r})\,\mathrm{d}z, \tag{34}$$

which, since $l/\lambda \ll 1$, radiates essentially as a current sheet. Here we have written

$$\begin{aligned}f_1(\boldsymbol{r}) &= f_1(z)\exp(\mathrm{i}\boldsymbol{k}_\parallel\cdot\boldsymbol{r}),\\ f_2(\boldsymbol{r}) &= f_2(z)\exp(\mathrm{i}\boldsymbol{K}_\parallel\cdot\boldsymbol{r}),\end{aligned} \tag{35}$$

for the first and second order fields respectively, where $\boldsymbol{k}_\parallel = (k_x, k_y, 0)$ is the wavevector of the linear fields in the plane of the surface, and $\boldsymbol{K}_\parallel = 2\boldsymbol{k}_\parallel$.

It is clear that, in this development, the current sheet is to be placed *outside* the bulk metal; all the shielding of the currents in the selvedge is taken into account in the equations of motion for the current in the selvedge. Restricting themselves to p-polarized light,

$$\boldsymbol{E}_1(z) = \hat{z}E_1^z(z) + \hat{k}_\parallel E_1^k(z), \tag{36}$$

etc., where $\hat{k}_\parallel = \boldsymbol{k}_\parallel/|\boldsymbol{k}_\parallel|$, Sipe et al. (1980b) found

$$Q^k = \frac{e}{2m\omega^2}\left[\frac{\epsilon(\omega)-1}{4\pi}\right]E_1^k(z=0^-)E_1^z(z=0^-). \tag{37}$$

This is precisely the result that follows from eq. (30) for the free electron theory; the agreement is expected, since the net induced surface charge, indicated by the difference between the normal components of $\boldsymbol{E}_1$ inside and outside the metal, is the same in both theories. Concerning the normal component of the selvedge current, Sipe et al. (1980b) did not solve for $P_2^z(z)$, but rather defined an "effective plasma frequency" of the second harmonic currents in the selvedge, ω_0, in terms of which they found

$$Q^z = -\frac{e}{m(\omega_0^2-\Omega^2)}\left[\frac{\epsilon(\omega)-1}{4\pi}\right]\left[\frac{\epsilon(\omega)+3}{2}\right][E_1^z(z=0^-)]^2. \tag{38}$$

Q^z contains information on the details of the motion of electrons normal to the surface, details which are undefined in the free electron theory (cf. discussion after eq. (31)) and which, even in the hydrodynamic theory, would require the solution of a complicated set of equations for their elucidation (Sipe et al. 1980b).

At this point we compare the results of Sipe et al. (1980b) with those of Rudnick and Stern (1971), who estimated $\boldsymbol{Q}$ from phenomenological arguments. They found, in the notation used here, that the components should be given by

$$Q^k = (e^3\bar{n}_0 b/2m^2\omega^4)\, E_1^k(z=0^-)E_1^z(z=0^-),$$

$$Q^z = (e\omega_p^2 a/16\pi m\omega^4)\,[E_1^z(z=0^-)]^2, \tag{39}$$

where a and b are of order unity. The two sets of coefficients agree if

$$b = -1,$$

$$a = -2(\omega_p^2-\Omega^2)/(\omega_0^2-\Omega^2), \tag{40}$$

and thus, if $\omega_0 \gtrsim \omega_p$ and $\Omega \ll \omega_p$, Sipe et al. (1980b) predict both an a and b of order unity. But if $\omega_0 \simeq \Omega \equiv 2\omega$, eq. (38) can lead to a very large $|a|$; since the phenomenological arguments of Rudnick and Stern (1971) employed the zero frequency response of the electron gas, they found no such resonance. Sipe et al. (1980b) stressed that the resonance denominator in eq. (38) should not be taken too seriously, since ω_0 may depend on Ω, and the hydrodynamic model itself may give an oversimplified description of the electron dynamics near the surface. Nonetheless, it is not unreasonable to expect a resonance in Q^z associated with an "effective plasma frequency" on general physical grounds, since the part of the second harmonic currents responsible for Q^z involve essentially longitudinal oscillations of charge density perpendicular to the surface; Q^k on the other hand, which results simply because the charge density accumulated at the surface is driven in second order parallel to the surface, should be free from such a resonance. The damping of this longitudinal oscillation might of course be very large, or it might occur sufficiently near ω_p so that, at optical and near UV frequencies where the theory developed here is valid, it could not be observed. However, these results indicate that further theoretical work on the dispersion of a, within more realistic models of the electron dynamics, would be worthwhile.

2.1.4. *More Realistic Calculations*

There are embarrassingly few calculations of the SHG of a bounded electron gas using descriptions of the electron dynamics more sophisticated than the hydrodynamic theory. Jha (1966) used time-dependent Hartree theory to calculate the SHG of a metal in the single-band approximation, and Jha and Warke (1967) extended the calculation to include interband contributions. However, in these calculations the response functions were evaluated in the long wavelenght limit; as pointed out by Rudnick and Stern (1971), this is insufficient because of the variation of the normal electric field at the surface over a few Fermi wavelengths. Using time-dependent Hartree theory, those authors calculated the SHG at a surface by using a current sheet in an infinite medium to mimic a surface; this approach includes the effect of the rapid variation of the normal electric field but, as they pointed out, does not include effects of the breaking of inversion symmetry at the surface on the dynamics of the electrons.

Bower (1976) returned to Jha and Warke's (1967) formalism, and included the large momentum components in the momentum–space Fourier expansion of the electromagnetic field, to correctly treat the rapid variation of the normal electric field. He found that the SHG was extremely sensitive to the details of the electronic structure at the surface, but did not calculate response parameters (such as a and b) for any model of the surface.

It seems clear that some model quantum mechanical calculations of the response parameters a and b would be a useful contribution to the field. Actual model calculations could of course also be made within the hydrodynamic theory, eliminating the need for the *ansatz* employed by Sipe et al. (1980b) in defining ω_0. Such calculations would be interesting, but because the hydrodynamic theory is a long wavelength approximation (cf. Griffin and Kranz 1977, Barton 1979), and because essentially all Fourier components up to the Fermi wavevector are important in determining the real charge density structure near the surface, it would not be clear what level of confidence should be assigned to hydrodynamic theory calculations of a. Quantum mechanical calculations, although more difficult and by necessity restricted to simple models, would include potentially significant processes, such as Landau damping (Eguiluz et al. 1976b), which are not contained in the hydrodynamic theory.

A simplification which would ease the computation difficulty is that the same splitting of the problem into "bulk" and "selvedge" parts employed by Sipe et al. (1980b) could also be applied to the quantum mechanical calculations; this has been discussed for the problem of linear response by Sipe (1980).

2.2. Geometries of Interest

We now turn to the calculation of frequency sum generation for geometries of interest. Calculations made in connection with early experiments will be mentioned in sect. 3; here we discuss a set of calculations made by Mills (1977), Fukui et al. (1978a, b, c, 1979) and Sipe et al. (1980a, b) in which bulk and surface frequency sum sources were included, the surface sources specified by parameters a and b. Sipe et al. (1980a, b), who considered SHG, took the bulk source polarization to be

$$\boldsymbol{P}_{\rm NL}(\boldsymbol{r}) = \gamma\boldsymbol{\nabla}(\boldsymbol{E}_1(\boldsymbol{r})\cdot\boldsymbol{E}_1(\boldsymbol{r})), \tag{41}$$

while for a metal–vacuum surface located at $z = z_0$ they took an effective current density

$$\boldsymbol{J}_{\rm NL}(\boldsymbol{r}) = -\,\mathrm{i}(2\omega)\boldsymbol{Q}\exp(\mathrm{i}\boldsymbol{K}_\parallel\cdot\boldsymbol{r})\delta(z - z_0^\pm), \tag{42}$$

with

$$\begin{aligned} Q^k &= \pm 4\gamma b E_1^z(z_0^\mp)E_1^k(z_0^\mp),\\ Q^z &= \pm 2\gamma a[E_1^z(z_0^\mp)]^2. \end{aligned} \tag{43}$$

The signs in eqs. (41)–(43) refer to the orientation of the surface: if $\hat{z}$ points from metal to vacuum, as in the derivation outlined in sect. 2.1.3, the upper sign was used, otherwise the lower sign was taken. This guaranteed that the current sheet was placed outside the "bulk metal" (see comment following eq. 35), and corrected eqs. (43) when the geometry for which eqs. (39) were derived was inverted. A value

$$\gamma = -\frac{e}{8m\omega^2}\left[\frac{\epsilon(\omega)-1}{4\pi}\right] \tag{44}$$

was adopted; if $\epsilon(\omega)$ is given by eq. (33), eqs. (41) and (43) reduce to the corresponding eqs. (32) and (39). However, to obtain the correct values for the Fresnel coefficients, Sipe et al. (1980a, b) used values of $\epsilon(\omega)$ and $\epsilon(2\omega)$, for silver metal at $\lambda = 1.06\,\mu$m and $0.53\,\mu$m respectively, obtained from experiment. Now, we note that Jha and Warke's (1967) calculations indicate that, for the bulk source polarization, nonresonant interband transitions are taken into account by using eq. (41) with

$$\gamma = -\frac{e}{2m\omega^2}\left[\frac{\epsilon(2\omega)-1}{4\pi}\right]. \tag{45}$$

Equations (44) and (45) are identical for a free electron $\epsilon(\omega)$, but differ by a non-negligible amount for silver at the frequencies considered; eq. (45)

would have been a better choice for the calculations of Sipe et al. (1980a, b), at least for the bulk source polarization, but a change from eq. (44) to eq. (45) would simply scale all their results for SHG from metal sources. Such a scaling does not affect their qualitative conclusions in the analysis of the experiment of Simon et al. (1977) (sects. 2.2.2, and 2.2.3). Mills (1977) and Fukui et al. (1978a, b, c, 1979), in their work on plasmon mixing, used the free electron model for $\epsilon(\omega)$ and $\epsilon(2\omega)$.

2.2.1. *Reflection from a Free Surface*

We consider first the traditional SHG experiment, where a p-polarized beam of light is incident at frequency ω on a vacuum–metal interface and, in addition to reflected light at frequency ω, light at 2ω is observed (fig. 1). We define the second harmonic reflection coefficient, R, by

$$I(2\omega) = R[I(\omega)]^2, \tag{46}$$

where $I(\omega)$ and $I(2\omega)$ are respectively the incident intensity and the second harmonic intensity. A straight-forward calculation (Sipe et al. 1980b) leads to $R = (2\pi/c)|A_{\rm vm}|^2$, where

$$A_{\rm vm} = \frac{4\pi \mathrm{i}\tilde{\Omega} k_\| t_{\rm vm}^2}{W_{\rm v}\mathscr{E}_{\rm m} + W_{\rm m}} \frac{e^2 F}{2m\omega^2} \frac{\epsilon_{\rm m} - 1}{4\pi}, \tag{47}$$

where

$$\tilde{\omega} = \omega/c,$$
$$w_j = (\tilde{\omega}^2 \epsilon_j - k_\|^2)^{1/2}, \tag{48}$$

and where the corresponding capital letters denote the values of the parameters at 2ω, $2k_\|$; v and m denote vacuum and metal respectively

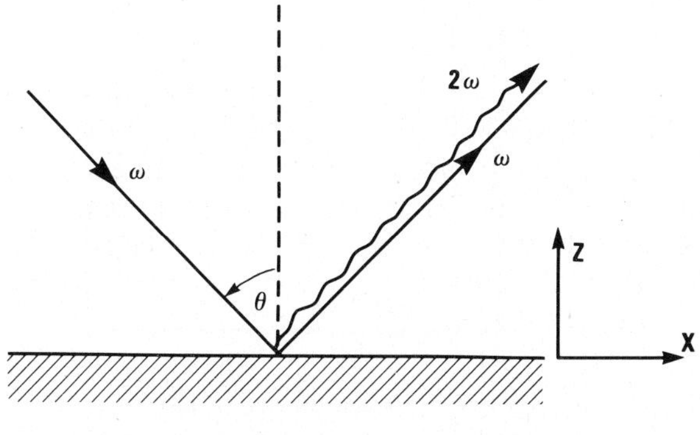

Fig. 1. Geometry for generating a second harmonic (2ω) on reflection of light (ω) from a metal surface with dielectric constant $\epsilon_{\rm m}$.

($\epsilon_v = \mathscr{E}_v = 1$), t_{vm} is the Fresnel transmission coefficient for p-polarized light, and

$$F = \left(\frac{k_\parallel^2}{\tilde{\omega}^2}\frac{\mathscr{E}_m}{\epsilon_m}a - \frac{w_m W_m}{\tilde{\omega}^2 \epsilon_m}b + \frac{1}{2}\right). \tag{49}$$

For $\epsilon_m \simeq 1 - \omega_p^2/\omega^2 \ll -1$, $\mathscr{E}_m \ll -1$, eq. (49) reduces to

$$F \simeq (\tfrac{1}{4}(k_\parallel^2/\tilde{\omega}^2)a - b + \tfrac{1}{2}), \tag{50}$$

where the $(\frac{1}{2})$ in both eqs. (49) and (50) is from the SHG in the bulk metal; the $k_\parallel$ in eq. (47) leads to the familiar $\sin^2\theta$ dependence of R at small θ (Bloembergen et al. 1968). The physical reasons for the relative sizes of the different terms in eqs. (49) and (50) are discussed by Sipe et al. (1980b); here we only note that, since $k_\parallel^2\tilde{\omega}^{-2} = \sin^2\theta$, only Q^k and the bulk SHG are important at small angles. Now both these terms are predicted essentially correctly by the free electron theory (cf. sect. 2.1); and, as a simple calculation shows, the placing of the second harmonic current sheet inside or outside the bulk leads to the same radiation *from* Q^k. Thus, the errors and *ad hoc* assumptions made in calculating the surface second harmonic current, in many early papers, do not affect their predictions for SHG at small θ (cf. sect. 3).

In fig. 2 we plot the predicted R's from silver, for incident light at

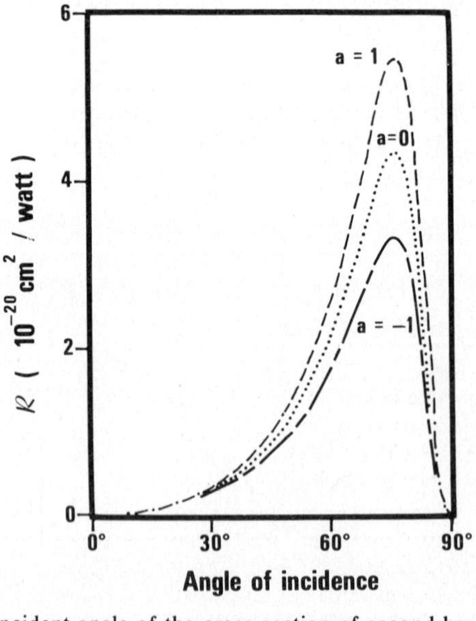

Fig. 2. Variation with incident angle of the cross section of second harmonic light generated on reflection of 1.06 μm light from a silver surface for three values of the surface parameter a (reproduced from Sipe et al. 1980).

1.06 μm (see table 1) and $a = -1, 0$ and $+1$ (all with $b = -1$); the proportionately larger contribution of Q^z at larger $k_\parallel$ is apparent. Since, comparing the $a = -1$ and $a = +1$ curves, the values of $R(40°)/R(\text{peak})$ differ by about 20% of that ratio, careful measurements of reflected SHG in the traditional geometry of fig. 1, although they would be at admittedly low intensities, could be used to determine a; the advantage of such an experiment is that its interpretation and dependence on optical constants and experimental parameters is rather more straightforward than some of the experiments discussed in sects. 2.2.2 and 2.2.3. For example, from eqs. (47) and (50) we see that, in the frequency region where $\epsilon_m, \mathscr{E}_m \ll -1$, we have $A_{vm} \sim \omega$ so $R \sim \omega^2$ as a function of incident frequency.

Because of the form of eqs. (49) and (50), we can expect more dramatic dependence on a for experiments in which incident fields can be directed with larger wavevectors parallel to the surface. Consider the experiments shown in fig. 3a, where we take a vanishingly small prism–metal separation, and assume there is no appreciable SHG in the prism. Since now $k_\parallel = n_p\tilde{\omega} \sin\theta$, where n_p is the refractive index of the prism, higher values of $k_\parallel$ can be reached. Sipe et al. (1980b) calculated R for this geometry with a rutile prism (see table 1) and found, for $b = -1$ and p-polarized light, the curves shown in fig. 4; as in fig. 2, the metal was taken to be silver and the incident beam at 1.06 μm. Note that the predicted signal in fig. 4 is both more intense than that shown in fig. 2, and there is larger difference between the curves with different values of a, the peaks in fact occurring for different

Table 1
Optical constants

	ω (1.06 μm)	2ω (0.53 μm)
Silver		
dielectric constant	$-67.03 + i2.44$[1]	$-11.9 + i0.33$[2]
Rutile[3]		
refractive index	2.479	2.671
Quartz[4]		
refractive index	1.536	1.542
nonlinear susceptibility:		
$d_\parallel = 0.39 \times 10^{-12}$ (m/V)		
$= 0.93 \times 10^{-9}$ (esu, cgs)		

References
[1] Mathewson et al. (1972), Dujardin and Theye (1971).
[2] Hollstein et al. (1977).
[3] American Institute of Physics, Handbook (1972).
[4] Handbook of Lasers (1971).

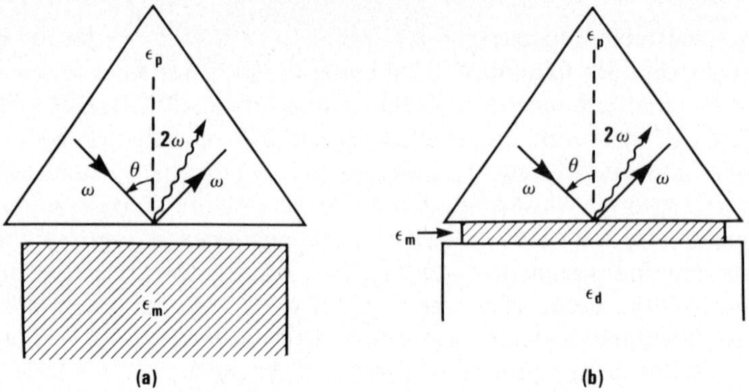

Fig. 3. ATR (attenuated total reflection) geometries for the generation of second harmonic light in a metal via the excitation of surface plasmons: (a) Otto geometry: (b) Kretschmann geometry.

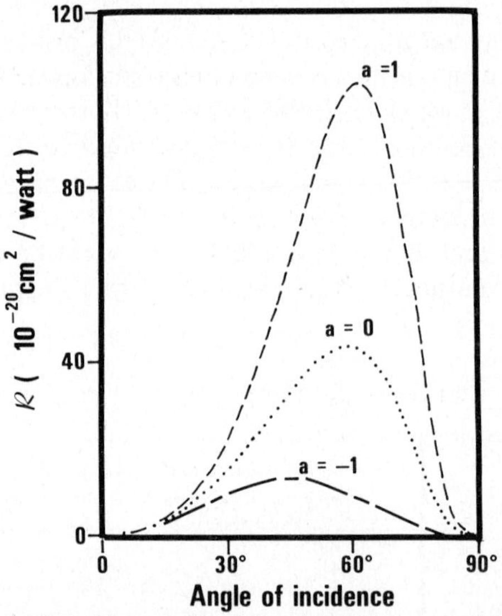

Fig. 4. The variation with incident angle of the theoretical second harmonic cross section obtained on the reflection of 1.06 μm light from silver in an Otto geometry (zero gap) for three values of the surface parameter *a* (reproduced from Sipe et al. 1980).

angles of incidence. Because of the comparatively large values of *R*, this experiment should be feasible; in fact, Sipe et al. (1980a, b) argued that it could be done as a simple extension of an experiment already performed, and we return to this point in the next section.

2.2.2. ATR Geometries

We now consider experiments using attenuated total reflection (ATR) geometries (Otto 1968, 1970) where the SHG can be enhanced due to surface polariton resonances (SPR) and second harmonic surface polariton resonances (SHSPR) (Simon et al. 1974, 1975, 1977). Sipe et al. (1980b) proposed the experiment shown in fig. 3a (Otto geometry), with a prism–metal gap of $d = 1\ \mu$m; they assumed prism refractive indices $n_p = N_p = 1.40$, with silver metal and an incident p-polarized beam at 1.06 μm. Their predicted results are shown in fig. 5, for $b = -1$ and three values of a. The enhancement (cf. fig. 2) is associated with the SPR at the vacuum–metal interface and with the vacuum light line, but is complicated by the fact that the gap is of the order of a wavelength; at larger gaps well-defined SPR peaks are predicted at higher angles, but they are smaller in intensity and very narrow. Enhancement due to the SHSPR at the vacuum–metal interface also occurs, but at a much lower intensity, since at those greater values of $k_\parallel$ the evanescent linear fields in the gap are more confined to the glass–vacuum surface, and much weaker SHG results. The predicted values of R shown in fig. 5 are fairly sensitive to d, but that distance could be determined from linear experiments; the predictions for smaller, but perhaps not experimentally reproducible, gaps are also dramatic.

We turn now to perhaps the most spectacular of the enhanced-SHG experiments, due to Simon et al. (1977), where p-polarized light at 1.06 μm is coupled through a rutile prism into a 550 Å silver film bounded by a quartz crystal (Kretschmann geometry, see fig. 3b). SHG occurs in the quartz, bulk metal and at both surfaces; it varies in intensity over about four orders of magnitude, as the incident angle is scanned over a few

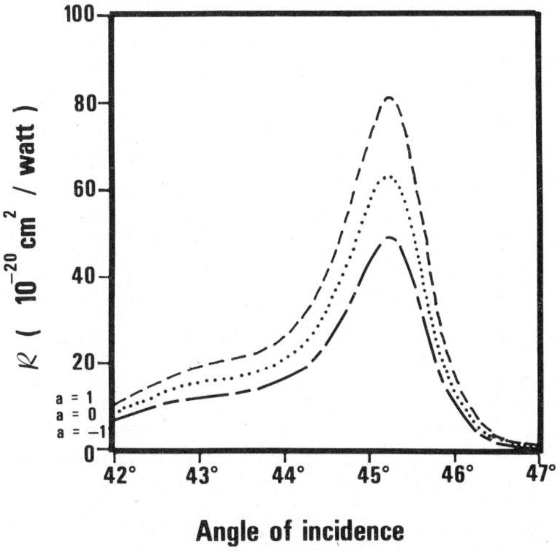

Fig. 5. The variation with incident angle of the theoretical second harmonic cross section obtained on the reflection of 1.06 μm light from silver in an Otto geometry ($d = 1\ \mu$m) for three values of the surface parameter a (reproduced from Sipe et al. 1980).

degrees, due to enhancement by passing through SPR and SHSPR at the metal–quartz interface.

The experiment was analyzed by Sipe et al. (1980a, b) using the parameters given in the table. A number of physical effects are important, and the predicted results are complicated by interference between the different sources of second harmonic light; we refer to the referenced papers for details of interpretation and postpone presenting the experimental results to sect. 3, but quote the predicted results here. In fig. 6 we show the reflection coefficients that would result if only SHG in the quartz were present (R_q), if only SHG in the bulk metal were present (R_m), and including SHG in both quartz and bulk metal (R_{qm}). Figure 7 gives the reflection coefficients predicted if SHG were present only at the rutile–metal selvedge (R_0), and if SHG were present only at the metal–quartz selvedge (R_d), for $b = -1$ and $a = 0$. Finally, fig. 8 gives the coefficient predicted including all sources, with $b = -1$ and $a = +2$, 0 and -2 at both selvedges; from fig. 7 it is clear that SHG from the metal–quartz selvedge is negligible.

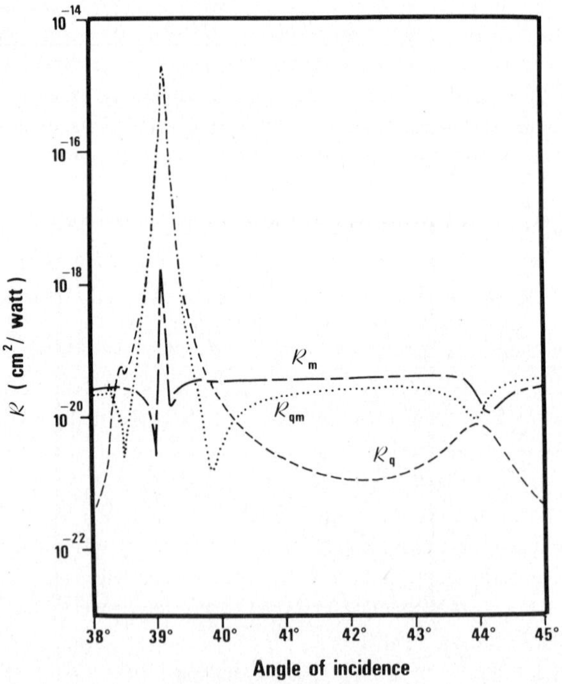

Fig. 6. The variation with incident angle of the theoretical second harmonic cross sections R_m (bulk metal), R_q (quartz) and R_{qm} (both) for reflection of 1.06 μm light from a rutile prism-silver film–air Kretschmann geometry (reproduced from Sipe et al. 1980).

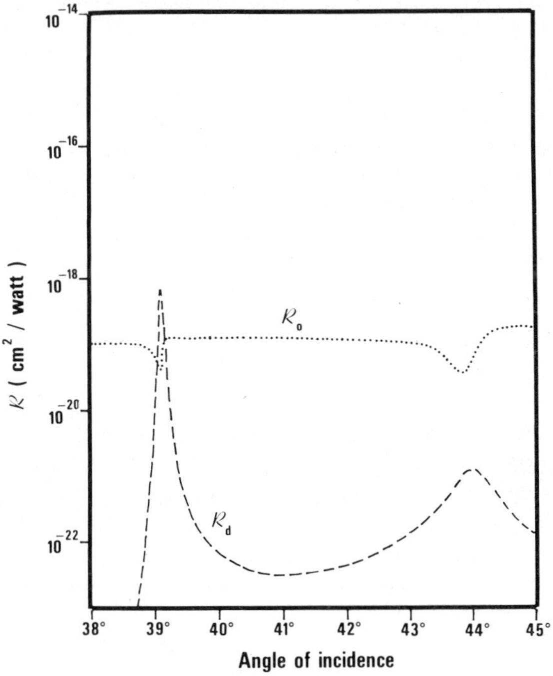

Fig. 7. Variation with angle of incidence of the selvedge (surface) second harmonic cross sections at both silver film surfaces for the reflection of 1.06 μm light in a Kretschmann geometry of rutile prism–silver film–air (reproduced from Sipe et al. 1980).

At the incident frequency considered the SPR at the metal–quartz interface is around 39°, where it is clear the SHG from the quartz dominates; however, SHG from the rutile–metal selvedge is crucial in determining the structure of R around 44° at the SHSPR, and both the bulk metal and rutile–metal selvedge SHG can be important between the resonances. These curves illustrate the necessity of taking in general all sources of second harmonic light into account, and the fact that physically different sources can, especially in an enhanced-SHG geometry, produce completely different reflection coefficient profiles.

We note from fig. 8 that the largest differences in the curves of different a occur at larger angles; this is expected from the discussion following eq. (50). To help determine a at the rutile–metal selvedge, it would thus be advantageous to extend Simon et al.'s (1977) experiment to larger angles, as suggested by Sipe et al. (1980a, b). At such large angles the results are really not characteristic of an ATR experiment but, since the SHG from the quartz and metal–quartz selvedge are here negligible, resemble the experiment discussed at the end of sect. 2.2.1; they are shown in fig. 9. The dip in the curve for $a = -2$ occurs when the second harmonic field from

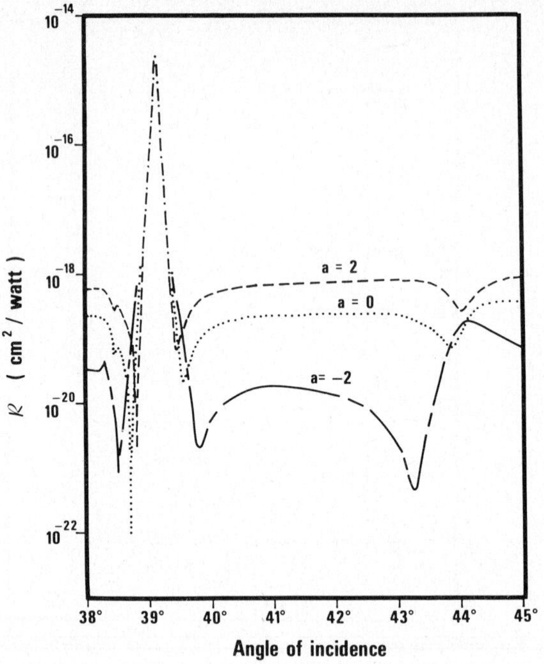

Fig. 8. The variation with angle of incidence (near the surface plasmon coupling angles) of the total theoretical second harmonic cross section for the reflection of 1.06 μm light from a rutile prism–silver film–air Kretschmann geometry for three values of the surface parameter a (reproduced from Sipe et al. 1980).

the rutile–metal selvedge, which is rapidly growing with increasing angle, reaches the same magnitude as the field from the quartz and metal, and adds *destructively*; it has nothing to do with plasmon enhancement. If a is in this range, its determination by such an experiment would be very simple indeed.

2.2.3. *Surface Polariton Interactions*

The third geometry that researchers have considered involves the interaction of freely propagating surface polariton modes. These waves can be excited with a coupling prism, as illustrated in fig. 10; light is directed at an appropriate angle of incidence at the base of a 90° prism near its corner, and surface polaritons are excited by the evanescent fields in the air gap. For this coupling method to be useful, the generated waves must propagate a few millimetres or more past the prism before they are damped out. Since the decay length L can be approximated by

$$L \simeq \left| \frac{c\epsilon_R^2(\omega)}{\omega\epsilon_I(\omega)} \right|, \tag{51}$$

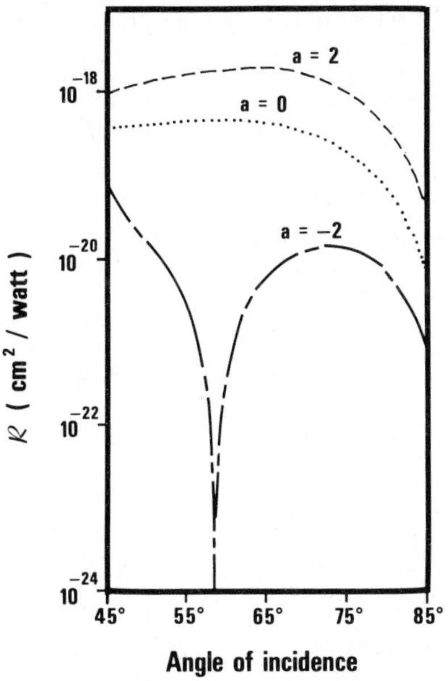

Fig. 9. The variation with angle of incidence (large angles) of the total theoretical second harmonic cross section for the reflection of 1.06 μm light from a rutile prism–silver film–air Kretschmann geometry for three values of the surface parameter a (reproduced from Sipe et al. 1980).

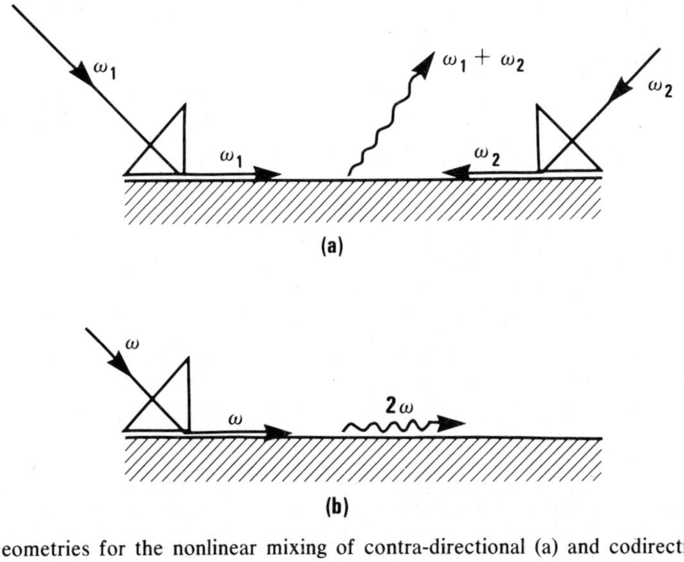

Fig. 10. Geometries for the nonlinear mixing of contra-directional (a) and codirectional (b) surface plasmons on a metal surface to produce (a) an air mode, and (b) a harmonic surface plasmon at the sum frequency.

where $\epsilon(\omega) = \epsilon_R(\omega) + i\,\epsilon_I(\omega)$, (cf. e.g. Schoenwald et al. 1973) the choice of material and frequency is important; in general, the decay length increases with decreasing frequency, and macroscopic distances can only be obtained for frequencies much less than the bulk plasma frequency.

2.2.3.1. Contrapropagating Incident Surface Plasmons. The geometry of contrapropagating incident surface plasmons is illustrated in fig. 10a; the two incident surface polaritons have fields in the metal of the form

$$\begin{aligned} \mathbf{E}(\omega_1) &= [\hat{x}E_x(\omega_1) + \hat{z}E_z(\omega_1)]\exp[-i(\omega_1 t - k_{1x}x) + \alpha_1 z], \\ \mathbf{E}(\omega_2) &= [\hat{x}E_x(\omega_2) + \hat{z}E_z(\omega_2)]\exp[-i(\omega_2 t + k_{2x}x) + \alpha_2 z], \end{aligned} \tag{52}$$

where

$$\begin{aligned} \alpha_j^2 &= k_{xj}^2 - \epsilon(\omega_j)\omega_j^2/c^2, \\ E_x(\omega_j)/E_z(\omega_j) &= (-1)^j \alpha_j/ik_{xj}, \end{aligned} \tag{53}$$

and j $(=1,2)$ labels the two waves, their frequencies and propagating constants; we have chosen $\pm\hat{x}$ to label the propagation directions. Mixing terms occur at the frequencies $\omega_1 \pm \omega_2$ and wavevectors $k_{x1} \pm k_{x2}$, but the only nonlinear source terms which can couple to normal mode fields, and thus radiate energy away from the overlap region, occur at the sum frequency and difference wavevector (cf. fig. 11). For $\omega/k_x > c$, where $\omega \equiv \omega_1 + \omega_2$ and $k_x \equiv k_{x1} - k_{x2}$, electromagnetic waves are radiated at the angle $\theta = \cos^{-1}(k_x c/\omega)$ relative to the normal to the surface; furthermore, if $\omega^2 > \omega_p^2 + k_x^2 c^2$, propagating waves can also be radiated into the bulk of the

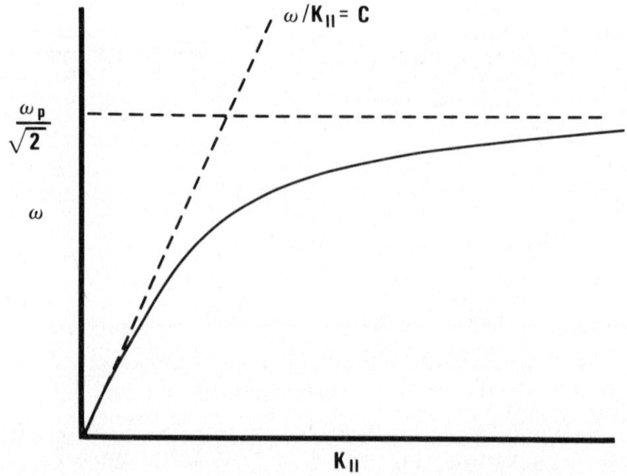

Fig. 11. The dispersion curve for surface plasmons.

metal. Here we review only the "free space" radiation fields evaluated by Fukui et al. (1978c, 1979).

Those authors followed the approach of Rudnick and Stern (1971) and found that, in the notation adopted in sect. 2.1 and with $\hat{z}$ direction from metal to vacuum, the components of the sum frequency surface source term $\boldsymbol{Q}$ are given by

$$Q^k = \frac{e\omega_p^2 b}{4\pi m^* \omega_1^2 \omega_2^2} \frac{[\omega_1 E_x(\omega_1)E_z(\omega_2) - \omega_2 E_x(\omega_2)E_z(\omega_1)]}{\omega},$$

$$Q^z = \frac{e\omega_p^2 a}{8\pi m^* \omega_1^2 \omega_2^2} E_z(\omega_1)E_z(\omega_2), \tag{54}$$

while the bulk source term is given by

$$\boldsymbol{P}_{\rm NL} = \frac{e\omega_p^2}{4\pi m^* \omega_1\omega_2(\omega_1+\omega_2)^2} \nabla[\boldsymbol{E}_1(\omega)\cdot \boldsymbol{E}_2(\omega)], \tag{55}$$

where the charge carriers are here given an effective mass m^*. The calculation of the radiation field is then straightforward, following the form of the calculation reviewed in sect. 2.2.1. For an arbitrary frequency the expressions for the generated fields are quite complicated; a simplification occurs in the limit $\omega_p \gg \omega_1$, ω_2, where $|\epsilon(\omega)| \gg 1$ and the bulk term is effectively shielded by the electron gas and may thus be neglected. Assuming $a = b = -1$, Fukui et al. (1978c, 1979) obtained a relatively simple expression for the total power radiated into the air,

$$P(\omega) = A_{\rm NL}(L/d)\, P(\omega_1)P(\omega_2), \tag{56}$$

where L is the overlap distance of the plasmon fields and d is their width, $P(\omega_1)$ and $P(\omega_2)$ are the input surface plasmon powers, and

$$A_{\rm NL} = \frac{\pi e^2 k_x^2(\omega_1^2+\omega_2^2+\omega_1\omega_2)^2 \cos\theta}{2(m^*)^2 c^5 \omega^2 \omega_p^2 |\Delta(\omega)|^2}, \tag{57}$$

with

$$\Delta(\omega) = \mathrm{i}\delta + \epsilon(\omega)\gamma,$$

$$\gamma^2 = \omega^2/c^2 - k_x^2,$$

$$\delta^2 = k_x^2 - \omega^2\epsilon(\omega)/c^2. \tag{58}$$

Numerical estimates were made for a silver surface with the surface plasmon frequencies (ω_1, ω_2) fixed at those available from $CO_2(\lambda = 10.6\,\mu m)$ and Nd : YAG ($\lambda = 1.06\,\mu$m) lasers. Estimating $\epsilon(\lambda = 1.06) \simeq -60 + \mathrm{i}$ and L from eq. (51), a power of $\sim 10^{-8}$ W was predicted for incident surface polariton powers of a few kW; the effect should be observable.

2.2.3.2. Second Harmonic Generation. The problem of second harmonic generation of surface polaritons on a free surface (fig. 10b) has been analyzed by Mills (1977) and Fukui and Stegeman (1978a). The nonlinear source terms are those discussed in sect. 2.1, and the calculation of the generated fields can be carried out using a Green function approach (Mills 1977) which yields the average field, or by solving for the total nonlinearly generated fields which propagate as waves of increasing amplitude. The Green function theory used by Mills (1977) is discussed by Maradudin and Mills (1975); the total field approach has been discussed by, e.g., Normandin et al. (1979), So et al. (1979), and reviewed by Sipe and Stegeman (1979).

Mills (1977) calculated the generation of second harmonic surface polaritons in copper excited at 10.6 μm, where all frequencies involved are far below the plasma frequency. Considering an input intensity of 50 MW/cm^2, a spot size of 100 μm and 10% efficiency at the input and output coupling prisms, he found that the predicted second harmonic power, 10^{-13} W, was just below the threshold of detectability. Fukui and Stegeman (1978a), on the other hand, considered generation in silver excited at 1.06 μm where the harmonic frequency falls close to the plasma frequency. Adopting slightly different excitation conditions (an input intensity of 10 MW/cm^2, a 200 μm diameter spot and 10% coupling efficiency), they calculated a readily detectable second harmonic power of 10^{-11} W, despite a small coherence length and large attenuation. The fact that their predicted cross section is larger than that of Mills (1977) is an indication of the importance of the frequency dependence of the linear fields and nonlinear sources in such geometries, a matter which has not been fully addressed in the literature.

A guided optical wave geometry has also been proposed (Fukui et al. 1978b) to mix freely propagating surface excitations; it is characterized by long decay lengths, variable coherence lengths and reasonable nonlinear cross sections in the visible. Surface polariton-like transverse magnetic (TM) modes guided by a thin dielectric film deposited on a metal (silver in the work of Fukui et al. 1978b) have electric fields in the metal which closely resemble those of surface plasmons (Kaminow et al. 1974). The presence of the film leads to a number of different TM modes (fig. 12), i.e., different transverse optical resonances across the film characterized by different values of $k_{\|}$; the resulting mode dispersion allows a phase-matching condition to be found. Fukui et al. (1978b) found that a TM_1 mode excited at 1.06 μm ($k_{\|} = 8.47 \times 10^4\ cm^{-1}$) can be phase-matched to a TM_3 mode at 0.53 μm for a glass film of thickness 2.6 μm; the mode crossing condition is shown in fig. 13. For details of the analysis, which is complicated by the number of interfaces present, we refer to the publication; we only note here that the phase-matched second harmonic power, which

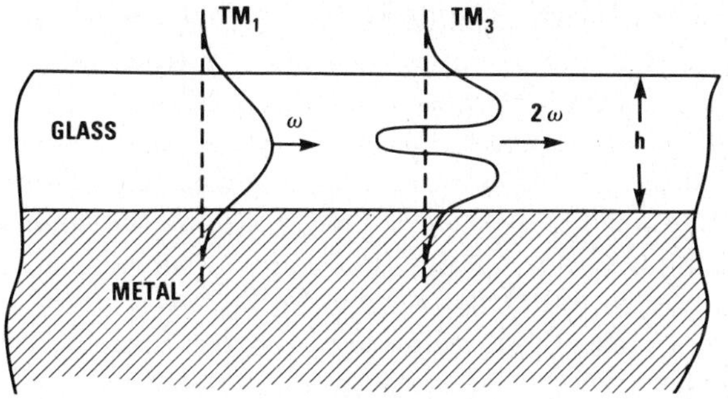

Fig. 12. Field distributions for TM_1 (ω) and TM_2 (2ω) optical modes guided by a glass film deposited on a metal surface.

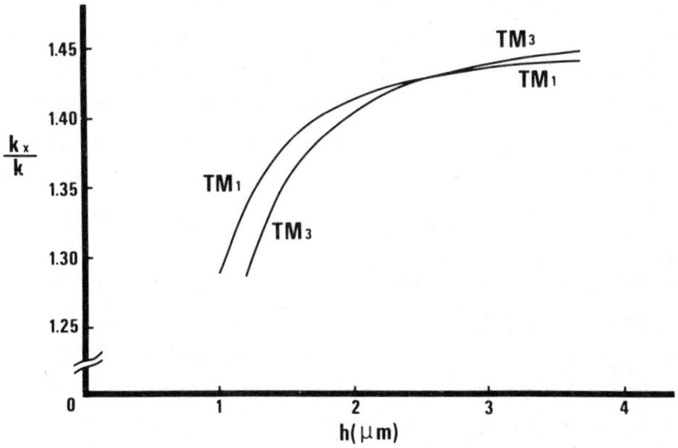

Fig. 13. Wavevectors for TM_1 (1.06 μm) and TM_3 (0.53 μm) optical modes guided by a glass film deposited on silver as a function of film thickness (reproduced from Fukui et al. 1978).

is limited by the 0.53 μm decay length of ~1 mm, was calculated to be about 5×10^{-6} W, assuming a 1 MW/cm^2 input over a 200 μm spot diameter with 10% prism coupling efficiencies. Since the dispersion curves are almost parallel at crossing, near phase-matching occurs for $2.4\ \mu m < h < 2.9\ \mu m$, where h is the thickness of the dielectric layer, with second harmonic powers $\sim 7\times10^{-8}$ W. These predicted values are very promising for future experimental investigations.

3. *Review of Experiments*

To date experiments have been reported which utilize only two of the three experimental geometries discussed in sect. 2.2. The simplest case, harmonic generation on reflection from a free metal surface, was studied first in the context of the evolution of nonlinear optics. In the last five years there have also been a limited number of ATR experiments utilyzing surface polariton resonances.

The metal surfaces studied to date have consisted of evaporated films of varying thickness; initially, silvered mirrors were used. The media were therefore isotropic and not representative of single crystal metal surfaces. Furthermore, preparation techniques played a crucial role since the dielectric constant of evaporated metal films is known to vary from sample to sample and hence should be measured in conjunction with the nonlinear experiments. This is an important (and frequently overlooked) aspect of the nonlinear problem, since the Fresnel coefficients are crucial in determining the harmonic intensity: they appear to the fourth power at the incident light frequency, and to the second power at the harmonic frequency. As indicated in the theory section, nonlinear surface currents are a dominant mechanism and hence the condition of the surface also plays an important role. Thus far no experiments have been reported on the ultraclean metal surfaces available only under very high vacuum conditions. There has, however, been some comparison between harmonic generation from silver surfaces maintained in a vacuum of $\sim 10^{-6}$ Torr and surfaces exposed to air.

The experiments on metal surfaces are amongst the most difficult reported in nonlinear optics. As discussed in sect. 2, the effects are much weaker than similar phenomena associated with non-centrosymmetric materials. Therefore high laser powers are required in order to obtain measureable signals, especially in the visible. On the other hand, large light intensities result in the generation of strong currents in the metal and dissipation effects (quantified by the imaginary component of the dielectric constant) can lead to strong local heating. For example, damage occurs experimentally in silver films for incident laser powers of 1–10 MW/cm^2 in the visible. (A theoretical treatment of surface damage by intense radiation fields is given in ch. 11.) Furthermore, the local heating leads to background thermal radiation being generated along with the second harmonic light.

An additional complication encountered in some of the initial investigations was the mode structure of the lasers used. Unless a laser oscillates in a single longitudinal and transverse mode of its cavity, the pulse envelope is not smooth and may be modulated, or even contain large intense spikes. Since the harmonic power at any instant in time varies quadratically with the instantaneous incident power, the ratio of the average harmonic to incident signals depends on the details of the pulse structure. In some experiments this problem was alleviated by monitoring the second harmonic produced in a non-centrosymmetric reference crystal.

3.1. Reflection Experiments

Experiments using this geometry (fig. 1) were all reported over the span 1965–70 and were directed at verifying the basic nonlinear mechanisms. They were performed before Rudnick and Stern (1971) showed that the nonlinear surface current sources were not shielded by the electron gas and effectively radiated from just above the metal surface. Hence the analyses used at that time for comparing experiment with theory incorrectly estimated the contribution of the nonlinear surface source directed normal to the boundary. However, assuming $b = -1$ and $a \simeq 1$ in eq. (50), it is evident that this error is minor for small angles of incidence and is probably not a serious mistake for $\theta \leq 45°$. In addition, as pointed out by Rudnick and Stern (1971) and others, there were errors in some calculations of the tangential surface currents. Hence, direct comparisons with experiments and the conclusions drawn from them should be re-examined in the light of recent developments.

A typical apparatus is shown in fig. 14. The two elements labelled A are polarizers which allow the polarization of the incident and harmonic fields to be adjusted relative to the plane of incidence. Copper sulphate in solution (B) was used to selectively attenuate the reflected and stray radiation at the incident laser frequency. The contribution of the diffuse thermal radiation (caused by local heating) to the observed signal was monitored by a detector. The second harmonic signal obtained from the metal surface on a pulse to pulse basis was calibrated by sampling the incident pulse with a beamsplitter and measuring the second harmonic generated, for example, in a z-cut quartz plate. Frequently photon counting techniques were used to process the weak signals obtained from the second harmonic detectors.

Two sets of parameters were measured in these experiments. Since the

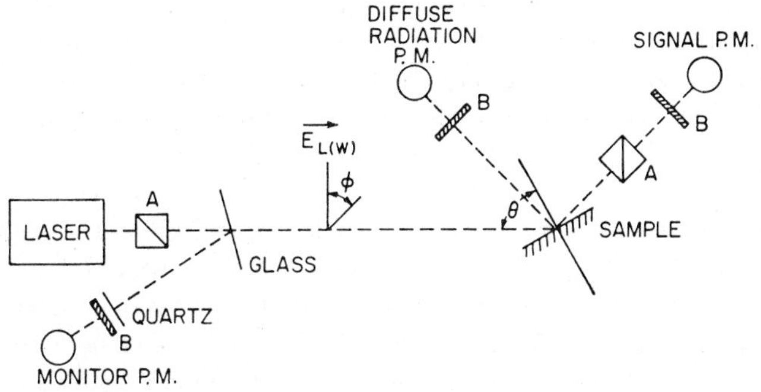

Fig. 14. Typical apparatus used for measuring second harmonic light generated on reflection from a metal surface (reproduced from Bloembergen et al. 1968).

surface nonlinear sources are in the simplest theory (sect. 2.1.2) proportional to $\boldsymbol{E}(\nabla \cdot \boldsymbol{E})$, then incident fields polarized only in the plane of the surface (normal to the plane of incidence) do not contribute to the harmonic signal. However the bulk terms proportional to $\boldsymbol{E} \times \boldsymbol{H}$ lead to harmonic generation for all incident polarizations. Neglecting these bulk sources a $\cos^4 \phi$ dependence on the polarization is predicted, where ϕ is the electric field angle relative to the plane of incidence. Deviations from this relation are indicative of contributions from the bulk term, and in particular the ratio

$$M = I_{2\omega}(\phi = \pi/2)/I_{2\omega}(\phi = 0) \tag{59}$$

yields information about the relative strengths of the surface versus bulk terms. Thus both the polarization angle dependence and the ratio M were objects of some initial studies. The second quantities of interest were the absolute harmonic signal levels and their variation with the angle of incidence from the surface normal. In both cases, the Fresnel coefficients (and the metal dielectric constants) played an important role.

The first experiments were reported by Brown et al. (1965) and Brown and Parks (1966). For 45° incidence onto a silver coated reflector, they measured the variation in harmonic signal with the polarization angle ϕ. The initial results exhibited a $\cos^4 \phi$ dependence as predicted for surface sources with a characteristic $\boldsymbol{E}(\nabla \cdot \boldsymbol{E})$ form. Subsequent improvements in their apparatus and techniques allowed them to show that $M \neq 0$ (0.03 for incident ruby laser light) and hence verify the existence of the so-called bulk magnetic dipole term: their results are shown in fig. 15. They also found that the angle at which the minimum signal occurred was in good agreement with the theory as proposed by Jha (1965).

The next group of experiments by Bloembergen and coworkers (1966, 1968), Sonnenberg and Heffner (1968) and Krivoshchekov and Stroganov (1969, 1970) dealt primarily with the magnitude of the second harmonic signal and its variation with the angle of incidence and film thickness. Of interest at that time was how accurately a free electron gas theory predicted the details of the second harmonic generation process and the contribution of the bound electrons. We again caution that the theories used for comparison were not complete as discussed previously. Except for the extensive work of Bloembergen and coworkers (1966, 1968), the experiments were carried out with ruby lasers and primarily on silver films (with the added exception of Krivoshchekov and Stroganov (1969)).

The importance of the interband transitions was first investigated by Bloembergen et al. (1966) on copper, gold, silver and silver–gold alloys. They found that the parameter M agreed reasonably well with a free electron theory only for silver with ruby laser excitation. At other wavelengths for silver, as well as the other metals studied, this ratio

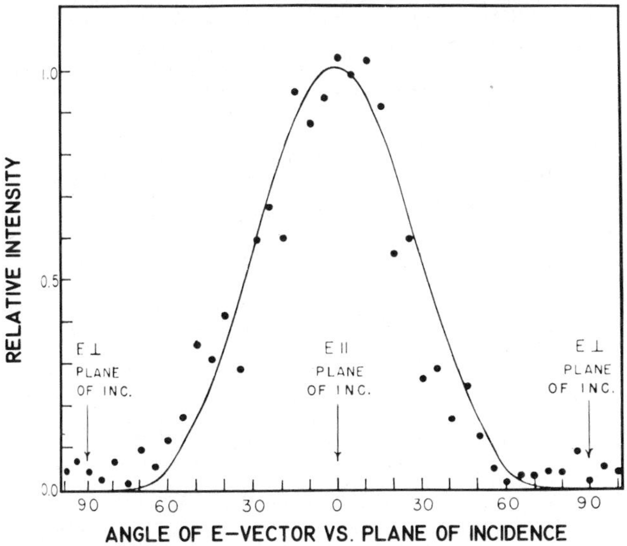

Fig. 15. Intensity of the second harmonic generated on reflection of ruby laser light from a silver surface for variable angle of polarization of the incident light (reproduced from Brown et al. 1966).

differed substantially from the free electron gas model predictions. Some improvement was obtained by including the effects of nonresonant interband transitions on the linear properties of the metals, i.e. $\mathscr{E}_m(\omega)$ and $\mathscr{E}_m(2\omega)$. However large discrepancies still remained which the authors attributed to contributions from the nonlinear interband transitions. The dispersion in harmonic signal with frequency (four values) was measured for silver and a variation of a factor of five was observed, also in substantial disagreement with the free electron theory. Two years later Bloembergen et al. (1968) summarized their work in a long theoretical and experimental paper. There they showed that the angle of incidence (θ) and polarization angle (ϕ) characteristics of the observed harmonic signal were not unique to conduction (free electrons) in metals and that they also described bound electron contributions in semiconductors.

Further experiments on very pure silver films were reported by Sonnenberg and Heffner in 1968. They made measurements on films stored in both air and vacuum ($\sim 10^{-6}$ Torr) and found no significant difference in the results. The signal was observed over a large range of incident powers (fig. 16) and the quadratic dependence of second harmonic on incident power was verified. Furthermore, the variation in harmonic signal with angle of incidence was measured over the largest range of angles reported to date (fig. 17). Their principal conclusion was that including the nonresonant contribution of the interband transitions via the measured values of $\mathscr{E}_m(\omega)$

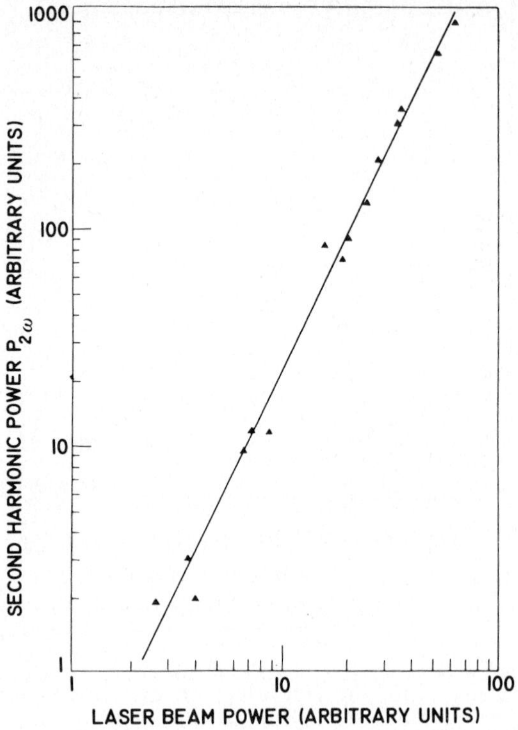

Fig. 16. Second harmonic power generated on the reflection of ruby laser light from a silver surface versus incident laser power (reproduced from Sonnenberg et al. 1968).

and $\mathscr{E}_m(2\omega)$ led to an improvement over the free electron gas case. Nevertheless, the measured powers were still approximately a factor of two larger than theory. A close examination of the shapes of the theoretical and experimental curves also favoured the model which included non-resonant interband contributions.

The dependence of the second harmonic power measured both on reflection from, and transmission through thin silver films was investigated by Krivoshchekov and Stroganov (1969). They studied films in the thickness range 150–1400 Å, both in air and in vacuum (10^{-7}–10^{-8} mm Hg). For a 45° angle of incidence, the reflected harmonic signal reached its asymptotic value at a thickness of ~300 Å, and the transmitted signal peaked at ~105 Å and then decayed exponentially with further increases in thickness. They found all of these results to be compatible with skin depth arguments at the fundamental and harmonic frequencies. Furthermore, no difference between the vacuum and air sample surfaces was found. Finally for the 200 Å thick film, the transmission signal varied with the angle of

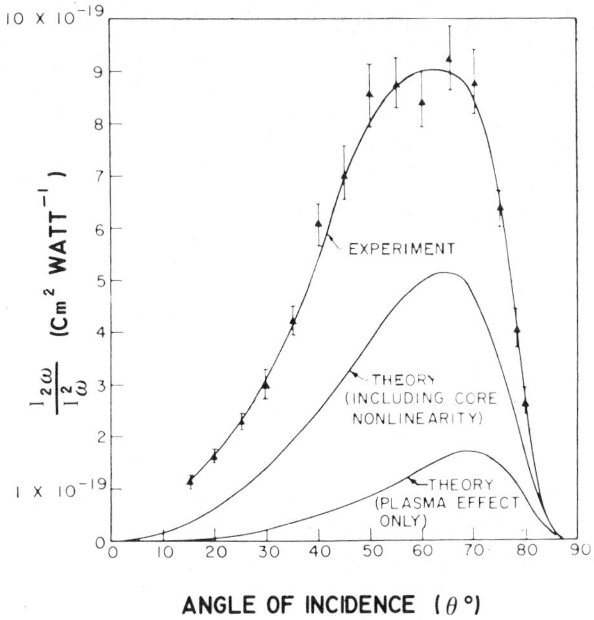

Fig. 17. Variation with angle of incidence of the second harmonic power obtained with 0.695 μm light incident on a silver surface and comparison with theory (reproduced from Sonnenberg et al. 1968).

polarization and incidence in a way similar to that reported previously in reflection experiments with thick films.

A year later these authors extended their measurements to the alkali metals sodium and potassium. These materials were chosen since there are no d electrons (as in silver and gold) and hence interband contributions are expected to be reduced significantly. In both cases the harmonic frequency is within approximately a factor of two of the plasma frequency. They found the harmonic signals obtained on reflection to be of comparable values, but in large disagreement (as much as an order of magnitude) with theory. Wang et al. (1973) attempted to reconcile this discrepancy with the free electron gas model but in so doing they neglected some very large nonlinear terms. At present the results of this experiment remain unexplained.

The experiments on nonlinear reflection from metal surfaces have as a group led to a great deal of insight into the nonlinear optics of metals. It was shown that the free electron gas model, as formulated by Jha (1965, 1966, 1967) provides a reasonable initial point for describing the nonlinear optical response of a metal. A number of experiments have indicated that the interband transitions play an important part for metals such as silver and their effects on the linear dielectric constant must be included. There is also strong evidence that the nonlinear contributions from the interband transitions are significant. Since in general the surface nonlinear current sources were not always analyzed properly, it is difficult to draw definitive conclusions about the magnitude of the disagreement between experiment and theory. Finally, we note that these experiments yielded essentially no information about the parameter a which is sensitive to the details of the surface region.

3.2. Attenuated Total Reflection Experiments

These experiments utilize the geometry discussed previously in sect. 2.2. They are characterized by the same material damage and low signal problems obtained in the reflection experiments; there are, however, some significant differences. The surface polariton modes are resonances at the air–metal interface and are only weakly coupled to radiation fields in the prism. Their amplitudes are typically an order of magnitude larger than the incident fields hence reducing the required laser powers relative to the reflection case. Furthermore, since surface polaritons are propagating waves, the coherence length for the nonlinear interaction can be many microns in size with the result that the harmonic fields can grow with propagation distance to be much larger than those in reflection experiments.

An important feature of this geometry is that the normal nonlinear surface current is comparable in magnitude to the other mechanisms. If the medium adjacent to the film (opposite side to the prism) has a refractive index n_d, then $(K_\parallel/\tilde{\omega})n_d \gtrsim 1$ with the result that the term proportional to a can now be studied experimentally. If this adjacent medium is not deposited but rather is placed against the metal surface, a coupling fluid of refractive index comparable to that of the material is used.

The first experiments were reported by Simon et al. in 1974. They used a 560 Å silver film and excited surface polariton resonances at the air–metal interface with a ruby laser. Dispersion in the prism caused the reflected fundamental and second harmonic to be separated by $\sim 1.5°$. An enhancement (relative to the free surface reflection case) of ~ 30 was obtained compared to the theoretical prediction of ~ 150. This discrepancy was associated with the film preparation techniques which may have led to film dielectric constants different from those reported in the literature. Furthermore, the angular divergence of the laser was comparable to the angular width of the measured peak with the result that the peak enhancement was probably larger than the quoted value.

The second experiment, also reported by Simon's group (1977), potentially contains a great deal of information about the nonlinear optical response of metals. The adjacent material was the nonlinear dielectric quartz and the nonlinearities in this material provided a local calibration for the metal nonlinearities via interference effects. Unfortunately, the metal film dielectric constants were not measured and a definitive interpretation of the results is not possible. In this case $n_d \simeq 1.54$ and the contribution of the normal surface current is comparable to the other nonlinear terms.

The geometry used is that of fig. 3b with 550 Å of silver deposited on a rutile prism and with l-bromonapthalene as the coupling fluid between the x-cut quartz plate and the metal. The large frequency dispersion in the prism caused a separation of 9.5° between the fundamental and harmonic fields. Although this large dispersion simplified the measurements, it also reduced the signal because the resulting coherence length was small. The results are shown in fig. 18. The peak magnitude is again reduced due to divergence limitations of the incident laser, a feature which does not allow the use of the peak value for normalization purposes.

The results show evidence of contributions from both the metal (actually near the rutile–metal interface) and the quartz. Nonlinearities in the quartz are responsible for the intense peak and the minima on either side are due to interference between it and the metal. The small minimum at ~44° is due to coupling of the second harmonic generated at the rutile–prism interface to surface plasmons at the quartz–metal boundary.

The complete angular spectrum has been calculated by Sipe et al. (1980) along the lines outlined in sect. 2. The calculated curves were shown previously in sect. 2.2 as figs. 5–8 and good agreement was obtained for

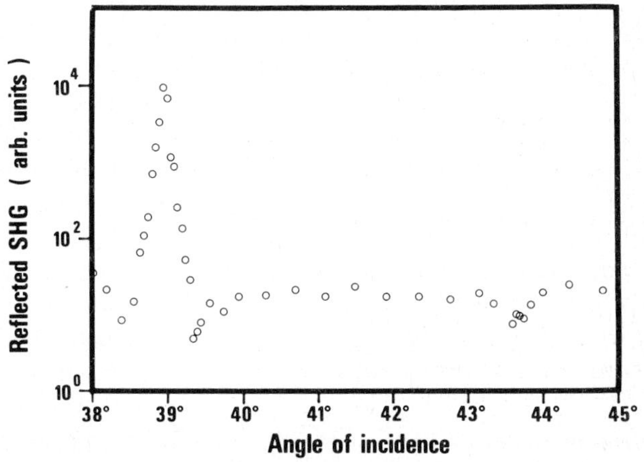

Fig. 18. Second harmonic power generated on reflection of 1.06 μm light from a prism–silver film–air Kretschmann sample geometry versus angle of incidence (reproduced from Simon et al. 1977).

$|a| \leq 2$. Extensions of this experiment which could lead to more definitive values of this parameter were discussed in sect. 2.

4. Concluding Remarks

In this chapter we have reviewed both the theoretical and experimental work on the nonlinear response of metal surfaces. Although this area of solid state physics has been studied by a number of workers over the course of about fifteen years, it is clear that much more work is required if the subject is to be understood at a quantitative level. However, we believe that the theoretical investigations have identified both the main physical effects and the parameters that should be the object of further experimental and theoretical studies. A convenient and physical way of representing the surface nonlinear sources is given by eqs. (42)–(43), and for the bulk source we suggest the expression

$$\boldsymbol{P}_{\mathrm{NL}}(\boldsymbol{r}) = \gamma d\, \nabla(\boldsymbol{E}_1(\boldsymbol{r}) \cdot \boldsymbol{E}_1(\boldsymbol{r})), \tag{60}$$

(cf. eq. 41), where d is a dimensionless parameter; in eqs. (42), (43), (60), the value of γ that should be used is given by eq. (45). With this characterization of the sources, the parameters to be determined are a, b and d. We note that:

(1) The free electron theory leads to $d = +1$ and $b = -1$, although the placement of the current sheet (42) is not unambiguously determined. If the free electron theory results are used to determine γ, the parameters a and b are those introduced by Rudnick and Stern (1971).

(2) The hydrodynamic theory leads to the free electron expression for γ, to values $d = +1$ and $b = -1$, and to the placement of the current sheet discussed in sect. 2 (cf. in particular the discussion at the start of sect. 2.2). That theory could be used to determine a value for a, but this has not yet been done.

(3) Quantum mechanical treatments based on the random phase approximation lead to a value $d = +1$, if there are no resonant interband transitions.

At a quantitative level, one would of course like agreement between theoretical predictions and experimental results for the frequency dependence of a, b and d, for different metals. Even at the qualitative level, however, there is a number of areas worthy of future study. We mention only three:

(1) Possible resonances in d. These would be due to interband transitions (Bower 1976) and, with the advent of high power lasers over a wide range of optical frequencies, should be observable if they exist. Since the quantum mechanical calculation of *bulk* SHG is much easier than the

calculation of *surface* SHG, and since the bulk SHG is not sensitive to surface conditions, this might be the area in which detailed comparison between theory and experiment would first be possible.

(2) Possible resonances in a. These would occur when the second harmonic frequency equalled the "effective plasma frequency" of the second harmonic current density at the surface (Sipe et al. 1980a, b). Because of possible interference between the different SHG sources (cf. sect. 2.2.2), such resonances could lead to dramatic changes in the angle dependence of SHG. Their observation would yield information on the electron dynamics within a few Fermi wavelengths of the surface.

(3) The effects of surface roughness on SHG. Although mentioned by Rudnick and Stern (1971), this has been neglected in most studies. It would probably be best investigated by determining the influence of surface roughness on b, since that parameter is relatively well–understood (cf. sect. 2.1.3) in the absence of roughness.

It is clear that, in addition to more theoretical work, *carefully designed* experimental studies are required to isolate different effects such as those mentioned above. In this context it is worth noting that the relative advantages of simple reflection experiments, ATR experiments, and surface polariton interaction experiments (sects. 2.2.1–2.2.3) should be weighed carefully in preparation for future work. In terms of second harmonic signal and signal analysis the respective advantages are simplicity in execution and interpretation, enhancement, and signal strength due to long coherence lengths. However, since these different types of experiments are sensitive to a, b and d in different ways, a complete comparison is not so straightforward. We also note that many of the performed and proposed experiments (sect. 2.3) involve metal–dielectric rather than metal–vacuum surfaces, and thus are not capable of studying the simpler and more fundamental interface. Finally, the fact that SHG is affected by the presence of adsorbed molecules (Brown and Matsuoka 1969) both opens up a new area of research which we have not discussed in this chapter, and points out that ultrahigh vacuum experiments will ultimately be necessary to study the details of at least the surface sources of second harmonic generation.

With as many qualitative, as well as quantitative questions still remaining, we feel that the nonlinear optical response of metals should be an interesting area of both experimental and theoretical work for a number of years to come.

Acknowledgment

One of us (J. E. S.) acknowledges financial assistance from the Killam Program of the Canada Council.

References

Adler, E., 1964, Phys. Rev. **134**, A728.

American Institute of Physics Handbook, 1972 (McGraw Hill, New York).

Ashcroft, N.W. and N.D. Mermin, 1976, Solid State Physics (Holt, Rinehart and Winston, New York).

Barton, G., 1979, Rep. Prog. Phys. **42**, 963.

Bennett, A.J., 1970, Phys. Rev. **B1**, 203.

Bloembergen, N., 1965, Nonlinear Optics (Benjamin, New York), p. 129.

Bloembergen, N., R.K. Chang and C.H. Lee, 1966, Phys. Rev. Lett. **16**, 986.

Bloembergen, N., R.K. Chang, S.S. Jha and C.H. Lee, 1968, Phys. Rev. **174**, 813.

Bloembergen, N. and Y.R. Shen, 1966a, Phys. Rev. **141**, 298.

Bloembergen, N. and Y.R. Shen, 1966b, Phys. Rev. **145**, 390(E).

Bonsall, L. and A.A. Maradudin, 1978, J. Appl. Phys. **49**, 253.

Bower, J.R., 1976, Phys. Rev. **B14**, 2427.

Brown, F., R.E. Parks and A.M. Sleeper, 1965, Phys. Rev. Lett. **14**, 1029.

Brown, F. and M. Matsuoka, 1969, Phys. Rev. **185**, 985.

Brown, F. and R.E. Parks, 1966, Phys. Rev. Lett. **16**, 507.

Chen, Y.J. and E. Burstein, 1977, Nuovo Cimento **39**, 807.

Chen, Y.J., W.P. Chen and E. Burstein, 1976, Phys. Rev. Lett. **36**, 1207.

Chen, C.K., A.R.B. deCastro and Y.R. Shen, 1980a, Optics Lett. **4**, 393.

Chen, C.K., A.R.B. deCastro, Y.R. Shen and F. DeMartini, 1980b, Phys. Rev. Lett. **43**, 946.

Cheng, H. and P.B. Miller, 1964, Phys. Rev. **134**, A683.

DeMartini, F., P. Ristori, E. Santamato and A.C.A. Zammit, 1981, Phys. Rev. **B** (in press).

Dujardin, M. and M. Theye, 1971, J. Chem. Phys. Solids **32**, 2033.

Eguiluz, A., 1979, Phys. Rev. **B19**, 1689.

Eguiluz, A. and J.J. Quinn, 1975, Phys. Rev. **B11**, 2118.

Eguiluz, A. and J.J. Quinn, 1976a, Phys. Rev. **B13**, 4299.

Eguiluz, A. and J.J. Quinn, 1976b, Phys. Rev. **B14**, 1347.

Fukui, M., J.E. Sipe, V.C.Y. So and G.I. Stegeman, 1978c, Solid State Commun. **27**, 1265.

Fukui, M., J.E. Sipe, V.C.Y. So and G.I. Stegeman, 1979, J. Phys. Chem. Solids **40**, 523.

Fukui, M. and G.I. Stegeman, 1978a, Solid State Commun. **26**, 239.

Fukui, M., V.C.Y. So and G.I. Stegeman, 1978b, Phys. Rev. **B18**, 2484.

Griffin, A. and H. Kranz, 1977, Phys. Rev. **B15**, 5068.

Handbook of Lasers, 1971 (Chemical Rubber Co., Cleveland).

Hollstein, T., U. Kriebig and F. Leis, 1977, Phys. Stat. Solidi **82b**, 545.

Jha, S.S., 1965, Phys. Rev. Lett. **15**, 412; Phys. Rev. **B140**, A2020.

Jha, S.S., 1966, Phys. Rev. **145**, 500.

Jha, S.S. and C.S. Warke, 1967, Phys. Rev. **153**, 751.

Kaminow, I.P., W.L. Mammel and H.P. Weber, 1974, Appl. Opt. **13**, 396.

Krivoshchekov, G.V. and V.I. Stroganov, 1969, Sov. Phys. Solid State **11**, 89.

Krivoshchekov, G.V. and V.I. Stroganov, 1970, Sov. Phys. Solid State **11**, 2151.

Landau, L.D. and E.M. Lifshitz, 1965, Quantum mechanics: Non-relativistic theory (Pergamon, Oxford).

Maradudin, A.A. and D.L. Mills, 1975, Phys. Rev. **B11**, 1392.

Mathewson, A.G., H. Aronsson and L.G. Bernland, 1972, J. Phys. **F2**, L39.

Mills, D.L., 1977, Solid State Commun. **24**, 669.

Normandin, R., V.C.Y. So., N. Rowell and G.I. Stegeman, 1979, J. Opt. Soc. Am. **69**, 1153.

Otto, A., 1968, Z. Physik, **216**, 398.

Otto, A., 1970, Phys. Status Solidi, **42**, K37.

Rudnick, J. and E.A. Stern, 1971, Phys. Rev. **B4**, 4274.

Rudnick, J. and E.A. Stern, 1974, Second Harmonic Radiation from a Metal Surface, in: Polaritons, ed. E. Burstein and F. DeMartini (Plenum, New York).
Schoenwald, J., E. Burstein and J.M. Elson, 1973, Solid State Commun. **12**, 185.
Simon, H.J., R.E. Benner and J.G. Rako, 1977, Opt. Commun. **23**, 245.
Simon, H.J., D.E. Mitchell and J.G. Watson, 1974, Phys. Rev. Lett. **33**, 1531.
Simon, H.J., D.E. Mitchell and J.G. Watson, 1975, Opt. Commun. **13**, 294.
Sipe, J.E., 1979, Surf. Sci. **84**, 75.
Sipe, J.E., 1980, Phys. Rev. **B22**, 1589.
Sipe, J.E. and G.I. Stegeman, 1979, J. Opt. Soc. Am. **69**, 1676.
Sipe, J.E., V.C.Y. So., M. Fukui and G.I. Stegeman, 1980a, Solid State Commun. **34**, 523.
Sipe, J.E., V.C.Y. So, M. Fukui and G.I. Stegeman, 1980b, Phys. Rev. **B21**, 4389.
So, V.C.Y., R. Normandin and G.I. Stegeman, 1979, J. Opt. Soc. Am. **69**, 1166.
Sonnenberg, H. and H. Heffner, 1968, J. Opt. Soc. Am. **58**, 209.
Wang, C.S., J.M. Chen and J.R. Bower, 1973, Opt. Commun. **8**, 275.

AUTHOR INDEX

Abelès, F. 204, 241, 253, 260, 263, 268, 269, 270
Abelès, F., *see* Chao, F. 318, 320
Abelès, F., *see* Lopez-Rios, T. 199, 200, 204, 258, 259, 263, 264, 294, 315
Abramowitz, M. 437
Abrikosov, A.A. 521, 597
Adams, A. 368
Adler, E. 671
Agarwal, G.S. 474
Agranovich, V.M. 7, 33, 42, 54, 59, 82, 95, 97, 99, 106, 107, 118, 130, 147, 149, 151, 168, 190, 191, 199, 200, 202, 205, 210, 211, 212, 213, 216, 218, 220, 221, 224, 225, 226, 228, 232, 257, 277, 279, 281, 284, 513, 514, 525, 559
Agranovich, V.M., *see* Bendow, B. 513
Aizu, K. 177
Akhmanov, S.A. 235
Alexander, R.W. 246, 279
Alexander, R.W., *see* Anderson, W.E. 43, 218
Alexander, R.W., *see* Begley, D.L. 103, 109, 112, 115, 117, 122, 251
Alexander, R.W., *see* Bell, R.J. 95, 106, 108, 130, 250, 294, 407
Alexander, R.W., *see* Bhasin, K. 130, 133, 264, 294
Alexander, R.W., *see* Bryan, D.A. 110, 122, 134, 264
Alexander, R.W., *see* Ward, C.A. 98, 100, 101, 103, 134
Alexander Jr., R.W., *see* Kovener, G.S. 29, 357
Alfaro Holbrook, J., *see* Hummel, R.E. 242
Anderman, G., *see* Kachare, A. 284, 286, 292
Anderson, J. 242
Anderson, J., *see* Rubloff, G.W. 242
Anderson, W.E. 43, 218
Arakawa, E.T. 29, 246, 279, 357
Arakawa, E.T., *see* Braundmeier Jr., A.J. 337
Arakawa, E.T., *see* Chung, M.S. 396
Arakawa, E.T., *see* Cowan, J.J. 391
Arakawa, E.T., *see* Gesell, T.F. 362, 363, 364, 365, 366
Arakawa, E.T., *see* Kretschmann, E. 396
Arakawa, E.T., *see* Loisel, B. 270
Arakawa, E.T., *see* Ritchie, R.H. 247, 391, 392, 393, 618
Arakawa, E.T., *see* Wheeler, C.E. 393
Arakawa, E.T., *see* Williams, M.W. 362
Aronsson, H., *see* Mathewson, A.G. 679
Aschenbach, B., *see* Trümper, J. 377
Ashcroft, N.W. 670
Ashley, I.C., *see* Kretschmann, E. 342, 396
Ashley, J.C., *see* Kretschmann, E. 323, 471, 472, 480, 495, 501, 502, 503
Aslangul, C., *see* Orrit, M. 225
Aspnes, D. 241
Aspnes, D.E., *see* McIntyre, J.D.E. 325
Aspnes, D.E., *see* Williams, M.D. 579
Aziza, A., *see* Ushioda, S. 368, 495, 580

Balkanski, M., *see* Proix, F. 148, 158
Banshchikov, A.G. 293
Banshchikov, A.G., *see* Novak, I.I. 293
Banshchikov, A.G., *see* Reshina, I.I. 31, 32, 33, 284
Barker, A.S. 170, 539, 541, 571, 598, 641, 643
Barker Jr., A.S. 24, 29, 37, 286
Barrick, D.E. 370
Barton, G. 667, 668, 675
Bäuerle, B., *see* Fischer, B. 27
Beaglehole, D. 19, 242, 370, 381
Beaglehole, D., *see* Hunderi, O. 370, 371
Beckmann, P. 380
Bedeaux, D. 359
Begley, D.L. 103, 109, 112, 115, 117, 122, 251
Begley, D.L., *see* Bell, R.J. 250
Begley, D.L., *see* Bryan, D.A. 110, 122, 134, 264
Begley, D.L., *see* Davarpanah, M. 111
Bell, M.I. 566
Bell, R.E., *see* Anderson, W.E. 43
Bell, R.I., *see* Kovener, G.S. 357
Bell, R.J. 95, 106, 108, 130, 250, 294, 407
Bell, R.J., *see* Alexander, R.W. 246, 279
Bell, R.J., *see* Anderson, W.E. 218
Bell, R.J., *see* Begley, D.L. 103, 109, 112, 115, 117, 122, 251
Bell, R.J., *see* Bhasin, K. 130, 133, 264, 294
Bell, R.J., *see* Bryan, D.A. 110, 122, 134, 264
Bell, R.J., *see* Kovener, G.S. 29
Bell, R.J., *see* Ward, C.A. 98, 100, 101, 103, 134

Bendow, B. 513, 555
Benner, R.E., *see* Simon, H.J. 589, 632, 650, 651, 652, 665, 666, 677, 681, 683, 696 697
Bennett, A.J. 667
Bennett, B.I. 469
Bennett, H.E. 380
Bennett, H.E., *see* Elson, J.M. 374, 380, 381
Bennett, H.E., *see* Seraphin, B.O. 569
Bennett, H.E., *see* Stanford, I.L. 362, 391
Bennett, J.M 422
Bennett, J.M., *see* Elson, J.M. 374, 380, 381
Berezinsky, V.L., *see* Blank, A.Ya. 218, 220
Bernard, J., *see* Turlet, J.M. 199, 230, 231
Bernland, L.G., *see* Mathewson, A.G. 679
Berreman, D.W. 148, 158, 356
Bewick, A. 325
Bhasin, K. 130, 133, 264, 294
Bhasin, K., *see* Bell, R.J. 250
Bhasin, K., *see* Bryan, D.A. 110, 122, 134, 264
Bhasin, K., *see* Ward, C.A. 134
Bird, V.M., *see* Hutley, M.C. 391
Birman, J.L. 513
Birman, J.L., *see* Bendow, B. 513
Blanchet, G.B. 242
Blank, A.Ya. 218, 220
Bloembergen, N. 593, 631, 632, 648, 649, 664, 668, 671, 678, 691, 692, 693
Blum, F.A. 44
Bockris, J.O'M. 303, 304
Bodesheim, J. 353, 408
Boersch, H. 18
Bösenberg, J. 250, 264, 265
Bonsall, L. 589, 666
Borburgh, J.C.M., *see* Van den Berg, P.M. 391
Borensztein, Y., *see* Abelès, F. 260
Born, M. 6, 21, 635
Borstel, G. 33, 39, 42, 631, 647
Borstel, G., *see* Falge, H.J. 41
Borstel, G., *see* Schuller, E. 29, 42, 279
Bower, J.R. 665, 675, 698
Bower, J.R., *see* Wang, C.S. 665, 671, 672, 695
Bradley, J.A., *see* Tsang, J.C. 589
Brambring, I. 360
Brantley, L.R., *see* Kachare, A. 284, 286, 292
Braundmeier, Jr., A.J. 337, 352, 384
Braundmeier, Jr., A.J., *see* Hall, D.G. 342, 352, 381, 471
Bräuninger, H., *see* Trümper, J. 377
Bray, R., *see* Mishra, S. 372
Brilliante, A., *see* Pockrand, I. 295, 315, 316
Brion, J.J. 48
Brion, J.J., *see* Hartstein, A. 33, 35, 53, 103, 644
Brion, J.J., *see* Wallis, R.F. 43, 49, 584
Brodin, M.S. 82, 232
Brown, F. 664, 692, 693, 699
Bruce, A.D., *see* Hisano, K. 16, 25
Bruns, R. 335, 341
Bryan, D.A. 110, 122, 134, 264
Bryan, D.A., *see* Begley, D.L. 109, 112, 115, 117, 251
Bryan, D., *see* Bhasin, K. 130, 133, 264, 294
Bryksin, V.V. 6, 7, 13, 16, 21, 22, 24, 25, 26, 27, 33, 37, 38, 43, 44, 45, 46, 147, 148, 151, 158, 168, 170, 284, 407
Buckel, W., *see* Perry, C.H. 31, 32
Buckel, W.J., *see* Fischer, B. 27
Burstein, E. 7, 242, 407, 411, 505, 561, 563, 583, 584, 589, 597, 619
Burstein, E., *see* Brion, J.J. 48
Burstein, E., *see* Chen, C.Y. 619
Burstein, E., *see* Chen, Y.J. 251, 321, 558, 559, 589, 600, 608, 609, 619, 622, 624, 665, 666
Burstein, E., *see* Hartstein, A. 33, 35, 46, 51, 52, 53, 103, 218, 644
Burstein, E., *see* Mills, D.L. 407, 411, 541, 556, 558, 559, 574, 589, 593
Burstein, E., *see* Pinczuk, A. 46
Burstein, E., *see* Schoenwald, J. 95, 108, 122, 190, 200, 250, 336, 545, 686
Burstein, E., *see* Talaat, H. 108, 127, 470
Burstein, E., *see* Wallis, R.F. 49, 584
Bush, I.A. 374, 375

Cabrera, N., *see* García, N. 442, 443, 444, 504
Cadilhac, M., *see* Petit, R. 387, 409
Callcott, T.A., *see* Chung, M.S. 396
Callcott, T.A., *see* Kretschmann, E. 396
Cardona, M. 553
Cardona, M., *see* Bell, M.I. 566
Cardona, M., *see* Weinstein, B.A. 566
Cavallini, M., *see* Gratton, L.M. 136
Celli, V., *see* Hill, N.R. 387, 409, 434, 447
Celli, V., *see* Marvin, A. 370
Celli, V., *see* Toigo, F.D. 342, 423, 424, 444, 447, 471, 472, 480, 486, 507
Chabal, Y.J. 109, 118, 133
Chabal, Y.L. 267
Chabrier, G. 250
Chang, R.K., *see* Bloembergen, N. 632, 648, 649, 664, 668, 671, 678, 691, 692, 693
Chao, F. 316, 318, 320
Chapet-Rousseau, M., *see* Reinisch, R. 589
Chauvineau, J.P., *see* Pariset, C. 267
Chen, C.K. 589, 619, 632, 650, 653, 654, 657
Chen, C.K., *see* Chu, K.C. 636
Chen, C.Y. 619
Chen, C.Y., *see* Burstein, E. 505, 583, 619
Chen, C.Y., *see* Chen, Y.J. 589, 609
Chen, J.M., *see* Chen, W.P. 265
Chen, J.M., *see* Wang, C.S. 665, 671, 672, 695
Chen, W. 98, 106, 293
Chen, W.P. 265
Chen, W.P., *see* Burstein, E. 242, 589
Chen, W.P., *see* Chen, Y.J. 251, 321, 589, 600, 608, 619, 622, 624, 667
Chen, W.P., *see* Talaat, H. 108, 127, 470

Chen, Y.J. 251, 321, 558, 559, 589, 600, 608, 609, 619, 622, 624, 665, 666
Chen, Y.J., *see* Burstein, E. 242, 589
Chen, Y.J., *see* Mills, D.L. 556, 558, 559, 574, 589
Cheng, H. 671
Chiaradia, P., *see* Chiarotti, G. 242
Chiarotti, G. 242
Chiu, K.W. 43, 48
Chu, K.C. 636
Chung, M.S. 396
Church, E.L. 392
Cohen, D.K., *see* Bush, I.A. 374, 375
Cohen, D.K., *see* Sari, S.O. 374, 376, 505, 506
Cohn, 408
Coleman, P.D., *see* Parsons, D.F. 566, 582
Colocci, M., *see* DeMartini, F. 84, 85, 86, 589, 631, 640, 643, 645
Conwell, E.M. 218, 219, 220, 258
Conwell, E.M., *see* Kao, C.C. 218, 219, 220
Costa, M., *see* Chao, F. 316, 318, 320
Cowan, J.J. 391
Cowan, J.J., *see* Ritchie, R.H. 247, 391, 392, 393, 618
Cowley, R.A. 16
Cox, F.T. 138, 139, 142, 292
Crowell, J. 364
Cummins, H.Z., *see* Birman, J.L. 513
Cunningham, J.A. 242
Cunningham, S.L. 146, 218, 219, 220, 258

Damen, T.C., *see* Scott, J.F. 566
d'Andrea, A. 90
Darmanyan, S.A., *see* Agranovich, V.M. 210, 211
Daudé, A. 364, 365, 422
Davarpanah, M. 111
Davarpanam, M., *see* Bell, R.J. 250
Davies, H. 380
Davis, L.C. 505
DeCastro, A.R.B., *see* Chen, C.K. 589, 619, 632, 650, 653, 654, 657
DeCastro, A.R.B., *see* Chen, Y.J. 667
De Crescenzi, M., *see* Abelès, F. 260
De Korte, P.A.I. 377, 378
Del Sole, R., *see* d'Andrea, A. 90
DeMartini, F. 23, 84, 85, 86, 589, 631, 637, 639, 640, 641, 643, 645, 651, 653, 667
DeMartini, F., *see* Chen, C.K. 589, 619, 632, 650, 653, 654
DeMartini, F., *see* Chen, Y.J. 667
Devanathan, M.A., *see* Bockris, J.O'M. 304
Dickertmann, D., *see* Schultze, J.W. 325
Dobberstein, A. 362
Dobberstein, P. 354, 369, 396, 408, 422
Dobberstein, P., *see* Sauerbrey, G. 354
Dobrzynski, L. 505
Dobrzynski, L., *see* Maradudin, A.A. 407
Dove, D.B., *see* Hummel, R.E. 242
Dresselhaus, G. 603
Drude, P. 191
Dubovsky, E.B. 213
Dubovskii, O.A. 33
Dubovskii, O.A., *see* Agranovich, V.M. 33, 202, 225, 226
Dujardin, M.M. 260, 679
Dzyaloshinskii, I.E., *see* Abrikosov, A.A. 521, 597

Eagen, C.F. 265
Economou, E.N. 54
Economou, E.N., *see* Ngai, K.L. 54, 257
Eguiluz, A. 505, 667, 668, 669, 670, 672, 675
Eldridge, H.B., *see* Ritchie, R.H. 73
Eliashberg, G.M. 524
Elson, I.M. 364, 366, 370
Elson, J.M. 374, 380, 381, 422
Elson, J.M., *see* Schoenwald, J. 190, 200, 250, 336, 545, 686
Elson, M., *see* Schoenwald, J. 95, 108, 122
Emerson, L.C., *see* Williams, M.W. 362
Endriz, I.G. 364, 365
Endriz, J.G. 523
Englman, R. 507
Englman, R., *see* Ruppin, R. 147, 148, 151, 155, 158, 168, 507, 559
Erlbach, E., *see* Beaglehole, D. 242
Ernst, S., *see* Gordon, J.G. 316
Erskine, J.L., *see* Cunningham, J.A. 242
Estrup, P.J., *see* Blanchet, G.B. 242
Evans, A.R., *see* Ushioda, S. 564
Evans, D.I. 366
Evans, D.J. 23, 29, 559, 568, 569, 580, 589

Falge, H.J. 35, 39, 40, 41
Falge, H.J., *see* Borstel, G. 39, 42, 631, 647
Falge, H.J., *see* Schuller, E. 29, 42, 279
Falicov, L.M., *see* Martin, R. 555
Fano, U. 242, 247, 391
Ferell, R.A., *see* Stern, E.A. 54
Ferrell, T.L., *see* Kretschmann, E. 323, 342, 396, 471, 472, 480, 495, 501, 502, 503
Feuerbacher, B.P. 362
Filipov, O.K., *see* Vinogradov, E.A. 160, 176
Firsov, E.I., *see* Zhizhin, G.N. 290
Firsov, Yu.A., *see* Bryksin, V.V. 6, 7, 13, 16, 21, 147, 151, 158, 168, 407
Fischer, B. 7, 20, 27, 43, 78, 79, 84, 148, 174, 507
Fischer, B., *see* Lagois, J. 76, 78, 79, 80, 81, 84, 88, 644
Fischer, B., *see* Marschall, N. 20, 27, 43, 148, 218, 566, 641
Fischer, B., *see* Perry, C.H. 31, 32
Flores, F., *see* García-Moliner, F. 80
Flynn, C.P., *see* Cunningham, J.A. 242
Forster, D. 539
Forstmann, F. 260, 307
Froitzheim, H. 82
Frolkov, J.A. 103
Fuchs, R. 12, 14, 147, 148, 168, 608

Fuchs, R., *see* Kliewer, K.L. 12, 14, 15, 242, 407, 549, 550
Fukui, M. 85, 589, 666, 676, 677, 687, 688, 689
Fukui, M., *see* Sipe, J.E. 666, 667, 669, 672, 673, 674, 675, 676, 677, 678, 679, 680, 681, 682, 683, 684, 685, 697, 699
Furtak, T.E. 325

Galantowicz, 397
Gammon, R.W. 28, 29, 43, 45, 46
Gammon, R.W., *see* Hartstein, A. 46, 51, 52
Gammon, R.W., *see* Palik, E.D. 45, 49, 50, 51
García, N. 442, 443, 444, 504
García, N., *see* Lopez, C. 505
García-Moliner, F. 80
Geiger, J., *see* Boersch, H. 18
Genzel, L. 31, 33, 507
Gerboshtein, Y.M. 589, 619
Gerbstein, J.M. 113
Gerbstein, Yu.M. 17, 55, 56, 57, 58, 148, 158, 281, 293, 294
Gerbstein, Yu.M., *see* Bryksin, V.V. 21, 22, 24, 25, 26, 27, 148, 170
Gerbstein, Yu.M., *see* Reshina, I.I. 43
Gerischer, H. 301, 304
Gerischer, H., *see* Kolb, D.M. 316
Gerson, R., *see* Bryan, D.A. 110, 122, 134, 264
Gervais, F. 140
Gesell, T.F. 362, 363, 364, 365, 366
Ginzburg, V.L. 101
Ginzburg, V.L., *see* Agranovich, V.M. 216, 218, 221, 559
Girlando, A. 321
Giuliani, G., *see* DeMartini, F. 23, 589, 631, 640
Glang, R. 123, 285
Glass, N.E. 452, 453, 458, 459, 504
Glockner, E. 199, 230, 231
Goben, C.A., *see* Begley, D.L. 109, 251
Goben, C.A., *see* Bell, R.J. 250
Goben, C.A., *see* Davarpanah, M. 111
Gordon, J.G. 253, 316
Gordon II, J.G., *see* Pockrand, I. 283
Gorkov, L.P. 521, 522
Gorkov, L.P., *see* Abrikosov, A.A. 521, 597
Grachev, V.L., *see* Vinogradov, E.A. 155, 162, 166, 168, 182
Gratton, L.M. 136
Greenlaw, D.K., *see* Cunningham, J.A. 242
Greenler, R.G. 294
Griffin, A. 667, 675
Griffith, S.L., *see* Davarpanah, M. 111
Grigos, V.I., *see* Zhizhin, G.N. 123, 131, 296
Grushevoi, G.V., *see* Vinogradov, E.A. 155, 162, 166, 168, 182
Guha, J.K., *see* Simon, H.J. 353, 384
Guidotti, D. 218, 219, 220, 258
Gurevich, L.E. 47

Hägglund, I. 391
Halevi, P. 7, 30, 53, 54, 100, 270, 280
Hall, D.G. 342, 352, 381, 471
Hall, D.G., *see* Braundmeier Jr., A.J. 352
Hamelin, A., *see* Valette, G. 314
Hamm, R.N., *see* Arakawa, E.T. 29, 246, 279, 357
Hamm, R.N., *see* Gesell, T.F. 362, 363, 364, 365, 366
Hamm, R.N., *see* Ritchie, R.H. 247, 391, 392, 393, 618
Hampe, A., *see* Dobberstein, A. 362
Hansen, W.N. 318
Hansma, P.K., *see* Adams, A. 368
Harrick, H.J. 22
Harris, L. 135
Harrison, J.A. 303
Harrison, M.J., *see* Melnyk, A.R. 260
Hartstein, A. 33, 35, 46, 51, 52, 53, 103, 218, 644
Hartstein, A., *see* Brion, J.J. 48
Hartstein, A., *see* Burstein, E. 7, 242, 407, 411, 584, 589
Hartstein, A., *see* Wallis, R.F. 49
Hartstein, H., *see* Wallis, R.F. 584
Hass, G., *see* Cox, F.T. 138, 139, 142, 292
Hayes, W. 553, 555, 561, 563
Heavens, O.S 61
Heffner, H., *see* Sonnenberg, H. 664, 692, 693, 694, 695
Heitmann, D. 369, 370, 371, 373, 374, 381, 390
Heitmann, D., *see* Girlando, A. 321
Henry, C.H. 566
Henvis, B.W., *see* Hartstein, A. 46, 51, 52
Heritage, J.P. 659
Hessel, A. 391
Hill, N.R. 387, 409, 434, 447, 471
Hill, N.R., *see* Toigo, F.D. 342, 423, 424, 444, 447, 471, 472, 480, 486, 507
Hillebrecht, F.U. 422, 494, 495
Hillebrecht, U. 349, 370, 371, 372, 373, 374, 381, 382, 384, 385
Hincelin, G. 250, 265
Hirabayashi, I. 84, 85
Hirabayashi, I., *see* Tokura, Y. 84
Hisano, K. 16, 25, 148, 158, 160
Hisano, K., *see* Placido, F. 16, 25
Hoffman, R.W., *see* Jansen, F. 374, 408, 422
Holah, G.D., *see* Hisano, K. 16, 25
Hollstein, T. 679
Holm, R.T. 46, 56, 57, 59
Holst, K. 263
Hopfield, J.J. 73, 75, 78, 79, 89, 208, 225, 296
Hopfield, J.J., *see* Henry, C.H. 566
Horing, N.J.M. 507
Hornauer, D.L. 337, 350, 351, 355, 380, 396
Hornauer, D.L., *see* Orlowski, R. 310, 351, 352, 353, 355, 356, 357, 384
Horstmann, C. 345, 349, 381, 382, 383, 422
Hua, Ch.T., *see* Rasigni, M. 369, 422
Huang, K. 73
Huang, K., *see* Born, M. 6
Hummel, R.E. 242

Hunderi, O. 370, 371
Hunderi, O., *see* Beaglehole, D. 370, 381
Hunter, W.R. 362
Hunter, W.R., *see* Cox, F.T. 138, 139, 142, 292
Huong, P.V. 147, 160
Hutley, M.C. 390, 391

Ibach, H. 18, 19, 537
Ibach, H., *see* Froitzheim, H. 82
Ikarashi, T. 55
Ivanov, I.A., *see* Vinogradov, E.A. 150, 181, 182
Ivchenko, E.L., *see* Zemski, V.I. 43

Jackson, J.D. 436
Jansen, F. 374, 408, 422
Jarovaya, R.G., *see* Schkl'arevskij, I.H. 126
Jasperson, S.N. 248, 354, 362, 363
Jeanmaire, D.L. 317, 505
Jha, S.S. 664, 671, 675, 676, 692, 695
Jha, S.S., *see* Bloembergen, N. 632, 648, 649, 664, 668, 671, 678, 691, 692, 693
Jogansen, L.V. 11?
Jones, H., *see* Mott, N.F. 241
Juranek, H.J. 359

Kachare, A. 284, 286, 292
Kagan, Yu.M.., *see* Dubovsky, E.B. 213
Kamada, A., *see* Fukui, M. 85
Kambe, K., *see* Takayamagi, K. 325
Kaminow, I.P. 397, 688
Kaneko, Y., *see* Hirabayashi, I. 84, 85
Kanstad, S.O., *see* Nordal, P.E. 135
Kao, C.C. 218, 219, 220
Kao, C.C., *see* Conwell, E.M. 218, 219, 220
Kapitza, H. 337, 338
Kapitza, H., *see* Hornauer, D.L. 337
Kaplan, H., *see* Palik, E.D. 45, 49, 50, 51
Kaplan, R., *see* Hartstein, A. 51, 52
Kaplan, R., *see* Palik, E.D. 45, 49, 50, 51
Kapusta, O.I., *see* Zhizhin, G.N. 61, 284
Kaspar, W. 354, 408, 422, 423
Keldysh, L.V. 212, 214, 218
Kelley, P.L. 235
Khokhlov, R.B., *see* Akhmanov, S.A. 235
King, W.T., *see* Zelano, A.J. 572
Kirtley, J.R., *see* Tsang, J.C. 589
Kizel, V.A. 191
Kleinman, D.W., *see* Spitzer, W.A. 140
Kliewer, K.L. 12, 14, 15, 242, 407, 549, 550
Kliewer, K.L., *see* Fuchs, R. 12, 14, 147, 148, 168, 608
Kloos, T. 315
Knox, R.S. 71, 72, 83
Koda, T. 84
Koda, T., *see* Hirabayashi, I. 84, 85
Koda, T., *see* Tokura, Y. 84
Kohn, S.E., *see* DeMartini, F. 84, 85, 86, 589, 631, 640, 643, 645
Kolb, D.M. 242, 301, 305, 306, 316, 322, 323, 325, 326
Kolb, D.M., *see* Gerischer, H. 301, 304
Kolb, D.M., *see* Kötz, R. 268, 312, 313, 324, 326
Kolb, D.M., *see* Pettinger, B. 318, 321
Kolb, D.M., *see* Tadjeddine, A. 309, 310, 311, 314, 318, 319, 325
Kolb, D.M., *see* Takayanagi, K. 325
Korsukov, V.E., *see* Banshchikov, A.G. 293
Korsukov, V.E., *see* Novak, I.I. 293
Kostyuk, V.P., *see* Schkl'arevskij, I.H. 126
Kottis, P., *see* Orrit, M. 225
Kottis, P., *see* Turlet, J.M. 199, 230, 231
Kötz, R. 268, 312, 313, 324, 326, 396, 502
Kötz, R., *see* Kolb, D.M. 305, 306, 322, 323, 325
Kötz, R., *see* Tadjeddine, A. 309, 311, 325
Kovacs, G.J. 263, 270
Kovener, G.S. 29, 357
Kovener, G.S., *see* Alexander, R.W. 246, 279
Kovener, G.S., *see* Bell, R.J. 95, 108
Kovener, G.S., *see* Ward, C.A. 98, 100, 101, 103
Krane, K.J. 308
Kranz, H., *see* Griffin, A. 667, 675
Kravtsov, V.E., *see* Agranovich, V.M. 224
Kreibig, U., *see* Genzel, L. 507
Kreibig, U., *see* Hollstein, T. 679
Kreibig, U., *see* Kaspar, W. 354, 408, 422, 423
Kretschmann, E. 242, 249, 294, 304, 305, 315, 323, 335, 339, 340, 342, 344, 345, 347, 351, 352, 353, 359, 360, 361, 364, 365, 380, 381, 384, 390, 396, 408, 470, 471, 472, 480, 495, 501, 502, 503, 589, 636
Kretschmann, E., *see* Chung, M.S. 396
Kretschmann, E., *see* Kötz, R. 396, 502
Kretschmann, E., *see* Kröger, E. 341, 342, 344, 354, 359, 361, 370, 387, 471, 494, 578
Krivoshchekov, G.V. 664, 665, 692, 694
Kröger, E. 341, 342, 344, 354, 359, 361, 370, 384, 387, 471, 494, 578
Kröger, E., *see* Kretschmann, E. 364, 365, 380
Kulik, I.O. 525

Lagois, J. 76, 78, 79, 80, 81, 84, 86, 87, 88, 89, 90, 644
Lagois, J., *see* Fischer, B. 78, 79, 84, 507
Lahar, W., *see* Maeland, A.J. 135
Lainé, R., *see* de Korte, P.A.I. 377, 378
Laks, B. 368, 445, 446, 447, 504
Lalov, I.I., *see* Agranovich, V.M. 199
Lambe, J., *see* McCarthy, S.L. 368
Lamprecht, G. 39
Landau, L.D. 9, 96, 424, 466, 519, 521, 529, 539, 668
Lang, N.D. 313

Laval, S., *see* Reinisch, R. 589
Lebedinskij, M.A. 136
Lee, C.H., *see* Bloembergen, N. 632, 648, 649, 664, 668, 671, 678, 691, 692, 693
Lee, Jr., R.G. 507
Lehmpfuhl, G., *see* Takayanagi, K. 325
Lehwald, S., *see* Ibach, H. 19
Leis, F., *see* Hollstein, T. 679
Leite, R.C.C., *see* Scott, J.F. 566
Lelyuk, L.G., *see* Schkl'arevskij, I.H. 126
Lemberg, H.L., *see* Guidotti, D. 218, 219, 220
Lemberg, M.L., *see* Guidotti, D. 258
Lenzen, R. 377
Leskova, T.A., *see* Agranovich, V.M. 59, 82, 190, 224, 228, 232, 277, 513
Leskova, T.A., *see* Vinogradov, E.A. 148
Leskova, T.M., *see* Agranovich, V.M. 559
Levenson, M.D. 654
Lewerenz, H.I., *see* Kötz, R. 396
Lewerenz, H.J., *see* Kötz, R. 502
Lichte, H. 378
Lifshitz, E.M., *see* Landau, L.D. 9, 96, 425, 466, 519, 521, 529, 539, 668
Llebaria, A., *see* Rasigni, M. 369
Loisel, B. 270
Lopez, C. 505
Lopez-Rios, T. 199, 200, 204, 254, 255, 256, 258, 259, 261, 262, 263, 264, 267, 271, 294, 315
Lopez–Rios, T., *see* Abelès, F. 260, 263, 268. 269, 270
Lopez-Rios, T., *see* Chao, F. 318, 320
Loudon, R. 540, 541, 560, 562, 566
Loudon, R., *see* Barker, A.S. 539, 598
Loudon, R., *see* Hayes, W. 553, 555, 561, 563
Loudon, R., *see* Nkoma, J.S. 12, 16, 31, 407, 538, 544, 556, 558, 559, 560, 589
Lozovik, Yu.E. 212, 213
Lozovik, Yu.E., *see* Agranovich, V.M. 212, 213
Lubeznikov, O.A., *see* Vinogradov, E.A. 150, 181, 182
Lubimov, J.N. 103
Lundquist, S., *see* Burstein, E. 583, 619
Lynch, D.W., *see* Furtak, T.E. 325
Lyubimov, V.N. 33

Macek, C.H. 250
Maeland, A.J. 135
Mahan, G.D. 76
Mal'shukov, A.G., *see* Agranovich, V.M. 54, 59, 210, 211, 212, 213, 257, 277, 281, 284
Mal'shukov, A.G., *see* Vinogradov, E.A. 148, 150, 155, 168, 173, 176, 294
Mammel, W.L., *see* Kaminow, I.P. 397, 688
Manasevit, H.M. 569
Maradudin, A.A. 16, 76, 78, 80, 342, 344, 370, 407, 471, 472, 473, 477, 478, 480, 481, 485, 488, 506, 507, 519, 526, 540, 578, 688
Maradudin, A.A., *see* Bennett, B.I. 469
Maradudin, A.A., *see* Bonsall, L. 589, 666
Maradudin, A.A., *see* Burstein, E. 7, 407, 411, 584
Maradudin, A.A., *see* Cunnigham, S.L. 46, 218, 219, 220, 258
Maradudin, A.A., *see* Dobrzynski, L. 505
Maradudin, A.A., *see* Eguiluz, A. 505
Maradudin, A.A., *see* Glass, N.E. 452, 453, 458, 459, 504
Maradudin, A.A., *see* Laks, B. 445, 446, 447, 504
Maradudin, A.A., *see* Lee Jr., R.G. 507
Maradudin, A.A., *see* Martin, B.G. 48
Maradudin, A.A., *see* Mills, D.L. 55, 366, 408, 422, 546, 547, 556, 558, 569, 593
Maradudin, A.A., *see* Rahman, T.S. 472, 495, 501, 502, 503
Maratoni, P., *see* DeMartini, F. 23
Marshall, N. 20, 27, 43, 148, 218, 566, 641
Marschall, N., *see* Fischer, B. 7, 20, 27, 43, 148, 174
Martin, B.G. 48
Martin, R. 555
Martin, T.P., *see* Genzel, L. 31, 33, 507
Marvin, A. 370
Marvin, A., *see* Toigo, F.D. 342, 423, 424, 444, 447, 471, 472, 480, 486, 507
Mataloni, P., *see* DeMartini, F. 589, 631, 640
Mathewson, A.G. 679
Matsko, M.G., *see* Brodin, M.S. 82, 232
Matsuoka, M., *see* Brown, F. 699
Mattei, G., *see* Ushioda, S. 495, 580
Mattei, G., *see* Valdez, J.B. 27, 563
Matumura, O., *see* Hisano, K. 148, 158, 160
Maystre, D. 390, 424
Maystre, D., *see* Hutley, M.C. 390
Maystre, D., *see* McPhedran, R.C. 391
McCarthy, S.L. 368
McCarthy, S.L., *see* Weber, W.H. 250
McIntyre, J.D.E. 322, 325
McIntyre, J.D.E., *see* Kolb, D.M. 242
McMullen, J.D. 102, 103, 122, 545
McMullen, J.D., *see* Evans, D.J. 23, 366, 559, 568, 569, 589
McMullen, J.D., *see* Ushioda, S. 31, 545
McPhedran, R.C. 391
McWhorter, A.L., *see* Mooradian, A. 569
Meecham, W.C. 439
Mekhtiev, M.A., *see* Agranovich, V.M. 54, 59
Melnik, N.N., *see* Vinogradov, E.A. 148, 160, 176
Melnyk, A.R. 260
Merkulov, M.A., *see* Gerbstein, J.M. 113
Merkulov, T.A., *see* Gerboshtein, Y.M. 589, 619
Mermin, N.D., *see* Ashcroft, N.W. 670

Merten, L., *see* Lamprecht, G. 39
Mertens, F.P. 135
Messiah, A. 444
Mettei, G., *see* Ushioda, S. 368
Millar, R.F. 387, 409, 434, 447, 452
Miller, P.B., *see* Cheng, H. 671
Miller, R., *see* Begley, D.L. 251
Mills, D.L. 55, 82, 127, 128, 229, 342, 366, 394, 407, 408, 411, 422, 424, 460, 465, 466, 467, 469, 472, 486, 493, 523, 538 541, 546, 547, 556, 558, 559, 569, 574, 578, 589, 593, 666, 676, 677, 688
Mills, D.L., *see* Burstein, E. 7, 407, 411, 584, 619
Mills, D.L., *see* Chen, Y.J. 558, 559, 589, 600
Mills, D.L., *see* Ibach, H. 537
Mills, D.L., *see* Laks, B. 368, 445, 446, 447, 504
Mills, D.L., *see* Maradudin, A.A. 76, 78, 80, 344, 370, 472, 473, 477, 478, 488, 506, 507, 519, 526, 540, 688
Mills, D.L., *see* Subbaswamy, K.R. 563, 574, 576
Mirlin, D.N. 55, 56, 57, 58, 59, 60, 62, 63, 277
Mirlin, D.N., *see* Bryksin, V.V. 6, 7, 13, 16, 21, 22, 24, 25, 26, 27, 33, 37, 38, 43, 44, 45, 46, 147, 148, 151, 158, 168, 170, 284, 407
Mirlin, D.N., *see* Gerbstein, Yu.M. 55, 56, 57, 58, 148, 158, 281, 293, 294
Mirlin, D.M., *see* Gerboshtein, Y.M. 589, 619
Mirlin, D.N., *see* Reshina, I.I. 31, 32, 33, 43, 284
Mirlin, D.N., *see* Zemski, V.I. 43
Mishra, S. 372
Mitchell, D.E., *see* Simon, H.J. 251, 589, 632, 648, 649, 665, 672, 681, 696
Möbius, D., *see* Waehling, G. 341
Mooradian, A. 569
Mooradian, A., *see* Blum, F.A. 44
Morosov, N.N., *see* Zhizhin, G.N. 296
Morozov, H.H., *see* Zhizhin, G.N. 123, 131
Moskalova, M.A., *see* Yakovlev, V.A. 61, 62, 284
Moskalova, M.A., *see* Zhizhin, G.N. 28, 29, 30, 56, 59, 61, 106, 109, 118, 119, 122, 123, 130, 131, 138, 205, 277, 279, 282, 284, 292, 294, 296
Mott, N.F. 241
Motulevich, G.P. 102
Motulevich, G.P., *see* Ginzburg, V.L. 101
Mühlschlegel, B., *see* Lee, Jr., R.G. 507
Mühlschlegel, B., *see* Rendell, R.W. 368
Muller, K., *see* Bockris, J.O'M. 304
Murata, J., *see* Hirabayashi, I. 84, 85
Musatov, M.I., *see* Yakovlev, V.A. 284, 288

Nazin, V.G., *see* Yakovlev, V.A. 56, 61, 62, 199, 284
Nazin, V.G., *see* Zhizhin, G.N. 30, 56, 59, 61, 106, 277, 279, 282, 284, 292
Ngai, K.L. 54, 257
Nannarone, S., *see* Chiarotti, G. 242
Nishanov, V.N., *see* Lozovik, Yu.E. 212, 213
Nkoma, J.S. 12, 16, 31, 407 538, 544, 550, 556, 558, 559, 560, 589, 600
Nordal, P.E. 135
Normandin, R. 688
Normandin, R., *see* So, V.C.Y. 688
Novak, I.I. 293
Novak, I.I., *see* Banshchikov, A.G. 293

Okamoto, Y., *see* Hisano, K. 148, 158, 160
Oliner, A.A., *see* Hessel, A. 391
Orlowski, R. 310, 351, 352, 353, 355, 356, 357, 384
Orrit, M. 225
Otto, A. 7, 21, 22, 23, 82, 108, 148, 242, 248, 250, 285, 304, 307, 308, 407, 553, 584, 589, 636, 680
Otto, A., *see* Bodesheim, J. 353, 408
Otto, A., *see* Borstel, G. 39, 42
Otto, A., *see* Falge, H.J. 35, 39, 40, 41
Otto, A., *see* Macek, C.H. 250

Paglia, S., *see* Gratton, L.M. 136
Pakhomov, V.I. 46, 48
Palange, E., *see* DeMartini, F. 23, 589, 631, 640
Palik, E.D. 45, 48, 49, 50, 51
Palik, E.D., *see* Gammon, R.W. 28, 29, 43, 45, 46
Palik, E.D., *see* Hartstein, A. 46, 51, 52
Palik, E.D., *see* Holm, R.T. 46, 56, 57, 59
Palmari, J.P., *see* Rasigni, M. 369
Palmer, R.E. 396, 502, 503
Paraire, N., *see* Reinisch, R. 589
Pardee, W.J., *see* Fuchs, R. 14, 148
Pariset, C. 267
Parks, R.E., *see* Brown, F. 664, 692, 693
Parks, W.F., *see* Bell, R.J. 95, 108
Parsons, D.F. 566, 582
Parsons, R. 302, 303
Passler, M.A., *see* Anderson, J. 242
Passler, M.A., *see* Rubloff, G.W. 242
Pastore, R., *see* Chiarotti, G. 242
Pekar, S.I. 78, 79, 80, 89, 207
Permien, V., *see* Heitmann, D. 369, 370, 371, 373, 374, 381
Perry, C.H. 31, 32
Pershan, P.S., *see* Bloembergen, N. 593, 631
Petit, R. 387, 409
Petit, R., *see* Maystre, D. 390
Pettinger, B. 318, 321
Philpott, M.R. 72, 199, 225, 232
Philpott, M.R., *see* Girlando, A. 321
Philpott, M.R., *see* Pockrand, I. 283, 295, 315, 316
Philpott, M.R., *see* Scherman, G. 295, 296
Philpott, M.R., *see* Sherman, P.G. 199
Philpott, M.R., *see* Syassen, K. 199, 230

Philpott, M.R., *see* Turlet, J.M. 199, 230
Pilipetsky, N.F., *see* Zeldovich, B.Ya. 236
Pinczuk, A. 46
Pinczuk, A., *see* Burstein, E. 561, 563
Pine, A.S., *see* Dresselhaus, G. 603
Piriou, B., *see* Gervais, F. 140
Placido, F. 16, 25
Placido, F., *see* Hisano, K. 16, 25
Pockrand, I. 263, 283, 295, 315, 316, 317, 386, 387, 388, 390, 391, 392, 394, 395, 396, 408
Pockrand, I., *see* Rosengart, E.H. 391, 392
Pokrowsky, P. 361, 362
Pons, J., *see* Rasigni, M. 369
Porteus, J.O., *see* Bennett, H.E. 380
Potter, P., *see* Stierwald, D. 147, 160
Pous, J., *see* Rasigni, M. 422
Prieur, J.-Y. 569, 571, 572, 589
Proix, F. 148, 158
Przasnyski, M., *see* Kolb, D.M. 316
Pschalek, N. 354
Puchkovskaya, G.A., *see* Frolkov, J.A. 103

Queisser, H.J., *see* Fischer, B. 7, 20, 27, 43, 148, 174
Queisser, H.J., *see* Marschall, N. 20, 43, 218
Quinn, J.J., *see* Chiu, K.W. 43, 48
Quinn, J.J., *see* Eguiluz, A. 667, 669, 670, 672, 675
Quinn, J.J., *see* Palik, E.D. 45, 49, 50, 51

Raether, H. 242, 250, 307, 308, 310, 335, 340, 343, 344, 360, 361, 366, 386, 392, 410, 584
Raether, H., *see* Brambring, I. 360
Raether, H., *see* Bruns, R. 335, 341
Raether, H., *see* Heitmann, D. 390
Raether, H., *see* Holst, K. 263
Raether, H., *see* Hornauer, D.L. 337, 396
Raether, H., *see* Krane, K.J. 308
Raether, H., *see* Kretschmann, E. 294, 339, 340, 470
Raether, H., *see* Kröger, E. 384
Raether, H., *see* Orlowski, R. 310, 357
Raether, H., *see* Pockrand, I. 387, 396, 408
Raether, H., *see* Pokrowsky, P. 361, 362
Raether, H., *see* Waehling, G. 341
Rahman, T.S. 472, 495, 501, 502, 503
Rako, J.G., *see* Simon, H.J. 589, 632, 650, 651, 652, 665, 666, 677, 681, 683, 696, 697
Randles, J.E.B., *see* Harrison, J.A. 303
Rasigni, G., *see* Rasigni, M. 369, 422
Rasigni, M. 369, 422
Rath, D.L., *see* Kolb, D.M. 326
Rayleigh Lord, 391, 409, 424
Rebane, K.K., *see* Birman, J.L. 513
Reddy, A.K.N., *see* Bockris, J.O'M. 303
Reinisch, R. 589
Rendell, R.W. 368
Reshina, I.I. 31, 32, 33, 39, 41, 43, 284
Reshina, I.I., *see* Bryksin, V.V. 33, 37, 38, 43, 44, 45, 46, 284
Reshina, I.I., *see* Mirlin, D.N. 55, 56, 57, 58, 59, 60, 62, 63, 277
Reshina, I.I., *see* Zemski, V.I. 43
Rice, S.A., *see* Guidotti, D. 218, 219, 220, 258
Richter, W. 555
Rimbey, P.R. 78
Ristori, P., *see* DeMartini, F. 651, 653, 667
Ritchie, R.H. 73, 247, 307, 391, 392, 393, 409, 618
Ritchie, R.H., *see* Arakawa, E.T. 29, 279, 357
Ritchie, R.M., *see* Arakawa, E.T. 246
Ritchie, R.H., *see* Crowell, J. 364
Ritchie, R.H., *see* Elson, I.M. 364, 370, 422
Ritchie, R.H., *see* Wheeler, C.E. 393
Ritchie, R.H., *see* Wilems, R.E. 361
Rittenhouse, R., *see* Maeland, A.J. 135
Robin, S., *see* Daudé, A. 364, 365, 422
Romano, P.V., *see* Maeland, A.J. 135
Romanov, Yu.A. 46
Rosengart, E.H. 387, 389, 391, 392, 397, 398
Rothballer, W. 391, 392
Rowell, N., *see* Normandin, R. 688
Rubinina, N.M., *see* Yakovlev, V.A. 284, 288
Rubloff, G.W. 242
Rubloff, G.W., *see* Anderson, J. 242
Rudnick, J. 663, 665, 666, 667, 671, 672, 674, 675, 687, 691, 698, 699
Rudolf, H.W. 251
Ruppin, R. 147, 148, 151, 155, 158, 168, 507, 559
Ruppin, R., *see* Englman, R. 507

Sannikov, D.G., *see* Lubimov, J.N. 103
Sannikov, D.G., *see* Lyubimov, V.N. 33
Santamato, E., *see* DeMartini, F. 651, 653, 667
Santo, R., *see* Girlando, A. 321
Santo, R., *see* Pockrand, I. 295, 315, 316
Sari, S.O. 374, 376, 505, 506
Sari, S.O., *see* Bush, I.A. 374, 375
Sass, J.K., *see* Gerischer, H. 301, 304
Sass, J.K., *see* Kötz, R. 268, 312, 313
Sauerbrey, G. 354
Sauerbrey, G., *see* Dobberstein, A. 362
Savary, A., *see* Daudé, A. 364, 365, 422

Scalpino, D.J., *see* Rendell, R.W. 368
Scattaglia, F., *see* Gratton, L.M. 136
Scherkoske, K.D., *see* Bush, I.A. 374, 375
Scherkoske, K.D., *see* Sari, S.O. 374, 376, 505, 506
Scherman, G. 295, 296
Schevchenko, V.V. 113
Schiffrin, D.J., *see* Harrison, J.A. 303
Schirmer, G., *see* Zhizhin, G.N. 292
Schkl'arevskij, I.H. 126
Schlesinger, Z. 205
Schnatterly, S.E., *see* Jasperson, S.N. 248, 354, 362, 363
Schnatterly, S.E., *see* Palmer, R.E. 396, 502, 503
Schoenwald, J. 95, 108, 122, 190, 200, 250, 336, 545, 686
Schoenwald, J., *see* Burstein, E. 7, 407, 411, 584
Schoenwald, J., *see* Talaat, H. 108, 127, 470
Schreiber, P. 361, 369
Schröder, E. 355, 361, 362, 366, 369
Schröder, U. 335, 339, 352, 355
Schuller, E. 29, 42, 279
Schultze, J.W. 325
Scott, G.D., *see* Kovacs, G.J. 270
Scott, J.F. 566
Sellborg, F., *see* Hägglund, I. 391
Septier, A., *see* Hincelin, G. 250, 265
Seraphin, B.O. 241, 569
Shen, Y.R., *see* Bloembergen, N. 664
Shen, Y.R., *see* Chen, C.K. 589, 619, 632, 650, 653, 654
Shen, Y.R., *see* Chen, Y.J. 667
Shen, Y.R., *see* Chu, K.C. 636
Shen, Y.R., *see* DeMartini, F. 23, 84, 85, 86, 589, 631, 639, 640, 643, 645
Sherman, P.G. 199
Sherman, P.G., *see* Philpott, M.R. 225
Shomina, E.V., *see* Zhizhin, G.N. 109, 118, 119, 122, 123, 130, 131, 138, 205, 279, 294, 296
Siddiqui, A.S. 137
Sievers, A.J., *see* Chabal, Y.J. 109, 118, 133, 267
Sievers, A.J., *see* Schlesinger, Z. 205
Sigarov, A.A., *see* Zhizhin, G.N. 123, 131, 296
Silfvast, W.T., *see* Scott, J.F. 566
Simon, H.J. 251, 353, 384, 589, 632, 648, 649, 650, 651, 652, 665, 666, 672, 677, 681, 683, 696, 697
Sipe, J.E. 666, 667, 669, 672, 673, 674, 675, 676, 677, 678, 679, 680, 681, 682, 683, 684, 685, 688, 697, 699
Sipe, J.E., *see* Fukui, M. 589, 666, 676, 677, 687, 689
Sivukhin, D.V. 190, 194
Sleeper, A.M., *see* Brown, F. 664, 692
So, V.C.Y. 688
So, V.C.Y., *see* Fukui, M. 85, 589, 666, 676, 677, 687, 688, 689
So, V.C.Y., *see* Normandin, R. 688
So, V.C.Y., *see* Sipe, J.E. 666, 667, 669, 672, 673, 674, 675, 676, 677, 678, 679, 680, 681, 682, 683, 684, 685, 697, 699
Sommerfeld, A. 408, 635
Sonnenberg, H. 664, 692, 693, 694, 695
Sparnaay, M.J. 303
Spicer, W.E., *see* Endriz, J.G. 523
Spicer, W.G.., *see* Endriz, I.G. 364, 365
Spitzer, W.A. 140
Spizzichino, A., *see* Beckmann, P. 380
Srivastava, V.K. 295
Stanford, J.L. 242, 362, 391
Stegeman, G.I., *see* Fukui, M. 85, 589, 666, 676, 677, 687, 688, 689
Stegeman, G.I., *see* Normandin, R. 688
Stegeman, G.I., *see* Sipe, J.E. 666, 667, 669, 672, 673, 674, 675, 676, 677, 678, 679, 680, 681, 682, 683, 684, 685, 688, 697, 699
Stegeman, G.I., *see* So, V.C.Y. 688
Stegun, I.A., *see* Abramowitz, M. 437
Steinmann, W., *see* Feuerbacher, B.P. 362
Steinmann, W., *see* Macek, C.H. 250
Steinmann, W., *see* Rudolf, H.W. 251
Stenschke, H., *see* Forstmann, F. 307
Stepanov, B.I. 147, 160
Stepanov, K.N., *see* Pakhomov, V.I. 46, 48
Stern, E.A. 54, 73, 343, 358, 370
Stern, E.A., *see* Rudnick, J. 663, 665, 666, 667, 671, 672, 674, 675, 687, 691, 698, 699
Stern, E.A., *see* Teng, Y.Y. 148, 174, 247, 391
Stickel, W., *see* Boersch, H. 18
Stierwald, D. 147, 160
Stiles, P.J., *see* Anderson, J. 242
Stiles, P.J., *see* Blanchet, G.B. 242
Stiles, P.J., *see* Rubloff, G.W. 242
Stoljarov, V.M., *see* Frolkov, J.A. 103
Strogaov, V.I., *see* Krivoshchekov, G.V. 664, 665, 692, 694
Subbaswamy, K.R. 563, 574, 576
Subbaswamy, K.R., *see* Mills, D.L. 538, 559
Sugakov, V.I. 199, 231, 232
Sukhorukov, A.P., *see* Akhmanov, S.A. 235
Sukhov, A.V., *see* Zeldovich, B.Ya. 236
Swalen, J.D., *see* Girlando, A. 321
Swalen, J.D., *see* Gordon, J.G. 253
Swalen, J.D., *see* Philpott, M.R. 72
Swalen, J.D., *see* Pockrand, I. 283, 295, 315, 316, 317
Swanson, L.R., *see* Bennett, B.I. 469
Syassen, K. 199, 230

Tabiryan, N.V., *see* Zeldovich, B.Ya. 236
Tada, O., *see* Fukui, M. 85
Tadjeddine, A. 309, 310, 311, 314, 318, 319, 325

Tadjeddine, A., *see* Abelès, F. 268
Tadjeddine, A., *see* Chao, F. 316, 318, 320
Tadjeddine, A., *see* Pettinger, B. 318, 321
Tajima, T. 584
Takayanagi, K. 325
Talaat, H. 108, 127, 470
Tarkhanian, R.G. 47
Tarkhanian, R.G., *see* Gurevich, L.E. 47
Teng, Y.Y. 148, 174, 247, 391
Theye, M.L., *see* Chao, F. 318, 320
Thèye, M.-L., *see* Dujardin, M.M. 260, 679
Thomas, B., *see* Bewick, A. 325
Thomas, D.G., *see* Hopfield, J.J. 75, 78, 79, 89, 208, 296
Thorsen, A.C., *see* Manasevit, H.M. 569
Tien, P.K. 113
Tilley, D.R., *see* Nkoma, J. 12, 16, 31, 407, 538, 544
Toigo, F.D. 342, 423, 424, 444, 447, 471, 472, 480, 486, 507
Toigo, F.D. *see* Marvin, A. 370
Tokura, Y. 84
Tokura, Y., *see* Hirabayashi, I. 84, 85
Tomaschke, H.E., *see* Braundmeier Jr., A.J. 384
Trasatti, S. 304, 314
Treherne, D.M., *see* Siddiqui, A.S. 137
Trümper, J. 377
Tsang, J.C. 589
Turlet, J.M. 199, 230, 231
Twersky, V. 391
Twietmeier, H. 350
Tyler, I.L., *see* Bell, R.J. 407
Tyler, J., *see* Ward, C.A. 134
Tyte, R.N., *see* Bell, M.I. 566

Uller, 408
Urner, P. 355, 356, 357, 381
Urner, P., *see* Orlowski, R. 310, 351, 352, 353, 355, 356, 357, 384
Ushioda, S. 31, 366, 367, 368, 407, 495, 545, 547, 564, 568, 580
Usioda, S., *see* Burstein, E. 561, 563
Ushioda, S., *see* Evans, D.J. 23, 29, 366, 559, 568, 569, 580, 589
Ushioda, S., *see* Prieur, J.-Y. 569, 571, 572, 589
Ushioda, S., *see* Tajima, T. 584
Ushioda, S., *see* Valdez, J.B. 27, 563, 565, 566

Vainshtein, L.A. 206, 221, 222, 223
Valdez, J.B. 27, 563, 565, 566
Valdez, J.B., *see* Ushioda, S. 368, 495, 564, 580
Valette, G. 314
Van den Berg, P.M. 391
Van Duyne, R.P. 619
Van Duyne, R.P., *see* Jeanmaire, D.L. 317, 505
Vetter, K.J. 303
Vincent, P., *see* Maystre, D. 424
Vinogradov, E.A. 148, 150, 155, 160, 162, 166, 168, 173, 176, 181, 182, 294
Vinogradov, E.A., *see* Vodopianov, L.K. 148, 158, 167
Vinogradov, V.S. 16
Vlieger, J., *see* Bedeaux, D. 359
Vodopianov, L.K. 148, 158, 167
Vuye, G. 254
Vuye, G., *see* Lopez-Rios, T. 199, 200, 204, 255, 258, 259, 263, 264, 294, 315

Waehling, G. 341
Waehling, R. 357
Wallis, R.F. 43, 49, 584
Wallis, R.F., *see* Brion, J.J. 48
Wallis, R.F., *see* Burstein, E. 7, 407, 411, 584
Wallis, R.F., *see* Cunnigham, S.L. 46, 218, 219, 220, 258
Wallis, R.F., *see* Hartstein, A. 33, 35, 53, 103, 644
Wallis, R.F., *see* Maradudin, A.A. 16, 407, 506
Wallis, R.F., *see* Martin, B.G. 48
Wallis, R.F., *see* Palik, E.D. 45, 49, 50, 51
Wang, C.S. 665, 671, 672, 695
Ward, C.A. 98, 100, 101, 103, 134
Ward, C.A., *see* Begley, D.L. 103, 122, 251
Ward, C.A., *see* Bell, R.J. 106, 130, 294, 407
Ward, W.H., *see* Ushioda, S. 564
Warke, C.S., *see* Jha, S.S. 664, 675, 676
Watson, J.G., *see* Simon, H.J. 251, 589, 632, 648, 649, 665, 672, 681, 696
Weber, H.P., *see* Kaminow, I.P. 397, 688
Weber, W.H. 250, 265, 266, 267
Weber, W.H., *see* Eagen, C.F. 265
Weinstein, B.A. 566
Welford, W.T. 380
Wenning, U., *see* Pettinger, B. 318
Weyl, H. 408
Wheeler, C.E 393
Wilems, R.E. 361
Williams, M.D. 579
Williams, M.W. 362
Williams, M.W., *see* Arakawa, E.T. 29, 246, 279, 357
Williams, M.W., *see* Gesell, T.F. 362, 363, 364, 365, 366
Wilson, A.H. 241
Wöckel, E., *see* Sauerbrey, G. 354
Wolf, E. 436
Wolf, E., *see* Born, M. 21, 635
Wolf, H.C., *see* Glockner, E. 199, 230, 231
Wood, R.W. 390
Wright, G.B., *see* Palik, E.D. 48
Wynne, J.J. 590
Wyss, J.C., *see* Adams, A. 368

Yakovlev, V.A. 27, 28, 56, 61, 62, 199, 284, 288
Yakovlev, V.A., *see* Zhizhin, G.N. 28, 29, 30, 56, 59, 61, 106, 109, 118, 119, 122, 123, 130, 131, 138, 205, 277, 279, 282, 284, 290, 292, 294, 296
Yamamoto, K., *see* Kolb, D.M. 325
Yeager, E. 303
Yndurian, F.J., *see* Lopez, C. 505
Yudson, V.A., *see* Vinogradov, E.A. 294
Yudson, V.I., *see* Vinogradov, E.A. 148, 150, 155, 162, 166, 168, 173, 176, 182

Zammit, A.C.A., *see* DeMartini, F. 651, 653, 667

Zavada, I.M., *see* Church, E.L. 392
Zelano, A.J. 572
Zeldovich, B.Ya. 236
Zemski, V.I. 43
Zenneck, J. 408
Zeyher, R. 90
Zhizhin, G.N. 28, 29, 30, 56, 59, 61, 106, 109, 118, 119, 122, 123, 130, 131, 138, 205, 277, 279, 282, 284, 290, 292, 294, 296
Zhizhin, G.N., *see* Vinogradov, E.A. 148, 150, 155, 160, 162, 166, 168, 173, 176, 181, 182, 294
Zhizhin, G.N., *see* Yakovlev, V.A. 27, 28, 56, 61, 62, 199, 284, 288
Zierau, W., *see* Maradudin, A.A. 342, 471, 472, 480, 481, 485,507, 578
Ziman, J.M. 162
Zolotarev, V.M. 139
Zolotarev, V.M., *see* Reshina, I.I. 39, 41

SUBJECT INDEX

additional boundary conditions 78–79
adsorbate layers, ordered 325–327
anharmonic damping 16
anharmonic effects 24–25, 27, 30–32
anisotropic crystals 33–42, 85–87
ATR method (see coupling techniques)
ATR spectra; explicit examples
  α-SiO_2 38, 58
  Al/Ag sandwiches 271
  Ag with Pd overlayer 254
  Au and Ag sandwiches 261, 262
  Au and Cd arachidate 253
  Au on MgF_2 270
  Au on SiO_2 60
  Bi on SiO_2 58
  LiF on rutile 286
  LiF on sapphire 286
  LiF on YIG 290
  MgF_2 30
  NaCl film 24
  n-InSb 44
  ZnO excitons 84, 85, 87, 88, 90
  ZnSe film 61
attenuation lengths
  effect of surface film 122–126, 129–133
  experimental values 121–122
  intrinsic 98, 100–104, 241–242
  Langmuir–Blodgett films 131–132
  natural oxide films 135–137
  SiO_x films 138–142
  surface discontinuities 120
  surface roughness; data 126–129; 339–340
  surface scratches 119

back bending 29, 49, 278–280, 289

chemisorbed overlayers 266–268
Coulomb frequency, definition 200
Coulomb modes, definition 149
Coupling techniques; linear
  ATR method 21–23, 82–83, 108, 248–249
  edge coupling 118
  experimental details 109–110
  Kretschmann method 249, 617–618, 636
  optimization of prism coupling 111–117
  ordered adsorbate layers 325–327
  Otto method 617–618, 636
  periodic gratings 19, 20, 107–108, 247

deformation potential Raman tensor 596
density of states; surface polaritons 228
detection techniques 249–250
dielectric films on metals (see also transition layer) 168–174, 257
dispersion curves for semi-infinite media; data
  Ag–electrolyte interface 312–314
  Ag single crystal faces 310
  effect of roughness 337
  $Gd_2(MoO_4)_3$ 179
  MgF_2 36
  NaCl 23
  sinusoidal gratings 385–391
  SiO_2 28, 39
  n-InSb 43, 50
  YIG 26
  ZnO 86
  ZnSe 179
dispersion curves; slabs and multilayer structures; data
  Ag with pyridine 318–319
  $GaAs/Al_2O_3$ 573
  LiF–sapphire 287
  LiF–YIG 289
  MgF_2–SiO_2 56
  NaCl 24–25
  ZnSe–LnSb 56

doped semiconductors 42–47
double interface surface polaritons
definition 546
examples 547–549
data for GaAs–Al_2O_3 573
double layer 303

electrooptic Raman tensor 596
electron energy loss method 18
ellipsometry 241
emissivity measurements; experimental details 159–161
exciton luminescence 225–232

field enhancement with surface polaritons 318–322, 619–624
fluctuation–dissipation theorem 541
Frankel excitons 72
free carrier effects 42–47

Green's functions for Maxwell equations
definition 473, 594
explicit form, semi-infinite media 476, 479

guided wave polaritons
definition 546
properties 550–553
Raman data, GaP 575–577

Helmholtz layer, definition 302

impedance steps 220–225
inhomogeneous surface (transition layers) 218–220
interface surface polaritons 52–63

Kretschmann configuration 617–618, 636

linewidth, intrinsic 544

magnetic field effects 45–52
multi-layered media, theory 632–639

non-linear excitation; data
GaP 644
ZnO exciton polariton 644–646
non-linear polarizability 560, 596
non-reciprocity 51

Otto configuration 317–618, 636

parametric mixing 598–599, 609
periodic gratings (see coupling techniques)
dispersion relation, experiment 384–392
dispersion relation, theory 425–432
phonon polariton 7
photoelastic coupling 460
polyatomic crystals 15, 27–28, 178–181

Rayleigh hypothesis 409, 426
reflectivity of rough surface; experiment 362–366
roughness correlation function
definition 347
experimental determination 350–354
roughness induced scattering
general discussion 370–377
data on Ag foils 375
limits of perturbation theory 381–389

scattering kinematics; surface Raman 556–558
second harmonics from surfaces
early data 691–696
surface polariton enhanced 647–653, 680–684, 696–698
self focusing 233–236
selvedge region, definition 417–418
semiconductor films; exciton effects 211–220
slab geometry
non-radiative modes 14, 26–27
radiative modes 14
splitting and transition layers 280–287, 291
stepped surfaces 324–325
surface active media, definition 52–53, 413, 589
surface CARS 653–658
surface damage; laser induced 523–525, 657

surface damping mechanisms 44
surface phonon polaritons 7
surface polaritons; polyatomic crystals 15, 27–28, 178–181
surface response of electrons; hydrodynamic picture 667–676
surface roughness correlation length, definition 334
structural phase transitions and surface polaritons 177–181

thermal emission by surface polaritons
  angle and polarization dependence 163–166
  theory 161–162
thin films
  absorption 155–157
  normal modes 149–154
transition layer (see also dielectric films on metal)
  definition 189
  effect on dispersion relation; experiment 253–254, 258–259
  effect on dispersion relation; theory 252–253
  effect on carrier mean free path 261–262
    non-local effects 260
transition layer; effective boundary conditions
  anisotropic case 194–196
  isotropic case 191–194
  impedance step 205–208
transition layer substrate combinations; specific examples and data
  Ag on Al 259
  Ag on Au 254
  Au on Ag 254
  Au on SiO_2 59–60
  Bi on SiO_2 58
  ZnSe on Al 173
  ZnSe on Ag 173
  ZnSe on Cu 254
transverse correlation length, definition 420, 422
twinned crystals; surface polaitons and emissivity 174

ultra high vacuum studies 263–268

Wannier excitons 72

X-ray scattering; roughness 377–378